Fracture and Damage of Concrete and Rock

FDCR-2

Fracture and Damage of Concrete and Rock

FDCR-2

Proceedings of the Second International Conference on Fracture and Damage of Concrete and Rock, Vienna, Austria, 9–13 November 1992

Edited by

H.P. ROSSMANITH

Fracture and Photomechanics Laboratory, Institute of Mechanics, Technical University of Vienna

CRC Press
Taylor & Francis Group
Boca Raton London New York

CRC Press is an imprint of the
Taylor & Francis Group, an **informa** business
A TAYLOR & FRANCIS BOOK

CRC Press
Taylor & Francis Group
6000 Broken Sound Parkway NW, Suite 300
Boca Raton, FL 33487-2742

First issued in paperback 2019

ISBN-13: 978-0-419-18470-6 (hbk)
ISBN-13: 978-0-367-86382-1 (pbk)

A catalogue record for this book is available from the British Library

Library of Congress Cataloging-in-Publication data available

Publisher's Note
The publisher has gone to great lengths to ensure the quality of this reprint but points out that some imperfections in the original may be apparent

Visit the Taylor & Francis Web site at
http://www.taylorandfrancis.com

and the CRC Press Web site at
http://www.crcpress.com

Contents

Preface

The Second International Conference on *Fracture and Damage of Concrete and Rock* (**FDCR-2**) was held in November 1992 following the very successful conference in 1988. It was organized by the Technical University Vienna and was dedicated to the impact of most recent research in fracture and damage of concrete and rock on improved understanding of the basic physical and mechanical principles with a strong view towards applications in construction engineering and mining engineering.

Over the past decades excessive damage by fatigue and ageing of complex concrete structures have been responsible for both loss of life and billions of dollars wasted. Despite worldwide intensive research in these fields many pressing problems urgently need to be solved.

Lack of virgin building space and prohibitive property values and costs in city centers and dense urban settlements require efficient demolition of (aged) concrete structures. Recycling and deposition of wrecked concrete structures and waste concrete products have already become one of the top environmental problems today and will become even more serious in the future.

Nowadays, an increasing number of engineering structures are being built on and in rock mass of low quality and difficult terrain. Safety requirements impose additional constraints and demand technically sound design. Rock fracture mechanics has matured to a fully recognized discipline within the field of rock mechanics and can routinely be applied to problems of excavation, fault-tunnel interaction, blasting and anchoring.

In anchorage engineering the life-time and performance of anchors subjected to varying loading and environmental conditions in rock and concrete structures is of primary concern. Numerical simulations and scale size as well as full size tests of anchors and fastening elements are being performed in a concerted worldwide research and testing program the results of which will be presented in this volume.

Technically feasible approaches and sound solutions to these problems are called for. New developments in the fields of material science, control and repair of concrete structures, advanced techniques in processing and recycling, and growing public awareness in conjunction with deposition of waste concrete and concrete-related materials have put another constraint and given a new direction to research in fracture and damage of concrete and rock.

FDCR-Conferences differ from other similar international conferences in their interdisciplinary character. Since many civil engineering structures are founded on rock one of the objects of these conferences is to bring together experts from the concrete and rock engineering communities to discuss problems of mutual interest. The conference participants share a common interest to learn from advances in the two areas of research, concrete mechanics and rock mechanics, and are fully aware of the public interest to find solutions to engineering problems thus avoiding or at least reducing the risk of failures to an acceptable limit.

It is the **FDCR**-Conferences which offer an international platform for this interdisciplinary cooperation and exchange of ideas and experience and promotes the cross-fertilization of ideas among scientists, engineers and practitioners.

Tremendous changes in politics and society at the dawn of a unified Europe and after the opening of the East have already created or are fostering a truly global field of science and technology. European partial integration and the creation of a single European market as well as the political and economic developments in various geographical regions offer a unique chance for an integrated interdisciplinary and global approach to failure prevention by intensified research in fracture and damage of concrete and rock. Relating thereto the excellent cooperation with RILEM and ESIS is kindly acknowledged.

FDCR-Conferences also aim at providing a comprehensive approach to important impacts and issues of fracture and fatigue failures at a local, national, regional and international level. It is mandatory for the enhancement of safe and reliable performance of engineering products that experts from the various disciplines intensify their cooperation and mutual understanding.

The scientific program was prepared by the International Committee:

Barr, B.	(UK)	Koyanagi, W.	(Japan)
Blight, G.E.	(RSA)	Li, V.C.	(USA)
Buchner, K.H.	(Austria)	Mang, H.	(Austria)
Carpinteri, A.	(Italy)	Mazars	(France)
Dougill, J.W.	(UK)	Mihashi, H.	(Japan)
Elfgren, L.	(Sweden)	Ouchterlony, F.	(Sweden)
Elices, M.	(Spain)	Rustan, A.	(Sweden)
Fourney, W.L.	(USA)	Shah, S.	(USA)
Hilsdorf, H.K.	(Germany)	van Mier, J.G.M.	(Netherlands)
Ingraffea, A.R.	(USA)	Widmann, R.	(Austria)
Karihaloo, B.L.	(Australia)	Wittmann, F.H.	(Switzerland)
Kasperkiewicz, J.	(Poland)	Zaitsev, Y.	(USSR)

More than 70 top researchers, engineers, practitioners, consultants and experts from other professional disciplines have contributed over 60 presentations on fracture modeling, mixed-mode fracture and damage, fatigue, demolition and recycling, fracture and damage of rock, various applications, anchorage engineering and innovative construction.

In particular, plenary lectures presented have addressed the areas of

* Nonlinear finite element analysis of reinforced and prestressed concrete structures
  H. Mang, Technical University, Vienna, Austria
* Modelling cracking in rocks and cementitious materials
  M. Elices, Universidad Politecnica, Madrid, Spain
* Micromechanics and localization in concrete and rock
  H. Horii, University of Tokyo, Japan
* Goals and barriers of recycling of concrete and masonry
  E. Lauritzen, DEMEX, Frederiksborg, Denmark
* Long-time tests of bonded anchors
  L. Elfgren, Luleå University of Technology, Sweden

The selection of papers gives an excellent and general state-of-the-art review of the many facets of research in the theory and applications of fracture and fatigue of concrete and rock.

Hopefully, the new results and experiences gained at the **FDCR**-Conferences will be taken into consideration and put to work in the rock and concrete engineering communities.

The Organizing Committee wishes to thank the following institutions and organisations for their assistance and cooperation during the preparation of this important meeting:

* Technical University Vienna
* The International Union of Testing and Research Laboratories for Materials and Structures (RILEM)
* WIFI of the Austrian Federal Economic Chamber (BWK)
* Austrian Society of Geomechanics (ÖGG = ISRM NG Austria)
* German Association for Materials Research and Testing (DVM)
* Swiss Society of Material Technology (SVMT)
* Austrian Association for Materials Testing (ÖVM)
* Austrian Federal Society of Engineers and Architects (ÖIAV)
* E&FN Spon (Chapman & Hall, London)

Vienna, November 1992 H.P.Rossmanith

PART ONE
FRACTURE MODELLING

1 MODELLING CRACKING IN ROCKS AND CEMENTITIOUS MATERIALS

M. ELICES, J. PLANAS and G.V. GUINEA
Departamento Ciencia de Materiales, Universidad Politécnica de Madrid, Madrid, Spain

Keywords: Concrete, Fracture Models, Cohesive Crack, Nonlocal Models, Softening Function, R curve, Equivalent Elastic Crack.

1 Introduction

Today, most experts accept that both the classical stress-strain formulations (plasticity and damage) and the classical linear elastic fracture mechanics (LEFM) are insufficient to deal with many practical structural problems involving fracture of cementitious materials and rocks. About 15 years ago, the pioneering works of Hillerborg on the single crack modelization and of Bazant on the band approach, led to the surge of the many models and approximations that emerged during these years. Most of the models, particularly the so called nonlocal, are very recent and there is not yet a comprehensive survey aimed at understanding their interrelation.

About 5 years ago, the authors contributed to the state of the art of RILEM Committee 90 with a chapter on material models (Elices and Planas, 1989) where they tried to display the essence of the models that were most extended at that time, with an aim tending to unify the different approaches rather than to strengthen the differences. The main conclusion was that Hillerborg's and Bazant's approaches could be conceived as approximations of each to the other, and, as far as symmetric modes were concerned, they could be considered as equivalent for engineering purposes. However, in that work the nonlocal approximations were not considered, and the treatment of equivalent elastic crack models was exceedingly brief.

Today things have evolved towards a more comprehensive understanding, and in this work we want to present a preliminary map of the modelling of fracture of concrete and quasi-brittle materials based on the relative position of the different models. To do so we divide the fracture models according to the dominant underlying theoretical framework: LEFM on one hand and classical continuum mechanics (using σ-ε relationships) on the other. The first group describes the material and the crack as in LEFM, but modifies the crack growth rules to take into account the nonlinear fracture behaviour: it is the group of the equivalent elastic crack models (EECM). The second group starts from a stress-strain formulation with softening, and modifies the classical formulation in such a way that strain localization into regions of measure zero is forbidden. In this paper we call them *regularized* models.

With this dichotomy, the cohesive crack model is somewhat transitional because it may be used with either LEFM or classical continuum mechanics as the background. Recent work (Planas, Elices and Guinea, 1993) has shown that the cohesive crack is a

Fracture and Damage of Concrete and Rock – FDCR-2.
Edited by H.P. Rossmanith. © 1993 Taylor & Francis. ISBN 0 419 18470 8.

particular case of a regularized model, and we shall include it in this group because of its wider scope.

The paper is made up of two main parts and a bridge. Section two contains the first part, which deals with *regularized models.* A very simple uniaxial model is first presented which contains the essence of the approach. It is shown how this model may be particularized to give (1) the (physically unacceptable) classical local theory; (2) general nonlocal average regularization; (3) an example of higher order gradient regularization; and (4) an example nonlocal regularization whose solution is a cohesive crack.

The bridging material is included in section 3, and is a summary of the essential properties of the *cohesive crack models* —the transitional model. It outlines the most usual hypotheses, global properties, computational methods currently available and recent experimental methods to estimate the softening function.

The fourth section deals with *effective elastic crack models.* They appear as an approximate procedure to solve cohesive crack problems, and bring a rational framework for methods based on crack growth resistance curves.

2 Regularized models

In this section we present a very simple uniaxial model, inspired in the recent work by de Borst *et al.*(1992), and show how it can be regularized. To be specific, only time-independent regularization is envisaged.

2.1 A uniaxial softening model in monotonic extension: local formulation

We consider the bar in Fig. 2.1, and assume that the uniaxial stress-strain curve of one of its material elements is that shown in Fig. 2.1a. We may represent this curve as a series coupling of an elastic element and an inelastic-softening element, as shown in the figure, so that the we write the total strain as

$$\varepsilon = \frac{\sigma}{E} + \varepsilon^{i} \tag{2.1}$$

The softening branch may be defined by expressing that for *monotonic* inelastic straining the stress is a decreasing function of ε^i:

$$\sigma = \phi(\varepsilon^i) \quad \text{for } \delta\varepsilon^i \geq 0 \tag{2.2}$$

where $\phi(\varepsilon^i)$ is the strain-softening function which may be independently plotted as in Fig. 2.1b. The stress at which the softening begins is the tensile strength f_t, and complete softening occurs when a critical strain ε_c is reached, as shown in the figure.

The energy dissipated in the complete breaking of a unit volume, γ_F, is the area under the stress-strain curve, which is easily shown to coincide with the area under the softening curve:

$$\gamma_F = \int_0^{\infty} \sigma \, d\varepsilon = \int_0^{\infty} \phi(\varepsilon^i) \, d\varepsilon^i \tag{2.3}$$

With this local model, a bifurcation takes place at the peak load (point *P* in Fig. 2.1). At this point, part of the specimen may continue straining (part *A,* say), while the remaining part of the specimen unloads (part *B*). The only condition is that equilibrium is satisfied, which in this case just means that the stress is uniform, and therefore the stress-strain states of parts A and B are respectively represented by points A and B in Figs. 2.1a and 2.1b. The width of the zone A is so far arbitrary; let us call it h.

To find out which is the physically probable value of h we just compute the total work required to break the specimen, and find its minimum (principle of minimum dissipation). Since no energy is dissipated in part B of the specimen, the fracture work W_F is just γ_F times the volume of part A:

$$W_F = \gamma_F \, S \, h \tag{2.4}$$

where S is the cross sectional area of the specimen.

Hence, the energy dissipated is minimum for $h = 0$, and thus, under the local hypothesis, the specimen will break with zero energy dissipation. Moreover, the inelastic displacement (contribution of ε^i to the total displacement) will also be zero. This means that the model predicts a zero-measure failure mode: no inelastic behaviour is measurable either as energy dissipation or as inelastic displacement. That is why regularization is essential.

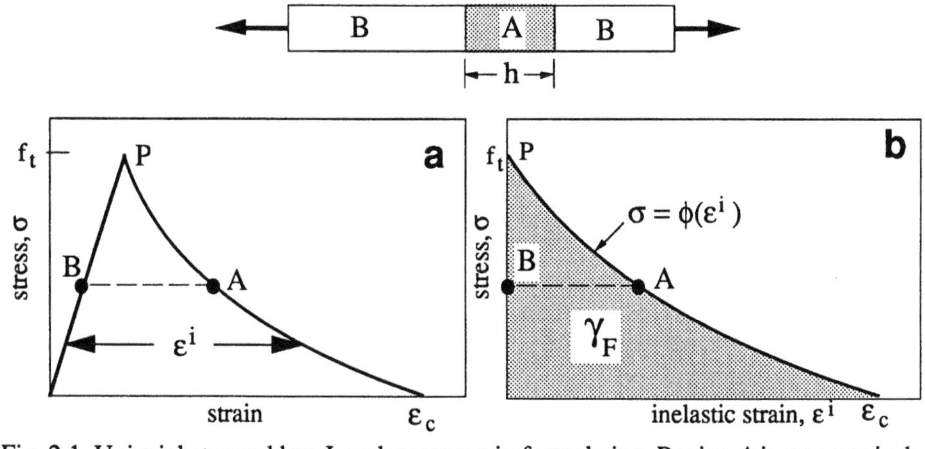

Fig. 2.1 Uniaxial stressed bar. Local stress-strain formulation. Region A is monotonicaly extended while regions B unload after peak.

2.2 Regularization technique

To regularize the local equations is, in our context, to introduce some conditions which forbid zero-measure solutions. Even better, a good regularization technique should make zero-measure solutions physically irrelevant: since they cannot be measured they must not influence any physical process; in that case the zero-measure solutions would be treated as avoidable discontinuities, avoided, and forgotten altogether.

The following presentation is a distillate of the most recent theories of regularization in which some inelastic variables are spatially coupled. In the early theories the coupled variables were ordinary variables (stress and strain) and they achieved regularization only with extreme pain, because "zero energy modes" appeared, i.e., the solutions included some strain distributions, usually alternate functions of arbitrary amplitude, which were physically unrealistic and which could not be excluded based on the model itself (see, e.g. Bazant, 1986). This problem was overcome a few years later when Bazant proposed to make nonlocal only irreversible variables, damage, cumulated plastic strain or the like (Bazant and Pijaudier-Cabot, 1987; Bazant and Lin, 1988; Pijaudier-Cabot and Bazant, 1987). The same idea of using a spatial regularization on irreversible variables within a gradient model (we prefer to call it differential for the reasons to follow) was used by de Borst *et al.* (1992). Now, it turns out that all these approaches may be unified in a single spatial regularization theory, and, moreover, a systematic treatment to explore the localization modes may be set. The following is the uniaxial formulation of the whole framework. Triaxial generalization is in progress.

A systematic way of regularizing is to define an intermediate nondimensional internal variable Ω, having a sense of damage (strength degradation), and postulate that it is determined by the spatial distribution of inelastic strain. Thus, Eq. (2.1) still holds, but Eq. (2.2) is substituted by the two equations

$$\sigma = \phi(\Omega) \quad \text{for } \delta\varepsilon^i \geq 0 \text{ and } \delta\Omega \geq 0 \tag{2.5}$$

$$\Omega(x) = \mathbf{F}[\varepsilon^i(x');x] \tag{2.6}$$

where now the softening function $\phi(\Omega)$ depends on Ω, and $\mathbf{F}[\varepsilon^i(x');x]$ denotes a spatial operator relating the distribution of inelastic strains to that of Ω. Here, $\phi(\Omega)$ plays exactly the same role as the local softening function and we plot it in Fig. 2.2. Its area still has the dimensions of energy per unit volume and we still call it γ_F.

Of course, when $\mathbf{F}$ is the identity operator we obtain the local formulation and no regularization is obtained. We may expect, therefore, that operators "close to" the identity operator will not regularize the problem. To our knowledge, no general analysis has been performed up to date to find the more general structure of the operators achieving regularization, so we will content ourselves with giving two examples: A high gradient (differential) regularization and a nonlocal (integral) regularization.

However, before doing that, we want to emphasize that the procedure is always the same as that used in the local approach:

(1) Analyze the bifurcation as before: part *A* of arbitrary width *h* softens while part *B* unloads;

(2) Impose the equilibrium condition: constant stress implies $\Omega = \Omega_A$ = constant in region *A* and $\Omega = 0$ in region *B*.

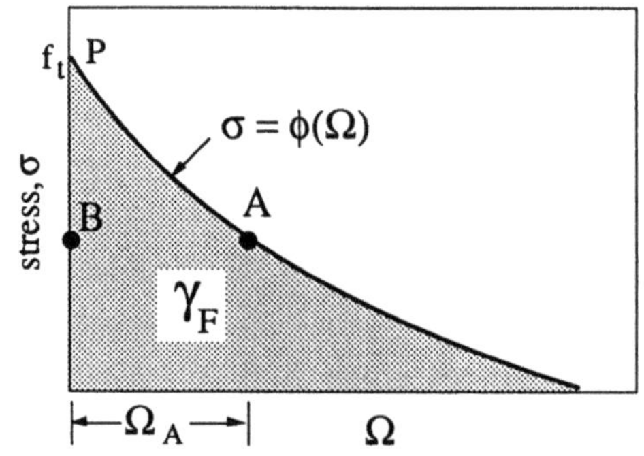

Fig. 2.2 Softening curve for the regularized model expressed as a function of the intermediate variable Ω.

(3) Substitute in Eq. (2.6) these conditions

$$\Omega(x) = \Omega_A \qquad \text{for } x \in A$$
$$\Omega(x) = 0 \qquad \text{for } x \in B \tag{2.7}$$

(4) Solve the resulting functional equation with the convenient boundary conditions and the essential condition that the inelastic strain is never decreasing, hence always positive everywhere:

$$\varepsilon^i \geq 0 \quad \text{everywhere} \tag{2.8}$$

If the regularization operator is properly designed, this conditions forces the width of the softening zone to be "quantified": only certain non-zero values of h allow the solution to satisfy the conditions.

(5) Among all the acceptable values of h, find the one for which fracture will take place with minimum energy dissipation.

Let us next outline two characteristic examples.

2.3 A high gradient (differential) model.

We apply the foregoing equations in the particular case where $\mathbb{F}$ is a (spatial) differential operator. We use a differential harmonic operator as the one used by de Borst *et al.* (1992) which, for convenience we write as:

$$\Omega(x) = \varepsilon^i + \left(\frac{\lambda}{2\pi}\right)^2 \frac{d^2\varepsilon^i(x)}{dx^2} \tag{2.9}$$

where λ is a material constant with dimensions of length which we call *internal length* and which will be related to the width of the softening localization. This differential equation is easily solved for the conditions (2.7). The solution is

$$\varepsilon^i(x) = \Omega_A + C_1 \cos\left(\frac{2\pi x}{\lambda} + \psi_1\right) \qquad \text{for } x \in A$$
$$\varepsilon^i(x) = C_2 \cos\left(\frac{2\pi x}{\lambda} + \psi_2\right) \qquad \text{for } x \in B \tag{2.10}$$

where C_1, C_2, ψ_1 and ψ_2 are integration constants. Condition (2.8) requires $C_2 = 0$ and $|C_1| \leq \Omega_A$. Taking the origin of x to coincide with the first interface between A and B (Fig 2.3) the continuity of the field at this interface requires $\psi_1 = 0$ and $C_1 = -\Omega_A$. Continuity at the second interface between regions A and B then requires

$$1 - \cos\left(\frac{2\pi h}{\lambda}\right) = 0 \quad \Rightarrow \quad h = n\lambda\,, \quad n = 1, 2, \ldots \tag{2.11}$$

which delivers the "quantization" condition. Fig. 2.3 shows the inelastic strain distribution for $n = 1$.

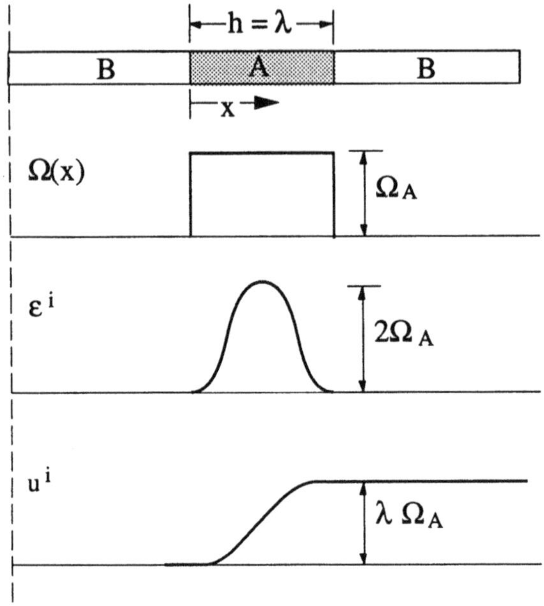

Fig. 2.3 Inelastic strain localization for harmonic regularization.

Now, the inelastic displacement contributed by the inelastic strain is:

$$u^i = \int_0^h \varepsilon^i(x)\,dx = \Omega_A \int_0^{n\lambda} \left[1 - \cos\left(\frac{2\pi x}{\lambda}\right)\right] dx = n\lambda\,\Omega_A \tag{2.12}$$

and the work of fracture is obtained as:

$$W_F = S \int_0^\infty \sigma\,du^i = n\,S\lambda \int_0^\infty \phi(\Omega_A)\,d\Omega_A = n\,S\,\lambda\,\gamma_F \tag{2.13}$$

and the minimum dissipation corresponds to $n = 1$, as expected. The fracture energy corresponding to this solution (energy per unit area of cross-section) is thus

$$G_F = \lambda\,\gamma_F \tag{2.14}$$

2.4 A nonlocal (integral) model.

Another case where the regularization is successful is that in which $\mathbf{F}$ is an integral equation. The authors have explored the simple constant weight averaging operator, where (2.6) is written as

$$\Omega(x) = \frac{1}{\lambda} \int_{-\lambda/2}^{\lambda/2} \varepsilon^i(x+s)\, ds \tag{2.15}$$

where λ is a material constant, which we call again the internal length.

Following the same systematic as before, the authors were able to deduce the following results, which are summarized in Fig. 2.4 (Planas, Elices and Guinea, 1993):

(1) The localization width h is "quantized" and the acceptable values of h are given, again, by

$$h = n\lambda, \qquad n = 1, 2, \ldots \tag{2.16}$$

(2) The valid inelastic strain distribution is an array of n Dirac's δ-distributions with spacing λ and intensity $\lambda\Omega_A$. The thermodynamically preferred mode is the one for n=1. In this case we have a δ–distribution centered in the softening band ($x = \lambda/2$, with the origin previously chosen):

$$\varepsilon^i(x) = \lambda\, \Omega_A\, \delta\left(x - \frac{\lambda}{2}\right) \tag{2.17}$$

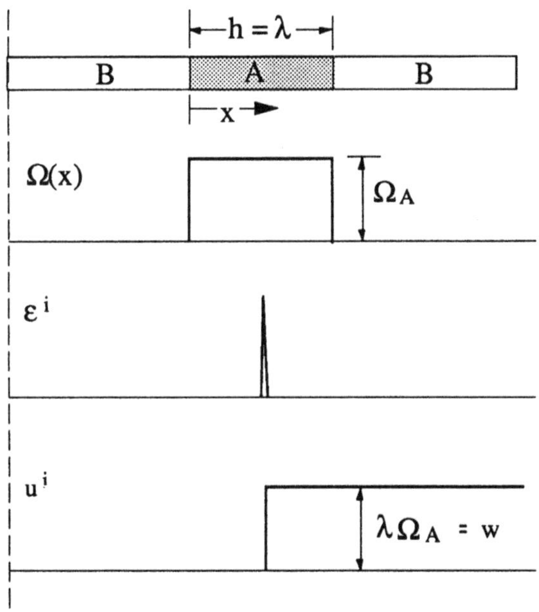

Fig. 2.4 Inelastic strain localization for integral regularization.

(3) The δ-function in inelastic strain produces a displacement jump which is interpreted as a stress-transferring crack. The displacement jump is the crack opening *w*, and its value is

$$w = \lambda\, \Omega_A \tag{2.18}$$

(4) The stress transferred for a crack opening *w* is obtained from (2.18) and the softening function:

$$\sigma = \phi(w/\lambda) = f(w) \tag{2.19}$$

(5) The surface fracture energy is related to γ_F by

$$G_F = \lambda\, \gamma_F \tag{2.20}$$

Therefore, as far as uniaxial models is concerned, the solution of the foregoing nonlocal model exactly coincides with a cohesive model as was developed by Hillerborg and will be used in the next sections. In this respect λ and Ω are eliminated from the equations and are not relevant for analyzing a single crack problem. However, it has not yet been investigated whether this holds true for triaxial generalizations.

If the average inelastic strain Ω is considered an observable strain, then this nonlocal models leads to a band model in Bazant's sense (constant "strain" across the band). However, the authors think that in the uniaxial framework the only really measurable magnitude is the inelastic strain ε^i, or, rather, its associated inelastic displacement *w*. To determine Ω given *w*, λ must be previously determined and there is no way to do that in the pure uniaxial case.

2.5 A tentative classification for uniaxial models

The essential feature of the foregoing spatial regularization scheme is that the spatial coupling functional $\mathbb{F}$ relates two irreversible variables $[\Omega(x), \varepsilon^i(x')]$, i. e., two non-negative and never-decreasing variables. This is what avoids the appearance of zero-energy modes that occurred in the early formulations of nonlocal models, where spatial coupling was applied to ordinary variables which can be positive or negative.

Fig. 2.5 displays a tentative classification of the uniaxial models. Two points are worth mentioning. First, we have not been able to find a particular regularization operator delivering as a solution a crack band in the restricted sense of Bazant, in which the observable strain is uniform throughout the band thickness. Differential regularization leads to a band-like solution, although the (observable) strain is not uniform across the band. The nonlocal model analyzed leads to a cohesive crack model rather than to a band model. However, from a practical point of view, they might be approximated by a band approach without much error. That is why, for the time being, we draw the crack band as an overlapping approach which cannot be obtained as a particular case of spatial regularization in our sense.

The second point refers to the usual claim that the differential harmonic regularization may be considered as an approximation to the integral formulation. This assertion is mathematically correct but difficult to apply in practice. The reason is that to obtain that relation, what one does is to expand in series the subintegral expression and then neglect all but the two first terms: these two terms are indeed a harmonic operator. However, it may be shown that for this approximation to be accurate it is necessary that the inelastic strain varies slowly over the integration interval. This may be valid in models where mild localization takes place. But in the kind of models where strong localization occurs, all the inelastic strain takes place over a zone of the same size as the integration interval, and the variation is so fast that the harmonic approximation is quite inaccurate. Therefore, in general, one cannot accurately approximate an integral operator by a harmonic one. That is why we clearly differentiate in Fig. 2.5 differential regularization from integral regularization.

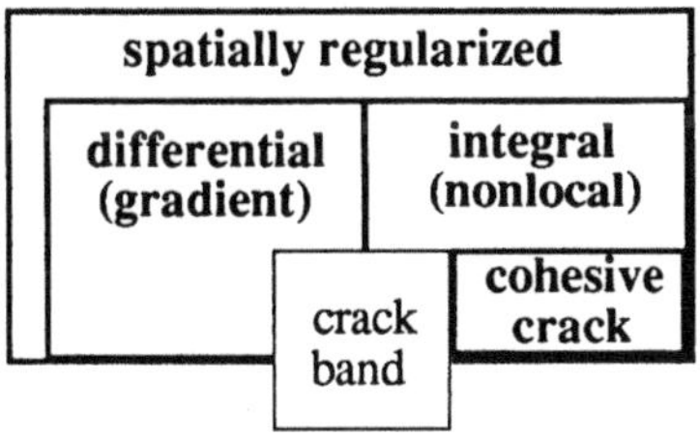

Fig. 2.5 Tentative classification of uniaxial spatially regularized models

3 Cohesive cracks

As pointed out in the introduction, cohesive crack models are transitional models which may be approached from three theoretical frameworks:

(1) As seen in the previous section they are a particular case of continuum mechanics formulation

(2) they can be set as a standalone model based on the interpretation of a tensile test, as done by Hillerborg (Hillerborg *et al.* 1976)

(3) and in their simplest formulation (linear elastic behaviour of the material outside the crack) they may be viewed as a modification of LEFM set to eliminate the stress singularity (as done by Barenblatt, 1962) or represent some kind of nonlinear material behaviour (as used, partly, by Dugdale, 1960).

A review of some of the more relevant properties of this models was done by the authors (Elices and Planas, 1989). A revised outline is given next, including new experimental methods to determine the softening curve, and a summary on numerical procedures.

3.1 Definition of a simple cohesive crack

In the traditional version, a cohesive crack model must be defined by the properties of the bulk material, the crack initiation condition and the crack evolution function. In the most usual version the assumptions regarding these aspects are much simplified, mostly to simplify the computations and to exploit well stablished methods in LEFM. However, this formulation is by no means the only one possible. Extensions of it will be outlined in a forthcoming section. The following are the most usual assumptions:

(1) The bulk material behaviour follows a linear elastic and isotropic stress-strain relationship, with elastic modulus E and Poisson ratio ν.

(2) The crack initiates at a point when the maximum principal stress σ^I at that point reaches the tensile strength f_t. Moreover, the crack forms normal to the direction of the major principal stress.

(3) After its formation, the crack opens while transferring stress from one face to another. The stress transferred, the *cohesive stress*, is assumed to be determined by the crack opening displacement history. For monotonic pure opening the stress transferred is normal to the crack faces and is a unique function of the crack opening:

$$\sigma = f(w) \tag{3.1}$$

The function f(w) is called the softening function –or softening curve– and it is a non-increasing function of the crack opening. It is the essential characteristic of the cohesive crack.

3.2 On the softening curve

The softening curve is, as already stated, the essential property of a cohesive crack. It has to be experimentally determined for a given concrete and this may be done, in principle, using stable tensile tests where a single crack is formed and opened keeping its faces parallel, as done by Petersson (1981). However, this kind of test is extremely difficult to perform, and in many cases its interpretation is not straightforward (Reinhardt *et al.*, 1986; van Mier, 1986). This is so mainly because the test tends to shift to asymmetric modes of failure, making it very difficult to achieve a crack of uniform opening as required. This is why most of the procedures to infer the softening function rely on indirect methods based on parametric fitting of a preselected family of softening functions, defined analytically. These procedures will be dealt with in a later section; here we analyze the essential aspects of an arbitrary softening function.

A softening function has two essential characteristics: the tensile strength f_t, which is, by definition, the stress transferred at the incipient opening; and the fracture energy G_F, which is the work supply required to bring completely apart the two faces of a unit surface of crack. Graphically speaking the fracture energy is the area under the softening curve, as depicted in Fig. 3.1. The analytical expressions are

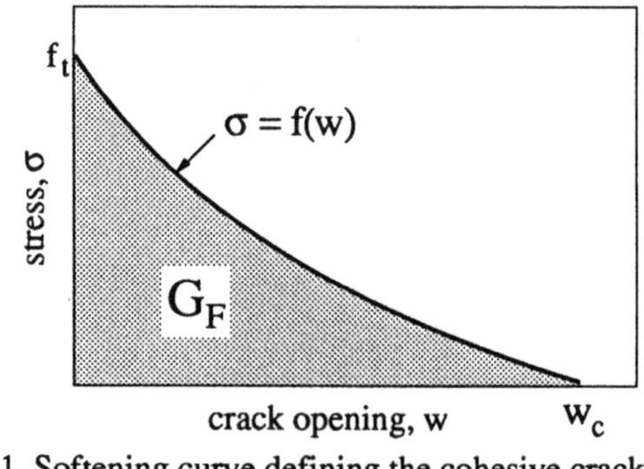

Fig. 3.1 Softening curve defining the cohesive crack model

$$f_t = f(0), \qquad G_F = \int_0^\infty \sigma \, dw = \int_0^\infty f(w) \, dw \tag{3.2}$$

These two parameters are in principle measurable in experiments without any explicit hypotheses about the shape of the softening curve.

Another characteristic property of the softening curve is the critical crack opening w_c, which is the crack opening for which the stress transferred becomes zero. The essential difference from the f_t and G_F is that its determination is very elusive. If direct tension tests are used, its value may be very dependent on the resolution of the measurement, and in indirect tests its determination relies on the assumed shape of the softening curve.

Apart from the foregoing parameters, the shape of the softening curve itself has an important influence on the predicted structural behaviour, and various shapes (analytical expressions) have been used in the literature to make computational simulations. One of the usual expedients used to present results comparing different softening shapes is to make a dimensionless representation of the softening function and write

$$\sigma = f_t \, f^*\!\left(\frac{f_t}{G_F} \, w\right) \quad \text{or} \quad \sigma^* = f^*(w^*) \tag{3.3}$$

where dimensionless stress and crack opening (σ^*, w^*) are defined as

$$\sigma^* = \frac{\sigma}{f_t}, \qquad w^* = \frac{f_t}{G_F} \, w \tag{3.4}$$

and $f^*()$ is a dimensionless function with unit vertical intercept and unit area:

$$f^*(0) = 1 \qquad \text{and} \qquad \int_0^\infty f^*(w^*) \, dw^* = 1 \tag{3.5}$$

The function $f^*(w^*)$ defines the shape of the softening. Once this function is fixed, the shape is fixed in the precise sense that the softening curves of all the materials with that shape can be made to coincide just by a change of scales, i. e. the softening curves are affine. Alternatively we could say that all the geometrical horizontal and vertical ratios are fixed.

The softening curves of a given shape constitute, in this restricted sense, a bi-parametric family of curves, the free parameters being f_t and G_F. Obvious examples of such curves are the rectangular softening, the linear softening and the exponential softening (Fig. 3.2). A widely used curve with this property is Petersson's bilinear softening, in which the relative position of the kink point is fixed. In dimensionless coordinates, the kink point (σ^*_k, w^*_k) is fixed at (1/3, 4/5) as shown in Fig. 3.2.

The most usual way of handling a fracture problem is to select, more or less a priori, one of those biparametric shapes for use in the numerical analysis. Then it is relatively simple to make predictions for a particular concrete, because only the tensile strength and the fracture energy have to be determined experimentally, and there are some international recommendations for the determination of these parameters.

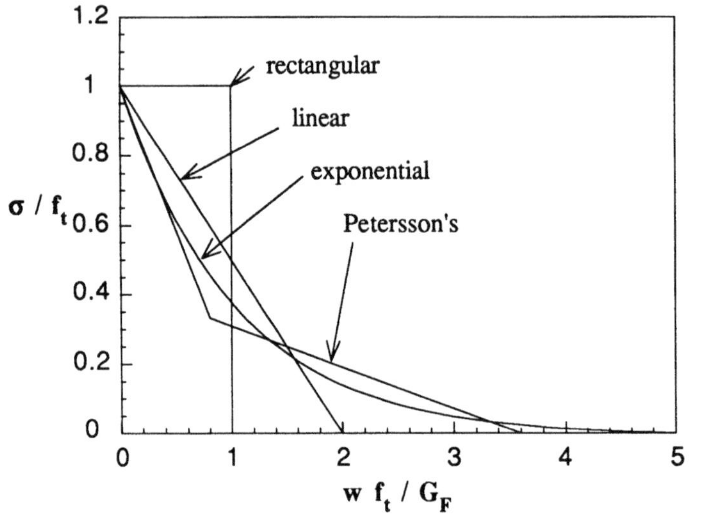

Fig. 3.2 Various widely used softening shapes

The selection of a particular analytical expression for the softening may depend very much on the kind of analysis one requires. If only trends are required, even the roughest rectangular softening may give a good idea of how the various parameters defining a particular problem influence the solution. It has the advantage that there are many analytical solutions in the literature, because it can be solved as a particular case of the Dugdale model (the Dugdale model includes constant cohesive stresses, but does not contain softening, i. e., the stress cut off is not part of the Dugdale formulation).

Linear softening has been extensively used in the literature. It has the advantage of making some equations linear, but the overall system is still non-linear, so that there is not a great gain in using this approximation for a particular problem. However, it is a good reference for comparison with other models.

Petersson's bilinear softening was designed to give a better fit of the softening curves observed in tensile tests. It is very much some kind of standard that has been used again and again in the literature to simulate the softening of concrete. However, it turns out that for some concretes, Petersson's curve may have too short a tail, as the comparison with the experimental results of Rokugo *et al.* (1989) show in Fig. 3.3. More convenient experimentally inferred curves may be obtained following the methods described in the forthcoming section 3.4.

3.3 Computational aspects

Assuming that a certain analytic expression for the softening curve has been selected the computational analysis of the fracture of any plain concrete structure should be possible. However, carrying out the corresponding computations is very difficult and many different expedients have been used.

One can consider three groups of computational approaches: (1) general purpose finite element codes, able, in principle, to analyze any structure subjected to any loading; (2) special computational strategies applicable only to cases where the cohesive crack path is known in advance; and (3) simplified approaches making use of more or less justified strong simplifications.

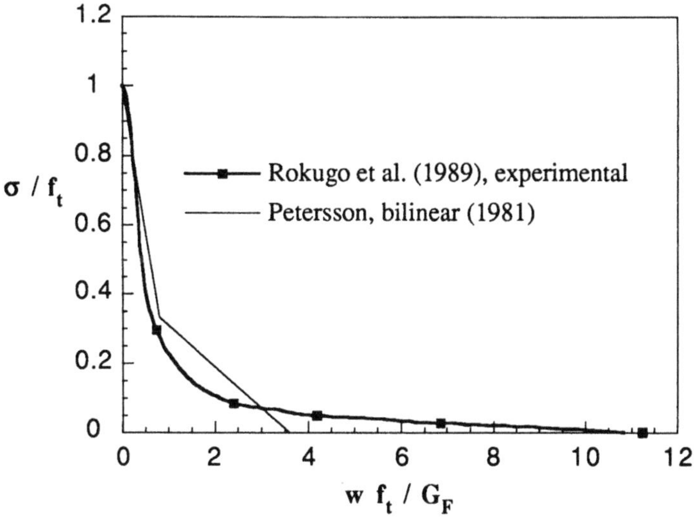

Fig. 3.3 Comparison of experimentally determined softening curve by Rokugo *et al.* (1989) and Petersson's bilinear softening function.

The first group is very scarce. The most extended approaches actually do not use a cohesive crack, but a smeared version of it similar to a crack band approach (mainly because its FE implementation is simpler). There are many codes in use, but the potential users must be aware of some ill-behaved characteristics when the crack propagates skew to the mesh. The analysis of this problem is outside the scope of this paper, and the reader is referred to the work of Rots (1992) and references therein. An interesting exception to the general trend is that led by Ingraffea at Cornell, who developed several versions of a finite element code with very powerful remeshing capacities, which is able to interactively follow the cohesive crack extension in 2–dimensional problems (for details, see Bittencourt *et al.*, 1992, and references therein).

The second group is essentially formed by special procedures focussed on solving symmetric cracking (mode I) which is the kind mainly found in the laboratory. There are two basic approaches in this group. One is to use a commercial finite element package, and situate interface elements with softening (usually of the strut or spring type) along the crack path. This kind of interface elements is available in most good finite element packages. The advantages of this approach are that it provides all the information about the stress, strains and displacement fields, and that it is applicable to any symmetric geometry. The main drawback is that if one wants an accurate computation, i. e., a fine mesh, the time to simulate a single test may be quite long. To get fast results (about 100 times faster) a special purpose program may be developed which is a special version of a boundary elements formulation.

The approach used by the authors is a modification of the influence method developed by Petersson (1981) in which a number of elastic solutions for the geometry used are superimposed to get a nonlinear solution for the crack propagation. Roughly speaking, if one has N elements along the crack path, one has to solve N+1 linear cases from which an "influence matrix" is constructed and from it a crack advance step is reduced to solve a problem with at most N degrees of freedom. This is to be compared with the previous approach where at each increment one has of the order of N^2 degrees of freedom. The authors have modified the Petersson algorithm to improve the performance and to accept nonlinear softening of any kind one may wish (Planas and Elices, 1991). A user-friendly program called SplittingLab has been developed by the authors and their collaborators. It runs on Macintosh computers and may simulate a complete fracture test in laboratory in only a few minutes. The program is menu driven and the most widely used specimen geometries and softening functions are selectable from a menu at a single mouse-click. A δ-version is available from the authors. All the solutions of cohesive crack growth included in this paper have been computed using this program.

The last group is that of special methods which reduce a problem of a large number of degrees of freedom to a minimum by using certain special simplifications. One simplification used is to assume that the crack opening profile is linear (Llorca and Elices 1990, Foote *et al.* 1986, Cotterel and Mai 1987). One entire family of approximations in which only two degrees of freedom are retained is based on the *equivalent elastic crack* concept, in which all the fields are approximated by linear elastic fields for a certain effective crack length. This kind of approximation may be quite systematized and is explained with plenty of details in the next section. The importance of this approximation is that it has been used in a completely different framework, and constitutes the link between the classical R-curve approach and the continuum mechanics based approaches, thus providing a global framework for the analysis and comparison of the different models.

3.4 Experimental determination of the softening curve

As pointed out in section 3.2, the experimental determination of the softening curve is far from easy. Very often, the softening curve is only partially determined. To do so one selects a convenient shape for the softening curve and, using (explicitly or implicitly) the nondimensional formalism explained in section 3.2, determines the tensile strength f_t and the fracture energy G_F.

Ideally, the tensile strength should be measured in direct tensile tests. However, the usual practice is to make an estimate of the tensile strength using the cylinder splitting strength (Brazilian test), which should give values within 20 percent of the actual f_t.

The fracture energy can, in principle, be determined easily using the procedure described in the RILEM draft recommendation (1985). However, it appears that direct application of such procedure delivers measured values of G_F which are strongly size-dependent. The authors have been working extensively on this problem and have shown that a number of spurious energy dissipation sources may be present during one of those tests: hysteresis of the testing arrangement, dissipation in the bulk of the specimen, dissipation by crushing at the supports, and neglect of the final portion of the load-displacement curve. When all these factors are properly taken into account, the size effect of the experimental values is strongly decreased (Guinea *et al.* 1992, Planas *et al.* 1992a, Elices *et al.* 1992).

To get information about the shape of the softening curve without recourse to direct tension tests, Wittmann and co-workers introduced an indirect fitting procedure based on making a parametrization of the softening curve and determining these parameters by best fitting to measured load-deflection curves of stable tests on notched specimens (Roelfstra and Wittmann, 1986; Wittmann *et al.*, 1987, 1988; Mihashi, 1992). To do so they used an unrestricted bilinear function, which has 4 independent parameters (4 degrees of freedom is the general bilinear softening which is a 4-parametric family of softening curves). They use an optimization algorithm to determine the four parameters, so that f_t and G_F are obtained as a part of the fitting procedure not from independent measurements.

In the same spirit, but with a very simple fitting procedure as an objective, the authors have developed a so called general bilinear fitting (GBF) procedure (Guinea, Planas and

Elices, 1993). An outline of the GBF method is given by Planas, Guinea and Elices (1992b).

In the GBF, four parameters completely defining the bilinear softening curve are determined quasi independently in four sequential operations. Fig. 3.4a illustrates the bilinear curve and the four parameters which are determined:

(1) If direct tension test results are not readily available, the tensile strength f_t is independently estimated as the cylinder splitting strength.

(2) The fracture energy G_F is evaluated by the total work of fracture method, using stable tests on notched specimens, with the modifications to the RILEM recommendation previously developed by the authors (Guinea *et al.* 1992, Planas *et al.* 1992a, Elices *et al.* 1992).

(3) The abscissa $\bar{w}$ of the center of gravity of the area under the softening curve is obtained by fitting a theoretical expression to the far end of the measured load-deflection (P-u) curve. The only special requisite is that the complete P-u curve must be experimentally obtained, which requires *complete specimen-weight compensation.*

(4) The horizontal intercept w_1 of the initial segment of the softening curve is obtained with a semi-graphical procedure which consists essentially in stating that for small size specimens, the peak load predicted by the bilinear softening curve must coincide with that predicted by the linear softening shown as a dashed line in Fig. 3.4a

Once the four parameters f_t, G_F, $\bar{w}$ and w_1 are determined, the bilinear softening is completely defined and the result may be used to compare numerical and experimental curves to check the fitting. Fig. 3.4b shows an example for a notched three point bend test (TPB) where it appears that the general bilinear softening obtained by the GBF procedure better reproduces the experimental curve than a Petersson's bilinear curve with identical f_t and G_F. Work is in process to extend this bilinear fitting to a 4-parameter smooth curvilinear softening.

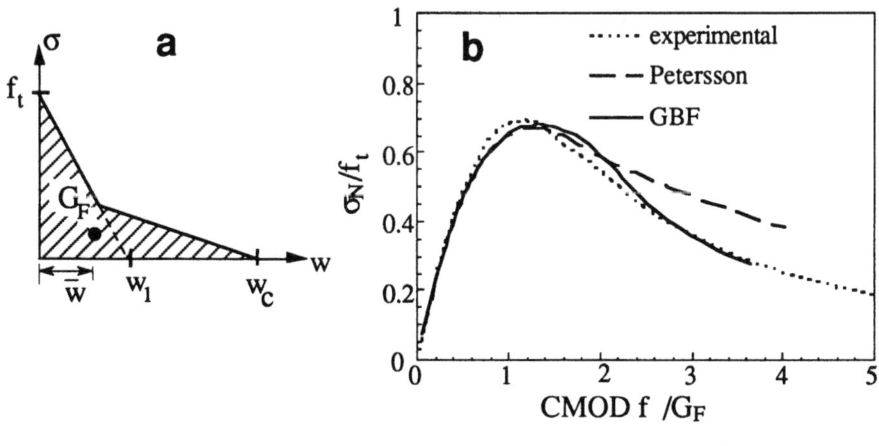

Fig. 3.4 (a) General bilinear softening curve. (b) Comparison of experimental load-CMOD curves of notched three point bend tests with predictions using Petersson's softening and general bilinear from GBF (Planas, Guinea and Elices 1992b).

3.5 Extensions of the cohesive crack

The foregoing formulation of the cohesive crack is the simplest version of it. But it is by no means the only one possible. There are at least three possible extensions that have been explored very little, mainly because they make the computations much more complex: (1) accept nonelastic behaviour in the material around the crack; (2) introduce dependence on triaxiality; and (3) introduce a fully consistent mixed-mode formulation.

The first issue was considered from the very beginning by Hillerborg as a very realistic possibility, and it has been implemented by the authors using for the material an elastoplastic hardening model with a Rankine criterion and associated flow rule (Planas, Elices and Guinea, 1992a). The computations using a perturbation approximate method showed that the influence of the inelastic strain around the cohesive crack was small. When expressed as dissipated energy, the influence was below 5 percent for notched specimen sizes. However its influence has not been investigated yet for unnotched specimens, where it is likely to be greater.

To the authors' knowledge, the second issue has never been considered at a *theoretical* level, and is one of the most frequently heard criticisms of the cohesive crack. However, there is no basic objection to introducing a dependence of the softening curve on the triaxial state. This may require further hypotheses regarding the loading-unloading behaviour, and may lead to a model which may be difficult to verify experimentally, because the only data so far available are about the influence on the strength, not on the post-peak (stable crack opening) behaviour. The first exception known to the authors is the research by Tschegg, Kreuzer and Zelezny (1992) on wedge-split cubes subjected to compression parallel to the crack plane. However their results have not yet led to a theoretical formulation, and further work in this direction is needed.

The last issue is a must. A complete vectorial formulation is required before general purpose programs may be confidently used. Even though there is at present a general agreement that the fracture initiates in pure opening mode (mode I, no shear), it is obvious that crack sliding displacement may take place after the crack opens and before complete fracture. The testing procedures in this field are tremendously difficult, but they go ahead of the theoretical formulation (van Mier *et al.*, 1991; Hassanzadeh, 1992). To the authors' knowledge a complete cohesive model relating the traction vector on the crack faces to the history of crack displacement vector is still lacking, as it is a sound theoretical interpretation of the experimental data already available.

4 The equivalent elastic crack models (EECM)

4.1 Introduction and objectives

The cohesive crack concept was proven a useful tool for modelling cracking of a large family of quasi-brittle materials. Nevertheless, its implementation is not straightforward, as is shown in the previous section. In this respect, the equivalent elastic crack concept emerges as a simplified procedure to evaluate crack growth in cohesive materials. Like all simplifications, it has its drawbacks and limitations that are worth examining.

There is another reason for considering the equivalent elastic crack. The use of crack growth resistance curves *(R-Δa)* to predict the behaviour of cracked specimens is a well established practice for ceramic and cementitious materials and for quasi-brittle materials. Assuming that the cohesive crack model can be applied to these materials, it will be shown that the use of R-Δa curves is equivalent to setting up an *elastic equivalence* between the actual (cohesive) specimen and a fictitious (linear elastic) one.

The equivalent elastic crack greatly simplifies the computations by shifting from a cohesive material (and hence, non lineal structural analysis) to a linear elastic material (which only requires linear analysis). But as already mentioned, this simplification is not

always acceptable and a discussion of its suitability and range of validity is the objective of this section.

A further reason for exploring how far a cohesive crack can be treated as a linear elastic one, is the use of the fracture toughness concept —K_{IC}— for these quasi-brittle materials. When design is based on K_{IC}, it is quite obvious that this value should be a *material property* in order to predict the behaviour of a component from a value extracted from a particular specimen tested in a laboratory. By using this parameter one is implicitly assuming that the specimen behaves in a linear elastic way because K_{IC} is a LEFM concept and, moreover, it is evaluated using expressions valid only for LEFM. Consequently, when these materials are characterized by K_{IC}, the validity and significance of this concept is by no means clear unless an acceptable degree of linear elastic behaviour is proven.

All the above mentioned procedures, the use of R-Δa curves and the validation of the fracture toughness parameters —concepts widely used for these quasi brittle materials— deserve a unified treatment to ascertain how far they are from linearity and to ensure a proper use of these parameters. This section reviews the concept of the *equivalent elastic crack* for cohesive materials —previously developed by the authors (Planas and Elices, 1990, 1992, 1993; Elices and Planas, 1993)— starting with the more familiar equivalences based on load i.e. P-Y equivalences and then the general X-Y equivalences. Finally, some comments and general trends for these equivalences are summarized. It should be remarked that all this section is valid only for materials that can be modelled with *cohesive cracks*, hence experimental values should coincide with numerical computations using a cohesive model. In what follows, "experimental" results and computed results from a cohesive model will be indistinguishable.

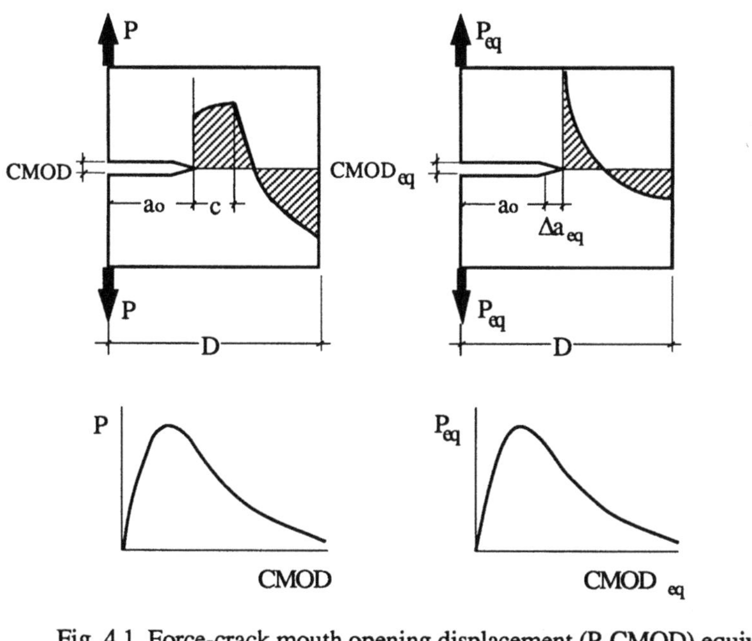

Fig. 4.1 Force-crack mouth opening displacement (P-CMOD) equivalence

4.2 Load-Y equivalences

The concept of equivalence is best illustrated by using a particular example; let us choose the force-crack mouth opening displacement equivalence or *P-CMOD* equivalence. As shown in Fig. 4.1, two geometrically identical cracked samples are loaded under *CMOD* control. One sample is made with a *cohesive material* and the other (the equivalent) with a *linear elastic material.* The measured loads in the two samples —P and P_{eq}— for the *same* crack mouth opening displacement *CMOD* will be in general different, but we can force the loads to match each other ($P = P_{eq}$) —at each value of *CMOD*— by choosing a suitable *equivalent crack length* a_{eq} and a suitable *equivalent crack growth resistance* R_{eq}.

In this procedure we force both samples to exhibit the same *P-CMOD* behaviour, but generally the equivalence ends here; stress fields, for example, need not be the same, as shown in Fig. 4.1. The price paid for the equivalence is that the linear elastic material has not a constant crack growth resistance (as is G_F for the cohesive case) but a changing value with crack length, or an *R-Δa* curve, as is usually known. Moreover —and this is worse— this *R-Δa* curve is not a material property; it depends on the geometry and specimen size.

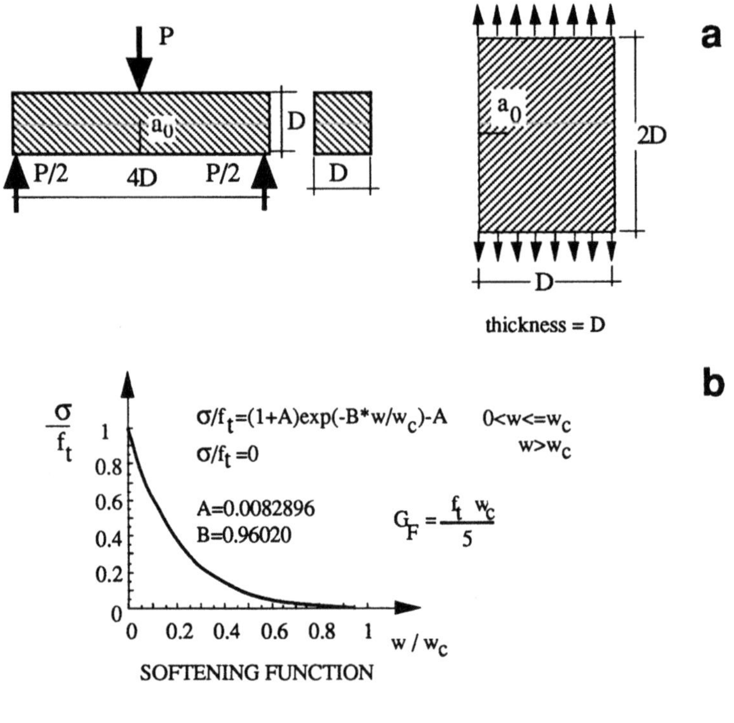

Fig. 4.2 Problem definition: (a) Reference beam and notched panel geometries ; (b) Softening function

For computing the R-Δa curve we proceed as follows: First, the equivalent elastic crack length is deduced. From the measured P and $CMOD$ values in the cohesive sample, the *equivalent elastic crack length* a_{eq} is derived from the expression of the compliance for the equivalent elastic sample

$$C_{eq}(a_{eq}) = \frac{CMOD}{P} \tag{4.1}$$

Then, the corresponding stress intensity factor is computed according to

$$K_{Ieq} = \frac{P_{eq}}{B\sqrt{D}} S\left(\frac{a_{eq}}{D}\right) \tag{4.2}$$

where B is the specimen thickness, D one of its characteristic in-plane dimensions and $S(a/D)$ is the geometrical shape factor. Finally, the R-Δa curve comes from the well known relation

$$R(\Delta a_{eq}) = \frac{1}{E'}\left[K_{Ieq}(P_{eq}, a_0+\Delta a_{eq})\right]^2 \tag{4.3}$$

where the equivalent crack was written explicitly as $a_{eq} = a_0 + \Delta a_{eq}$ in order to remark the dependence on the crack increment, and E' is the generalized elastic modulus. Notice that this R-Δa curve is size and specimen geometry *dependent* because of the implicit size and geometry dependence of both members of Eq. (4.1).

This procedure and the use of the R-Δa curve to predict the behaviour of other specimens are exemplified in the following exercise: Starting with measurements of *P-CMOD* done in a three point notched beam (TPB) made of a cohesive material, ascertain the *P-CMOD* behaviour of a single edge notched panel in tension (SENT) made of the same cohesive material by using the *P-CMOD* equivalence, or, in other words, using the R-Δa_{eq} curve corresponding to the TPB sample. Fig. 4.2a shows the geometrical definition of the specimens.

In this example the "measured" values were computed using a cohesive material with a quasi-exponential softening as defined in Fig. 4.2b. The resulting *P-CMOD* curve is shown in Fig. 4.3a for $D = l_{ch}$ (where $l_{ch} = EG_F/f_t^2$) and $a_0 = 0.5\ D$. The resistance curve, R-$\Delta a^{P\text{-}CMOD}$, of the equivalent elastic beam is depicted in Fig. 4.3b. This R-Δa curve is then used to predict the *P-CMOD* curves for the SENT specimens. Fig. 4.4a shows the results for two different sizes and notch-to-depth ratios. For $D= l_{ch}$ and $a_0= 0.5\ D$ the prediction is very close to the "experimental" result (obtained by using the cohesive model). The reason for this agreement is that the R-curves of the TPB and SENT specimens are very similar for the sizes and relative notch depths considered, as shown in Fig. 4.4b. However when a SENT specimen of sharply different size and relative notch depth is considered ($D=3\ l_{ch}$, $a_0= 0.1D$), the agreement is not so good (Fig. 4.4a). This is so because as Fig. 4.4b shows, in this case the R-curve based on a *P-CMOD* equivalence is sharply different from those for the specimens previously considered.

Moreover, when the relative ligament $(D-a_0)/l_{ch}$ is larger than the reference one —as happens with this second example— the equivalent crack stops after having advanced a distance equal to the length of the reference ligament. This unphysical behaviour is because at this length the equivalent crack feels an infinite resistance and can no longer propagate. (The R curve tends to infinity when the crack approaches to the free surface). This behaviour is shown in Fig. 4.4a and need not be taken into consideration for practical purposes.

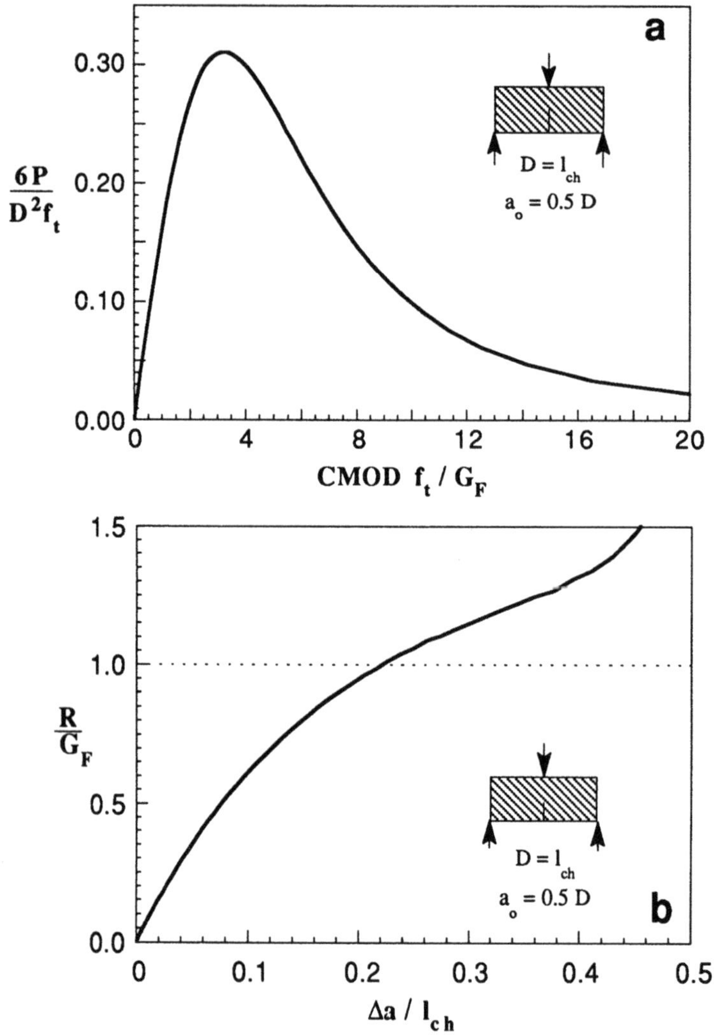

Fig. 4.3 Reference beam: (a) Measured P-CMOD values; (b) Corresponding R-Δa curve (for P-CMOD equivalence)

For large specimens, the linear elastic equivalence may be useful for the following reasons: In previous papers (Planas and Elices, 1990, 1992, 1993), the authors have shown that, for cohesive materials, as the specimen size increases the crack resistance curve tends towards an asymptotic curve that is independent of the specimen geometry (and of size, of course) and depends solely on the shape of the softening function. In particular, for the quasi-exponential softening chosen for these examples, the asymptotic R-Δa curve is shown in Fig. 4.5a. The tangency point of the rising R curve with the horizontal branch ($R/G_F = 1$) happens at $\Delta a_{c\infty}{}^{P\text{-}CMOD} = 2.48\, l_{ch}$ (approx.) for this particular softening (Planas and Elices, 1990, 1992). Here, $\Delta a_{c\infty}{}^{P\text{-}CMOD}$ is the equivalent crack extension for infinite specimens at peak load. Also shown in the same figure are two R-Δa curves corresponding to two large specimens of quite different geometry; a notched beam TPB ($D = 6\, l_{ch}$ and $a_0 = D/6$) and a notched panel SENT ($D = 6\, l_{ch}$ and $a_0 = 0.1\, D$). It is easy to realize that both curves are very similar (except at the very end) and tend to merge with the asymptotic one.

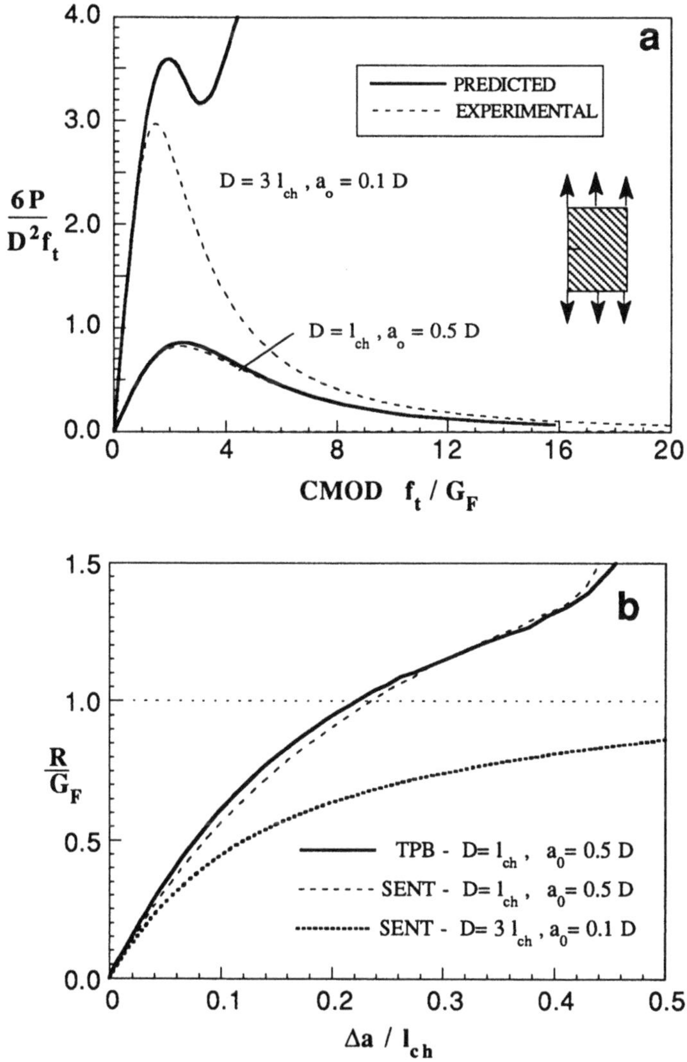

Fig. 4.4 Predictions for notched panels using a P-CMOD equivalence: (a) P-CMOD curves for the SENT specimens; (b) Reference R-Δa curve (TPB) and corresponding R-Δa curves of SENT specimens

Results of *P-CMOD* curves for the SENT panel are shown in Fig. 4.5b; these results came from the reference beam (TPB) and from the asymptotic *R* curve. Also, experimental values are drawn for comparison. These results show that it is possible to compute accurate values of the maximum loads and of the shape of the *P-CMOD* curve using the equivalent elastic crack concept when samples are large enough.

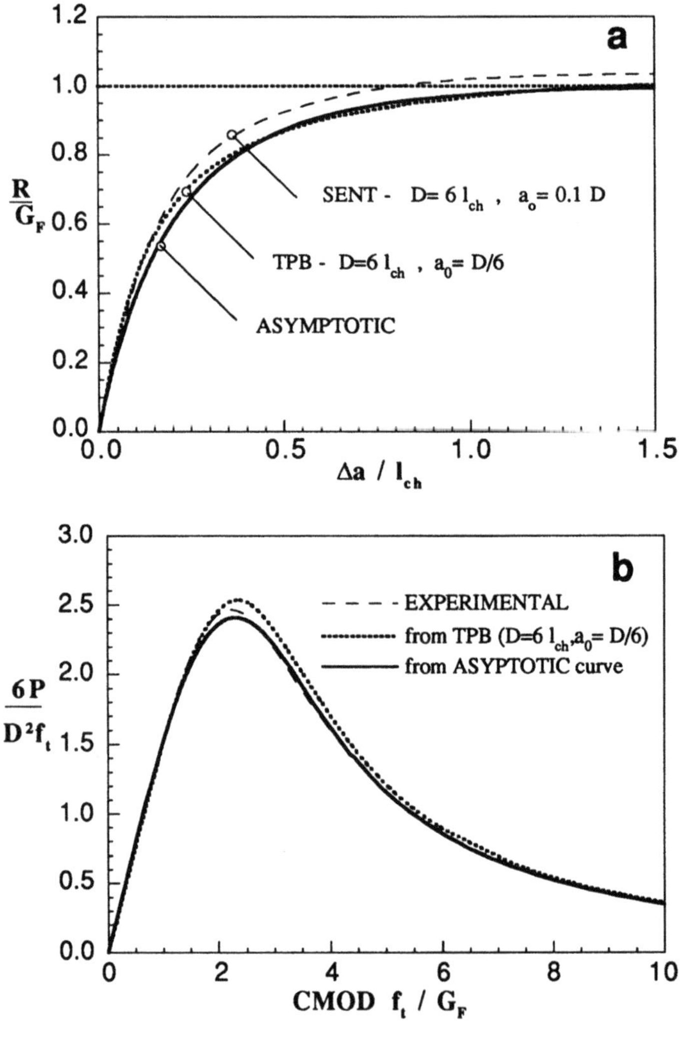

Fig.4.5 Predictions for large notched panels, using a P-CMOD equivalence: (a) R-Δa curves for large specimens; (b) P-CMOD curves for a SENT specimen predicted from the TPB R-Δa curve, from the asymptotic R-Δa curve and experimental values

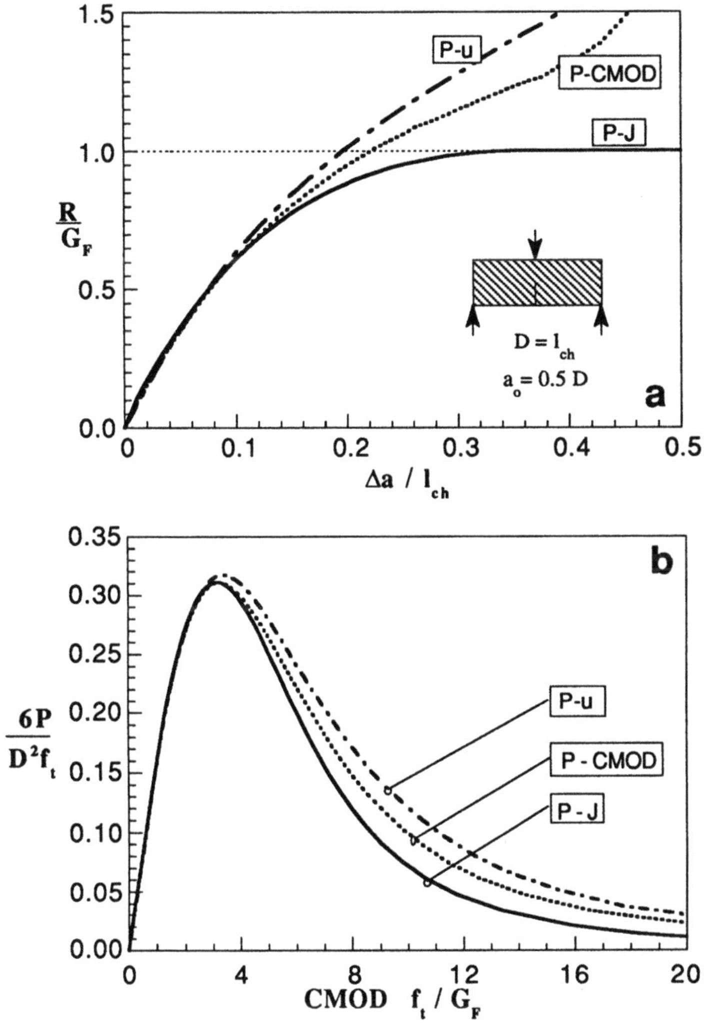

Fig. 4.6 Comparison of different equivalences: (a) R-Δa curves for TPB ($D=l_{ch}$, $a_0=0.5$ D) for three equivalences; (b) P-CMOD curves for the three equivalences. (Notice that , by definition, the P-CMOD curve "predicted" by the P-CMOD equivalence exactly coincides with the "experimental" P-CMOD curve)

So far, only the *P-CMOD* equivalence was considered but other *P-Y* equivalences may be sought (where *Y* stands for other possible magnitudes). For example; when the displacement *u* in the load-displacement record is chosen as a second variable one arrives at the load-displacement *(P-u)* equivalence. If the second variable is the crack tip opening displacement, the *P-CTOD* equivalence results. Also, a P-J equivalence may be considered when information about the softening behaviour of the cohesive material is known. All these equivalences are analyzed in detail in a forthcoming paper (Elices and

Planas, 1993). The corresponding crack growth resistance curves need not be the same, as found in numerical computations; Fig. 4.6a shows the *R*-Δa curves for a notched beam TPB ($D = l_{ch}$, $a_0 = 0.5\ D$) for the three equivalences *P-CMOD*, *P-u* and *P-J*. Although they do not coincide, they show similar shapes for not too large crack extensions and similar outputs should be expected regardless of the type of equivalence, as shown in Fig. 4.6b. The main difference between the mentioned equivalences and the *P-J* equivalence appears at the very end of the *R*-Δa curves, because the *P-J* equivalence does not induce stopping of the crack when it reaches the free surface. Nevertheless this fact should not be taken into consideration in practice, as already mentioned.

The asymptotic behaviour of the *P-CMOD* equivalence may serve as a guide line for discussing the asymptotic behaviour of other equivalences. In this respect it is convenient to gather the equivalences based on variables associated with fields far away from the cohesive zone (Planas and Elices, 1990). The variables *P*, *CMOD*, *u* and *J* used to define equivalences are far field variables. In previous papers (Planas and Elices, 1990, 1992, 1993) the authors have shown that the equivalences based on far field variables merge into a single one for very large sizes. In particular, the far field crack extensions for the different equivalences (but for the same softening material) are the same, i.e.:

$$\Delta a_\infty^{FF} = \Delta a_\infty^{P\text{-}CMOD} = \Delta a_\infty^{P\text{-}u} = \Delta a_\infty^{P\text{-}J} \tag{4.4}$$

This means that all *R*-Δa curves will tend towards a unique *R*-Δa curve for very large sizes, independently of the considered geometry. This curve is still depending on the softening properties of the cohesive material (Planas and Elices, 1990). On the basis result one can predict that computations for large sizes and from different equivalences will bring almost the same results, provided they are based on far field equivalences. When variables not related with far fields are considered—such as the crack tip opening displacement *CTOD*— the general trend is still similar, but the asymptotic *R*-Δa curve is not coincident with the previous one, as discussed in Planas and Elices (1992, 1993) and in the next paragraph.

4.3 X-Y Equivalences

The concept of *P-Y* equivalence can be generalized when the load *P* is no longer the controlling variable. The same procedure used for the *P-Y* equivalences can be extended to any couple of variables *X-Y*. This generalization will allow us to consider under the same framework apparently unrelated procedures such as Bazant *R*-Δa approach (Bazant, Gettu and Kazemi, 1989) or the *J-CTOD* method.

Now, the cohesive specimen and the equivalent (linear elastic) specimen are not bearing the same load, in general, and the equivalent load $P^{X\text{-}Y}$ and equivalent crack length $a^{X\text{-}Y}$ corresponding to the equivalent specimen can be computed by equating *X* and *Y* in both specimens.

$$X_{eq}\,(P^{X\text{-}Y}, a^{X\text{-}Y}) = X \tag{4.5}$$

$$Y_{eq}\,(P^{X\text{-}Y}, a^{X\text{-}Y}) = Y \tag{4.6}$$

where the left hand members are the elastic expressions for magnitudes *X* and *Y* in the equivalent specimen and the right hand members the actual values of *X* and *Y* (either measured or computed using the cohesive model).

Eqs. (4.5) and (4.6) determine the equivalent load and crack length from which any other magnitude can be found for the equivalent specimen. In particular, the equivalent crack growth resistance curve $R\text{-}\Delta a$ is obtained from Eqs. (4.2) and (4.3) with $P_{eq} = P^{X\text{-}Y}$ and $a_{eq} = a^{X\text{-}Y}$.

As in the previous paragraph, working with a particular equivalence will help in clarifying the concepts. Let us consider the *J-CTOD* equivalence. It is well known —see for example Rice, 1968— that J can be written as

$$J = W_F(CTOD) = \int_0^{CTOD} f(w)\, dw \tag{4.7}$$

where $f(w)$ is the softening function of the cohesive material and w the corresponding crack opening. Notice that w for the crack tip is the variable *CTOD*. For the non cohesive sample, J is equal to K_I^2/E'.

The couple of Eqs. (4.5) and (4.6) for this particular equivalence become:

$$\left[K_{Ieq}\left(P^{J\text{-}CTOD}, a^{J\text{-}CTOD}\right)\right]^2/E' = W_F(CTOD) \tag{4.8}$$

$$w_{eq}\left(P^{J\text{-}CTOD}, a^{J\text{-}CTOD}, a_0\right) = CTOD \tag{4.9}$$

where $w_{eq}(P, a, a_0)$ is the crack opening at the initial crack tip location a_0 of the equivalent specimen subjected to load P when the crack length is a.

The set of Eqs. (4.8) and (4.9) are a parametric representation of the $R\text{-}\Delta a$ curve for this particular equivalence. After rearranging its terms this curve can be written more explicitly in parametric form, with parameter *CTOD*, as (Planas and Elices, 1990):

$$R^{J\text{-}CTOD} = W_F(CTOD) \tag{4.10}$$

$$\Delta a^{J\text{-}CTOD} \cdot L^2\left(\frac{\Delta a^{J\text{-}CTOD}}{D}\right) = \frac{\pi}{32}\,\frac{E'\, CTOD^2}{W_F(CTOD)} \tag{4.11}$$

where $L(\Delta a/D)$ is a purely geometric function which can be computed using LEFM for any given specimen geometry.

The first equation is the *R-CTOD* relation associated with the material softening function. It is a material property and hence size and geometry *independent*. The second equation gives the *J-CTOD* equivalent crack extension. Notice that this $R\text{-}\Delta a$ curve is size and geometry *dependent* because of the presence of the function $L(\Delta a/D)$. Only for very large sizes, when $\Delta a/D$ becomes very small and L tends to 1, does the $R\text{-}\Delta a$ curve become geometry independent (Planas and Elices 1990).

To explore the predictive capacity of this equivalence, let us work on the same example already used for *P-Y* equivalences. Starting with "experimental" results from a notched beam made with a cohesive material (with quasi-exponential softening in this example) compute *P-CMOD* curves for similar beams of different sizes (to explore the size effect) and also *P-CMOD* curves for notched panels subjected to tension with different sizes and initial notch depths (to explore the geometry influence).

The selected reference beam was the same as that shown in Fig. 4.2a, a TPB ($D = l_{ch}$; $a_0 = 0.5\, D$) and the size effect was explored by computing *P-CMOD* values —using the *J-CTOD* equivalence— for two other homothetic TPB specimens of respective depths $0.3\, l_{ch}$ and $3\, l_{ch}$. The results are plotted in Fig. 4.7a, with the experimental values for comparison. Again, it seems that predictions for smaller beams are on the safe side and the reverse is the case for larger sizes.

The influence of the geometry was checked by considering two SENT panels, one with a ligament and size similar to the notched beam ($D = l_{ch}$; $a_0 = 0.5\ D$) and another much larger and with a small relative initial notch SENT ($D = 3\ l_{ch}$; $a_0 = 0.1\ D$). The computed *P-CMOD* curves, jointly with the experimental ones, are drawn in Fig. 4.7b. It seems that results are more sensitive to the relative ligament length $(D - a_0)/l_{ch}$ than to the geometry. Results for similar ligaments are quite good —$(P_{eq}/P_{ex}) = 1.05$—. When the ligament is larger than the reference one, predictions for maximum load are not as accurate and are unsafe as happens with beams —$(P_{eq}/P_{ex}) = 1.16$—.

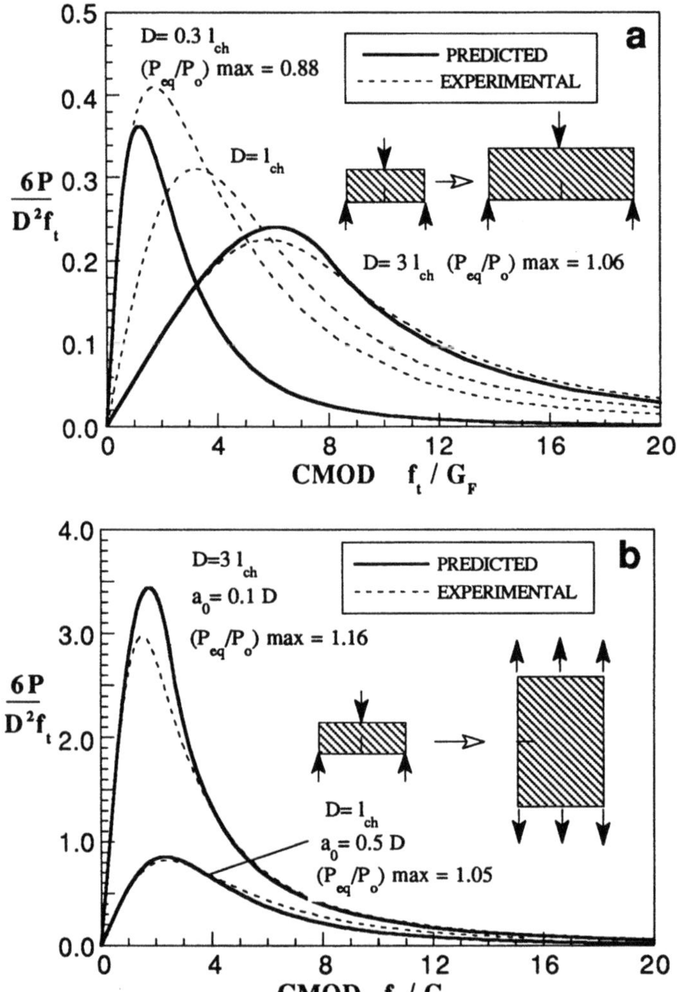

Fig.4.7 Predictions based on the J-CTOD equivalence: (a) P-CMOD values for homothetic beams; (b) P-CMOD values for notched panels.

As already mentioned, the accuracy of the predictions depends on how close are the R-Δa curves —the reference curve and the actual curve—. The trends found in Fig. 4.7 match the trends of the R-Δa curves; the nearest they are to the reference curve, the better the approximation.

This reasoning brings us to the consideration of the asymptotic behaviour for this equivalence. Here, one of the chosen variables, the *CTOD*, is related to the cohesive zone, and the far field property (discussed in section4.2) cannot be directly exploited. This fact does not guarantee that the equivalent crack extension $\Delta a_\infty^{J\text{-}CTOD}$ will be the same as Δa_∞^{FF} already found, and in fact they are not. The value of $\Delta a_\infty^{J\text{-}CTOD}$ can be easily obtained from Eq. (4.11) by using the fact that when $\Delta a/D$ tends to zero, the function L approaches unity, then:

$$\Delta a_\infty^{\text{J-CTOD}} = \frac{\pi}{32} \cdot \frac{E' \, CTOD^2}{W_F(CTOD)} \tag{4.12}$$

In a previous work the authors published a lower bound theorem for the far field equivalent crack extension (Planas and Elices, 1992). It was found that for a cohesive material and a general geometry under proportional mode I loading, the far-field effective crack extension —up to order c/D— always verifies:

$$\Delta a_\infty^{FF} \geq \frac{\pi}{32} \frac{E' \, CTOD_C^2}{G_F} \tag{4.13}$$

where c and D are respectively the size of the cohesive zone and the specimen size. It happens that the right hand member of Eq. (4.12), when *CTOD* reaches its critical value, coincides with the right hand member of Eq. (4.13). Consequently, the lower bound theorem states that

$$\Delta a_\infty^{FF} \geq \Delta a_\infty^{\text{J-CTOD}} \tag{4.14}$$

The far field equivalent extension is larger than the *J-CTOD* equivalent crack extension.

In the *J-CTOD* equivalence and in the *P-Y* equivalences considered in section 4.2, all the computed R-Δa curves were *size* and *geometry* dependent. Nevertheless, Bazant proposed a method to compute a size-independent —but *geometry* dependent— R-Δa curve from the knowledge of the size-effect curve (or variation of peak load P_{peak} with size D) for geometrically similar precracked structures (Bazant, Kim and Pfeiffer, 1986).

Bazant's approach is a general X-Y equivalence in the sense that two conditions are imposed; the first condition is that the peak load must be the same on the actual and on the equivalent elastic specimen for every size D. The second condition is that the R-Δa curve must be size independent. The only input is the size effect curve, which may be written as a function $P_{peak}(D)$, known from experiments or by computation. The unknowns are the equivalent resistance and the equivalent crack extension at peak load for every size, i.e.: the R_{eq}-Δa_{eq} curve. This curve was obtained by Bazant (see the quoted references) as the envelope of the following uniparametric family of functions, with parameter D,

$$R^{B\text{-}G} = \frac{1}{E'}\left[\frac{P_{peak}(D)}{B\sqrt{D}} \, S\left(\frac{a_0 + \Delta a^{B\text{-}G}}{D}\right)\right]^2 \tag{4.15}$$

The superindex B-G stands for Bazant and General, because no special size effect curve is postulated here. The size effect may come from a cohesive model, from Bazant's size effect Law, or from any other imaginable model. This $R^{B\text{-}G}$-$\Delta a^{B\text{-}G}$ curve depends on the chosen size effect [i.e.; on $P_{peak}(D)$] and on specimen geometry through $S(a^{B\text{-}G}/D)$, although it is size independent *by construction.*

This equivalence, when based on a cohesive model, is also a far field equivalence because it depends on the knowledge of the peak loads over a certain range of sizes. Consequently, for sizes large enough

$$\Delta a_{\infty}{}^{B\text{-}C} = \Delta a_{\infty}{}^{FF} \tag{4.16}$$

where B-C stands for Bazant-Cohesive, to distinguish between other procedures to obtain the size effect curve, as already indicated.

4.4 Final Comments

It was shown that the equivalent elastic crack concept is an approximate procedure to handle non linear cohesive crack problems. Furthermore, the equivalent elastic crack provides a rational framework for methods based on crack growth resistances curves (R-Δa) (Elices and Planas, 1993; Planas and Elices, 1993). This equivalence is a valuable tool and the examples developed in this chapter show that in some circumstances, the use of the *equivalence concept* provides acceptable results.

Some general trends, inferred from the above examples, are briefly summarized: For homothetic samples, predictions for maximum loads are on the safe side if the specimen is smaller than the reference one, and the contrary is true for larger samples. Specimens with different geometry or different initial notch size give almost the same *P-CMOD* response provided the relative ligament size is similar. In this sense, the relative ligament size seems to be one of the most relevant geometrical parameters. Finally, the type of equivalence does not cause great differences, except for the fictitious crack arrest that is not present in *P-J*, *CTOD-J* and similar equivalences.

The analysis of the behaviour of large samples provides further insights into the elastic equivalence. In this chapter it was found that every specimen of a cohesive material has an associated R-Δa curve for a particular equivalence and that this curve was dependent on the specimen *size* and *geometry*. The authors have proven (Planas and Elices, 1990, 1992, 1993) that when size increases, the R-Δa curves merge into a single one that depends only on the softening properties of the material and on the type of equivalence. Moreover, when the equivalences are based on far field variables (P, u, *CMOD*, J, ...) the asymptotic R-Δa curves merge into the same crack growth resistance curve. All the equivalences considered in this chapter converge into two —slightly different— R-Δa asymptotic curves; one corresponding to the far-field variables and the other to the *J-CTOD* equivalence (notice that *CTOD* is not a far field variable). These findings are useful for approximate computations of large specimens when the asymptotic curve is used as the reference curve.

The asymptotic crack growth resistance curve can be obtained in the following way: The plateau corresponds to the value of the specific fracture energy G_F. A procedure to measure this basic parameter was described in a recent series of papers (Guinea *et al.* 1992, Planas *et al.* 1992a, Elices *et al.* 1992). The initial rising part of the curve was computed by the authors for far field equivalences and for the *CTOD-J* equivalence (Planas and Elices, 1990). This suffices for drawing the R-Δa curve, provided the softening properties of the cohesive material are known. A reasonable approximation of the softening function, —based on a four parameter fitting— can be obtained using a procedure developed by the authors (Guinea, Planas, and Elices 1993) or according to other methods (Roelfstra and Wittmann, 1986; Wittmann *et al.*, 1987, 1988; Mihashi, 1992).

The following exercise exemplifies the use of an asymptotic *R-Δa* curve: Ascertain the *P-CMOD* response of a large notched beam TPB ($D = 3\ l_{ch}$; $a_0 = 0.1\ D$) and of a notched panel SENT ($D = 6\ l_{ch}$; $a_0 = 0.1\ D$) made with a cohesive material whose properties are modelled with the quasi-exponential softening function of Fig. 4.2b, as used in previous examples. By selecting a *J–CTOD* equivalence, the asymptotic *R-Δa* curve can be drawn according to the above procedures. Results of *P-CMOD* curves for both samples are shown in Fig. 4.8, jointly with experimental results for comparison. The agreement is quite good. This exercise shows the advantage of *equivalent elastic crack* method when applied to large enough specimens .

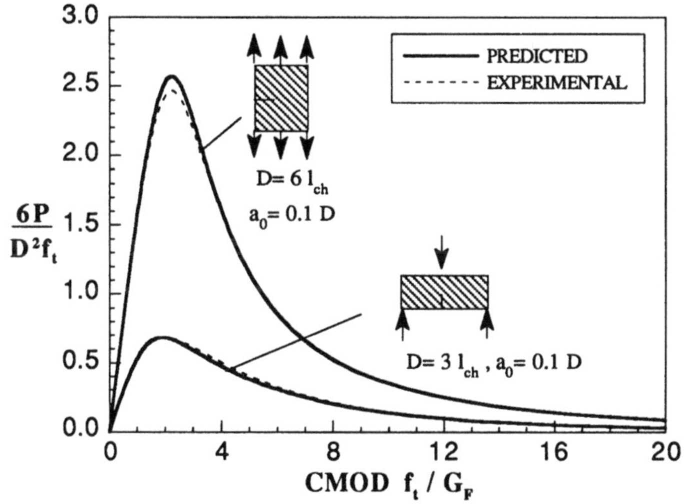

Fig.4.8 Predictions based on the asymptotic R-Δa curve (J-CTOD equivalence, quasi- exponential softening).

Acknowledgement. The authors acknowledge financial support for this research provided by the Dirección General de Investigación Científica y Técnica, Spain, under grants PB90-0276 and MAT92-0031. Computational help from Dr. Gonzalo Ruiz is gratefully acknowledged.

5 References

Barenblatt, G.J. (1962) The mathematical theory of equilibrium crack in brittle fracture. **adv. in Applied Mechanics**, 7, pp. 55-125.

Bazant, Z. P. (1986) "Mechanics of Distributed Cracking," **Appl. Mech. Rev., ASME**, 39, 5, pp. 675-705.

Bazant, Z. P. and Lin, F.-B (1988) Nonlocal Yield Limit Degradation. **Int. J. for Num. Meth. in Engng.**, 26, pp. 1805-1823.

Bazant, Z. P. and Pijaudier-Cabot, G. (1987) Modeling of Distributed Damage by Nonlocal Continuum with Local Strain. Prep.*4th Int. Conf. on Num. Meth. in Fract. Mech.,* A. R. Luxmore, D. R. J. Owen and M. F. Kanninen, eds., pp. 432-441.

Bazant, Z.P., Gettu, R. and Kazemi, M.T. (1989) Identification of Nonlinear Fracture Properties from Size Effect Tests and Structural Analysis based on Geometry-Dependent R-Curves. Report No. 89-3/498p, Center for Advanced Cement-Based Materials, The Technological Institute, Northwestern University, Evanston, Illinois 60208.

Bazant, Z.P., Kim, J.K. and Pfeiffer , P.A. (1986) Nonlinear Size Effect Properties from Size Effect Tests. **J. of Structural Engineering, ASCE**, Vol. 112, pp. 289-307.

Bittencourt, T.N., Ingraffea, A.R. and Llorca, J. (1992) Simulation of arbitrary cohesive crack propagation. In **Fracture Mechanics of Concrete Structures** (Z. P. Bazant, Ed.) Elsevier Applied Science, London, pp. 339-350.

de Borst, R., Mühlhaus, H.-B. and Pamin, J. (1992) A gradient continuum model for mode-I fracture in concrete and rock. In **Fracture Mechanics of Concrete Structures** (Z. P. Bazant, Ed.) Elsevier Applied Science, London, pp. 251-257.

Cotterel, B. and Mai, Y.W. (1987) Crack grwth resistance curve and size effect in the fracture of cement paste, **J. of Materials Science**, 22, pp. 2734-2738.

Dugdale, D.S. (1960) Yielding of steel sheets containing slits. **J. Mech. Phys. Solids**, 8, pp.100-104.

Elices, M. and Planas, J. (1989) Material Models. Chapter 3 in **Fracture Mechanics of Concrete Structures** (L. Elfgren, Ed.) Chapman and Hall, London, pp. 16-66.

Elices, M. and Planas, J. (1993) The Equivalent Elastic Crack: 1. Load-Y Equivalences. Submitted to **Int.J. of Fracture**.

Elices, M., Guinea, G.V. and Planas, J. (1992) Measurement of the fracture energy using three-point bend tests: 3. Influence of cutting the P-δ tail. **Materials and Structures**, 25, pp. 327-334.

Foote, R.M.L., Mai, Y.W. and Cotterel, B. (1986) Crack growth resistance curves in strain softening materials, **J. Mech. Phys. Solids**, 34, pp. 593-1607.

Guinea, G.V., Planas, J. and Elices, M. (1992) Measurement of the fracture energy using three-point bend tests: 1. Influence of experimental procedures. **Materials and Structures**, 25, pp. 212-218.

Guinea, G.V., Planas, J. and Elices, M. (1993) A General Bilinear Fitting for the Softening Curve of Concrete. Submitted to **Materials and Structures**.

Hassanzadeh, M. (1992) Behaviour of fracture process zones in concrete influenced by simultaneously applied normal and shear displacements. Div. of Building Mat., Lund Inst. of Technology, Report TVBM-1010.

Hillerborg, A., Modéer, M. and Petersson, P.E. (1976) Analysis of crack formation and crack growth in concrete by means of fracture mechanics and finite elements. **Cement and Concrete Research**, 6, pp.773-782.

Llorca, J. and Elices, M. (1990) A simplified model to study fracture behaviour in cohesive materials, **Cement and Concrete Research**, 20, pp. 92-102.

van Mier, J.G.M. (1986) Fracture of concrete under complex stress. **Heron**, 31, pp. 2-90.

van Mier, J.G.M., Nooru-Mohamed, M.B. and Schlangen, E. (1991) Experimental analysis of mixed mode I and II behaviour of concrete. In **Analysis of Concrete Structures by Fracture Mechanics** (L. Elfgren and S.P. Shah, Eds.) Chapman and Hall, London, pp. 32-43.

Mihashi, H. (1992) Material structure and tension softening properties of concrete, in **Fracture Mechanics of Concrete Structures** (ed. Z.P. Bazant), Elsevier, London, pp. 239-250.

Petersson, P.E. (1981) Crack growth and development of fracture zone in plain concrete and similar materials. Div. of Building Mat., Lund Inst. of Technology, Report TVBM-1006.

Pijaudier-Cabot, G. and Bazant, Z. P. (1987) Nonlocal Damage Theory. **J. Eng. Mech., ASCE,** 113, pp. 1512-1533.

Planas, J. and Elices, M. (1990) Asymptotic Analysis of Cohesive Cracks and its Relation with Effective Elastic Cracks. In **Toughening Mechanisms in Quasi-Brittle Materials**, (S.P. Shah, Ed.), pp. 217-230.

Planas, J. and Elices, M. (1991) Nonlinear fracture of cohesive materials. **Int. J. of Fracture**, 51, pp. 139-157.

Planas, J. and Elices, M. (1992) Asymptotic analysis of a cohesive crack: 1. Theoretical background. **Int. J. of Fracture,** 55, pp. 153-177.

Planas, J. and Elices, M. (1993) The Equivalent Elastic Crack: 2. X-Y Equivalences and Asymptotic Analysis.Submitted to **Int. J. of Fracture**.

Planas, J., Elices, M. and Guinea G.V. (1992a) Measurement of the fracture energy using three-point bend tests: 2. Influence of bulk energy dissipation. **Materials and Structures**, 25, pp. 305-312.

Planas, J., Guinea G.V. and Elices, M. (1992b) Softening curves for concrete and structural response. Int. Conf. on Fracture and Damage of Concrete and Rock, FDCR-2, Vienna, Austria, November 9-13.

Planas, J., Elices, M. and Guinea G.V. (1993) Cohesive cracks versus nonlocal models: Closing the gap. Dep. Ciencia de Materiales, Universidad Politécnica de Madrid, Spain. Report 92-jpr-04, June. Revised version submitted to **Int. J. Fracture**.

Reinhardt, H.W., Cornelissen, H.A.W. and Hordijk, D.A. (1986) Tensile tests and failure analysis of concrete. **J. of Structural Engineering, ASCE,** 112, pp. 2462-2477.

Rice, J. (1968) Mathematical Analysis in the Mechanics of Fracture. In **Fracture**, Vol. 2 (H. Liebowitz, Ed.), Academic Press, pp. 192-311.

RILEM Draft Recommendation (1985) Determination of fracture energy of mortar and concrete by means of three-point bend test on notched beams. **Materials and Structures**, 18, pp.285-290.

Roelfstra, P.E. and Wittmann, F.H. (1986) Numerical method to link strain softening with failure of concrete. In **Fracture Toughness and Fracture Energy of Concrete** (F. H. Wittmann, Ed.) Elsevier Science, Amsterdam, pp. 163-175.

Rokugo, K., Iwasa, M., Suzuki, T. and Koyanagi, W. (1989) Testing methods to determine tensile strain softening curve and fracture energy of concrete. In **Fracture Toughness and Fracture Energy** (Mihashi et al., Eds.) Balkema, Rotterdam, pp. 153-163.

Rots, J.G. (1992) Removal of finite elements in strain-softening analysis of tensile fracture. In **Fracture Mechanics of Concrete Structures** (Z. P. Bazant, Ed.) Elsevier Applied Science, London, pp. 330-338.

Tschegg, E. Kreuzer, H. and Zelezny, M. (1992) Fracture of concrete under biaxial loading-Numerical evaluation of wedge splitting test results. In **Fracture Mechanics of Concrete Structures** (Z. P. Bazant, Ed.) Elsevier Applied Science, London, pp. 455-460.

Wittmann, F.H., Roelfstra, P.E., Mihashi, H., Huang, Y.Y., Zhang, X.H. and Nomura, N. (1987) Influence of age of loading, water-cement ratio, and rate of loading on fracture energy of concrete. **Materials and Structures**, 20, pp.103-110.

Wittmann, F.H., Rokugo, K., Bruhwiler, E., Mihashi, H. and Simoni, P. (1988) Fracture energy and strain softening of concrete as determined by means of compact tension specimens. **Materials and Structures**, 21, pp. 21-32.

2 THE ROLE OF MICROCRACKS AND MICROVOIDS IN THE TOUGHENING OF QUASI-BRITTLE MATERIALS

X. HUANG and B.L. KARIHALOO
School of Civil and Mining Engineering,
The University of Sidney, Australia

Abstract
The tensile/flexural response of quasi-brittle materials, such as concrete, rocks, and most ceramics is characterized by moderate strain hardening before, and significant tension softening after, the attainment of their ultimate tensile/flexural strength. These non-linear phenomena which are a result of the growth of existing flaws (microcracks and microvoids), and of the nucleation of fresh flaws, enhance the fracture toughness of the quasi-brittle materials. Based on a micromechanical model this paper estimates the contributions of microcracks and microvoids to the toughening of these materials and shows that the dominant contribution is made by the microvoids. The theoretical results are shown to compare favourably with the available test results for concretes.
Keywords: Quasi-brittle Materials, Concrete, Rocks, Fracture Toughness, Tension Softening, Microvoids, Microcracks, Toughening

1 Introduction

The last decade or so has improved our understanding of the role of flaws in the tensile response not only of metallic materials (e.g. Tvergaard, 1988) but also of materials, such as concrete, rocks, and most ceramics that have traditionally been regarded as brittle. It is now generally known that concrete (Cornelissen *et al.*, 1986), rocks (Labuz *et al.*, 1985), and most ceramics exhibit moderate strain hardening prior to the attainment of their ultimate tensile/flexural capacity (region AB in Fig. 1), reminiscent of the response of high strength metallic materials. Beyond the peak capacity, these materials are characterized by an increase in deformation with decreasing tensile carrying capacity, i.e. by tension softening (region BCD in Fig. 1). They may therefore be called quasi-brittle materials. This paper will attempt to quantify the contributions of the pre-peak nonlinearity and the post-peak tension softening to the fracture toughness of quasi-brittle materials. It will identify the roles of microcracks and microvoids in this toughening process and show that the dominant contribution to toughening is made by the microvoids.

Fracture and Damage of Concrete and Rock – FDCR-2.
Edited by H.P. Rossmanith. ISBN 0 419 18470 8.

The theoretical results will be shown to compare favourably with the available test results for a range of concrete mixes.

The pre-peak nonlinear response has been satisfactorily described from the point of view of damage mechanics by Karihaloo and Fu (1989). The limit of applicability of the micromechanical damage model coincides with the onset of damage localization along the eventual fracture planes. The critical flaw density at this instant, designated p_c, corresponds to what is commonly called the ultimate effective strain $\hat{\epsilon}^{ult}$ (or ultimate effective stress $\hat{\sigma}^{ult}$). However, the failure of quasi-brittle materials under tension/flexure does not occur at the attainment of p_c. This is because the localization of damage results in the formation of a discontinuous macroflaw rather than of a single macrocrack of critical size due to bridging by interlocking second phase particles and by unbroken ligaments.

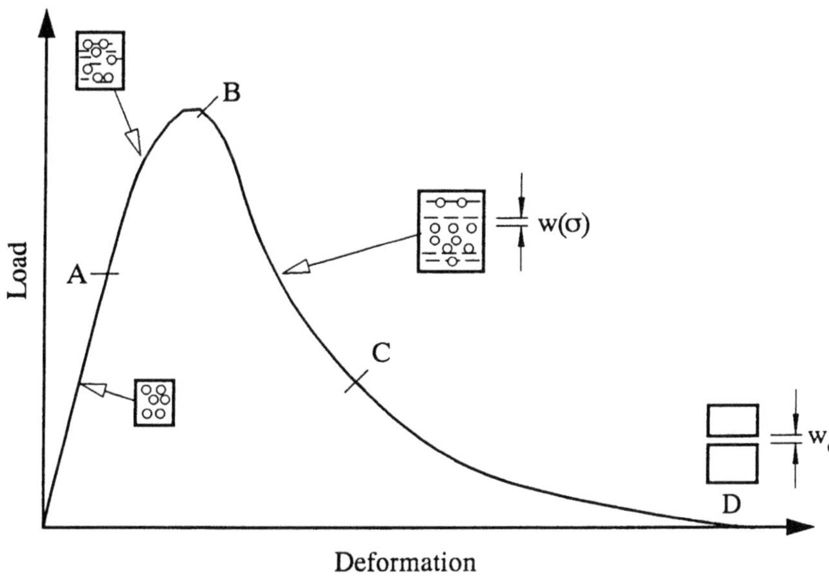

Fig. 1. Load-deformation response of a typical quasi-brittle material under tensile loading.

2 Tension softening

As a starting point for the description of the tension softening behaviour we adopt the model proposed recently by Karihaloo *et al.* (1991) that treats the discontinuous macroflaw as a collinear array of microvoids with edge cracks separated by unbroken ligaments. In the limit, when no microvoids are present, this model coincides with that of Horii *et al.* (1987) and of Ortiz (1988). It allows evaluation of the roles of microvoids and microcracks in the toughening process.

The discontinuous macroflaw is assumed to consist of an infinite row of collinear circular micropores of radius a_o with edge cracks of length ℓ_o. The neighbouring segments are separated by unbroken ligaments of length $[d - 2(\ell_o + a_o)]$, so that $2(\ell_o + a_o)/d$ may be regarded as equal to the accumulated damage per unit area (volume) at the onset of strain localization, i.e. equal to p_c, which will henceforth be designated as γ_o.

Without going into detail it may be shown that the tension softening relation between the residual stress carrying capacity σ and the inelastic deformation w in terms of the material constants at the onset of localization of the deformation, namely the matrix fracture toughness K_{Ic}^m and the accumulated damage at peak load γ_o $(= p_c)$ is given by

$$\frac{\sigma}{f_t} = \left(\frac{\gamma_o}{\gamma}\right)^{\frac{1}{2}} \frac{f(\beta_o, \gamma_o)}{f(\beta, \gamma)}, \tag{1}$$

$$\frac{wE'}{\pi(a_o + \ell_o)f_t} = \frac{g(\beta_o, \gamma_o)}{\pi\gamma_o} \left(\frac{\sigma}{f_t}\frac{g(\beta, \gamma)}{g(\beta_o, \gamma_o)} - 1\right). \tag{2}$$

Here $\beta = a/(a+\ell), \gamma = 2(a+\ell)/d$, and f_t is the microscopic ultimate tensile strength of the material. The polynomials $f(\beta, \gamma)$ and $g(\beta, \gamma)$ are given in Karihaloo *et al.* (1991). The tension softening curves predicted by the present model are shown in Fig. 2 for two values each of γ_o and β.

3 Effective fracture toughness

The curves corresponding to β=1/11 are characterized by a distinct maximum inelastic deformation beyond which the condition $K_I = K_{Ic}^m$ on the tension softening curve can only be met if w decreases with decreasing σ (broken lines). However, this unstable behaviour has nothing to do with the material response but is a result of the collinear crack model. It is known (Melin, 1983) that in tensile mode I tip to tip coalescence of collinear cracks cannot take place. Instead, when the tips come so close to each other that the straight path would be unstable, they avoid each other by curving and in the process form 'eyelets'.

The observed material behaviour, namely a tension softening branch with a pronounced tail (Cornelissen *et al.*, 1986), is therefore better described not by an array of collinear microcracks ($\beta = 0$) but by the present model with $\beta > 0$. In fact, for $\beta > 0$ it is easily verified that $w \to \infty$, as $\gamma \to 1$ at complete rapture. For $\beta = 0$, $g(0, \gamma)/f(0, \gamma) \to 0$, as $\gamma \to 1$, thereby confirming the instability inherent to the collinear microcrack model. It is therefore important to realise that tension softening models cannot predict the exact value of the critical crack opening displacement w_c at rupture ($\gamma = 1$). A notional w_c may be calculated by limiting

γ to $1 - \delta$ (with $\delta = 10^{-4}$, say).

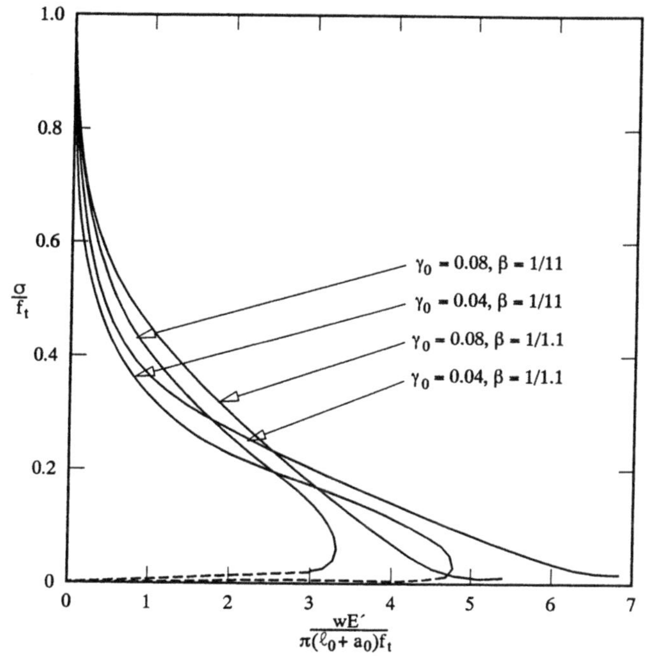

Fig. 2. Tension softening curves.

On the other hand, it is easily verified that the area under the tension softening curve between the limits $\gamma_o \leq \gamma \leq 1$ is a finite quantity

$$W_d = \int_{f_t}^{0} w(\sigma)\, d\sigma \equiv \int_{\gamma_o}^{1} w(\gamma)\, \frac{\partial \sigma}{\partial \gamma}\, d\gamma, \tag{3}$$

where for a fixed β, Eq. (1) gives

$$\frac{\partial \sigma}{\partial \gamma} = -\frac{1}{2}\, f_t\, \frac{f_o}{f}\, \sqrt{\gamma_o/\gamma^3}\, \left(1 + 2\frac{\gamma}{f}\, \frac{\partial f}{\partial \gamma}\right), \tag{4}$$

with

$$\frac{1}{f}\,\frac{\partial f}{\partial \gamma} = \frac{\pi}{2}\left(\frac{1}{\sin \pi\gamma} - \frac{1}{\pi\gamma}\right) + \frac{\sum_{r=1}^{3}\sum_{s=1}^{3} s\, C_{rs}\, \beta^r\, \gamma^{s-1}}{\left(1 + \sum_{r=1}^{3}\sum_{s=0}^{3} C_{rs}\, \beta^r\, \gamma^s\right)}. \tag{5}$$

For brevity, the abbreviations $f_o = f(\beta_o, \gamma_o)$ and $f = f(\beta, \gamma)$ have been introduced in Eqs. (4)–(5). The coefficients C_{rs} are given in Karihaloo *et al.* (1991).

The knowledge of W_d is useful in several ways. It may be used, for instance to calculate an effective fracture toughness K_{Ic}^e or an effective toughness (energy release rate) G_{Ic}^e of the quasi-brittle material in terms of the matrix fracture toughness

K_{Ic}^m. The effective Griffith toughness G_{Ic}^e is defined as the critical energy release rate for an equivalent brittle material that would consume the same energy (W_d) as used in the creation of traction-free crack surfaces of length $(1 - \gamma_o)$ in the actual quasi-brittle material

$$W_d = G_{Ic}^e\,(1-\gamma_o) = \frac{(K_{Ic}^e)^2}{E'}\,(1-\gamma_o). \tag{6}$$

The introduction of an effective toughness allows one to estimate the toughening provided by the bridging mechanism in the quasi-brittle material. For this estimation, it is convenient to rewrite Eq. (3) as

$$W_d = \frac{(K_{Ic}^m)^2}{\pi E' \gamma_o f_o}\left(W_d^* - \frac{g_o}{f_o}\right), \tag{7}$$

where, in view of Eqs. (4)–(5)

$$W_d^* = \frac{\gamma_o f_o}{2}\int_{\gamma_o}^{1}\frac{g}{f^2}\,\frac{1}{\gamma^2}\left[\frac{\pi\gamma}{\sin\pi\gamma} + \frac{2\sum_{r=1}^{3}\sum_{s=1}^{3}C_{rs}\,\beta^r\,\gamma^s}{\left(1+\sum_{r=1}^{3}\sum_{s=0}^{3}C_{rs}\,\beta^r\,\gamma^s\right)}\right]d\gamma, \tag{8}$$

and for brevity, $g = g(\beta,\gamma)$ and $g_o = g(\beta_o,\gamma_o)$.

An examination of Eq. (7) shows that the energy represented by W_d^* includes the energy consumed in the pre-peak nonlinear stage, i.e. in the pre-peak slow crack growth. The toughening provided by the bridging mechanism can now be formally expressed as the ratio (using Eqs. 6 and 7)

$$\frac{K_{Ic}^e}{K_{Ic}^m} = \sqrt{(W_d^* - \frac{g_o}{f_o})\ /\ [\pi\gamma_o f_o(1-\gamma_o)]}. \tag{9}$$

The integral in Eq. (8) (the integrand has an integrable singularity) was calculated numerically for various values of γ_o and β in the ranges $0.02 \le \gamma_o \le 0.25$ and $0 \le \beta \le 1$. The numerical results were substituted in Eq. (9) and the resulting values of the toughening ratio K_{Ic}^e/K_{Ic}^m were fitted with a polynomial approximation (error $< 1\%$)

$$\frac{K_{Ic}^e}{K_{Ic}^m} = \sum_{r=0}^{1} Y_{r0}\,\gamma_o^r + \sum_{r=0}^{1}\sum_{s=1}^{3} Y_{rs}\,\gamma_o^r\,\beta^s. \tag{10}$$

The coefficients Y_{rs} are given in Table 1.

Table 1. Coefficients Y_{rs} in the analytical approximation (10)

r	s	Y_{rs}
0	0	1.003944
0	1	1.408276
0	2	-3.878354
0	3	3.201900
1	0	0.326490
1	1	0.928382
1	2	-2.502774
1	3	1.921063

The knowledge of the energy consumed to create a traction-free crack of length (1 – γ_o) also comes in handy when an approximate estimate of the influence of the pre-peak nonlinearity on the effective fracture toughness K^e_{Ic} is desired. For this calculation, it is sufficient to compare the K^e_{Ic} obtained above with that resulting from the full W^*_d, i.e. when the energy consumed in the pre-peak nonlinear zone is included in the calculation of the energy consumption. Of course, this estimate is valid only for small values of γ_o, i.e. for moderate pre-peak slow crack growth, in view of the definition of the effective fracture toughness (Eq. 6). It was for this reason that γ_o was restricted to the range $0.02 \leq \gamma_o \leq 0.25$ in the above calculations (Eq. 10). Denoting by Δ the relative contribution from the pre-peak nonlinearity to K^e_{Ic}, it follows from Eqs. (7) and (9) that

$$\Delta = 1 - \left(1 - \frac{g_o}{\gamma_o W^*_d}\right)^{\frac{1}{2}}. \tag{11}$$

4 Results and discussion

Fig. 3 shows the ratio of the effective fracture toughness K^e_{Ic} to the matrix fracture

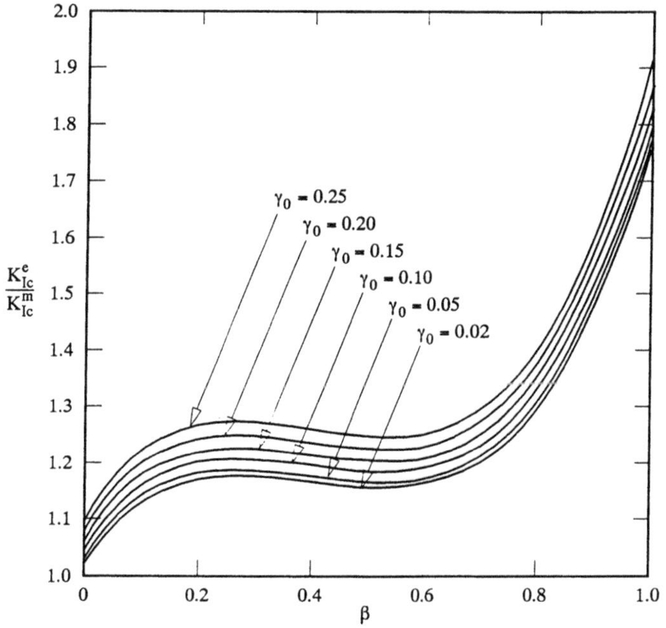

Fig. 3. Variation of the toughening ratio with β for several levels of the accumulated damage at peak load γ_o.

negligibly small edge microcracks), using the appropriate coefficients from Table 1

$$K_{Ic}^e/K_{Ic}^m = \begin{cases} 1.0039 + 0.3265\,\gamma_o & (\beta = 0) \\ 1.7360 + 0.6732\,\gamma_o & (\beta \approx 1) \end{cases} \tag{12}$$

Thus, for a material such as a high strength concrete, even when the accumulated damage at peak load is large ($\gamma_o = 0.25$, say) the toughening provided by this bridging mechanism ($\beta = 0$ in Eq. (12)) would be just 8.5%. Concretes of normal strength, though can be expected to have a β value in the range of $0.15 \leq \beta \leq 0.6$, so that the toughening provided by the bridging mechanism would range between 15% and 20% (Fig. 4) depending on the damage present at peak load. This result is consistent with that of Horii and Nirmalendran (1990).

The relative contribution of the pre-peak slow crack growth to the toughening given by Eq. (11) is shown in Fig. 5 for the full range of values of β ($0 \leq \beta < 1$) as a function of γ_o. The relative contribution in the range $0.15 \leq \beta \leq 0.75$ remains practically constant. It is clear from Fig. 5 that the pre-peak nonlinearity makes but a marginal contribution to the toughening of quasi-brittle materials, the bulk of toughening being a result of the bridging mechanism. This is consistent with the results of Horii and Nirmalendran (1990).

To complete the discussion, it may be pointed out that for concrete, Eq. (10) shows that K_{Ic}^e can be expressed in terms of only K_{Ic}^m and V_f (in the range $0.15 \leq \beta \leq 0.6$). A similar dependence of K_{Ic}^e on K_{Ic}^m and V_f is also predicted by the tension softening model for concrete proposed by Li and Huang (1990), in which the discontinuous macroflaw is treated as a series of curved cracks separated by the largest aggregate particles (assumed as circular disks).

Finally, the effective fracture toughness K_{Ic}^e calculated above from a micromechanical point of view may be used to estimate the *brittleness* of a quasi-brittle material from tests using conventional test specimens (compact tension, three or four point bend) of characteristic dimension W containing an pre-crack of depth c_o. To do this, one needs to apply linear elastic fracture mechanics to an equivalent brittle material containing an effective crack $c_e = c_o + \Delta c$

$$K_{Ic}^e = \sigma_n \sqrt{c_e}\, F(\frac{c_e}{W}), \tag{13}$$

where σ_n is the nominal failure stress of the test specimen, and $F(c_e/W)$ is the well-known compliance function that depends on only the shape of the test specimen. A measure of the *brittleness* of the quasi-brittle material is then given by the supplementary fictitious traction-free crack Δc. The smaller its value, the more brittle the material. From the discussion in the preceding paragraph it follows that Δc will also depend only on K_{Ic}^m and V_f. The concept of the effective Griffith crack is already widely used in the fracture testing of concrete (Jenq and Shah, 1985, Karihaloo and Nallathambi, 1989) and of rocks (Labuz *et al.*, 1985). The toughness K_{Ic}^m (Eq. 10) for several levels of the accumulated damage at peak load γ_o as a function of the parameter β. The latter, it will be recalled is a measure of the relative fractions of microvoids and microcracks. Thus, $\beta = 0$ denotes the absence of any microvoids, whereas $\beta \approx 1$ implies microvoids with negligibly small edge microcracks. The curves on Fig. 3 are reminiscent of the classical *creep* curves, with an initial *primary* region of increasing K_{Ic}^e/K_{Ic}^m with increasing $\beta < 0.15$, an intermediate *secondary* or *steady state region* for $0.15 \leq \beta \leq 0.6$ in which K_{Ic}^e/K_{Ic}^m remains practically constant, and a *tertiary* region in which this ratio increases sharply with increasing $\beta > 0.6$. The variation of K_{Ic}^e/K_{Ic}^m with γ_o for a fixed β is fairly moderate (Fig. 4). Thus, it appears that microvoids with small edge microcracks have a far larger toughening effect than microcracks alone. This may explain the well-known fact that the thermal shock resistance (related to fracture toughness) of refractory ceramics increases with increasing porosity. It may also explain why microcracking alone cannot make a large contribution to the toughening of quasi-brittle materials.

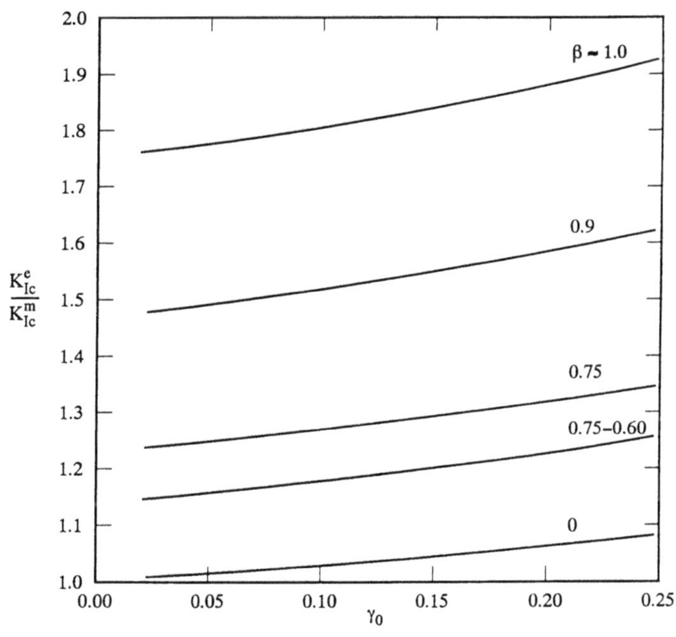

Fig. 4. Variation of the toughening ratio with γ_o for various relative fractions of microvoids and microcracks (β).

To illustrate these two points, we write explicitly the toughening ratio from Eq. (10) for the two extreme cases of $\beta = 0$ (no microvoids) and $\beta \approx 1$ (microvoids with micromechanical description reported in this paper puts it on a sound physical foundation.

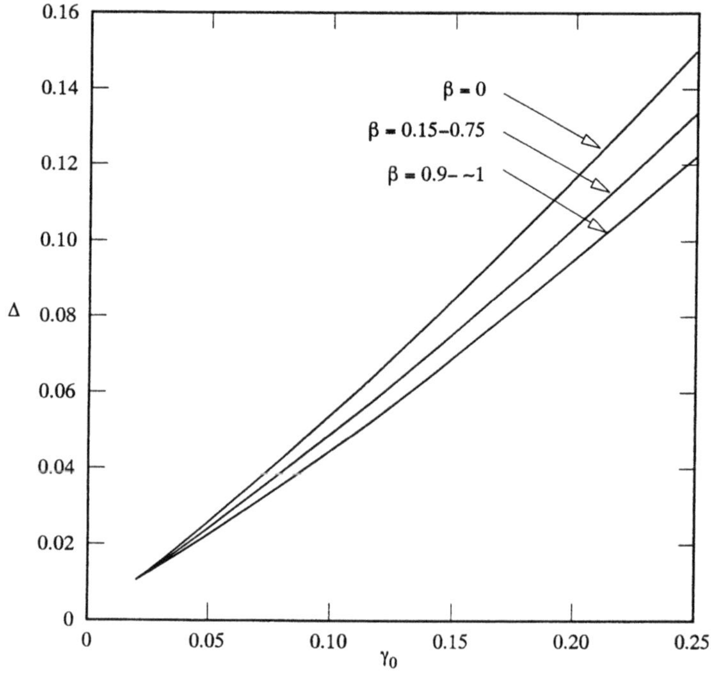

Fig. 5. Relative contribution (Δ) of the pre-peak nonlinearity (γ_o) to the toughening.

5 References

Cornelissen, H.A.W., Hordijk, D.A. and Reinhardt, H.W. (1986) Experimental determination of crack softening characteristics of normal weight and lightweight concrete. **Heron**, 31, 45–56.

Horii, H., Hasegawa, A. and Nishino, F. (1987) Fracture process and bridging zone model and influencing factors in fracture of concrete, in **Fracture of Concrete and Rock** (eds S.P. Shah and S.E. Swartz), Springer-Verlag, New York 1989, pp 205–219.

Horii, H. and Nirmalendran, S. (1990) Roles of microcracking and bridging in fracture of quasi-brittle materials, in **Micromechanics of Failure of Quasi-Brittle Materials** (eds S.P. Shah, S.E. Swartz and M.L. Wang), Elsevier Appl. Sci, London, pp. 569–578.

Jenq, Y.S. and Shah, S.P. (1985) Two parameter fracture model for concrete. **J. Eng. Mech., ASCE**, 111, 1227–1241.

Karihaloo, B.L. and Fu, D. (1989) A damage-based constitutive law for plain concrete in tension. **Eur. J. Mech. A/Solids**, 8, 373–384.

Karihaloo, B.L. and Nallathambi, P. (1989) Fracture toughness of plain concrete from three-point bend specimens. **Mater. & Struct.**, 22, 185–193.

Karihaloo, B.L., Fu, D. and Huang, X. (1991) Modelling of tension softening in quasi-brittle materials by an array of circular holes with edge cracks. **J. Mech. Mater.**, 11, 123–134.

Labuz, J.F., Shah, S.P. and Dowding, C.H. (1985) Experimental analysis of crack propagation in granite. **Int. J. Rock Mech. Min. Sci. & Geomech. Abstr.**, 22, 85–98.

Li, V.C. and Huang, J. (1990) Relation of concrete fracture toughness to its internal structure. **Engng. Fracture Mech.**, 35, 39–46.

Melin, S. (1983) Why do cracks avoid each other? **Int. J. Fracture**, 23, 37–45.

Ortiz, M. (1988) Microcrack coalescence and macroscopic crack growth initiation in brittle solids. **Int. J. Solids Struct.**, 24, 231–250.

Tvergaard, V. (1988) 3D-analysis of localization failure in a ductile material containing two size scales of spherical particles. **Engng. Fracture Mech.**, 31, 421–436.

3 NON-LOCAL PLASTICITY SOFTENING MODEL FOR BRITTLE MATERIALS

A. SCARPAS and J. BLAAUWENDRAAD
Department of Civil Engineering, Delft University of Technology, Delft, The Netherlands

Abstract
A fracture energy based constitutive formulation is presented which, by augmenting the classical non-associated plasticity theory with a material lenght parameter and a mixed softening law, enables the development of a unified formulation for the smooth transition from the plasticity to the fracture domains of material response.
Keywords: Finite elements analysis, fracture, non-associated plasticity, mixed softening.

1 Introduction

Since the pioneering work of Bazant and co-workers [1979], [1983], fracture mechanics based softening models are routinely used at present for finite element simulations of the uniaxial tensile behaviour of brittle materials like concrete, rock, masonry etc. In most of these formulations, the degradation of response is related to the physically meaningful crack opening displacement w by means of a decaying function in the form :

$$\frac{\sigma}{f_t'} = g\left(\frac{w}{w_{ft}}\right) \tag{1.1}$$

in which w_{ft} corresponds to the crack opening at the time of complete stress annihilation.

In more complex loading situations, this same formulation is used to simulate the behaviour of the tensile components for all stress regions over which a tension cut-off type criterion is valid. All other regions are treated by means of plasticity or some other type of constitutive models. Appealing as it may sound, this type of implementation has two drawbacks :

(i) it does not provide adequately for the response of the transverse stress component if this happens to be compressive. The standard practise of assuming an elastic modulus along its direction, allowing micro-crack development to occur and then, **posteriori**, reducing any excess state of stress by means of a plasticity, or other, reduction algorithm, has been known to result to non-converging iterations between the smeared and the reduction algorithms, Wang [1989].

(ii) there exists by now an abundance of experimental evidence attesting to the fact that, with the exception of very highly confined triaxial states of compressive loading,

Fracture and Damage of Concrete and Rock – FDCR-2.
Edited by H.P. Rossmanith. © 1993 Taylor & Francis. ISBN 0 419 18470 8.

the response of brittle materials, in all other compressive loading ranges, is highly influenced by the magnitude of the maximum principal tensile strain.

Indeed, the experimental work of van Mier [1986] has cast doubts on the applicability of classical plasticity based formulations for simulating the post-fracture characteristics of concrete-like materials. In a series of carefully planned and adequately instrumented tests he demonstrated that, similar to uniaxial tension fracture, strain can **not** be considered as **a material property** for compressive fracture under low to medium ranges of lateral confinement.

Kotsovos [1983] has established that in uniaxial compression test, the mode of fracture is axial splitting running parallel to the applied compressive load, along planes perpendicular to the maximum principal tensile strain. However, the resulting splitting is of a much more distributed nature compared with the single plane splitting mode typically observed in fracture under tensile loading, Liu et al. [1972], Willam et al. [1986].[1]The more distributed nature of compressive splitting results to larger overall crack opening displacements than those typically observed in uniaxial tension. A ratio of 10 : 1 has been noted by Willam et al. [1986] and can be also verified from the uniaxial tests of van Mier [1986].

Axial splitting along planes perpendicular to the direction of maximum tensile strain is also the fracture mode in case of biaxial or triaxial tension-compression tests and, for compression tests under low lateral confinement, e.g. Horii and Nemat-Nasser [1985]. With increasing levels of confinement, fracture under compressive loading manifests itself in one or more shear bands inclined to the direction of load application. These have not only been observed experimentally, Nemat-Nasser and Horii [1982], Vardoulakis and Graf [1985], van Mier [1986], but have also been reproduced numerically, Leroy and Ortiz [1990], Droz and Bazant [1989]. The material within the shear band can be assumed to be in a state of pure shear, Vardoulakis and Graf [1985], which gives rise to the development of a diagonal tension micro-cracking field, Bazant and Pfeiffer [1985].

From the above short overview it can be concluded that, fracture in brittle materials is almost always associated with splitting. It appears therefore reasonable to postulate that, it may be possible to formulate a criterion, for the post-fracture response of brittle solids, on the basis of the **maximum tensile deformation** which occurs in the material after fracture. In the following sections, the formulation of an elasto-plastic-softening model based on this assumption will be presented.

2 Fracture Criterion

Many brittle engineering materials portray some type of anisotropy. Common examples are layered rock, soil deposits and masonry. To allow for the constitutive modelling of these materials as well, an anisotropic formulation was chosen as the basis of material

(1)Only in the presence of friction between the loading plates and the material specimen, an hour-glass fracture mode can develope, Kotsovos [1983].

formulation. Reduction to the isotropic case is of course trivial. Without loss of generalization it will be assumed in the following that :

(i) the physical composition of the material allows adoption of a state of anisotropy possessing three mutually orthogonal material axes,

(ii) a pressure sensitive Hoffman type fracture criterion is adequate in describing the onset and the development of fracture in the material.

Following standard notation, the fracture surface can be expressed in general form as :

$$f = f\,[\,\underline{\sigma}\,,\kappa_0\,] = 0 \tag{2.1}$$

in which : $\underline{\sigma}$ = the stress vector,
κ_0 = some monotonically varying scalar measure controlling the evolution of the surface after fracture

3 Fracture Characteristics in Uniaxial Tension

Various expressions have been proposed for the load-deformation relation of brittle materials like concrete and rock. In the present work, the formulation advocated by Hurlbut [1985] will be adopted. It reads :

$$\frac{\sigma}{f_t'} = \exp\left(-\frac{c\cdot w}{w_{ft}}\right) \tag{3.1}$$

and the shape coefficient c is of the order of 5 for concrete.

The fracture energy needed to separate the opposite crack faces can be computed as :

$$G_f = \int_0^{w_{ft}} \sigma\cdot d\,w \tag{3.2}$$

For finite element computations, the crack opening w_{ft} is smeared over a certain width within the finite element which, has become known in the literature as the characteristic length l_c. Following Rots et al [1985], the resulting smeared crack strain ε_{cr} is defined from :

$$w_{ft} = l_c\cdot \varepsilon_{cr} \tag{3.3}$$

By means of this artifice, Eq. 3.1 may now be re-written as :

$$\frac{\sigma}{f_t'} = \exp\left(-\alpha\cdot\varepsilon_{cr}\right) \tag{3.4}$$

with : $\alpha = \dfrac{f_t'\cdot l_c}{G_f}$.

Eq. 3.4, or similar, constitutes the basis of most smeared cracking methodologies for the simulation of the degrading response of semi-brittle materials in uniaxial tension.

4 Softening Formulation for Brittle Materials

The fracture surface of Eq. 2.1 corresponds to the locus of states of stress (s.o.s.) delimiting the onset of fracture. Upon further straining, damage of the material manifests itself in load response degradation. Response degradation is equivalent to a contraction of the initial fracture surface in stress space. The multitude of subsequently resulting loci of s.o.s. have been termed in the literature "**unloading surfaces**".

The continuously diminishing size of the unloading surfaces implies that the monotonically varying scalar measure κ_0 of Eq. 2.1 must diminish as well. It is a natural choice to associate κ_0 to the overall degree of **damage** η sustained by the material after initial fracture. In the model this interelationship is defined as:

$$\kappa_0 = \eta \cdot \kappa_{0i} + (1 - \eta) \cdot \kappa_{0r} \tag{4.1}$$

in which κ_{0i} and κ_{0r} are some limiting values of κ_0 whose role will be explained in the following. In similarity to the uniaxial load-deformation expression, Eq. 3.4, the degree of damage is postulated to be a function of the **maximum-principal-tensile-smeared** crack strain ε_{cr}^{m} :

$$\eta = \exp(-\alpha \cdot \varepsilon_{cr}^{m}) \tag{4.2}$$

in which the coefficient α was defined in Eq. 3.4.

As pointed out in Sect. 1, for compressive loading, the magnitude of the tensile splitting displacements is much larger than in uniaxial tension. To account in the model for the less detrimental effects of cracking strain with increasing confinement, w_{ft} is substituted in Eq. 3.1 by an effective crack opening w_{eff} which is defined as follows :

$$w_{eff} = \begin{cases} w_{ft} & \text{for } I_1 \geq 0 \\ (1-\nu) \cdot w_{ft} + \nu \cdot w_{fc} & \text{for } -|f_c| \leq I_1 < 0 \\ w_{fc} & \text{for } I_1 < -|f_c| \end{cases} \tag{4.3}$$

in which $\nu = \dfrac{|I_1|}{f_c'}$ and w_{fc} corresponds to the ultimate, uniaxial compression, splitting crack opening displacement ($w_{fc} \approx 10 \cdot w_{ft}$ for concrete, van Mier [1986]).

The definition of w_{eff} is analogous to the definition of "effective fracture energy" utilized in the fracturing model of Willam et al. [1986]. However, as pointed out by Bazant [1986], "... the fracture energy for shear also includes energy obtained when the area under the compression stress-strain diagram ... is multiplied by the width of the fracture process zone..." hence, it appears that effects other than tensile splitting are included in Willam's definition of effective fracture energy.

5 Mixed Hardening

From the classical plasticity theory point of view, the softening implied by Eq. 2.1 is of "isotropic" nature. Even though this type of fracture surface evolution can, and, has been utilized successfully by many researchers for modelling the hardening behaviour of various materials, it is not always adequate to describe the softening behaviour of anisotropic materials. For realistic modelling of the softening response in all stress domains, a "mixed" softening rule is much more flexible.

In the present investigation, the existence of **two congruent nested limiting surfaces** of material response is postulated, Fig. 3.1. The first corresponds to the **initial fracture surface** introduced in Sect. 2. The other corresponds to the locus of s.o.s. beyond which the response stabilizes to the residual strength characteristics of the material. It will be named in the following the **residual strength surface**.

Because of their definition, the latter surface is contained within the initial fracture surface. During softening, the initial fracture surface progressively translates and contracts until, in the limit, coincides with the residual strength surface.

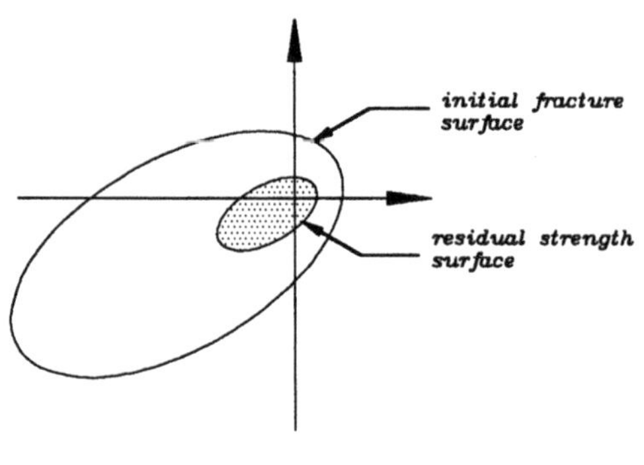

Fig. 5.1 Limit surfaces of material response

In agreement with the mixed nature of the proposed softening law, the development of the subsequent softening surfaces, Fig. 5.2(a), consists of a translation component, Fig. 5.2(b) and, an isotropic contraction component, Fig. 5.2(c). Translation is postulated to take place in such a manner that the center K of the resulting softening surfaces always lies on the line OO' joining the centers of the limiting surfaces. As this results to a locus

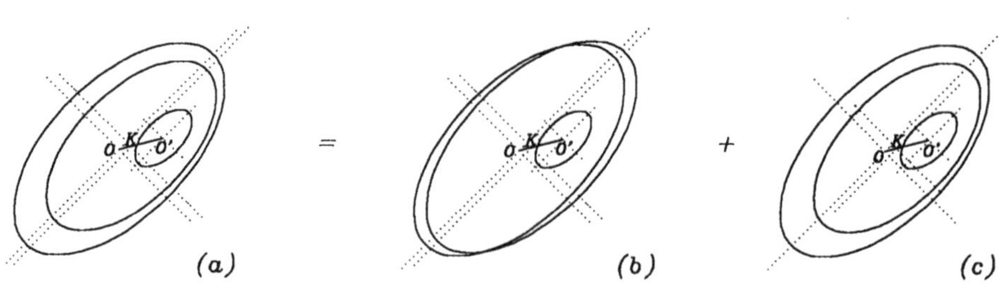

Fig. 5.2 Proposed mixed hardening law

of s.o.s. which may lie outside the initial fracture surface, isotropic contraction is called upon to reduce its size.

If s is defined as the scalar measure of the distance between the center $\underline{\alpha}^n$ of the current softening surface and the center $\underline{\alpha}_r$ i.e. :

$$s = \left| \underline{\alpha}_r - \underline{\alpha}^n \right| \tag{5.1}$$

it is natural to expect that s varies as a function of κ_0 in the form $s = s(\kappa_0)$. The actual form of functional interelationship can be determined, for the specific material under consideration, on the basis of experimental evidence regarding the size and the translation of the softening surface.

6 Flow Rule and Plastic Strain Increments

The mixed nature of the softening law dictates expressing the fracture surface in terms of the coordinates of its center. In this **canonical** form, the quadratic of Eq. 2.1 can be expressed as :

$$f(\underline{\tilde{\sigma}}) = f(\underline{\sigma} - \underline{\alpha}, \kappa_0) = 0 \tag{6.1}$$

in which : $\underline{\alpha}$ = the center of the fracture ellipsoid in stress space,
$\underline{\tilde{\sigma}}$ = $\underline{\sigma} - \underline{\alpha}$

For numerical computation, Eq. 6.1 can be conveniently rewritten in matrix form as :

$$f(\underline{\tilde{\sigma}}) = (\underline{\tilde{\sigma}})^T \cdot \underline{A} \cdot \underline{\tilde{\sigma}} - \tilde{\kappa}_0 \tag{6.2}$$

with $\underline{A}$: the matrix of anisotropy coefficients in the canonical coordinate system.

The truncated Taylor series expansion of $f(\underline{\tilde{\sigma}}, \tilde{\kappa}_0)$ around the trial state of stress $(\underline{\sigma}^t - \underline{\alpha}^n, \kappa_0^n)$ is :

$$\begin{aligned} f^{n+1} &\approx f^t + \left(\frac{\partial f}{\partial \underline{\sigma}}\right)^T \cdot (\underline{\sigma}^{n+1} - \underline{\sigma}^t) + \frac{\partial f}{\partial \kappa_0} \cdot (\kappa_0^{n+1} - \kappa_0^n) + \frac{\partial f}{\partial \underline{\alpha}} \cdot (\underline{\alpha}^{n+1} - \underline{\alpha}^n) \\ &\approx 0 \end{aligned} \tag{6.3}$$

The return mapping of the trial state of stress $\underline{\sigma}^t$ to the fracture surface results to :

$$\underline{\sigma}^{n+1} = \underline{\sigma}^t - \Delta\lambda \cdot \underline{D} \cdot \frac{\partial g}{\partial \underline{\sigma}} \tag{6.4}$$

while the translation of the softening surfaces can be defined as :

$$\begin{aligned} \underline{\alpha}^{n+1} - \underline{\alpha}^n &= \Delta\underline{\alpha} \\ &= -\underline{\hat{u}} \cdot \left(s^{n+1} - s^n\right) \\ &= -\underline{\hat{u}} \cdot \Delta s \end{aligned} \tag{6.5}$$

in which : $\hat{\underline{u}}$ = the unit vector along the direction OO'.

Assuming $s = s(\kappa_0)$, the motion of the center of the softening surface can be expressed in terms of the softening parameter κ_0 as :

$$d\,\underline{\alpha} = -\hat{\underline{u}} \cdot \left(\frac{\partial s}{\partial \kappa_0}\right) \cdot d\,\kappa_0 \tag{6.6}$$

Also :

$$\left(\frac{\partial f(\tilde{\underline{\sigma}})}{\partial \underline{\sigma}}\right)^T = \frac{\partial f}{\partial \tilde{\underline{\sigma}}} \cdot \frac{\partial \tilde{\underline{\sigma}}}{\partial \underline{\sigma}} = \frac{\partial f}{\partial \tilde{\underline{\sigma}}} = 2\,\underline{A} \cdot \tilde{\underline{\sigma}} \tag{6.7}$$

$$\left(\frac{\partial f(\tilde{\underline{\sigma}})}{\partial \underline{\alpha}}\right)^T = \frac{\partial f}{\partial \tilde{\underline{\sigma}}} \cdot \frac{\partial \tilde{\underline{\sigma}}}{\partial \underline{\alpha}} = -\frac{\partial f}{\partial \tilde{\underline{\sigma}}} \tag{6.8}$$

Substituting the above into Eq. 6.3 it results in differential form :

$$f^t + \left(\frac{\partial f}{\partial \tilde{\underline{\sigma}}}\right)^T \cdot \left[-d\lambda \cdot \underline{D} \cdot \frac{\partial g}{\partial \tilde{\underline{\sigma}}}\right] + d\,\kappa_0 \cdot \left[\frac{\partial f}{\partial \kappa_0} + \left(\frac{\partial f}{\partial \tilde{\underline{\sigma}}}\right)^T \cdot \frac{\partial s}{\partial \kappa_0} \cdot \hat{\underline{u}}\right] = 0 \tag{6.9}$$

Also, from the discussion in Section 4, the size κ_0 of the unloading surfaces is related to the damage η , hence :

$$d\,\kappa_0 = \left(\frac{\partial \kappa_0}{\partial \eta}\right) \cdot \frac{\partial \eta}{\partial \varepsilon_{cr}^m} \cdot < d\,\varepsilon_{cr}^m > \tag{6.10}$$

Limiting for demonstration purposes the presentation to the plane stress situation, the increment of maximum principal cracking strain ε_{cr}^m can be computed on the basis of the other plastic strain increments as :

$$< d\,\varepsilon_{cr}^m > = d\lambda \cdot \left\{\frac{\partial g/\partial\sigma_x + \partial g/\partial\sigma_y}{2} + \sqrt{\left(\frac{\partial g/\partial\sigma_x + \partial g/\partial\sigma_y}{2}\right)^2 + \left(\frac{\partial g/\partial\tau}{2}\right)^2}\right\}$$

$$= d\lambda \cdot < \mu > \tag{6.11}$$

In the above expressions, the < > brackets indicate that only the tensile values are to be retained. Substituting successively, the value of the indeterminate parameter $d\lambda$ can be computed from Eq. 6.9 as :

$$d\lambda = \frac{f^t}{-<\mu> \cdot \frac{\partial \kappa_0}{\partial \eta} \cdot \frac{\partial \eta}{\partial \varepsilon_{cr}^m} \cdot \left[\frac{\partial f}{\partial \kappa_0} + \left(\frac{\partial f}{\partial \underline{\sigma}}\right)^T \cdot \hat{\underline{u}} \cdot \frac{\partial s}{\partial \kappa_0}\right] + \left(\frac{\partial f}{\partial \underline{\sigma}}\right)^T \cdot \underline{D} \cdot \frac{\partial g}{\partial \underline{\sigma}}} \tag{6.12}$$

The first term of the sum in the denominator of Eq. 6.12 indicates that, response degradation occurs only for tensile values of the maximum principal cracking strain in the material.

7 Performance Tests

The performance of the proposed strain softening formulation is demonstrated by plane-stress tests on a masonry material.

The initial fracture surface and the residual surface, Fig. 7.1, have been determined from the test data of Spanos [1987]. The Young modulus was estimated at 2000. MPa, the Poisson ratio as 0.1 and the fracture energy in uniaxial tension as 0.05 MPa.

For verification of the material-constitutive-law, a driver program was developed. In all tests, uniform lateral pressure (compressive or tensile) of the desired magnitude was first applied. Then, the vertical strain was increased monotonically.

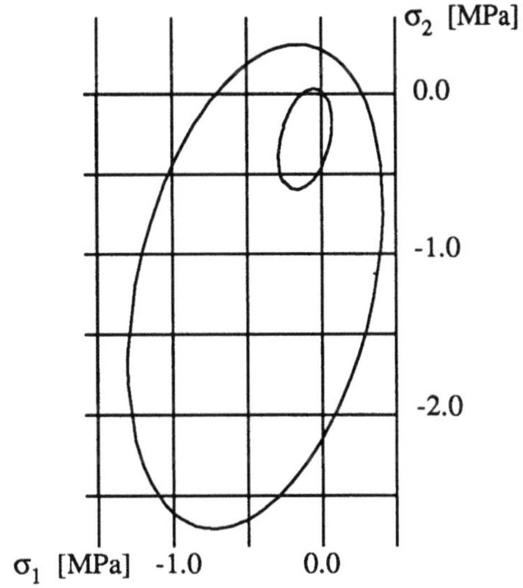

Fig. 7.1 Biaxial test data of Spanos

The influence of the characteristic length on the uniaxial tensile stress-strain response of the material is portrayed in Fig. 7.2. The numerical legends refer to multiples of the basic characteristic length l_c = 140 mm. As expected, an increase in l_c results to a more brittle material behaviour in uniaxial tension.

As mentioned earlier, one of the main features of the model is its capability to treat the transition from the plasticity to the fracture domain in a unified consistent formulation. For this reason a number of validation tests were performed spanning the biaxial tension-compression and compression-compression regions.

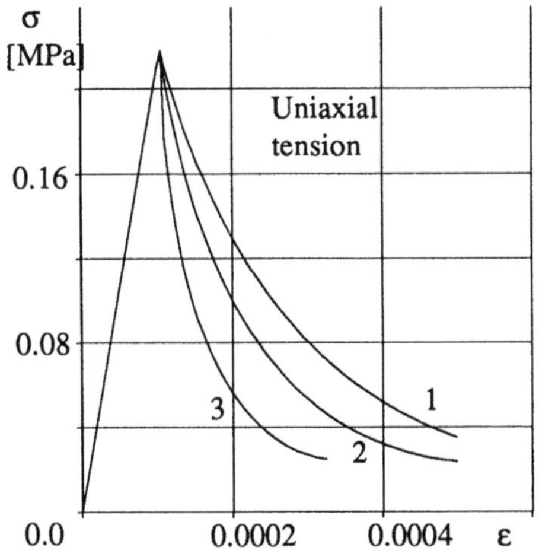

Fig. 7.2 Influence of characteristic length

Various degrees of lateral confinement were applied. The legends in Fig. 7.3 correspond to the following conditions :

ht–c :	high lateral tension (0.10 MPa)	+	compression
lt–c :	low lateral tension (0.01 MPa)	+	compression
c :	no lateral pressure	+	compression
lc–c :	low lateral compression (0.20 MPa)	+	compression

Despite the lack of any additionally imposed tension cut-off criterion, the model can

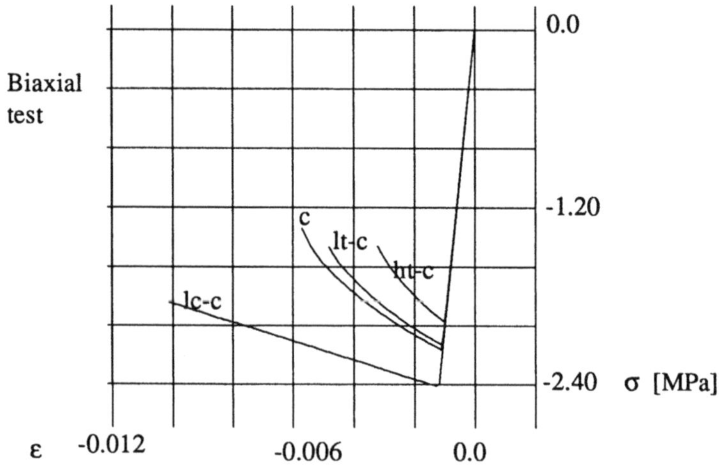

Fig. 7.3 Compression softening

simulate softening in both, the tension-compression and the compression-compression regions, ensuring, at the same time that, the correct amount of fracture energy is dissipated. These are significant advantages over commonly available models.

In agreement with Sect. 4, higher confinement and hence smaller magnitudes of maximum principal tensile strain, result to a more ductile response accompanied with diminished magnitudes of strength degradation. In the limit, for pure triaxial compression s.o.s., the behaviour will resemble that of an ideal elasto-plastic material with no softening.

From the above it can be concluded that, the maximum principal tensile strain criterion can be utilized as a versatile descriptor of damage in fracturing materials.

8 References

Bazant Z.P. and Cedolin L. (1979) Blunt crack band propagation in finite element analysis. **Journal of the Engineering Mechanics Division, ASCE**, 105, 297-315.

Bazant Z.P. and Oh B. H. (1983) Crack band theory for fracture of concrete. **Materials and Structures, RILEM**, 16, 155-177.

Bazant Z.P. and Pfeiffer P. A. (1986) Shear fracture tests of concrete. **Materials and Structures, RILEM**, 19, 111-121.

Droz P. and Bazant Z.P. (1988) Nonlocal analysis of stable states and stable paths of propagation of damage shear bands, in **Cracking and damage : strain localization and size effect**, eds. Mazars J. and Bazant Z.P., Elsevier Applied Science, N.Y., pp. 415-425.

Horii H. and Nemat-Nasser S. (1985) Compression-induced micro-crack growth in brittle solids : axial splitting and shear failure. **Journal of Geophysical Research**, 90, 3105-3125.

Hurlbut B. (1985) Experimental and computational investigation of strain-softening in concrete. M.S. Thesis, CEAE Dept., Univ. of Colorado, Boulder.

Kotsovos M. (1983) Effect of testing techniques on the post-ultimate behaviour of concrete in compression. **Materials and Structures, RILEM**, 16, 3-12.

Leroy Y. and Ortiz M. (1990) Finite element analysis of transient strain localization phenomena in frictional solids. **International Journal of Numerical and Analytical Methods in Geomechanics**, 14, 93-124.

Liu T.C.Y., Nilson A.H. and Slate F.O. (1972) Stress-strain response and fracture of concrete in uniaxial and biaxial compression. **ACI Journal**, 69, 291-295.

van Mier J. (1986) Fracture of concrete under complex stress. **HERON**, 31, No.3.

Nemat-Nasser S. and Horii H. (1982) Compression-induced nonplanar crack extension with application to splitting, exfoliation and rockburst. **Journal of Geophysical Research**, 87, 6805-6821.

Rots J.G., Nauta P., Kusters G.M. and Blaauwendraad J. (1985) Smeared crack approach and fracture localization in concrete. **HERON**, 30, No. 1.

Spanos C. (1987) Failure characteristics of masonry walls in compression. M. Eng. Thesis, Fac. of Civil Engineering, TU Athens, Athens.

Vardoulakis I. and Graf B. (1985) Calibration of constitutive models for granular materials using data from biaxial experiments. **Geotechnique**, 35, 299-317.

Wang Q.B. (1989) Shear strength of reinforced concrete beams and slabs without shear reinforcement, Research Rept. 25.2-89-2-10, Faculty of Civil Engineerimg, TU Delft, Delft.

Willam K., Hurlbut B. and Sture S. (1986) Experimental and Constitutive aspects of concrete failure, in **Proc. US-Japan Seminar on FE Analysis of RC Structures**, Tokyo, ASCE, Special Publication, pp. 226-254.

4 MICROMECHANICS AND LOCALIZATION IN CONCRETE AND ROCK

H. HORII
Department of Civil Engineering, University of Tokyo,
Bunkyo-ku, Tokyo, Japan

Abstract
Localization phenomenon is a challenging subject of solid mechanics, and is of importance in many engineering problems since it governs the total behaviors or ultimate loads in various kinds of problems. It is also of great concern for development and expansion of fracture mechanics of concrete and rock. In this paper, firstly a thermomechanics-based theory of localization phenomena is introduced. A unified theory for energy dissipation processes and a general condition for localization are presented. The theory is considered to provide a framework for the algorithm in the analysis of distributed cracks and their localization in *e.g.* concrete structures. Secondly, a micromechanics-based continuum theory is introduced which can reproduce localization phenomena. The information important for localization is lost through usual homogenization processes. It is necessary to account for the interaction effect between microstructures directly in the evolution law. In the present theory a new field quantity is introduced which represents the interaction effect. The additional governing equation associated with the new field quantity is shown to be an integral equation. Numerical results show that the softening behavior following a peak point and bifurcation from homogeneous deformation to localized one are reproduced by the present theory.
Keywords: Localization, Micromechanics, Thermomechanics, Homogenization

1 Introduction

After more than ten-year extensive studies on fracture mechanics of concrete and rock, the mode one crack growth phenomena in concrete and rock have been captured and general agreement on governing mechanisms, their modelling, material characterization, and analysis method has been attained to some extent, while discussions are still required for details. Studies on fracture mechanics of concrete and rock entered the second stage. What is the next thing to do? The following is the list of topics which the author considers to be of importance in the second stage of studies on fracture mechanics of concrete and rock.

1) Establishment of general analysis tool
   It is time to unify or classify the existing models of crack growth in concrete and rock and to establish a standard model. The model should reproduce localization of cracking, *i.e.* the transition from distributed crack mode to a mode with a single crack growing and other cracks closing. One of the final goals is to establish a general analysis tool which can predict or reproduce mechanical behaviors of concrete structures under arbitrary conditions. It is required to combine fracture mechanics of concrete and FEM of concrete structures.

Fracture and Damage of Concrete and Rock – FDCR-2.
Edited by H.P. Rossmanith. © 1993 Taylor & Francis. ISBN 0 419 18470 8.

2) Extension of fracture mechanics

The prediction of the ultimate load of concrete structure is one of the important tasks expected from analytical studies. It is important to predict initiation and propagation of cracks because it determines the geometry of the final state. However, the prediction of crack growth alone cannot lead to the prediction of the ultimate load of concrete structures. Fracture mechanics of concrete will be much more attractive if it is extended to failure under compression.

3) Size effect

It has been reported that existence and growth of cracks are responsible for size effect of strength of concrete structures. However, there are many factors which affect the size effect. It is necessary to identify and classify the mechanisms of the size effect.

4) Application for design

Recently, a new design provision for SFRC concrete tunnel lining was published in Japan. It is based on the concept of fracture mechanics of concrete. To estimate the maximum resultant forces of the critical cross section, existence of a crack and transmitted stress by fibers are considered. This is an example of application of fracture mechanics for design.

5) Development and design of new materials

The knowledge obtained in the study of fracture mechanics of concrete should be utilized for the development and design of new cementitious composites which possesses epoch-making properties.

Localization is the main issue in this paper. To capture localization phenomena is one of important tasks in the second stage of fracture mechanics of concrete and rock. For example, it is important to establish a general analysis tool for fracture behaviors. It is known that numerical analysis of three point bending test of concrete with no reinforcement results in distributed cracks as shown in Fig. 1.

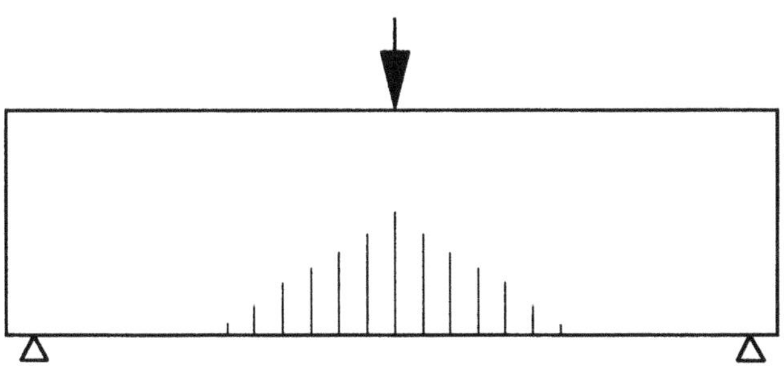

Fig.1. Numerical result of distributed cracks in concrete specimen.

In reality, crack growth is localized and a single crack continues to grow while other cracks start to close. It is known that the unloading path as shown in Fig. 2 is necessary to reproduce the localization phenomena. However, the introduction of the unloading path is not sufficient and an appropriate numerical algorithm is required in the analysis method. The numerical algorithm should be based on a theoretical framework. Hence, the establishment of theoretical basis for localization phenomena is of great importance.

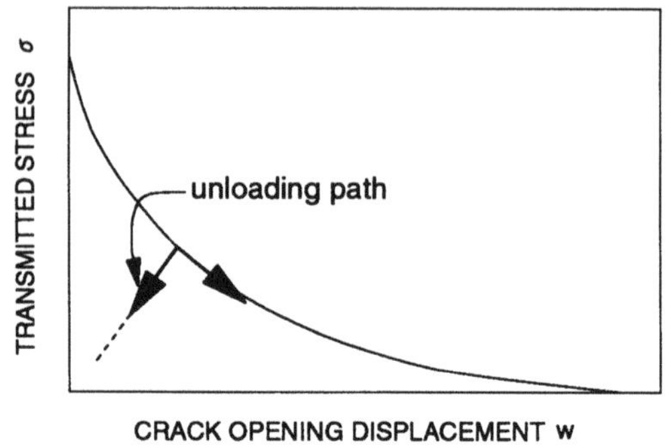

Fig.2. Tension-softening curve and unloading path.

Another important aspect related to localization is the prediction of ultimate load in compression. To extend fracture mechanics of concrete and rock from tensile fracture to compressive fracture and to predict the critical state under general conditions, it is necessary to construct a continuum theory which can reproduce localization phenomena. Empirical laws are limited for certain conditions and cannot be applied for general conditions.

In the present paper, firstly a theory based on thermomechanics is introduced for general description of energy dissipation process and localization phenomena. Secondly, a micromechanics-based continuum theory is presented which can reproduce localization phenomena as a result of numerical solution of boundary value problems.

2 Thermomechanics for energy dissipation processes and localization phenomena

Localization phenomena are observed in many materials, such as metal, concrete, soil, and rock, and govern the global behaviors in most cases. Since the localization itself is common in different materials with different mechanisms of deformation and failure, a unified theory is expected which captures localization phenomena in different materials.

In this section a theory of thermomechanics is introduced for a unified treatment of energy dissipation processes and localization phenomena after Sakata (1992). The formulation is based on Nguyen (1987); for basic background see a text book by Maugin (1992).

Consider a system whose total potential energy E is given as a function of a reversible parameter u, an irreversible parameter α, and a loading parameter λ as $E(u,\alpha,\lambda)$. The displacement field serves as the reversible parameter. The irreversible parameter represents the amount of energy dissipation and takes different forms in different materials depending on the form of energy dissipation. It can be a scaler quantity or a tensor quantity and is defined as a field quantity or takes a finite number of values for a system. For plasticity, plastic strain serves as α. For crack growth phenomena, crack lengths are taken for α.

The equilibrium equation is expressed as

$$E,_u(u,\alpha,\lambda)\cdot\delta u = 0 \,. \tag{1}$$

The solution of this equilibrium equation is obtained as $u = u(\alpha,\lambda)$. Substituting the solution of the equilibrium equation into $E(u,\alpha,\lambda)$, the total potential energy is expressed as $W(\alpha,\lambda) = E(u(\alpha,\lambda),\alpha,\lambda)$.

Then irreversible force associated with α is defined by

$$A = -W,_{\alpha}(\alpha,\lambda) . \tag{2}$$

In the case of crack growth phenomena where crack length c serves as α, the irreversible force A is found to be the energy release rate G. The irreversible force is understood to represent a force which promotes the energy dissipation.

The behavior of materials involving energy dissipation is described by providing a so-called evolution law. The evolution law generally provides relationship between the irreversible parameter and the irreversible force from which the variation of the irreversible parameter is determined as a function of the loading parameter.

The standard material is an idealized material for which the evolution law is provided by the following normality rule; see Fig. 3. In the space of irreversible force, the irreversible force takes values in a convex region C of admissible forces. Inside C, the increment of the irreversible parameter is zero; $\dot{\alpha} = 0$. On the boundary ∂C, the increment of the irreversible parameter takes a non-zero value and the increment is directing outward normal to the boundary of C. It is shown that the energy dissipated in the process described by the normality rule is always non-negative, *i.e.*

$$A \cdot \dot{\alpha} \geq 0 , \tag{3}$$

and the second law of thermodynamics is always satisfied.

The localization phenomena and the condition for localization are studied by considering the following simple example. We consider a two-dimensional problem of an infinite plane containing two cracks subjected to uniform stresses and concentrated forces as shown in Fig. 4. Two cracks have an equal initial length $c_1 = c_2 = c_0$ which is small enough compared with their distance d. The forces increase proportionally with $P = kc_0\sigma$ where k is a given constant. We assume LEFM for crack growth, *i.e.* a crack starts to grow when the energy release rate G reaches the fracture toughness G_c and keeps $G = G_c$ during crack growth. Note that two cracks grow equally when $k = \infty$ *i.e.* $P \neq 0$, $\sigma = 0$; homogeneous solution. When $k = 0$, only one crack extends while another crack remains; localized solution.

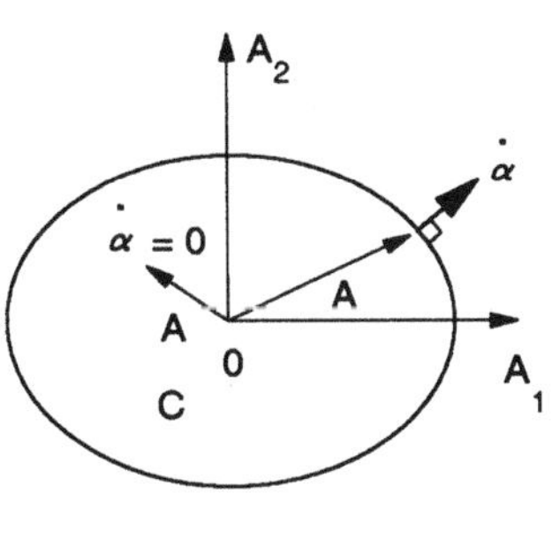

Fig.3. Convex domain of admissible forces and normality rule.

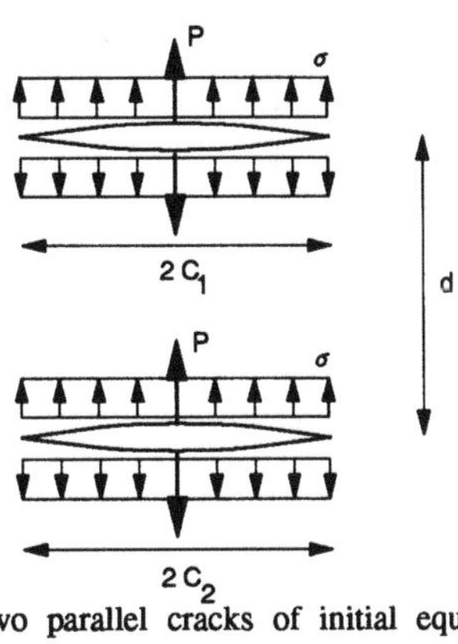

Fig.4. Two parallel cracks of initial equal length.

Our interest is the condition for bifurcation from the homogeneous solution to the localized solution. According to the theory of thermomechanics introduced above, the irreversible parameter in this problem is $\boldsymbol{c} = (c_1,c_2)$ and the associated irreversible force is $\boldsymbol{G} = (G_1,G_2)$. The region C of admissible values of $\boldsymbol{G}$ is as shown in Fig. 5. Inside C, $\dot{\boldsymbol{c}} = \boldsymbol{0}$. On AB, $\dot{\boldsymbol{c}} = (\dot{c}_1,0)$ since $G_1 = G_c$ and $G_2 < G_c$. On BC, $\dot{\boldsymbol{c}} = (0,\dot{c}_2)$. At the corner point B, $\dot{\boldsymbol{c}}$ is not unique and the bifurcation from the homogeneous solution to the localized solution can occur at the corner point.

The solutions of the problem, *i.e.* variation of crack lengths with the increasing loading parameter, are presented as solution paths in the crack length space; see Fig. 6. It is shown that the two cracks grow equally (homogeneous solution) until critical crack length is attained. After the critical point, only single crack continues to grow while the other crack stops to grow. To identify the solution path to be selected, the stability of the solution path (not the stability of the crack growth) is defined as follows.

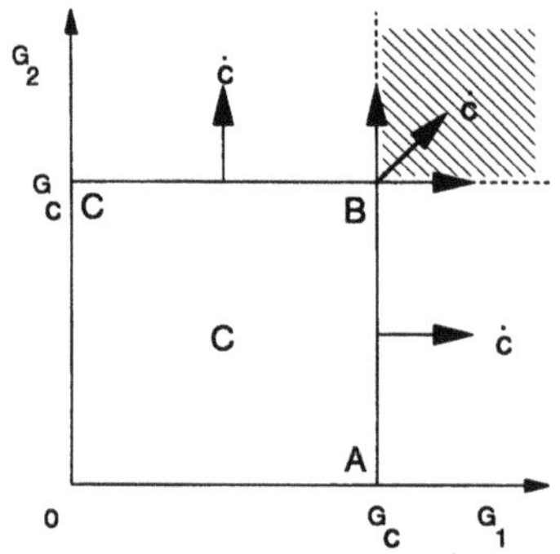

Fig.5. Space of energy release rate and normality rule.

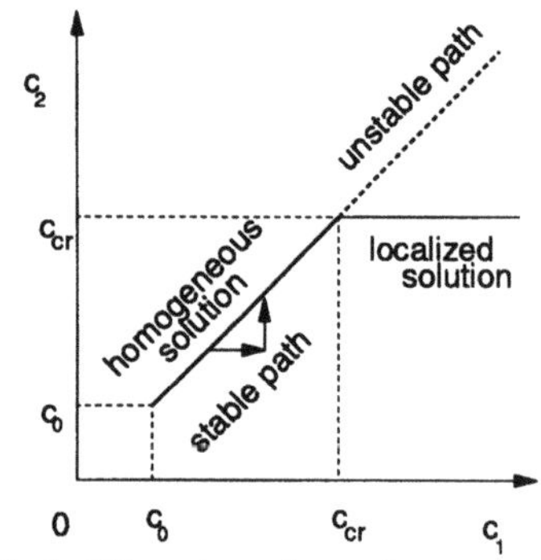

Fig.6. Bifurcation from homogeneous solution to localized solution.

The homogeneous solution path is stable if the solution point returns back to the original path when small disturbance in the crack length is prescribed. Otherwise the homogeneous solution path is said to be unstable; see Fig. 6.

The stability of the homogeneous solution path is investigated directly by evaluating the stress intensity factor (or energy release rate) with, for example, method of pseudotractions by Horii and Nemat-Nasser (1985) which is a general method to evaluate the interaction effect between microstructures. It can be shown that a homogeneous solution path is stable if all of the eigen values of the Hessian, the matrix which consists of second derivatives of the total potential energy at equilibrium,

$$\begin{bmatrix} \dfrac{\partial^2 W}{\partial c_1^2} & \dfrac{\partial^2 W}{\partial c_1 \partial c_2} \\ \dfrac{\partial^2 W}{\partial c_2 \partial c_1} & \dfrac{\partial^2 W}{\partial c_2^2} \end{bmatrix} . \tag{4}$$

In this example, the maximum and minimum eigen values correspond to the stability of crack growth along the homogeneous solution path and the stability of the homogeneous solution path, respectively. When the maximum eigen value is positive, the crack growth along the homogeneous path is stable, which is the case under consideration here. After the minimum eigen value becomes negative the homogeneous solution path is unstable. The bifurcation of the solution occurs when the minimum eigen value vanishes.

Then, it is seen that the examination of the stability of the homogeneous solution path is equivalent to the examination of the principal curvature, *i.e.* the convexity, of the energy surface. It can also be shown that the path with minimum potential energy is selected even in the case of energy dissipation process.

Figure 7 shows the homogeneous solution path and the localized solution paths on the energy surface, and the cross section of the energy surface presenting how the convexity of the surface is lost when the bifurcation occurs.

In this paper it is only mentioned that the results can be extended to the cases of N cracks and distributed cracks by changing the dimension of the problem from two to N or infinity.

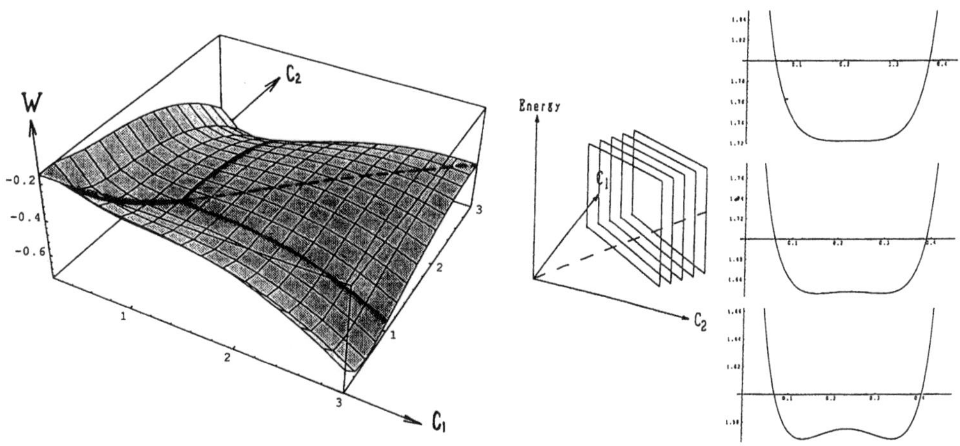

Fig.7. Homogeneous solution path and localized solution path on energy surface, and cross section of energy surface showing loss of convexity at bifurcation point.

3 Micromechanics-based continuum theory for localization phenomena

The mechanism of localization phenomena is attributed to the interaction effect between microstructures. The bifurcation from the homogeneous solution path to localized solution path discussed in the previous section result from the interaction effect.

Micromechanical models which directly take into account of the interaction effect can reproduce localization phenomena. For example, Shi and Horii (1989) showed that strain localization in sand is reproduced by the proposed microslip model. Although micromechanical models clarify the mechanism of localization phenomena, models which treat microstructures individually are not suitable for the analysis of engineering problems. To provide answers to engineering problems, a continuum theory and an associated analysis tool which can reproduce localization phenomena are required.

Since the localization phenomena are governed by the interaction effect between microstructures, the continuum theory must be based on micromechanics. To derive a continuum theory from a micromechanical model, a homogenization process is employed. In the homogenization process information on the interaction effect is most likely to be lost. As a result the continuum theory derived through the homogenization process cannot reproduce localization phenomena. Here, a new continuum theory based on micromechanics is introduced and is compared with a conventional micromechanics-based continuum damage mechanics; see Okui, Horii, and Akiyama (1992) for details.

In this paper, we consider crack growth under compression which is the main mechanism of deformation and failure behavior of rock. The crack growth due to frictional sliding or other mechanisms at initial defects is modelled as shown in Fig. 8; see Horii and Nemat-Nasser (1986).

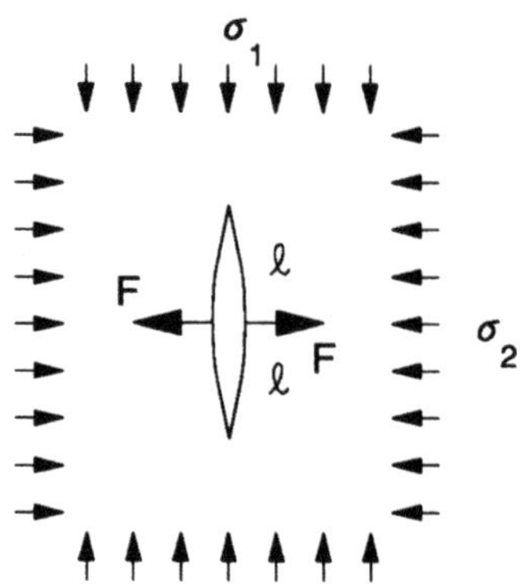

Fig.8. Model of crack growth under compression.

The effect of initial defects to drive the crack growth is modelled by the concentrated forces. The crack density (number of cracks per unit volume) is assumed to be constant and is prescribed. The inelastic strain due to growth and opening of the crack is calculated by integrating the crack opening displacement as

$$\varepsilon_{ij}^{IN} = \frac{1}{V}\int_S \frac{1}{2}([u_i]n_j + [u_j]n_i)dS = f(l,\sigma_1,\sigma_2) \,. \tag{5}$$

The evolution law is obtained by adopting LEFM and equating the stress intensity factor with the fracture toughness as

$$K_I = g(l,\sigma_1,\sigma_2) = K_c \,. \tag{6}$$

Note that the crack length l serves as the damage parameter.

The relationship (5) between the inelastic strain and the damage parameter (crack length) l, and the evolution law (6) are derived by considering a single crack in an infinite solid, which implies to neglect the interaction effect and to assume equal crack growth in a representative volume element. This is the simplest and usual homogenization method.

Then, a micromechanics-based continuum damage mechanics is formulated as follows,

$$\sigma_{ij,j} + p_i = 0 \,,$$

$$\varepsilon_{ij} = \frac{1}{2}(u_{i,j} + u_{j,i}) \, ,$$

$$\sigma_{ij} = C_{ijkl}(\varepsilon_{kl} - \varepsilon_{kl}^{IN}) \, , \qquad (7a\text{-}e)$$

$$\varepsilon_{kl}^{IN} = f_{kl}(\alpha, \sigma_{mn}) \, ,$$

$$\phi(\alpha, \sigma_{mn}) = 0 \, .$$

All quantities in the above equations are understood to be average quantities over a representative volume element. The last two equations are derived from the micromechanical model.

As shown later localization phenomena cannot be reproduced by this theory because the interaction effect is not directly accounted. A change in the damage parameter at a point x results in the change in inelastic strain ε_{ij}^{IN} at the point x which affects the stress at a different point x' through equilibrium equation. Then it results in the change in the damage parameter at the point x' through the evolution law. The evaluation of the interaction between the damage at different points is indirect and involves equations whose derivation have already employed homogenization processes. To reproduce localization phenomena, direct evaluation of the interaction effect is necessary in the evolution law.

To account the interaction effect in the evolution law, we generalize the method of pseudotractions by Horii and Nemat-Nasser (1985). The method of pseudotractions is for the evaluation of interaction effect between discrete microstructures. The problem is decomposed into subproblems and pseudotractions are introduced for each microstructures; see Fig. 9. The pseudotractions are determined by the consistency equation which assures the satisfaction of the boundary condition in the original problem.

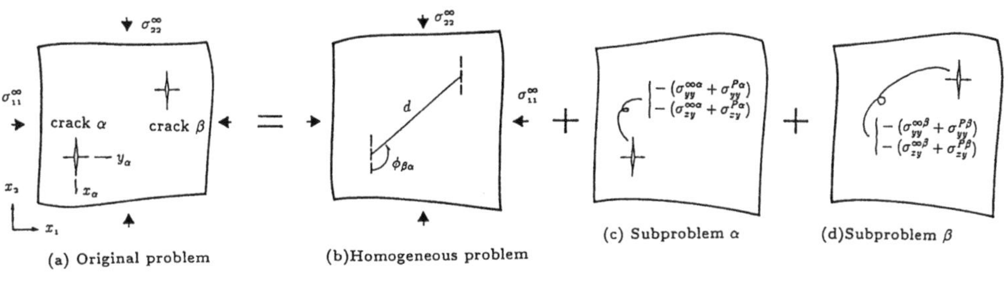

Fig.9. Decomposition of original problem in method of pseudotractions.

We consider continuously distributed cracks. Cracks exist everywhere with certain crack density. This concept itself is homogenization. Then the pseudotraction defined for each crack becomes field quantity $\sigma_{ij}^{P}(x)$ called interaction field. The interaction field represents the interaction effect at a point from all other distributed cracks. The consistency condition that the boundary condition is to be satisfied when all subproblems are superimposed leads to an integral equation for the interaction field,

$$\sigma_{ij}^{P}(x) = \int_V K_{ijkl}(\alpha, x, x')[\sigma_{kl}(x') + \sigma_{kl}^{P}(x')] dV(x') \, . \qquad (8)$$

The kernel of the integral represents stress at point $\boldsymbol{x}$ due to a crack at other point $\boldsymbol{x}'$.

With this new field quantity and new governing equation, we have a new continuum theory called interaction field theory (IFT):

$$\sigma_{ij,j} + p_i = 0 \; ,$$
$$\varepsilon_{ij} = \frac{1}{2}(u_{i,j} + u_{j,i}) \; ,$$
$$\sigma_{ij} = C_{ijkl}(\varepsilon_{kl} - \varepsilon_{kl}^{IN}) \; , \qquad \text{(9a-f)}$$
$$\varepsilon_{kl}^{IN} = f_{kl}(\alpha,\sigma_{mn}) \; ,$$
$$\phi(\alpha,\sigma_{mn},\sigma_{mn}^{p}) = 0 \; ,$$
$$\sigma_{ij}^{p}(\boldsymbol{x}) = \int_V K_{ijkl}(\alpha,\boldsymbol{x},\boldsymbol{x}')[\sigma_{kl}(\boldsymbol{x}')+\sigma_{kl}^{p}(\boldsymbol{x}')]dV(\boldsymbol{x}') \; .$$

In the evolution law, the interaction effect is taken into account through the interaction field. The additional governing equation is an integral equation.

It is interesting to notice that if we eliminate the damage parameter and the interaction field from the last three equations, we obtain a nonlocal constitutive equation in which the material behavior at a point depends on the state at other points.

By introducing the finite element discretization and assuming constant damage (crack length) and interaction field σ_{ij}^{p} in each element, the problem is reduced to usual FEM equations and additional nonlinear algebraic equations. By solving the total system of algebraic equations numerically, distribution of damage as well as stress, strain, and displacement are obtained.

Figure 10 shows the relationship between stress and damage (crack length) for a specimen under biaxial compression. Eight elements are used. Results by the Continuum Damage Mechanics

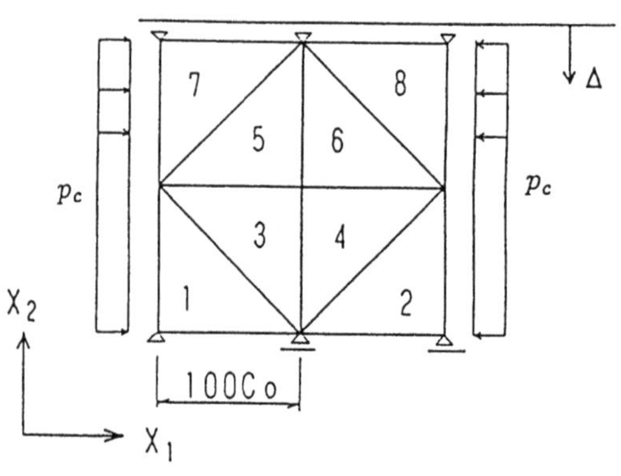

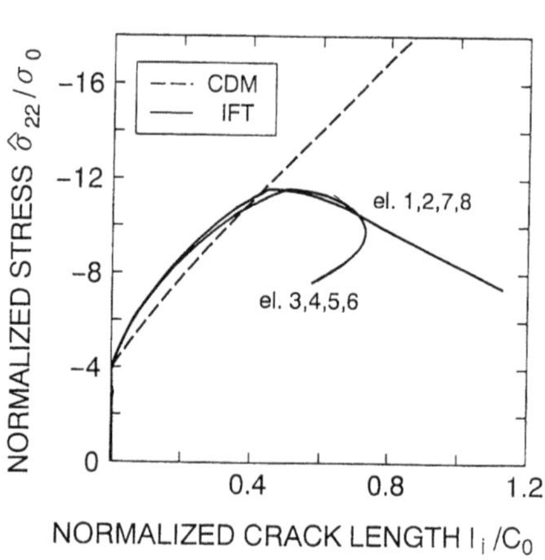

Fig.10. Axial stress vs. crack length with p_c/σ_0 = -2.0.

(CDM) show that deformation is always uniform and stress increases monotonously. On the other hand the present Interaction Field Theory (IFT) reproduces a peak and following softening behavior. The deformation is localized around the peak point and the softening behavior is not a material behavior but a structural behavior. Note that the ultimate stress is predicted as a result of numerical analysis without any failure criterion. No information is required for the prediction of the failure load; the critical value of the damage parameter is not used for this prediction.

Figure 11 shows a relationship between the applied stress and the damage (crack length) in the element 6. The bifurcation of the solution occurs before the peak point of the original path. The original path corresponds to the homogeneous crack growth and the bifurcated solution corresponds to the localized crack growth. It is known that the shear failure in rock under compression is due to the localization of microcracking, and the present result agree with the experimental facts; see Horii and Nemat-Nasser (1986).

Similar results are obtained with more elements. The effect of confining pressure is investigated with 24 elements. As is seen from Fig. 12, the peak stress increases with increasing confining pressure.

It is interesting to notice that the transition of the failure mode from axial splitting to shear failure is naturally reproduced by the present analysis. This result agree with experimental observation in many kinds of rock.

To compare with experimental results, input data are selected based on the reported data for Westerly granite. The reported values of the Young's modulus E, Poisson's ratio ν, and fracture toughness K_c are used. The density of cracks and size of initial defects are determined based on the direct observation with SEM by Hadley (1975). Note that input data are only those five parameters.

The ultimate axial stress is plotted as a function of the confining pressure in Fig. 13. The solid curve is the prediction by the present theory and is compared with experimental data reported by Brace *et al.* (1966) and Mogi (1966). The agreement between experimental data and numerical results of the present method is satisfactory.

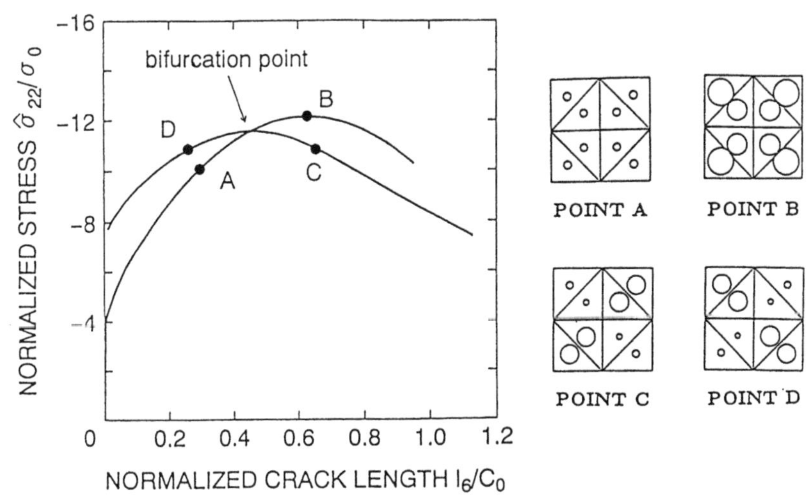

Fig.11. Axial stress vs. crack length in element 6 and crack growth modes.

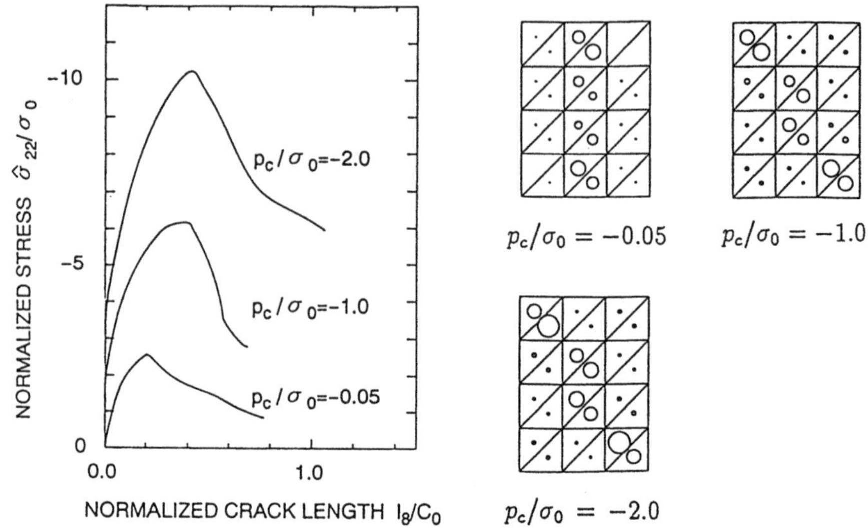

Fig.12. Effect of confining pressure in stress vs. crack length curve and distribution of crack length.

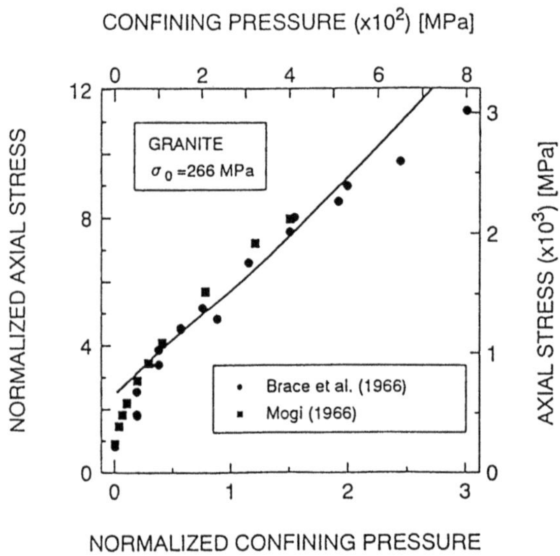

Fig.13. Axial compressive strength as a function of confining pressure.

4 Concluding remarks

Localization phenomenon is an attractive subject of solid mechanics, and is of importance in many engineering problems since it often governs the total behavior of the structures or materials. It is also of great concern for development and expansion of fracture mechanics of concrete and rock. The thermomechanics-based theory of localization phenomena introduced in Section 2 would provide a theoretical framework for the algorithm in the analysis of distributed cracks and their localization. The micromechanics-based continuum theory based on an appropriate micromechanical model for concrete deformation would be employed for the prediction of failure in compressive region which is required for the estimation of ultimate loads in many concrete structures.

Acknowledgments

The present study has been supported in part by the Grant-in-Aid of Scientific Research from the Japanese Ministry of Education, Science and Culture and by TEPCO Research Foundation.

References

Brace, W.F., Paulding, B.W. and Scholz, C. (1966) Dilatancy in the fracture of crystalline rock. **J. Geophys. Res.**, 71, 3939-3953.

Hadley, K. (1975) Comparison of calculated and observed crack densities and seismic velocities in Westerly granite. **J. Geophys. Res.**, 81, 3484-3494.

Horii, H. and Nemat-Nasser, S. (1985) Elastic fields of interacting inhomogeneities. **Int. J. Solids Structures**, 21, 731-745.

Horii, H. and Nemat-Nasser, S. (1986) Brittle failure in compression: splitting, faulting and brittle-ductile transition. **Phil. Trans. R. Soc. Lond.**, A319, 337-374.

Maugin, G.A. (1992) **The Thermomechanics of Plasticity and Fracture.** Cambridge University Press.

Mogi, K. (1966) Some precise measurements of fracture strength of rocks under uniform compressive stress. **Rock Mech. and Eng. Geology**, IV, 41-55.

Nguyen, Q.S. (1987) Bifurcation and post-bifurcation analysis in plasticity and brittle fracture. **J. Mech. Phys. Solids**, 35, 303-324.

Okui, Y., Horii, H. and Akiyama, N. (1992) A continuum theory for solids containing microdefects. **Int. J. Engrg. Sci.**; in print.

Sakata, K. (1992) **Thermomechanics of strain localization in solids with microstructures.** Master Thesis, University of Tokyo.

Shi, Z.-H. and Horii, H. (1989) Microslip model of strain localization in sand deformation. **Mechanics of Materials**, 8, 89-102.

5 SOFTENING CURVES FOR CONCRETE AND STRUCTURAL RESPONSE

J. PLANAS, G.V. GUINEA and M. ELICES
Departamento Ciencia de Materiales, Universidad Politécnica de Madrid, Madrid, Spain

Abstract
The softening function is the main input needed to model fracture of concrete when using a cohesive crack approach. The influence of different aspects of the softening function on the structural response are analyzed to show that four parameters are needed to describe reasonably well the softening function: the tensile strength f_t, the specific fracture energy G_F, and two more parameters characterizing the shape of this function. A recently developed method to determine the four parameters from stable tests on notched specimens and simple fitting procedures is outlined and discussed.
Keywords: Concrete, Cohesive Crack, Softening Function, Fracture Energy.

1 Introduction

Modelling concrete cracks by using cohesive cracks has been a successful procedure since its proposal by Hillerborg and co-workers (Hillerborg *et al.*,1976). This success is partly due to its physical meaning and to its simplicity. A basic ingredient of the model is the *softening* curve, a material property. This function relates the stresses —the cohesive stresses— on the crack to the corresponding crack openings.

In principle, the soundest method to obtain the softening function is by means of direct tensile tests. Unfortunately this procedure has many drawbacks (Reinhardt *et al.*, 1986; van Mier, 1986), and it is difficult to achieve a crack of uniform opening, as required. This is why most of the procedures to infer the softening function rely on indirect methods based on the parametric fitting of the experimental results of bending beams or compact specimens (Wittmann *et al.*, 1987, 1988; Mihashi, 1992). These methods are, however, very complex in their present version because the parametric fit is nonlinear and requires an iteration loop which includes a full finite element computation of the load-displacement (or load-CMOD) curve.

In most practical situations, the users of a cohesive crack model do not rely on direct or indirect determination of the actual softening curve, but they rather select an analytically defined curve from those available in the literature and use the values of only two parameters, namely the tensile strength and the fracture energy. This paper discusses this approach by showing, in section 2, how one may expect the different parts of the softening curve to affect the various aspects of the structural behavior, when the structure size is taken into account. It is shown that at least 4 parameters of the softening curve are expected to play a role in the structural behavior.

Section 3 outlines a recent procedure —the general bilinear fitting (GBF)— which delivers reasonable estimations of the 4 independent parameters of a bilinear softening function from tests performed on notched beams and relatively simple fitting procedures.

Section 4 discusses the results and draws the main conclusions.

Fracture and Damage of Concrete and Rock – FDCR-2.
Edited by H.P. Rossmanith. © 1993 Taylor & Francis. ISBN 0 419 18470 8.

2 Softening curves and structural response

2.1 Softening curves

As far as monotonic mode I is concerned, a cohesive crack is defined by its softening curve. Two characteristic (and essential) parameters of this curve are the tensile strength f_t, and the fracture energy G_F, which are identified, respectively, with the transferred stress at zero opening and with the area below the softening curve. These two parameters are, in principle, measurable in experiments without any explicit hypotheses about the shape of the softening curve.

Another characteristic property of the softening curve is the critical crack opening w_c, which is the crack opening for which the stress transferred becomes zero. The essential difference with the f_t and G_F is that its determination is very elusive. In direct tension tests its value may be very dependent on the resolution of the measurement, and in indirect tests its determination relies on the assumed shape of the softening curve.

Bi-parametric softening curves are commonly used, taking f_t and G_F as free parameters. Obvious examples of such curves are the rectangular softening, the linear softening and the exponential softening. A widely used curve is Petersson's bilinear softening, in which the relative position of the kink point is fixed. In fact, the coordinates of the kink point (σ_k, w_k) are ($f_t/3$, $0.8G_F/f_t$), hence fixed given f_t and G_F (Petersson, 1981).

In some circumstances further variable parameters are included, as in the general bilinear softening which is a 4-parametric family of softening curves because the relative position of the kink point (2 parameters) is left free, in addition to the tensile strength and the fracture energy (Wittmann *et al.*, 1987, 1988; Mihashi, 1992).

In this work we also have chosen the general bilinear softening to approximate the softening function, but before showing the GBF technique used to fit this kind of softening to particular experimental results, it is appropriate to explore the influence of the different aspects of the softening curve on the mechanical response of a structure, because this exploration was what induced the GBF method.

2.2 Influence of softening on the structural response

Consider a notched plain concrete specimen which is progressively deformed up to complete fracture, and assume that the softening curve is well behaved: monotonically decreasing and concave upwards. How do the different parts of the softening function influence the different parts of the response? The answer is relatively complex because it depends on the size of the specimen. However, a number of general rules may be first obtained from the analysis of the fracture process as the fracture zone grows, and then the size dependence analyzed for some limiting cases: very large size, very small size and small size.

General rules: Whatever the size, at the early stages of loading the cohesive zone is small. Thus the maximum crack opening is small too, less than the critical crack opening. Fig.1a represents such a situation and the corresponding state of the points of the cohesive zone on the softening curve. Only the arc PA of the softening curve plays a role at this stage.

The arc PA grows as the cohesive crack extends until the initial crack tip opening reaches the critical value w_c (Fig.1b). This is called the critical state and does not correspond, in general, with the peak load situation.

After the critical state, the cohesive zone detaches from the initial crack tip and situations such as that depicted in Fig.1c are produced where the states of the different points of the cohesive zone span the whole softening function.

The point is that at early stages, well before the critical state, only a short arc of the softening curve is in play for the determination of the overall behavior. How this is related to the observed structural response depends, however, on the size of the structure. This dependence is analyzed next for the limiting cases.

Very large sizes: When the size is very large, the cohesive crack zone is of negligible extent relative to *all* the linear dimensions of the specimen. If one limits oneself to measuring remote field values (load, displacement, CMOD), LEFM approximately applies. Then the pre-peak deviation from linearity is negligible and the peak load situation coincides with the critical condition (Planas and Elices, 1991, 1992). This means that any state giving measurable deviation from linearity is of the type shown in Fig.1c, and the whole softening curve is in play. However, as is well known, in this limiting case the influence of the softening curve is only through its area: only G_F matters.

Note that the "size" of the specimen is an elusive term which is only mildly related to the dimensions of an external envelope of the structure. To define it precisely an intrinsic size was introduced by the authors (Planas and Elices, 1988, also 1991). This intrinsic size is defined as

$$\frac{1}{D_i} = \frac{\partial \ln K_I^2}{\partial a} \tag{1}$$

where a is the non-stress-transferring portion of the crack.

At the onset of loading the value of a is the notch depth, a_0. If the notch depth is very small then $K_I \propto \sqrt{a_0}$, and from (1) $D_i \approx a_0$ and the intrinsic size is also small. At the end of loading, when the crack is close to the free surface, the remaining ligament, $l = D-a$ is small and then $K_I \propto l^{-3/2}$, and from (1) $D_i \approx l/3$ (Fig.1c). Therefore, the initial stages of loading in unnotched specimens and the final stages of loading in any specimen correspond to very small intrinsic size specimens, whatever the observable physical size. Of course when the physical (enveloping) size is itself small, so is the intrinsic size.

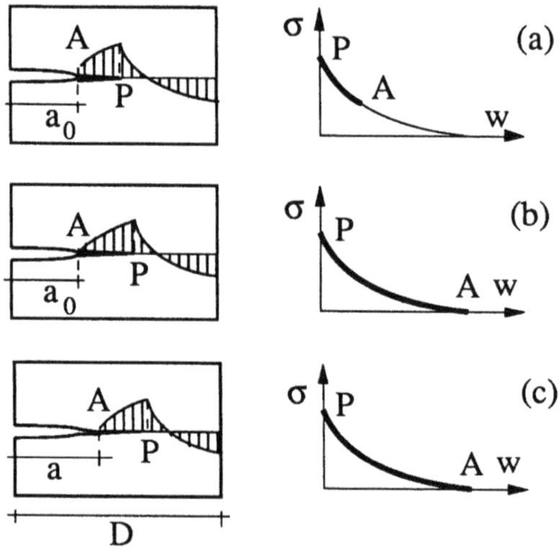

Fig.1. Evolution of a cohesive zone ahead of an existing notch. (a) Early stages of loading; (b) critical situation; (c) advanced state of fracture

Very small sizes: When the sizes are very small the analysis may be performed using a rigid-softening analysis. This is so because to decide whether the (intrinsic) size of the specimen is small or not, D_i is to be compared with the characteristic material size l_{ch}:

$$l_{ch} = \frac{EG_F}{f_t^2} \tag{2}$$

An obvious way to obtain a relative small size is to use a large l_{ch}, and that may be done using a virtual very large value of *E*, the modulus of elasticity, hence a rigid material.

We must distinguish now the beginning of loading and the far end of the loading. In a rigid-softening system the peak load occurs right at the beginning, because the cohesive zone has to spread completely across the specimen to allow displacements to take place. Hence, at peak load only the upper point of the softening curve (point P in Fig.1) matters and the only relevant parameter to determine the peak load is the tensile strength f_t.

At the other end of the test, when the ligament is very small, the rigid body kinematics shown in Fig.2 may be used to determine the moment-rotation curve. The result (Elices, Guinea and Planas, 1992) is that this function depends only on two global properties of the softening function, namely, the fracture energy G_F, and the opening $\overline{w}$ corresponding to the center of gravity of the area defined by the softening curve. The expression is

$$M \approx \frac{1}{\theta^2}\ \overline{w}\ G_F \tag{3}$$

where *M* is the bending moment at the central section and θ the rotation angle (Fig.2).

Because of the previous reasoning showing that the intrinsic size of a specimen is always small when the ligament is small, Eq. (3) must apply to the far end of any test whatever the physical size of the specimen, or, better still, it must apply for large enough rotation angles.

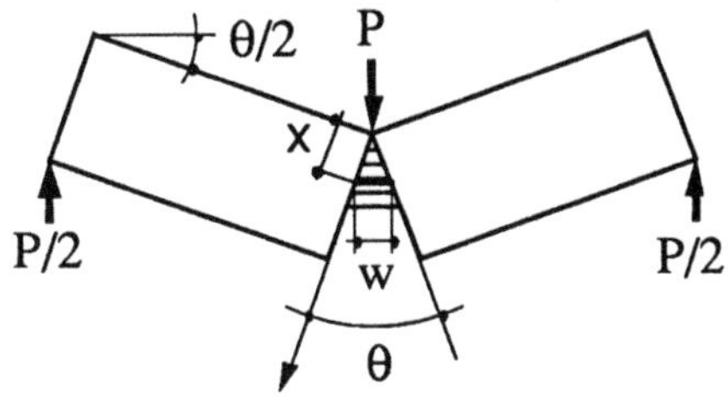

Fig.2. Rigid body kinematics used to describe the far-end portion of a test.

<u>Small sizes:</u> When the sizes are small, but not so small that the rigid-softening analysis applies, one may make an extrapolation of the very small size case. In that case we said that the peak load takes place before any measurable softening occurs at any point. If we relax this limitation, we may expect that for small sizes the peak load will take place at a situation similar to the one in Fig.1a. Hence, we may expect that for small sizes the peak load for the actual material (softening) will not appreciably differ from the peak load predicted for a linear softening with the same tensile strength and a slope determined to fit the initial portion of the actual softening curve (Fig.3). This is an essential property that explains why so many numerical computations using linear softening give an excellent picture of the near peak behavior of concrete specimens of usual laboratory size, even though the softening function of concrete is known to be far from linear.

The interdependence of structural behavior and softening curve just analyzed is the basis of the indirect fitting procedure developed by the authors which is summarized next.

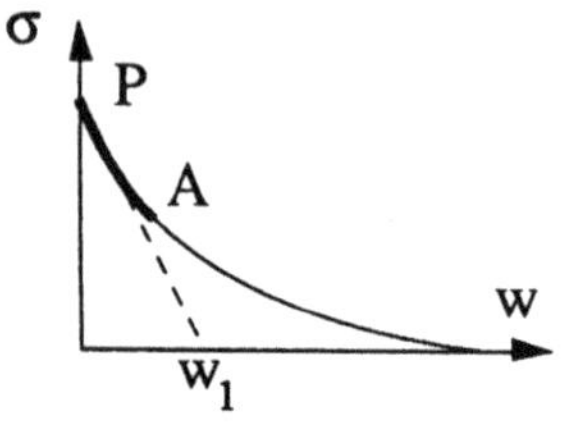

Fig. 3. Linear softening approach: arc PA may be approximated by the mean initial tangent with horizontal intercept w_1.

3 Outline of the general bilinear fitting (GBF)

3.1 The basis

In order to make an estimate of the softening curve we remark that the limiting cases analyzed in the previous section were controlled essentially by 4 parameters: (1) The tensile strength f_t which controls the peak load for very small specimens; (2) the fracture energy G_F which controls the large size behavior; (3) the abscissa $\bar{w}$ of the center of gravity of the area defined by the softening curve, which controls (together with G_F) the far end of the load-deflection curve; and (4) the initial mean tangent, defined by its horizontal intercept w_1 which, together with f_t, controls the peak load of small specimens.

Therefore, we may select a 4-parameter softening curve and make a fit by determining (1) an approximate value of the tensile strength in independent tests, (2) G_F is measured using stable bending tests on small (relatively) notched beam specimens from the area of the P-δ curve, (3) $\bar{w}$ from the far end of the P-δ or P-CMOD curves, and (4) w_1 from the peak load.

A general bilinear softening is considered, as depicted in Fig.4. The GBF procedure is summarized next, in parallel with its application to a particular example. The tests on which the example is based are documented elsewhere (Maturana, Planas and Elices, 1990). However, the geometry and sizes of the specimens used are summarized in Fig.5 for the sake of completeness.

3.2 Tensile stress: f_t

In this simplified method, the tensile strength is estimated from cylinder splitting tests (Brazilian test, ASTM C-495). This may not be very accurate, but it is simple and may be enough for most practical purposes and for normal strength concrete, as discussed by Mihashi (1992). For the concrete of the example, the splitting test yielded f_t = 2.8 MPa.

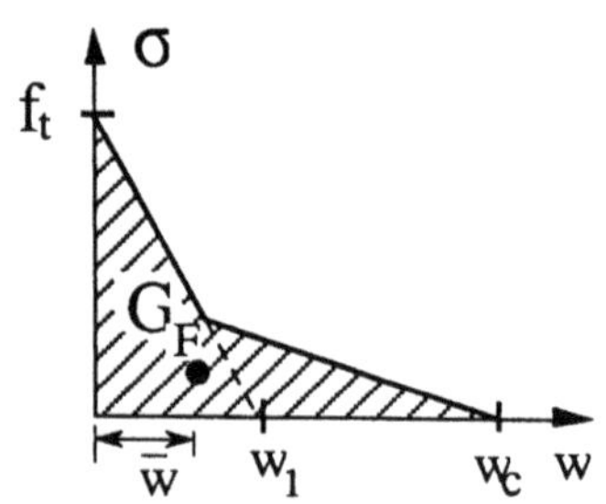

Fig. 4. Bilinear softening function. The heavy dot is the center of gravity of the area.

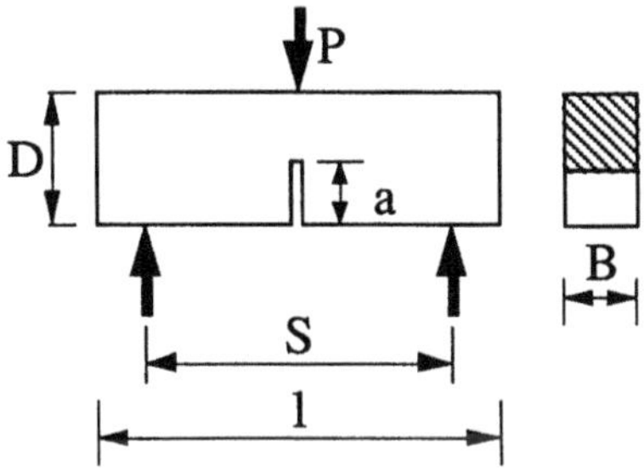

Size	l	S	D	a	B
IV	800	750	300	100	100
III	550	500	200	67	100
II	550	250	100	33	100
I	300	125	50	17	100

in mm

Fig. 5. Specimen geometry and dimensions

3.3 Specific fracture energy: G_F

The area enclosed under the softening function is the specific fracture energy G_F. This parameter was measured according to the RILEM draft recommendation (1985) with some improvements (Guinea *et al.*, 1992; Planas *et al.*, 1992; Elices *et al.*, 1992). Basically, G_F was obtained by dividing the measured work of fracture W by the ligament area:

$$G_{FMeas} = \frac{W}{Bb} \tag{4}$$

where the ligament b is $b = D - a$, according to Fig.5. These G_{FMeas} results displayed a clear size effect, increasing with specimen size as shown in the first row of Table 1. When sources of spurious energy dissipation —hysteresis of the testing equipment, volume dissipation and energy dissipation at the supports—, and the effect of interrupting the test at some fixed rotation angle are taken into account, a corrected value of G_F is obtained (second row of Table 1) which is fairly size independent. The average corrected value, for our concrete, was $G_F = 81$ N/m.

Table 1. Specific Fracture Energy Results

(J/m^2)	Size I	Size II	Size III	Size IV
G_{FMeas}	57 ± 2	75 ± 13	82 ± 2	94 ± 5
G_F	71 [13]	85 [23]	79[7]	89 [11]

Values in [ ] are estimated variation intervals.

3.4 Abscissa of the center of gravity of the softening curve: $\bar{w}$

To determine the abscissa $\bar{w}$ of the center of gravity of the area under the softening curve, the far end of the load-displacement (P - δ) curve is transformed into a moment-rotation diagram (writing $M = PS/4$, $\theta = 4\delta/S$ according to the notation in Fig.5). Since G_F has already been independently determined as described in the previous section, $\bar{w}$ may be found by fitting Eq. (3) to the experimental results.

Note that this requires the determination of the true far end of the P-δ curve, so that the self weight of the specimen *must* be compensated.

For the concrete of our example, the fitting described by Elices *et al.* (1992) led to a value of $\bar{w}$ = 61 μm.

3.5 Initial tangent intercept: w_1

As discussed in section 2.2, for small specimen sizes the peak load for the bilinear softening is the same as that for an imaginary linear softening having the same horizontal intercept w_1, as represented by the dashed line in Fig.4. Hence, the value of w_1 may be found by finding out what linear softening yields the same peak loads as those found in the experiments.

To do so, an easy semi-graphic procedure has been found, as shown elsewhere. When a log-log plot is used, as depicted in Fig.6, to represent the peak load versus the size of geometrically similar specimens (using the dimensionless variables shown in the figure) a unique "master" curve may be proved to exist representing the size effect curve for a linear softening (solid curve in Fig.6). This curve is obtained once and for all for a given specimen geometry using a suitable numerical computation such as the influence method (Planas and Elices, 1991). The curve shown in Fig.6 corresponds to the geometry in Fig.5 (notch-to-depth ratio equal to 0.3). Similar "master" curves may be obtained for other geometries by cross-plotting size effect results found in the literature (for example those of Petersson, 1981). If the actual concrete did soften linearly, the experimental points would lie on this "master". However, if the concrete softening is bilinear the experimental plots are horizontally shifted towards the left. For small sizes, the horizontal displacement is constant so that the experimental points should lie on a line obtained by translating the master curve a horizontal distance Δx towards the left, as shown by the dashed line in Fig.6. This horizontal displacement in the log-log plot is related to the initial tangent intercept w_1 by

$$\Delta x = \log\left(\frac{2G_F}{w_1 f_t}\right) \quad \text{or} \quad w_1 = \frac{2G_F}{f_t}\, 10^{-\Delta x} \tag{5}$$

In principle, a single size experimental result is enough to determine Δx and w_1, but a range of sizes is required to check that the specimen size is small enough to guarantee that the experimental results do lie on a parallel to the master curve. For our concrete, the experimental results in Fig.6 verify this condition reasonably well, and the value $w_1 = 37$ μm is found.

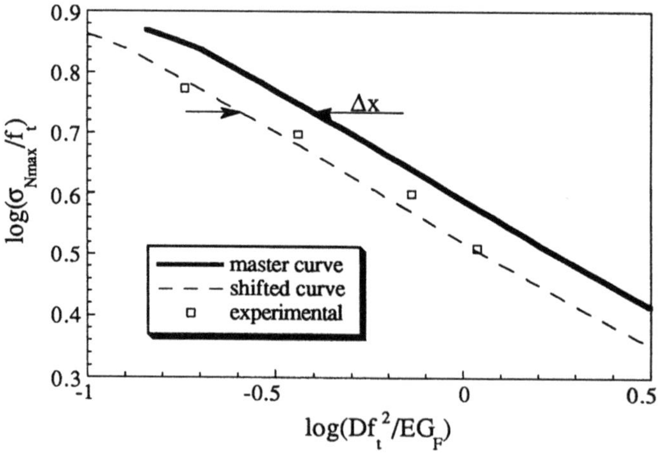

Fig.6. Size effect plot, and graphic method to determine w_1. The master curve is for the geometry depicted in Fig.5.

3.6 Final fit to bilinear softening curve

From f_t, G_F, $\bar{w}$ and w_1, the characteristic points of the bilinear softening may be computed using simple geometrical relations. For our concrete, the critical crack opening w_c is found to be equal to 367 μm, and the coordinates of the kink point are (35 μm, 0.176 MPa).

4 Discussion and conclusions

The behavior predicted using the softening curve obtained with the foregoing GBF procedure is worth comparing with the experimental results and the prediction of a widely used bi-parametric bilinear softening, that defined by Petersson (1981).

Fig.7 shows the maximum loads as a function of specimen size. The agreement between the experimental results and the numerical predictions found when using a cohesive model with the GBF softening function, is quite good. Predictions based on the bilinear softening function proposed by Petersson, using the same G_F and the same f_t, are also quite good. This coincidence reflects the fact, already mentioned in 2.2, that for normal sizes the relevant part of the softening function for predicting the maximum load is the initial part. When values of w_1 for the two softening functions are compared, one realizes that they are similar (w_1 = 37 μm for GBF and w_1 = 35 μm for Petersson's function). This means that the slope of the first segment part is practically the same for both functions.

Fig.8 shows the dimensionless load-CMOD curves for samples with D = 100 mm (size II in Fig.5). The experimental curve is the average value of two tests. Up to the maximum load, both models, Petersson's and ours (GBF), give the same values and agree quite well with the experimental results. This behavior was expected according to previous comments. After the peak load the two models differ: GBF predictions still fit the experimental results, while Petersson's model predicts higher loads.

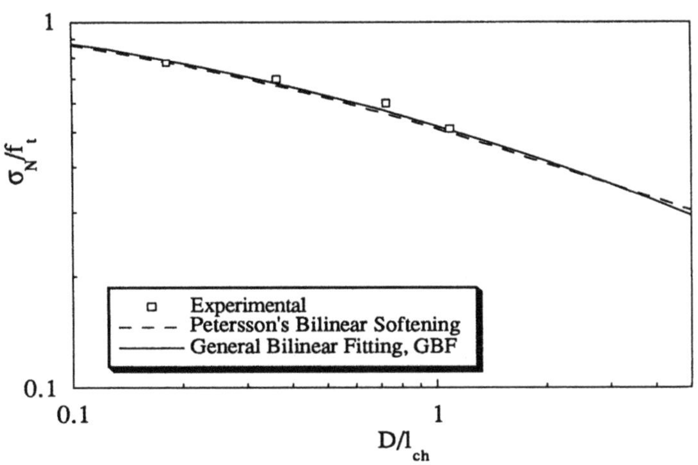

Fig.7. Maximum load; experimental data and model predictions ($\sigma_N=3P_{max}S/2BD^2$, and $l_{ch}=EG_F/f_t^2$)

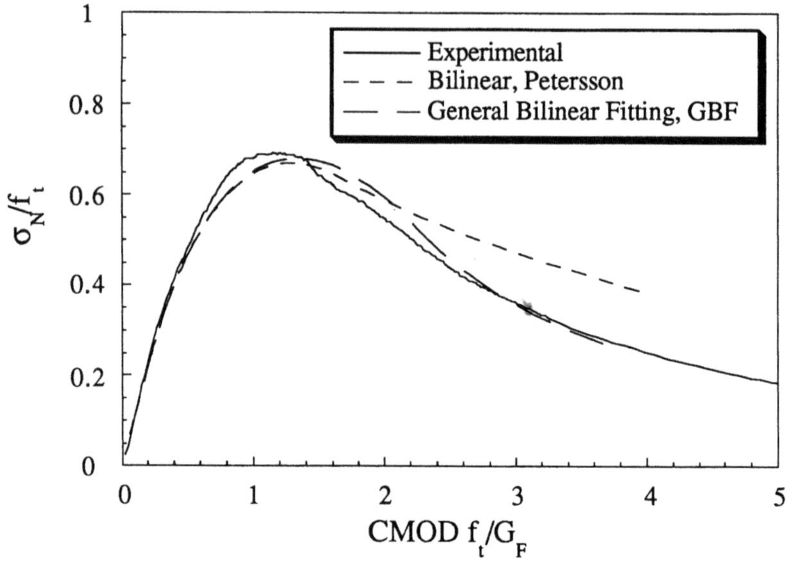

Fig.8. Load-CMOD curves for D = 100 mm specimen

At this point it is worth noting that the essential difference between these two softenings is that the length of the softening tail, hence the critical crack opening, is much larger for the GBF than for Petersson's softening. In the case of Petersson's softening $w_c = 3.6\ G_F/f_t$, while for the GBF (and for this specific concrete) $w_c = 12.7\ G_F/f_t$.

To conclude, some remarks about the proposed procedure to obtain the four parameters are worth consideration. First, we assumed that the maximum value of the softening function coincides with the tensile stress f_t measured by standard procedures; this assumption needs further support. Second, to measure G_F it was necessary to refine the RILEM draft recommendation by taking into account the improvements developed by the authors when weight compensation was used (Guinea *et al.*, 1992; Planas *et al.*, 1992; Elices *et al.*, 1992). To measure the abscissa of the centroid of the area of the softening curve, careful measurements of the tail of the load-displacement curve are needed. Finally, to determine w_1, it must be ensured that the specimen sizes are small enough for the linear approximation to be valid up to the peak load. This may be done by using several specimen sizes to check that the translational property in Fig. 6 holds (as done in our example) or, alternatively, a single size may be used and the validity of the assumption verified *a fortiori* by numerical computation.

To sum up, 4-parameter softening curves appear to be necessary in order to describe the most important aspects of the structural behavior. A relatively easy procedure for a general bilinear fitting has been described and is available to obtain a reasonable approximation of the softening for most engineering purposes.

Acknowledgment. The authors gratefully acknowledge financial support for this research provided by the Dirección General de Investigación Científica y Técnica, Spain, under grants PB90-0276 and MAT90-1153-E.

5 References

Elices, M., Guinea, G.V. and Planas, J. (1992) Measurement of the fracture energy using three-point bend tests: 3. Influence of cutting the P-δ tail. **Materials and Structures**, 25, pp. 327-334.

Guinea, G.V., Planas, J. and Elices, M. (1992) Measurement of the fracture energy using three-point bend tests: 1. Influence of experimental procedures. **Materials and Structures**, 25, pp. 212-218.

Hillerborg, A., Modéer, M. and Petersson, P.E. (1976) Analysis of crack formation and crack growth in concrete by means of fracture mechanics and finite elements. **Cement and Concrete Research**, 6, pp.773-782.

Maturana, P., Planas, J. and Elices, M. (1990) Evolution of fracture behavior of saturated concrete in the low temperature range. **Engineering Fracture Mechanics**, 35, pp. 827-834.

Mihashi, H. (1992) Material structure and tension softening properties of concrete, in **Fracture Mechanics of Concrete Structures** (ed. Z.P. Bazant), Elsevier, pp. 239-250.

Petersson, P.E. (1981) Crack growth and development of fracture zone in plain concrete and similar materials. Div. of Building Mat., Lund Inst. of Technology, Report TVBM-1006.

Planas, J. and Elices, M. (1988) Size effect in concrete structures: Mathematical approximations and experimental validations, in **Cracking & Damage, Strain Localization and Size Effect** (eds. J. Mazars and Z.P. Bazant), Elsevier, pp. 462-476.

Planas, J. and Elices, M. (1991) Nonlinear fracture of cohesive materials. **Int. J. of Fracture**, 51, pp. 139-157.

Planas, J. and Elices, M. (1992) Asymptotic analysis of a cohesive crack: 1. Theoretical background. **Int. J. of Fracture**, 55, pp. 153-177.

Planas, J., Elices, M. and Guinea G.V. (1992) Measurement of the fracture energy using three-point bend tests: 2. Influence of bulk energy dissipation. **Materials and Structures**, 25, pp. 305-312.

Reinhardt, H.W., Cornelissen, H.A.W. and Hordijk, D.A. (1986) Tensile tests and failure analysis of concrete. **J. of Structural Engineering, ASCE**, 112, pp. 2462-2477.

RILEM Draft Recommendation (1985) Determination of fracture energy of mortar and concrete by means of three-point bend test on notched beams. **Materials and Structures**, 18, pp.285-290.

van Mier, J.G.M. (1986) Fracture of concrete under complex stress. **Heron**, 31, pp. 2-90.

Wittmann, F.H., Roelfstra, P.E., Mihashi, H., Huang, Y.Y., Zhang, X.H. and Nomura, N. (1987) Influence of age of loading, water-cement ratio, and rate of loading on fracture energy of concrete. **Materials and Structures**, 20, pp. 103-110.

Wittmann, F.H., Rokugo, K., Bruhwiler, E., Mihashi, H. and Simoni, P. (1988) Fracture energy and strain softening of concrete as determined by means of compact tension specimens. **Materials and Structures**, 21, pp. 21-32.

6 TENSION SOFTENING DIAGRAM OF STEEL FIBER REINFORCED CONCRETE AND SIZE EFFECT ON THE FLEXURAL STRENGTH

W. KOYANAGI, K. ROKUGO and Y. UCHIDA
Department of Civil Engineering, Gifu University, Japan

Abstract
This paper deals with the tension softening diagram and the size effect on flexural strength of steel fiber reinforced concrete (SFRC, 1% vol. of fiber) beams. Load-displacement curves of SFRC beams in various sizes were examined both experimentally and analytically. Tension softening diagram was determined through modified J-integral based method. Load-displacement curves of SFRC beams (depth: 10, 20, 40 cm) are simulated and the results are in good accordance with the test results. There are two peak loads in the load-displacement curve of SFRC beams and the second peak load becomes greater than the first one when the size of specimens becomes greater. Numerical analysis can fairly explain that the first peak load of SFRC beams depends on the matrix resistance and that the second peak load appears to depend on the resistance due to the fiber reinforcement. Size effect was recognized for flexural strength of SFRC beams. Bending strength corresponding to the first peak load decreased with the increase of size and the gradient was similar to that for normal concrete. The strength to the second peak load also decreased with the increase of size, though the slope became gentle.
Keywords: Steel fiber reinforced concrete, Fracture mechanics, Tension softening, Flexural strength, Size effect, Load-displacement curve.

1 Introduction

Steel fiber reinforced concrete (SFRC) has various outstanding properties, such as high flexural strength, high toughness, high impact resistance and so on and is widely used in the field of construction. Yet, the fracture properties of SFRC has to be investigated more in detail in order to make the structural design method for SFRC as precise as that for ordinary concrete.

The characteristics regarding to the tension softening diagram and the size effect are the most important subjects relating to the fracture mechanics of concrete. Experimental and analytical works were done on the flexural behavior of SFRC beams and the shapes of load-displacement curves were investigated. The detailed information of this study and the further discussions are described in Japanese journal by Rokugo et al. (1993).

2 Outline of experiment

2.1 Specimens and concrete

Four types of beam specimens were made. All the specimen had square cross section. The height (depth) of the section was 5, 10, 20 and 40 cm. These

Fracture and Damage of Concrete and Rock – FDCR-2.
Edited by H.P. Rossmanith. © 1993 Taylor & Francis. ISBN 0 419 18470 8.

specimens are denoted as S5, S10, S20 and S40, hereafter. The length of the specimens was four times the height.

Ready mixed concrete was used to cast the specimens. Base concrete was conformed to JIS 5308 with proof compressive strength of 255 kgf/cm^2 (22.1MPa) and the slump value of 15 cm. The maximum size of coarse aggregate was 15 mm and air content was 4%. The steel fiber was indented type as shown in Fig. 1 and the size was ø0.6x30mm. Volume content of the fiber was 1.0 %. When the fiber was added to the base concrete, the slump value of SFRC became about 3 cm. High range water reducing air entraining agent was used.

2.2 Loading manner

Loading span length in the flexural test of the beam specimens was three times the height. The third point loading test was made. The loading manner with measuring system is shown in Fig. 2. More than four specimens were provided for each test condition. The relation between load versus deflection at loading points was measured except for S5 specimens. Only the maximum load was measured for S5 specimens.

In order to obtain the fracture energy G_f of SFRC, three-point bending test on notched beam specimens was made, following the RILEM recommendation (1985). The size of the specimen was 10x10x84 cm, the notch depth was 5 cm and the span length was 80 cm. The set up of the loading and the measuring system are illustrated in Fig. 3. Five specimens were provided for this test.

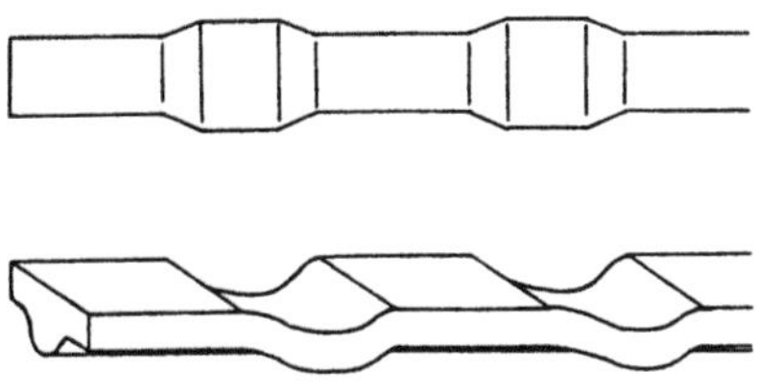

Fig.1 Shape of steel fiber.

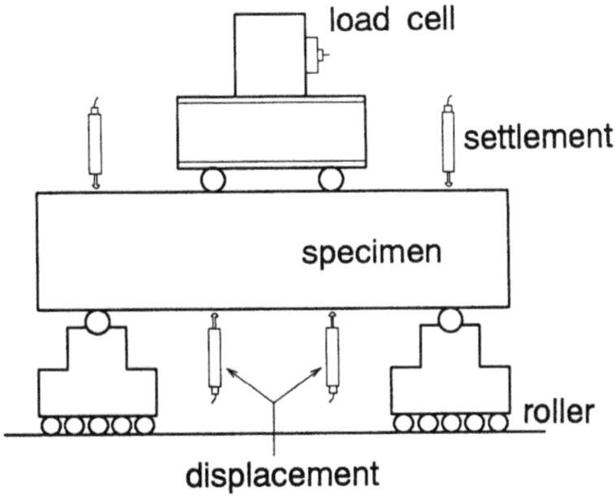

Fig.2 Configuration of third-point bending test.

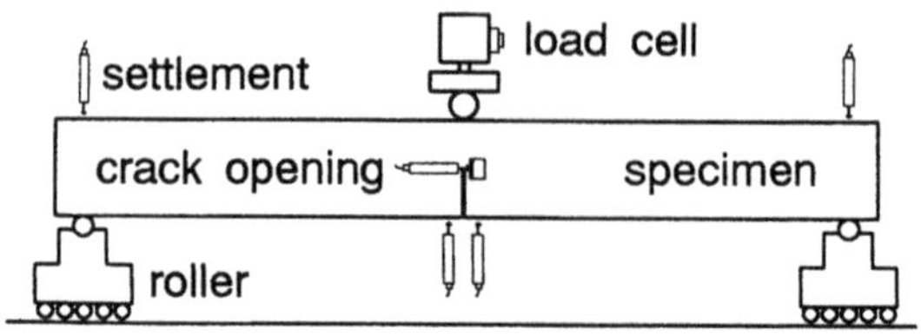

Fig.3 Configuration of three–point bending test on notched beam.

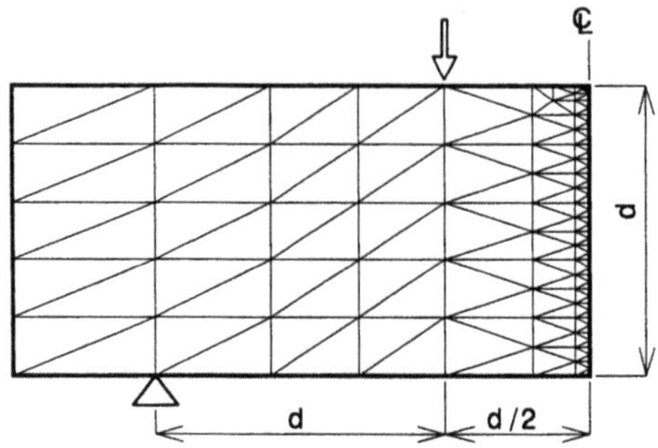

Fig.4 FE mesh.

3 Analytical method

Finite element analysis was made where fictitious crack model was introduced at the center of the specimen. In the analysis, the extension of the crack was modeled so that each node was separated in turn and the equivalent nodal forces were transmitted through the nodes, where the forces were derived from tension softening diagram. The tension softening diagram of the SFRC was estimated from the relation between load and displacement at loading points and the relation between crack width of the crack tip (COD) and displacement at loading points, using the modified J–integral based method proposed by Uchida et al. (1991), which was originated from Li's J–integral based method (1989). FE mesh adopted in the analysis is shown in Fig. 4. The crack was assumed to start from the tension face and to develop upwards straightly along the center of the specimen. Only a half portion of FE mesh is shown because of its symmetricity.

4 Experimental results

Obtained load–deflection curves for S10, S20 and S40 are shown in Fig. 5(a)–(c). Two peak loads were observed when the specimen sizes were large (S20 and S40). When the specimen size was small (S10), there was only one peak load just like in case of normal concrete. In case of S20 the first peak load was higher than the second one, but in case of S40, the second peak load was higher than the first one. The first peak load is named P1 and the second, P2. Nominal flexural strengths corresponding to the averaged load P1 and P2 are denoted as σ_{b1} and σ_{b2}, respectively, and are shown in Fig. 6. The nominal flexural strength was calculated from peak load divided by its section modulus, where the effect of the specimen's own weight was taken into account. σ_{b1} decreased with the increase of the specimen height and so–called size effect was observed just as in case of normal concrete. σ_{b2} increased slightly when the specimen height increased from 20 cm to 40 cm.

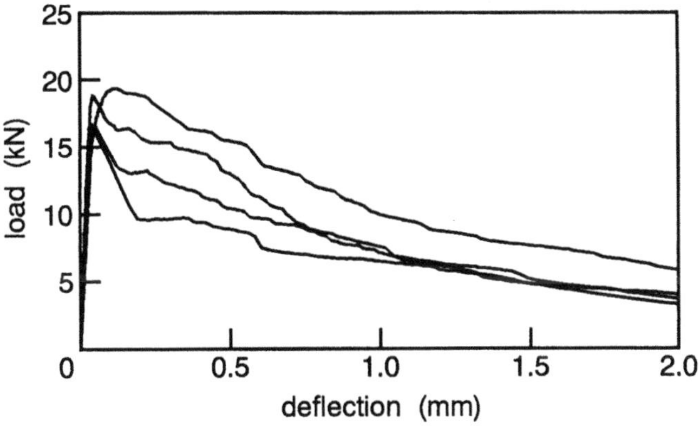

(a) specimen size : 10×10×40cm (span 30cm)

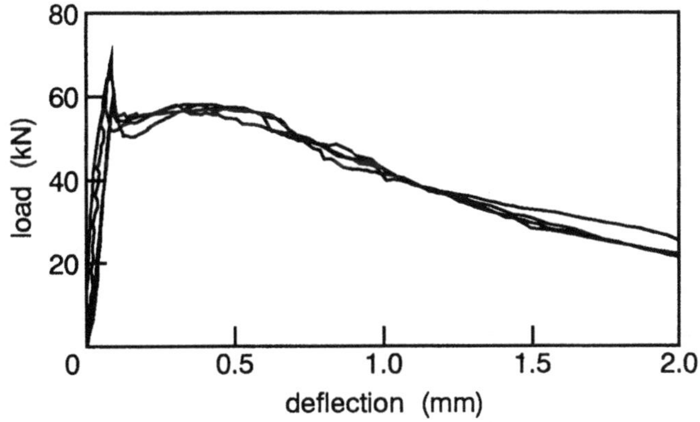

(b) specimen size : 20×20×80cm (span 60cm)

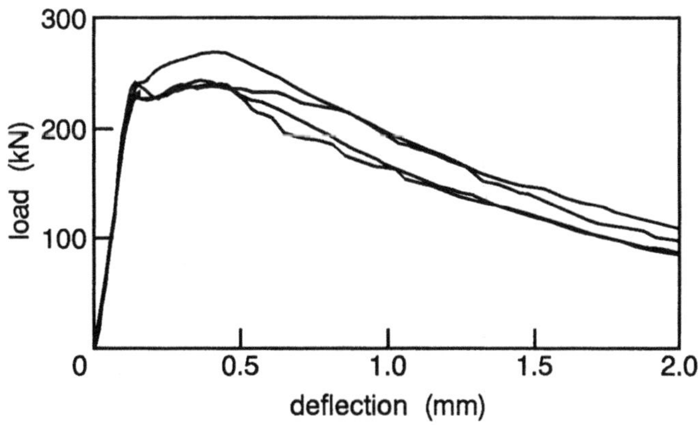

(c) specimen size : 40×40×180cm (span 120cm)

Fig.5 Observed load–deflection diagrams.

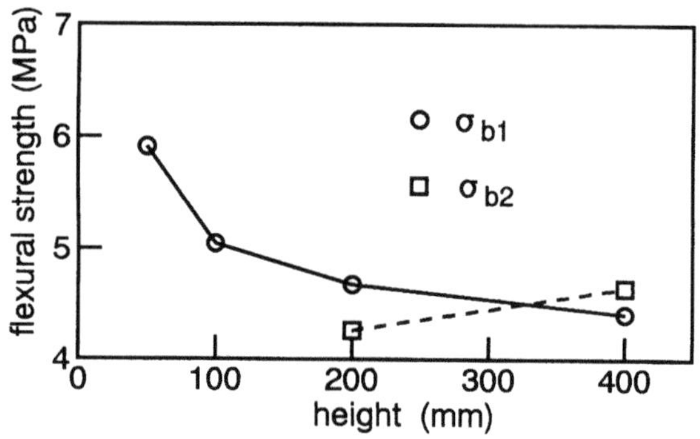

Fig.6 Flexural strength vs. height of beam (experiment).

Strengths of SFRC were obtained by standard tests. The compressive strength was 38.2 MPa (ø10x20 cm specimen), the splitting tensile strength was 3.05 MPa (ø15x15 cm specimen), the flexural strength was 5.05 MPa (10x10x40 cm specimen) and the elastic modulus was 31 GPa.

5 Fracture energy G_f and tension softening diagram

Fracture energy G_f obtained from the load–deflection curve of notched beams by three point RILEM bending test was 3.4 kN/m as an average. Not only the fracture energy in tensile cracking zone but also a certain small amount of fracture energy in compressive zone was included in G_f when the RILEM method is applied to SFRC.

The tension softening diagram of SFRC was obtained through the modified J–integral based method and is shown in Fig. 7. The diagram is called J–curve, hereafter. Just after the softening initiation point (tensile strength failure point), the transmitted stress decreased suddenly with the increase of crack width. When crack width reached about 0.03 mm, the transmitted stress increased gradually because the reinforcing effect of steel fiber began to work.

6 Analysis of load–deflection curves and of size effect

6.1 Load–deflection curve

The load–deflection curve of notched beam was analytically calculated using the J–curve and the result is shown in Fig. 8 together with the load–deflection curve obtained experimentally. The analytical result was well correlated to the experimental curve except for the magnitude of the first peak load, where analytical one was higher by about 15%. Load–deflection curves for the specimens of various depths were calculated using J–curve and illustrated in Fig. 9. In the analysis, J–curve was taken to be composed of united short straight lines. They coincide fairly well with the experimental results and represent two kinds of peak loads.

The tension softening diagram derived from the modified J–integral based method is considered to be only approximation. J–curve for SFRC can be modified further by the union of smooth lines with the same G_f as shown in Fig. 10 and the curve is named model–I. Analyzed load–deflection diagrams for each specimen size based on the model–I are also shown in Fig. 9. They agreed well with experimental results rather than that from J–curve does. Especially, the relation of two peak loads, P1 and P2, for various specimen sizes was properly estimated.

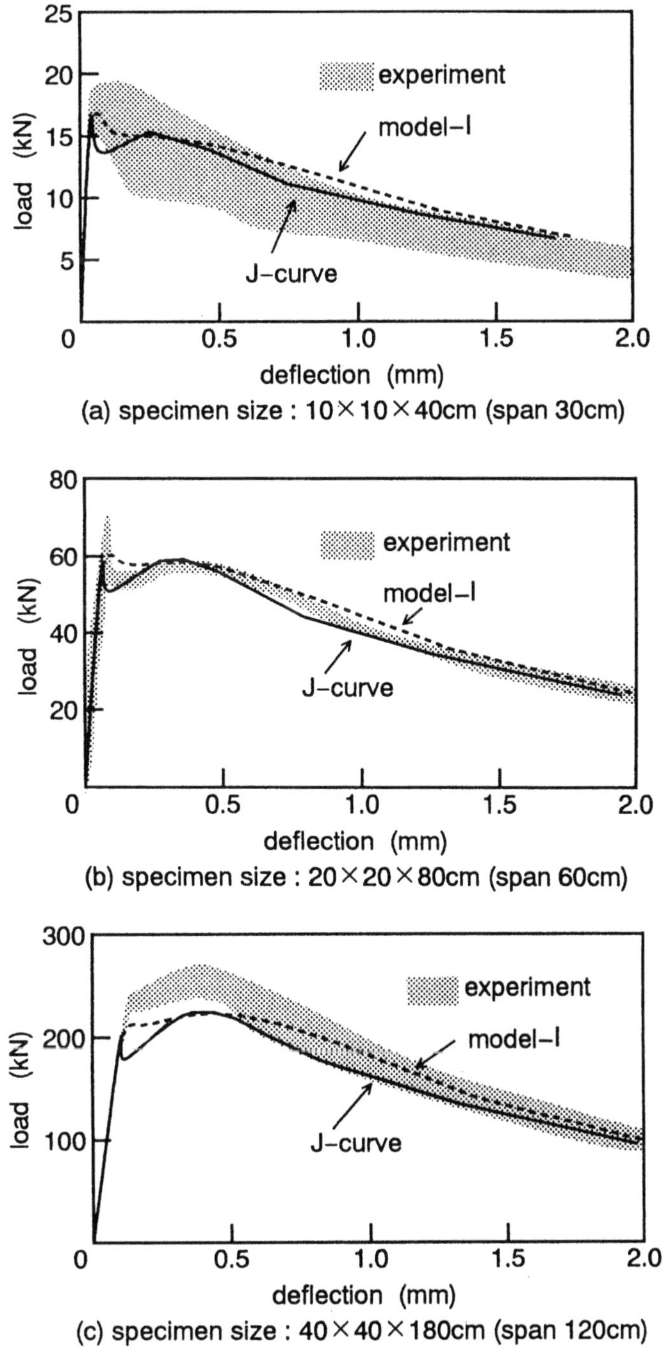

Fig.7 Load-deflection diagrams.

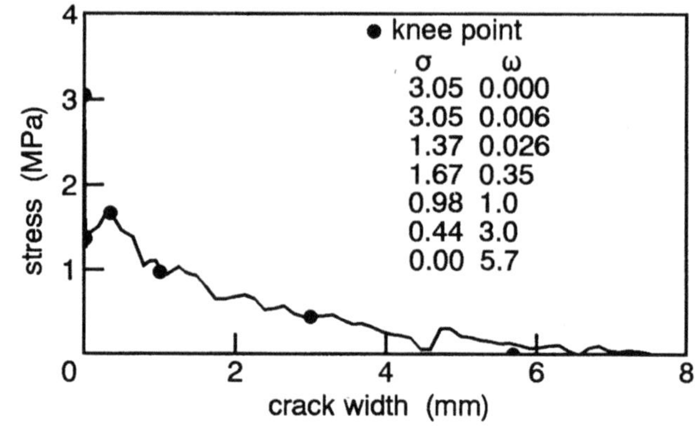

(a) whole shape

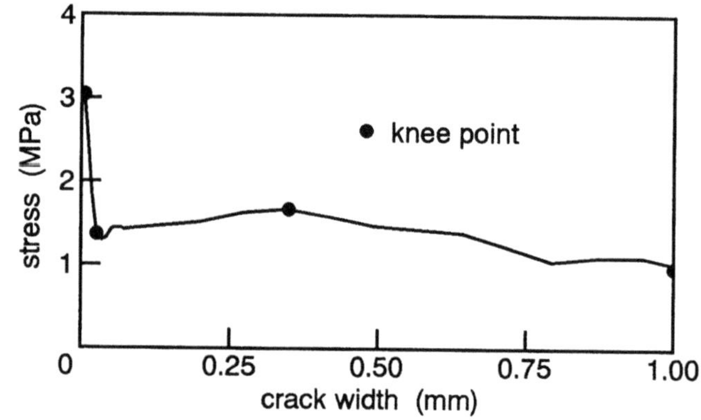

(b) initial part

Fig.8 Strain softening diagram by modified J-integral based method.

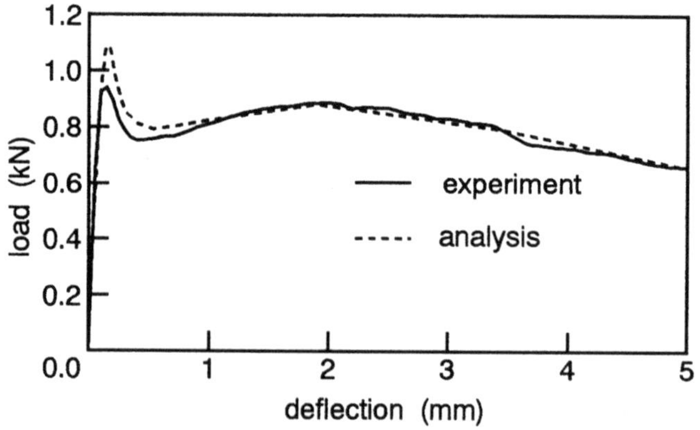

Fig.9 Load-deflection curves of notched beam.

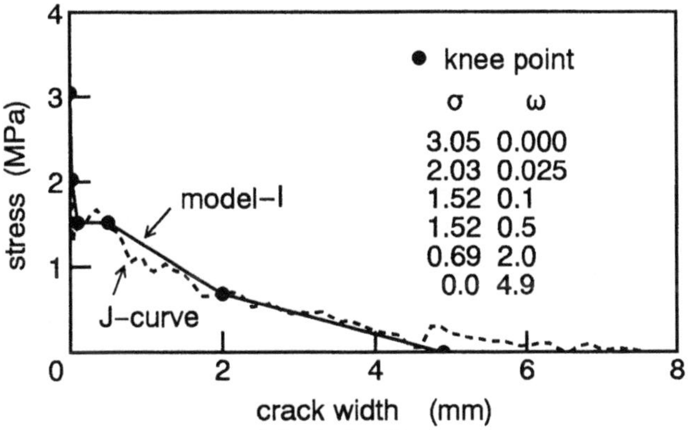

(a) whole shape

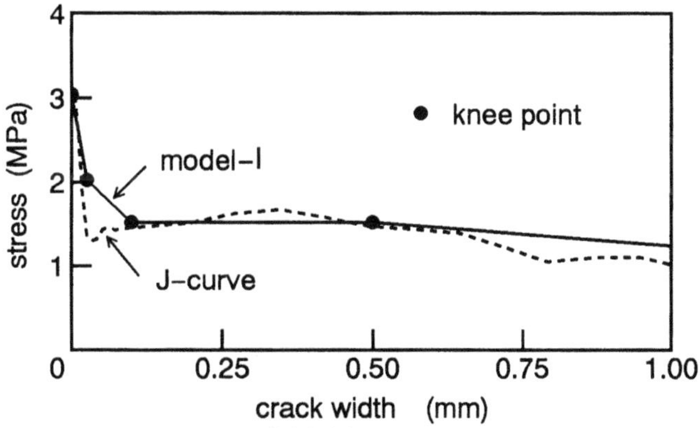

(b) initial part

Fig.10 Models of tension softening diagram.

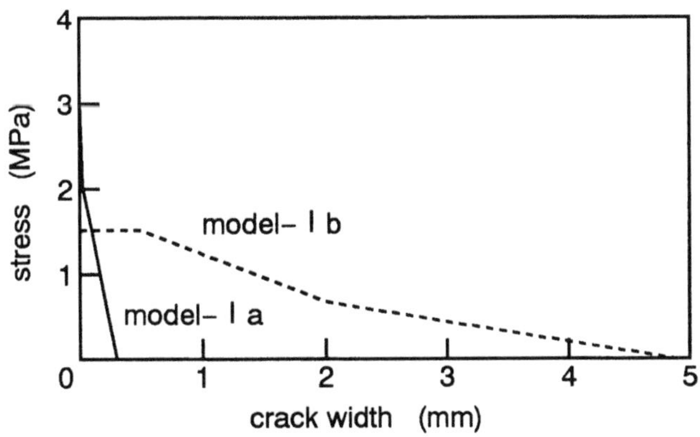

Fig.11 Components of the model-I.

The model–I is considered to be composed of two portions; model–Ia and –Ib. The first part is considered to be basically corresponding to the resistance of normal concrete, and the second part, to the resistance of fiber reinforcement as shown in Fig. 11. The load–deflection curves derived from the model–I and its components are shown in Fig. 12.

6.2 Size effect on flexural strength

The relation between calculated flexural strength and the height of specimen is shown in Fig. 13. Model–I was used in the analysis as the tension softening diagram. The calculated results for normal concrete reported by Uchida et al. (1992) is also shown in Fig. 13, where tensile strength of concrete f_t was 2.79MPa, fracture energy G_f was 0.1 kN/m and so–called 1/4 model for tension softening diagram proposed by Rokugo et al. (1989) was adopted. As shown in Fig. 13, the flexural strength of SFRC decreases with the increase of specimen size. The slope of the decrement of σ_{b1} for SFRC is almost the same for that of normal concrete, though the strength itself is higher for SFRC than for normal concrete. The gradient of the decrement for σ_{b2} is smaller than that for σ_{b1}. It means that the size effect becomes smaller when the

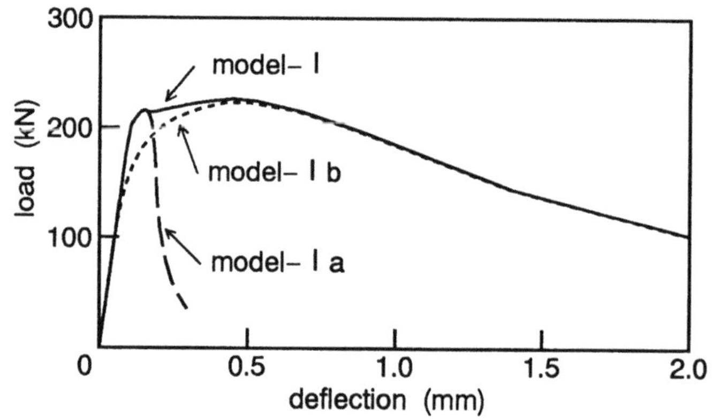

Fig.12 Load–deflection curves derived from the components of model–I.

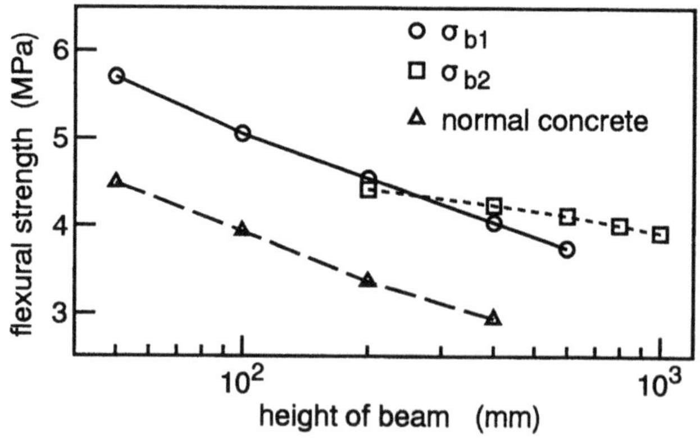

Fig.13 Flexural strength vs. height of beam (analysis).

effect of steel fiber reinforcement becomes dominant, which means the behavior of material becomes ductile.

7 Conclusions

Studies on the tension softening diagrams and the size effect of SFRC with 1% vol. content of indented type fiber were made both analytically and experimentally. Principal conclusions derived are as follows;

(1) In the load–deflection curves, two peak loads were observed when the specimen sizes were large, though only one peak load was observed for the small size specimens. The second peak became higher than the first one when the size became large.

(2) Calculated load–deflection curves for each size specimen were agreed well with the experimental data including the relation between the two peak loads and specimen height, where model–I (further modified J–curve) was used as the softening diagrams.

(3) The model–I is composed of two parts. The first part is considered to be basically corresponding to the resistance of normal concrete, and the second part, to the fiber reinforcement.

(4) The size effect of SFRC on the flexural strength was recognized. The slope of the decrement of the flexural strength corresponding to the first peak load is almost the same for that of normal concrete. The slope of the decrement of that to the second peak load is smaller than that for the first one.

(5) The size effect becomes smaller when the effect of steel fiber reinforcement becomes dominant, which means the behavior of materials becomes ductile.

8 Acknowledgement

The authors acknowledge financial supports for this study by the Kajima Foundation's Research Grant and by the Grant–in–Aid of the Ministry of Education.

9 References

Li, V.C. and Ward, R.J. (1989) A novel testing technique for post–peak tensile behavior of cementitious materials. **Fracture Toughness and Fracture Energy.** Balkema, pp.183–195

RILEM Draft Recommendation (50–FMC)(1985) Determination of the fracture energy of mortar and concrete by means of three–point bend tests on notched beam. **Materials and Structures.**, 106, 285–290

Rokugo, K., Iwasa, M., Suzuki, T. and Koyanagi, W. (1989) Testing methods to determine tensile strain softening curve and fracture energy of concrete. **Fracture Toughness and Fracture Energy** (eds. H.Mihashi, H.Takahashi and F.H.Wittmann) Balkema, Rotterdam, pp. 153–163

Rokugo, K., Uchida, U., Katoh, H. and Koyanagi,W. (1993), Tension softening diagram and flexural failure behavior of steel fiber reinforced concrete. **Proc. Japan Soc. Civil Eng.,** (in Japanese, to be published)

Uchida, Y., Rokugo, K. and Koyanagi, W. (1991) Determination of tension softening diagrams of concrete by means of bending tests. **Proc. Japan Soc. Civil Eng.,** 426, 203–212 (in Japanese)/ **Concrete Library of JSCE,** 18, 171–184

Uchida, Y., Rokugo, K. and Koyanagi, W. (1992) Application of fracture mechanics to size effect on flexural strength of concrete. **Proc. Japan Soc. Civil Eng.,** 442, 101–107 (in Japanese)

7 APPARENT TENSILE STRENGTH AND FICTITIOUS FRACTURE ENERGY OF CONCRETE: A FRACTAL GEOMETRY APPROACH TO RELATED SIZE EFFECTS

A. CARPINTERI and G. FERRO
Department of Structural Engineering,
Politecnic of Turin, Italy

Abstract

Aim of the present paper is the study of the influence of the structural size on the apparent tensile strength and on the fictitious fracture energy. It is well known how these parameters are varying with the structural size. The fictitious fracture energy increases when the structural size increases, while the apparent tensile strength decreases.

The study is focused onto two aspects. The first one is experimental and concerns a set of tensile tests on concrete specimens. Specimens of different sizes are tested and for each size the fictitious fracture energy and the apparent tensile strength are evaluated. Tensile tests require a very careful execution: a particular instrumentation is realized to avoid the load eccentricity, due to the non-simmetry in the softening zone development. In the second part of the paper a fractal geometry explanation for related size effects is presented.

1 Introduction

The study of tensile properties of heterogeneous materials (concrete, mortar, rocks, etc.) has involved numerous researchers in recent years. Different test techniques were developed and several theories were formulated in support of the results obtained. The substantial difficulties that the tension test had to overcome are the following:

- to maintain always the tensile load perfectly centered so to avoid the presence of bending moments which can alterate the results;
- to obtain a stable and complete post-peak softening response;
- to resolve the grip problems.

Gopalaratnam and Shah (1985) proposed a test technique with a universal joint to minimize in-plane and out-of-plane bending effects and a particular wedge-type frictional grip system to assure that fracture occurred away from the grips. With

Fracture and Damage of Concrete and Rock – FDCR-2.
Edited by H.P. Rossmanith. ISBN 0 419 18470 8.

this instrumentation they performed tests on constant size specimens and observed that the notches do not influence the average response of the specimens and that the tangent modulus of elasticity in tension is comparable to that in compression.

Reinhardt and co-workers (1986) performed cyclic tests on notched specimens and proposed softening constitutive laws on the basis of the obtained experimental results.

Phillips and Zhangh (1991) designed a stiff frame which proved highly effective in order to obtain stable and complete stress-deformation curves on both unnotched and notched plain concrete. They also established a substantial homogeneity of the notched specimens behaviour compared with that of the unnotched ones. Moreover they studied the effect of water-cement ratio on fracture energy and proposed a brittleness index as the ratio of the maximum elastic deformation to the deformation at failure.

Van Mier (1989) faced the problem of the influence of boundary rotation stiffness and indicated some conditions on the test stability. Moreover he presented a vacuum impregnation experiment using a low viscosity fluorescing fluid which allows monitoring of the complete crack growth process. He could observe self-similar crack patterns, as well as Saouma et al.(1990) emphasized the fractal nature of the concrete fracture surfaces.

Bažant and Cedolin (1992) justified why in direct tension test the specimens flex and break at midlength, but this explanation presupposes the presence of an eccentricity in the test procedure.

This paper deals with the size effect fenomena on apparent tensile strength and on fictitious fracture energy of concrete. It was necessary to realize an instrumentation to avoid any load eccentricity and to obtain a very stable test. In order to explain the size effects on the apparent tensile strength and on the fictitious fracture energy, the concepts of self-similarity of the defect size distribution and of fractal dimension of the fracture surface are introduced, respectively.

2 Description of test equipment

The first problem to solve was the choice of the shape and the dimension of the specimens. The choice of the shape depends on the grip system and on the convenience to create a failure preferential zone. On the other hand, the specimen dimensions are limited by the dimension of the available equipment as well as by the aggregate size. In the present study relatively small aggregate sizes are used to allow tests on small size specimens.

The clepsydra shape for specimens, see Fig.1, permits to dispose of a central zone with a cross-sectional area equal to one third of that of the grip zones. The specimen thickness was kept constant for all sizes and equal to 10 cm, whereas the width of the central section (characteristic specimen size) was assumed equal to 0.5, 1, 2, 4 times the thickness, respectively. The specimens were fixed to the testing machine by a bicomponent glue of epoxy resin.

The load application was realized employing three servocontrolled jacks to annul in-plane and out-of-plane bending stress. The central hydraulic jack was suspended from contrast structure by a hinge while the connection to the movable crosshead was executed by a universal joint. That jack applied the traction load to the centre of the crosshead and the load control was put into practice through the signal sum from the four extensometers applied to the specimen.

The other two hydraulic jacks were disposed the former on the end of the contrast crosshead and the latter on the ground and connected to the movable crosshead through an auxiliary bar normal to it. The peculiar position of these two jacks consents to generate two normal couples able to interact and to annul any load eccentricity. The test set-up is illustrated in Fig.2.

The specimens were loaded by the central jack so as to maintain a constant rate of the specimen elongation while the other two jacks interacted with the central one.

The tests were carried out with the lower crosshead fixed to the principal frame. During the test execution horizontal forces due to the non perfect alignment of central jack with the specimen axis arose. The presence of these horizontal forces in the upper extremity of the specimen, whose structural arrangement is given by a cantilever loaded by a concentrated force at the free extremity, produces a linear bending moment along the specimen. This bending moment undergoes a traslation so much that it is zero at midlenght of the specimen, due to the action of the jack linked to the auxiliary crosshead. This moment being proportional

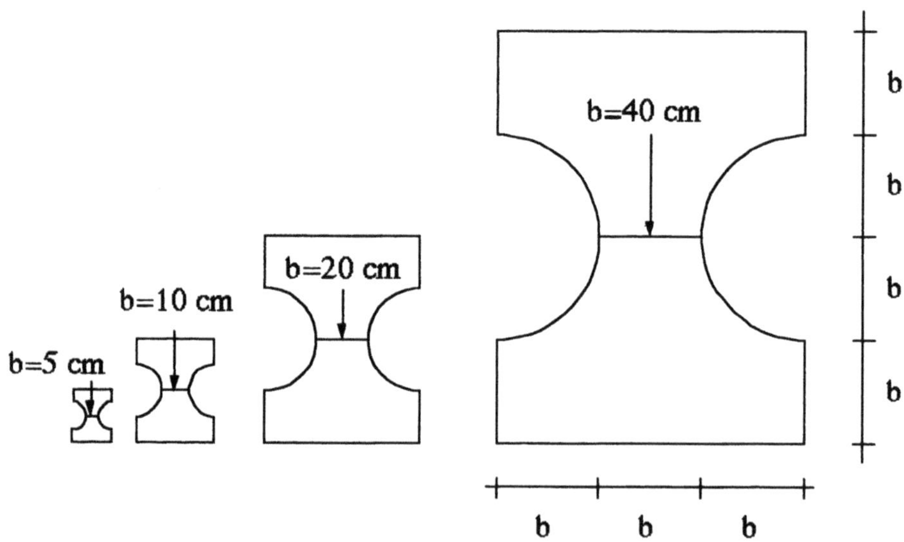

Fig. 1. Geometries of the specimens tested.

to the length of the specimen, it originated stresses at the extremities that, in the cases of the large specimens, caused the separation of the anchorage plates. Therefore two hinges were arranged at the extremities of the lower crosshead to permit specimen rotations. In this way the specimen axis aligns to the central jack and the horizontal forces vanish during the test.

3 Experimental results

The tests were carried out on four specimen sizes. The dimensions of the minimum cross-section of the smallest specimen were 5x10 cm, 10x10 for the second size, 20x10 for the third and 40x10 for the largest size (Fig.1). Four specimens for each size were cast. For the smallest size only one complete diagram is available, in that the other three were used to set the sensibility of the three servocontrolled machines. For the subsequent size two complete diagrams are available. For the specimens of length 80 cm the hinges were introduced to bind the lower crosshead, since in the first two specimens a failure near the grip plates had occurred. For this size only a complete curve is disposable. On the other hand, for the largest specimens only the values of the tensile strength are available, because the length of the extensometers (40 cm), was greater than the critical one and therefore snap-back behaviour occured. The results of the tests are reported in Table 1.

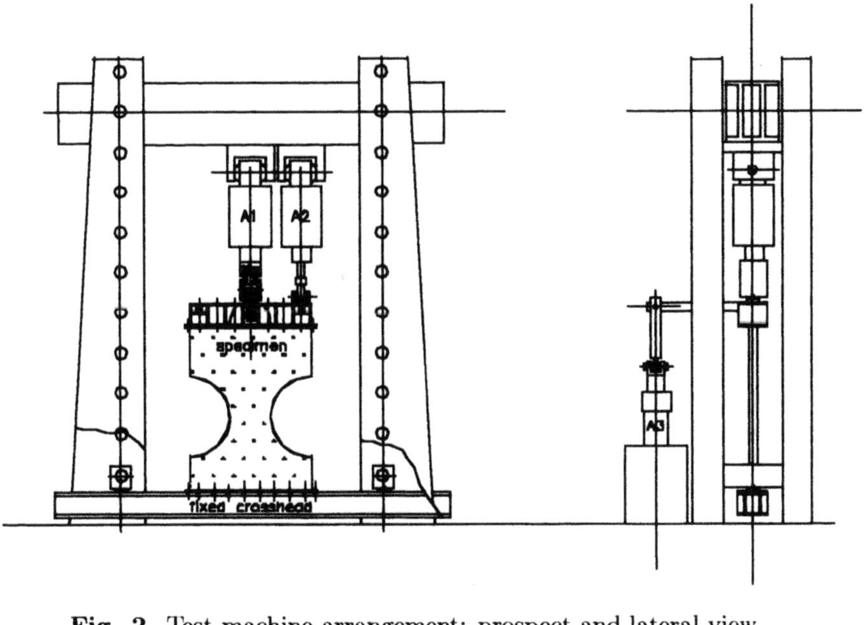

Fig. 2. Test machine arrangement: prospect and lateral view.

Table 1. Experimental average values for tensile strength and fracture energy

Specimen Width(cm)	*Cross Sectional Area(cm^2)*	*Tensile Strength* σ_u (kg/cm^2)	*Fracture Energy* $\mathcal{G}_F(kg/cm)$
5	50	43.3	0.085
10	100	38.5	0.104
20	200	37.1	0.145
40	400	31.7	-

The experimental load versus extension curves for the three specimen sizes are showed in Fig.3. It is important to note how the slope of elastic portion is the same for all sizes. This is due to the fact that the lengths of the extensometers were proportional to the size of the specimens (5 cm for the smallest ones, 10 cm and 20 cm for the other two).

Moreover the shape of these curves is in any case the same and this fact means that this kind of test procedure is independent of the machine stiffness. In fact,

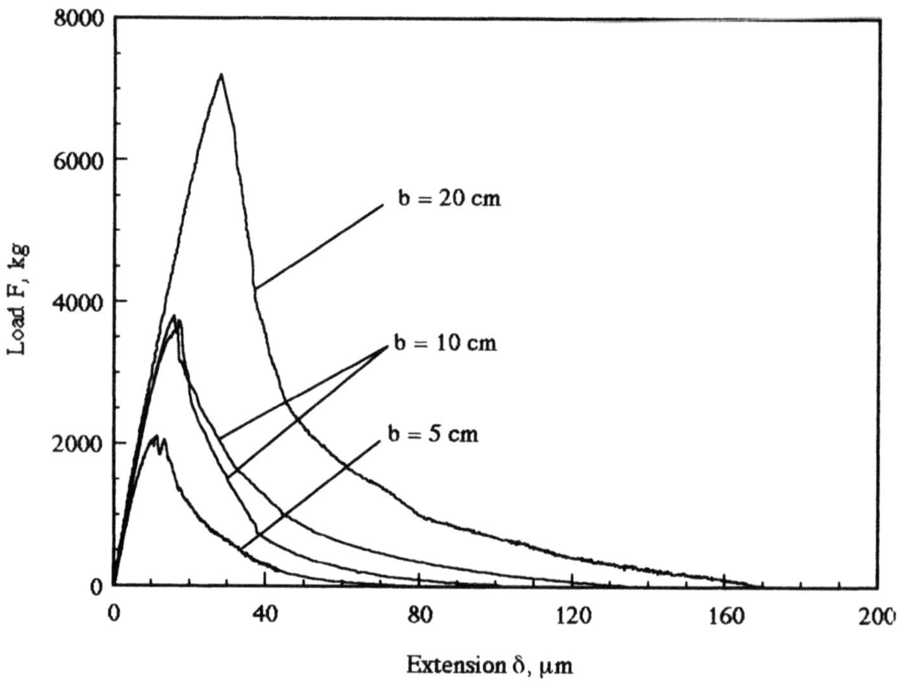

Fig. 3. Experimental load versus extension curves obtained from three different specimen sizes.

unlike what underlined by van Mier (1989) for the specimens tested with fixed grip plates, no bumps were found. This is due to the load which is always centred, when the control takes place both in-plane and out-of-plane. In the tests discussed by van Mier, on the contrary, the check was only in the specimen plane, and no check was performed for the out-of-plane eccentricity.

The values of the apparent tensile strength are plotted in Fig.4 against the specimen size in a bilogarithmic plane. The size effect is represented by the slope of the linear regression of the points of the diagram. It is evident how the tensile strength decreases with increasing specimen size.

The fracture energy instead shows the opposite trend; i.e., it increases with specimen size. The values obtained experimentally are reported in Fig.5 on the bilogarithmic plane $\ln \mathcal{G}_F$ versus $\ln b$. The variation of $\mathcal{G}_F$ as a function of size is represented by the slope of the linear regression diagram.

4 Fractal interpretation of size effects

The fracture energy $\mathcal{G}_F$, extracted from the $F - \delta$ curves, does not appear to be constant when varying the specimen size, but it shows a growing trend when the specimen size increases.

Herein an interpretation of the size effect is presented. The fracture energy

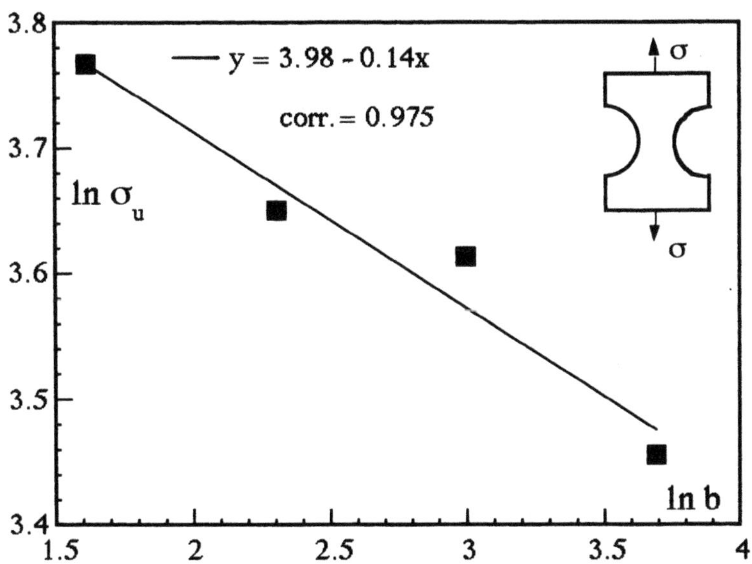

Fig. 4. Size effect on apparent tensile strength.

is defined as the absorbed energy per unit projected area of fracture zone after the complete fracture process. On the other hand, the fracture surface for heterogeneous materials like concrete is not contained in one plane. It is possible to suppose that this fracture surface has a dimension larger than 2, which is characteristic of bidimensional domains. Numerous researchers tried to estimate the fractal dimension of these fracture surfaces (Mandelbrot et al. 1984, Saouma et al. 1990). It is clear how, assuming this fractal dimension equal to a real number included between 2 and 3, the increase of the fracture surface follows an exponential law, when increasing the specimen size. The exponent is equal to the dimension of the same surface and, this number being greater than 2, that fracture surface increases more quickly than the projected area. The expression of fracture energy is the following:

$$\mathcal{G}_F = \frac{1}{A}\int_0^{\delta_c} F(\delta)d\delta, \tag{1}$$

where A is the projected area previously defined. Replacing A with the fractal measure of the fracture zone a constant value of fracture energy with varying the specimen size is obtained. Obviously the problem consists in finding the fractal dimension of the fracture surface. It should be noted that, assuming a fractal dimension of the fracture surface equal to 2.38, any size effect on fracture energy disappears.

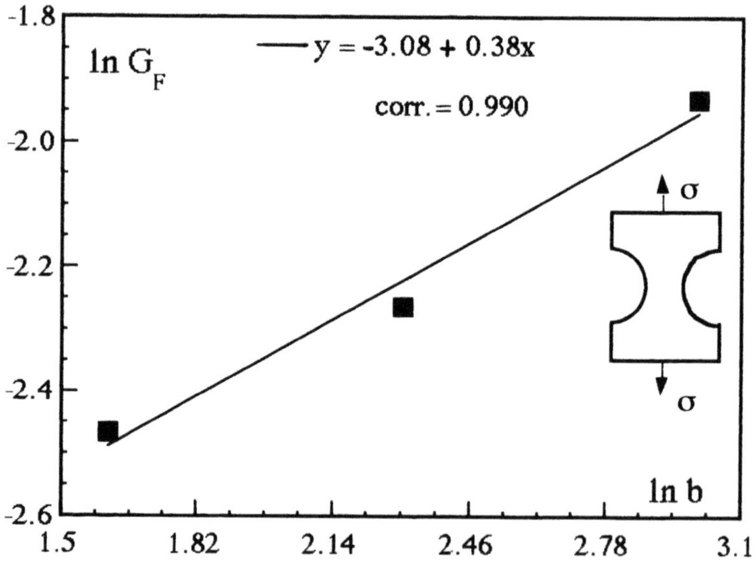

Fig. 5. Size effect on fictitious fracture energy.

Even the value of tensile strength is not constant when varying the specimen size. The strength decreases when the size increases. To explain this effect a defect size distribution of self-similarity is defined, for which the maximum defect size turns out to be proportional to the size of the specimen. That distribution can be expressed in the following form (Carpinteri 1986, 1989):

$$p(a) = \frac{C}{a^2}, \quad \forall a \gg \overline{a} \tag{2}$$

where C is a constant with the physical dimension of a length and $\overline{a}$ is the defect size mode. This equation will be referred to as the defect size distribution of self-similarity. A Monte Carlo simulation on an ideal material with that statistical distribution of defects was executed (Carpinteri and Ferro 1992). A law for the tensile strength decrease with size is obtained. The size effect on tensile strength is represented by a linear $\ln \sigma_u$ versus $\ln b$ diagram where the slope is the power of the stress singularity produced by the defect. The slope ranges between zero and 1/2 in any case.

5 Conclusions

A tensile test carried out in deformation-control that avoids the load eccentricity both in the specimen plane and out-of-plane is presented. The advantages of this testing technique are the following.

1. Total elimination of load eccentricity and, for this reason, the possibility to carry out tests up to total separation of the specimens into two parts when the load vanishes.

2. Possibility to carry out tests on specimens of any shape and size. The shape here adopted permits to dispose of a preferential zone for the crack initiation.

3. Elimination of the influence of boundary rotation stiffness of test machine and presence of tensile stress also at the end of the specimen, due to the particular grip system adopted.

From the tests comes out a size effect both on the apparent tensile strength, that decreases with increasing size of the specimen, and on the fictitious fracture energy, that instead increases.

The fractal dimension concept for the fracture surface is invoked, i.e., the dimension that cancels the size effect on experimental fracture energy results.

Eventually a defect size distribution of self-similarity is assumed. The size effect on tensile strength is represented by a linear $\ln \sigma_u$ versus $\ln b$ diagram where the slope coincides with the power of the stress-singularity produced by the defect. This slope is always comprised between zero and 1/2.

References

Bažant, Z.P.and Cedolin L. (1992) Why direct tension specimens flex and break at midlength, in **Fracture Mechanics of Concrete Structures** (ed Z.P. Bažant), Elsevier, London and New York, 1992, pp. 443-448 (Proc. of FRAMCOS1, Colorado, USA, 1-5 june 1992).

Carpinteri, A. (1986) **Mechanical Damage and Crack Growth in Concrete.** Martinus Nijhoff, Dordrecht.

Carpinteri, A. (1989) Decrease of apparent tensile and bending strength with specimen size: two different explanations based on fracture mechanics. **Int. J. Solids and Structures**, 25, 407-429.

Carpinteri A. and Ferro G. (1992) Effetti di scala sulla resistenza dei materiali dovuti alla dispersione statistica dei difetti, **XI Congresso A.I.M.E.T.A.**, Trento, Italia, 28 september-2 october 1992.

Gopalaratnam, V.S. and Shah S.P. (1985) Softening response of plain concrete in direct tension. **J. ACI**, 82, 310-323.

Mandelbrot, B.B. Passoja, D.E. and Paullay, A. (1984) Fractal character of fracture surfaces of metals. **Nature**, 308, 21-722.

Phillips, D. V. and Zhang, B. (1991) Fracture energy and brittleness of plain concrete specimens under direct tension, in **Fracture Behaviour and Design of Materials and Structures** (ed D. Firrao), EMAS, Warley,1990, pp.646-652 (Proc. of the 8th European Conference on Fracture, Torino, Italia, 1-5 october 1990).

Reinhardt, H. W. Cornelissen, A. W. and Hordijk, D. A. (1986) Tensile tests and failure analysis of concrete. **J. Struct. Eng.**, 112, 2462-2477.

Saouma, V.E. Barton, C.C. and Gamaleldin, N.A. (1990) Fractal characterization of fracture surfaces in concrete. **Engineering Fracture Mechanics**, 35, 47-53.

Van Mier, J. G. M. (1989) Scaling in tensile and compressive fracture of concrete, in **Applications of Fracture Mechanics to Reinforced Concrete** (ed A. Carpinteri), Chapman and Hall, London, pp. 19-31.

8 SIMULATION OF CONCRETE CRACKING: A NUMERICAL EVALUATION OF THE ROTATING CRACK MODEL

Y.-J. LI and Th. ZIMMERMANN
LSC, Swiss Federal Institute of Technology,
Lausanne, Switzerland

Abstract
Among existing concrete crack simulation models the rotating crack model is certainly one which lends itself to a relatively simple implementation. It is therefore of interest to evaluate its simulation capabilities. This is the purpose of this article. Two questions are adressed :

- Is the model capable of simulating mode I and mode II type cracking ?
- Size effect : does the model perform correctly over the whole range of structural sizes ?

Results show satisfactory results for mode I cracking, but poor results for mode II.
Keywords : Finite elements, numerical simulation, concrete, cracking

1. Introduction

The rotating crack model (RCM) is basically a smeared crack model in which the crack rotates in order to stay perpendicular to the maximum tensile stress. This rotation offers the possibility to describe the softening process associated with a crack submitted to multiaxial stresses directly from the one-dimensional softening curve.

A description of the model can be found in [Willam et al. 1987, Crisfield and Wills 1987, Rots 1988, de Borst 1991]. The details of the implementation used here are given in [Li & Zimmermann, 1992]. Attention is focused, in this article, on performance tests.

The evaluation of the model is performed on a number of test problems which include : the double edge cracked plate tension specimen, the single edge cracked three point bending specimen, the edge slant cracked rectangular plate tension test, and the four point bending test. All tests are performed on a variety of specimen sizes covering the whole range of size effects. Results obtained with program FRACT (see Li & Zimmermann, 1992) are described next.

Fracture and Damage of Concrete and Rock – FDCR-2.
Edited by H.P. Rossmanith. © 1993 Taylor & Francis. ISBN 0 419 18470 8.

2. Tests

2.1 The double edge cracked plate tension specimen

This problem is used to test if the RCM can be applied to the Mode-I fracture problem of LEFM (linear elastic fracture mechanics). The geometry of the specimen is shown in figure 1.

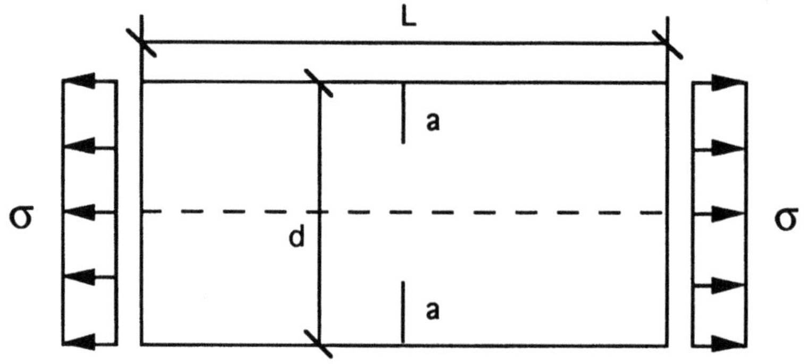

Fig. 1. Geometry of specimen

The shape of the softening diagram is bilinear, as shown in figure 2.

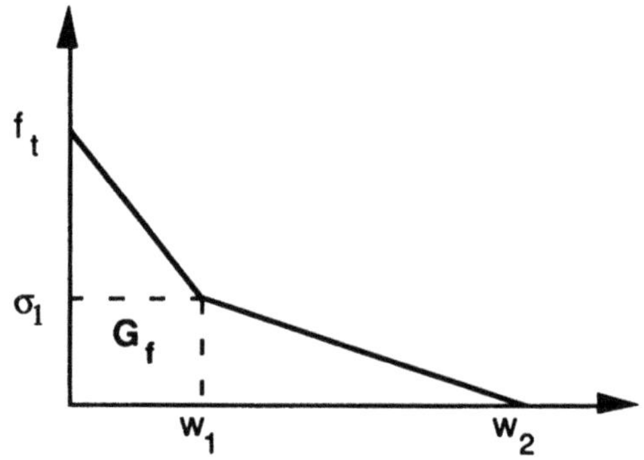

Material data

Young's modulus	E	$= 20000$ N/mm^2
Poisson's ratio	ν	$= 0$
Tensile strength	f_t	$= 2.5$ N/mm^2
Fracture energy	G_f	$= 0.1$ N/mm
	σ_1	$= 0.83$ N/mm^2
	w_1	$= 0.032$ mm
	w_2	$= 0.15$ mm

Fig. 2. Softening diagram and material data

According to the theory of LEFM the fracture toughness of the material K_{Ic} can be obtained directly from the fracture energy G_f as follows

$$K_{Ic} = \sqrt{\frac{G_f\ E}{1-\nu^2}} = 44.72\ \text{N/mm}^{3/2} \qquad (1)$$

On the other hand, the maximum applicable load is related to the fracture toughness by (2) [Murakami, 1987], so that the maximum load computed by FRACT can be used to generate a computed value of K^c_{Ic} to compare with the theoretical value obtained from (1)

$$K^c_{Ic} = \sigma_{max}\sqrt{\pi a}\, f(\alpha), \quad \alpha = \frac{2a}{d}$$
$$f(\alpha) = 1.112 - 0.154\alpha + 0.807\alpha^2 - 1.894\alpha^3 + 2.494\alpha^4 \qquad (2)$$

The mesh used for the computation is shown in figure 3.

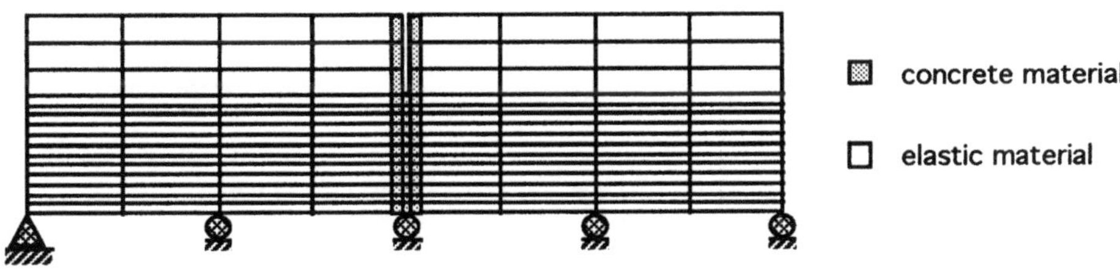

Fig. 3. Finite element dense mesh

The calculation is performed on a plane strain mesh with concrete (nonlinear) material only in the central part and dimensions ratios as follows

$d/L = 0.5$
$2a/d = 0.4$
$h/d = 1/20$ ($d < 3000$ mm)
$h/d = 1/30$ ($d \geq 3000$ mm)

where h is the mesh size at crack tip.

Results are shown in figure 4. It is observed that, as desirable, the numerical results tend towards the fracture results for large specimen sizes.

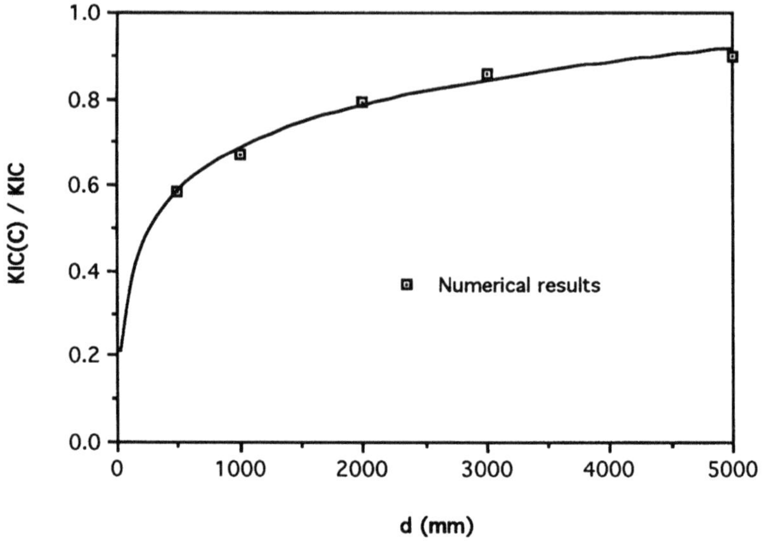

Fig. 4. Curve (K_{Ic}^{c}/K_{Ic}) as a function of d (the specimen size)

2.2 Single edge cracked three point bending test
This problem is used to test if the RCM can be applied to the indirect tension mode I fracture problem. The geometry and loading of the specimen are shown in figure 5. The loading was applied via two concentrated forces in the mi-point of the upper edge of the beam, in the Y direction.

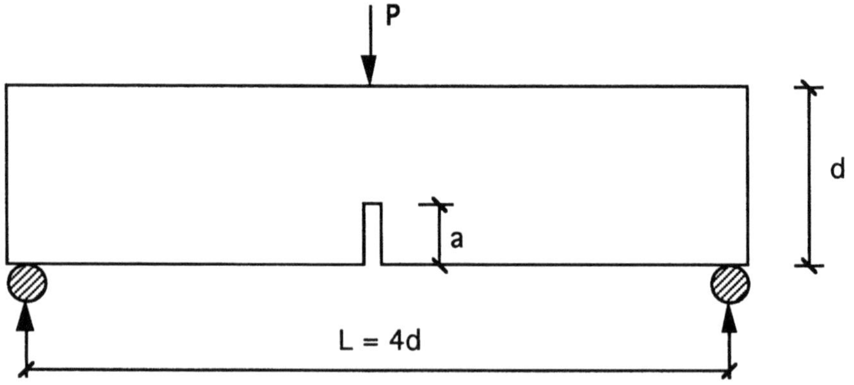

Fig. 5. Geometry of specimen

Material data are

Young's modulus	E	$= 20000\ N/mm^2$
Poisson's ratio	ν	$= 0$
Tensile strength	f_t	$= 2.5\ N/mm^2$
Fracture energy	G_f	$= 0.1\ N/mm$

The shape of the softening diagram remains the same as for the previous test.

According to Carpinteri [1985.], the brittle and plastic cases can be treated as two extreme modes of failure corresponding to LEFM and L.A. (limit analysis), respectively.

For the three point bending tests the stress intensity factor is [Murakami, 1987].

$$K_I = \frac{P\ L}{t\ d^{3/2}}\ f\left(\frac{a}{d}\right) \tag{4}$$

where

$$f\left(\frac{a}{d}\right) = 2.9\left(\frac{a}{d}\right)^{1/2} - 4.6\left(\frac{a}{d}\right)^{3/2} + 21.8\left(\frac{a}{d}\right)^{5/2} - 37.6\left(\frac{a}{d}\right)^{7/2} + 38.7\left(\frac{a}{d}\right)^{9/2} \tag{5}$$

The critical load P_{cr} for crack propagation can be obtained from

$$\frac{P_{cr}\ L}{t\ d^2\ t} = \frac{S}{f(a/d)} \qquad \text{with } S = K_{Ic}/\left(\sigma_y \sqrt{h}\right) \tag{6}$$

where K_{Ic} is the plane strain fracture toughness, σ_y the yield stress and h the characteristic size of the specimen.

Plastic flow collapse by limit analysis corresponds to

$$\frac{P_{cr}\ L}{\sigma_y\, d^2\ t} = \left(1 - \frac{a}{d}\right)^2 \tag{7}$$

The following notations are used in order to simplify comparisons

$$A = (P_{cr}\ L)/(\sigma_y\ d^2\ t) \tag{8.1}$$

$$B = S/f(a/d) \tag{8.2}$$

$$C = (1 - a/d)^2 \tag{8.3}$$

where A is obtained through numerical simulation and B and C result from material data.

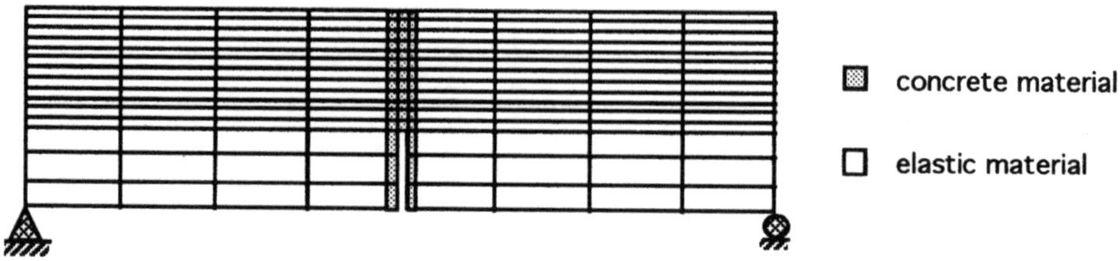

Fig. 6.Finite element dense mesh

Six specimens of different sizes were calculated, the height of the specimen ranging from 200 to 5000 mm, with the following dimensions ratios :

d/L = 1/4 (constant)
a/d = 0.4 (constant)
h/d = 1/20 $\quad$ d ≤ 3000 mm
h/d = 1/30 $\quad$ d > 3000 mm

where L, d a, h are the specimen height, the crack length and the mesh size at the crack tip (Fig. 6), respectively.

Comparisons of results are based on the following "error" estimates :

Er1 = ((B - A)/B) [%]
Er2 = ((C - A)/C) [%]

The results of the calculation and analyses of the relative error are listed in table 1.

d mm	S	a/b	f(a/b)	B	C	Pcr (N)	A	Er1	Er2
200	1.26	0.4	1.98	0.64	0.36	45	0.360	44%	0%
300	1.03	0.4	1.98	0.52	0.36	65	0.347	33%	4%
500	0.80	0.4	1.98	0.40	0.36	100	0.320	20%	11%
1000	0.57	0.4	1.98	0.29	0.36	150	0.240	17%	33%
3000	0.32	0.4	1.98	0.16	0.36	290	0.155	3.3%	57%
5000	0.25	0.4	1.98	0.13	0.36	420	0.134	3.1%	63%

Table 1. The results of calculation

The results illustrate that the RCM is capable of simulating situations corresponding both to plastic collapse (d = 200 mm) and to LEFM type failure (d = 5000 mm), with a smooth transition from the first to the second situation.

2.3 Edge slant rectangular plate subjected to uniform uniaxial tension stress

This problem is used to test if the RCM can be applied to the mixed mode fracture problem of LEFM. The geometry of the specimen is shown in figure 7. The loading is applied via uniform distributed forces along the left and right edges, in the horizontal direction

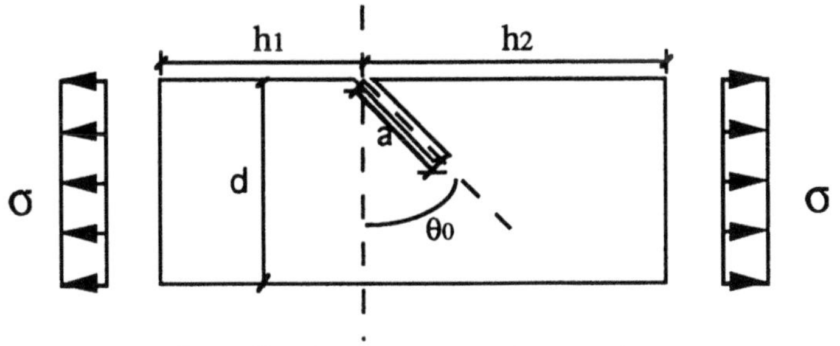

Fig. 7. Geometry of specimen

The fracture material parameters are assumed to be

Young's modulus $E = 20000\ N/mm^2$

Poisson's ratio $\nu = 0$

Tensile strength $f_t = 2.5\ N/mm^2$

The fracture energy and the shape of the softening diagram are the same as before, i.e. $G_f = 0.1$ N/mm, and again

$$K_{Ic} = \sqrt{\frac{G_f\ E}{1-\nu^2}} = 44.72\ N/mm^{3/2}$$

The stress intensity factors K_I and K_{II} resulting from LEFM theory are

$$K_I = \alpha\ \sigma\sqrt{\pi a}$$

$$K_{II} = \beta\ \sigma\sqrt{\pi a}$$

where $\alpha = 1.2$, $\beta = 0.57$ [Murakami, 1987].

Using the theory of maximum tangential stress of mixed mode LEFM theory, the angle of propagation of crack and fracture toughness K_{Ic} can be calculated as follows

$$K_I \sin\theta + K_{II}\ (3\cos\theta - 1) = 0$$

$$K^*_{Ic} = \cos\ (\theta/2)\left[K_I \cos^2\ (\theta/2) - \frac{3}{2}\ K_{II} \sin\theta\right]$$

Comparing K_{Ic} with K^*_{Ic} a relative error analysis can be performed with

$$Er = \left[\frac{K_{Ic} - K_{Ic}^{*}}{K_{Ic}}\right] \%$$

Five specimen of different sizes were calculated in this case. The height of these specimen ranges from 1000 mm to 15000 mm. The following geometry is adopted (see Fig. 7)

h_1 = d (constant)

h_2 = 1.5 d (constant)

a/d = 0.5 (constant)

θ_0 = 45° (constant)

h/d = (1/20 ~ 1/50, see table 2)

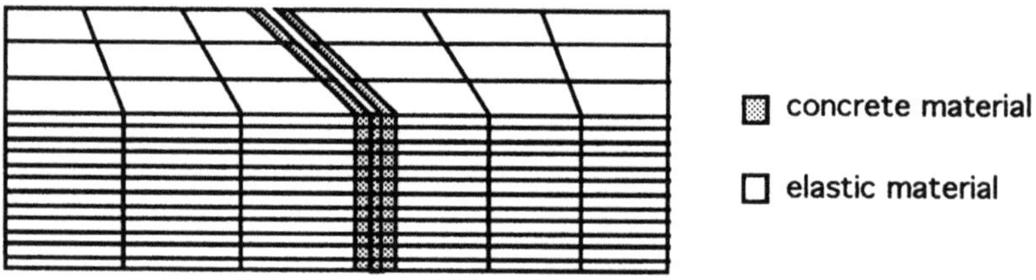

Fig. 8. Finite element dense mesh

a, θ_0, and h are the crack length, the crack angle relative to the vertical direction and the mesh size at the crack tip, respectively.

The results of the calculation are listed in table 2 and illustrated in figure 9.

d (mm)	h/d	σ_{max} N/mm^2	K_I $N/mm^{1.5}$	K_{II} $N/mm^{1.5}$	K_{IC}^{*} $N/mm^{1.5}$	K_{IC} $N/mm^{1.5}$	$\frac{K_{IC}^{*}}{K_{IC}}$
1000	1/20	0.6	28.54	13.55	11.93	44.72	0.27
3000	1/20	0.45	37.07	17.60	15.49	44.72	0.35
5000	1/30	0.45	47.86	22.73	20.00	44.72	0.45
10000	1/30	0.45	67.68	32.14	28.28	44.72	0.63
15000	1/50	0.45	82.88	39.37	34.62	44.72	0.77

Table 2. Results of calculation

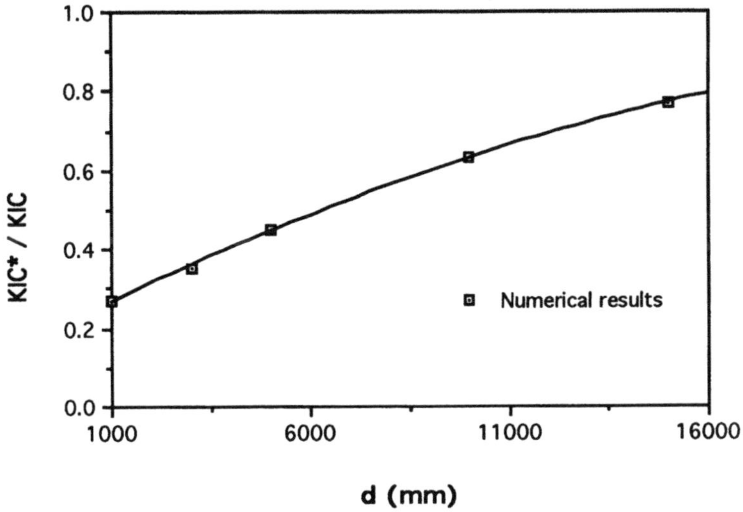

Fig. 9. Curve K_{IC}^{c}/K_{Ic} vs. d

From figure 9, it is observed that K_{Ic}^{*} tends slowly towards K_{Ic} with an increase of specimen height. When the specimen height reaches 15000 mm the relative error is however still 23%. This result is insensitive to the value of Poisson's ratio.

2.4 The four point bending test

The notched shear beam tested by [Arrea and Ingraffea, 1982.] is analyzed next using the RCM model. The geometry of the problem and the finite element mesh are shown in the figure 10.

The material properties are the following

Young's modulus	$E = 24800$ N/mm^2
Poisson's ratio	$\nu = 0.18$
Tensile strength	$f_t = 4.55$ N/mm^2
Compressive strength	$f_c = 45.5$ N/mm^2

Fracture energy $G_f = 100$ J/m $= 0.1$ N/mm. The shape of the softening diagram is the same as before.

In this case the support and the loading condition are non-symmetric with respect to the notch. The boundary conditions and the loading system are shown in figure 10. The dense mesh in the central part is shown in figure 11.

Again nonlinear (concrete) material is used only in the central part.

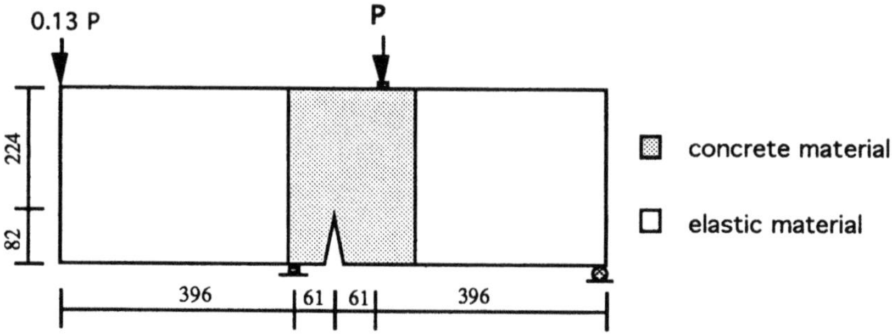

Fig. 10. Loading system and boundary conditions

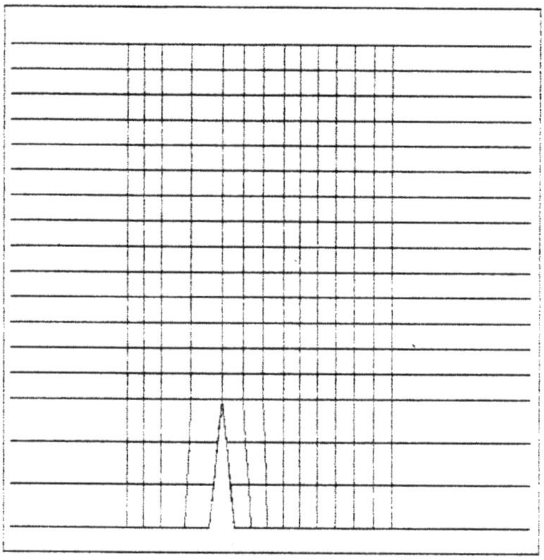

Fig 11 Dense mesh in the central part

Comparison with the test yield the following results

	Experiment	Simulation
Maximum load	160 kN	152 kN
Fracture initiation angle	67°	50° - 60°
Crack trajectories	see figure 12	

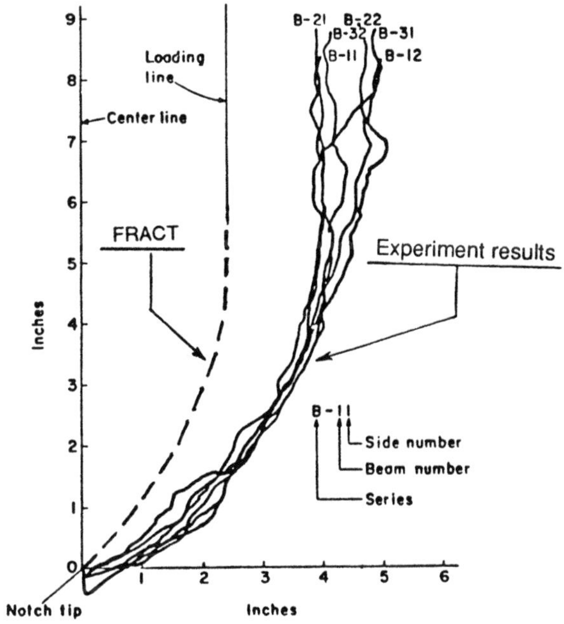

Fig. 12. Crack trajectories

It is observed that simulated crack trajectories differ significantly from the experiment. This was already observed by [Rots, 1988.].

Concluding remarks

In this article a numerical evaluation of the rotating crack model (RCM) is performed, based on four common fracture tests.

Results show that the RCM performs well in Mode I type fracture, over the whole range of sizes covered by Linear Elastic Fracture Mechanics and Limit Analysis. Under mixed mode fracture, results are poor.

All tests were performed under plane strain conditions with Poisson's ratio equal to zero, in order to compare with plane stress situations. Conclusions drawn appear insensitive to the value of Poisson's ratio.

Presently, more tests are being performed and improvements of the RCM model are being attempted by the authors.

References

Arrea M., Ingraffea A. R., 1982, Mixed mode crack propagation in mortar and concrete. Report No. 81-13, Cornell University

Carpinteri A.,1985, Scale effects in fracture of plain and reinforced concrete structures, Fracture mechanics of concrete (Edited by G.C. SIH, A. DITOMMASO)

Crisfield M. A. and Wills J., 1987, Numerical comparisons involving different 'concrete-models', IABSE Reports 54, Coll. Comp. Mech. of Reinforced Concrete, Delft Univ. Press

de Borst R., 1991, Computational aspects of failure of reinforced concrete, colloquium on Numerical analysis of massive concrete structures, Lausanne

Li Y.-J., Zimmermann Th., 1992, Simulation of concrete cracking. A numerical evaluation of the rotating crack model, Internal Report, No. 92/21, LSC, EPFL

Murakami Y, Editor-in Chief, 1987, Stress Intensity Factors Handbook, Pergamon press

Rots J. G., 1988, Computational modeling of concrete fracture, Dissertation, Delft Univ. of Technology, Dept. of Civil Engrg., Delft

Willam K., Pramono E. and Sture S., 1987, Fundamental issues of smeared crack models, Proc. SEM-RILEM Int. Conf. on Fracture of Concrete and Rock (S. P. Shah and S. E. Swartz eds.), SEM, Bethel, pp. 192-207

9 ENVIRONMENTAL FACTORS FOR POST-TENSIONING MATERIALS IN CONCRETE STRUCTURES

E.G. SEGAN
US Army Construction Engineering Research Laboratories, Champaign, Illinois, USA
P. KURATH
University of Illinois at Urbana-Champaign, Urbana, Illinois, USA

Abstract
High strength steels are typically used for post-tensioning systems in concrete structures. These steels are typically thought not to be subject to corrosion due to the high pH (typically 10 to 12) present in the concrete environment. However, Environmentally Assisted Cracking (EAC) can occur even at high bulk pH and can result in reduction in mechanical properties of steels while leaving the base material generally unaffected. This process can result in geometrical stress raisers and reduction of materials properties including crack growth rates and critical stress concentration factor for fracture (K_{IEAC}).

Previous failures of post-tensioning systems have illustrated the potential for EAC failures. In this study, published EAC resistance of high strength steels in various environments are reviewed to determine which materials properties have a direct effect on EAC resistance and determine whether EAC is a potential problem in steel-concrete systems. It is shown that in many cases, materials properties generally used to specify materials (e.g. ultimate tensile strength, yield strength, percent reduction in area (%RA), hardness etc.) render little predictive information for resistance to EAC. Thus, there is a need to specify material resistance to EAC when selecting these high-strength post-tensioning materials. The results also illustrate that concrete environments can lead to reduction in material resistance to EAC.
Keywords: Environmentally Assisted Cracking, High Strength Steels, Stress Corrosion Cracking, Hydrogen Embrittlement, Post-Tensioning Systems.

1 Introduction

Post-tensioned systems have been used for concrete structures since the 1930's but the majority of this type of construction was completed in the past thirty years. Failures of these systems, although infrequent, have occurred in the past. Schupack reviewed known failures before 1978 and stated: "In the 28 structures with known corrosion incidents cited, some 200 of the 30 million tendons estimated to have been used to date were involved." This statistic has been be interpreted to suggest that further investigation into

Fracture and Damage of Concrete and Rock – FDCR-2.
Edited by H.P. Rossmanith. © 1993 Taylor & Francis. ISBN 0 419 18470 8.

this phenomenon is unwarranted. Further, most failures were traceable to design deficiencies or improper workmanship, and there were no catastrophic occurrences. However, there is continuing controversary about the effects of corrosion on the behavior of post-tensioned steels.

More recent failures of post-tensioned steel in structures such as the Berlin Congress Hall (Isecke, Elices) and Grundfors hydropower station (Jurell) occurred 23 and 26 years after construction, respectively. In both instances, there was evidence of Environmentally Assisted Cracking (EAC) of the steel reinforcing. This suggests that other existing structures that are nearing twenty years of age may be candidates for possible corrosion-related problems. Other recent failures of post-tensioned systems have been discussed by Segan et. al. 1985,Isecke 1982, Elices 1983, and Jurell 1985. These incidents have contributed to concerns regarding the long-term reliability of such systems.

Materials used for this type of construction are high strength steels with ultimate strengths often well in excess of 825 MPa. They are documented to have lower fracture toughness and ductility than lower strength materials of similar chemical composition. Project specifications dictate that considerable care to be taken ensure that no materials degradation occurs at the construction site and also deal with installation procedures. Specifications intended to ensure material integrity (e.g. ASTM A-416, ASTM A-421,and ASTM A-722) address minimum macroscopic material properties such as percent elongation, and yield and ultimate strengths. Due to the nature of the structural arrangement of the aforementioned materials, present technology provides limited possibilities to ensure material integrity after installation.

Corrosion of metals may take many different forms which range from obvious damage due to general corrosion and pitting to subtle damage in the form of environmentally assisted cracking. General corrosion, pitting corrosion, and other types of corrosion of steels can occur in many environments. In these cases although material properties remain unchanged, load bearing cross-section is reduced to the point where the remaining cross-section can no longer sustain the load and failure occurs. Corrosion by-products (i.e., hydrated forms of iron oxide) are usually visible during this type of corrosion.

Steel is often used as a reinforcing material in concrete environments because it develops a passive film in the high pH (often pH over 12) environment of the water, sand, aggregate, and cement mixture. This film substantially reduces the corrosion rate of the steel. Since the 1960's, much research on general corrosion and pitting of steel in concrete has been conducted. Slater summarized some of the proposed mechanisms of this type of damage. It should be noted that in this type of corrosion damage, the steel exhibits visible corrosion products and often results in spalling of concrete. This is the result of tensile stresses placed on the concrete due to the formation of hydrated iron oxide which has a larger volume than the base metal.

The subject of this paper is EAC which encompasses stress corrosion cracking (SCC) and hydrogen embrittlement (HE). The three factors necessary for EAC are:

1. a sensitive material,
2. the presence of stress, and
3. the presence of a corrosive environment.

EAC reduces a materials resistance to cracking. Cracks can nucleate and/or propagate through a material at applied stresses well below the yield stress. It should be noted that EAC typically causes little or no general corrosion or pitting attack. This is in contrast to general and pitting corrosion phenomena, where corrosion products are readily apparent. Although SCC and HE mechanisms differ, the result of either can be catastrophic material failure with little or no warning.

The environments in which EAC can occur are complex, not well understood and as a result, not easily predictable. Further, the bulk environment is not necessarily indicative of the local environment at a crack or crevice. Brown, et al., noted that the pH in a crack or crevice is about 3 and independent of the bulk pH from 3 to above 12. This implies that although the pH of the bulk environment in concrete is in the safe range of about 12, the environment in a pit of crevice may be in the range where EAC can occur. Another effect which is not well understood is the lowering of the local pH by external means such as allowing air to enter the post-tensioned system.

In general, materials which are highly susceptible to EAC are those with high strengths and low ductility. In the presence of stress and an aggressive environment for EAC, these materials can exhibit significantly reduced resistance to cracking. There are few data available on the EAC behavior of post-tensioning steels in concrete environments but data are available on the behavior of these steels in various other types of environments. These studies are helpful in identifying trends in material behavior. The goal of this work is to evaluate the effect of materials properties on the resistance of high strength steels to EAC.

Present material specifications for post-tensioned structures consider only minimum percent elongation, yield, and ultimate strength allowable. Post-tensioned structures are unique within the realm of civil engineering in that the usual factor of safety of between 1.8 and 3 below the yield stress for design stresses is not incorporated into tendon calculations. Design stresses routinely exceed 70% of the yield strength. A literature review of past failures in this type of construction was conducted to ascertain the cause of failure and its relation to the preloading tendons. Strength, ductility, microstructural properties, corrosion resistance, and environmentally assisted cracking (EAC) resistance of materials that could be employed in these structures were examined with the intent of improving current specifications and identifying material limitations.

2 Background

2.1 Post-Tensioned Systems

The basic premise of post-tensioned and pre-tensioned construction is that a low cost material, concrete, can be used to build massive structures, with the preloaded steel providing superior results and cost savings when compared to conventional reinforcing. Post-tensioning is often employed to upgrade structural performance (Anon, 1973) or to remedy unforeseen design shortcomings. In either case, high strength steel bars or cables are in tension. To optimize the design, stress levels in this steel are often greater than 70 percent of the ultimate strength of the material.

Putting aside corrosion considerations for a moment, these preloading materials are quite susceptible to manufacturing defects (inclusions, voids, etc.) and installation defects at high stress levels, due to their lower fracture toughness (K_{IC}). Therefore, specifications exist which prohibit welding, which could likely cause inclusions or voids, and mandate protective sheathing to prevent damage in handling. These measures are intended to ensure that no defects exist in the reinforcement at the time of installation. Visual inspection techniques prior to installation are presently mandated (i.e. John Day Lock and Dam Specifications, Old River Control Auxiliary Structure Specifications). However, present technology offers more rigorous means (i.e. ultrasonic, x-ray, etc) of ensuring initial material integrity with regard to the presence of defects. While such precautions are costly, they could be required on structures where failure could have serious consequences. For less critical structures, life cycle cost analysis could be employed to determine if the implementation of more rigorous inspection procedures are justified.

2.2 Laboratory Investigations into Environmentally Assisted Cracking

The susceptibility of materials employed in post-tensioned construction have been investigated by various researchers. The results of selected studies are summarized in the following section. An ASTM-A416 grade 270 was tested by Hahin, Gambell, and Scott. This material has an ultimate strength requirement of 1865 MPa. Tests were conducted in HCl and $Ca(OH)_2$ solutions with pH values from 1 to 12.5. Failures were noted after one month at loads up to 90 percent of the ultimate strength of the material for pH values greater than 3. A sharp decrease in performance was noted at lower pH values for similar stress levels, leading to the researchers conclusion that pH level dominates the failure mechanism.

Segan, et al. investigated the susceptibility of post-tensioning wire conforming to ASTM A-416-80 in pH = 3 HCl solution. The value of pH was chosen to simulate the expected condition in a pit or crevice. It was found that the critical stress intensity factor for EAC (K_{IEAC}) was less than 50 percent of K_{IC} observed in an air atmosphere for this material.

Ellyin and Matta (1981) tested two high strength steels, σ_u = 1680 MPa and σ_u = 1620 MPa, which correspond closely to ASTM-A586 and ASTM-A421, respectively. Tests were conducted at stress levels ranging from 25 percent to 75 percent of the ultimate strength of the material in distilled water, sodium chloride, and calcium nitrate solutions. No specimens fractured during the duration of the tests (30 days). However, subsequent tensile testing revealed a degradation in properties such as ultimate strength and percent elongation. At load hold levels greater than 25 percent of the ultimate strength of the material, a decrease in monotonic properties was reported. The amount of this decrease was dependent on the environment. Stress level was cited by the authors as a major factor, along with environment, leading to the failure of post-tensioning materials. A decrease in monotonic properties was also noted for specimens that did not experience corrosion failures in Hahin et al.

Klodt tested 1240 MPa ultimate strength steel in calcium hydroxide, sodium chloride, and hydrogen sulfide solutions. Stress levels ranged from 0 to 90 percent of the ultimate strength of the material. The longest test lasted 49 days. The results indicated chlorides were detrimental, and hydrogen sulfide caused hydrogen embrittlement failures. An extensive section of the paper was devoted to the electrochemical response of the material in the various environments, as in a paper by Lewis and Copenhagen. Seven wire strands, ASTM-A416 grade 270, were tested in the presence of chloride, nitride, and sulfide ions (Griess and Naus). The test procedure involved exposure to the corrosive environment and macroscopic examination for signs of material degradation, followed by slow strain rate monotonic testing. The longest exposure time was 130 days. Sulfides and nitrides were found to degrade material performance while chlorides, regardless of pH, were discounted as a damaging medium.

The effect of chloride ions was further investigated by McGuinn and Griffiths on a 1630 MPa ultimate strength steel. Precracked specimens were tested at constant load or constant deflection to obtain fracture toughness data (i.e. K_{ISCC}). The authors concluded that, at pH values greater than 12, the presence of the chloride ions has little effect on material properties. Treadway also investigated the presence of chloride ions at higher pH in prestressing wire stressed to 70 to 80 percent of its ultimate tensile strength. Some specimens were protected by concrete containing various additives of chloride ions, and others were tested in hot aqueous chloride solutions. Though failure occurred in some specimens after 200 hours exposure, the cause was attributed to pitting and subsequent ductile failure, not environmentally assisted cracking.

The presence of NO_3^- and NO_4^+ ions in solution was investigated for a low strength steel (DeSanta et al.). Both ductile and brittle failures were attributed to the presence of corrosive ions based on microscopic examination of the failed samples. A 1800 MPa ultimate strength material was tested at pH 10, 12, and 14 in the presence of chloride ions by Cherry and Price (1980). Under these conditions, pitting or crevice corrosion was noted. Stress level had no influence on the depth of penetration of corrosive layer in tests lasting up to one year, EAC was not observed. Even in specimens precharged with hydrogen, no microscopic evidence of hydrogen embrittlement failures was detected.

These data indicate that while a majority of the materials tested are classified as prestressing or post-tensioning, variations in experimental data exist with regard to the effect of chloride ions. Many investigators have tried to recreate field situations where concrete surrounds the reinforcing bars or cables; thus, the pH for their tests are often greater than 10. Various researchers have shown concrete can help prevent tendon corrosion (Beeby, 1978, Scott, 1985, Hausmann, 1964, Szilard, 1969). Macroscopic examination of unstressed steel specimens are employed by these authors to draw their conclusions, with no consideration of changes in material properties where EAC is most likely to manifest itself. There is a consensus that sulfide and nitride ions cause environmentally assisted cracking in tendon materials.

One factor worth noting is that very few tests have been conducted lasting longer than one month. Many are slow strain rate tests that last a few days at most. Sanchez-Galvez, Caballero, and Elices have shown that, while slow strain rate tests render more consistent results for a given strain rate, a sensitivity to strain rate was identified. Depending on the imposed potential level, some slow strain rate data are easily correlated with constant load data. At other strain rates appreciable discrepancies exist. Their results indicate comparison between previous data may not be feasible. The structures in question are in the 20 year age category, requiring data to be extrapolated more than two orders of magnitude. While accelerated testing is certainly advantageous, no clear correlation exists to substantiate its extrapolation to two orders of magnitude. Evidence exists for the harmful effects of sulfide and nitride ions in the accelerated testing and for these environments the extrapolation of the data is probably appropriate. The lack of consensus on the harmful effects of chloride ions may be due to the short duration of most tests.

3 Discussion

Specifications for post-tensioning materials are vague regarding chemical composition and microstructure (ASTM-A416 and ASTM-A722). Limitations are placed on a few trace elements, such as sulfur and phosphorous, and minimum mechanical properties of yield and ultimate strength are dictated. Previously discussed data have dealt with different materials and experimental techniques making comparison between different investigations difficult. McGuinn and Elices made comparisons regarding production variables, such as the method of stress relieving. Twelve material states were tested with failure in the presence of NO_4^+ ions serving as a baseline. Chlorides were deemed to cause stress corrosion cracking in some of the materials even at pH values greater than 12, whereas other materials displayed no sensitivity to the chloride ions. These mirror discrepancies existing in the prior data and suggest that unspecified production variables may be responsible for the observed results. Elices, et al. have investigated residual stresses resulting from manufacturing processes and concluded that compressive surface residual stresses suppress EAC for initially uncracked specimens. Fatigue precracked specimens displayed a diminished effect of manufacturing variables. Further substantiation of these observations was presented by Felip'ev, et al.(1972).

Accelerated stress corrosion tests have been conducted in the presence of sulfide ions. Cracking in this environment is not necessarily indicative of a material's performance in others, but does offer some insight into the material's susceptibility to similar corrosion phenomena. The results of Shape, in which twenty-six samples of different chemical compositions were subjected to various heat treatments, are reviewed with the intent of establishing overall tendencies for these materials. A hydrogen sulfide environment was used for tests which were terminated after 28 days. The threshold strength, (σ_{th}) is defined as the highest stress level where no stress corrosion cracking occurred after a 28 day exposure to the sulfide ions. These data are compared to baseline yield strength($\sigma_{y0.2\%}$) values for the material.

Figures 1 through 6 show the authors' interpretation of various experimental results for quenched steel with retained martensite (square data points) and those with no retained martensite (Shape 1968: circle data points). Figure 1 illustrates that there is little correlation between yield strength and corrosion threshold for quenched steels with retained martensite (square data points). For lower strengths (i.e. < 700 MPa), one could crudely estimate a design criteria stipulating 25 percent of the yield strength. At higher yield strengths, susceptibility to stress corrosion is evident. Figure 2 illustrates that there is also limited correlation of ultimate strength with the corrosion threshold for these steels.

Figure 3 illustrates that, for steels with retained martensite (square data points), correlation between the yield strength and the ultimate strength is marginal. If fracture toughness can be interpreted from %RA, the data indicated for this type of heat treatment, a correlation exists between σ_u and %RA (Figure 4). Some of these materials display ductility to a degree far in excess of the ASTM specifications; however, they perform poorly with regard to their susceptibility to stress corrosion cracking (Figure 1). These results seem to indicate that untempered martensites in the microstructure, while resulting in admirable baseline properties, are sensitive to corrosive environments. Indeed, ASTM-416 and ASTM-572 specify cold work (i.e. cold drawing) to achieve material strength properties, with subsequent stress relieving heat treatment, rather than thermal methods to achieve high strength in conventional steels.

Examination of the data for quenched and tempered steels with no retained martensite (Shape 1968, circle data points) lends further insight into the importance of microstructure. When yield strength (Figure 1) and ultimate strength (Figure 2) are compared to corrosion threshold for this class of material (circle data points), it is evident that a spherodized carbide microstructure offers superior resistance to corrosive environments compared to other possible microstructures.

In contrast to the lack of correlation between yield and ultimate strength for the non-spherodized microstructures, there appears to be a linear relation between yield and ultimate strength for the spherodized microstructures (Figure 3, circle data points). Figure 4 shows ultimate tensile strengths versus %RA for the spherodized microstructures.

For both microstructures there appears to be an optimum heat treatment that achieves the highest possible threshold strength. This value is approximately 620 MPa and 500 MPa for spherodized and other

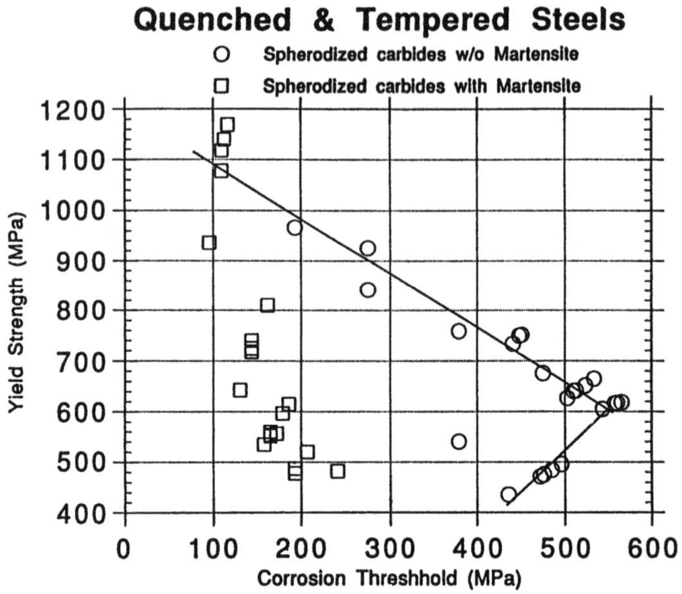

Figure 1
Comparison of Macroscopic Materials Properties

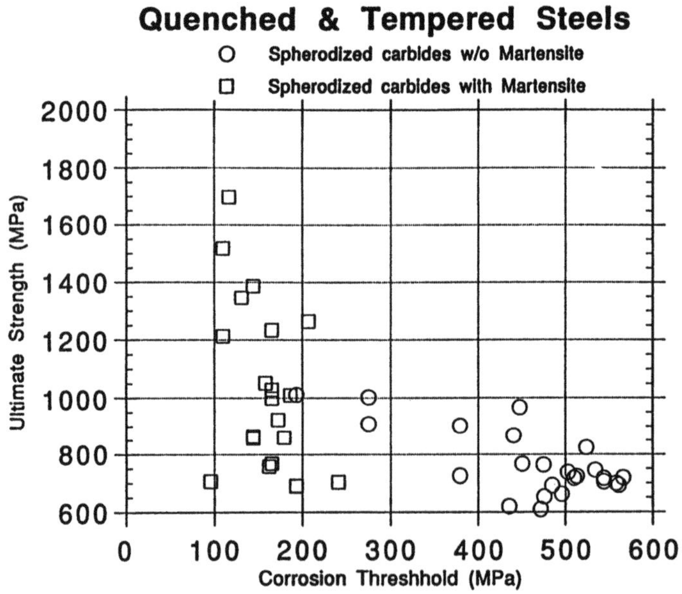

Figure 2
Comparison of Macroscopic Materials Properties

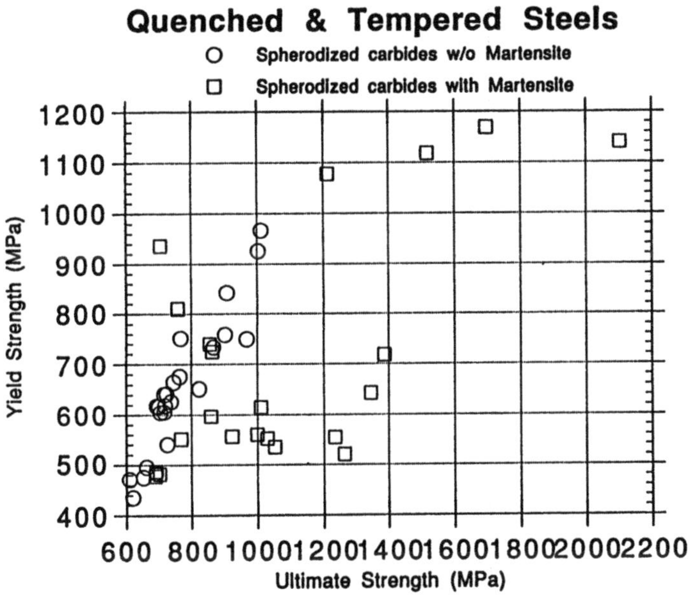

Figure 3
Comparison of Macroscopic Materials Properties

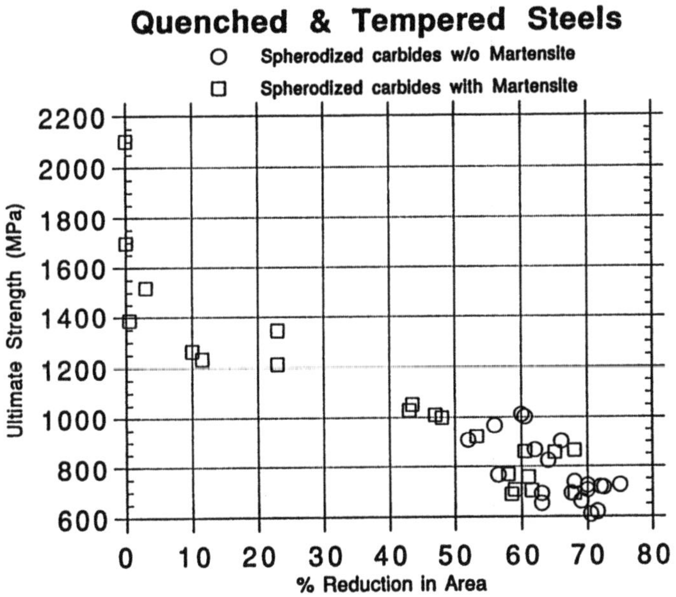

Figure 4
Comparison of Macroscopic Materials Properties

microstructures, respectively. At higher yield strengths, a decrease in corrosion performance is expected (Figure 1).

Figure 5 illustrates that regardless of strength level the spherodized microstructure displays appreciable ductility. In addition, if %RA is taken to be indicative of the material's ductility, it is evident yield strength is not well correlated to ductility. Figure 6 shows that for both microstructures, % elongation and %RA are relatively well correlated.

These observations tend to pose the problem that, while one could design systems using a 620 MPa yield strength material with an 550 MPa working stress, present structures often require stress levels twice as large. Designing twice the steel reinforcement into some structures is physically impossible. An alternate hypothesis postulates corrosion problems, and impeccable corrosion protection must be provided upon installation, and subsequently maintained. Gilmour and Anon (1978) review protection techniques and procedures involved. Exotic materials have been suggested (Ault et al., Anon, 1979, Cauchois) to resist corrosion; however, they violate the basic constraint of keeping the structure at a low cost. One could speculate that since the spherodized microstructures without retained martensite possess exceptional ductility (Figure 5) they could be cold worked to higher yield strength levels, but Figure 3 discounts this premise since a marginal increase in ultimate strength is anticipated for the higher yield strength material.

Other researchers have proposed alternate corrosive environments for laboratory testing, such as $Ca(OH)_2$ modified by NaCl and HCl to provide chloride ions yet maintain various pH levels (Parkins, et al.). An NH_4CNS testing environment has been proposed by the Prestressing International Federation (Brachet). Material microstructural features, heat treatment, and chemical composition, along with test methodology (i.e. slow strain rate or static load) are among the other variables also influencing material performance in these environments. The Shape data have been reviewed in depth due to the large variation of materials tested with an identical testing technique. At present there is no consensus on the corrosive environment that most accurately simulates the corrosion occurring in post-tensioned structures. However, the concepts forwarded should be applicable to EAC in other media.

Experimental data are lacking to ascertain the optimal material configuration of mechanical and microstructural characteristics resistant to corrosive environments. It is doubtful that a material will surface that will eliminate the need for corrosion protection within the cost constraints of this type of construction. However, it is feasible that an optimal material can be identified and, in conjunction with protective measures, render more reliable service than if the aforementioned factors are not considered. With regard to both new and existing structures, two possible maintenance procedures could be implemented. Tendons could be replaced periodically at a lower bound life expectancy, a procedure analogous to that being performed in the aircraft turbine engine industry. In practice this would prove expensive and highly impractical. Alternately, inspection techniques could be developed, and methodologies similar to the Air Force's "retirement for cause" (Annis, Hyzak) could be invoked.

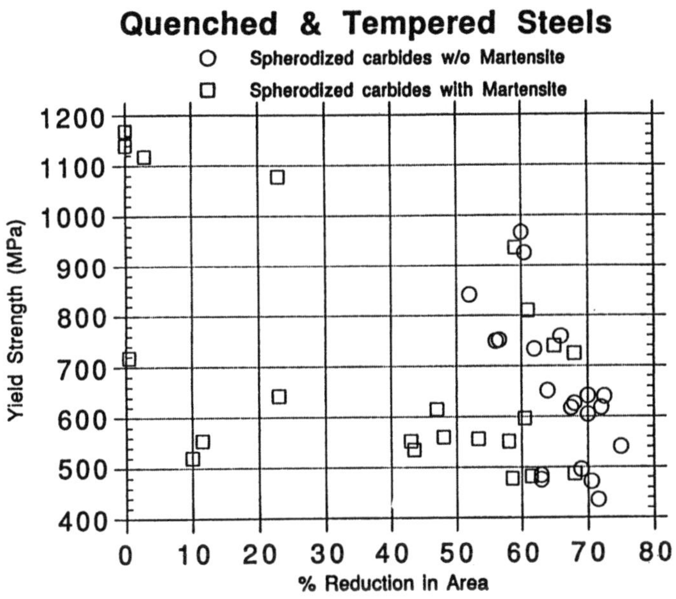

Figure 5
Comparison of Macroscopic Materials Properties

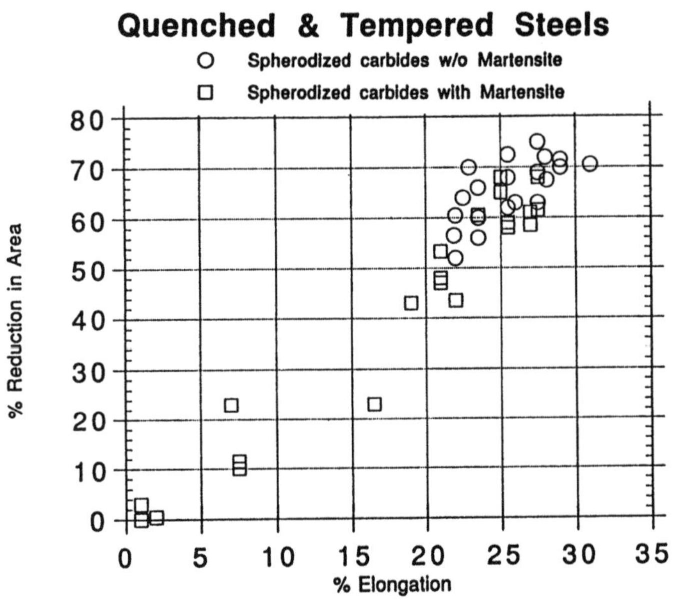

Figure 6
Comparison of Macroscopic Materials Properties

4 Conclusions

Environmentally assisted cracking (EAC) has been a documented cause of failures for post-tensioning structures. Laboratory investigations of post-tensioning materials reveal widely varying sensitivities to EAC. Corrosive environment, pH, corrosion potential, material microstructure and material processing have been identified as important testing variables. One existing data set allowed the comparison of baseline macroscopic material properties (ultimate strength, yield strength, and %RA) and performance in a hydrogen sulfide corrosive environment. An optimal microstructure (spherodized) and yield strength level (~ 600 MPa) were proposed based on the review of this data. The optimal strength levels in hydrogen sulfide corrosive media are lower than current design requirements. Therefore, corrosion protection methods are vital to ensure safe structures. While one may argue the applicability of a hydrogen sulfide environment to evaluate post-tensioning materials, the concept of optimal material strength levels and microstructure are anticipated for other corrosive environments. In the authors' opinion, the hydrogen sulfide testing technique offers a short duration test to identify materials that warrant further testing in a more representative environment.

Present material specifications, while adequate in some regards, need to be refined to more adequately specify microstructure, chemical composition and thermal treatment procedures. On a macroscopic material properties level, prescribing both lower bound strength levels (established in ASTM-A416 and ASTM A772) as well as upper bounds on yield and ultimate strength, would tend to crudely bound microstructure. Percent elongation is a material property the authors view with extreme skepticism. %RA more accurately reflects the localized deformation associated with static load or slow strain rate tests. Neither %RA nor percent elongation have been shown to be indicative of a material's fracture toughness in a corrosive environment. Experimental data are insufficient at the present time to advance exact modifications to present design methods, and experimental work to provide a sound basis for these decisions ought to be initiated.

REFERENCES

Annis, C. G., Jr., VanWanderham, M. C., Harris, J. A., Jr., and Sims, D. L., "Gas Turbine Engine Disk Retirement-For-Cause: An Application of Fracture Mechanics and NDE," **ASME 80-GT-127.**

Anon., "Old Dam Tied Down to Bedrock to Stand up to Greater Flows," **Engineering News Record**, Vol. 199, No. 9, 1973, pp. 18-19.

Anon., "Sulfide Stress Cracking Resistant Metallic Material for Oil Field Equipment," **NACE Standard MR-0175**, National Association of Corrosion Engineers, 1979, 17 pp.

Anon., "Grouting of Tall Vertical Prestressed Ducts," **Concrete**, London, Vol. 12, No. 5, 1978, pp. 63-64.

Ault, R. T., Holtmann, R. B., and Myers, J. R., "Heat Treatment of a Cr-Mo-Co Martensitic Stainless Steel for Optimum Combinations of Strength, Toughness, and Stress-Corrosion Resistance," **Transaction of the ASM**, Vol. 61, 1968, pp. 75-84.

Beeby, A. W., "Corrosion of Reinforcing Steel in Concrete and its Relation to Cracking," **Structural Engineer**, Vol 56A, No. 3, 1978, pp. 77-81.

Brachet, M., "Stress Corrosion Resistance Test for Prestressing Tendons," **Report on Prestressing Tendons: 5**, Prestressing International Federations, and Cement and Concrete Association, London, 1980.

Brown, B. F., "Stress Corrosion Cracking of High Strength Steels," **The Theory of Stress Corrosion Cracking in Alloys**, ed., J. C. Scully, NATO Scientific Affairs Div., Brussels, 1971, pp. 186-204.

Cauchois, L., Didier, J., and Herzog, E., "A Special N-80 Steel Tubing Developed in France to Resist Sulfide Stress Corrosion in Sour Gas Wells," **Corrosion**, Vol. 13, No. 3, 1957, pp. 263t-269t.

Cherry, B. W., and Price, S. M., "Pitting, Crevice, and Stress Corrosion Cracking Studies of Cold Drawn Eutectoid Steels," **Corrosion Science**, Vol. 20, No. 11/12, 1980, pp. 1163-1184.

Cherry, B. W., and Price, S. M., "The Corrosion of Steel Prestressing Tendons," **Australian Corrosion Association Conference**, Melbourne, Australia, 1978, 17 pp.

DeSanta Maria, M. S., Verdeja, J. I., and Perosanz, J. A., "Failure Analysis of a Low-Carbon Steel Structure: A Stress-Corrosion Problem," **Microstructural Science**, Vol. 9, 13th Annual Technical Meeting, International Metallographic Society, Brighton, England, 1980, pp. 398-405.

Elices, M., "Problems de Corrosion n bajo tension n en el hormig n pretensado," **I Congreso Iberoamericano de Corrosion y Protecci n**, 1983, pp. 21-31.

Elices, H., Maeder, G., and Sanchez-Galvez, V., "Effect of Residual Stress on Hydrogen Embrittlement of Prestressing Steels," **British Corrosion Journal**, Vol. 18, No. 2, 1983, pp. 80-81.

Ellyin, F., and Matta, R. A., "An Approach to the Evaluation of Stress-Corrosion of High Strength Steels," **Canadian Journal of Civil Engineering**, Vol. 8, No. 4, 1981, pp. 416-424.

Ellyin, F., and Matta, R. A., "Bonding and Corrosion Protection Properties of Two Coatings for Prestressing Steels," **Journal of the American Concrete Institute**, Vol. 79, No. 5, 1982, pp. 366-371.

Felip'ev, A. A., Alekseev, S. N., and Krasouskaya, G. M., "The Effect of Manufacturing Parameters on the Tendency to Corrosion Failure of Prestressed Wire Used in Reinforced Concrete Construction," **Zashchita Metallov**, Vol. 8, No. 4, 1972, pp. 458-461.

Gilmour, R. S., and Walker, A. L., "A Fracture Mechanics Approach to the Stress-Corrosion Susceptibility of Prestressing Tendons," **Proc., Effects of Chemical Environment on Fracture Processes**, University of Melbourne, Australia, 1974, pp. 261-273.

Griess, J. C., and Naus, D. J., "Corrosion of Steel Tendons Used in Prestressed Concrete Pressure Vessels," **Corrosion of Reinforcing Steel in Concrete**, ASTM STP 713, D. E. Tonini and J. M. Gaidis, Eds., American Society for Testing and Materials, 1980, pp. 32-50.

Hahin, C., Gambell, J., and Scott, J. K., "An Investigation of theSusceptibility of Post Tensioning Cables to Stress-Corrosion Cracking," CERL Technical Report M-199, U.S. Army Corps. of Engineers, Jan. 1977, 11 pp.

Hausmann, D. A., "Electrochemical Behavior of Steel in Concrete," **Journal of the American Concrete Institute**, Vol. 61, No. 2, 1964, pp. 171-188.

Hyzak, J. M., Allosin, J. E., and Reimann, W. H., "Development of Quantitative NDI for Retirement-For-Cause," AFML-TR-78-198, Feb. 1979.

Isecke, B., "Collapse of the Berlin Congress Hall Prestressed Concrete Roof," **Materials Performance**, Dec. 1982, pp. 36-39. [2]

Jurell, G., "Investigation into the Failure of a Prestressed Rock Anchor," **Water Power and Dam Construction**, Feb. 1985, pp. 45-47.

Klodt, D. T., "Studies of Electrochemical Corrosion and Brittle Fracture Susceptibility of Prestressing Steel in Relation to Prestressed Concrete Bridges," **Proc. 25th Conference, National Association of Corrosion Engineers**, Houston, TX, March 10-14, 1964, pp. 78-87.

Lewis, D. A., and Copenhagen, W. J., "Corrosion of Reinforcing Steel in Concrete in Marine Atmospheres," **Corrosion**, Vol. 15, No. 7, 1959, pp. 382t-388t.

McGuinn, K. F., and Elices, M. "Sorting Tests for Stress Corrosion Resistance of Cold Drawn Prestressing Tendon," **British Corrosion Journal**, Vol. 16, No. 3, 1981, pp. 132-139.

McGuinn, K. F., and Griffiths, J. R., "Stress Corrosion Cracking of Cold-Drawn Eutectoid Steel Wire," **Tewksbury Symposium on Fracture, 3rd**, University of Melbourne, Australia, 1974, pp. 274-285.

Parkins, R. N., Elices, M., Sanchez-Galvez, V., and Caballero, L., "Environment Sensitive Cracking of Prestressing Steels," **Corrosion Science**, Vol. 22, No. 5, 1982, pp. 379-405.

Sanchez-Galvez, V., Caballero, L., and Elices, M., "The Effect of Strain Rate on the Stress Corrosion Cracking of Steels for Prestressing Concrete," Lab. Corrosion Tests and Standards, **ASTM STP 886**, G. Haynes and R. Baboian, Eds., Amer. Soc. for Testing and Materials, 1985, pp. 428-436.

Schupack, U., "A Survey of the Durability Performance of Post-Tensioning Tendons," **Journal of the American Concrete Institute**, Vol. 75, No. 11978, pp. 501-510.

Scott, G. N., "Corrosion Protection Properties of Portland Cement Concrete," **Journal of American Water Works Association**, Vol. 57, No. 8, 1965, pp. 1038-1051.

Segan, E. G., Blackmon, D. R., Morrow, D., Socie, D., and Kurath, P., "Investigation of Susceptibility of Post Tensioning Systems to Environmentally Assisted Cracking," Corrosion Cracking, **Proc., Int. Conf. Exhibition on Fatigue, Corrosion Cracking, Fracture Mechanics and Failure Analysis, Salt Lake City**, UT, 1985, pp. 73-80.

Shape, E., "Roles of Composition and Microstructure in Sulfide Cracking of Steel," **Corrosion**, Vol. 24, No. 9, 1968, pp. 261-288.

Slater, J. E., "Corrosion of Metals in Association with Concrete," ASTM STP 818, Amer. Soc. for Testing and Materials, 1983.

Szilard R., "Corrosion and Corrosion Protection of Tendons in Prestressed Concrete Bridges," **Journal of the American Concrete Institute**, Vol. 66, No. 1, 1969, pp. 42-59.

Treadway, K. W. J., "Corrosion of Prestressed Steel Wires in Concrete," British **Journal of Corrosion**, Vol. 6, No. 2, 1971, pp. 66-72.

U.S. Army Corps of Engineers, Old River Control Auxiliary Structure Specifications, Contract No. DACW29-82-B-0072, 1982.

U.S. Army Corps of Engineers, John Day Lock and Dam Specifications, Contract No. DACW57-81-C-0057, 1981.

Wu, S. T., and Clifton, J. R., "Analysis and Modeling of Corrosion of Steel in Prestressed Concrete," Report No. NBSIR81-2390, National Bureau of Standards, Washington, DC, 1981, 24 pp.

10 EMPIRICAL ANALYSIS OF THE FRACTURE PROCESS IN HIGH STRENGTH CONCRETE LOADED IN UNIAXIAL COMPRESSION

L. TAERWE
Magnel Laboratory for Reinforced Concrete,
University of Ghent, Belgium

Abstract
The fracture process of high strength concrete (HSC), submitted to uniaxial compression, is analysed by means of loading tests on cylinders under special closed loop control. Conclusions are drawn from axial and circumferential strain curves, cross sections of specimens impregnated with a fluorescent dye and visual observations. The evolution of the internal crack extension is revealed and it is shown that under stable progressive fracture, predominantly aggregate bond failure and crack branching occur with the cracks passing around the coarse aggregates. The onset of damage is explained in terms of elementary force transfer concepts and a volumetric damage parameter is introduced.
Keywords : High strength concrete, Fracture, Strain-softening, Crack propagation, Closed loop testing, Stress strain diagram.

1 Introduction

As concrete strength increases, the descending branch of the stress-strain curve gradually becomes steeper and for HSC an almost vertical softening curve is obtained. When no particular precautions are taken during the compression test, a very explosive failure occurs. Typically, the fracture surface crosses the coarse aggregates and a fairly smooth fracture surface results. For these reasons, HSC is generally considered as a brittle material. This brittleness causes some concern with respect to applications in structural elements. However, HSC beams and slabs show a higher flexural ductility than corresponding normal strength concrete (NSC) specimens. This apparent contradiction between material brittleness and member ductility was discussed in detail by Taerwe (1991).

Hence, it is of great practical importance to gain fundamental insights into the fracture process of HSC and the factors that influence it. This can result in appropriate measures to improve material toughness or in rational ways to account for it in design procedures.

Fracture and Damage of Concrete and Rock – FDCR-2.
Edited by H.P. Rossmanith. ISBN 0 419 18470 8.

2 Test specimens

The loading tests on cylinders discussed in this paper are part of a comprehensive research program that includes normal, medium and high strength concrete, the latter with and without fibre addition. A general survey of the characteristics of the plain high strength concrete cylinders (designated as HS followed by a number) is given in table 1. More details of the complete program are given by Taerwe (1992). Concrete composition is as follows : river gravel 4/14 (1300 kg), river sand 0/5 (550 kg), Portland cement (400 kg), water (120 kg), high-range water-reducer (10 ℓ). The water-cement ratio equals 0.30. The concrete shows a mean slump of 200 mm and a mean compressive strength of 93 MPa at an age of 4 months. The specimens were cured at 20°C in a humid environment with relative humidity of at least 90 %. Eight specimens have a diameter of 155 mm, while the other three are 192 mm in diameter. This corresponds to a height/diameter ratio of 2.45 and 2.52 respectively.

TABLE 1. Survey of test specimens

Designation	Diameter, mm	Height, mm	Compressive strength, MPa	ε_1^*, per thousand
HS1	155	380	87.5	2.67
HS2	155	380	89.7	2.54
HS3	155	380	93.0	2.66
HS4	155	380	93.0	2.78
HS5	192	483	96.6	2.76
HS6	192	483	96.6	2.71
HS7	192	483	94.8	2.77
HS8	155	380	90.9	2.67
HS9	155	380	95.2	2.82
HS10	155	380	94.1	2.70
HS11	155	380	94.1	2.74

*ε_1 is the axial strain at peak stress

3 Experimental procedure

At an age of 4 months, the specimens were submitted to axial compression in a stiff closed loop testing machine with a capacity of 6000 kN (axial stiffness about 2500 kN/mm). No special precautions were taken to prevent friction between the end faces of the specimens and the loading platens. It is deemed that due to the height/diameter ratio of 2.5 the extent of the disturbed end zones is sufficiently limited as shown in fig. 1. However, this figure does not represent the volumetric ratio which is smaller than the ratio of the indicated areas. No steel columns or bars were loaded in parallel with the specimens, as described by Glavind, Stang (1991) and others.

When an inappropriate control signal is used to command the testing machine, a sudden, explosive failure occurs when a high-strength specimen reaches the peak stress. The occurrence of explosive failures is to some extent related to the stiffness of the testing machine relative to the stiffness of the specimen. For HSC this problem is of minor importance as it can generally be overcome by a closed loop system with high speed response. The most important factor that influences the nature of specimen failure is the feed back signal that is used by the closed loop system.

Directly controlling the cross-head movement or the rate of axial deformation does not avoid explosive failure because these quantities show snap-back behaviour as will be discussed further on. Hence, there is a need for a signal that shows a continuous increase in course of time. Shah, Gokoz and Ansari (1981) put forward the idea of using the circumferential strain as control signal. Due to longitudinal splitting, the rate of increase of circumferential strain in the descending part is greater than that of axial strain and thus provides a more sensitive control signal.

To measure the average expansion of the specimen, Shah, Gokoz and Ansari (1981) developed a special measuring device, consisting of a piano wire wrapped around the cylindrical specimen. The wire was fixed at one end and was attached

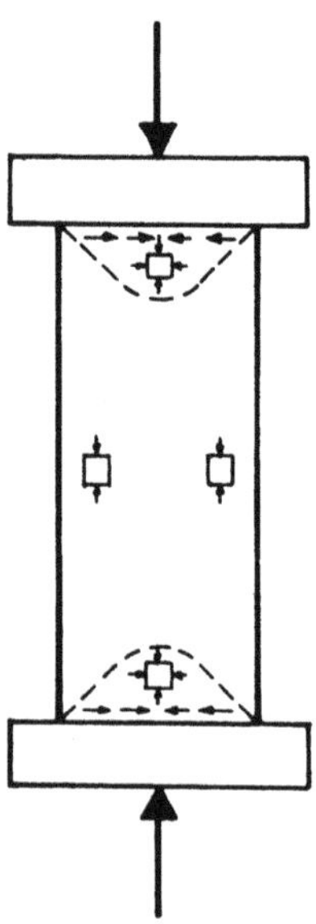

Fig. 1 Disturbed zones in specimens with h/d = 2.5

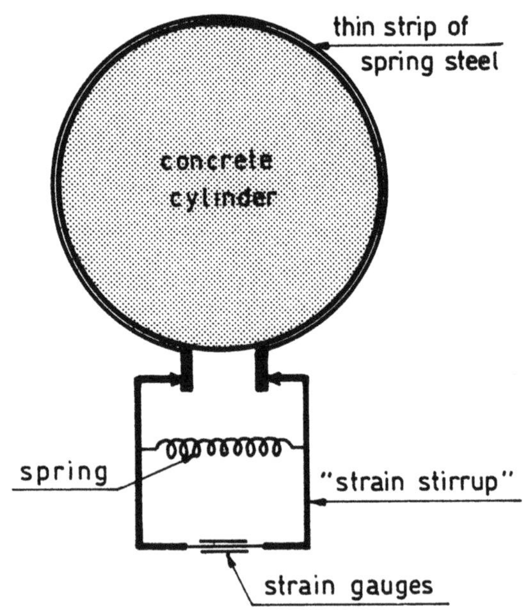

Fig. 2 Strain stirrup for measurement of circumferential strain

through a spring to a linear variable differential transformer (LVDT). During loading, the expansion of the specimen caused the free end of the wire to move, producing a proportionate change in the voltage of the LVDT signal. In this way, a feedback signal that steadily increased in course of the loading process was obtained as input for the regulation of the hydraulic system.

In this research program, the same testing principle was adopted. However, instead of a piano wire, a thin strip of spring steel was wrapped around the specimen, its length being somewhat shorter than the cylinder's circumference (fig. 2). Two small steel blocks, fixed at both ends of the strip, were fitted in between the extremities of a "strain-stirrup". This device, in fact a small portal frame, consisted of two legs and a connecting beam element. The section of this small beam was reduced over the central zone to increase its flexibility. In this zone, four electrical strain gauges were applied in full bridge connection. The measured strain signal is proportional to the change in distance between the legs' extremities, which in turn represents the integral of the circumferential strain. The strain-stirrup was kept in place by means of a spring connecting both legs (fig. 2). The device yielded a reliable control signal as it is almost insensitive to friction effects. Moreover, it proved to be extremely robust. The control technique was further improved by taking a combination of axial and circumferential strain as feed back signal. This aspect is discussed in more detail by Glavind and Stang (1991).

4 Stress-strain curves

The stress-strain curves of four HSC specimens are shown in fig. 3. The axial strain is defined as $\Delta h/h$ with h the specimen's height. Axial compressive stress is normalized with respect to peak stress f_c. The specimens were unloaded deliberately at different stress levels along the descending branch to allow investigation of the progression of the fracture process in more detail by means of sections obtained after resin impregnation.

The stress-strain curve of specimen HS11 (fig. 3) exhibits some characteristics typical for the HSC specimens. After the peak stress, the curve shows a fairly steep part followed by a general snap-back. At a stress equal to about 70 MPa, a spontaneous unloading in the σ-ε path occurs, which can only be obtained with the type of control signal used. Axial strain control would not allow this phenomenon, and explosive failure would result. At a stress of about 55 MPa, a kind of intermediate plateau occurs. Subsequently, before an axial strain of about 0.003 is reached, a steep part with slightly positive slope and local snap-backs results again. Below 20 MPa, a more gradual decrease can be seen. At a strain of about 0.0035, the specimen was unloaded deliberately.

In fig. 4 the circumferential strain $\varepsilon_t = \Delta d/d$, with d the specimen's diameter, is shown in function of the axial stress. This signal shows a monotonous increase in time and no snap-back behaviour is observed. Hence it yields a suitable control signal. The typical features of the curve will be discussed further on.

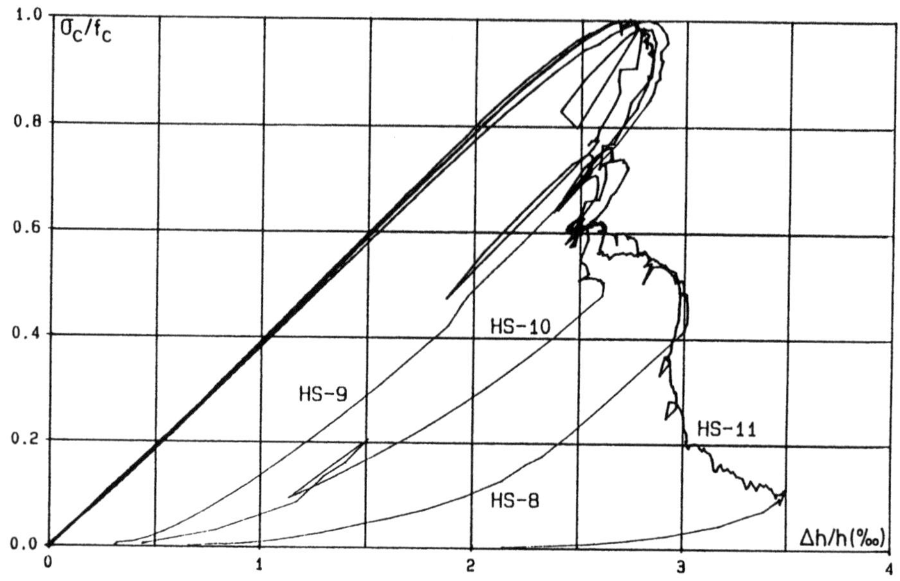

Fig. 3 Stress-axial strain curves of four specimens

5 Internal crack pattern

After unloading, a selected number of specimens was impregnated with a low viscosity epoxy resin to which a fluorescent dye was added. After hardening of the resin, the specimens were sawn in two halves along a meridian plane. The crack distribution in this plane can clearly be observed under ultraviolet light. The impregnation technique reveals the structure of the continuous internal crack pattern that is connected to outer surface cracks. Internal localized microcracking cannot be detected by this technique.

Specimen HS9 was unloaded in the descending branch after the first lobe with snap-back behaviour (fig. 3). On the outer surface, only slight scaling over about half of the perimeter and at variable height was visible (fig. 5). The impregnated section revealed a small crack in the neighbourhood of a coarse aggregate, extending to a depth of about 15 mm from the surface. The picture with the internal crack pattern cannot be reproduced with sufficient quality and contrast to reveal the details of interest.

Specimen HS10 was unloaded at the small plateau mentioned before, that occurs between 0.5 and 0.6 f_c. At each side of the specimen, an inclined branching crack can be observed (fig. 6). This, and the following photographs are reproduced in inversed mode with the cracks showing up as black lines. From the more detailed picture in fig. 7, it can be seen that successive branching of cracks occurs, mainly at

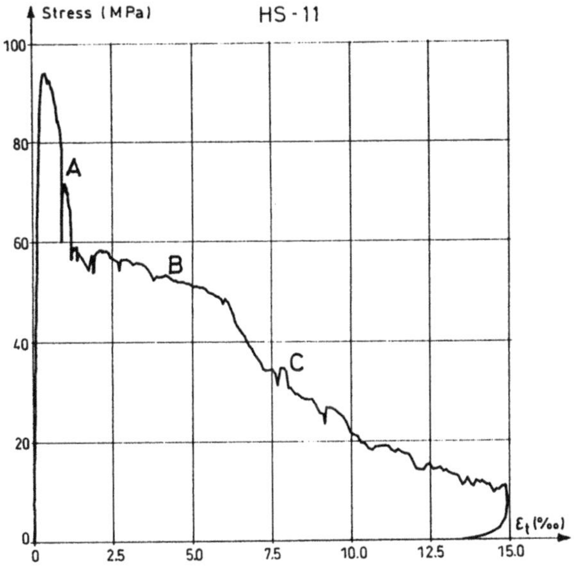

Fig. 4 Circumferential strain for specimen HS11

Fig. 5 Specimen HS9 after unloading

coarse aggregates, resulting in a rather extensive finely distributed crack pattern. Further, these saw cuts indicate that fracture lines predominantly pass along coarse aggregates as in normal strength concrete (except for flat shaped and weaker particles), and rarely run through the aggregates, as is observed in the case of explosive failure. Hence, the smoothness of the fracture surface is mainly related to the speed of crack propagation. Van Mier (1991) arrived at similar findings for the case of uniaxial tension. He also concluded that stable failure of HSC is mainly caused by debonding of aggregates and matrix.

Specimen HS11 (fig. 8) additionally shows a transverse crack, connecting both damage zones that occur at diagonally opposite locations. This crack points to a kind of shear band which can be observed at the cylinder's surface.

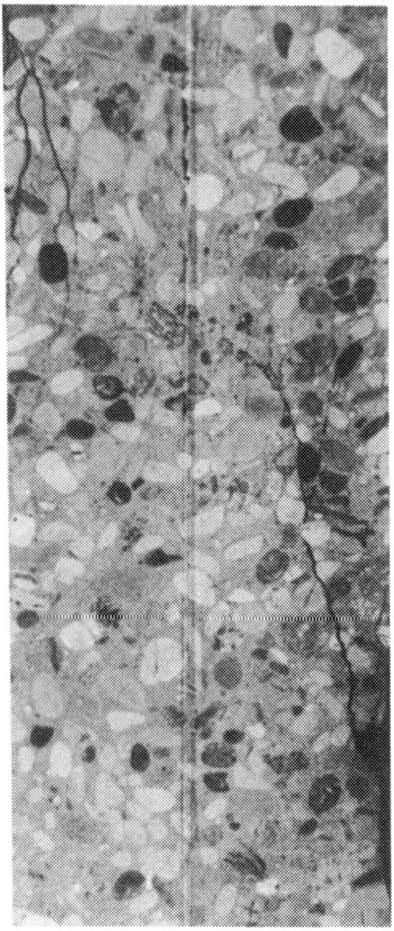

Fig. 6 Cross-section of specimen HS10

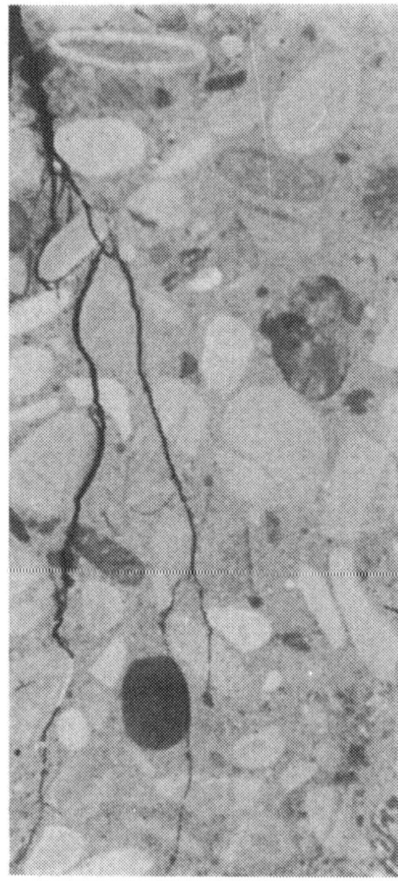

Fig. 7 Detail of the upper left corner of fig. 6

6 Fracture process and damage distribution

In this section, a basic description of the fracture process and the damage distribution is presented. First, attention is drawn to fig. 4, where three different stages can be distinguished in the descending branch. Stage A, with fairly steep descent, corresponds to the snap-back lobe in the axial strain curve. Subsequently, a fairly extensive plateau occurs (stage B), corresponding to the smaller one in the axial strain curves. Finally stage C shows a gradual decrease of the σ-ε_t curve.

From the previous observations (including section 5) the following fracture sequence is derived

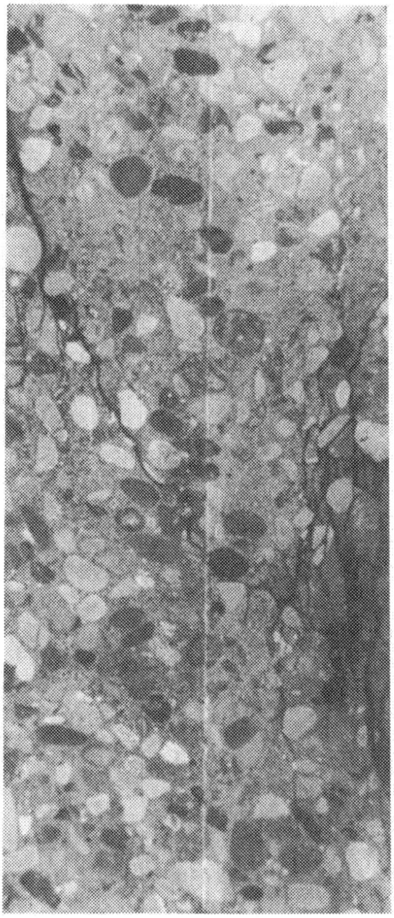

Fig. 8 Specimen HS11

Fig. 9 Specimen HSR5

- The fracture process is initiated by the occurrence of very localized damage at the specimen surface, gradually extending around the major part of the perimeter (stage A).
- Extension of the cracks, mainly in height but also in depth, which yields an important expansion of the specimen and causes a small increase in axial shortening due the more or less intact core (stage B).
- Formation of a shear band, crossing the core of the specimen, connecting opposite damage zones and yielding a finely distributed crack pattern with the coarse aggregates acting as crack flank bridges (stage C).

The fact that in fig. 4, the slope of the curve in stage C is steeper than in stage B, is partly attributed to the fact that the shear band mainly affects the core of the specimen, which has a smaller cross section. It is suggested that the inflexion point in the softening branch of the stress-strain curve of NSC can be associated with the transition of two fracture processes i.e. on the one hand the gradual extension of surface cracks over the height and to the core of the specimen and on the other hand shear of the more or less intact core.

7 Onset of damage process

7.1 Extent of damage zone

As mentioned already for specimen HS9, the onset of damage occurs in a very small zone at the outer surface. There is reasonable experimental evidence that this occurs around a coarse aggregate. This is confirmed by specimen HSR5 that is shown in fig. 9 and that belongs to another test series than the one mentioned in table 1. This specimen shows pop-outs at coarse aggregates in the vicinity of the side face.

The fact that the initial damage zone is small, not only holds with respect to the height but also with respect to the cross-section. Hence, it is necessary to consider the volume of the damaged zone and not only the extension in one or two dimensions. Due to the small volume of the initial damage zone, snap-back behaviour at the onset of the softening branch can be explained by the elastic extension of the large non-damaged zone that dominates the axial shortening in the damage zone. It is deemed that a one-dimensional series model with softening band, as recently considered for compression by way of extension of the more obvious representation in tension, has to incorporate the extension of damaged zones both in height and cross section. Hence, it is proposed to consider a volumetric damage parameter defined as

$$\omega = V_d/V = (V - V_n)/V \qquad (1)$$

where V is the nominal volume, V_d the volume of the damage zone, V_n the volume of the non-damaged zone. Formula (1) is an extension of the similar concept used for damaged areas in a cross-section (see e.g. Løland, 1980).

7.2 Damage initiation

As pointed out in sub-section 7.1, it appears reasonable to state that damage is initiated by the heterogeneity caused by one (or more) of the coarse aggregates. It is often claimed that HSC is a more homogeneous material than NSC and hence is prone to a more brittle behaviour. This is argued by the fact that the mechanical properties of the aggregates and the mortar matrix are less different than is the case for NSC. However, for the type of coarse aggregate used in the research program considered so far, Van Nieuwenburg (1973) found a mean compressive strength of 260 MPa and a modulus of elasticity of 69 GPa. The difference with respect to the mortar properties still remains reasonably large. Hence we may conclude that the coarse aggregates still act as rigid inclusions in HSC, at least for the strength level considered.

Under this hypothesis it can be shown that damage and cracking initiate at the edge of a specimen by the following simplified model. Consider a regular array of circular inclusions, and a load transfer between the coarse aggregates as indicated in fig. 10. An aggregate at the edge needs a horizontal tensile force for equilibrium whereas for an inward aggregate both tensile forces cancel each other. When, due to bond or matrix failure, a vertical crack occurs, the horizontal equilibrium of the edge aggregate can only be assured by a compression exerted by the outer concrete layer. When this fails, pop-out occurs and a local damage zone initiates.

However, this two-dimensional model is not sufficient for explaining the occurrence of vertical radial cracks. For this purpose, a spatial load transfer mechanism is considered (fig. 11). Consider a cylindrical specimen and three aggregates, two of them being located along the edge of the specimen and in the same horizontal plane. Due to the divergency of the compressive forces that are transmitted from A to B and C, a hoop tension results that causes vertical cracks at the cylinder's curved side face. It follows that consideration of a 3D-model instead of a 2D-model results in an additional contribution to the volumetric damage. This aspect should also be considered in numerical simulations. The previous conclusion is in accordance with Vonk (1992), who suggests, on the basis of a two-dimensional simulation : *"It is not unlikely that a part of the necessary increase in heterogeneity can be found in extending the model to the third dimension* (p. 190)".

Further to the more homogeneous nature of HSC, it was found that less micro-cracks were observed in HSC before peak stress than in similar NSC specimens. This results in a more linear ascending branch of the stress-strain curve. The fact that less bond cracks were found was attributed to a higher bond strength between the aggregates and the mortar matrix. However, when it is assumed that bond strength is directly proportional to tensile strength and as the latter approximately follows a $\sqrt{f_c}$-law, it follows that bond strength increases less than proportional with peak stress. Hence, it is suggested that the phenomena that cause micro-cracks do not have the same magnitude in HSC, so that bond strength is not exhausted too early. In the case of silica fume addition, the aggregate-matrix interface is undoubtedly of better quality.

8 Size effect

The size effect concept in compression is mainly supported by experimental observations from Van Mier (1984), and was explained later on by a hypothetical localized crack band. Further tests conducted by Vonk (1992) do not clearly confirm the size effect. The crack band hypothesis is very reasonable for tension, but for compression, generally no such localized band can be found in a vertical cross section. As pointed out previously, extension of the damage zones occurs both in height and cross-section.

The volume of specimens HS5, HS6 and HS7 is about twice as large as that of the other specimens. Specimens HS5 and HS6 exhibited explosive failure at about 0.65 f_c. Specimen HS7 failed in a stable way and no significant difference with the

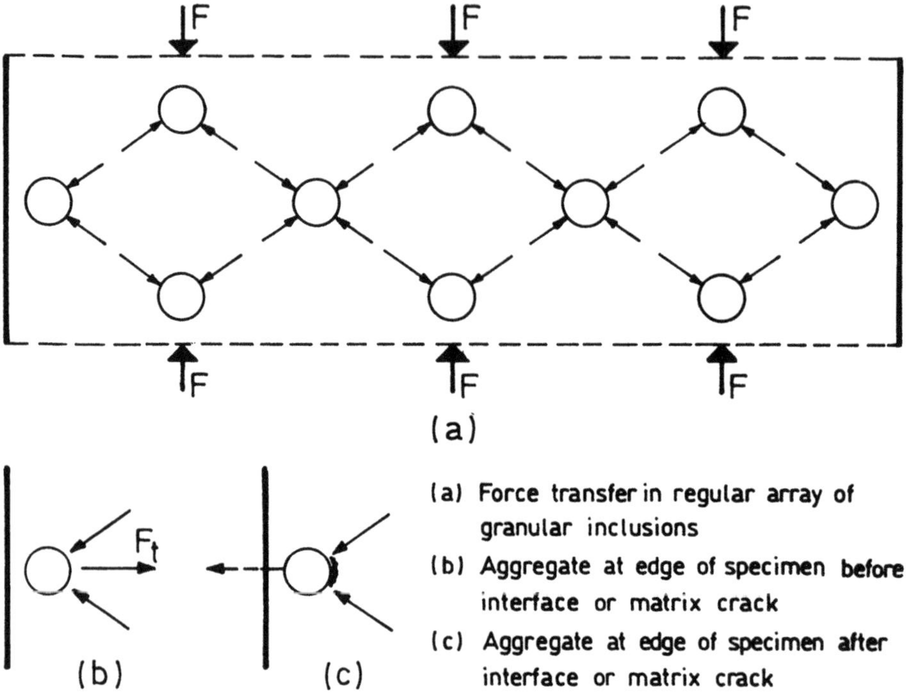
F
F
F
F
F
F
(a)
F_t
(b)
(c)
(a) Force transfer in regular array of granular inclusions
(b) Aggregate at edge of specimen before interface or matrix crack
(c) Aggregate at edge of specimen after interface or matrix crack

Fig. 10

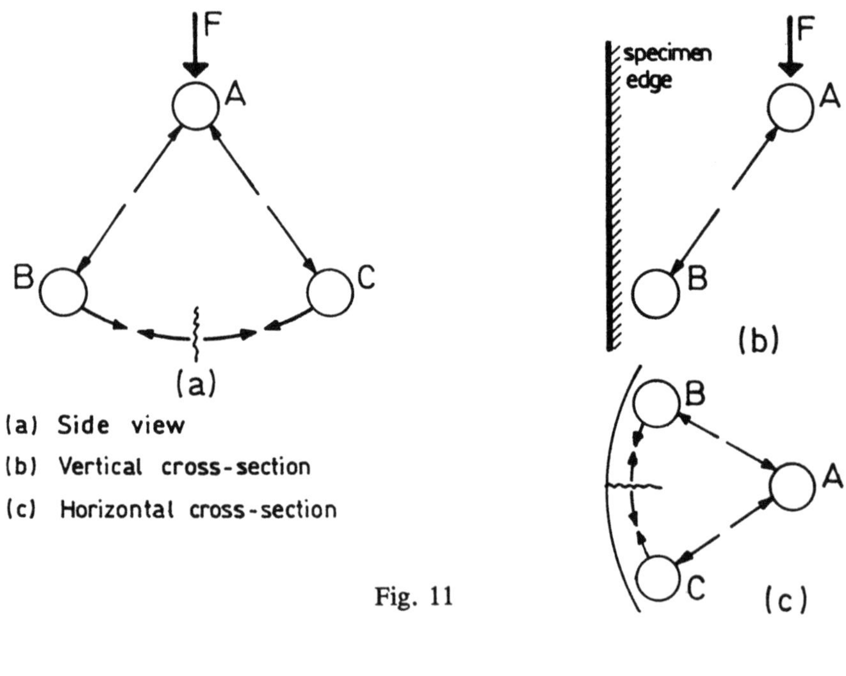
F
A
B
C
(a)
(a) Side view
(b) Vertical cross-section
(c) Horizontal cross-section
specimen edge
F
A
B
(b)
B
A
C
(c)

Fig. 11

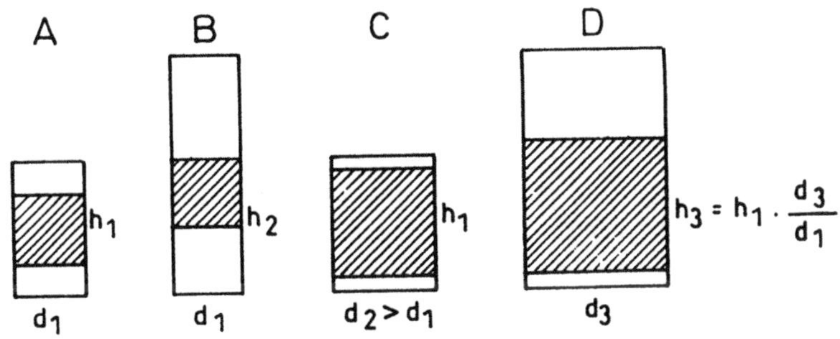

Fig. 12 Characterization of size effect by h/d-ratio

softening part of the other σ-ε curves could be observed. Taerwe (1992) tested HSC with steel fibre addition and found a remarkable increase in toughness. In that case, with smoothly decreasing softening branch, the size effect was apparent. Also a more finely distributed crack pattern could be observed, especially in the shear band.

From observations of crack patterns in cross-sections of test specimens, the following hypothesis is proposed. The height of the cumulated damage zone is roughly equal to the transverse dimension i.e. the diameter in case of a cylinder. Hence, the final volumetric damage parameter ω is the same for specimens with the same h/d-ratio. If the hypothesis can be extended to the evolution of ω also, then it can be concluded that for the specimens shown in fig. 12, the magnitude of the volumetric damage as defined by (1) decreases in the following order : C, A (= D) and B. The approach also allows extension to real structural members (columns, compression zone of beams, ...).

It is not necessary that all damage zones are closely connected, although this is mostly the case. The previous formulation supports the size effect concept introduced by Van Mier, at least if the volumetric damage evolution is directly related to the axial strain decrease. As explained previously, several damage mechanisms simultaneously or consecutively contribute to the progression of damage. Apparently the core is subject to another failure mechanism but has a smaller volume and hence a smaller damage content.

9 Conclusions

- Stable testing of the softening branch of HSC is only possible by making use of the circumferential strain as part of the command signal for the closed loop testing . For the purpose of this research program a robust and reliable device was developed.
- The stress-axial strain curve of HSC shows snap-back behaviour.
- The circumferential strain shows a continuous increase in function of time.
- In the case of stable progressive fracture, the internal cracks mainly pass along the coarse aggregates and debonding and branching are the main phenomena that determine crack extension. The fracture aspect is very similar to NSC.

- The first stage of the fracture process is characterized by an extension of cracks both in height and in cross-section of the specimen. In the final stage a shear band type of failure occurs.
- The onset of damage occurs at coarse aggregates, located in the vicinity of the specimen's outer surface, that act as rigid inclusions. Both a two dimensional as a three-dimensional force transfer model allow to explain certain damage aspects that are observed in experiments.
- Damage should be characterized by a volumetric damage parameter that tends to a final value which is closely related to the ratio d/h, as it is proposed to take the height of the damage zone equal to the transverse dimension of the specimen. The size effect is formulated in function of h/d.

10 Acknowledgment

This research was partly financed by the Belgian National Fund for Basic Research (grant 1992/1993). The author gratefully acknowledges the support of F. Burick for the practical realization of the testing equipment.

11 References

Glavind, M. and Stang, H. (1991) Evaluation of the complete compressive stress-strain curve for high strength concrete. **"Fracture Processes in Concrete, Rock and Ceramics"**, Edited by Van Mier J., Rots J. and Bakker A., E & F.N. Spon, London, 749-759.

Løland, K. (1980) Continuous damage model for load-response estimation of concrete. **Cement and Concrete Research**, 10, 395-402.

Shah, S.P. ; Gokoz, U. and Ansari, F. (1981) Experimental technique for obtaining complete stress-strain curves for high-strength concrete. **Cement, Concrete and Aggregates**, Vol. 3, 1, Summer 1981, 21-27.

Taerwe, L. (1991) Brittleness versus ductility of high strength concrete. **Structural Engineering International (IABSE)**, 4/91, 40-45.

Taerwe, L. (1992) Influence of steel fibers on strain-softening of high-strength concrete. **ACI Materials Journal**, January/February, 54-60.

Van Mier, J. (1984) Strain-softening of concrete under multiaxial loading conditions. **Doctoral Thesis**, Eindhoven University of Technology, The Netherlands.

Van Mier, J. (1991) Crack face bridging in normal, high strength and Lytag concrete. **"Fracture Processes in Concrete, Rock and Ceramics"**, Edited by Van Mier J., Rots J. and Bakker A., E & F.N. Spon, London.

Van Nieuwenburg, D. (1973) Study of the deformation behaviour of structural lightweight concrete (in Dutch). **Doctoral thesis**, University of Ghent, Magnel Laboratory for Reinforced Concrete, Ghent, Belgium.

Vonk, R. (1992) Softening of concrete loaded in compression. **Doctoral Thesis**, Eindhoven University of Technology, The Netherlands.

11 INDIRECT DETERMINATION OF THE σ-w RELATION OF HSC THROUGH THREE-POINT BENDING

J.P. ULFKJÆR and R. BRINCKER
University of Aalborg, Aalborg, Denmark

Abstract
The results from an experimental investigation are presented. The purpose of the investigation has been to determine the constitutive parameters in the fictitious crack model and to examine size effects in high strength concrete. Eight different beam geometries and 40 beams in all have been tested. The beams were submitted to three-point bending in a servo-controlled materials testing system. The constitutive parameters in the fictitious crack model were determined by using three different indirect techniques, which are described and evaluated. The results show that the fracture energy is increasing with the beam depth, the uniaxial tensile strength is both increasing and decreasing with beam depth and the kink coordinates of the softening relation are decreasing with beam depth and are smaller than those used for NSC.
Keywords: Experiments, Fictitious crack model, Modulus of rupture, Fracture energy, Fracture toughness, Optimization, Size Effects.

1 Introduction

It is very difficult to determine the constitutive parameters in the fictitious crack model and for high-strength concrete, so far only the fracture energy has been determined, and only by using the RILEM recommendation developed by Hillerborg and co-workers, RILEM 50-FMC (1985). Instead indirect methods can be applied, e.g. the fracture parameters are determined through three-point bending experiment, Roelfstra, P.E. and Wittmann F.H. (1986). Here the experimental results, in form of load displacement curves, are compared with numerical results obtained by using the direct sub-structure method, Brincker and Dahl (1989), and the fracture parameters are then assumed to be those who minimize a certain function, which in some way describes the difference between the experiment and the numerical result. Three of such functions are tested on 8 different beam geometries, and two different optimization techniques have been used. This method of determining the constitutive parameters is evaluated, and conclusions are drawn.

Fracture and Damage of Concrete and Rock – FDCR-2.
Edited by H.P. Rossmanith. © 1993 Taylor & Francis. ISBN 0 419 18470 8.

2 Experiments

2.1 Materials

A high strength concrete similar to the one used at the Great Belt project in Denmark has been used. The largest aggregate size in this concrete is, however, only 8 mm.

Table 2. Mix proportions.

Cement	Fly ash	Silica fume	Water	Plast. (*)	Super Plast.	Sand	Gravel
312	44	29	122	1.56	8.58	614	1238

(*) Melamine-Naphthalene based plasticizer; Units are [kg/m^3]

Table 3. Mechanical properties of tested concrete.

Cylinder compressive strength		Cylinder splitting strength		Young's modulus	
Mean	S.Dev	Mean	S.Dev.	Mean	S.Dev.
81,7	5.0%	5.55	8.1%	42660	2.5%

Units are [MPa]

Table 2. shows the mix of the concrete and Table 1. shows the mechanical properties.

The experiments were carried out on 8 different beam geometries (all combination of width and depth: 50 mm 100 mm and 200; length = 800 mm except 200 x 50 x 800). These geometries were chosen for two reasons. Firstly, in order to be able to study statistical and structural size effects separately, and secondly because of the limitations of the testing equipment. A total amount of 40 beams were cast over a period of 4 weeks. The beams were named after size, cast day and batch number (e.g. A-3a is beam size A cast on day 3 from the first batch). The day after casting the beams were striped from their molds and cured 7 days in water (37°C). Then a notch of rectangular cross section was saw cut in the beams with a depth of 1/10 of the beam depth. The beams were then further cured under water (20°C) for 7 days. Hereafter, the beams were stored in the laboratory at relative humidity of 100% at 20°C until the day of testing.

2.2 Testing equipment and procedure

The beams were submitted to three-point bending in a servo-controlled materials testing system. In order to measure the true beam deflection a reference bar was placed on each side of the beam, and the beam deflection was measured as the distance from the load point to the reference bar using two LVDT's with a base of 20.0 mm and a sensitivity of 0.5 V/mm. The piston displacement was measured using the build-in LVDT with a base of 5.0 mm and a sensitivity of 2.0 V/mm. The crack opening displacement was measured using a clip-gauge with a base of 2.0 mm and with a sensitivity of 5.0 V/mm. The load was measured using a 50.0 kN load cell with a sensitivity of 0.2 V/kN. A schematic of the test set-up is shown in Fig. 1.

All signals together with the time, t, were recorded using an analog to digital converter and an AT Personal Computer. The test was controlled by a feedback signal that included contributions from both the piston displacement and the COD. The feed back signal, δ, was created by analog addition of the corresponding signals:

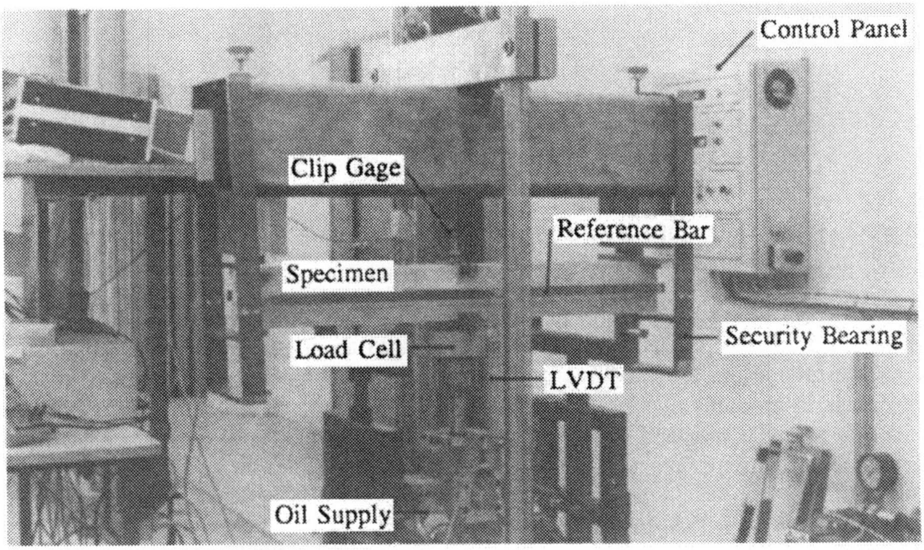

Fig. 1. Photo of test set-up.

$$\delta=\alpha\delta_{COD}+\beta\delta_p \tag{1}$$

where, δ_{COD}, is the crack opening displacement and, δ_p, is the piston displacement. The weight factors, α, and, β, were chosen to $\alpha=10.0$, 5.0 and 2.5 and $\beta=1.0$, where α is decreasing with the beam depth.

The reference signal, a linear ramp, was generated with the same AT PC using a digital to analog converter.

3 Fracture Parameter Results

3.1 The Modulus of Rupture

The modulus of rupture is calculated as the bending tensile strength of a Bernoulli-Euler beam. In Fig. 2. the mean value of the modulus of rupture is drawn as a function of the beam depth, for the different beam thickness. It is seen that except for the smallest beam

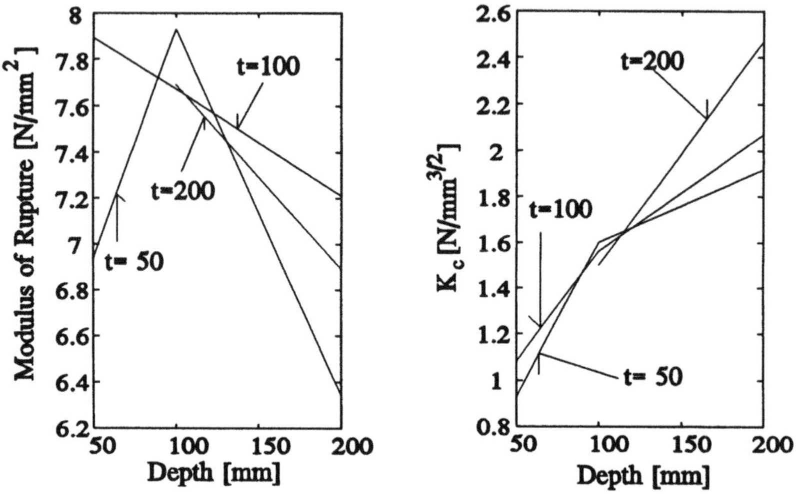

Fig. 2. Mean values of modulus of rupture and fracture toughness as function of the beam depth for the three beam thicknesses.

(d = 50 mm, t = 50 mm) the same trend is seen as in the Reagel and Willis (1931) experiments, thus, the modulus of rupture is decreasing with the beam depth. The decrease in modulus of rupture by doubling the beam depth is 4% for the thicknesses 50 mm and 100 mm and 10% for the beam thickness 200 mm which is of the same order as in the experiments by Reagel and Willis (1931).

3.2 The Fracture Toughness

The fracture toughness, K_c is calculated according to ASTM:

$$K_c=\frac{3}{2}\frac{F_{max}l\sqrt{a}}{bd^2}f\left[\frac{a}{d}\right] \tag{2}$$

where

$$f\left[\frac{a}{d}\right]=1.93-3.07\left[\frac{a}{d}\right]+14.53\left[\frac{a}{d}\right]^2-25.11\left[\frac{a}{d}\right]^3+25.8\left[\frac{a}{d}\right]^4 \tag{3}$$

F_{max} is the peak load and d,t and a is the beam depth, thickness and notch depth, respectively.

The mean values of the critical stress intensity factor are plotted in Fig. 2. It is seen that the fracture toughness is strongly increasing with the beam depth, and is almost independent of the beam depth. The increase by doubling the beam depths is 53% for $t=50$ mm, 46% for $t=100$ mm and 65% for $t=200$ mm. Thus, a distinct size effect is observed, when LEFM is used.

3.3 Fracture Energies According to RILEM

According to the RILEM recommendation, RILEM TC-50 (1985), the fracture energy should be determined on a specimen with the dimensions l = 800 mm, d = 100 mm, t = 100 and initial notch depth a_i = 50 mm. In this test series the initial notch depth ratio was chosen to be only 1/10 of the normalized beam depth, because LEFM is a better approximation for small notch depth. By using this reduced notch depth the amount of energy dissipated outside the fracture surface will increase. Therefore it can be expected that a certain size effect will be measured by using the RILEM method.

The mean value of the fracture energies for each geometry is shown in Fig. 3. It is seen that the fracture energy is increasing with the beam depth and is almost independent of the beam thickness. The increase by doubling the beam depth is for $t= 50$ mm 19%, for $t= 100$ mm 8% and for $t= 200$ mm 16%. The reason for the size dependency of the fracture energy is attributed to the fact that energy dissipation in the bulk is not taken into consideration.

3.4 Fracture Parameters according to the Fictitious Crack Model and Data fitting

The fracture energy, the tensile strength, the modulus of elasticity, the coordinates of the kink on a bi-linear σ-w relation were estimated by solving the minimizing of an objective function which in some manner describes the difference between the measured load displacement relation and one calculated using a numerical method. Three different objective functions and two different optimization procedures were tested in order to obtain the best estimate of the constitutive parameters. The functions will be presented in the order they were tested by the authors, and the short comings of each function will be given.

Objective function I:

The first objective function to be tested were

$$\min_{\bar{z}} \frac{\sum_{i=1}^{N}\left[(\hat{F}_i^I(\bar{z})-F_i^I)^2\ (F_i^I)^2\right]}{F_{max}^4} \tag{4}$$

$$\bar{z} = [\hat{\sigma}_u\ \hat{w}_c\ \hat{E}\ ]^T$$

where F_i^I is the experimental load deflection curve at the applied displacement δ_i, which were calculated at equidistant displacements and $\bar{z}$ is the state vector. No bounds were imposed on the elements in the state vector. $\hat{F}_i^I(\bar{z})$ is a load deflection curve determined using the direct sub-structure method, Brincker and Dahl (1988). The σ-w relation was assumed to be bi-linear with the normalized kink-coordinate at (0.16,0.31). $\hat{\sigma}_u$, $\hat{w}_c$ and $\hat{E}$ are the estimated uniaxial tensile strength, critical crack opening displacement and the modulus of elasticity respectively. The term $(F_i^I)^2$ is a weight that emphasizes the importance of the peak.

The optimization problem was solved using the Nelder and Mead algorithm on a 386 20 MHz Personal Computer and it took about 1 hour for each optimization. However,

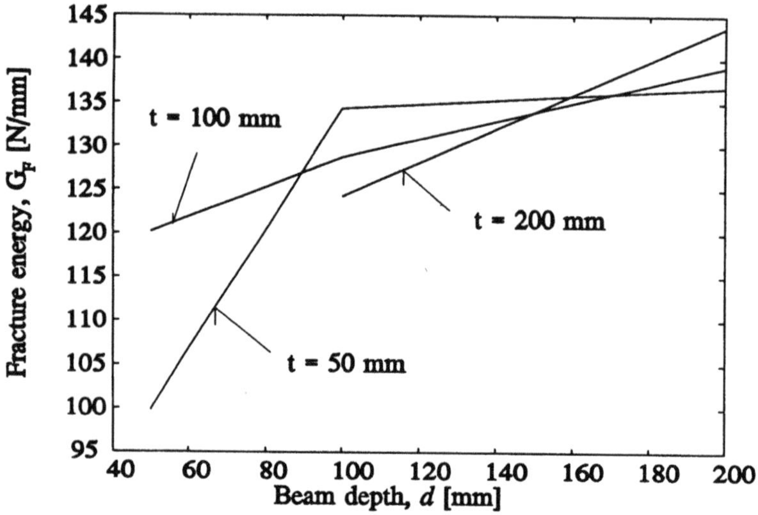

Fig. 3. Fracture energies determined according to the RILEM recommendation.

in order to avoid local minima several runs were necessary. Using this objective function initially rather good results were obtained and it was concluded that the method was rather good. However, as the results became more and more scattered, and as it was observed that the fitting results in some cases became rather poor, especially at the descending branch, another objective function was formulated.

<u>Objective function II:</u>

The second objective function is similar to the first, however, the normalized coordinates of the kink point $(\hat{w}_1/\hat{w}_c, \hat{\sigma}_1/\hat{\sigma}_u)$ of the softening relation are taken into account

$$\min_{\vec{z}} \frac{\sum_{i=1}^{N} \left[(\hat{F}_i^{II}(\vec{z}) - F_i^{II})^2 \; (F_i^{II})^2 \right]}{F_{\max}^4} \tag{5}$$

$$\vec{z} = [\; \hat{\sigma}_u \; \hat{G}_F \; \hat{E} \; \hat{w}_1/\hat{w}_c \; \hat{\sigma}_1/\hat{\sigma}_1 \;]^T$$

and the state vector is subjected to the following constraints

$$
\begin{array}{lcl}
0 < & \hat{\sigma}_u & < 10.0 \text{ MPa} \\
0 < & \hat{E} & < 80{,}000 \text{ MPa} \\
0 < & \hat{w}_c & < 0.5 \text{ mm} \\
0 < & \hat{w}_1/\hat{w}_c,\ \hat{\sigma}_1/\hat{\sigma}_u & < 0{,}5
\end{array} \tag{6}
$$

Another important difference is that the loads F_i^{II}, were calculated at equidistance points at the load-displacement curve and not just for equidistant displacements. By doing so the descending branch was weighted equally as much as the ascending branch, and even snap back is taken into account by this method. The number of points, N, were set to 350. In order to increase the calculation speed the objective function was implemented using the C programming language, and instead of the Nelder and Mead scheme the NLPQL scheme developed by Schittkowski (1980) implemented in the programming language FORTRAN was used.

In this way a more complicated objective function with two additional parameters is used with almost the same calculation time as the previous function. However, as in the case of the first objective function it is necessary to try more than one start guess, in order to find the true minimum.

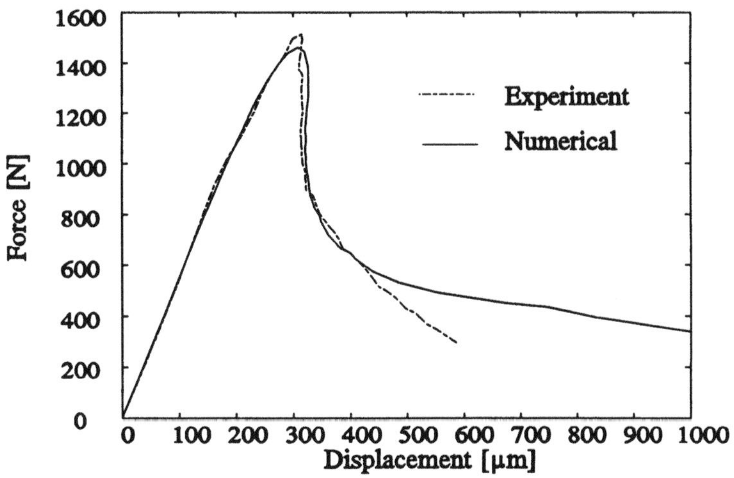

Fig. 4. Optimization result where the second objective function fails.

In most cases this object function yields very satisfactory results, however, it is in some cases not possible to obtain an acceptable agreement at the tail at the descending branch, which usually results in a too large fracture energy. An example of such an optimization result is seen in Fig. 4.

Objective function III:
In order to solve the above outlined problem a third objective function was designed

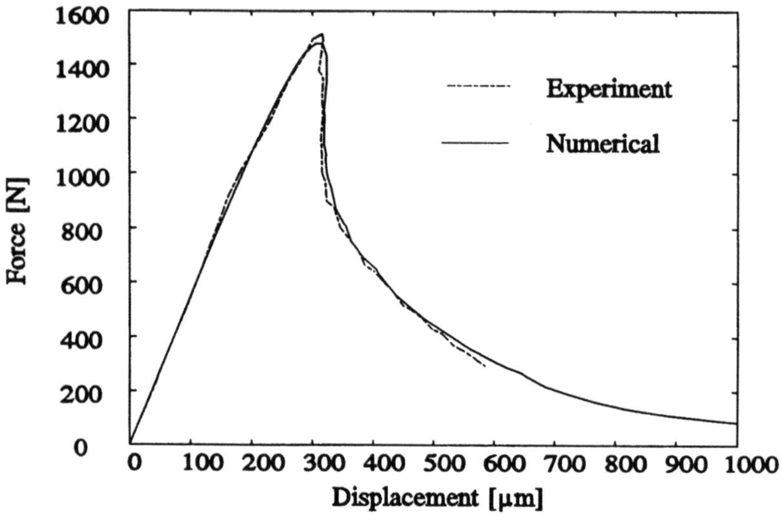

Fig. 5. Optimization result from using the third objective function.

$$\min_{\vec{z}} \frac{\sum_{i=1}^{N} \left[\left(\hat{F}_i^{II}(\vec{z}) - F_i^{II}\right)^2 \left[\left(\frac{F_i^{II}}{F_{max}}\right)^2 + \left(\frac{\delta_i^{II}}{\delta_{max}}\right)^2 \right] \right]}{F_{max}^2} \qquad (7)$$

$$\vec{z} = [\ \hat{\sigma}_u\ \hat{G}_F\ \hat{E}\ \hat{w}_1/\hat{w}_c\ \hat{\sigma}_1/\hat{\sigma}_u\ ]^T$$

subjected to the constraints in (6). By using this objective function the optimization result in Fig. 4. changes to Fig. 5.

The mean value of the estimated fracture energies is shown in Fig. 6., and the trend observed in the RILEM method is also observed here, which is due to the fact that the fitting is made so that the curve will be as equal as possible and that the energy dissipation in the bulk has not been taken into consideration in this method. The increase in the fracture energy by doubling the beam depth is 15% for t = 50 mm and t = 200 mm and 10% for t = 100.

The mean values of the uniaxial tensile strength, σ_u, are shown in Fig. 6. No distinct trend is observed from the estimated uniaxial tensile strength. The tensile strength is smallest for the small specimens, are then increasing and then decreasing a little. It is observed that the uniaxial tensile strength is smaller than the modulus of rupture.

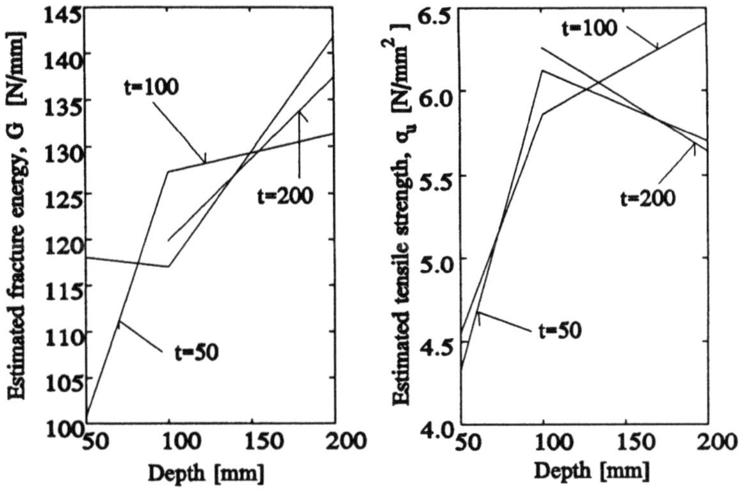

Fig. 6. Estimated fracture energies and uniaxial tensile strength as a function of the beam depth.

The mean values of the normalized kink point of the softening relation are shown in

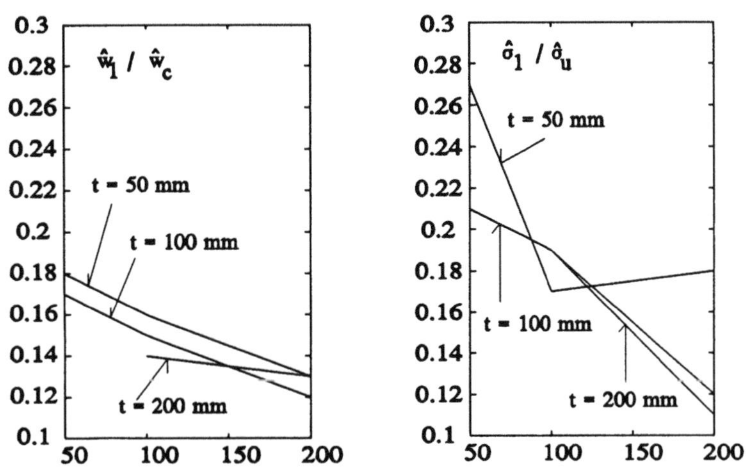

Fig. 7. Estimated results for the normalized kink coordinate of the softening relation.

Fig. 7. It is also seen that the normalized kink coordinate is decreasing with the beam size and that the values are lower than those of NSC ((0,16,0.31)), indicating that HSC is more brittle than NSC.

4 Conclusions

For all the three models it was seen that there were a significant size effect. The size effect on the modulus of rupture was of the same order as that of experiments performed by Reagel and Willis (1931). The increase on the fracture toughness with the increasing beam depth was very large and it is concluded that LEFM, is not applicable for HSC concrete of laboratory size. The size effect on the fracture energies was also significant, and the RILEM method for determining the fracture energy seems inappropriate. By using the data fitting technique the uniaxial tensile strengths and the kink coordinate on the descending branch were determined and it was observed that the tensile strength was both decreasing and increasing with the beam depth. The normalized kink coordinates were decreasing with the beam depth. For all parameters it was concluded that they were independent of the beam thickness.

5 Acknowledgements

Thanks are due to Dr. H.H. Bache of Aalborg Portland, Denmark, for fruitful discussions and for technical assistance with the production of the high-strength concrete specimens. The financial support from the Danish Council for Scientific and Industrial Research is gratefully acknowledged.

6 References

ASTM (1974), E399-74: Method of test for plane strain fracture toughness of metallic materials.

Brincker, R. and Dahl, H. (1989), Fictitious Crack Model of Concrete Fracture, *Magazine of Concrete Research*, 41, No. 147, pp. 79-86.

Hillerborg, A.(1985), Influence of Beam Size on Concrete Fracture Energy Determined According to a Draft RILEM Recommendation; Report to RILEM TC50-FMC.Rapport TVBN-3021, pp. 1-15.

Reagel, F.V. and Willis, T.F. (1931), The Effect of the Dimensions of Test Specimens on the Flexural Strength of Concrete, *Public Roads*, Vol. 12, No. 2, pp.37-46.

RILEM 50-FMC (1985), Determination of the Fracture Energy of Mortar and Concrete by Means of Three-Point Bend tests on notched beams, *Materials and Structures*, Vol. 18, No. 106, pp. 285-290.

Roelfstra, P.E. and Wittmann F.H. (1986), Numerical Method to Link Strain Softening with Failure of Concrete, *in Fracture Toughness and Fracture Energy of Concrete, Edited by F.H. Wittmann*, Elsevier, pp. 163-175.

Schittkowski, K. (1985), A FORTRAN subroutine Solving Constrained Non-Linear Programming Problems, *Annals of Operation Research*, Vol. 5, pp. 485-500.

12 ASYMPTOTIC TIP FIELDS FOR A STEADILY GROWING CRACK IN PRESSURE-SENSITIVE MATERIALS

D. BIGONI and E. RADI
Istituto di Scienza delle Costruzioni, University of Bologna, Italy

Abstract
Mode I steady-state, quasi-static crack propagation is analysed in elastoplastic pressure-sensitive solids obeying the Drucker-Prager yield condition. The asymptotic crack-tip fields are numerically obtained with reference to the incremental small strain theory, in the case of linear isotropic hardening, under plane-stress and plane strain conditions. The determination of the stress and strain near-tip fields is fundamental for the understanding of fracture in ceramics, amorphous rocks at low temperature and concrete.
Keywords: Crack Propagation, Incremental Elastoplasticity, Linear Hardening, HRR Fields, Drucker-Prager Yield Condition, Frictional Materials.

1 Introduction

The determination of asymptotic stress and strain fields in the plastic zone near a crack tip is extremely important for the understanding of the crack propagation mechanism. The case of J_2-flow theory with linear hardening was initiated by Amazigo and Hutchinson (1977) on the basis of the HRR singularity of the fields ahead a stationary crack (Hutchinson, 1968 a, b; Rice and Rosengren, 1968). The problem was further developed by Ponte-Castañeda (1987 a) who considered plastic reloading on crack flanks, by Achenbach et al. (1981) and Ostlund and Gudmundson (1988), who considered dynamic propagation.

The Drucker-Prager yield surface may be appropriate for the description of the behaviour of ceramics (Chen and Reyes-Morel, 1986), amorphous rocks at low temperature and concrete. Moreover, this criterion may be roughly adequate in modelling the appearance of inelastic deformation in concrete (Kotsovos and Newman, 1978), soils and rocks (Rudnicki, 1977). When used in connection with the associative flow-rule, the Drucker-Prager yield function predicts a plastic volumetric dilatancy, which is generally an overestimation of the experimental results. However, the dilatancy measurements are generally mediated over entire samples and thus a global measure of dilatancy can underestimate the dilatancy in a fracture process zone. The behaviour of frictional materials as concrete may be also described by non associative constitutive relation (Palaniswamy and Shah, 1974) and this case is studied in Radi and Bigoni (1992).

Fracture and Damage of Concrete and Rock – FDCR-2.
Edited by H.P. Rossmanith. © 1993 Taylor & Francis. ISBN 0 419 18470 8.

In the present paper numerical results are given for the near tip stress and velocity fields of a steadily propagating crack as functions of the hardening parameter α and the pressure-sensitivity parameter μ. In the case $\mu = 0$ (J_2-flow theory) the result reduces to those of Ponte-Castañeda (1987 a).

2 Basic relationship

Reference is made to an isotropic elastoplastic model with linear hardening based on the Drucker-Prager yield function:

$$f(\sigma, k) = \frac{\mu}{3} tr\, \sigma + \sqrt{J_2} - k = 0, \tag{2.1}$$

where σ denotes the stress tensor, J_2 the second invariant of deviatoric stress, k is the hardening parameter which governs the size and the evolution of the yield surface and μ is the measure of the pressure-sensitivity. In perfect plasticity μ and k are related to the yield strengths in tension f_t and compression f_c through the following relations:

$$\mu = \sqrt{3}\frac{f_c - f_t}{f_c + f_t} \qquad k = \frac{2}{\sqrt{3}}\frac{f_c\, f_t}{(f_c + f_t)} \tag{2.2}$$

The elastoplastic stress-strain relationships, relating the stress rate $\dot{\sigma}$ to the velocity of deformation $\dot{\varepsilon}$, can be deduced, via Prager consistency, in the form:

$$\dot{\varepsilon} = \frac{1}{2G}\left[\dot{\sigma} - \frac{\nu}{1+\nu}(tr\, \dot{\sigma})\mathbf{I} + 2\left(\alpha^{-1} - 1\right) < \mathbf{Q} \bullet \dot{\sigma} > \mathbf{Q}\right], \tag{2.3}$$

where ν is the Poisson ratio, α the ratio between tangent modulus G_t and elastic shear modulus G, the operator < > denotes the McAulay brackets, $\mathbf{Q}$ is the gradient of the yield function (2.1). Let $\mathbf{S}$ denote the deviatoric stress, then:

$$\mathbf{Q} = \frac{\mu}{3}\mathbf{I} + \frac{\mathbf{S}}{2\sqrt{J_2}}, \tag{2.4}$$

A cylindrical co-ordinate system (r, θ, x_3) is employed with the origin attached to the moving crack tip, which is propagating in the $\theta = 0$ direction. The x_3-axis is assumed along the direction of zero stress or strain. For this choice of the reference frame, the steady-state propagation condition implies:

$$\dot{(\;)} = V\left[(\sin\theta/r)(\;)_{,\theta} - \cos\theta\,(\;)_{,r}\right], \tag{2.5}$$

where V denotes the (constant) crack-tip speed and the comma indicates differentiation.

The compatibility conditions

$$\dot{\varepsilon}_{rr} = v_{r,r} \qquad \dot{\varepsilon}_{\theta\theta} = \left(v_{\theta,\theta} + v_r\right)/r \qquad \dot{\varepsilon}_{r\theta} = \left(r\, v_{\theta,r} + v_{r,\theta} + v_\theta\right)/(2r) \tag{2.6}$$

allow for the corresponding substitutions. Therefore the problem will be formulated considering the stress components and the two in-plane velocities v_1 and v_2 as unknowns. The equilibrium equations in cylindrical co-ordinates write:

$$(r\,\sigma_{rr})_{,r} + \sigma_{r\theta,\theta} - \sigma_{\theta\theta} = 0 \qquad\qquad (r\,\sigma_{r\theta})_{,r} + \sigma_{\theta\theta,\theta} + \sigma_{r\theta} = 0 \tag{2.7}$$

Constitutive equations (2.3) and equilibrium equations (2.7) form a system of first order PDEs which governs the problem of the near crack-tip fields determination. It is worth noting that all equations of the mentioned system of PDEs are homogeneous in r, and therefore solutions are sought in the HRR separable form (Hutchinson, 1968 a,b; Rice and Rosengren, 1968):

$$\begin{aligned} v_r(r,\theta) &= (V/s)\,\underline{r}^s\,y_1(\theta), & v_\theta(r,\theta) &= (V/s)\,\underline{r}^s\,y_2(\theta),\\ \sigma_{r\theta}(r,\theta) &= E\,\underline{r}^s\,y_3(\theta), & \sigma_{rr}(r,\theta) &= E\,\underline{r}^s\,y_4(\theta),\\ \sigma_{\theta\theta}(r,\theta) &= E\,\underline{r}^s\,y_5(\theta), & \sigma_{33}(r,\theta) &= E\,\underline{r}^s\,y_6(\theta), \end{aligned} \tag{2.8}$$

where $E = 2G(1+\nu)$, s is the fields singularity coefficient and $\underline{r}$ denotes the non-dimensional ratio between r and any characteristic dimension of the plastic zone. Therefore the solution is determined except for an amplitude factor (Ponte-Castañeda, 1987 b), as expected from asymptotic analysis of a homogeneous problem.

Assuming a straight path motion of particles, a generic particle ahead the crack tip experiences unloading when $\mathbf{Q}\bullet\dot{\sigma} \le 0$. Plastic reloading on crack flanks occurs when the same particle reaches a stress state on the yield surface left at the beginning of unloading. Introducing the effective stress σ_e:

$$\sigma_e = \frac{\sqrt{3}}{\mu - \sqrt{3}}\left(\frac{\mu}{3}\,tr\,\sigma + \sqrt{J_2}\right) = E\,\underline{r}^s\,y_e(\theta), \tag{2.9}$$

reloading takes place at the angle θ_2, if

$$y_e(\theta_1)/(\sin\theta_1)^s = y_e(\theta_2)/(\sin\theta_2)^s \tag{2.10}$$

where θ_1 is the angle corresponding to the elastic unloading.

It has been proved that (Drugan and Rice, 1984) for quasi-static problems in elastoplasticity with (positive) isotropic hardening, all stress components must be continuous if the material is stable in the Drucker sense. Thus, for the present model, full continuity of stress components, i. e., of functions y_i, should be imposed.

3 Plane stress (mode I)

The substitution of (2.8) into (2.7) and (2.3) with the plane stress condition $\sigma_{33} = 0$, yields a system of five first order ODEs in the standard form (Bigoni and Radi, 1992):

$$\mathbf{y}' = \mathbf{f}(\theta, \mathbf{y}, s, \underline{\alpha}, sign(\mathbf{Q}\bullet\dot{\sigma})). \tag{3.1}$$

The system (3.1) depends on the conditions of plastic loading, i.e. on *sign*($\mathbf{Q}\bullet\dot{\sigma}$) and *sign*[$f(\sigma)$]. Therefore, the parameter $\underline{\alpha}$ in equation (3.1) is equal to 1 when $f(\sigma) < 0$ or equal to α when $f(\sigma) = 0$.

The symmetry of the mode I problem restricts the study to the interval $0 \le \theta \le \pi$. The mode I symmetry and regularity of the stress functions ahead the crack tip require:

$$y_2(0) = y_3(0) = y_4'(0) = y_5'(0) = 0. \tag{3.2}$$

Moreover, on the crack surface, the tensions $\sigma_{\theta\theta}$ and $\sigma_{r\theta}$ must vanish, hence:

$$y_3(\pi) = y_5(\pi) = 0. \tag{3.3}$$

Equations (3.2) and (3.3) are five independent homogeneous boundary conditions. In order to solve the system of ODEs the standard Runge-Kutta procedure (IMSL library, subroutine DIVPRK) is used. This approach requires all the initial values of the functions y_i. Therefore, the normalisation condition $y_4(0) = 1$ is employed to avoid the trivial solution and the values of $y_5(0) = q$ and s are initially guessed. Further, equations (3.1) are evaluated in $\theta = 0$, and an algebraic system is obtained, where the unknowns are the components of $\mathbf{y}(0)$ and $\mathbf{y}'(0)$ which are not specified by the initial boundary conditions (3.2). Therefore, the following auxiliary condition is obtained:

$$y_1(0) = -s\left\{1 - \nu q + \frac{1}{3h}\left[\mu + \frac{2-q}{2\sqrt{\underline{J_2}(0)}}\right]\left[\frac{\mu}{3}(1+q) + \sqrt{\underline{J_2}(0)}\right]\right\} \tag{3.4}$$

where $h^{-1} = 2(1+\nu)(\underline{\alpha}^{-1} - 1)$ and $\underline{J_2}(0) = (1 + q^2 - q)/3$. The integration is then performed and a check is made on the values $y_3(\pi)$ and $y_5(\pi)$. On the basis of this check, the guessed values for q and s are reassigned and the process is iterated using a modified Powell hybrid method (IMSL library, subroutine DNEQNF), until $y_3(\pi)$ and $y_5(\pi)$ are found to be zero. All the results are re-normalised assuming $y_e(\theta_1) = 1$. In Tables 1-2 the values of the singularity s and of the unloading angle θ_1 (in degrees) are reported for ν = 1/2 for the cases $\alpha = 0.001$ and $\alpha = 0.1$.

Table 1. Values of s, q and θ_1 for $\alpha = 0.001$ and $\nu = 1/2$

μ	s	q	θ_1
0.00	-0.02866	1.96890	53.202
0.20	-0.02656	2.45930	47.770
0.30	-0.02578	2.81838	45.502
0.40	-0.02515	3.31713	43.445
0.50	-0.02467	4.06674	41.559
0.60	-0.02430	5.33904	39.816
0.80	-0.02394	17.65462	36.692

The system (3.1) depends on the conditions of plastic loading, i.e. on $sign(\mathbf{Q}\bullet\dot{\sigma})$ and $sign[f(\sigma)]$. Therefore, the parameter $\underline{\alpha}$ in equation (3.1) is equal to 1 when $f(\sigma) < 0$ or equal to α when $f(\sigma) = 0$.

The symmetry of the mode I problem restricts the study to the interval $0 \le \theta \le \pi$. The mode I symmetry and regularity of the stress functions ahead the crack tip require:

$$y_2(0) = y_3(0) = y_4'(0) = y_5'(0) = 0. \tag{3.2}$$

Moreover, on the crack surface, the tensions $\sigma_{\theta\theta}$ and $\sigma_{r\theta}$ must vanish, hence:

$$y_3(\pi) = y_5(\pi) = 0. \tag{3.3}$$

Equations (3.2) and (3.3) are five independent homogeneous boundary conditions. In order to solve the system of ODEs the standard Runge-Kutta procedure (IMSL library, subroutine DIVPRK) is used. This approach requires all the initial values of the functions y_i. Therefore, the normalisation condition $y_4(0) = 1$ is employed to avoid the trivial solution and the values of $y_5(0) = q$ and s are initially guessed. Further, equations (3.1) are evaluated in $\theta = 0$, and an algebraic system is obtained, where the unknowns are the components of $\mathbf{y}(0)$ and $\mathbf{y}'(0)$ which are not specified by the initial boundary conditions (3.2). Therefore, the following auxiliary condition is obtained:

$$y_1(0) = -s\left\{1 - \nu q + \frac{1}{3h}\left[\mu + \frac{2-q}{2\sqrt{\underline{J_2}(0)}}\right]\left[\frac{\mu}{3}(1+q) + \sqrt{\underline{J_2}(0)}\right]\right\} \tag{3.4}$$

where $h^{-1} = 2(1+\nu)(\underline{\alpha}^{-1}-1)$ and $\underline{J_2}(0) = (1+q^2-q)/3$. The integration is then performed and a check is made on the values $y_3(\pi)$ and $y_5(\pi)$. On the basis of this check, the guessed values for q and s are reassigned and the process is iterated using a modified Powell hybrid method (IMSL library, subroutine DNEQNF), until $y_3(\pi)$ and $y_5(\pi)$ are found to be zero. All the results are re-normalised assuming $y_e(\theta_1) = 1$. In Tables 1-2 the values of the singularity s and of the unloading angle θ_1 (in degrees) are reported for $\nu = 1/2$ for the cases $\alpha = 0.001$ and $\alpha = 0.1$.

Table 1. Values of s, q and θ_1 for $\alpha = 0.001$ and $\nu = 1/2$

μ	s	q	θ_1
0.00	-0.02866	1.96890	53.202
0.20	-0.02656	2.45930	47.770
0.30	-0.02578	2.81838	45.502
0.40	-0.02515	3.31713	43.445
0.50	-0.02467	4.06674	41.559
0.60	-0.02430	5.33904	39.816
0.80	-0.02394	17.65462	36.692

Table 2. Values of s, q and θ_1 for $\alpha = 0.1$ and $\nu = 1/2$

μ	s	q	θ_1
0.00	-0.23726	1.60583	73.646
0.20	-0.21810	1.92533	69.296
0.30	-0.21043	2.13341	67.468
0.40	-0.20379	2.39237	65.826
0.60	-0.19302	3.18331	63.016
0.80	-0.18497	4.92100	60.732
1.00	-0.17914	12.61329	58.894
1.10	-0.17694	101.54746	58.129

The effects of pressure sensitivity consist in a decreasing of the fields singularity and of the plastic sector amplitude. The functions $y_i(\theta)$, evaluated for $y_e(\theta_1) = 1$, are reported in Figs. 1-2, for the values of $\alpha = 0.001$ and $\alpha = 0.1$. These plots represent (unless an amplitude factor) the stress and velocity fields for different values of μ. At the increasing of the parameter μ, a remarkable reduction is observed in the radial stress ahead the crack tip, which corresponds to a reduction in the mean normal stress ahead the crack tip. This conclusion is in good agreement with the observations of Li and Pan

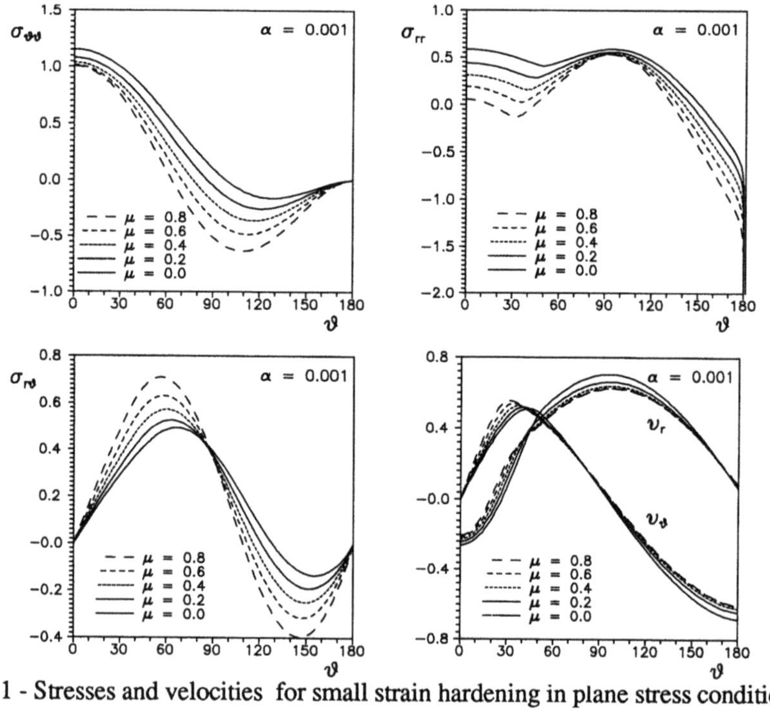

Fig. 1 - Stresses and velocities for small strain hardening in plane stress condition.

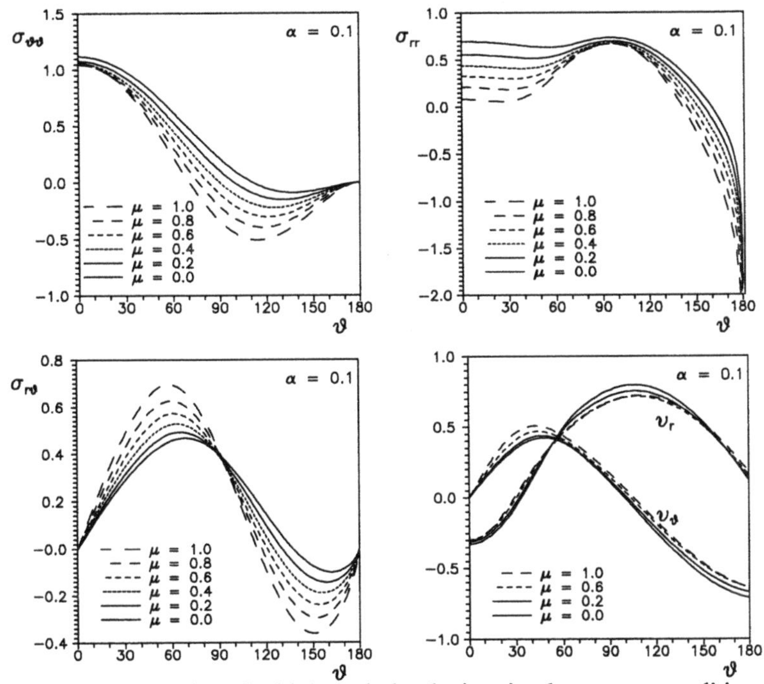

Fig. 2 - Stresses and velocities for high strain hardening in plane stress condition.

(1990 b). In the perfect-plasticity case, these Authors found a decrease in $\sigma_{rr}(0)$ with μ ($\sigma_{rr}(0) \rightarrow 0$, for $\mu = \mu_{lim} = 3^{1/2}/2$). For small values of α, an analogous behaviour is observed (see Table 2). However, for high values of hardening, $\sigma_{rr}(0)$ vanishes when μ approaches values greater than μ_{lim} (see Table 2). Further, for high values of α the effects caused by the pressure sensitivity decrease and all fields tend to approach the elastic solution.

4 Plane strain (mode I)

The plane strain condition $\dot{\varepsilon}_{33} = 0$, with the compatibility relations (2.6) may be substituted in the constitutive relations (2.3) to obtain with the equilibrium equations (2.7) a system of six first order PDEs in the six unknown functions σ_{rr}, $\sigma_{\theta\theta}$, $\sigma_{r\theta}$, σ_{33}, v_r, and v_θ, whose solutions are sought in the HRR form (2.8) as in the plane stress case. Under this hypothesis, the governing system of six first order ODEs may be reduced to the explicit form (Bigoni and Radi, 1992):

$$\mathbf{y}' = \mathbf{f}(\theta, \mathbf{y}, s, \underline{\boldsymbol{\alpha}}, sign(\mathbf{Q} \bullet \dot{\boldsymbol{\sigma}})). \tag{4.1}$$

The system depends on whether plastic loading, elastic unloading or the subsequent plastic reloading occurs, similarly to the plane stress case.

The vanishing of the tractions on the crack surface requires

$$y_3(\pi) = y_5(\pi) = 0, \tag{4.2}$$

while mode I symmetry and regularity in $\theta = 0$, require:

$$y_2(0) = y_3(0) = y_4'(0) = y_5'(0) = y_6'(0) = 0, \tag{4.3}$$

The scheme for solving the system (4.1) is similar to the plane stress case, except for the determination of the initial boundary condition. In this case, the normalisation $y_5(0) = 1$ is adopted and the values of $q = y_4(0)$ and s are initially guessed. Further, the following conditions on $y_1(0)$ and $y_6(0)$ are obtained, by using equations (4.1) evaluated in $\theta = 0$:

$$y_1(0) = -s\left\{q - \nu\left(1 + y_6(0)\right) + \frac{1}{3h}\left[\mu + \frac{2q - 1 - y_6(0)}{2\sqrt{\underline{J_2}(0)}}\right]\left[\frac{\mu}{3}\left(1 + q + y_6(0)\right) + \sqrt{\underline{J_2}(0)}\right]\right\} \tag{4.4}$$

$$y_6(0) - \nu\left(1 + q\right) + \frac{1}{3h}\left[\mu + \frac{2\,y_6(0) - 1 - q}{2\sqrt{\underline{J_2}(0)}}\right]\left[\frac{\mu}{3}\left(1 + q + y_6(0)\right) + \sqrt{\underline{J_2}(0)}\right] = 0 \tag{4.5}$$

where $\underline{J_2}(0) = \left(1 + q^2 + y_6^2(0) - q - y_6(0) - q\,y_6(0)\right)/3$.

The numerical value for $y_6(0)$ corresponding to the trial value of q is obtained by solving the non linear equation (4.5) (subroutine DZBREN in the IMSL library). When $y_6(0)$ is known, the corresponding value for $y_1(0)$ is furnished by (4.4).

The values of the singularity s and the unloading and reloading angles θ_1 and θ_2 have been reported in Table 3, for $\nu = 0.3$ and $\alpha = 0.001$. The effect of pressure sensitivity consist in a decreasing of the angles of unloading and subsequent reloading as well as of the stress singularity s, similarly to the plane stress case.

Table 3. Values of s, q, θ_1 and θ_2 for $\nu = 1/3$ and different values of α and μ

α	μ	s	q	θ_1	θ_2
0.1	0.00	-0.20956	1.09031	122.012	175.318
0.1	0.10	-0.20186	1.01006	107.416	179.845
0.01	0.00	-0.08242	0.79740	134.998	146.242
0.01	0.10	-0.07840	0.88486	131.477	144.584
0.01	0.20	-0.07415	0.99017	127.663	143.727
0.001	0.00	-0.05640	0.75379	136.965	138.444
0.001	0.10	-0.05367	0.82806	133.816	135.458
0.001	0.20	-0.05002	0.89693	130.549	132.531
0.001	0.30	-0.04561	0.96363	127.095	129.741
0.001	0.33	-0.04405	0.98837	125.999	128.962

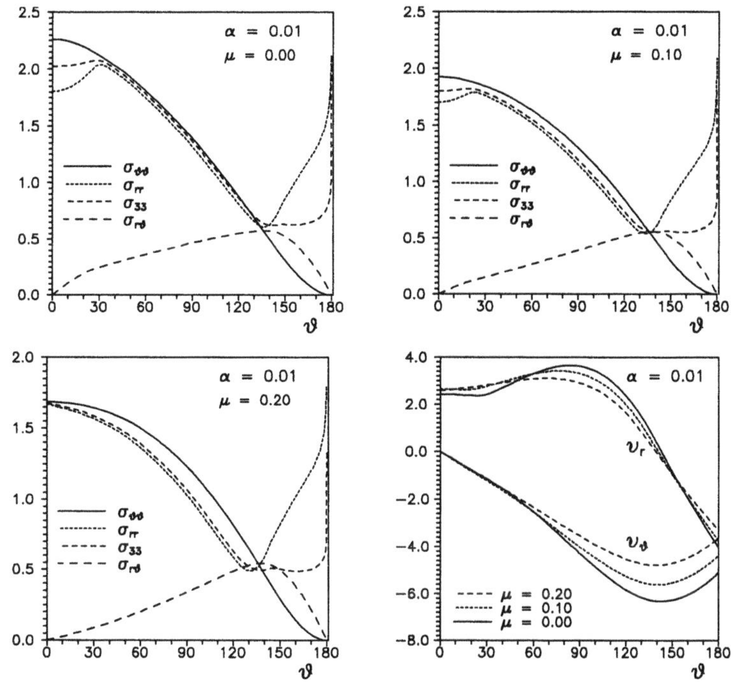

Fig. 3 - Stresses and velocities for different values of μ, in plane strain condition

At the increasing of the pressure sensitivity parameters μ, the values of q tend to 1 and the stress state near θ = 0 tends to a hydrostatic state of tension. The plots of the stresses and velocity functions y_i , evaluated for $y_e(\theta_1) = 1$, are reported in Fig. 3.

5 Conclusions

In the paper, the near tip fields of a steadily propagating crack under mode I condition have been numerically obtained for the Drucker-Prager incremental elastoplastic model in plane-stress and plane strain conditions. In both cases, the increasing of μ yields a the lowering in the strength of the singularity and a reduction of the plastic sectors size. Therefore the pressure sensitivity acts with a stabilising effect in crack propagation. Moreover, in the plane stress case, a lowering is observed in the ratio between the radial and hoop stresses ahead the crack tip, related to a decreasing of the mean normal stress ahead the crack tip, while under plane strain condition the pressure sensitivity reduces the components of the deviatoric stress ahead the crack tip. These results fit coherently with those of the perfectly plastic and deformation-theory models of Li and Pan (1990a,b), who studied the plane stress and plane strain static problem. From the above, it can be concluded that the pressure sensitivity of the model sensibly modifies the stress and velocity patterns and therefore has to be considered an important parameter for crack-propagation in frictional material as ceramics, amorphous rocks at low temperature and concrete.

6 References

Achenbach, J.D., Kanninien, M.F. and Popelar, C.H. (1981). Crack tip fields for fast fracture of an elastic-plastic material. **J. Mech. Phys. Solids**, 29, 211-225.

Amazigo, J. and Hutchinson, J.W. (1977). Crack-tip fields in steady crack-growth with linear strain hardening. **J. Mech. Phys. Solids**, 25, 81-97.

Bigoni, D. and Radi, E. (1992). Mode I crack propagation in elastoplastic pressure-sensitive materials. Accepted on **Int. J. Solids Structures**.

Chen, I.W. and Reyes Morel, P.E. (1986). Implications of transformation plasticity in ZrO_2-containing ceramics: I, shear and dilatation effects. **J. Am. Ceram. Soc.**, 69, 181-189.

Drugan, W.J. and Rice, J.R. (1984). Restrictions on quasi-statically moving surfaces of strong discontinuity in elastic-plastic solids. In **Mechanics of Material Behavior**, (eds G.J. Dvorak and R.T. Shield), Elsevier, Amsterdam, pp. 59-73.

Hutchinson, J.W. (1968 a). Singular behaviour at the end of a tensile crack in a hardening material. **J. Mech. Phys. Solids**, 16, 13-81.

Hutchinson, J.W. (1968 b). Plastic stress and strain fields at a crack tip. **J. Mech. Phys. Solids**, 16, 337-347.

Kotsovos, M.D. and Newman, J.B. (1978) Generalized stress-strain relations for concrete. **J. Eng. Mech. Div. ASCE**, 104, 845-856.

Li, F.Z. and Pan, J. (1990 a). Plane-strain crack-tip fields for pressure-sensitive dilatant materials. **J. Appl. Mech.**, 57, 40-49.

Li, F.Z. and Pan, J. (1990 b). Plane-stress crack-tip fields for pressure-sensitive dilatant materials. **Engnr. Fracture Mechanics**, 35, 1105-1116.

Ostlund, S. and Gudmundson, P. (1988). Asymptotic crack tip fields for dynamic fracture of linear strain-hardening solids. **Int. J. Solids Structures**, 24, 1141-1148.

Palaniswamy, R. and Shah, S.P. (1974). Fracture and stress-strain relationship of concrete under triaxial compression. **J. Struct. Eng., ASCE**, ST5, 100, 901-916.

Ponte Castaneda, P. (1987 a). Asymptotic fields in steady crack growth with linear strain-hardening. **J. Mech. Phys. Solids**, 35, 227-268.

Ponte Castaneda, P. (1987 b). Plastic stress intensity factors in steady crack growth. **J. Appl. Mech.**, 54, 379-387.

Radi, E. and Bigoni, D. (1992). Asymptotic fields of mode I steady-state crack propagation in non-associative elastoplastic solids. Accepted on **Mechanics of Materials**.

Rice, J.R. and Rosengren, G.F. (1968). Plane strain deformation near a crack tip in a power-law hardening material. **J. Mech. Phys. Solids**, 10, 1-12.

Rudnicki, J.W. (1977). The inception of faulting in a rock mass with a weakened zone. **J. Geophys. Res.**, 82, 844-854.

PART TWO
MIXED-MODE FRACTURE AND DAMAGE

13 ACOUSTIC EMISSION AND SOURCE LOCATION IN CONCRETE SUBJECTED TO MIXED-MODE LOADING

W. SUARIS
Department of Civil and Architectural Engineering,
University of Miami, Coral Gables, Florida, USA
J.G.M. van MIER
Stevin Laboratory, Delft University of Technology,
The Netherlands

Abstract
The paper contains results of acoustic emission studies conducted on mortar and concrete. The loading was conducted both in tension and shear. The results of hit-rate, energy etc. indicate that the amount of AE activity is strongly dependent on the aggregate size. The located events are found to lie adjacent to the surface cracks.
Keywords: Acoustic Emission, Source Location, Concrete, Aggregate Size, Mixed-Mode Loading, Tension, Shear.

1 Introduction

Concrete Structures subjected to loading undergo continuous deterioration due to the cracking of concrete. Surface crack growth in concrete may be traced by optical techniques (van Mier, 1991). Impregnation techniques may also be used to determine the extent of crack growth in a specimen which has been subjected to a certain amount of load (van Mier, 1990).

The cracking of concrete may also be monitored using acoustic methods. In the active method, acoustic energy is transmitted through the specimen. The cracking of the specimen has been found to be related to certain characteristics of the received signal, such as the amplitude (Suaris and Fernando, 1987). In the passive method, the acoustic emission (AE) generated by the cracking of concrete is monitored by using AE transducers mounted on the surface of the specimen. AE data obtained during mixed-mode (tension-shear) loading of a 8mm concrete, 2 mm mortar and hardened cement paste (hcp) is presented here. The location of the more pronounced events are also deduced from the AE data and are compared with visual observations.

2 Loading Equipment

The mixed-mode testing machine developed at the Stevin Laboratory was used for the tests. The equipment contains two individually controlled hydraulic actuators, which apply vertical loads (tension or compression) and horizontal loads (shear). Details of the

Fracture and Damage of Concrete and Rock – FDCR-2.
Edited by H.P. Rossmanith. © 1993 Taylor & Francis. ISBN 0 419 18470 8.

test-set-up can be found in Nooru-Mohamed, 1992.

Two load paths were investigated in the present tests. In load path I, the specimens were precracked in tension up to a crack opening of 100 μm, unloaded, then loaded in shear under displacement control, maintaining zero vertical load. In load path II, an axial load and a lateral shear load were applied to the specimen such that the ratio of the axial to lateral deformation remained at a constant value of 2.

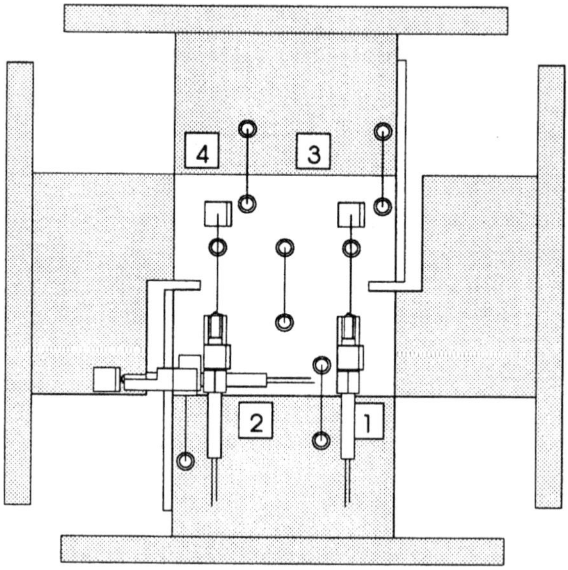

Fig. 1a. Placement of AE transducers.

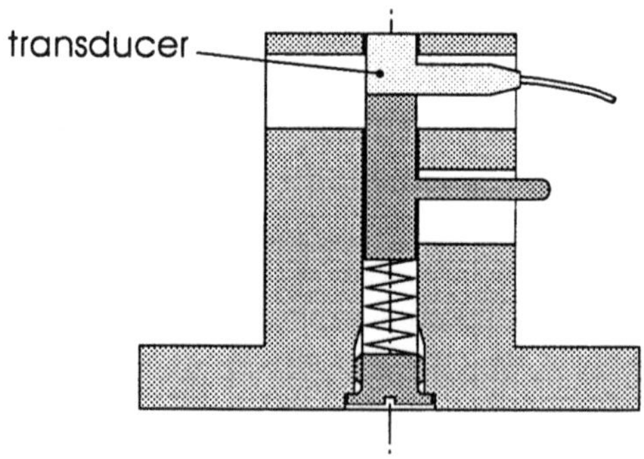

Fig. 1b. Detail of transducer placement in loading platen (cross-sectional view).

3 Acoustic System

The specimens tested were 100x100x50 mm, double-edge-notched to a depth of 12.5 mm as shown in Fig. 1a. Four AE transducers arranged in a triangular format were used for AE monitoring. They were all placed in the mid-plane of the specimen as shown in Fig 1a. The transducers were mounted directly on the specimens through slots made in the top and bottom loading platens. The transducers were spring loaded against the platens to assure a uniform contact pressure (Fig. 1b).

The transducers used had a resonant frequency of approximately 150 kHz. A LOCAN-AT manufactured by Physical Acoustics Corp. was used for AE data acquisition. The AE signals were amplified with a 40 db gain of a preamplifier and a 20 db gain in the system. The threshold was set at 45 db. To limit the amount of data however, a filter was applied and data for events with a maximum amplitude greater than 65db only were recorded. For the mortar specimens the filter level was reduced to 60db as the acoustic signals were weaker than in the concrete.

Although the LOCAN-AT system was capable of source location, it would function satisfactorily only if the AE waves from the four transducers initially go all positive or all negative. This is because the threshold in the system is set at a certain positive voltage and not at an absolute value. However for shear type events, it is expected that the response of some transducers would be different from that of the others. Two, 2-channel Nicolet oscilloscopes were therefore used to record the AE waves. The oscilloscopes were coupled to trigger at the same instant and the trigger level was adjusted to record AE waves having a maximum amplitude greater than approxima-tely 70db. Each channel in the oscilloscope had a data acquisition frequency of 2 MHz.

4 Results/AE Characteristics

The load vs. time curves for a load path I test in concrete are shown in Fig. 2. Three different phases of the test can be identified: tension loading, unloading and shear loading. The horizontal axis can be considered to correspond to the displacement in these loading stages, as a constant rate of 1.6 μm/min was specified for both the tensile and shear displacement.

The cumulative energy of the AE signals is plotted vs. time in Fig. 3. The energy recorded by the transducer, which initially receives the wave from each event, was used for drawing the graph. All AE hits which follow within 25 μs of an initial hit were considered to be from the same event and were not included in the graph. The energy dissipation in the tension phase starts gradually and increases to a maximum in the softening region just beyond the peak load. The energy dissipation in the shear phase also started gradually and kept on increasing linearly upto the time the test was stopped due to reaching the limit of vertical displacement (400 μm) for the equipment.

A histogram of the number of events vs. time is given in Fig. 4. The time scale is divided into 50 bins, with each bin corresponding to 200 μs. Each bin contains the sum of all events during the corresponding time period. It can be observed that the hit-rate in the tension phase is a maximum in the region just beyond the peak load. The maximum hit-rate in the shear phase occurs in the region of the peak of the load-deformation curve.

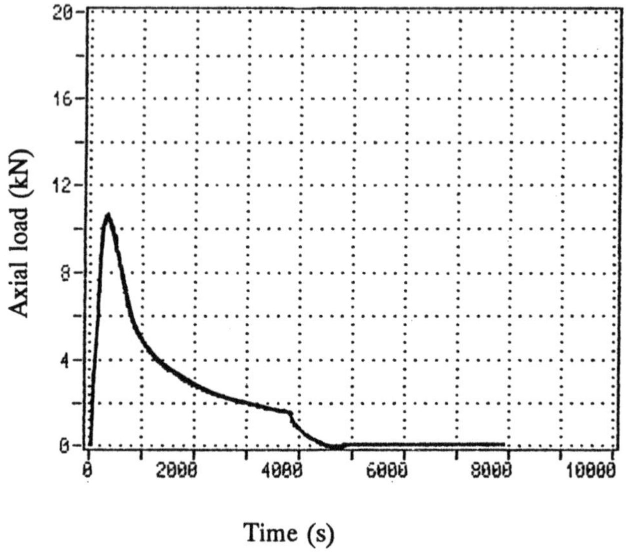

Fig. 2a. Axial load vs. time for load path I test on concrete.

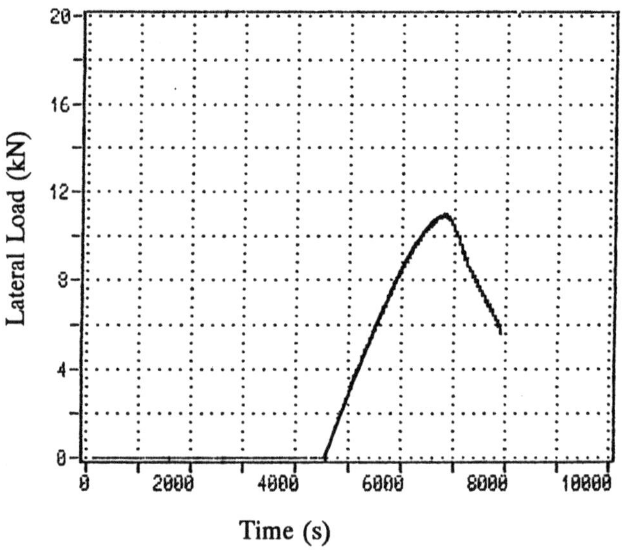

Fig. 2b. Lateral load vs. time for load path I test on concrete.

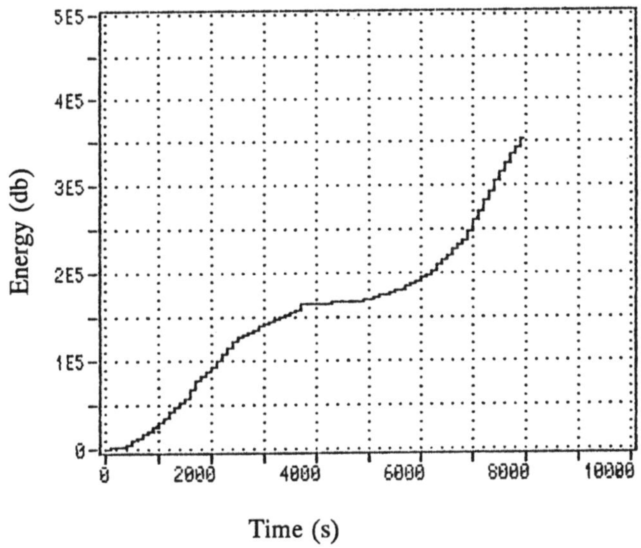

Fig. 3. Cumulative energy dissipation vs. time for load path I test on concrete.

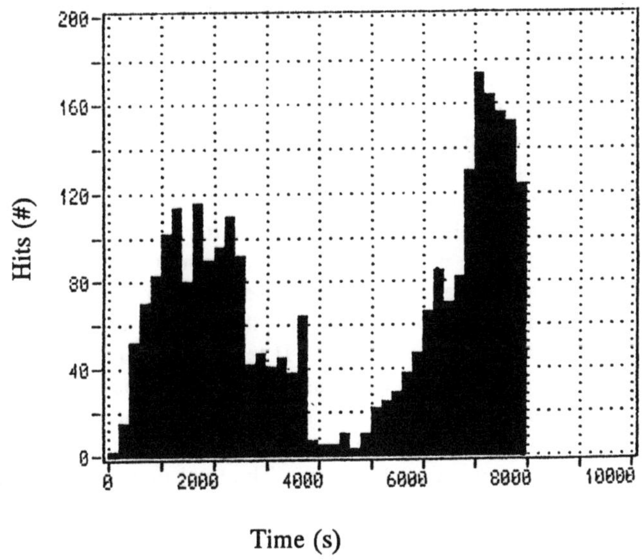

Fig. 4. Histogram of hits vs. time for a load path I test on concrete.

Table 1. AE Characteristics

	hcp	Mortar	Concrete
Max. Aggregate size (mm)	-	2	8
Energy (db)	9750	106000	350000
Peak load (kN)			
Tension	7.7	7.6	10.5
Shear	-	7.7	11.0
# of events (> 65 db)	63	512	1900
Max hit-rate (/s)			
Tension	0.18	0.07	0.58
Shear	-	0.30	0.87

A large proportion of the events recorded were at the lower end (65db) of the amplitude spectrum. Of the hits having a maximum amplitude greater than 65db, only 7% of the events had a maximum amplitude greater than 80db.

The mortar specimens had similar trends as the concrete except that the hit-rates, cumulative energy etc., were all lower than for the concrete. The failure of the cement-paste specimens occurred suddenly during the tension phase and were uncontrollable. One large acoustic event accompanied fracture. Table 1 summarizes the data for the three different materials. The cumulative energy dissipation in concrete is about 3 times that of the mortar specimen and about 30 times that of the cement-paste specimen, for the same test duration of 8000s. This clearly displays the toughness of the concrete despite its strength being only marginally higher than the mortar. The number of events in concrete which are greater than 65db is about 4 times that in mortar. The hit-rate is also higher in concrete in both the tension and the shear phase. Another noteworthy point is that the maximum hit-rate in the shear phase is higher than in the tension phase.

The maximum hit-rate in the load path II test on mortar was 0.1/s which was close to that in the tension phase of the load path I tests. The maximum hit-rate also occurred in a region just beyond the tension peak. The axial force then went into compression as a constant ratio was maintained between the axial deformation and the lateral deformation. The hit-rate continued to decrease thereafter. The total number of events was approximately half that for a load path I test, corresponding to a similar time period.

AE waves received by the four transducers, for a typical event during the tension phase are shown in Fig. 5a. All the waves have an initial negative peak. In contrast, AE waves recorded during the shear phase show that some waves have an initial positive peak (Fig. 5b). This is characteristic of a shear type event, for the present geometry and transducer placement.

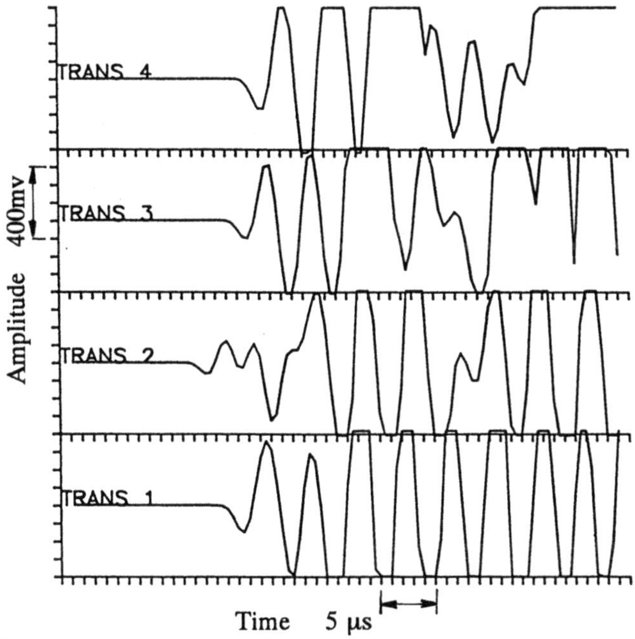

Fig. 5a. Typical AE waves recorded during tensile loading.

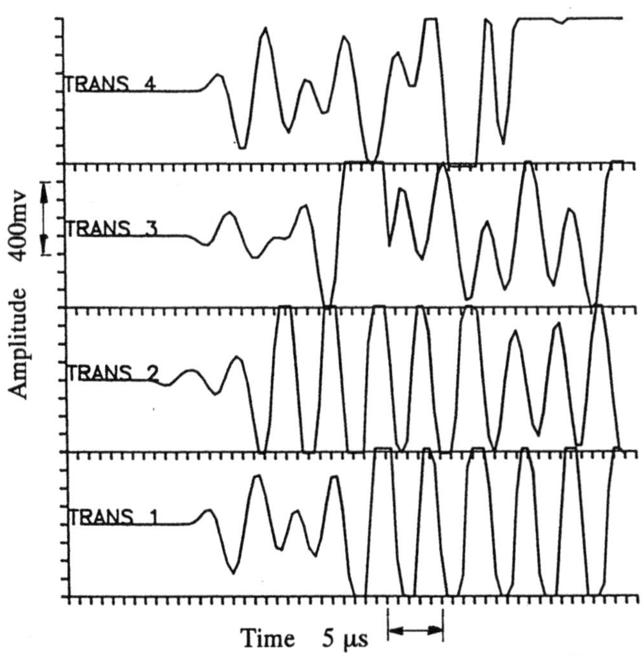

Fig. 5b. Typical AE waves recorded during shear loading.

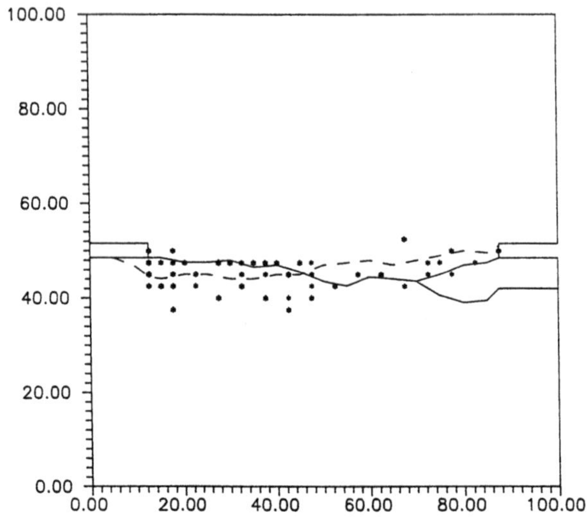

Fig. 6. AE sources and crack trajectories
for mortar in load path I.

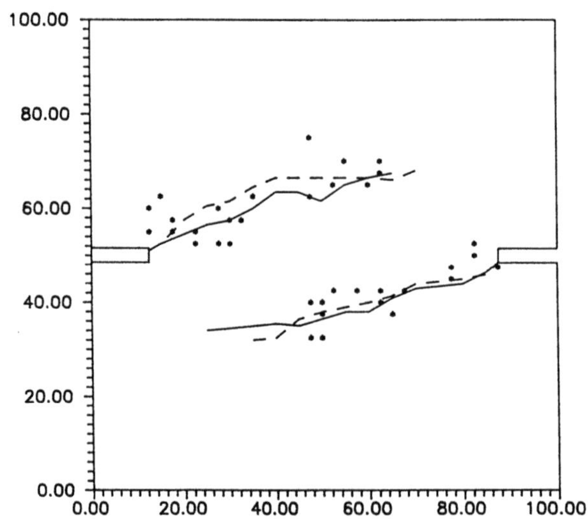

Fig. 7. AE sources and crack trajectories
for mortar in load path II.

5 AE Source Location

The location of the AE sources was determined from the arrival time differences of the AE waves. As the location was determined in 2-D, only the first three arrivals were necessary. The P-wave velocity in both the concrete and mortar were found to be 4200 m/s. The events that were located had a maximum error in arrival time of 0.5 μs. Considering this threshold, the events were located on a grid of 2.5mm. Approximately 85% of the events recorded were located, subjected to this limitation.

Fig. 6 displays all the located events for a load path I test in mortar. The surface cracks in the front and the rear of the specimen are also plotted for reference. It can be seen that most of the sources are located in the vicinity of the observed cracks.

The sources located in a load path II test is shown in Fig. 7. A set of overlapping cracks developed in these tests and the events are found to lie along the crack trajectories. Note that events occuring between the two overlapping cracks

6 Conclusions

The AE activity in concrete was found to increase substantially as the aggregate size was increased. The events during the shear phase were also found to differ from those during the tension phase. The events thus located were found to follow the crack trajectories very closely. The amplitude data from the events show good prospects for studying the source characteristics.

Acknowledgements

The research was conducted while the principal author was on sabbatical leave from the University of Miami. The financial support provided by the National Science Foundation under the US-Netherlands Cooperative research program and the fellowship awarded by the Delft University of Technology are gratefully acknowledged.

References

Nooru-Mohamed, M.B. (1992), "Mixed-Mode Fracture of Concrete: An Experimental Approach", *PhD-Thesis*, Delft University of Technology, The Netherlands.

Suaris, W. and Fernando V. (1987), "Ultrasonic Pulse Attenuation as a Measure of Damage Growth During Cyclic Loading of Concrete", *ACI Journal*, 84(3), 185-193.

Van Mier, J.G.M. (1990), "Internal Crack Detection in Single Edge Notched Concrete Plates Subjected to Uniform Boundary Displacement", in *Micromechanics of Failure of Quasi-Brittle Materials*, eds. S.P Shah, S.E Swartz and M.L.Wang, Elsevier Science, pp. 33-42.

Van Mier, J.G.M. (1991), "Crack Face Bridging in Normal, High Strength and Lytag Concrete", in *Fracture Processes in Concrete, Rocks and Ceramics*, eds, J.G.M. van Mier, J.G. Rots and A. Bakker, Chapman & Hall, London/New York, pp. 27-40.

14 MIXED-MODE FRACTURE PROPAGATION: A COMBINED NUMERICAL AND EXPERIMENTAL STUDY

E. SCHLANGEN and J.G.M. van MIER
Stevin Laboratory, Delft University of Technology, The Netherlands

Abstract
In this paper a mixed mode fracture study for five different concretes is presented. Experimental results of four-point-shear tests are compared with the outcome of uniaxial tensile tests. It is found that even when the loading is shear the fracture growth takes place perpendicular to the maximum tensile strength. For comparing the fracture process in the different concretes it is important to be aware of differences in hardening and drying. Next to this also the differences in microstructure, which result in differences in mechanical and physical properties, lead to deviations in fracture behaviour. With a two dimensional lattice model the four-point-shear experiments are simulated for three concretes. Although the results look already promising, three dimensional simulations are needed to describe the observed fracture mechanisms correctly.
Keywords: Mixed Mode Fracture, Concrete, High Strength Concrete, Lightweight Concrete, Fibre Concrete, Cement Paste, Lattice Model.

1 Introduction

In the past years many researchers studied the mixed mode fracture process in concrete. The main question is: does mode II fracture exist? Recent results of mixed mode experiments on normal weight concrete, Schlangen & Van Mier (1991a), indicate that the only influence of mode II loading is a rotation of the principal stresses. However, fracture takes place perpendicular to the maximum tensile stress only. The same behaviour is observed by Carpinteri et al. (1991).

In the present study the influence of mode II loading on the fracture process is investigated for five types of concrete, i.e. normal weight concrete with a maximum aggregate size of 8mm (*con8*), lytag lightweight concrete (*lytag*), high strength concrete (*hsc*), fibre concrete (*fcon*) and hardened cement paste (*hcp*). Mixed mode experiments carried out in the four-point-shear test set-up in the Stevin Laboratory are compared with uniaxial tensile tests performed on specimens of the same material.

The mixed mode experiments are simulated with a lattice type model, originally developed in theoretical physics, Herrmann (1988). In this investigation such a lattice model is made suitable for simulating concrete fracture. The properties of the individual components in the concrete composite can be varied. Following this procedure simulations are carried out for three different sets of input parameters representing normal weight, lytag and high strength concrete.

Fracture and Damage of Concrete and Rock – FDCR-2.
Edited by H.P. Rossmanith. © 1993 Taylor & Francis. ISBN 0 419 18470 8.

2 Experiments

2.1 Test set-ups and measurements

Two different test set-ups are used in the investigation. For the uniaxial tensile tests the equipment described in Hordijk (1991) is used. Prisms are loaded in deformation control using a closed loop hydraulic system. The specimens are glued in the machine between fixed end platens. In figure 1a the specimen dimensions are shown. Test control was over four LVDTs mounted at the front and rear of the specimen as shown in figure 1a.

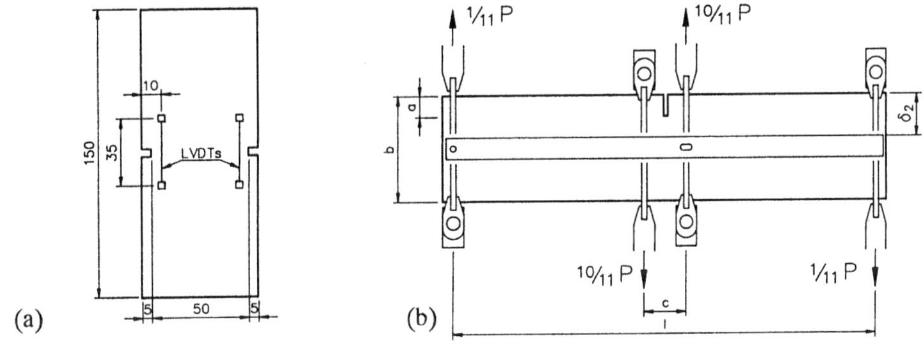

Fig.1. Uniaxial tensile specimen (a); Four-point-shear specimen (b).

The four-point-shear experiments are carried out in the machine shown in figure 2. To overcome friction in the supports the loads are introduced in the specimen via tensile bars as described in Schlangen & Van Mier (1991a). Two specimen sizes are tested, large (a=40, b=200, c=80, l=800 and thickness=100 mm) and small beams (a=20, b=100, c=40, l=400 and thickness=100 mm). The beams are tested under deformation control using the average signal of two LVDTs measuring the deformation δ_2 (Fig. 1b) as feed back in a closed-loop servo control.

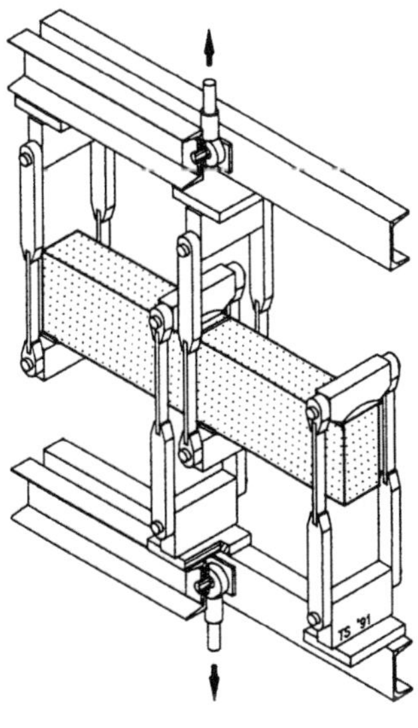

Fig.2. Loading configuration for four-point-shear tests.

2.2 Materials and specimen preparation

As mentioned before five different concretes are used in the investigation. The composition of the five materials tested is given in Table 1. The materials have been used before in other research projects in the Stevin Laboratory, see Van Mier (1991).

Table 1. Concrete Compositions

	con8	*lytag*	*hsc*	*fcon*	*hcp*
Cement PC-B[1)]	375	330	350	375	979
Microsilica[2)]	-	-	35	-	-
w/c-ratio	0.5	0.34	0.37	0.5	0.4
Superplast. Melment	-	5	14	7.5	12
River gravel					
16-8 mm	-	-	609	-	-
8-4 mm	540	-	431	540	-
2-4 mm	363	79	304	363	-
1-2 mm	272	158	216	272	-
.5-1 mm	272	158	151	272	-
.25-.5 mm	234	237	107	234	-
.125-.25 mm	127	158	95	127	822
Lytag[3)] 12-8 mm	-	484	-	-	-
Lytag 8-4 mm	-	264	-	-	-
Steel-fibres					
Dramix OL 25/.5	-	-	-	157[4)]	-
28 day strength					
f_c [MPa]	46.6	55.8	81.9	46.5	66.9
f_{spl} [MPa]	3.1	3.8	4.6	3.9	-
f_b [MPA]	-	-	-	-	11.8

1) all quantities are given in kg/m^3, except the w/c-ratio which is the ratio by weight.
2) solid solution only
3) the Lytag aggregates were saturated in water before casting. The given values are inclusive of 15% and 14% water for the 4-8 mm and 8-12 mm grading respectively.
4) quantity = 2% of total volume

In each batch two large beams of size 840*260*100 mm^3 and two small beams of size 440*160*100 mm^3 were casted. After two days of casting the specimens were demoulded and cured under water. After 14 days the beams were sawn to the required size as given in Figure 1b. After sawing, the specimens were again placed in a fresh water basin. After 28 days of casting the specimens were taken out of the water and left in the laboratory to dry until testing. The age at testing varied between 76 and 106 days. The tensile specimens were sawn from plates (300*300*50 mm^3) that were casted in each batch in addition to the beams. The specimen size is shown in Figure 1a. The curing procedure was the same as for the beams. From each batch three tensile specimens were tested at the same age as the beam specimens. Also included in each batch were six cubes (150 mm) for

determining the compressive and splitting tensile strength of the material, except for the *hcp*. The results are given in Table 1. For *hcp*, the strength was determined from small prismatic specimens of size 40*40*160 mm^3. After failing these beams in three-point-bending, the two remaining specimen halves were loaded in compression (loading area 40*40 mm^2). The flexural strength and compressive strength are included in Table 1.

3 Experimental results

The differences in uniaxial tensile stress-deformation response for the various concretes, is shown in figure 3. The load-displacement curves of the four-point-shear tests are plotted in figures 4a,b. Of all the materials and beam sizes four experiments are performed. The curve of only one test is plotted. In the tests on the large fibre concrete beams the maximum load capacity of the machine was reached. Therefore the test had to be stopped.

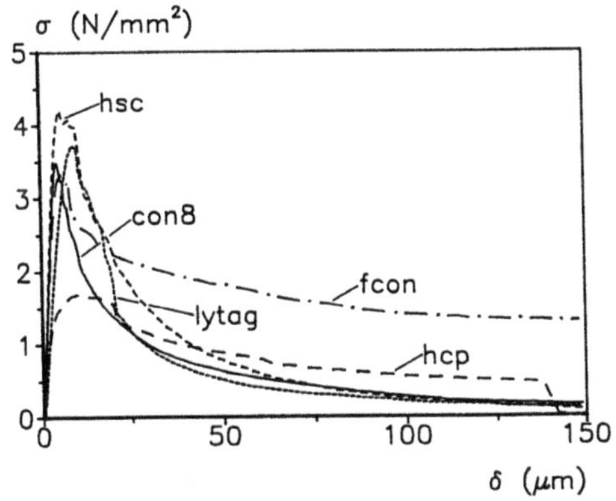

Fig.3. Stress-deformation curves of uniaxial tensile tests.

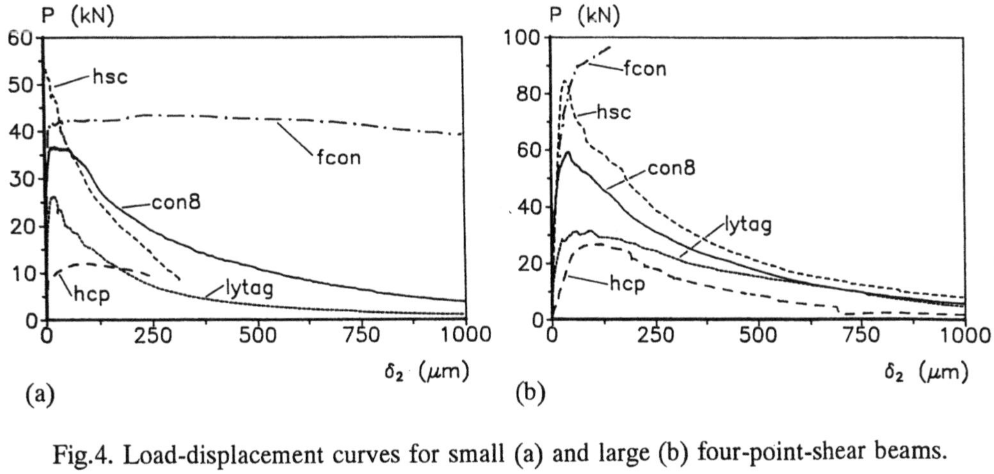

Fig.4. Load-displacement curves for small (a) and large (b) four-point-shear beams.

The cracks in the four-point-shear beams have all the same patterns for the different materials. The crack starts at the notch and continues to the opposite support as shown in figure 5a. In the small fibre concrete beams a more distributed crack pattern was observed as shown in figure 5b.

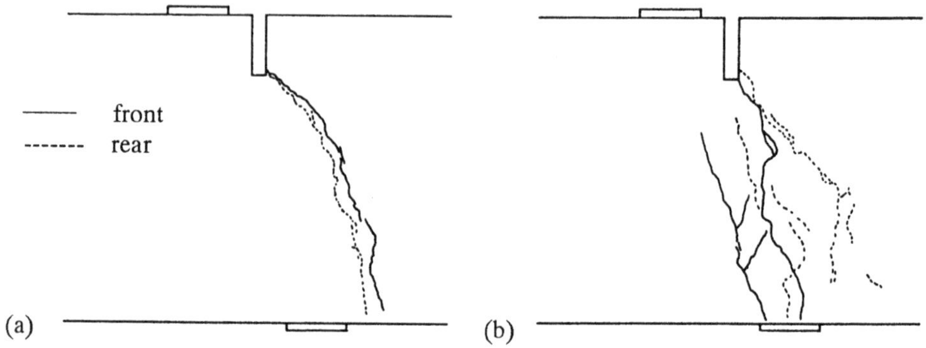

Fig.5. Crack patterns in four-point-shear beams; concrete (a) and fibre concrete (b).

4 Discussion of results

From the present and earlier (Schlangen & Van Mier (1991a)) results it must be concluded that crack growth takes place perpendicular to the maximum tensile stress. The shear loading leads to a rotation of the direction of the principal stresses, which forces the crack to start under an angle. With a propagating crack through the material the stress distribution will change which results in a curved crack pattern. By changing the boundary conditions of the beams the loading condition is changed, which results in a different stress state and therefore different crack patterns as shown in Van Mier et al. (1992).

In the discussion below of the behaviour of the different materials under tension and shear, always a comparison with the normal weight concrete will be made.

lytag:
In the tensile tests the *lytag* gives a peak stress which is a few percent higher compared to the *con8*. However in the four-point-shear tests the peak load is much lower for the *lytag*. For the small beams the difference is 30% and for the large beams even 45%. An explanation for this decrease in maximum load could be a difference in moisture content. To check this the specimens were put into the oven to dry after the test. After 30 days, when there was no further decrease in weight, the tensile specimens lost 4% of their weight and the large beams 6.7%. Because of a larger thickness of the beam specimens they contain more water at the time of testing. This probably leads to a stress gradient over the cross section and therefore to a lower peak load, as indicated earlier by Wittmann & Roelfstra (1987).

high strength concrete:
Comparing the tensile strength of *hsc* with that of *con8* shows that the increase is approximately 20%. In contrast, the compressive strength increases with 75%. A reason for this could be that the tensile specimens are all pulled in the direction of casting. Due to bleeding under the aggregate particles this is maybe the weakest direction.

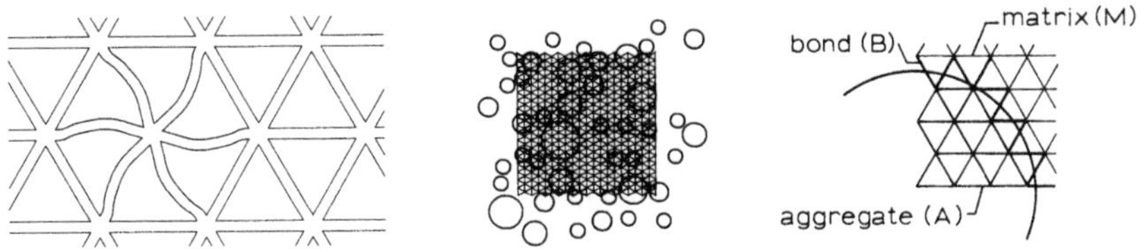

Fig.6. Triangular lattice and projection on generated grain structure.

For introducing the heterogeneity of concrete a grain structure is generated. For the distribution of aggregate particles in the concrete mix a Fuller curve has been chosen. Using statistical formulas the distribution of circle diameters in a certain cross section can be generated. The triangular lattice is projected on top of the generated grain structure, and different strength and stiffness are assigned to the respective beam elements in the matrix, the aggregates and the bond zone, see Figure 6.

5.2 Simulations

First a simulation of the uniaxial tensile test on normal weight concrete is carried out. This simulation is used to determine the input parameters for the model. The parameters that gave the best fit with the experiments are shown in Table 2. For more details about the influence of the individual parameters see Schlangen & Van Mier (1992a). The simulated load displacement response is always too brittle compared to the experimental outcome. Possible explanations for this can be found in the fact that the three dimensional fracture process is simulated in two dimensions and because the smaller aggregates are excluded from the generated grain structure, Schlangen & Van Mier (1992b).

Table 2. Input parameters for simulations.

General:	beam elements: l=5/3 mm, b=specimen thickness, h=0.68*l aggregates in grain structure: 3-8 mm α = 0.005 ß = 2.0			
I	$\sigma_A = 10$ $E_A = 70$	$\sigma_M = 5.0$ $E_M = 25$	$\sigma_B = 1.25$ $E_B = 25$	MPa GPa
II	$\sigma_A = 1.25$ $E_A = 30$	$\sigma_M = 5.0$ $E_M = 30$	$\sigma_B = 2.5$ $E_B = 30$	MPa GPa
III	$\sigma_A = 10$ $E_A = 70$	$\sigma_M = 5.0$ $E_M = 25$	$\sigma_B = 2.5$ $E_B = 25$	MPa GPa

Using the same input parameters the four-point-shear experiments are simulated. Only the part in the mesh where cracks are expected to grow contains the triangular lattice. The remainder of the specimen is modelled using isoparametric eight-noded plane stress elements available in DIANA. Only the small specimen size is simulated. The input parameters are varied in order to simulate the different concretes. The parameters of simulation I represent the normal weight concrete. The high strength concrete is simulated by increasing the strength of the bond zone (simulation III). For the lytag lightweight

In the four-point-shear tests an increase of approximately 43% of the maximum load is observed for both beam sizes. Here the cracks start under an angle with the casting direction. At failure the direction of the crack is perpendicular to the casting direction. This supports the above mentioned explanation for the relatively small increase in tensile strength compared to the compressive strength increase.

fibre concrete:
In the *fcon* the cracks are more distributed as compared to *con8*. Due to the fibres the stress will be redistributed, which leads to a wider crack branch as shown in figure 5b. The same tendency was found by Arslan et al. (1991) in shear tests on fibre concrete. From the stress-deformation curves of the tensile tests it can be seen that the maximum stress is almost equal for *fcon* and *con8*, but the descending branch is higher for *fcon*. There is some crack opening needed before the fibres start carrying load. In the beam test the crack starts opening at one side (at the notch). If the maximum load is reached in the small beams the crack opening at the notch has already reached such a value that the fibres start transferring stress. Therefore almost no stress drop after the peak load takes place. In the large beams, before the peak load is reached, the crack opening at the notch is already so large that the fibres start transferring stress which increases the peak load. The capacity of the machine was not enough to break the large beams.

hardened cement paste:
In the *hcp* specimens, both in the prisms for the tensile tests and in the four-point-shear beams, a large amount of shrinkage cracks could be observed before any loading was applied. During the hardening process the temperature in the material will increase, because of the large cement contents. After demoulding the specimens are put in a water basin. Probably the temperature difference between the specimens and the water was too large which caused most of the shrinkage cracks. Alternatively the strength development of the *hcp* was not large enough at the time when the specimen were taken from the water to withstand the hygral stresses. Due to these cracks the maximum load that was measured in the tests was considerably lower than for *con8*.

5 Numerical simulations

5.1 Lattice model

The four-point-shear experiments are simulated with a lattice model, Schlangen & Van Mier (1991b). In the model the material is schematized as a lattice of brittle breaking beam elements. The model is based on a linear elastic finite element model (DIANA). Fracturing of the material takes place by removing in each load step the beam element with the highest stress (relative to its strength). The stress is calculated from the fracture law:

$$\sigma_t=\beta*(F/A+\alpha*(|M_i|,|M_j|)_{max}/W)$$

where F is the normal force in the beam element, M_i and M_j are the bending moments in nodes i and j of the beam element respectively, A=b*h is the cross-sectional area of the beams, and $W=b*h^2/6$. The coefficient α is introduced in order to select a failure mode where bending plays either a dominated or a restricted role. The coefficient ß is a scaling factor for the global maximum stress.

concrete the strength of the aggregates is decreased and the strength of the bond zone higher than for normal weight concrete (simulation II). The aggregates are now the weakest link. The values of strength and stiffness are listed in Table 2.

5.3 Numerical results

The final crack patterns for the three simulations of the four-point-shear beams are plotted in figure 7. The load displacement curves of the simulations are compared in figure 8 with the experimental curve for the normal weight concrete.

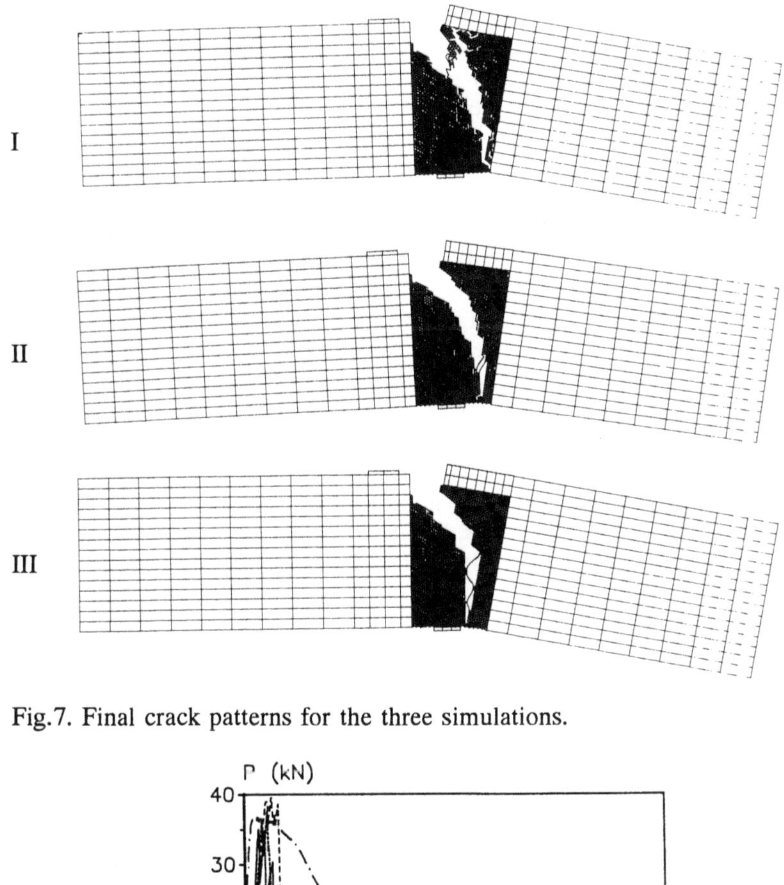

Fig.7. Final crack patterns for the three simulations.

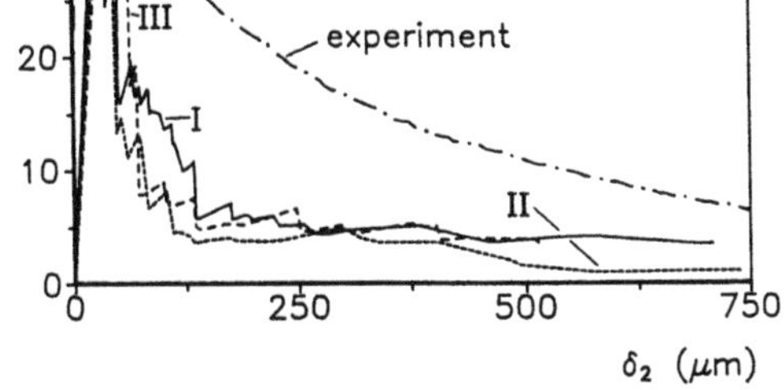

Fig.8. Load displacement curves for the three simulations compared with the experimental curve for normal weight concrete.

In figure 7 it can be seen that if the bond strength is increased (III) the crack path is more straight and there is less crack branching and bridging as compared to the simulation for normal weight concrete. In the simulation of the lytag concrete (II) first the beam elements inside the aggregate are removed from the mesh. Thereafter the matrix elements reach their maximum stress. If the calculated load-displacement curves are compared with the experiments it can be seen that the response is again too brittle. Possible explanations for this are discussed above. If the bond strength is doubled (III) the maximum load of the specimen is only increased with about 10%. This means that the maximum load is mainly governed by the strength of the matrix. The same tendency is observed in a large survey of experiments by Mindess (1989).

Here only an indication of the possibilities of the model is given. However, to capture the real fracture process it is necessary to do three dimensional simulations. Yet, this leads to an enormous increase in computer time.

6 Conclusions

It is shown that fracture grows under mode I, even when the external loading is mixed-mode. The influence of the shear loading is a rotation of the principal stress.

To compare the fracture process in different concretes, it is important to include in the discussion influences due to differences in manufacturing, drying and hardening.

Different fracture behaviour is often the result of the difference in microstructure, which is obvious in the case of fibre concrete, but lytag and high strength concrete also have a different fracture mechanism than normal weight concrete as shown by Van Mier (1991).

The lattice model has proved to be a promising tool to study the fracture behaviour in concretes with different microstructures. However to capture influences of gradients due to moisture, shrinkage and temperature three dimensional simulations are necessary.

7 Acknowledgement

The authors are greatly indebted to Mr. A.S. Elgersma for his substantial assistance in carrying out the experiments. The work was sponsored through a grant from the Dutch Technology Foundation (STW).

8 References

Arslan, A. Hughes, T.G. and Barr, B.I.G. (1991) Mixed mode fracture - including torsion in a new compact test specimen geometry, in *Fracture Processes in Concrete, Rock and Ceramics* (eds. J.G.M. van Mier, J.G. Rots and A. Bakker), Chapman & Hall Publishers, London/New York, 1991, pp. 737-746.

Carpinteri, A. Valente, S. Ferrara, G. Melchiorri, G. (1991) Is mode II fracture toughness a real material property?, in *Proceedings International Conference on "Mixed Mode Fracture and Fatigue"*, July 15-17, 1991, Vienna, Austria.

Herrmann, H.J. (1988) Introduction to modern ideas on fracture patterns, in *Random Fluctuations and Pattern Growth: Experiments and Models* (eds. H.E. Stanley and N. Ostrowsky), Kluwer Academic Publishers, Dordrecht, pp. 149-160.

Hordijk, D.A. (1991) Local approach to fatigue of concrete. *Doctoral Thesis*, Delft University of Technology, Delft, pp. 210.

Mindess, S. (1989) Interfaces in Concrete, in *Materials Science of Concrete I* (ed. J.P. Skalny), The American Ceramic Society, Inc., Westervill (OH), pp. 163-180.

Schlangen, E. and Van Mier, J.G.M. (1991a) Boundary effects in mixed mode I and II fracture of concrete, in *Fracture Processes in Concrete, Rock and Ceramics* (eds. J.G.M. van Mier, J.G. Rots and A. Bakker), Chapman & Hall Publishers, London/New York, 1991, pp. 705-716.

Schlangen, E. and Van Mier, J.G.M. (1991b) Lattice model for numerical simulation of concrete fracture, in *Proceedings Int'l. EPRI Conference on Dam Fracture* (eds. V.E. Saouma, R. Dungar and D. Morris), EPRI, pp. 511-527.

Schlangen, E. and Van Mier, J.G.M. (1992a) A new lattice model for predicting localized fracture in concrete, in *Localized Damage II, Vol I: Fatigue and Fracture Mechanics* (eds. M.H. Aliabadi, D.J. Cartwright and H. Nisitani), Computational Mechanics Publications, Southampton/Boston and Elseviers Applied Science, London/New York, pp 425-438.

Schlangen, E. and Van Mier, J.G.M. (1992b) Lattice model for simulating fracture of concrete, in *Proceedings 1st Bolomey-Workshop*, Zürich, July 16-17, ETH-Zürich, pp. 164-174.

Van Mier, J.G.M. (1991) Crack face bridging in normal, high strength and lytag concrete, in *Fracture Processes in Concrete, Rock and Ceramics* (eds. J.G.M. van Mier, J.G. Rots and A. Bakker), Chapman & Hall Publishers, London/New York, 1991, pp. 27-40.

Van Mier, J.G.M. Schlangen, E. and Nooru-Mohamed, M.B. (1992) Shear fracture in cementitious composites, part I: experimental observations, in *Proceedings FraMCoS I* (ed. Z.P. Bažant), Breckenridge, Colorado, June 1-5, 1992, Elsevier Publishers, London/New York, pp. 659-670.

Wittmann, F.H. and Roelfstra, P.E. (1987) Constitutive Relations for Transient Conditions, in *Proceedings IABSE Colloquium Delft 1987*, pp. 239-259.

15 APPLICATION OF FRACTURE MECHANICS TO THE STUDY OF CRACK PROPAGATION IN CONCRETE STRUCTURES USING A GRANULAR MICROCRACKED MODEL

J. WANG, P. NAVI and C. HUET
Laboratoire de Matériaux de Construction, Département des Matériaux, École Polytechnique Fédérale de Lausanne, Switzerland

Abstract
Fracture process in concrete structures is studied by applying the principles of linear elastic fracture mechanics with taking the material's heterogeneity and pre-existing microcracks into consideration. At the present stage the hard inclusions in the concrete body are represented by circles, around which microcracks are introduced. Crack propagation in such granular microcracked concrete body are investigated numerically by the finite element method. Two examples concerning concrete specimens subjected to the direct tension are demonstrated. The numerical results that are presented include the load-displacement curves, the trajectories of the crack propagation and the K-a curve. These results are discussed by comparison with the case of the same specimen made of a homogeneous material.
Keywords: Crack propagation, Fracture mechanics, Numerical simulation, Material's heterogeneity, Microcracks, Direct tension.

1 Introduction

Fracture process of concrete structures deviates from linear elastic fracture mechanics (LEFM) due to the development of a relatively large fracture process zone in front of the crack. Within the fracture process zone there exists the accumulation of microcracks. In contrast to the real concrete behavior, LEFM considers infinitely small fracture process zone in front of the crack. This is the reason that to study the fracture of concrete structures macroscopically, nonlinear approaches are needed. But with nonlinear approaches it is difficult to obtain the results that are independent of the geometry of the specimen and the loading conditions. Since the nonlinear behavior of concrete may be attributed to its granular structure, it may be assumed that LEFM is adequate to examine the fracture of concrete if its microscopic structure is considered. On the other hand the application of the energy principles which allows the study of stability of crack propagation is not confined to the linear behavior of the material. By applying the energy principles of LEFM, Huet [1973, 1974] has shown that the stability of crack propagation and the post-peak part of load-displacement curve for a homogeneous material are highly dependent on the geometry of specimen, the mode of loading control and the crack size, and not on the material itself. It is very important to investigate what happens to stability and post-peak behavior in a granular material from this point of view.

In this investigation a two dimensional numerical model of concrete considered as a granular material with pre-existing microcracks given in Wang et al. [1992a] is applied to the study of the fracture process in the concrete specimens subjected to the direct tension. The crack trajectories, the global load-displacement curves and the curve of

Fracture and Damage of Concrete and Rock – FDCR-2.
Edited by H.P. Rossmanith. © 1993 Taylor & Francis. ISBN 0 419 18470 8.

stress intensity factor - crack length (K-a curve) obtained for two examples are presented. The results are discussed mainly from the stability viewpoint and by comparison with the homogeneous case with one central crack.

2 Numerical model

2.1 Numerical generation of pre-microcracked concrete structure

Concrete is generally treated as a multi-phase composite which is composed of coarse and fine aggregates, hardened cement paste and an interface layer of about 0.1 mm thickness. In our model the aggregates are taken to be circles. Their diameters and positions are calculated with an algorithm similar to that used by Roelfstra [1989]. The diameters of inclusion follow the Fuller curve, whereas the coordinates of the center of the inclusion are generated randomly. Microcracks are introduced only in the interface layers, which is the weakest phase in normal concrete. Two random variables are used to determine the length and location of the microcracks. One is the opening angle relative to the center of the inclusion which gives the length of the microcrack. The other is the bisection of the opening angle which determines the position of the microcrack. Fig. 1 shows one example of computer generated concrete structure with 21 inclusions and pre-existing microcracks.

2.2 Automatic mesh generation process

A computer code of automatic mesh generation has been developed to triangulate the microcracked composite described in 2.1. This computer code of mesh generation creates the points representing the geometry of the inclusions, the microcracks and the other points within the specimen. These points are then connected to each other into triangular elements. To decrease the computational error an algorithm for improving the element shape has been developed. In order to calculate accurately the stress intensity factors, the elements around the crack tip are always kept with the optimal shapes. Fig. 2 gives the resultant mesh generation corresponding to the concrete structure shown in Fig. 1. After each incremental crack propagation, the structure is remeshed automatically.

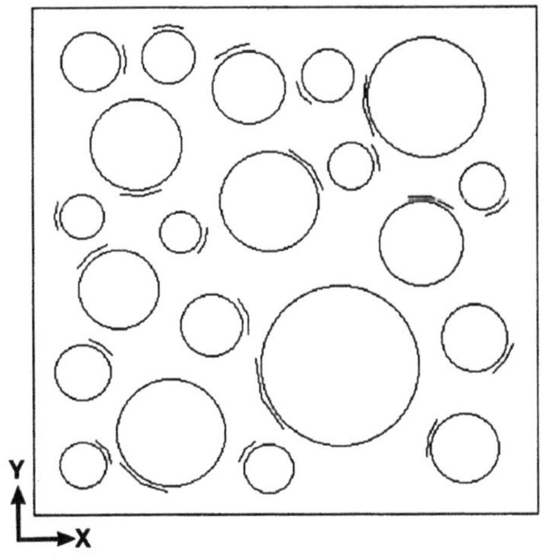

Fig. 1 Computer generated concrete structure.

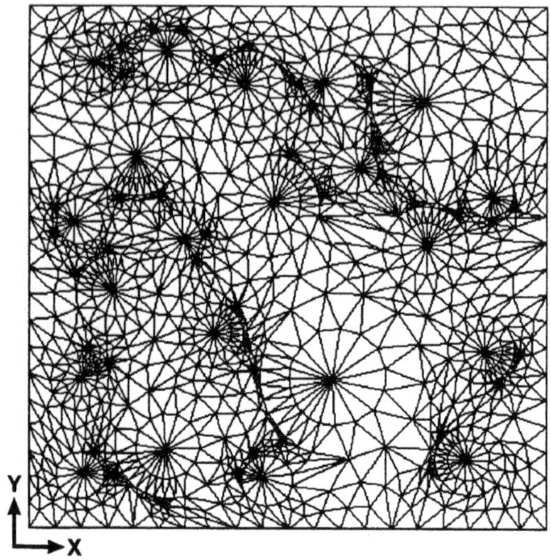

Fig. 2 Resultant mesh generation corresponding to the concrete structure shown in Fig. 1.

2.3 Numerical calculation of stress intensity factors and crack propagation
The stress intensity factor K_I (mode I) and K_{II} (mode II) are calculated from the values of the node displacements of quarter-point singular elements located around each crack tip, Freese et al. [1976] and Ingraffe et al. [1985]. This particular element produces the stress singularity with the order of (−1/2). When the values of K_I and K_{II} are known, the angle of crack propagation θ_m and the stress intensity factor $\bar{K}$ of mixed mode are calculated according to the maximum principal stress criterion given by Erdogan et al. [1963] as follows:

$$\tan\frac{\theta_m}{2} = \frac{1}{4}\frac{K_I}{K_{II}} \pm \frac{1}{4}\sqrt{\left(\frac{K_I}{K_{II}}\right)^2 + 8} \tag{1}$$

$$\bar{K} = K_I\left(\cos\frac{\theta_m}{2}\right)^3 - 3K_{II}\left(\cos\frac{\theta_m}{2}\right)^2\sin\frac{\theta_m}{2} \tag{2}$$

The corresponding necessary conditions for crack initiation and instability are given by:

$\bar{K} \geq K_{IC}$ (crack propagation) (3)

$(\partial\bar{K} / \partial a) > 0$ (unstable crack propagation) (4) where a is the crack length and K_{IC} is the critical stress intensity factor of the material. Equation (1) for the propagating angle is only valid when the crack tip is far from the interface between the hardened cement paste and the inclusions. It is known from analytical solutions (Murakami [1987]) that when the crack tip approaches the interface between two different perfectly bonded materials, the stress intensity factors decrease if the material in front of the crack has a greater shear modulus. The stress intensity factors become zero when the crack terminates at the interface. To consider this, it is assumed that there is a small distance between the circumference of the inclusion and the crack tip whose value is chosen to be 0.6 mm, and that crack is not allowed to enter the inclusion. This assumption is close to the reality of crack propagation in normal concrete, for which the aggregates are generally stiffer and tougher than the hardened cement paste.

When a crack approaches the interface, there are two possible paths for its further propagation. One is that the crack goes around the inclusion to the right hand of its original position, the other to the left, as shown in Fig. 3. In order to decide which path the crack takes we apply the analytical solution given by Wang [1992b] for the semi-infinite crack terminating at the material's interface. The crack configuration shown in Fig. 4 is composed of an inclined semi-infinite crack terminating at the interface between two different materials. According to our analytical solution for the case where the shear modulus of material 1 is less than that of material 2, the cleavage stress ($\sigma_{\theta\theta}$) is much larger on the plane of interface with an obtuse angle between the crack and the

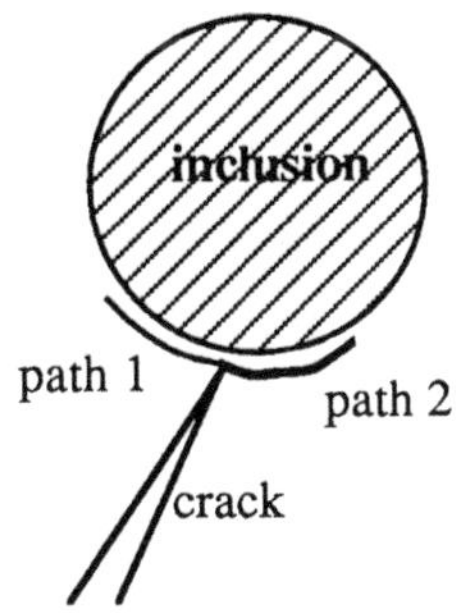

Fig. 3 Two possible propagation paths of crack approaching an inclusion.

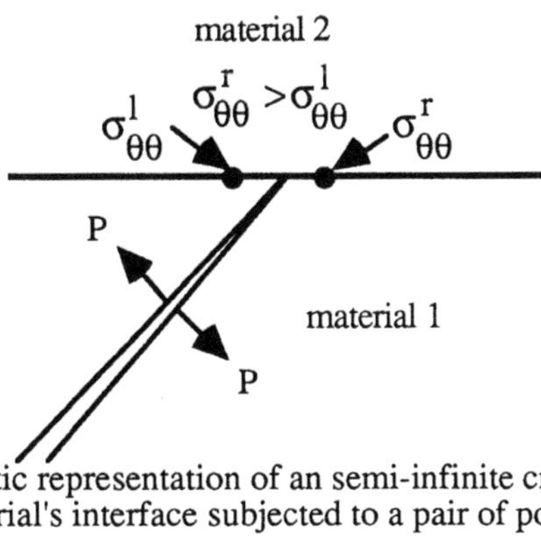

Fig. 4 Schematic representation of an semi-infinite crack terminating at the material's interface subjected to a pair of point forces P.

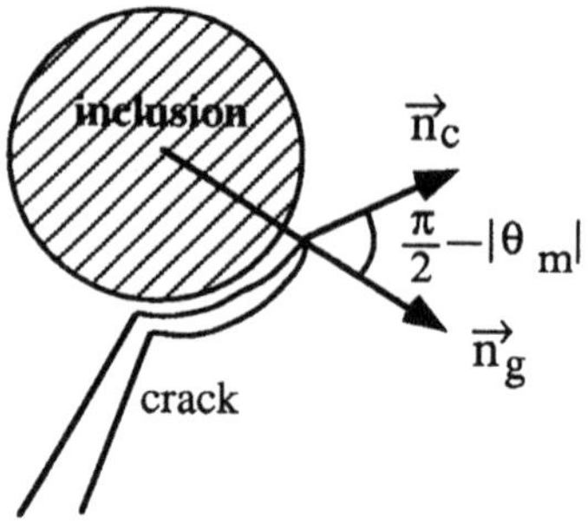

Fig. 5 Two unit direction vectors n_c and n_g at crack tip.

interface than the acute one. For the case shown in Fig. 4, $\sigma_{\theta\theta}$ at the interface on the right hand side of the crack is much larger than that on the left hand side, namely, $\sigma^r_{\theta\theta} > \sigma^l_{\theta\theta}$. For this reason we assume that the crack propagates around the inclusion in the direction which makes an obtuse angle between the original crack direction and the tangent to that inclusion, which is path 2 in Fig. 3.

For the case that a crack is around an inclusion, two unit direction vectors, as shown in Fig. 5, will determine whether the crack still propagates around the inclusion or deviates into the matrix. One of these unit vectors is the direction vector of crack propagation n_c calculated from equation (1). The other is the outer normal vector of the inclusion n_g. For $n_c \cdot n_g \geq 0$ the crack will deviate into the matrix and for $n_c \cdot n_g < 0$ the crack will still go around the inclusion.

During the fracture process the stress intensity factor K_I of mode I may become negative for some microcracks . The negative K_I means that the crack closes. For the case of crack closing, two nodes situated on the opposite crack surface are jointed together and become only one node. In this way the discontinuity of the displacement normal to the crack surface is eliminated. Thus, for the present stage the glide and the friction between the two crack surfaces are not taken into consideration. When negative K_I occurs, K_I and K_{II} are both forced to be zero.

To follow the fracture process in the numerical simulation, the incremental length of crack propagation is given prior to the analysis. Either external displacement or force can be used as the mode of loading control. The steps of numerical simulation in terms of displacement control are explained as follows:

a. An arbitrary initial displacement $\delta^{(0)}$ is imposed along the specimen boundary.

For this displacement the stress intensity factors of mixed mode $\bar{K}_i$ ($i = 1,\cdots n$ | n: the total number of crack tips) for all the crack tips are computed. The maximum value $\bar{K}_{max}$ is selected and the ratio of $(\bar{K}_{max} / K_{IC}) = A$ is calculated, where K_{IC} is the critical stress intensity factor of the material.

b. From equation :

$$\delta^{(1)} = \frac{\delta^{(0)}}{A} \tag{5}$$

the value of the displacement $\delta^{(1)}$ corresponding to the onset of the crack propagation is obtained. The external forces (reactions) which correspond to the displacement $\delta^{(1)}$ are computed.

c. All the cracks satisfying the relation:

$$K_{IC} \leq \frac{\overline{K}_i}{A} \leq K_{IC}\ (1+\ \varepsilon) \tag{6}$$

are forced to propagate simultaneously by a given increment. The amplitude interval ε is taken as some small percentage value.

d. For the new configuration of crack position, remeshing is performed. Then the above steps a, b, c are repeated.

The introduction of ε in equation (6) allows the possibility of simultaneous propagation of several cracks. In the following numerical examples, ε is given the value 4%.

The above mentioned numerical procedure of determination of displacement and force provides us with the *equilibrium curve* (or *Griffith locus*) of Berry [Berry, 1960], which is valid for all types of loading. The analysis of the stability of crack propagation under different types of loading can be made from this equilibrium curve directly.

3 Numerical analysis for direct tension

The specimen shown in Fig. 1 is loaded with a uniform displacement along its two opposite sides in the x direction. The corresponding tensile reaction forces are computed for each loading step. The Young's modulus and Poisson's ratio of the inclusion and the matrix are taken as 60 GPa, 0.2 and 20 GPa, 0.2 respectively. The critical stress intensity factor K_{IC} of the matrix is given as 20.0 $N/mm^{-3/2}$. Plane stress condition is assumed in the calculation.

Figs. 6 and 7 give the crack trajectory and the nominal stress-displacement curve respectively. The original positions of microcracks are represented by the bold lines in Fig. 6. The numbers along the crack trajectory illustrated in Fig. 6 are the successive steps of crack propagation during the simulation. For example, number 1 indicates the first step of propagation. The corresponding values of stress and displacement are shown with the same number indicated in Fig. 7. The stress-displacement curve for the homogeneous matrix material with one crack at the center of the same specimen is also shown by dashed line in Fig. 7 for the purpose of comparison. For all the possible loading programs, the stability domain in Fig. 7 for the heterogeneous microcracked body corresponds to the area encompassed by the loading line before the peak load, the equilibrium curve and the displacement axis. A loading program is globally stable if the representing point starting from the equilibrium curve enters into the stability domain and remains inside it to the end. The local stability means that the representing point having entered the stability domain leaves it by crossing the equilibrium curve. No stable propagation can be achieved for the homogeneous case under fixed displacement condition. Crack propagation in heterogeneous microcracked body from peak load to step 4 is similar to that of the homogeneous body with one central crack, except for step 2. After step 4 the crack branch II shown in Fig. 6 becomes a single-edged crack. It is known that the stability of single-edged crack propagation in homogeneous body under fixed displacement is dependent on its length. For a short single-edged crack, propagation is unstable. Whereas for a longer one, propagation becomes stable. This seems to be the reason for which the crack propagation after step 4 shown in Fig. 7

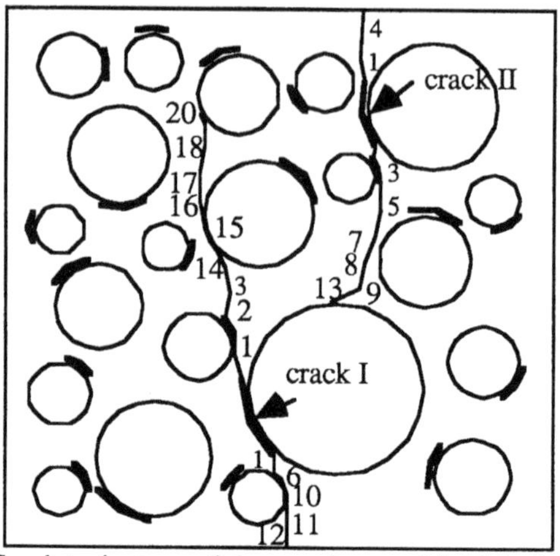

Fig. 6 Crack trajectory of the specimen with 21 inclusions.

turns out to be stable under fixed displacement conditions, especially after step 8, for which the length of the crack branch II is longer than that at step 4. The presence of inclusions generates several local stable propagation regions corresponding to the indentations on the equilibrium curve indicated by number 2, 5, 14 indicated in Fig. 7. Within these local stable regions, stable propagation can be obtained under either fixed load or fixed displacement condition. Therefore the stability of crack propagation in a granular body is not only dependent on the type of loading, the crack size and the specimen geometry, but also on the material's structure itself.

In order to examine more deeply the stability of crack propagation, the variation of relative stress intensity factor ($\bar{K} / K_{IC}$) in terms of crack length a for a

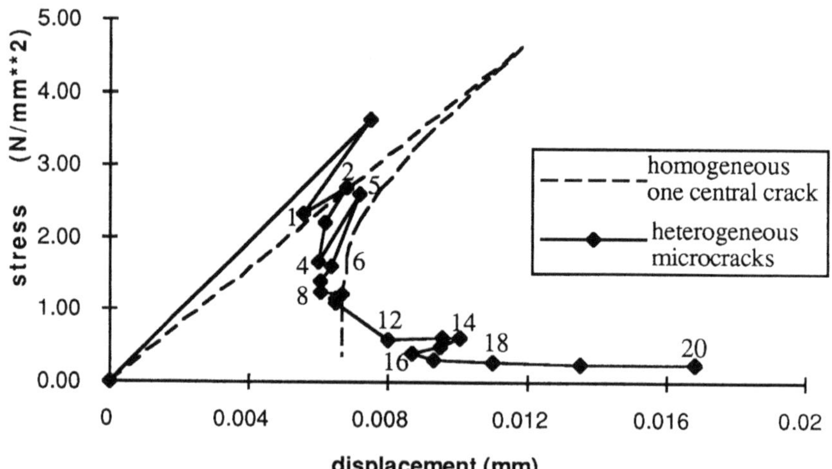

Fig. 7 Nominal stress-displacement curve for the specimen with 21 inclusions.

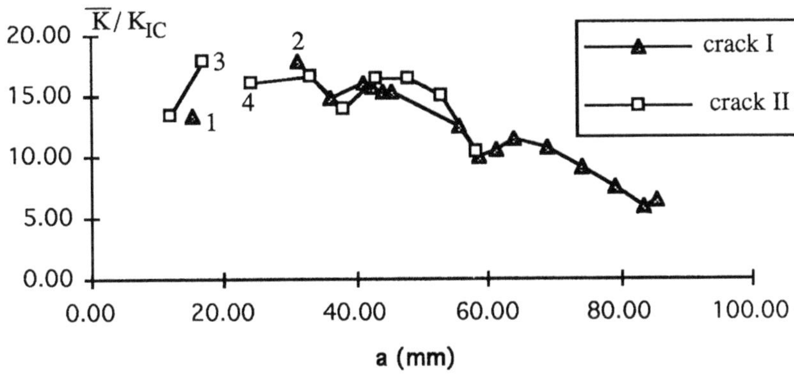

Fig. 8 Relation between relative stress intensity factor of mixed mode $\overline{K} / K_{IC}$ and crack length a for the value of the imposed displacement equal to 0.1 mm.

constant displacement (K-a curve) is shown in Fig. 8, where crack I and crack II are the two crack branches indicated in Fig. 6. We note that when the tip of a propagating crack meets an existing microcrack in the material, a jump occurs in the crack length and the values of the stress intensity factors cannot be evaluated. Therefore as shown in Fig. 8 the values of the stress intensity factors before the jump (point 1 for crack I and point 3 for crack II) and after the jump (point 2 for crack I and point 4 for crack II) cannot be connected to each other. It is recognized from this figure that crack I and crack II experience several stable and unstable propagations during the whole fracture process. The unstable propagation corresponds to the branches of the curve with positive slope, while the stable propagation corresponds to the ones with negative slope. It has to be pointed out that other K-a curves for other types of loading can also be obtained from the same set of numerical data.

The influence of material's heterogeneity was studied by comparison with another numerical example. For this example there are only 12 inclusions and the volume

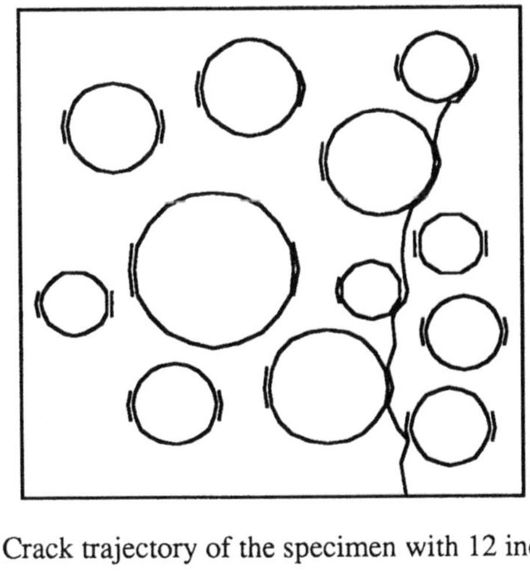

Fig. 9 Crack trajectory of the specimen with 12 inclusions.

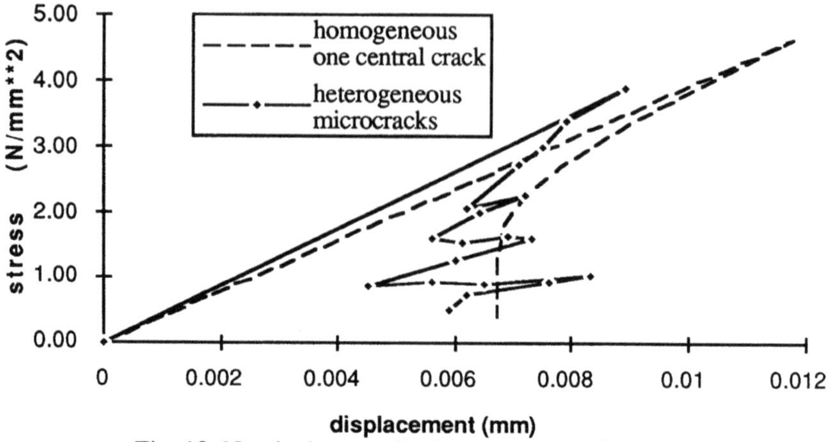

Fig. 10 Nominal stress-displacement curve for the specimen with 12 inclusions.

content of inclusion is 33%. Each inclusion has two microcracks. All the microcracks are perpendicular to the direction of the external load. Fig. 9 gives the crack trajectory and Fig. 10 the stress-displacement curve. It appears that for this case the crack propagation at fixed displacement corresponding to the peak load is totally unstable and that the resultant crack trajectory has only one branch. Nevertheless, the influence of the hard inclusions provides a succession of local stability regions under either fixed displacement or fixed load condition. This local stability might be reached, for instance, by a fast unloading after peak load and then reloading.

4 Conclusions

Crack propagation in concrete is highly influenced by its granular microstructure. Numerical simulation is a useful tool to study this influence. A granular microcracked model has been developed in order to take into account the material's heterogeneity and pre-existing microcracks. This model takes concrete as a two-phases material: hard inclusions embedded in the matrix. It is assumed that the principles of linear elastic fracture mechanics can be applied to the matrix. The equilibrium load-displacement curve is obtained with the finite element method. This equilibrium curve provides us with the possibilities to study the stability of crack propagation under different loading schemes, which might be difficult to achieve in experiments. We can also study the influences of several other parameters on the stability of crack propagation. These parameters can be the volume content, distribution and maximum size of inclusions, the density and size of microcracks, the presence of notch, shrinkage in the matrix, etc. The examples studied for direct tension show that the global trends of the behavior of the homogeneous material are conserved in the heterogeneous one. But the presence of the inclusions results in local stable regions on the equilibrium curve which can hinder crack propagation. It also provides the possibility for crack multiplication and/or branching when the volume content of inclusion is large enough. The distribution of microcracks may result in a transition from a central crack regime to a single-edged one. The former gives instability under the condition of both fixed load and fixed displacement corresponding to peak load, while the latter gives a global stability for fixed displacement if the crack length is long enough. Our numerical tool provides also the possibility to explain and study the fast increase of specific fracture energy during the beginning of crack propagation (R-curve) observed in experiment. An application of this numerical tool to the analysis of anchor bolts is presented in another contribution to this conference, Wang et al. [1992c].

5 Acknowledgment

Financial support from Swiss National Science Foundation is gratefully acknowledged.

6. References

J.P. Berry (1960), Some kinetic considerations on the Griffith criterion for fracture, I:, **J. Mech. Phys. Solids**, 8, 194-206, II: ibid, p. 207-216.

F. Erdogan, G.C. Sih (1963), On the crack Extension in Plates under Plane Loading and Transverse Shear, **J. Basic Engng.**, 85, pp. 519 - 527.

C.E. Freese, D.M. Tracey (1976), The Natural Isoparametric Triangle versus Collapsed Quadrilateral for Elastic Crack Analysis, **Int. J. Fracture**, 12, pp. 767 - 770.

C. Huet (1973), Méthode de détermination de l'énergie spécifique de rupture et application aux céramiques et à divers matériaux, I: Bases théoriques, **Cahiers du Groupe Français de Rhéologie**, novembre, t. 3, n° 3, pp. 128-142.

C. Huet (1974) Méthode de détermination de l'énergie spécifique de rupture et application aux céramiques et à divers matériaux, II: Conséquences pratiques et expérimentation , **Cahiers du Group Français de Rhéologie**, septembre, t. 3, n° 4, pp164-174.

A.R. Ingraffe, V. Saouma (1985), Numerical Modeling of Discrete Crack Propagation in Reinforced and Plain Concrete, in **Fracture Mechanics of Concrete** (Editors, G.C. Sih, A. Ditommaso), Martinus Nijhoff Publishers, 171-225.

Y. Murakami (Editor-in-chief) (1987), **Stress Intensity Factors Handbook**, Pergamon Press, pp. 429-640.

P. Roelfstra (1989), A numerical approach to investigate the properties of concrete: numerical concrete, Ph D dissertation No 788, Ecole Polytechnique Fédérale de Lausanne.

J. Wang, P. Navi, C. Huet (1992a), Numerical Study of Granule Influences on the Crack Propagation in Concrete, in **Fracture Mechanics of Concrete Structure** (Ed. Z. P. Bazant), pp.373-378.

J. Wang (1992b), Stress Distribution of an Inclined Semi-Infinite Crack Terminating at Interface between Dissimilar Materials, accepted for **Eighth International Conference on Fracture**, Kiev, June 8-14, 1993.

J. Wang, P. Navi, C. Huet (1992c), Finite Element Analysis of Anchor Bolts Based on Fracture Mechanics, accepted for **International Conference on Fracture and Damage of Concrete and Rock**, Vienna, Austria, November 9-13, 1992.

16 EXPERIMENTAL CRACK IDENTIFICATION OF MIXED-MODE BY ACOUSTIC EMISSION

M. OHTSU and M. SHIGEISHI
Department of Civil and Environmental Engineering,
Kumamoto University, Japan

Abstract
Source inversion technique of acoustic emission (AE) is developed as a SiGMA procedure (simplified Green's function for moment tensor analysis), which determines crack location, crack type, and crack orientation from AE waveforms recorded. For the easy use in experimental crack identification, the procedure is modified to be available for two-dimensional models (SiGMA-2D). In the present paper, crack identification in uniaxial compression tests of PMMA plates with inclined slits is performed by AE waveform analysis. In the experiments, cracks of mixed mode are nucleated and observed. The analysis shows that crack locations (AE sources located) are closely distributed around cracks propagated from the slit. In these sources, both types of tensile cracks and shear cracks are observed. The crack orientations of both cracks are in remarkable agreement with crack trajectories observed. The results confirm that the SiGMA-2D analysis is readily applicable to elucidating fracture mechanisms of plate models.
Keywords: Acoustic Emission, Source Inversion, Mixed-Mode Crack, PMMA Plates.

1 Introduction

Acoustic emission (AE) technique is recently applied to experiments on fracture mechanics by Berthaud et al.(1991), Nomura et al.(1991), and Chen et al.(1992). They locate AE sources to identify the crack front and further the fracture process zone. Due to the limited accuracy of AE location technique, the detection of the fracture process zone is marginally attained (Mindes, 1991).

A theoretical AE waveform analysis has been studied extensively (Ohtsu, 1982; Ohtsu and Ono, 1986). It is clarified by Ohtsu (1987) that a moment tensor

Fracture and Damage of Concrete and Rock – FDCR-2.
Edited by H.P. Rossmanith. © 1993 Taylor & Francis. ISBN 0 419 18470 8.

analysis can provide information on crack kinematics. By using a source inversion based on the moment tensor, crack types and crack orientations are determined, in addition to crack locations. When the source inversion procedure is applied to experiments on fracture mechanics, further information on microfracturing zone could be obtained and thus the limitation of the location accuracy is compensated.

An application of the moment tensor analysis to damage assessment in concrete was attempted by Ouyang et al. (1991), although theoretical treatment was not well-developed. Their procedure proposed was neither practical nor available for general AE tests. In order to determine the moment tensor components from multi-channel AE waveforms and to analyze crack kinematics, a computer software is developed by Ohtsu (1991), and named as a 'SiGMA' software (simplified Green's function for moment tensor analysis). The procedure was successfully applied to AE waveforms detected at in situ hydrofracturing tests.

In the present paper, the SiGMA procedure is modified into the two-dimensional case (SiGMA-2D). Compared with visual observation, crack identification by AE source inversion is performed in the 2-D PMMA (poly-methyl methacrylate) model experiments.

2 Experiment of PMMA plate

A two-dimensional (2-D) PMMA (poly-methyl methacrylate) model with an internal slit is employed. Under uniaxial compression, mixed-mode crack propagation from the slit is observed. A set-up for AE detecting system is shown in Fig. 1. In uniaxial compression

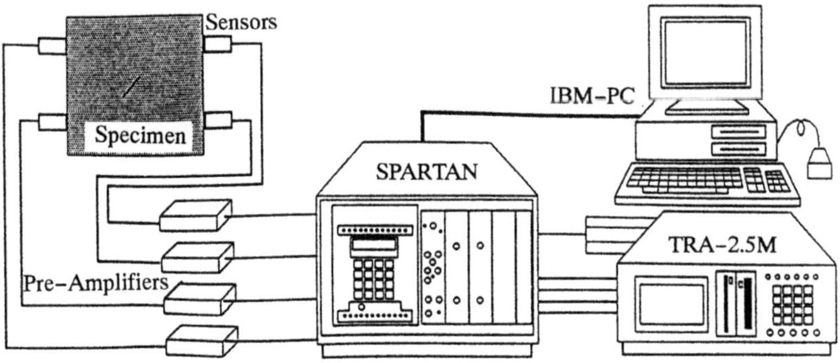

Fig. 1 A setup for AE detecting system.

tests of PMMA plates, four AE sensors are attached to both sides of the specimen. The sensor is of flat response in the selected frequency band between 50 kHz and 300 kHz for recording AE waveforms. Total gain of the measurement is 60 dB. One AE waveform is digitized into 2 kilo-word in length at 2 MHz sampling frequency. To monitor AE activity during the test, AE event counting is performed at channel 1. The configuration of the specimen and sensor locations are shown in Fig. 2. A PMMA plate of dimension 100 mm

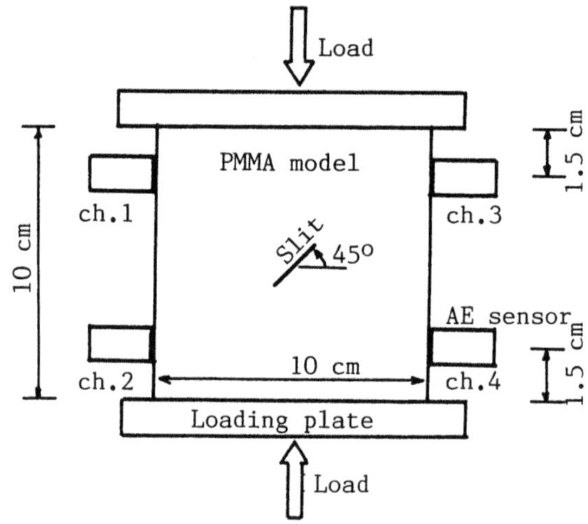

Fig. 2 A PMMA model and AE sensor locations.

x 100 mm x 10 mm contains a through-thickness slit of 2 mm wide and 20 mm long. The inclined angle of the slit is 45°. P-wave velocity measured is 2941 m/s and Poisson's ratio is 0.35.

In the experiments, the uniaxial compression tests of the PMMA plates were conducted, detecting AE waveforms. To decrease the friction between the specimen and the loading plates, teflon sheets of 0.5 mm thickness were inserted along with silicone grease. Load was applied at the rate of 150 kg/min. in both loading and unloading. When high AE activity was observed at channel 1, load was either kept constant or slightly decreased until few AE events were observed.

3 Moment tensor analysis

A waveform analysis starts with AE location procedure, determining the coordinates of a crack. In the 2-D case, the AE location is performed by recording AE waveforms at more than four channels. From the arrival time differences of the first motions at the four channels, eventually a set of two linear algebraic equations is obtained with two unknowns on crack locations in the x-y coordinates. Then, a converged solution on crack location is determined from the iteration procedure (Ohtsu and Ono, 1988).

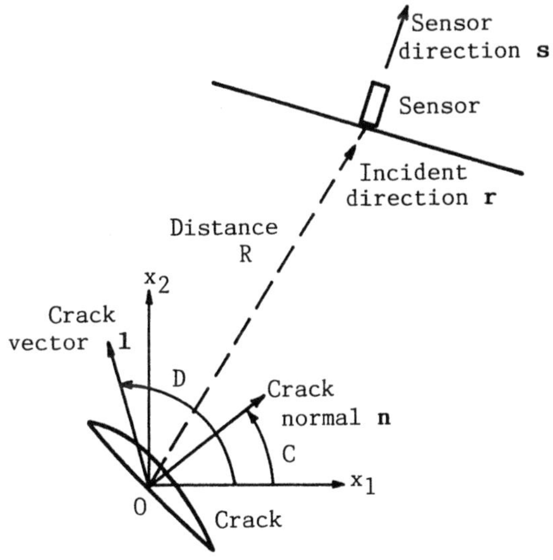

Fig. 3 Crack geometry.

As shown in Fig. 3, one AE waveform due to a crack at a known location is detected at one sensor, of which vibrating direction is denoted by vector **s**. The distance from the crack to the sensor is R and its direction vector is $\mathbf{r}(r_1,r_2)$. From the theory (Ohtsu, 1987), the amplitude of the first motion, A, at the sensor is expressed, as follows;

$$A = C\ \mathrm{Re}(\mathbf{s},\ \mathbf{r})(r_1r_1m_{11} + 2r_1r_2m_{12} + r_2r_2m_{22})/R, \quad (1)$$

where C is material constant, $\mathrm{Re}(\mathbf{s}, \mathbf{r})$ is the reflection coefficient, and m_{ij} are moment tensor components.

When AE waveform due to one AE event is recorded at one sensor, the amplitude of the first motion is represented by eq. 1. After the crack location, all values are known except for three independent elements of the moment tensor m_{ij} in the 2-D problem. As discussed, the four-channel recording system is needed for the crack location. Thus, a set of four linear algebraic equations with three unknown m_{ij} is obtained from eq. 1, by substituting the amplitudes of the first motion into A at four sensors. Solving this overdetermined system, moment tensor components are determined for one AE event.

In the configuration of Fig. 3, theoretical moment tensor components are calculated. The direction of crack normal is **n**:(cosC, sinC) and crack motion is b**l**:(bcosD, bsinD), where the direction of crack motion is indicated as crack vector **l**. The moment tensor components are,

$$m_{ij} = bG\begin{pmatrix} 2\nu\cos(D-C)/(1-2\nu)+2\cos C\cos D, & \sin(C+D) \\ \sin(C+D), & 2\nu\cos(D-C)/(1-2\nu)+2\sin C\sin D \end{pmatrix}, \qquad (2)$$

where G is the shear modulus of elasticity and ν is Poisson's ratio. From eq. 2, it is obvious that moment tensor components contain information on crack vectors **l** and **n**, namely the angles between C and D, although these two directions are not readily separable. In this respect, the eigenvalues of the moment tensor present particular values, depending on the angle between vector **l** and vector **n**, namely cos(D-C). In the 2-D case, the following relation is obtained by Ohtsu and Shigeishi(1992),

$$\cos(D-C) = (1 - 2\nu)(1 + e)/(1 - e), \qquad (3)$$

where e is the ratio of the maximum eigenvalue to the minimum. In the case of a tensile crack, the direction of crack vector is identical to that of crack normal. As a results, angle D = angle C and cos(D-C) = 1. For a shear crack, the direction of crack vector is perpendicular to that of crack normal, and cos(D-C) = 0.

Based on these cosine values, the classification of crack type was attempted. Then, it was found that the classification was rather ambiguous, except for the extreme cosine values either close to unity or zero. In addition, the cosine values in eq. 3 were too sensitive to Poisson's value.

In the three dimensional (3-D) problem, the classification of crack type is rationally performed, by applying the unified decomposition of the eigenvalues (Ohtsu, 1991). Even in the present case where theoretical treatment is of the 2-D, elastic waves (AE) detected are essentially of 3-D. Taking into account this fact, the moment tensor is projected onto the 3-D space by assuming $l_3 = n_3 = 0$,

$$m_{ij} = \begin{pmatrix} m_{11} & m_{12} & 0 \\ m_{12} & m_{22} & 0 \\ 0 & 0 & m_{33} \end{pmatrix}, \tag{4}$$

where tensor components m_{11}, m_{12}, and m_{22} are indicated in eq. 2. m_{33} is obtained theoretically, as follows (Shigeishi and Ohtsu, 1992);

$$m_{33} = \nu(m_{11} + m_{22}). \tag{5}$$

Since the eigenvalue analysis and the unified decomposition are already established in the 3-D case, the crack classification is readily performed as the same as reported elsewhere (Ohtsu, 1992). Based on the results of numerical experiments (Shigeishi and Ohtsu, 1992), AE events of which the shear ratio is greater than 75% are classified into shear cracks, while those less than 25% are classified into tensile cracks. Otherwise, both types and orientations are determined.

From the eigenvalue analysis of the tensor in eq. 2, two eigenvectors **e1** and **e2** are obtained,

$$\mathbf{e1} = (\cos D+\cos C,\ \sin D+\sin C),$$
$$\mathbf{e2} = (\cos D-\cos C,\ \sin D-\sin C). \tag{6}$$

Thus, the direction of crack opening in the tensile crack is identical to the first vector **e1**. For the shear crack, the direction of motion is recovered from the sum of two vectors **e1** and **e2**.

4 SiGMA procedure

After uniaxial compression tests of PMMA plates were completed, AE waveforms were displayed on a CRT screen as shown in Fig. 4. From one waveform, two parameters of the arrival time (P1) and the amplitude of the first motion (P2) were read on the screen and were stored onto a file. In the next step, the crack location analysis was performed, based on the arrival time differences (P1). Then, the moment tensor components were determined from the first-peak amplitudes (P2). Based on eq. 1, an overdetermined set of four linear algebraic equations with three unknown; m_{11}, m_{12}, and m_{22} was obtained. Solutions were obtained by the least-square error procedure. The classification of crack types and the determination of crack orientations followed the procedure mentioned above.

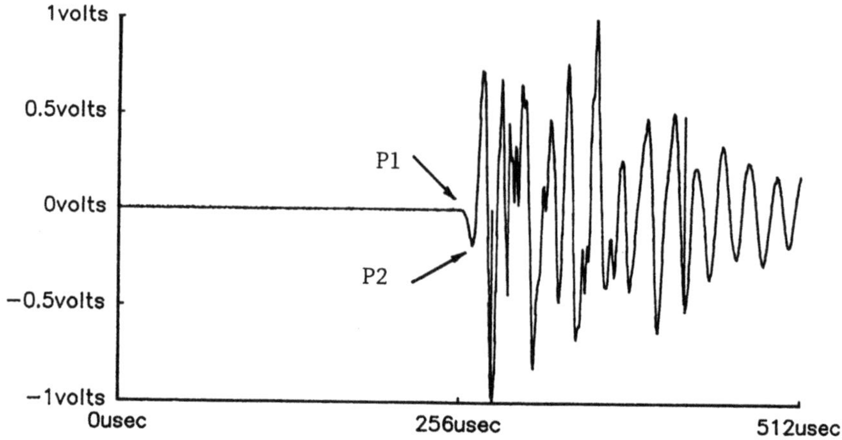

Fig. 4 An AE waveform recorded.

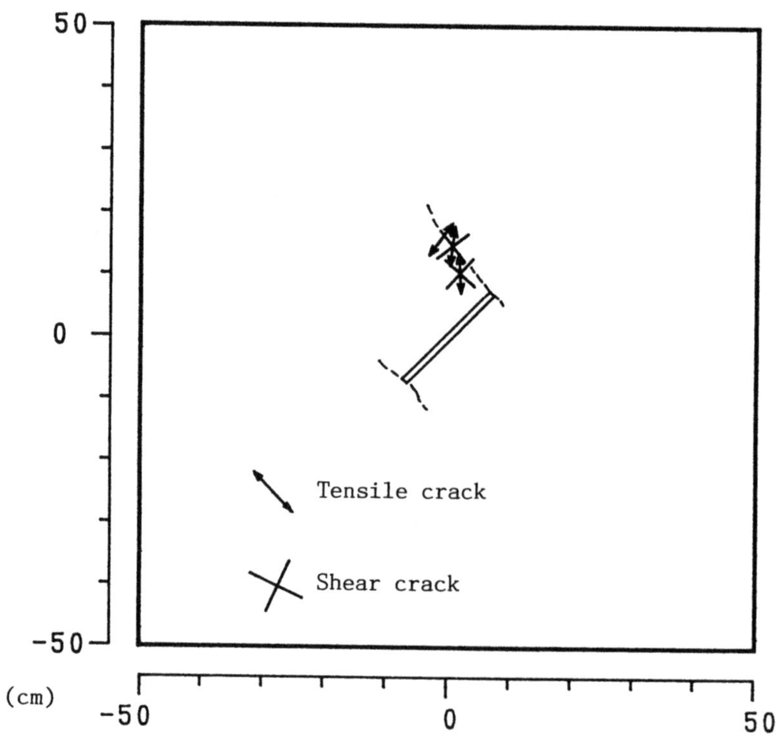

Fig. 5 Crack locations, types, and orientations in one PMMA model.

5 Results and Discussion

Two results of the PMMA model experiments are shown in Figs. 5 and 6. Since the limited number of AE waveforms are of the clear first motions, total number of AE events analyzed are quite a few. Still from both figures, it is obviously observed that AE sources are located along visible crack trajectories. A tensile crack is indicated by an arrow symbol, of which direction corresponds to crack opening. A cross symbol denotes a shear crack. Two directions of the cross correspond to either crack normal or crack vector.

In the experiment of Fig. 5, compressive force was immediately unloaded as visible cracks were found. As a result, only three events are analyzed and plotted. One AE source far from the slit is classified into a tensile crack, and the direction of crack motion is in remarkable agreement with the opening direction of visible crack trajectory. For the other two, both directions of tensile and shear cases are shown, because the shear ratios are bigger than 25%

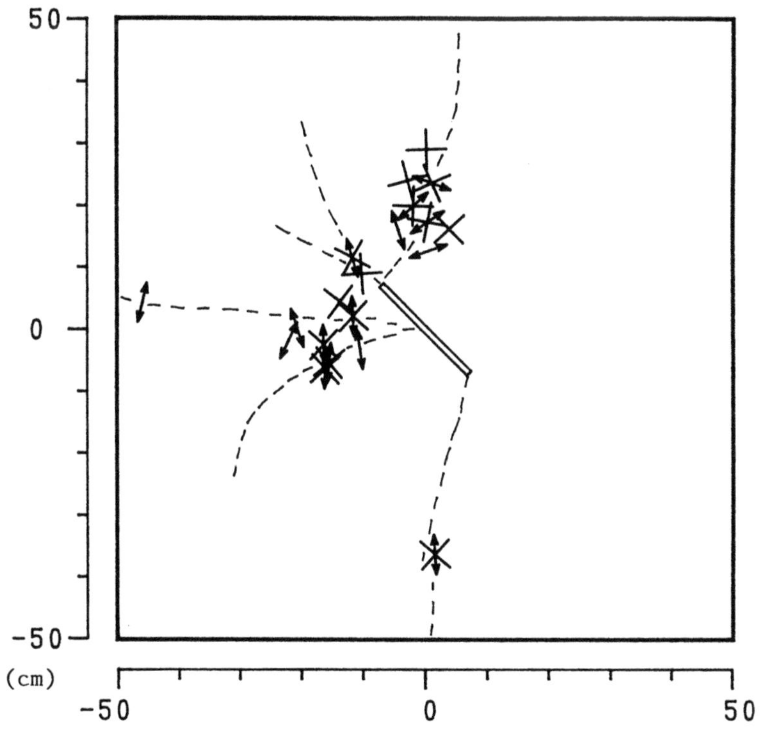

Fig. 6 Crack locations, types, and orientations in another PMMA model.

but smaller than 75% and thus cracks are classified into mixed mode. It is clearly observed that the directions of shear mode are parallel to the crack trajectory.

In the experiment of Fig. 6, load was applied up to the final failure due to out-of-plane buckling. Thus, many crack trajectories and AE sources are observed. Some tensile cracks are observed far from the slit. Shear cracks are mainly observed in AE cluster, although almost of all cracks are classified into the mixed mode. It is observed that the directions of shear cracks are almost parallel to directions of visible cracks. In the cracks classified into tensile cracks, the directions of crack opening are vertical to the visible crack surface. These results show that the shear cracks and the mixed mode cracks are located behind the tensile cracks.

6 Conclusion

A new SiGMA-2D software is developed and applied to the uniaxial compression tests of the PMMA plates. The following conclusions are obtained.

(1) Under uniaxial compression, crack propagation from the 45° inclined slit is observed in the transparent PMMA plate. Cracks propagate tortuously, directing the loading axis. AE sources analyzed are classified into the tensile cracks, the shear cracks and the mixed mode cracks.

(2) A few tensile cracks are observed far from the slit. The directions of tensile motion are in remarkable agreement with the visible crack trajectories. crack-opening directions are observed as vertical to the crack trajectories.

(3) Most AE sources are classified into either the shear cracks or the mixed mode cracks. In these sources, the directions of shear cracks are almost parallel to the crack trajectories.

(4) Distribution of these tensile cracks and shear cracks suggests the mechanisms, where the tensile cracks are nucleated first and the shear cracks follow on the crack surface created.

7 References

Berthaud, Y., Ringot, E. and Schmitt, N. (1991) Experimental Measurements of Localization for Tensile Tests on Concrete, in **Fracture Processes in Concrete, Rock and Ceramics** (eds. J. G. M. van Mier, J. G. Rots and A. Bakker), E. & F. N. Spon, London,

pp. 41-50.
Chen, L. H., Cheng, C. T. and Chen, S. E. (1992) Determination of Fracture Parameters of Mortar and Concrete Beams by Using Acoustic Emission, **Materials Evaluation**, Vol. 50, No. 7, pp. 888-894.
Mindes, S. (1991) Fracture Process Zone Detection, in **Fracture Mechanics Test Methods for Concrete** (eds. S.P. Shah and A. Carpinteri), Chapman and Hall, London, pp. 231-261.
Nomura, N., Mihashi, H. and Izumi, M. (1991) Proper ties of Fracture Process Zone and Tension Softening Behavior of Concrete, in **Fracture Processes in Concrete, Rock and Ceramics** (eds. J. G. M. van Mier, J. G. Rots and A. Bakker), E. & F. N. Spon, London, pp. 51-60.
Ohtsu, M.(1982) Source Mechanism and Waveform Analysis of Acoustic Emission in Concrete, **J. Acoustic Emission**, Vol. 2, No. 1, pp. 103-112.
Ohtsu, M. and Ono, K.(1986) The Generalized Theory and Source Representations of Acoustic Emission, **J. Acoustic Emission**, Vol. 5, No. 4, pp. 124-133.
Ohtsu, M. (1987) Determination of Crack Orientation by Acoustic Emission, **Materials Evaluation**, Vol. 45, No. 9, pp. 1070-1075.
Ohtsu, M. and Ono, K. (1988) AE Source Location and Orientation Determination of Tensile Cracks from Surface Observation, **NDT International**, Vol. 21, No. 3, pp. 143-150.
Ohtsu, M. (1991) Simplified Moment Tensor Analysis and Unified Decomposition of Acoustic Emission Source: Application to In Situ Hydrofracturing Test, **J. Geophysical Research**, Vol. 96, No. B4, pp. 6211-6221.
Ohtsu, M. and Shigeishi, M. (1992) Inverse Analysis in the Two-Dimensional PMMA Model with a Slit, **Proc. of the Fourth International Symposium on Acoustic Emission From Composite Materials**, ASNT, pp. 428-435.
Ohtsu, M. (1992) AE Inversion in Pull-Out Test, presented in **Seminar on Anchorage Engineering**.
Ouyang, C., Landis, E. and Shah, S. P. (1991) Damage Assessment in Concrete Using Quantitative Acoustic Emission, **J. Engineering Mechanics**, ASCE, Vol. 117, No. 11, pp. 2681-2698.
Shigeishi, M. and Ohtsu, M. (1992) A SiGMA Analysis of the 2-Dimensional PMMA Model, **Progress in Acoustic Emission VI**, JSNDI (in press).

17 MULTI-DOMAIN BEM IMPLEMENTATION FOR MIXED-MODE CRACKING IN CONCRETE

A.H. CHAHROUR and M. OHTSU
Department of Civil and Environmental Engineering,
Kumamoto University, Japan

Abstract
A multi-domain boundary element method (BEM) approach is applied to mixed-mode cracking in concrete, based on linear elastic fracture mechanics (LEFM) principles. Automatic remeshing of the boundary is accomodated with the increment of crack growth, and the prediction of crack trajectory is accomplished by just generating new boundaries. Mixed-mode crack propagation in center-notched plain concrete beams is investigated based on the criterion of the maximum circumferential tensile stress. Mixed-mode experiments were performed to compare with the theoretical predictions. Reasonable agreement between the experimental and analytical results is obtained. The analytical results suggest that crack initiation and propagation in the early stage is governed by the tensile mode-I fracture, the shear mode-II fracture taking over in the later stages.

1 Introduction

The applicability of fracture mechanics principles to analyze crack initiation and propagation in concrete under tensile mode-I loading has encouraged researchers for their application to mixed-mode fracture investigation. Crack propagation in concrete structures normally takes place under combined action of both tensile (mode-I) and shear (mode-II) types of loading. In theoretical treatment, such a mixed-mode propagation of fracture poses a complicated problem to predict crack trajectory and crack instability. Accordingly, the development of a numerical analysis technique for mixed-mode cracking is of significant importance in research on fracture mechanics of concrete.

In this respect, finite element method (FEM) implementation in the analysis is widely reported by Ingraffea and Gerstle (1985), John and Shah (1990), and Swartz and Taha (1991). Ohtsu (1988) shows that employing the boundary

Fracture and Damage of Concrete and Rock – FDCR-2.
Edited by H.P. Rossmanith. © 1993 Taylor & Francis. ISBN 0 419 18470 8.

element method (BEM) comes as a natural choice because BEM deals with the discretization of only the boundary and holds a privileged applicability to the discrete crack model. It is more advantageous to use BEM than FEM for the analysis of mixed-mode crack propagation, because BEM needs fewer input data and allows much simpler automatic remeshing of the boundary to accomodate crack propagation.

In the present paper, crack propagation in center-notched concrete beams subjected to mixed-mode loading is analyzed by a two-domain BEM approach, based on the criterion of maximum circumferential stress. Crack propagation is modeled by creating new boundaries in the numerical analysis with automatic remeshing to accomodate crack extension. Stress intensity factors K_I and K_{II} at the crack tip are computed from the displacements of nodes on the crack tip element. Based on the computation of crack direction, a new boundary is created with the two-domains stitched at the common interface boundary.

The analyzed crack trajectories, load-CMOD curves and relative values of the stress intensity factors K_I and K_{II} are discussed for the cases of 20, 30, and 40 mm center-notched plain concrete beams.

2 Boundary element method formulation

2.1 Two-dimensional BEM

Fig. 1 describes a boundary value problem of an elastic body with boundary surface S surrounding domain D. The boundary is classified into stress-prescribed boundary S_t and displacement-prescribed boundary S_u.

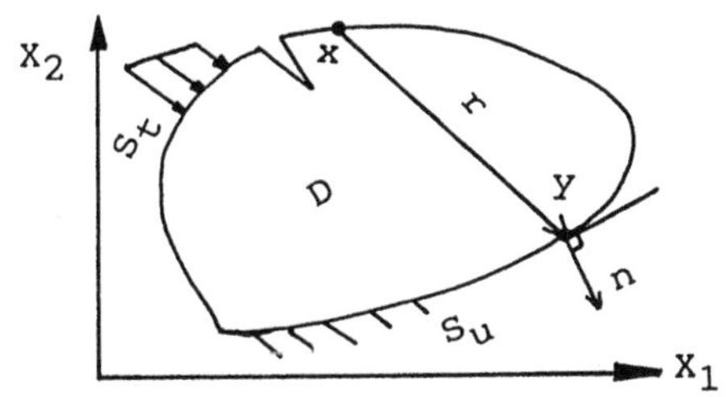

Fig. 1 Boundary value problem of an elastic body

The governing boundary integral equation is given, as follows:

$$C_{ij}u_i(x) = \int_S [G_{ij}(x,y)t_j(y) - T_{ij}(x,y)u_j(y)]dS, \qquad (1)$$

where C_{ij} is the configuration coefficient determined by eliminating rigid body motion in Eq. 1. $G_{ij}(x,y)$ is the fundamental singular displacement solution and $T_{ij}(x,y)$ is the associated traction solution. For the two-dimensional linear elastic plain strain field, they are given by,

$$G_{ij}(x,y) = (1+\nu)\{(3-4\nu)\delta_{ij}\log(1/r)+r_{,i}r_{,j}\}/4\pi E(1-\nu), \quad (2)$$

$$T_{ij}(x,y) = \{[(1-2\nu)\delta_{ij}+2r_{,i}r_{,j}]\,\partial r/\partial n\}/4\pi(1-\nu)r - \{(1-2\nu)(r_{,i}n_j-r_{,j}n_i)\}/4\pi(1-\nu)r, \quad (3)$$

where $r=|x-y|$ and the comma denotes the spatial derivative. δ_{ij} is the Kronecker delta, E is Young's modulus, and ν is Poisson's ratio.

The boundary of the domain is discretized into elements L_j $(j=1,...,N)$ and thus Eq. 1 becomes,

$$C_{ij}u_i(x) + \sum_{j=1}^{N} \int_{L_j} T_{ij}(x,y)u_j(y)dS = \sum_{j=1}^{N} \int_{L_j} G_{ij}(x,y)t_j(y)dS. \quad (4)$$

2.2 Two domain BEM

Shown in Fig. 2 is the case of two elastic domains D_1 and D_2 joined along common interface S_C. S_1 and S_2 represent the external boundaries of D_1 and D_2, respectively. Hence, $S_{1T}=S_1+S_C$ constitutes the boundary of domain D_1, and $S_{2T}=S_2+S_C$ designates the boundary of D_2.

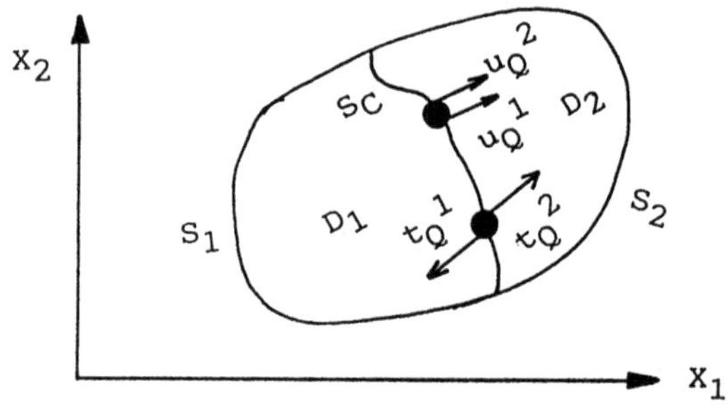

Fig. 2 Subdivision of computational domain

The boundary integral equation for each region α is given by Cruse and Polch (1986), as follows:

$$C_{ij}{}^{\alpha}u_i{}^{\alpha}(x) + \int_{S^\alpha} T_{ij}{}^{\alpha}(x,y)u_j{}^{\alpha}(y)dS = \int_{S^\alpha} G_{ij}{}^{\alpha}(x,y)t_j{}^{\alpha}(y)dS, \quad (5)$$

where S^α is equal to S_{1T} for D_1, and S_{2T} for D_2.

The boundary conditions are those imposed seperately on S_1 and S_2, while the interface conditions on S_C are given by,

$$u_j{}^1(Q) = u_j{}^2(Q), \quad t_j{}^1(Q) = -t_j{}^2(Q). \qquad (Q \in S_C) \quad (6)$$

Discretizing the boundaries of the two domains and the

interface boundary, a constant element is adopted for traction and a linear element is assigned for the displacement on the boundary elements (Chahrour and Ohtsu, 1991), as follows:

$$\{t(\xi)\} = t_j(yy_k) = \text{Constant},$$
$$\{u(\xi)\} = (1 - \xi)u_j(y_k)/2 + (1 + \xi)u_j(y_{k+1})/2, \tag{7}$$

where ξ is the variable of local coordinates as shown in Fig. 3.

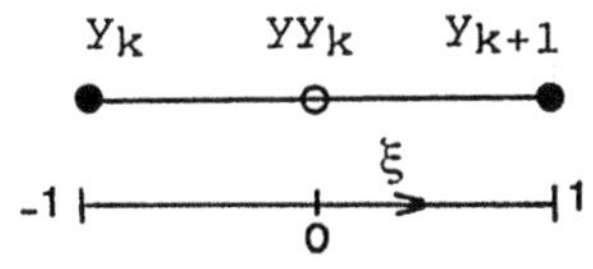

Fig. 3 Linear boundary element

Carrying out the numerical integration over the elements using 6-point Gaussian integration and taking care of the diagonal terms, Eq. 5 is written in the following matrix-form,

$$[H_{ij}{}^{\alpha}]\{u_j{}^{\alpha}(y_k)\} = [G_{ij}{}^{\alpha}]\{t_j{}^{\alpha}(yy_k)\}. \quad (\alpha = 1,2) \tag{8}$$

To make the numerical scheme comparable to the conventional FEM code, Eq. 8 is converted into the equation, not of traction on the element, but of nodal forces, as follows:

$$N_{km}{}^{\alpha}[G_{ml}{}^{-1}H_{lj}]^{\alpha}\{u_j{}^{\alpha}\} = \{f_k{}^{\alpha}\}, \quad (\alpha = 1,2) \tag{9}$$

where N_{km} is the shape function for obtaining nodal forces from tractions. Making use of Eqs. 6, the matrix Eqs. 9 for (α = 1,2) are coupled together, and thus the resulting equation is solved for the nodal forces and displacements.

3 Analytical model and crack propagation analysis

3.1 Analytical model and procedure

The analytical model for the analysis of mixed-mode fracture of center-notched concrete beams is given in Fig. 4, where double nodes are assigned along the notch, and the broken line with nodes represents the stitching interface. It joins the notch tip to point R as shown in the figure.

When a crack propagates, the node at the tip is seperated into two nodes, creating two new crack tip elements. This results in the increase of the total number of nodes. A new stitching boundary is defined as the straight line joining the new crack tip elements, defined based on the

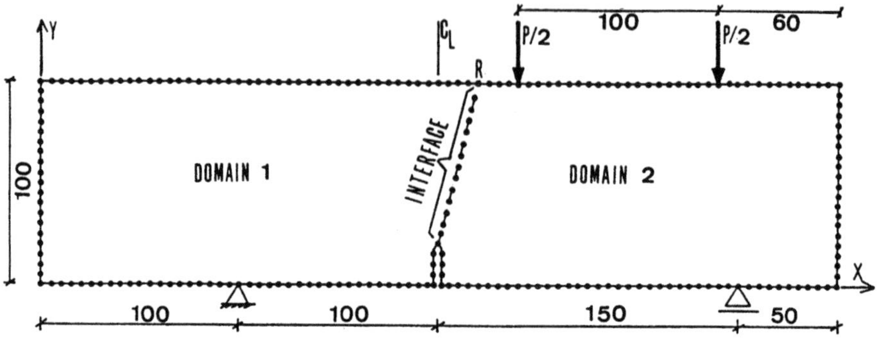

Fig. 4 Two-domain BEM model

crack direction and increment, to point R on the top of beam surface. Automatic remeshing is accomodated to take care of the newly created stress-free crack tip elements. The same element length of 5 mm was considered for domains 1 and 2, and for the crack increment during crack propagation.

3.2 Crack propagation analysis

The stress intensity factors at the crack tip are determined by Smith's one point formulae (Smith and Mason, 1982). K_I and K_{II} are determined by using displacement u_i in the X-direction and displacement v_i in the Y-direction, as shown in Fig. 5.

$$K_I = (2\pi/L)^{1/2}E[v_i(B) - v_i(A)]/4(1 - \nu^2), \quad (10)$$

$$K_{II} = (2\pi/L)^{1/2}E[u_i(B) - u_i(A)]/4(1 - \nu^2), \quad (11)$$

where L denotes the length of the crack tip element.

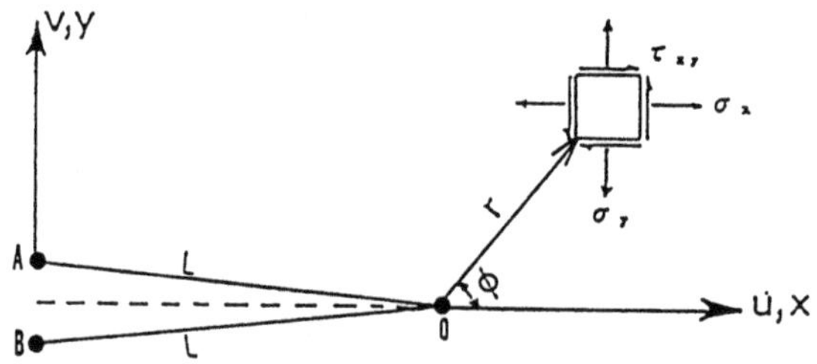

Fig. 5 Crack tip element

The direction of crack propagation is determined from the criterion of maximum circumferential stress given by Eq. 12, while Eq. 13 states the adopted criterion for crack nucleation,

$$K_I^* \sin \phi + K_{II}^* (3\cos \phi - 1) = 0, \quad (12)$$

$$\cos(\phi/2)[K_I^* \cos^2(\phi/2) - 3K_{II}^* \sin(\phi/2)] >= 1, \quad (13)$$

where $K_I^* = K_I/K_{IC}$ and $K_{II}^* = K_{II}/K_{IC}$. ϕ is taken to be the angle corresponding to the bigger maximum tensile stress as shown in Fig. 6.

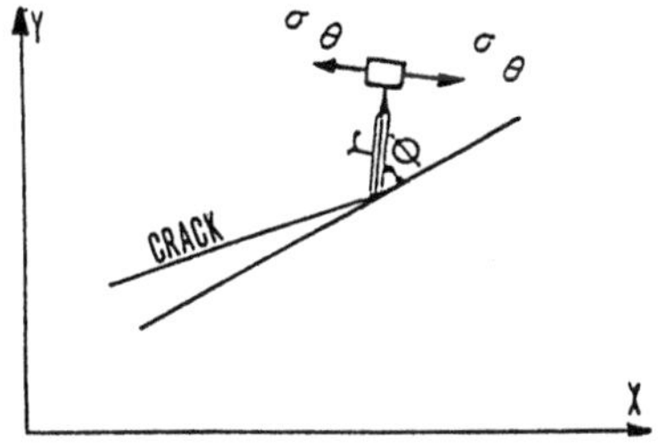

Fig. 6 Crack growth direction

4 Experimental study

4.1 Specimens and test set-up

Center-notched plain concrete beams of 400 mm length, 100 mm depth, and 100 mm width were cast. 20, 30, and 40 mm notches were inserted into the specimens by using steel plates of 1 mm thickness fixed to the mould face. The specimens were submerged in water till the time of testing which was carried out after 40 days. Concrete was of the following mechanical properties: unit weight of 2390 Kg/m^3, compressive strength of 25.46 Mpa, tensile strength of 1.93 Mpa, Young's modulus E of 26.66 Gpa, and Poisson's ratio of 0.18. These values comprised the average results of three cylindrical specimens of 100 mm diameter and 200 mm height, tested after 32 days of moisture curing. The bending strength of the concrete was 4.50 Mpa.

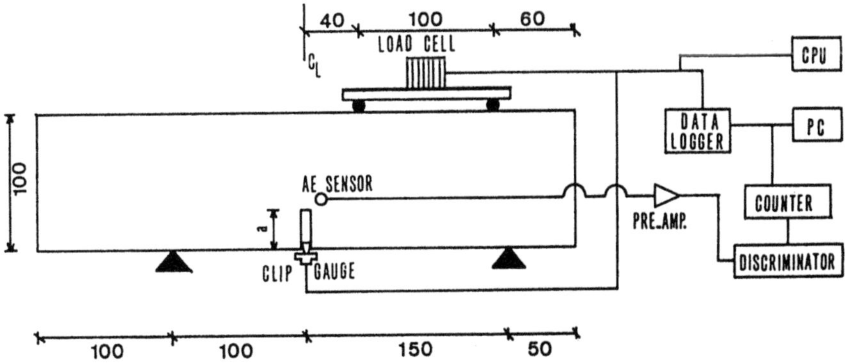

Fig. 7 Experimental set-up for mixed-mode fracture

Fig. 7 shows the experimental set-up used for studying mixed-mode crack propagation in center-notched beams subjected to anti-symmetric loading. A servo-valve controlled loading machine was employed and complete load-CMOD curves were obtained using a X-Y plotter. Load was measured by a load cell and CMOD was measured by a clip gauge inserted at the notch bottom. Load was applied at the rate of 0.0010 mm/s and load-CMOD data was stored in a processing unit. Acoustic emission (AE) events were monitored by an AE sensor, a pre-amplifier, a discriminator, and an event counter. Load- cumulative AE count data was processed through a data logger and stored in a personal computer. AE events were recorded with the aim of assessing the damage induced in the concrete during crack propagation (Ohtsu, 1989). This damage is observed as the sudden accelerated increase in the recorded AE events before reaching the maximum load.

The crack started from the notch tip, for all the tested beams, and propagated upwards till reaching the top surface of the beam. Tracing the obtained crack trajectories after the tests revealed that the crack propagated along different paths on the front and back sides of the beam, suggesting that the crack path is tortuous, curved and wavy along the beam width with some branching observed in some of the specimens. It took about 11, 10, and 4 minutes to reach the peak load in testing 20, 30, and 40 mm notched beams, respectively.

4.2 Critical stress intensity factor (K_{IC})

The critical stress intensity factor or fracture toughness of concrete K_{IC} normally refers to the ultimate load P_U at which the crack becomes unstable. In this research, the fracture toughness is determined based on P_{AE} which represents the load at which high AE activity is observed in a four-point bending test. It is shown by Chahrour and Ohtsu (1991) that P_{AE} corresponds to the load at which the stress at the notch tip becomes equal to the tensile strength of concrete. This procedure does not differ in principle from the recommendations of RILEM report 5, by Karihaloo and Nallathambi (1991). Here, P_{AE} is associated with the original notch depth, while in RILEM procedure K_{IC} is computed based on P_U and the effective crack length.

K_{IC} is determined from P_{AE} using the equation,

$$K_{IC} = 3P_{AE}(S_1 - S_2)a^{1/2}Y(a/H)/2BH^2, \text{ where} \qquad (14)$$

$Y(a/H)=1.99-2.47(a/H)+12.97(a/H)^2-23.17(a/H)^3+24.80(a/H)^4$, for mode-I fracture of center-notched concrete beams under four-point bending.

Mode-I tests for 20, 30, and 40 mm center-notched beams, of the same mechanical properties as described above, were carried out to determine K_{IC}. The fracture toughness of concrete was found to be fairly independent of the notch depth and the computed value was 0.321 $Mpam^{1/2}$.

5 Analytical results and discussion

5.1 Crack trajectories

Figs. 8(a), (b), and (c) show the analytical and experimental results for crack trajectories of mixed-mode fracture in center-notched beams with 20, 30, and 40 mm notch depths, respectively. The shaded area reveals the crack front across the beam width. The simulated BEM results showed good agreement with the actual crack trajectories. In numerical aspects, considering different meshes for domains 1 and 2 in the analytical model did not produce good results. It demonstrates to be important to keep the same mesh division on the boundary surfaces of the two domains and along the interface. Choosing the stitching interface is another important issue. For the given loading configuration, the case of point R located between the center-line of the beam and the left loading point produced the best results. The interface should end up at a point on the top surface somewhere near the actual termination point of the crack.

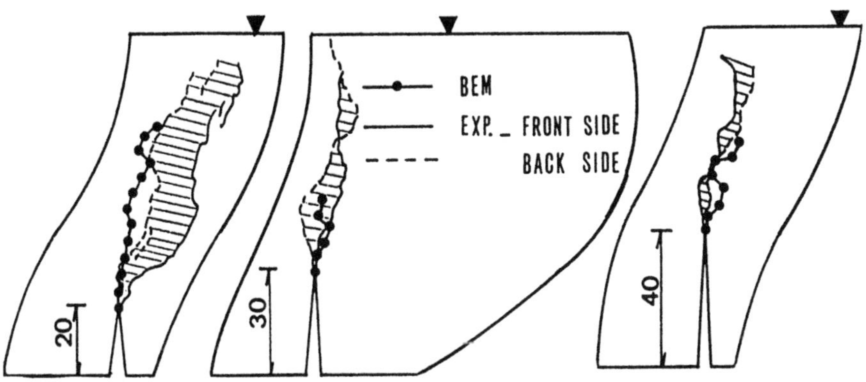

(a) 20 mm notch (b) 30 mm notch (c) 40 mm notch

Fig. 8 Actual and simulated crack trajectories

5.2 Load-CMOD curves

Load-CMOD curve for the center-notched beam of 20 mm notch depth is shown in Fig. 9. Even though the abrupt decrease in the analytical load is observed, the obtained curve reproduces the essential feature of the experimental load-CMOD curve. In all cases, the peak load was overestimated. This may be attributed to the size effect law(Carpinteri and Swartz, 1991).

5.3 Analysis of crack propagation mechanism

K_I/K_{IC} and $|K_{II}/K_{IC}|$ values versus crack extension for the 20 mm notched beam is shown in Fig. 10. Absolute values of K_{II}/K_{IC} are considered because negative values of K_{II}/K_{IC} are obtained at several stages in the analysis. The obtained results reveal an important illustration on the mechanism of crack propagation. Crack starts from the notch with K_I/K_{IC} value almost equal to unity and corresponding K_{II}/K_{IC} less than one tenth. This indicates that the crack starts with almost pure mode-I fracture. As the crack propagates, K_I/K_{IC} decreases while K_{II}/K_{IC} increases and gradually mode-II fracture becomes more dominant. The beam ends up failing in shear with K_{II}/K_{IC} approaching unity. The analysis was terminated when K_I became negative, which means that inward overlapping of the opening displacements at the crack tip element occurred.

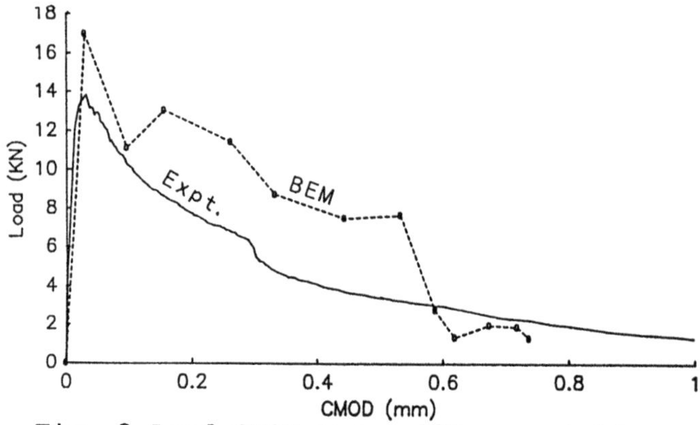

Fig. 9 Load-CMOD curve (20 mm notch)

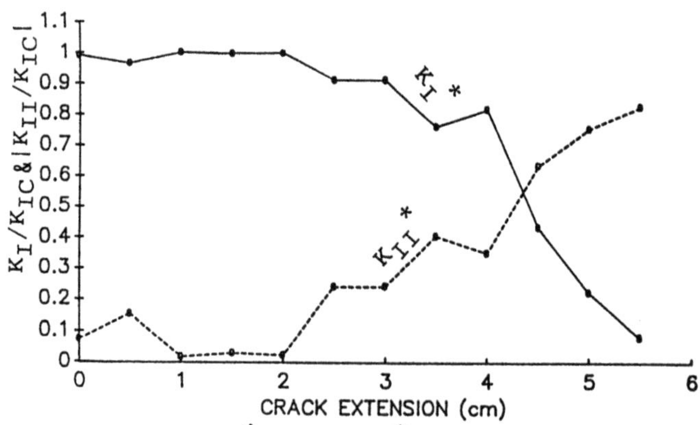

Fig. 10 Relative K_I^* and K_{II}^* relation (20 mm notch)

6 Concluding remarks

As a numerical technique, multi-domain BEM is very promising for implementation in the analysis of mixed-mode fracture of concrete, based on linear elastic fracture mechanics. The method is quite easy to apply and allows simple automatic remeshing to model crack propagation in comparison with FEM. Although the overestimation of the peak load is obtained, the simulated load-CMOD curves reproduce essential feature of the experimental curves. Crack initiation seems to be solely of the mode-I type with mode-II dominating at the later stages, resulting in the interaction of both modes as the crack propagates.

7 References

Carpinteri, A. and Swartz, S. (1991) Mixed-mode crack propagation in concrete, Fracture mechanics test methods for concrete (eds S.P. Shah and A. Carpinteri), RILEM Report 5, pp. 129-197.

Chahrour, A.H. (1991) BEM analysis of crack propagation based on fracture mechanics, Master thesis, Kumamoto U.

Chahrour, A.H. Ohtsu, M. and Fukuchi, S. (1992) BEM analysis of mixed-mode fracture, Proceedings of the Japan concrete institute, Vol. 14, No. 2, pp. 1011-1016.

Cruse, T.A. and Polch, E.Z. (1986) Proceedings of 3rd Japan symposium on boundary element methods, JASCOME, pp. 111-133.

Ingraffea, A. and Gerstle, W.H. (1985) Non-linear fracture models for discrete crack propagation, Application of fracture mechanics to cementitious composites (editor S.P. Shah), Martinus Nijhoff publishers, pp. 247-285.

John, R. and Shah, S.P. (1990) Mixed-mode fracture of concrete subjected to impact loading, Journal of structural engineering, Vol. 116, No. 3, pp. 585-602.

Karihaloo, B.L. and Nallathambi, P. (1991) Notched beam test:Mode-I fracture toughness, Fracture mechanics test methods for concrete, RILEM Report 5, pp. 1-86.

Ohtsu, M. (1988) Crack propagation in concrete: Linear elastic fracture mechanics and boundary element method, Theoretical and applied fracture mechanics, Elsevier science publishers, Vol. 9, No. 1, pp. 55-60.

Ohtsu, M. (1989) AE application to fracture mechanics, Proceedings of the Japan concrete institute, Vol. 11, No. 2, pp. 673-678.

Smith, R.N.L and Mason, J.C (1982) A boundary element method for curved crack problems in two dimensions, Boundary element methods in engineering (ed. Brebbia), Springer, Berlin, pp. 472-484.

Swartz, S. and Taha, N.M. (1991) Crack propagation and fracture of plain concrete beams subjected to shear and compression, ACI Structural Journal, Vol. 88, No. 2, pp. 169-177.

18 QUANTITATIVE ASSESSMENT OF FRACTURE OF CONCRETE BY USING BEM FOR INHOMOGENEOUS MATERIALS

ZHOU RUIZHONG and LI TONGMING
Department of Civil Engineering, Fuzhou University,
Fujian, China

Abstract
The effect of inhomogeneity of concrete on fracture behavior and quantitative assessment method are presented in this paper by using BEM for inhomogeneous materials. Considering two opposite factors reinforced by aggregate and weakened by interface crack between aggregate and matrix, we make a computer program to calculate the real process of failure. Some examples show the fluctuating range of fracture index is about 0–18. %.
Keywords: Inhomogeneity, Fracture, Concrete

1 Criterion for local fracture of a quasibrittle material

Concrete is inhomogeneous material, and it is composed of multiphase media, As a quasibrittle material, the behaviors of fracture have obviously the local characteristics. Either the relationship of stress–strain near the tip of crack, or the propagation of a cracking is directly influenced by the location, size, elastic modulus of aggregate and interface microcracking. The most difficulty of fracture mechanics applied to design for a structure is the inhomogeneity and nonlinearity of concrete, because the linear fracture is satisfactory only in the framework of an ideally elastic and homogeneous body. The prediction of linear fracture sometimes is far from the fact of real process.

There were two reasons obstructing its rapid development:[1]

1. Before unstable expansion, the crack's stable and slow expanding is found. This process is obstructed by aggregate in front of main crack, while it is stimulated by microcracking that exists in matrix or on the jointed surface between aggregate and mortar. Even if we use the advanced equipment, we can't exactly find the crack tip and its real length, especially at the moment of last stage of the failure.
2. When crack propagation occurs, the multiple microcracking system will be found in the region near the tip of this crack. This is the result from concentration of tension strain that exceeds the limit of tension strain in the principal direction. The microcrackings make this region as an uncontinuous area. It will cause following consequences:

(1) The boundary between propagation of main crack and continuous damage area becomes unclear, because it is very difficult to figure out the crack tip and its length.

(2) This strain–softened area is influenced by many factors, such as size and geometry of specimen, rate of the loading, character of aggregates and initial distribution of cracks. Therefore, many test data shows[2] the toughness K_c or critical energy G_c of concrete is not a constant, but a certain range of fluctuating value.

There are two possibilities for failure process of the structure with a main crack in its edge:

Fracture and Damage of Concrete and Rock – FDCR-2.
Edited by H.P. Rossmanith. © 1993 Taylor & Francis. ISBN 0 419 18470 8.

1. Whole front region near the tip of crack reaches up to yielding point, the exhausting irreversible work is consumed mainly with the growth of micro– cracking zone and increment of the free surface of crack. (Most of metals were found to suit this case, the material can exhaust the capability for plastic strain.)
2. After the nonlinear area has grown into a certain size, the local crack propagation occurs as soon. The new microcracking zone is formed as the increment of crack length and the translation of the stable nonlinear zone. The shape and size of the microcracking zone are basically not changed. Therefore, the character of quasibrittle materials is that the nonlinear zone in front of a crack is a constant of the material. It is independent of the loading history and the geometry of the specimen. This rule was observed by concrete and rock to a certain extent. When the crack propagates through the surface of aggregates, it has a phenomenon of obvious 'jump'.

From the point of view of theoretical analysis, when the equivalent radius of plastic zone is $r \to 0$, the brittle fracture occurs. As opposed to this case, when $r \to \infty$, the ductile fracture will occur. A real–life material is the combination of two limiting cases mentioned above.

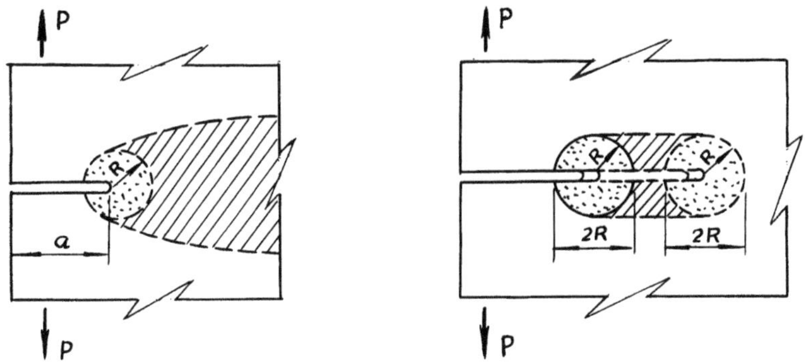

Fig.1 Two possibilities of crack propagation

In order to estimate the influence of inhomogeneity of concrete on the fracture behaviors, meanwhile to use the existing theory and formulae for homogeneous body, current study approaches to this problem usually with the different ways:

1. Equivalent method.
   Typical theories are damage mechanics, elastoplastic theory, synthetic deformation method etc.[3] The common point of these methods is that all changes reflect in the deformation modulus. Its advantage is simple. The deformation modulus can be used conveniently in engineering practice. Otherwise, they are too crude for a control index. The result from these methods is only a compatible or essential condition, but isn't a complete or full condition, so it can't exactly give the expression of real failure process.
2. Fracture mechanics for inhomogeneous multiphase media.
   Based on real compositions of concrete, this method considers three compositions —— aggregate, matrix (cement mortar) and interface microcracking. Then it was calculated by computation program and determined whether the main crack is in propagation, or the propagation is in stable. Other problem is the growth of interface microcracking. The stress state of interface may be fallen into slide or rotation

between the aggregate and cement–mortar. The advantage of this method is that it can express the real process of crack propagation in structure. The result from this method shows good agreement with the test, but its computer procedure is somewhat complex, and the high accuracy of calculation is also required.

The boundary integral equation and boundary element method are employed in this paper. The effect of inhomogeneity of concrete on the fracture behaviors is taken into account. The numerical solution can be applied to reference of engineering.

2 Subregion integral equation and boundary element technique

The fictitious stress method (indirect method) or the direct boundary integral method is deduced from Somigliana's formulas. They are formulas for displacement and stress.

$$u_i(P)=\int_\Gamma \tau_j(Q)U_{ij}(P,Q)d\Gamma(Q)-\int_\Gamma u_j(Q)T_{ij}(P,Q)d\Gamma(Q)$$
$$+\int_\Omega x_j(q)U_{ij}(p,q)d\Omega_q \qquad p,q\in\Omega,\quad Q\in\Gamma \tag{1}$$

It shows the relationship between displacement $u_j(Q)$, traction $\tau_j(Q)$ on a boundary point Q and body force term $x_j(q)$.
Where $U_{ij}(p,q)$ and $T_{ij}(p,q)$ are solved by Kelvin's solution. i. e.

$$U_{ij}(p,q)=1/[8\alpha\pi G(1-v)\cdot r^{\alpha-1}]\cdot\{(3-4v)\delta_{ij}[(\alpha-2)lnr+\alpha-1]+r_{,i}r_{,j}\} \tag{2}$$

where $\alpha=1$ (for 2–D problem) $\alpha=2$ (for 3–D problem) and $r=r(p,q)$, $r_{,i}=\dfrac{\partial r}{\partial x_i}$.

$$T_{ij}(p,q)=-1/[4\alpha\pi(1-v)\cdot r^{\alpha}]\cdot\{\frac{dr}{dn}[(1-2v)\delta_{ij}+\beta r_{,i}r_{,j}]$$
$$-(1-2v)(n_jr_{,i}-n_ir_{,j})\} \tag{3}$$

$U_{ij}(p,q)$, $T_{ij}(p,q)$ denote respectively the displacement, traction in i direction at point p caused by unit force at point q in j direction.

The Somigliana's stress formula is:

$$\sigma_{ij}(P)=\int_\Gamma \tau_k(Q)V_{ijk}(P,Q)d\Gamma(Q)-\int_\Gamma U_k(Q)S_{ijk}(P,Q)d\Gamma(Q)$$
$$+\int_\Omega X_k(q)V_{ijk}(p,q)d\Omega_q \tag{4}$$

It shows the relationship between stress $\sigma_{ij}(P)$ of internal point P and boundary displacement $U_k(Q)$, boundary traction $\tau_k(Q)$. There are:

$$V_{ijk}(p,q)=1/[4\alpha\pi(1-v)\cdot r^{\alpha}]\cdot[(1-2v)(\delta_{ik}r_{,j}-\delta_{jk}r_{,i}+\delta_{ij}r_{,k})+\beta r_{,i}r_{,j}r_{,k}] \tag{5}$$

$$S_{ijk}(p,q)=G/[2\alpha\pi(1-v)\cdot r^{\beta}]\cdot\{\beta\frac{dr}{dn}[(1-2v)\delta_{ij}r_{,k}+v(\delta_{ik}r_{,j}+\delta_{jk}r_{,i})$$
$$-\gamma r_{,i}r_{,j}r_{,k}]+\beta v(n_ir_{,j}r_{,k}+n_jr_{,i}r_{,k})$$
$$+(1-2v)(\beta n_kr_{,i}r_{,j}+n_j\delta_{ik}+n_i\delta_{jk})-(1-4v)n_k\delta_{ij}\} \tag{6}$$

$$\begin{cases}\alpha=1, & \beta=2, & \gamma=4 & for\ 2-D\ problem\\ \alpha=2, & \beta=3, & \gamma=5 & for\ 3-D\ problem\end{cases}$$

Employed quadratic interpolating, we have:

$$\phi_1=\frac{1}{2}\xi(\xi-1),\qquad \phi_2=(1-\xi)(1+\xi),\qquad \phi_3=\frac{1}{2}\xi(1+\xi),$$

$$d\Gamma = \left\{\sqrt{\left(\frac{dx}{d\xi}\right)^2 + \left(\frac{dy}{d\xi}\right)^2}\right\} d\xi = |G| d\xi,$$

$$d\Omega = det\begin{bmatrix} \frac{\partial x}{\partial \xi} & \frac{\partial y}{\partial \xi} \\ \frac{\partial x}{\partial \eta} & \frac{\partial y}{\partial \eta} \end{bmatrix} d\xi d\eta = |J| d\xi d\eta$$

The direct boundary integral equations can be written as follows,

$$C^i U^i + \sum_{j=1}^{NE}\{\sum_{k=1}^{L} W_k(\mathbf{P}^* \phi)_k |G|\}\mathbf{U}^j = \sum_{j=1}^{NE}\{\sum_{k=1}^{L} W_k(\mathbf{U}^* \phi)_k |G|\}\mathbf{P}^j + \sum_{s=1}^{M}\{\sum_{p=1}^{r} W_p(\mathbf{U}^* \mathbf{B})_p |J|\} \quad (7)$$

When $i = j$, $u^i = u^j$, the system of algebraic equations is formed by considering the boundary condition for each element. It can be written as:

$$\sum_{j=1}^{NE} \mathbf{H}^{ij}\mathbf{U}^j = \sum_{j=1}^{NE} \mathbf{G}^{ij}\mathbf{P}^j + \sum_{s=1}^{M} \mathbf{B}^{is} \quad (8)$$

For the study of influence of aggregate, jointed surface and microcracking on main crack's propagation, we consider a specimen of concrete as a subregion elastic domain. It shows in figure 2.

To simply the process in following work, we remove the body force term away and assume that the interface between aggregate and matrix is good contact besides the microcracking. The juncture– surface has initial continuity of the displacement and traction under loading.

From this condition, we obtain the expression in the global coordinate system as:

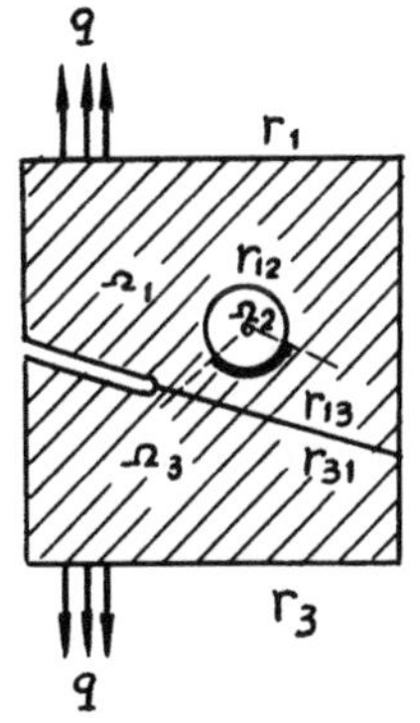

Fig.2 Subregion media

$$U_{12} = U_{21}, \quad U_{31} = U_{13}, \quad P_{12} = -P_{21}, \quad P_{31} = -P_{13} \quad (9)$$

We replace these equations by the algebraic system for every subregion and have:

$$\begin{bmatrix} \mathbf{H}_1 & 0 & \mathbf{H}_{12} & -\mathbf{G}_{12} & \mathbf{H}_{13} & -\mathbf{G}_{13} \\ 0 & 0 & \mathbf{H}_{21} & G_{12} & 0 & 0 \\ 0 & \mathbf{H}_3 & 0 & 0 & \mathbf{H}_{31} & \mathbf{G}_{31} \end{bmatrix} [\mathbf{U}_1 \quad \mathbf{U}_3 \quad \mathbf{U}_{12} \quad \mathbf{P}_{12} \quad \mathbf{U}_{13} \quad \mathbf{P}_{13}]^T = \begin{bmatrix} \mathbf{G}_1 & 0 \\ 0 & 0 \\ 0 & \mathbf{G}_3 \end{bmatrix} \begin{bmatrix} \mathbf{P}_1 \\ \mathbf{P}_3 \end{bmatrix} \quad (10)$$

After solving the linear system (10), we obtain the unknowns and Stress Intensity Factors. Either main crack's propagation or microcracking's stability is easily solved. Considering the change of each parameter in the figure 2, we note the SIF or COD should be a range of certain value, when the distance from aggregate to main crack tip or the location of curved microcracking on the interface is varying.

Here we point out two important problems:

1. There is microcracking on the bottom of aggregate, which may be caused by water

separation or shrinkage of cement–mortar. Its state is one of the follows: slide, joint separation, elastic contact. When the equilibrium loses, the new distribution of stress will occur. Therefore, the stress–strain relation on the interface is nonlinear. Obviously, the linear Mohr–coulomb principle is an approximate method.

In the real calculation, we divide the states of interface into following cases[4]:

(1) When $|\sigma_n| - P_w \geqslant 0$, the normal pressure σ_n applied to interface is greater than penetrating water head P_w, while $|(\sigma_n - P_w) \cdot f + C_0| \geqslant |\tau|$, the state is in pressure–shear condition. On the other hand, when $|(\sigma_n - P_w) \cdot f + C_0| < |\tau|$, the state is in relative slide condition. The direction of slide depends upon the sign of shear stress.

(2) When $|\sigma_n| - P_w < 0$, the surface is in joint separating. Once the sliding or separating state occurs, the excessive stress will be translated to neighbor element. This step is repeatedly done until the stable condition in each contact element is satisfactory.

2. For microcracking on the bimaterial joint surface, (aggregate and mortar) the singularity of stress distribution near the crack tip is no longer equal $r^{\lambda} = r^{-1/2}$, but depends on the rate of two elastic modulus,[5] i.e. $\lambda = -\frac{1}{2} \pm \beta i$, $\beta = \frac{1}{2\pi} \cdot ln\frac{\alpha + æ_1}{1 + \alpha æ_2}$, $\alpha = G_1 / G_2$ is shear modulus rate.

$$
æ_i = \begin{cases} 3 - 4\nu_i & \textit{for plane stress problem} \\ \dfrac{3 - \nu_i}{1 + \nu_i} & \textit{for plane strain problem.} \end{cases}
$$

Strictly say, here when the number λ is complex, the function $sin(\beta \cdot lnr)$ in any sector near the end of the crack undergoes an infinite number of changes of sign. The stress will undergo an infinite number of changes of sign when $r \to 0^+$. This result indicates the solution with complex λ, will make no physical sense and incorrectness of the mathematical statement.

However, if the complexity of λ is weak in a certain sense, the solution can be given an exact physicalmeaning. As regards the microcracking on the juncture surface between aggregate and matrix, the limiting case is $|\beta|_{max} = ln3 / 2\pi$. Therefore, $Exp(-1 / |\beta|) \leqslant 0.0007 \ll 1$, the mathematical and physical conditions mentioned above are all satisfactory, it's result will be correct.

When $y \geqslant 0$, $\lambda = -\frac{1}{2} + \beta_i$, the stress formulas can be obtained as follows:[6]

$$
\begin{aligned}
\sigma_x = {} & K_1 / A \cdot \left\{e^{-\beta(\theta-\pi)}(3cos\frac{\theta}{2} - 2\beta sin\theta cos\frac{3}{2}\theta - sin\theta sin\frac{3}{2}\theta) - e^{\beta(\theta-\pi)}cos\frac{\theta}{2}\right\} \\
& - K_2 / A \cdot \left\{e^{\beta(\theta-\pi)}sin\frac{\theta}{2} + exp^{-\beta(\theta-\pi)}(-3sin\frac{\theta}{2} + sin\theta cos\frac{3}{2}\theta - 2\beta sin\theta sin\frac{3}{2}\theta)\right\}
\end{aligned} \tag{11}
$$

$$
\begin{aligned}
\sigma_y = {} & K_1 / A \cdot \left\{e^{-\beta(\theta-\pi)}(cos\frac{\theta}{2} + 2\beta sin\theta cos\frac{3}{2}\theta + sin\theta sin\frac{3}{2}\theta) + e^{\beta(\theta-\pi)}cos\frac{\theta}{2}\right\} \\
& - K_2 / A \cdot \left\{-e^{\beta(\theta-\pi)}sin\frac{\theta}{2} + e^{-\beta(\theta-\pi)}(sin\frac{\theta}{2} - sin\theta cos\frac{3}{2}\theta + 2\beta sin\theta sin\frac{3}{2}\theta)\right\}
\end{aligned} \tag{12}
$$

$$
\begin{aligned}
\tau_{xy} = {} & K_1 / A \cdot \left\{e^{-\beta(\theta-\pi)}(sin\frac{\theta}{2} + sin\theta cos\frac{3}{2}\theta - 2\beta sin\theta sin\frac{3}{2}\theta) - e^{\beta(\theta-\pi)}sin\frac{\theta}{2}\right\} \\
& - K_2 / A \cdot \left\{-e^{\beta(\theta-\pi)}cos\frac{\theta}{2} + e^{-\beta(\theta-\pi)}(-cos\frac{\theta}{2} + sin\theta sin\frac{3}{2}\theta + 2\beta sin\theta cos\frac{3}{2}\theta)\right\}
\end{aligned} \tag{13}
$$

where $A = 2ch\pi\beta \cdot \sqrt{2\pi r}$ and the K_1, K_2 are Stress Intensity Factors in the global coordinate system. Thus

$$
\sigma_y + i\tau_{xy} = (K_1 + iK_2) \cdot (2\pi r)^{1/2}, \quad \theta = 0^\circ \tag{14}
$$

Based on Boundary Integral Equation and Boundary element method, we have written computer program "MFAC–91"(Multiphase Fracture Analysis of Concrete), which used quadratic isoparametric element and considered the characteristic of aggregates and interface microcracking.

3 Calculated examples

3.1 Example 1.

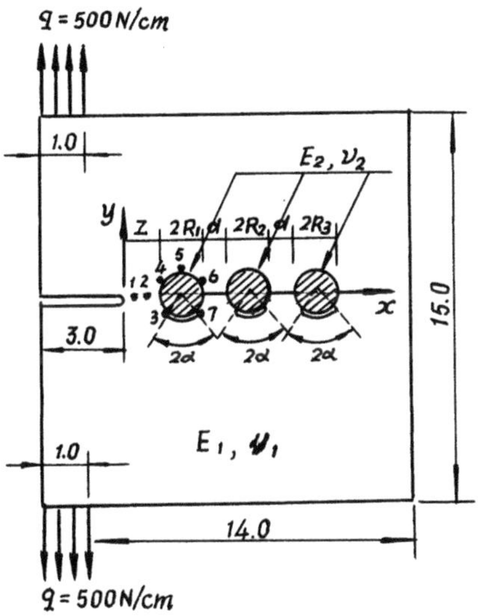

Fig.3 The specimen with a main crack in its edge

C20# concrete, E = 2.0 × 10^4MPa, $\nu = 0.167$, aggregate: E2 = 2.0~20.0 × 10^4MPa, strength against compression is 20MPa, and strength against tension is 2.5MPa.

On the joint surface, besides microcracking 2d = 90 °, we suppose the rest has friction coefficient f = 1.1, the locations of measuring points 1–7 are shown in figure 3. Among them, the point 1 is r = x = 0.05 cm and point 2 is r = x = 0.2 cm. when $E_2 = E_1 = 2.0 \times 10^4$MPa, the specimen is a homogeneous material.

The SIF calculated by computer program is $K_1 = \sqrt{2\pi r} \cdot \sigma_y = \sqrt{2\pi \times 0.05} \times 1187.5 = 66.59 \text{kgf} / \text{cm}^{3/2}$. Otherwise, the SIF calculated by formula $K_1 B\sqrt{W} / P = 29.6(a/W)^{1/2} - 185.5(a/W)^{3/2} + 655.7(a/W)^{5/2} \cdots$ is $K_1 = 5.166 \times 500 \times (14.5)^{-1/2} = 67.84 kgf/cm^{3/2}$, so their relative error is just 1.88%. The result shows sufficient accuracy. List the stresses τ_y, σ_y of points 1–7 and SIF of main crack on the table 1.

Table 1 shows: when the aggregate in front of the main crack propagation, which has the higher elastic modulus than the matrix, will reinforce the capability of concrete against the fracture failure. The quantitative increment depends on the distance from crack tip to aggregate. The shorter is distance, the bigger is increment.

Table 1 The stresses near the crack tip, SIF of main crack and stresses on the juncture surface
($2\alpha = 90°$, $E_2 / E_1 = 3$, $R_1 = R_2 = R_3 = 1.0$cm, $d = 0.5$cm)

No.	distance from tip toaqqre. Z(cm)	stress near the crack tip ($\times 10^{-2}$MPa)		SIF (kgf / $cm^{3/2}$)	compared with homo. body %	stress on the joint surface ($\times 10^{-2}$MPa)				
		P. 1 $\frac{\tau_{xy}}{\sigma_y}$	P. 2 $\frac{\tau_{xy}}{\sigma_y}$			P. 3 $\frac{\tau_{xy}}{\sigma_y}$	P. 4 $\frac{\tau_{xy}}{\sigma_y}$	P. 5 $\frac{\tau_{xy}}{\sigma_y}$	P. 6 $\frac{\tau_{xy}}{\sigma_y}$	P. 7 $\frac{\tau_{xy}}{\sigma_y}$
1	0.6	−5.2 1054.5	311.2 388.1	59.11	88.7	−63.1 −624.7	−268.9 152.4	98.4 105.1	141.0 331.4	318.7 −359.6
2	1.0	−27.3 1082.2	283.7 406.2	60.66	91.1	−91.7 −451.3	−236.4 108.1	83.0 82.2	131.5 268.4	303.4 −278.2
3	2.0	−22.5 1119.5	282.3 428.3	62.75	94.2	−108.6 −82.2	−139.0 22.7	40.4 13.3	90.2 1.38	210.5 −48.2

Moreover, to study the influence of microcracking at the aggregate bottoms, we reduce the angle from $2\alpha = 90°$ to $2\alpha = 5°$, while other conditions are completely same, the calculated results show in table 2.

Table 2 The calculated results from reducing the microcracking's length
($2\alpha = 5°$, $E_2 / E_1 = 3$, $R_1 = R_2 = R_3 = 1.0$cm, $d = 0.5$cm)

No.	distance from tip toaqqre. Z(cm)	stress near the crack tip ($\times 10^{-2}$MPa)		SIF (kgf / $cm^{3/2}$)	compared with table 1 %	stress on the joint surface ($\times 10^{-2}$MPa)				
		P. 1 $\frac{\tau_{xy}}{\sigma_y}$	P. 2 $\frac{\tau_{xy}}{\sigma_y}$			P. 3 $\frac{\tau_{xy}}{\sigma_y}$	P. 4 $\frac{\tau_{xy}}{\sigma_y}$	P. 5 $\frac{\tau_{xy}}{\sigma_y}$	P. 6 $\frac{\tau_{xy}}{\sigma_y}$	P. 7 $\frac{\tau_{xy}}{\sigma_y}$
1	0.6	−64.0 861.1	232.2 297.7	48.26	81.7	−200.7 −370.5	−259.5 10.9	23.90 188.6	70.4 −47.7	188.7 169.6
2	1.0	−19.6 1051.7	287.4 399.2	58.95	97.2	75.8 −206.5	−288.4 14.4	84.3 137.2	154.0 −4.1	139.6 −188.9
3	2.0	−21.4 1115.0	282.4 426.7	62.49	99.6	23.5 −34.6	−163.5 5.87	36.8 20.7	103.1 −4.2	80.4 −39.1

Table 2 shows: the length of interface microcracking has very sensible influence on the main crack's propagation. The longer is length of microcracking, the easier is propagation of the main crack. When the distance between microcracking and main crack is more short, the action is more obvious.

The stresses of round points of aggregate also have obvious changes. General tendency is the first: the tension stress and the possibility of failure are more as the microcracking longer and closer; the second: the stress is even as the microcracking shorter. The maximum tension will occur at the top symmetrical point 5, then it moves toward point 6 when microcracking is longer. Obviously, the location of the maximum tension depends upon the micro–cracking's length.

Other important factor is the elastic modulus of the aggregate. It is shown in table 3.

Table 3 Tthe stresses at points 1–7 as the change of elastic modulus rate of aggregate with matrix
($z = 0.6$cm, $2\alpha = 90°$, $R_1 = R_2 = R_3 = 1.0$cm, $d = 0.5$cm)

No.	E_2 / E_1	stress near the crack tip ($\times 10^{-2}$MPa)		SIF (kgf / $cm^{3/2}$)	compared with homs. body %	stress on the joint surface ($\times 10^{-2}$MPa)				
		P. 1 $\frac{\tau_{xy}}{\sigma_y}$	P. 2 $\frac{\tau_{xy}}{\sigma_y}$			P. 3 $\frac{\tau_{xy}}{\sigma_y}$	P. 4 $\frac{\tau_{xy}}{\sigma_y}$	P. 5 $\frac{\tau_{xy}}{\sigma_y}$	P. 6 $\frac{\tau_{xy}}{\sigma_y}$	P. 7 $\frac{\tau_{xy}}{\sigma_y}$
1	1.0	−136.8 1187.5	290.3 440.5	66.59	100	−83.0 −442.2	−190.0 108.8	68.0 74.4	97.2 30.8	284.5 −293.8
2	3.0	−5.2 1054.5	311.2 388.1	59.11	88.7	−63.1 −624.7	−268.9 152.4	98.4 105.1	141.0 331.4	318.7 −359.6
3	10.0	16.8 978.6	325.2 360.7	54.85	82.4	−34.2 −732.0	−318.0 176.8	119.1 127.5	168.1 289.3	316.1 −376.2

The result shows: the elastic modulus of aggregate is greater, the capability of resistance of concrete against the crack propagation also is higher. No doubt this conclusion agrees with our experience.

3.2 Example 2.

The microcracking occurs usually at the bottom of aggregates, but sometimes it occurs also at the other part of aggregate. Joint separating can be caused by smooth surface, oil pollution or inevitable shrinkage. For these reasons we calculate the cases of microcracking, which rounds the aggregate. (ref. to figure 4) The microcracking can be described in three different locations: AB, CD, EF. The calculated result by computer is listed in table 4. It shows the closer is microcracking from main crack tip, the easier is propagation of main crack. The microcracking at the AB and EF, the changed range of SIF is about 12.6%.

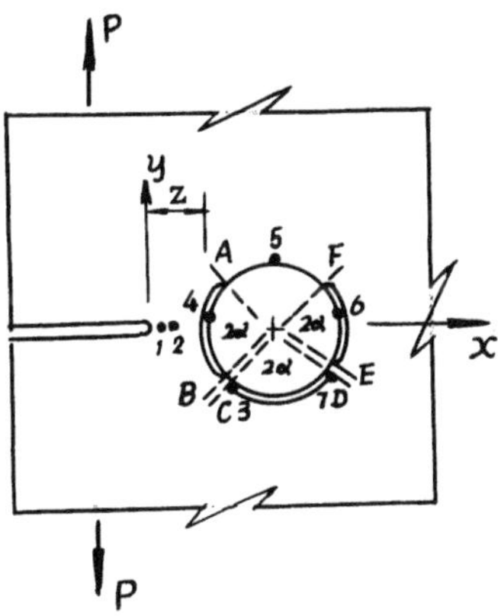

Fig.4 Different locations of microcracking rounded the aggregate

Table 4 Stresses at measuring points as the different locations of microcracking ($2\alpha=90°$, $R_1=R_2=R_3=1.0$cm, $d=0.5$cm)

No.	the locations of micro-cracking	stress near the crack tip ($\times 10^{-2}$MPa)		SIF (kgf/cm$^{3/2}$)	relative rate %	stress on the joint surface ($\times 10^{-2}$MPa)		
		P. 1 $\frac{\tau_{xy}}{\sigma_y}$	P. 2 $\frac{\tau_{xy}}{\sigma_y}$			P. 3 $\frac{\tau_{xy}}{\sigma_y}$	P. 5 $\frac{\tau_{xy}}{\sigma_y}$	P. 7 $\frac{\tau_{xy}}{\sigma_y}$
1	AB	−106.9 1152.3	129.9 354.3	64.59	100.0	−576.7 329.8	67.5 −1.03	−548.5 −300.7
2	CD	−5.2 1054.5	311.2 388.1	59.11	91.5	−63.1 −624.7	98.4 105.1	318.7 −359.6
3	EF	−8.6 1008.7	302.8 380.2	56.54	87.5	398.9 −89.4	−366.7 −4.48	385.7 68.0

In addition, we also calculate different ranks of aggregates, (ref. to figure 5) the result shows that most important factor is length and location of interface microcracking, especially the distance from main crack tip to neighboring aggregate. The rank of aggregates has a little influence. It is restricted by pages of this paper, they will be discussed in other place.

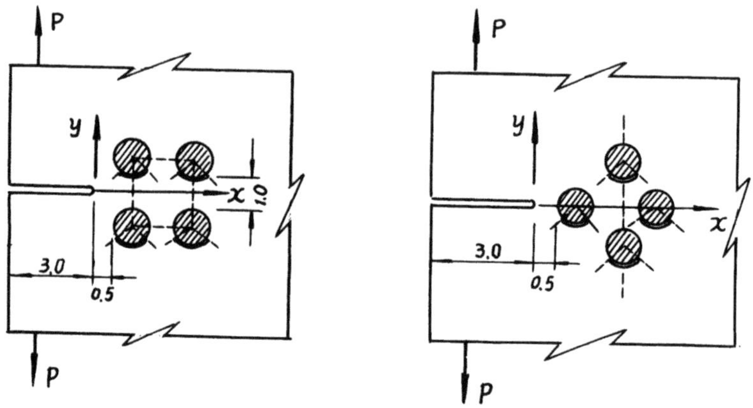

Fig.5 Different ranks of aggregates

Acknowledgement: this paper is supported by Natural Science foundation.

References

〔1〕 Alberto.Carpinteri: ASCE. Structure Diversion Vol.108, No.4, 1982.4, PP833–848

〔2〕 S.Mindess: Fracture toughness testing of Cement and Concrete ", Chapter 3, ISBN 90–247–2959–9, PP67–102

〔3〕 Zhou Rui–Zhong: The Synthetic Deformation Modulus and Synthetic Poisson's Ratio of Multiple Crack Body ", Engineering Fracture Mechanics Vol.35, No.4 / 5, 1990

〔4〕 Zhou Rui–Zhong & Lui Guan–Ting: "Numerical–Experimental Analysis of the Cracks in Massive Concrete Under Compression–Shear Condition", Proceedings of the Inter national Conference on the Development of Pumped Storage Power Station, Beijing, 1990.10.

〔5〕 Zhou Rui–Zhong: "Fracture Analysis for Curved Cracks on the Juncture–Surface between Aggregate and Cement–Mortar" Micro Mechanics of Failure of Quasi–Brittle Materials, Edited by S.P.Shah etc. ISBN 1–85166–511–0, 1990.6.

〔6〕 G.P.Cherepanov: "Mechanics of Brittle Fracture", McGraw–Hill Inc. 1979, ISBN 0–07–0107394.

PART THREE
FATIGUE

19 PROPERTIES OF CONCRETE RELATED TO FATIGUE DAMAGE

G. PETKOVIC
SINTEF Structural Engineering – FCB, Trondheim, Norway
R. LENSCHOW
Division for Structural Engineering, Norwegian Institute of Technology, Trondheim, Norway

Abstract
This paper gives an overview of the changes of strength and strain related properties of concrete under the influence of repeated loading. Changes in strain, strain rate, modulus of elasticity, and Poisson's ratio have been followed during cyclic loading, and in static tests after cyclic preloading. Loading situations resulting and not resulting in fatigue failure have been compared, so as to relate the criteria for an endurance limit to material characteristics. Special attention is paid to properties in the direction perpendicular to the application of loading. The relationship between transverse strain rate and the fatigue life was used for obtaining a formulation for the average damage rate as a function of the stress level. This article summarizes the contents of a thesis, and due to the lack of space includes only its most important parts. For more detailed information please see the full thesis (Petkovic, 1991).
Keywords: Fatigue, Compression, Strain, Damage, High strength concrete, Lightweight aggregate concrete.

1 Introduction

The expressions used for fatigue design are based on experimentally obtained S-N diagrams, which are in the usual form relatively well known and are similar in the existing codes of practice. The fact that this usual form is not satisfactory, and that not much of today's research in fatigue has yet found its way to design codes makes fatigue of concrete be still looked upon as a new research field. No experimental procedures are standardized, which makes results obtained in different laboratories difficult to compare. Due to the costs connected to such experiments one lacks data at the load levels which are most interesting for practice and has abundant data at less interesting load levels. Finally, which is most important, we do not have enough knowledge about the phenomenon, the effect of cyclic loading on measurable material properties. The objective of the work described in this article was formed based on the belief that a better understanding of the changes in the material structure caused by repeated loading would provide a better basis for determining fatigue damage and for the forming of design rules.

Fracture and Damage of Concrete and Rock – FDCR-2.
Edited by H.P. Rossmanith. © 1993 Taylor & Francis. ISBN 0 419 18470 8.

The experimental basis of this work is threefold:

- A series of tests on cube specimens of normal density ND65 concrete, exposed to cyclic preloading and static reloading.
- A series of constant amplitude fatigue tests on two types of normal density concrete, ND65 and ND95, and a lightweight aggregate concrete, LWA75, where the numbers present the planned compressive strength in MPa. The results belong to a large fatigue investigation on high strength concrete carried out at SINTEF FCB in the period from 1987 to 1990.
- A series of fatigue tests with a variable amplitude, from the same investigation of fatigue on high strength concrete.

2 The structure of concrete - the basis for material behaviour

Fatigue failure occurs at load levels lower than those causing failure under static loading. Two characteristics of the loading are important:

- The repetitive characteristic of loading implies that each new load cycle acts on a slightly changed material structure. A crack opened by one cycle will be the location of a stress concentration in the next cycle.
- Fatigue loading is usually a long term loading, revealing viscous characteristics of concrete. A low cyclic loading repeated many times can cause a rearrangement of the concrete material structure not necessarily resulting in crack propagation.

One can assume that two phenomena are occurring simultaneously under fatigue loading: crack propagation, and a form of material consolidation. There have been presented several hypotheses (e g Cook and Chindaprasirt, Sawko and Saha, Shah and Chandra) about what the consolidation is and all of them are connected to the inhomogeneity of concrete.

In an experimental series on normal density concrete with strength 65 MPa, cube specimens were firstly exposed to cyclic loading up to 30 or 60% of their expected fatigue life, and after that tested statically to failure. The duration of the cyclic preloading was determined by calculating the expected fatigue life from the relationship between strain rate and fatigue life obtained in an introductory test series. Keeping the strain rate in the preloading at a certain value meant "prescribing" a certain life time to the specimen. In static reloading, the compressive and splitting strength were measured, both in the direction of the cyclic preloading and in the perpendicular direction. Fig. 1 shows the results of strength and modulus of elasticity measurements.

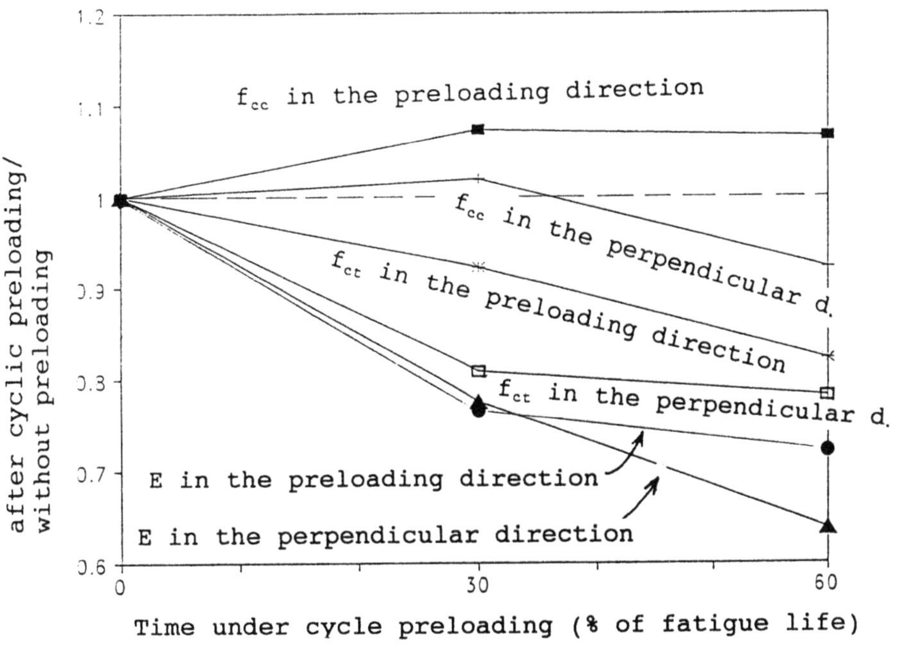

Fig.1. Strength and modulus of elasticity measured on specimens previously exposed to a cyclic preloading of duration either 30 or 60% of the expected fatigue life, compared to initial values

The following could be noted:

The compressive strength both in the same direction as the cyclic loading and in the perpendicular direction was higher than the initially measured strength after an exposure to cyclic preloading lasting 30% of the expected fatigue life.

After cyclic preloading up to 60% of the expected fatigue life, the compressive strength in the same direction as the cyclic loading remained higher than the initial strength and the compressive strength in the perpendicular direction was reduced.

The tensile strength was reduced in both directions, and already after 30% of the fatigue life.

The modulus of elasticity was also reduced, the reduction being higher for the perpendicular direction.

The changes in the stress-strain relationship under static reloading dependent on the duration of the cyclic preloading and the direction are shown in Fig. 2. Fig. 2a) shows the stress-strain relationship registered in reloading in the same direction as the cyclic preloading, and Fig. 2b) shows the same in the perpendicular direction.

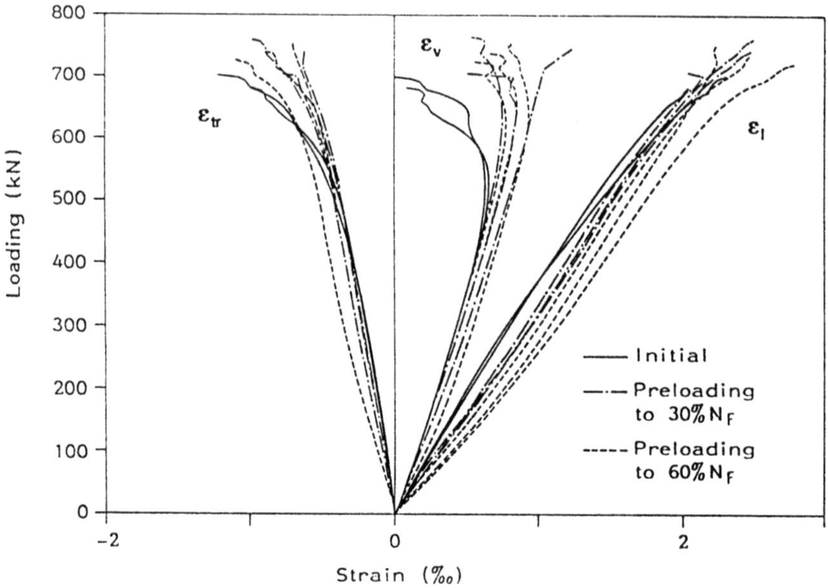

a) Static reloading in the same direction as cyclic preloading

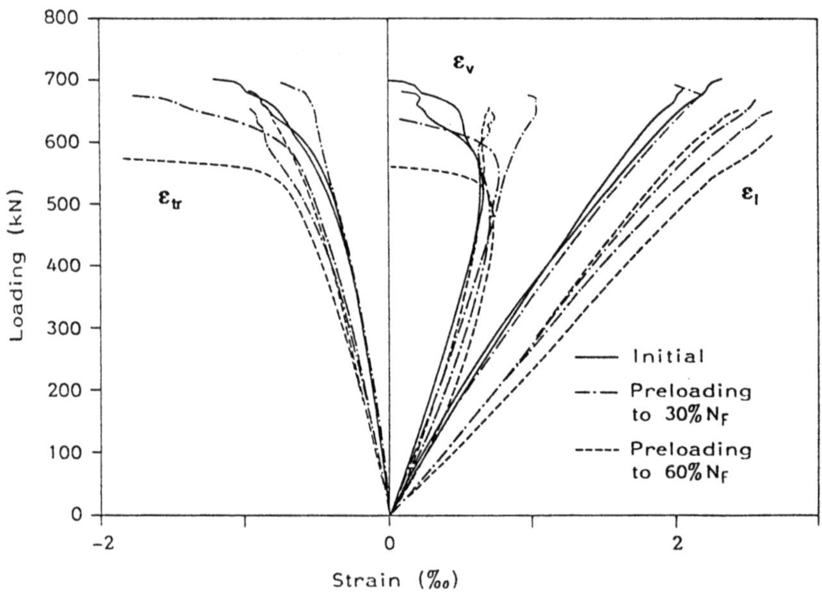

b) Static reloading in the perpendicular direction

Fig.2. Stress-strain relationship measured on specimens exposed to cyclic preloading

It can be noted that the shape of the stress-strain diagram changes towards more concave towards the stress axis for specimens reloaded in the same direction as the cyclic preloading. This change of shape has influence on the value of the modulus of elasticity and Poisson's ratio. The modulus of elasticity may be higher and Poisson's ratio lower if calculated from strain values corresponding to a higher load level.

The experiment showed that fatigue damage has a direction. This loss of isotropy is the most general indication of fatigue damage.

3 Strain parameters indicating fatigue damage

The changes of strain related parameters under cyclic loading was followed on specimens of two types of normal density concrete and one type of lightweight aggregate concrete were included: ND65, ND95, and LWA75, where the numbers present the planned compressive strength in MPa. These specimens were exposed to cyclic loading of minimum relative load level from 5 to 60% of the static strength, and maximum load level from 55 to 95% of the static strength.

It was assumed that it would be possible to establish a reliable mathematical relationship between the strain and fatigue life. This was, however, not the case. Fig. 3 shows strain development curves typical for the fatigue tests. The last, rapid, phase of strain development, indicating oncoming fatigue failure, starts approximately at 75% of the fatigue life. A relatively reliable relationship could be found between the transverse strain at 75% of the fatigue life (or strain at the start of the failure phase) and the fatigue life, but it should be verified on an larger number of specimens.

It can be seen from the same figure that lightweight aggregate concrete exhibits the largest part of the strain at the very beginning and the end of the fatigue life. Strain development both under static and cyclic loading are dependent on the combinations of stiffness of aggregate and cement pasta. For normal density high strength concrete, the stiffness of the cement paste is closer to the aggregate stiffness. For high strength lightweight aggregate concrete, this difference is even smaller. Similarly to the absence of transverse dilatation before failure in static tests on lightweight aggregate concrete, in fatigue tests of lightweight aggregate concrete no clear indications of oncoming fatigue failure can be found in the strain development curves.

Concerning strain rate, the existence of a linear relationship between the logarithm of the strain rate and the logarithm of the number of loading cycles to failure has already been found in earlier investigations (Cornelissen and Siemes, Sparks, Weigler and Rings), and was confirmed in this test series. The scatter around the relationship established between the average transverse strain rate and the fatigue life was found to be smaller for the transverse direction. The relationship obtained for the transverse direction was used later on in developing the expression for the average damage rate as a function of the stress levels. Tests on variable load levels have shown that the strain rate is very much dependent on the

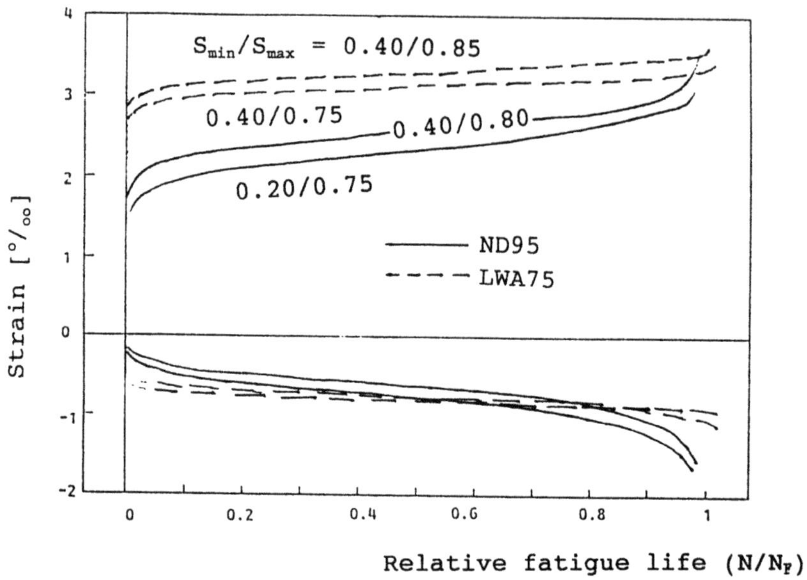

Fig.3. Typical strain development curves for NC and LWA concrete under fatigue loading

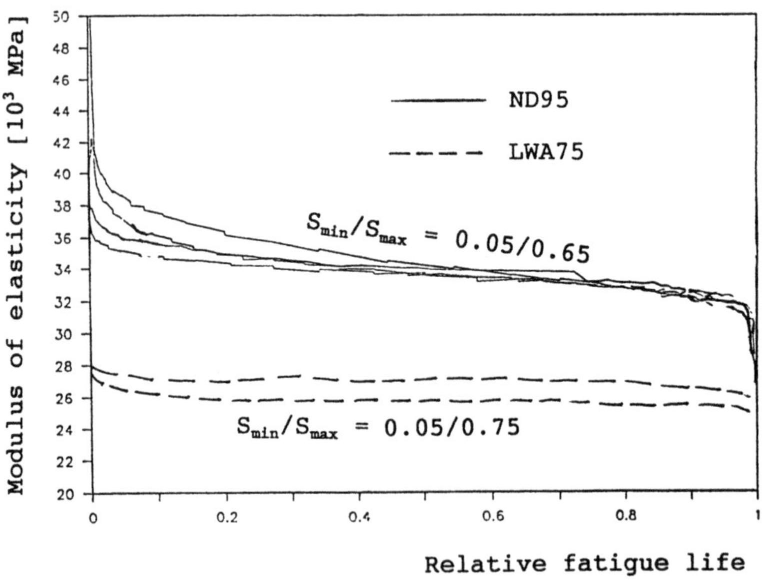

Fig.4. The development of the modulus of elasticity under fatigue loading, calculated within each cycle

previously applied loading. Relationships between strain rate and fatigue life are, therefore, relevant only for constant amplitude tests.

The **modulus of elasticity** decreases under fatigue loading, and Fig. 4 shows some curves typical for the test results. Also here there seems to be a stage of almost linear decrease. The rate of decrease is found to be mostly dependent on the stress amplitude. Tests with variable stress amplitude have shown that this factor is also dependent on the loading history. Lightweight aggregate concrete exhibits almost no change in the modulus of elasticity.

It was also interesting to look into the **relationship between longitudinal and transverse strains**. Two such parameters were calculated:

The relationship between the transverse and longitudinal strain within one loading cycle; and

The relationship between the totally developed transverse and longitudinal strain.

The total strain ratio has been shown to be higher for specimens with a shorter fatigue life, in that way indicating that loads resulting in a shorter fatigue life cause a relatively smaller consolidation compared to transverse dilatation, while the loads resulting in longer fatigue life cause higher consolidation.

All the investigated material parameters have shown to be more indicative for fatigue life if expressed by their rate, change in time. Even then they express the fatigue damage only in qualitative terms. Quantitatively, the relationship between material parameters and fatigue life is very much dependent on the load level for constant amplitude loads and the loading history for variable amplitude loads.

In the fatigue tests with variable amplitude, some specimens were exposed to low fatigue loads: the minimum load was 5% of the static strength and the maximum load was 45 or 55% of the static strength. These tests did not end in fatigue failure. Using the usual term, this loading could be said to be under the endurance limit of concrete. In such tests it was noted that both the strain rate, the modulus of elasticity, and Poisson's ratio reach a certain level, this level, however, being dependent on the loading history. One could conclude from the results that as long as these parameters do not undergo changes fatigue failure is not likely to occur. Indications of oncoming fatigue failure are an increasing strain rate, a more rapid modulus of elasticity reduction, and an in creasing total strain ratio. For lightweight aggregate concrete, however, it is much more difficult to establish such conclusions.

4 Damage presentation model and ideas for future research

A effort was put into establishing a relationship between the measured parameters and fatigue life, resulting in the diagram shown in Fig. 5. The fatigue life enters this diagram through the parameter, Δ, called "average damage rate", which is the reciprocal value of the fatigue life, $1/N_F$, where N_F is the number of cycles to failure. N_F is here calculated based on the relationship with the transverse strain rate, which enables specimens not resulting in fatigue failure to be included. Fig. 5 shows the relationship between the average damage rate and the minimum and maximum stress level.

As in the S-N curves, a solution was not found that would enable elimination of the logarithmic scale. However, contrary to the S-N curves, in this diagram the area of low stress levels and corresponding long fatigue lives is relatively small compared to the whole diagram area.

In fatigue of concrete it is usual to use the Palmgren-Miner hypothesis of linear damage development. What makes this method unsuitable for concrete is its inability to take into account the sequence of loading and that it gives no one answer to how one should take into account the low loads not resulting in fatigue failure, but in combinations with higher loads contributing to a higher fatigue life. Approaching the problem from the aspect of the structure of concrete, it is reasonable to assume that damage development curves will not be the same independent on the stress level. A lower load may cause a slower damage development in the beginning, ending with a rather sudden failure after the consolidation capacity of the material is used up. A higher load may, on the contrary, cause a steeper damage development in the beginning and continue with an even damage growth until failure.

Instead of using the stress levels, one can define "low" or "high" load by the help of the average damage rate, Δ. If the change of slope is defined as a function of the average damage rate, a simple possibility of which is shown in Fig. 6, a set of non-linear damage development curves can be developed, which enables taking into account the sequence of loading levels. This is the direction in which this work is planned to be continued.

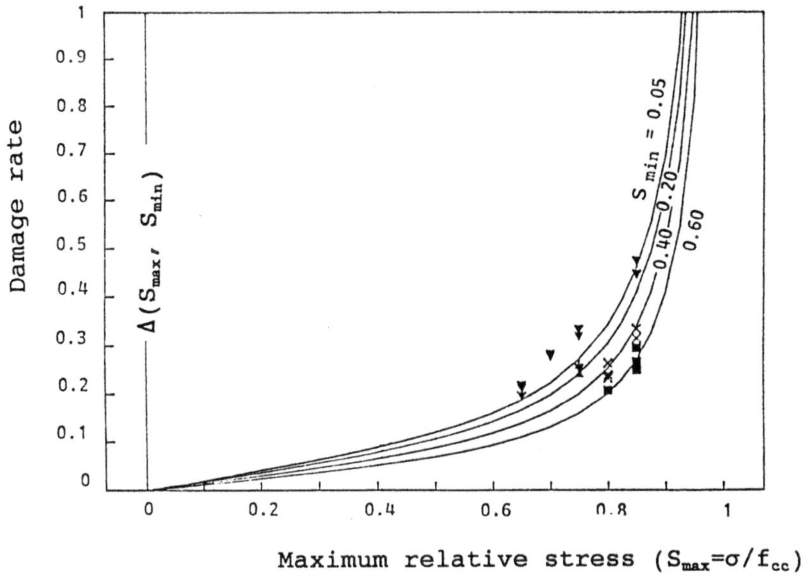

Fig.5. The relationship between the stress levels and the fatigue life expressed through a parameter Δ, average damage per cycle

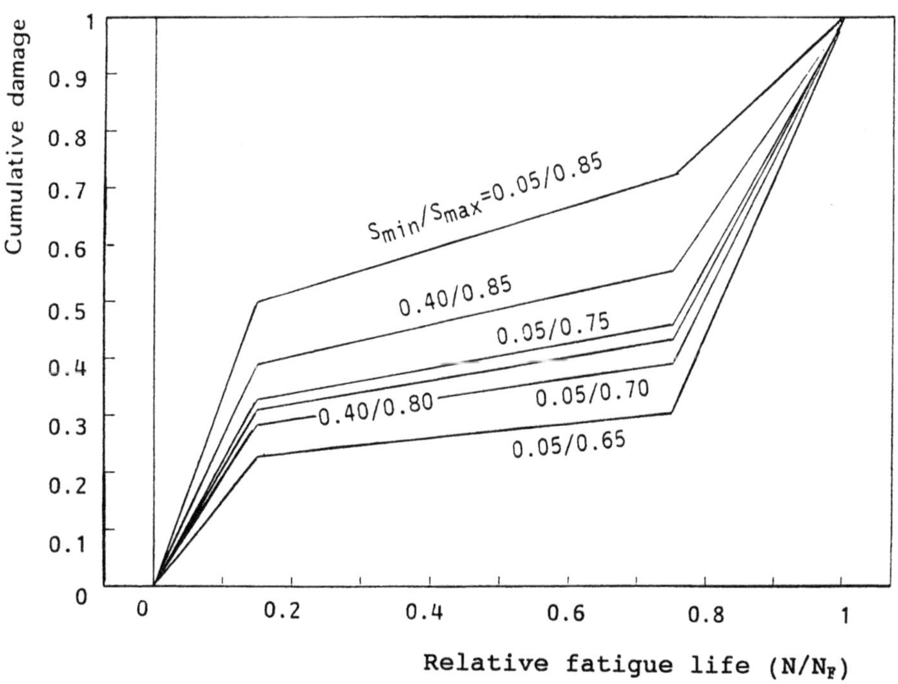

Fig.6. Hypothetical non-linear damage development curves

5 REFERENCES

Cook, D.J. and Chindaprasirt, P.: "Influence of loading history upon the compressive properties of concrete", **Magazine of Concrete Research**, Vol.32, No.111, June 1980, pp 89-100.

Cornelissen, H.A.W. and Reinhardt, H.W.: "Effect of static and fatigue preloading on residual strength and stiffness of plain concrete", **6th Biennal European Conference on Fracture** - ECF6, Amsterdam, The Netherlands, June 1986, Vol.III, pp 2087-2095.

Cornelissen, H.A.W. and Siemes, A.J.M. (1985) "Plain concrete under sustained or tensile and compressive fatigue loadings", **BOSS Conference Amsterdam**, Elsevier Science Publishers B.V., pp 487-498.

Petkovic, G. (1991) **Properties of concrete related to fatigue damage with emphasis on high strength concrete**, Dr.Ing thesis, The Norwegian Institute of Technology, Trondheim, Norway, 1991.

Petkovic, G., Rosseland, S., and Stemland, H. (1989) **Fatigue of high strength concrete**, SINTEF Report STF65 F89025.

Shah, S.P., and Chandra, S. (1970) "Fracture of concrete subjected to cyclic and sustained loading", **ACI Journal**, pp 816-825

Sawko, F. and Saha, G.P. (1968) "Fatigue of concrete and its effect upon prestressed concrete beams", **Magazine of Concrete Research**, Vol 20, No 62, pp 21-30.

Sparks, P.R. (1982) "The influence of rate of loading and material variability on the fatigue characteristics of concrete", **ACI publication SP-75**, pp 331-343.

Weigler, H. and Rings, K-H. (1988) "Unbewehrter und bewehrter Beton unter Wehselbeanspruchung", **Duetscher Ausschuss fur Stahlbeton**, Heft 383.

20 CYCLIC LOADING OF REINFORCED CONCRETE CONTINUOUS BEAMS: FINITE ELEMENT ANALYSIS AND EXPERIMENTAL INVESTIGATION

D. BRUNO, G. PORCO, G. SPADEA and R. ZINNO
Department of Structural Engineering, University of Calabria, Cosenza, Italy

Abstract

The object of the present work is an experimental investigation for the determination of post-cracking behaviour of reinforced concrete continuous beams under cyclic loading. In addition a FEM procedure is developed in order to model the non-linear behaviour of hyperstatic reinforced concrete structures.

The modelling of concrete material includes non-linear stress-strain behaviour, tensile cracking, tension-stiffening, compression crushing and strain-softening, whereas steel material is assumed to have an elasto-plastic piece-wise behaviour which simplifies the Ramberg-Osgood equation for uniaxial reverse loads.

The experimental data to determine the moment-curvature relationships are discussed in terms of the gage length and its position with respect to the location of longitudinal reinforcement.

Keywords: Reinforced Concrete, Load Redistribution.

1 Introduction

Modern computational techniques like the finite element method have been used in the non-linear analysis of reinforced concrete structures for more than a decade.

Reinforced concrete has a very complex behaviour, involving phenomena like inelasticity, cracking, time-dependency and interactive effects between concrete and reinforcement. The development of a material model for uncracked and cracked reinforced concrete is a particularly challenging field. The special properties of reinforced concrete structures must be considered for all stages of the computational process, including the variational formulation, the discretization techniques and the solution strategy for the non-linear equations.

Fracture and Damage of Concrete and Rock – FDCR-2.
Edited by H.P. Rossmanith. © 1993 Taylor & Francis. ISBN 0 419 18470 8.

The finite element approach requires more computer equipment and engineer expertise but enables a more general and consistent analysis. If a more realistic assessment of the behaviour of reinforced concrete structures is carried out, the cost is frequently reduced and higher safety is usually gained.

Generally the overcoming of the first crack moment in a bent beam provokes, as a consequence of crack growth, the reduction of its flexural rigidity.

Suitable moment-curvature relationships, must take into account both the finite increment of curvature due to the crack opening and the transferring of the tensile loads both to the reinforcing steel and to the concrete included between two consecutive cracks ("tension stiffening effect"). In this way it is possible to develop finite element models which give results sufficiently close to available experimental ones relative to the monotonically loaded beams. Furthermore, as is well known, the behaviour of reinforced concrete structures under cyclic loading is influenced both by the hysteretic response of materials and by the loss of the bond and tension stiffening contribution under repeated loading.

With this in mind, the present work develops an experimental investigation for the evaluation of moment redistribution in reinforced concrete continuous beams monotonically loaded or in a regime of low cycle fatigue and also presents a numerical analysis carried out by using the finite element method.

The results are then checked against experimental results obtained from tests on full-scale beams.

2 Moment curvature relationships

The evaluation of the actual deflections in reinforced concrete structures requires the knowledge of an effective moment-curvature relationship of a generic beam element.

If a beam element is subjected to flexure its behaviour varies between two limit states, determined by the position of the neutral axis. In the first stage, called state I, and characterized by the maximum depth of the neutral axis, the total cross section (steel and both tensed and compressed concrete) reacts to the bending moment which is lower than the first cracking moment M_{cr}. In the second stage, state II, because of the crack, the depth of the neutral axis reaches a minimum and only steel and compressed concrete react to equilibrate the bending moment.

Moreover, the actual behaviour of bent reinforced concrete beams under service loads evidences average values of inertia (J_m) comprised between that of state I (J_1) and those of state II (J_2). Fig. 1 shows, schematically, the moment-curvature diagram of a reinforced concrete beam: the linear behaviour of state I is acted on when $M < M_{cr}$, while for $M > M_{cr}$ the actual behaviour is intermediate between the states I and II because of the tension stiffening effect, i.e. the increase of rigidity with respect to the state II behaviour due to the reacting of tensed concrete comprised between two consecutive cracks. In addition in fig.1 J_g indicates the inertia of the gross section, without considering the contribution of steel reinforcement.

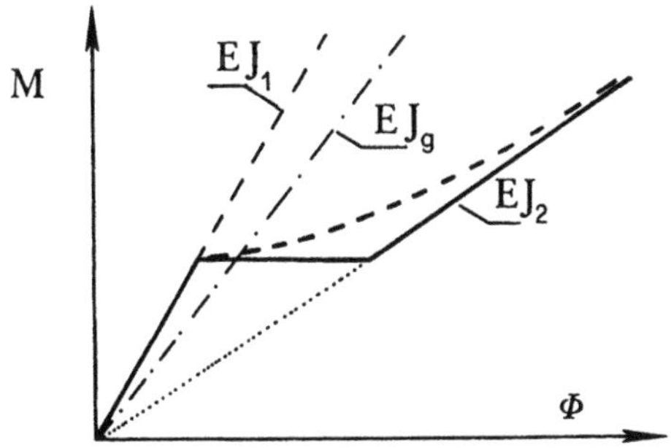

Fig.1. Moment-curvature relationships of a reinforced concrete section

Several formulations suitable for modelling the tension stiffening effect have been proposed. Among these, that proposed by CEB has been adopted in the present work.

In this formulation the actual moment of inertia is given by:

$$J_m = (J_1 J_2)/[J_1\alpha + J_2(1 - \alpha)] \tag{1}$$

where the coefficient α assumes the following values:

$$\begin{aligned} \alpha &= 1 - \beta_1\beta_2\ (M_{cr}/M)^2 && \text{when } M \geq M_{cr} \\ \alpha &= 0 && \text{when } M < M_{cr} \end{aligned} \tag{2}$$

In the eqn.(2) β_1 is a coefficient which characterizes the bond properties of the steel bars and β_2 is a coefficient which takes into account the duration of the application and repetition of the loads.

3 Formulation of the problem

3.1 Kinematics and constitutive equation

According to the beam model proposed by Timoshenko, which includes the shear deformability effects, the displacement field in the point (x,y,z) of a beam can be put as:

$$\begin{aligned} u(x,y,z) &= u(x) - z\phi(x) \\ v(x,y,z) &= 0 \end{aligned} \tag{3}$$

$$w(x,y,z) = w(x)$$

where u, v and w are the displacement components in the x, y and z direction respectively.

From eqns. (3) we obtain the strain components:

$$\varepsilon_{xx} = u' - z\phi'; \qquad \gamma_{xz} = 1/2\ (w' - \phi) \tag{4}$$

Furthermore, the constitutive equations are:

$$\sigma_{xx} = E\varepsilon_{xx}; \qquad \tau_{xz} = 2GK\gamma_{xz} \tag{5}$$

where E is the Young's modulus, G is the transverse shear modulus and K is the shear correction factor.

3.2 Variational formulation and FEM discretization

The behaviour of hyperstatic reinforced concrete structures subject to monotonically increasing loads is now analysed by using a FEM discretization .

With reference to the Timoshenko beam theory above, we consider the total potential energy functional.

In particular we have:

$$\pi = 1/2 \int_o^l \int_A \sigma_{ij}\varepsilon_{ij}\ dV \ - \int_o^l qw\ dx - \Sigma_i F_i u_i \tag{6}$$

where the first term on the right hand of equation (6) represents the strain energy of the beam, while the sum of the second and the third ones gives the work of the external loads.

The stationary principle applied to the functional (6) gives :

$$\delta\pi = EA\int_o^l u'\delta u'dx + EJ\int_o^l \phi'\ \delta\phi' dx + KGA\int_o^l (w'\delta w' + \phi\delta\phi + \\ -w'\delta\phi - \phi\delta w')dx - \int_o^l q\delta w\ dx - \Sigma_i F_i \delta u_i \tag{7}$$

In the previous equations l denotes the lenght of the beam, A is its cross section area and J is its moment of inertia.

The beam kinematics (3) allows us the use of lagrangian C^o - elements. Therefore, the following interpolation of displacements can be used:

$$u = \Sigma_i u_i f_i \qquad \phi = \Sigma_i \phi_i f_i \qquad w = \Sigma_i w_i f_i \qquad (i= 1,N) \tag{8}$$

where u_i, ϕ_i, w_i are the nodal values of the unknown functions, f_i are the global interpolation functions and N is the number of elements of discretization. We obtain the following equation:

$$K\ U = F \tag{9}$$

The solution to the problem depends on the cracking pattern of the entire structure.

In fact, under service loads, a beam element can be in the cracked or in the uncracked state. Consequentely its contribution to the flexural rigidity can be different.

Therefore, a step by step procedure must be adopted: according to the bending force applied to each beam element, its flexural rigidity must be assumed with reference to the actual moment of inertia (eqn.(1)).

Obviously, at the first step, for all elements it is assumed the inertia relative to the gross section of concrete: $J_g = J_1$.

4 Behaviour of reinforced concrete beams under repeated loads

The analysis of reinforced concrete structures under cyclic loadings is particularly complicated both for the hysteretic behaviour of the materials and for the loss of bond under repeated loading.

Reinforced concrete structures under cyclic loading beyond the elastic limit can be studied taking into account the non-linear behaviour of the materials and by using a simplified approach.

The mechanical model adopted in this work is based on the following assumptions:

- Idealization of steel behaviour with a piece-wise linearization of the classic Ramberg-Osgood relationship;
- Stress-strain relationship of cycling loaded concrete according to Smith and Young's proposal;
- Confining effect of web reinforcement;

As proposed by (Bruno et al., 1990) a step by step procedure, suitable in identifying the sectional behaviour under cyclic loads has been adopted.

The envelope curve of stress-strain relationship for concrete in compression under repeated load is represented by the equation (Smith et al., 1955; Ahmad et al., 1982):

$$\sigma_c/\sigma_{co} = \varepsilon_c/\varepsilon_{co} \exp(1 - \varepsilon_c/\varepsilon_{co}) \tag{10}$$

where $(\sigma_{co},\varepsilon_{co})$ are the coordinates of the peak point for the envelope curves and:

$$\begin{aligned} \sigma_{co} &= f_{ck}+K_1F_{rp} \\ \varepsilon_{co} &= \varepsilon_{ck}+K_2F_{rp} \end{aligned} \tag{11}$$

The terms F_{rp}, K_1 and K_2 take into account the confining effect of lateral reinforcement. Their explicit expression are reported in (Bruno et al., 1990) together with the definition of envelope curves according to the loading-unloading condition. The simplification of the Ramberg-Osgood equation for cyclically loaded steel, consists of sets of straight lines parallel to the elastic slope and inclined to it. In the analysis, the tension stiffening effect and the loss of bond is also taken into account.

5 Experimental Investigation

The models outlined for the analysis of force redistribution in hyperstatic reinforced concrete structures and for the determination of their sectional behaviour under cyclic loading have been validated both by a numerical and an experimental investigation.

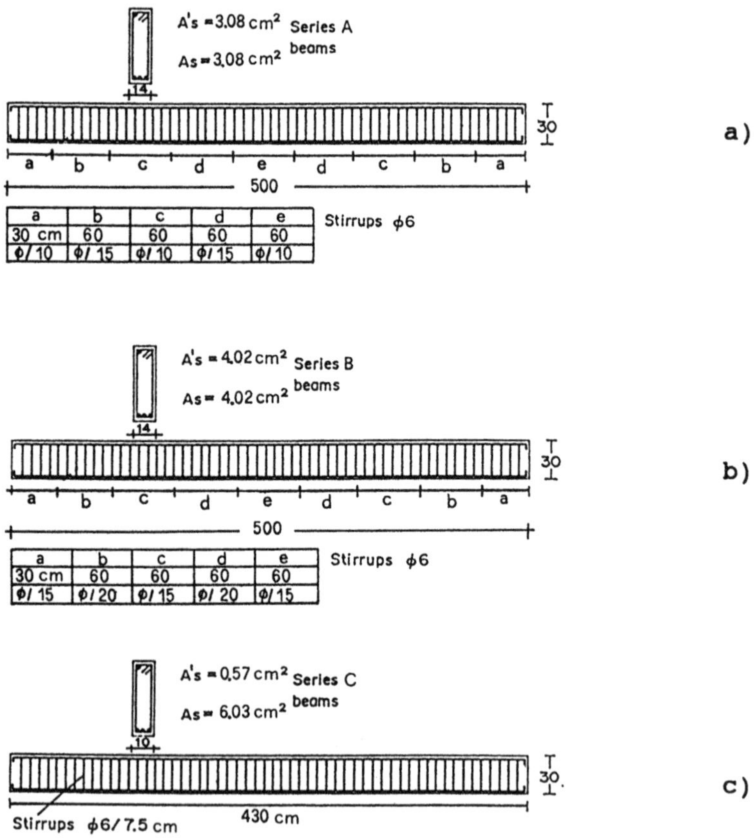

Fig. 2 Full-scale beam samples.

The experimental program has been carried out by testing a set of 12 full scale beams.

The proportions by weight of concrete mix prepared (cement: fine aggregate: coarse aggregate:water), were: 1:2:3:2:0.5.

Type III ASTM Portland cement, fine and coarse aggregates, with a maximum size respectively of 1 mm and 15 mm were employed. The grading curve of aggregates was chosen in order to have a good fit with the Fuller's curve.

Three series of beams have been manufactured with the steel reinforcement (longitudinal and lateral) as indicated in fig.2

The beams of Series A and B were tested by using the framework and testing apparatus represented in fig.3. The scheme of continuous beam was achieved by using, as supports, three strain-gage load cells. In this way the reaction's redistribution could be measured.

The loading was applied by using a 300 KN worm gear actuator: so that controlled dispacement tests could be carried out. This was of fundamental importance in the cracked or in the unloading stages.

The measuring apparatus set up is indicated in the same fig.3.

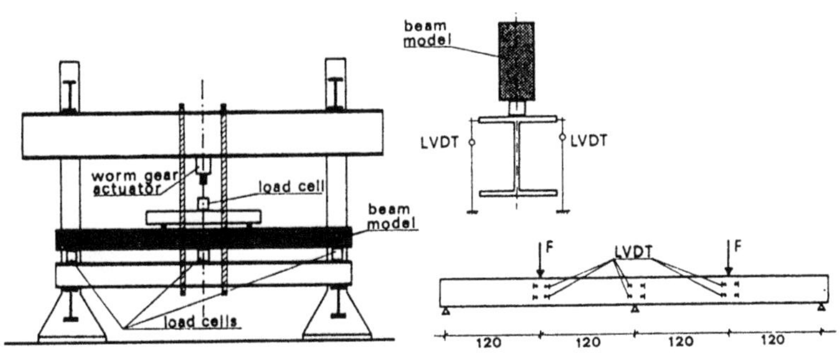

Fig. 3 Testing apparatus: loading scheme and instruments.

In particular the curvatures have been determined by using pairs of LVTD inductive displacement transducers with a gage-length to beam-height ratio equal to 1.

This choice allowed us to measure the average curvature of the beam element under consideration, which is lower compared to the local curvature in those zones which include a crack. More details of the experimental apparatus set up can be found in (Spadea, 1990).

The moment-curvature diagrams relative to the series A beams, both for the central support section and for the midspan section are given in fig.4

It can be noted that the finite element approach gives results close to those of the experimental investigation.

Moreover the cracking evolution represented by passing

from state I to state II is well evidenced.

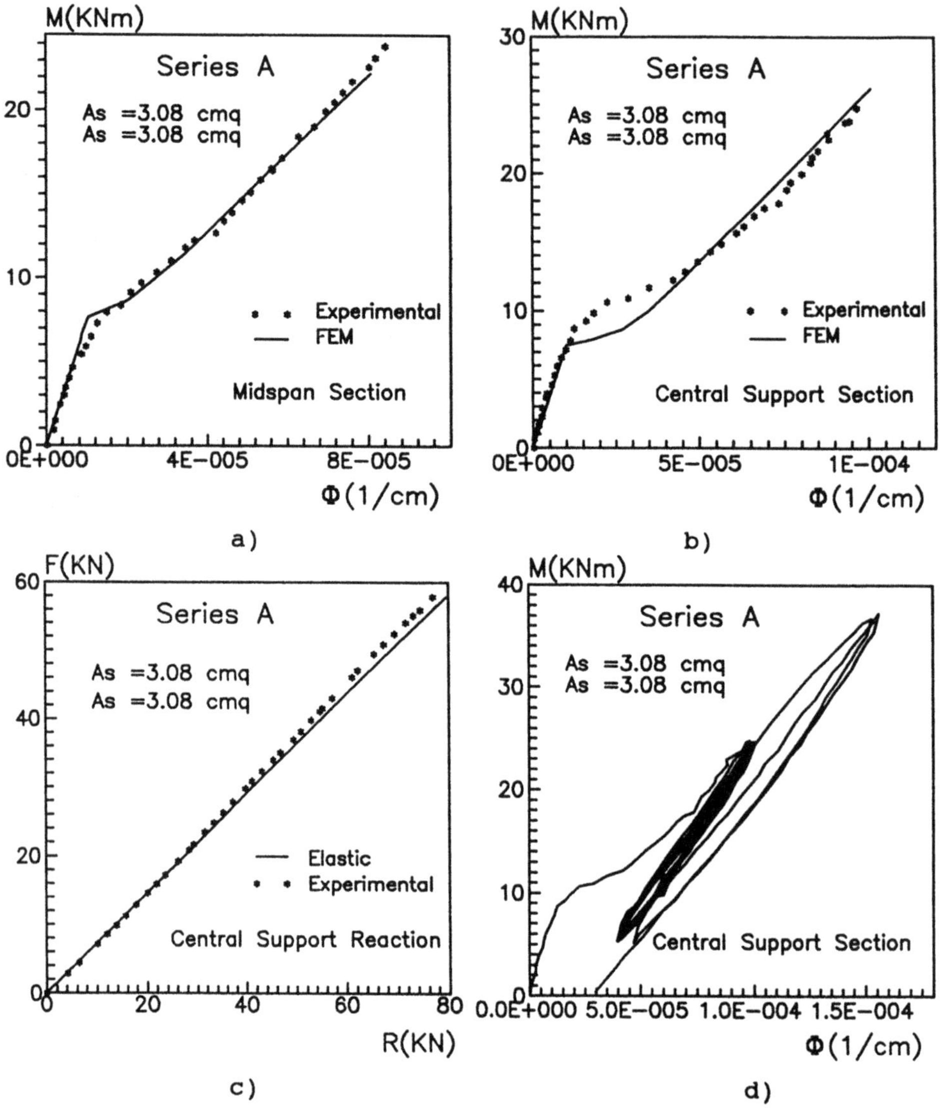

Fig. 4 a,b,c,d. Main results of series A beams.

Fig. 4c shows the variation of the reaction in the central support, as the external load increases, compared with the linear behaviour. It can be noted that, for the model tested, the force redistribution is almost negligible. A slight higher redistribution occurs when the reinforcement ratio is lower, expecially in the central support sections (Spadea, 1990).

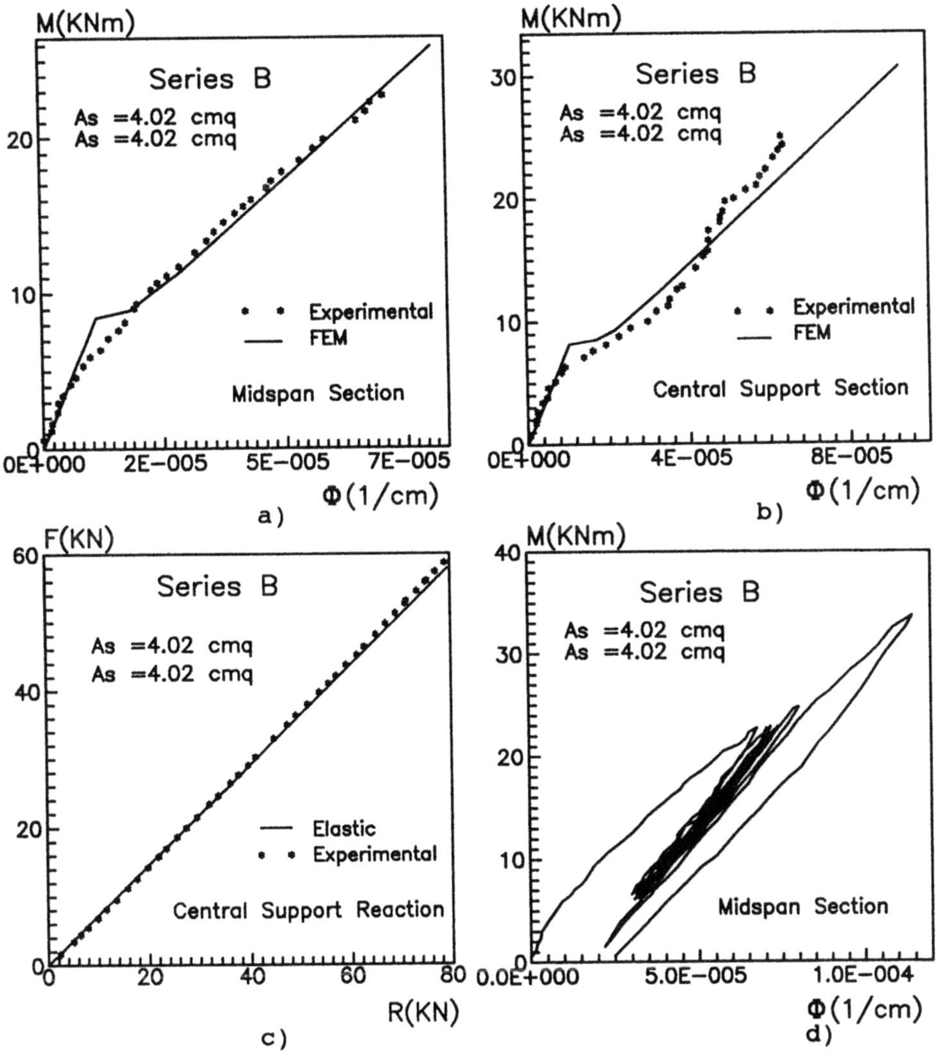

Fig 5 a,b,c,d Main results of series B beams.

Fig. 4d shows the moment-curvature diagram in the central support section, experimentally determined, with cyclic loading. After the first step the unloading and reloading branches are about linear: a little increase of curvature with cycling can be observed.

Fig. 5 a,b,c,d, give the analogous diagrams for the series B beams except for the cyclic loading diagram which is that of the midspan section beam element.

Finally, Fig. 6 shows the moment-curvature diagram determined for a simply supported reinforced concrete beam of series C, subject to cyclic reversed loading. The testing apparatus (loading scheme and instruments' position) is that illustrated with more details in (Bruno et al., 1990).

The experimental results are comparated with those obtained by using the numerical approach outlined in the previous sections. It can be noted that the curvatures calculated by using the described step by step procedure close enough to those experimentally determined.

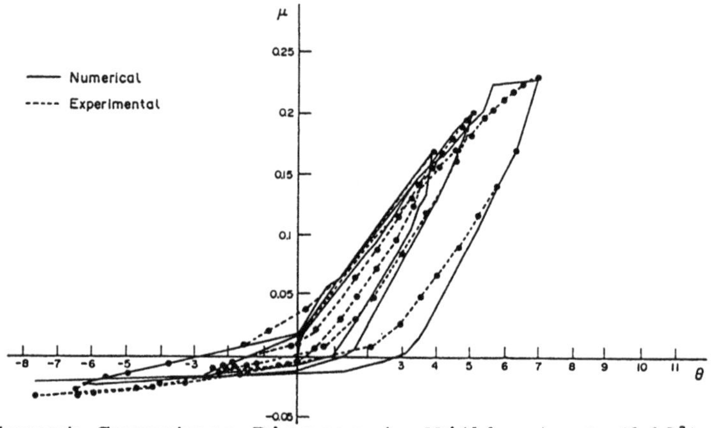

Fig. 6. Moment-Curvature Diagram ($\mu=M/(bh\sigma_{co})$; $\Theta=\phi h10^3$)

6 References

Ahmad, S.H. Shah, S.P. (1982) Stress - Strain Curves of Concrete Confined by Spiral Reinforcement. **ACI Journal**, 79-46.

Bruno, D. Olivito, R.S. and Spadea, G. (1990) Fracture in Reinforced Concrete Beams Under Repeated Loads. **Engineering Fracture Mechanics**, Vol 35, N.1/2/3, pp.269-275.

C.E.B. Manual (1985) **Cracking and deformation**. Ecole Politechnique Federale de Lausanne.

Cosenza, E. (1990) Finite Element Analysis of Reinforced Concrete Elements in a Cracked State. **Computers & Structures**, Vol 36, N.1, pp.71-79.

Smith, G.M. and Young, L.E. (1955) Ultimate Theory in Flexure by Exponential Function. **ACI Journal**, 349.

Smith, G.M. and Young, L.E. (1956) Ultimate Flexural Analysis Based on Stress - Strain Curves of Cylinders. **ACI Journal**, 597.

Spadea, G. (1990) Experimental Results on Moment Redistribution in Reinforced Concrete Continuous Beams. **Applied Stress Analysis**, (T.H. Hyde and E. Ollerton in Eds.). Elsevier Applied Scienze, London & New York..

21 CYCLIC BEHAVIOR OF REINFORCED CONCRETE BEAMS: DIFFERENT RESPONSES DEPENDING ON MECHANICAL AND GEOMETRICAL PROPERTIES

A. CARPINTERI and R. MASSABÓ
Department of Structural Engineering,
Politecnic of Turin, Italy

Abstract
A Linear Elastic Fracture Mechanics Model suitable for the analysis of different kind of collapses in reinforced concrete members is proposed. A concrete beam element, with a steel layer and a propagating through–thickness edge crack, subjected to unidirectional cyclic bending moment, is analyzed. Moment versus localized rotation relationships for the beam cross section are obtained. Yielding of steel, slippage between steel and concrete, elastic shake–down, plastic shake–down and stable and unstable crack propagation in concrete are examined. The influence of mechanical and geometrical properties on the behavior of the beam is discussed.
Keywords: Reinforced Concrete, Fracture Mechanics, Cyclic Loading, Shake–down.

1 Introduction

Reinforced concrete is often used for structures subjected to repeated loading such as traffic loads or earthquakes. Innumerable experimental tests have shown that reinforced concrete members under cyclically varying loading exhibit both stiffness and strength degradation principally due to flexural cracks propagation as well as steel yielding and slippage and inelastic shear deformations (Popov 1980).

Various hysteresis models have been introduced to simulate this behavior and among them the Takeda Model is the most sophisticated (Takeda, Sozen and Nielsen 1970). These empirical models were developed based on the results of several experimental studies. They define a primary load–deformation curve for monotonic loading (trilinear in the Takeda Model) and a series of unloading and reloading curves according to rules defining global stiffness and strength degradation. In the model proposed by Wang and Shah (1987) accumulated damage depending on the number of cycles is also taken into account.

Object of this paper is the simulation of the global response under cyclic loading of a reinforced concrete section with a Linear Elastic Fracture Mechanics Model which considers different elementary collapse mechanisms such as

Fracture and Damage of Concrete and Rock – FDCR-2.
Edited by H.P. Rossmanith. ISBN 0 419 18470 8.

crack propagation and steel yielding or slippage. In particular a precracked concrete beam, reinforced by a steel layer, under unidirectional cyclic bending moment is considered. The model follows a previous one proposed by Carpinteri Al. and Carpinteri An. (1984) with the introduction of some essential modifications regarding the global scheme, the imposed congruence conditions, the stress intensity factors, etc.

The behavior of the r.c. cross section is examined neglecting the phenomena relating to the surrounding concrete such as distributed compliances, diffused damage and the presence of other macrocracks. The model has been previously used to simulate monotonic tests on r.c. beams in bending and the numerical results turned out to be in agreement with the experimental ones (Bosco, Carpinteri and Debernardi 1990; Bosco and Carpinteri 1992; Carpinteri and Massabó 1992).

Moment–localized rotation response is evaluated by simulating low cycle fatigue tests. Different kinds of collapse and the ductile–brittle transition in the beam behavior are examined when varying geometrical and mechanical properties.

2 Theoretical Model

Let us examine the reinforced concrete beam element subjected to a bending moment as shown in fig.1. The beam is reinforced by a steel layer and initially precracked with a through–thickness edge crack.

The beam geometrical characteristics are illustrated in fig.1: h and b are respectively the beam depth and thickness, a is the crack length and c is the distance of the reinforcement from the bottom of the beam.

An elastic–perfectly plastic law is assumed for steel while a linear–elastic law together with a L.E.F.M. crack propagation condition is assumed for concrete. Moreover steel–concrete bond slips can be taken into account through an elastic–plastic law which relates the bond stress with the slip–displacement.

In analyzing the cracked cross section the unknown reinforcement action is represented by indeterminate closing forces directly applied on the crack faces. During loading and unloading cycles the statically indeterminate reaction of the bar is evaluated by applying a congruence condition to the crack opening displacement in correspondence to the reinforcement.

The crack opening displacement in correspondence to the bar is defined by the following relation:

$$\delta = \delta_M + \delta_P = \lambda_{PM} M - \lambda_{PP} P \tag{1}$$

in which δ_M e δ_P are the crack opening displacements respectively due to the bending moment M and to the force P; λ_{PM} and λ_{PP} are the localized compliances due to the crack. A method to evaluate localized compliances is summarized below.

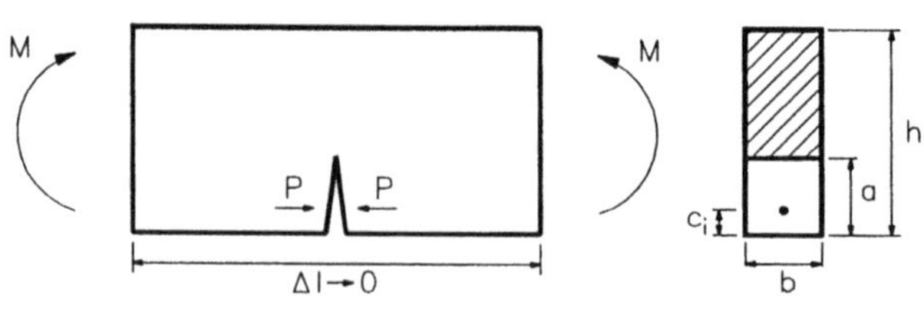

Fig.1. Beam element.

According to Linear Elastic Fracture Mechanics the intensity of the singular stress field at the crack tip is measured by a stress intensity factor K_I depending on the applied loads and on the crack length. The stress intensity factors due to bending moment M and to opening forces P applied on the crack faces (fig.1) take the form (Tada, Paris and Irwin 1973; Rooke and Cartwright 1976):

$$K_{IM} = \frac{M}{h^{1.5}b} Y_M(\xi) \tag{2}$$

$$K_{IP} = \frac{P}{h^{0.5}b} Y_P(\xi, \zeta) \tag{3}$$

where: $\xi = \frac{a}{h}$, $\zeta = \frac{c}{h}$, $Y_M(\xi)$ and $Y_P(\xi,\zeta)$ are polynomial functions given by:

$$Y_M(\xi) = \begin{cases} 6(1.99\xi^{0.5} - 2.47\xi^{1.5} + 12.97\xi^{2.5} - 23.17\xi^{3.5} + 24.8\xi^{4.5}), \xi \le 0.6 \\ \dfrac{3.99}{(1-\xi)^{1.5}}, \quad \xi \ge 0.6 \end{cases} \tag{4}$$

$$\begin{aligned} Y_P(\xi,\zeta) &= \frac{2}{(\pi\xi)^{0.5}} \left[\frac{3.52(1-\frac{\zeta}{\xi})}{(1-\xi)^{1.5}} - \frac{4.35 - 5.28\frac{\zeta}{\xi}}{(1-\xi)^{0.5}} + \right. \\ &\left. + \left(\frac{1.3 - 0.3(\frac{\zeta}{\xi})^{1.5}}{(1-(\frac{\zeta}{\xi})^2)^{0.5}} + 0.83 - 1.76\frac{\zeta}{\xi} \right)(1 - \xi + \zeta) \right] \end{aligned} \tag{5}$$

The total stress intensity factor at the crack tip K_I can be readily obtained by using the superposition principle:

$$K_I = K_{IM} - K_{IP} \tag{6}$$

As well known, the stress intensity factor K_I is connected with the strain energy release rate $\mathcal{G}_I$ through the following expression:

$$\mathcal{G}_I = \frac{K_I^2}{E} \tag{7}$$

in which E is the Young's Modulus of concrete.

Furthermore, by means of an energy balance approach, one can demonstrate the equivalence between $\mathcal{G}_I$ and the total potential energy W released during the formation of a unitary increment of the crack surface A:

$$\mathcal{G}_I = -\frac{dW}{dA} = -\frac{dW}{da}\frac{1}{b} \tag{8}$$

By applying Betti's theorem (from which $\lambda_{PM} = \lambda_{MP}$) and Clapeyron's theorem, the total potential energy of the beam element takes the form:

$$W = -\frac{1}{2}\lambda_{PP}P^2 + \lambda_{PM}PM - \frac{1}{2}\lambda_{MM}M^2 \tag{9}$$

Finally, the expressions of the localized compliances are achieved through the integration of eq.(8) and by equating the result to the right–hand side of eq.(9):

$$\lambda_{PP} = \frac{2}{bE}\int_{\zeta}^{\xi} Y_P^2(\xi,\zeta)d\xi \tag{10}$$

$$\lambda_{PM} = \frac{2}{hbE}\int_{\zeta}^{\xi} Y_P(\xi,\zeta)Y_M(\xi)d\xi \tag{11}$$

$$\lambda_{MM} = \frac{2}{h^2bE}\int_{0}^{\xi} Y_M^2(\xi)d\xi \tag{12}$$

Hence, by making use of a proper congruence condition, one can obtain the unknown reaction P in the different situations of loading and unloading, accompanied by crack propagation, steel yielding and/or slippage between steel and concrete.

2.1 Monotonic Loading

Under monotonic loading or during the first loading ramp in a cyclic process, the unknown reaction of the bar is evaluated by assigning zero value to the crack opening displacement δ until yielding or slippage of the steel bar is reached. In nondimensional form it results:

$$\frac{Ph}{M} = \frac{\lambda_{PM}h}{\lambda_{PP}} = \frac{1}{r^{II}(\xi,\zeta)} \tag{13}$$

The moment of plastic flow of the steel bar M_P can be evaluated by equating P to the yield force $P_P = A_s\sigma_y$, A_s being the steel area and σ_y the yield stress:

$$\frac{M_P}{P_P h} = r^{II}(\xi,\zeta) \tag{14}$$

If bond slip takes place before yielding of steel, the slippage moment can be evaluated by making use of the pull out force instead of the yielding force in eq.(13).

After yielding or slippage of the steel bar, the crack opening displacement δ, which is a priori unknown, can be evaluated by substituting the yield force in eq.(1).

The crack propagation condition is given by:

$$K_I = K_{IC} \tag{15}$$

in which K_{IC}, the critical value of the stress intensity factor, represents the concrete toughness.

Then the fracture moment can be found by substituting eq.(15) in eq.(6), yielding:

$$\frac{M_F}{K_{IC}h^{1.5}b} = \frac{1}{Y_M(\xi)}[1 + N_P\alpha(\xi,\zeta)Y_P(\xi,\zeta)] \tag{16}$$

in which $\alpha(\xi,\zeta)$ is a parameter given by:

$$\alpha(\xi,\zeta) = \begin{cases} 1, & if \;\; M_F \geq M_P \\ \dfrac{P}{P_P}, & if \;\; M_F < M_P \end{cases} \tag{17}$$

and N_P is a nondimensional number depending on the mechanical and geometrical properties of the beam (Carpinteri 1984):

$$N_P = \frac{P_P}{K_{IC}h^{0.5}b} = \rho\frac{\sigma_y h^{0.5}}{K_{IC}} \tag{18}$$

where $\rho = \frac{A_s}{bh}$ is the steel percentage.

The nondimensional fracture moment $\frac{M_F}{K_{IC}h^{1.5}b}$ against the nondimensional crack length ξ is plotted in fig.2 for $\zeta = 0.1$, when varying geometrical and mechanical properties by means of the single parameter N_P.

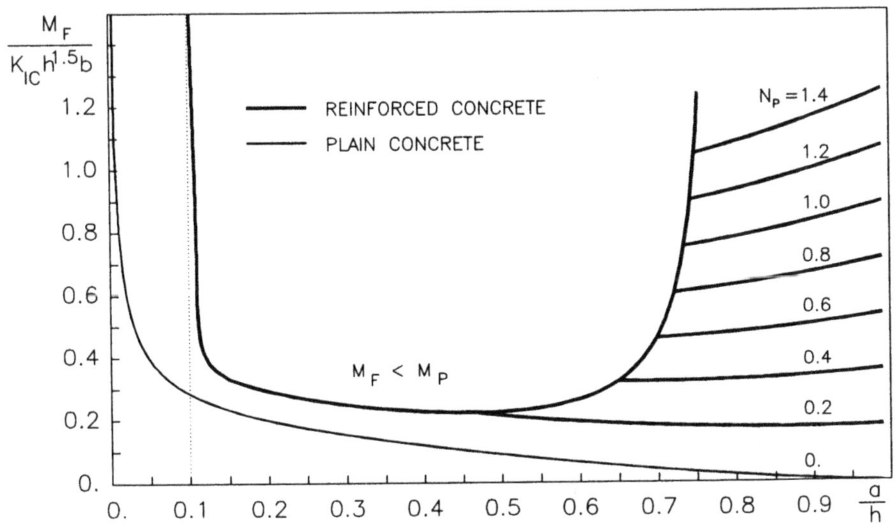

Fig.2. Nondimensional fracture moment versus relative crack length.

It can be seen in the diagram that, for $\xi \le \zeta$, the behavior of the beam is the one characteristic of plain concrete the crack being unstable, while, for $\xi > \zeta$, the reinforcement increases the fracture toughness of the member. When the initial crack depth and the brittleness number N_P of the beam are given, the diagram makes it possible to follow the evolutive phenomenon of crack propagation. In particular, it can be shown that, for N_P less than a limit transitional value N_{PC}, the crack propagation is always unstable, while for N_P greater then N_{PC} the reinforcement makes the behavior of the beam stable, after a phase of unstable propagation.

2.2 Cyclic Loading

Let us examine now the reinforced concrete beam subjected to a bending moment cyclically varying between $M = 0$ and $M = \overline{M}$.

In order to analyze the behavior of the beam during the unloading process one needs to know the situation of the beam at the peak load and then to impose suitable congruence conditions.

In particular, three different situations are possible for a given ξ_0:
(a) if the peak moment is lower of both the plastic flow (or slippage) moment and the fracture moment, the behavior of the beam is linear–elastic;
(b) if the peak moment is greater than the fracture moment but lower of both the plastic flow moment and the slippage moment, since during the loading phase the crack remains closed in correspondence of the bar, the congruence condition has to assure that δ will remain zero during the unloading phase. In particular, when the applied moment is zero, the reaction of the bar will vanish. Nevertheless, since during the loading phase the crack propagates, the stiffness of the beam will decrease;
(c) if the peak moment is greater than the moment of plastic flow, it is necessary to impose a congruence condition for which the crack opening displacement δ will remain constant during unloading up to reverse yielding. In particular, when the applied moment is put equal to zero the application of eq.(1) leads to:

$$\lambda_{PM}\overline{M} - \lambda_{PP}P_P = -\lambda_{PP}P_{SC} \tag{19}$$

in which P_{SC} is the unknown unloading reaction of the bar. Moreover if $\overline{M}$ is lower than the fracture moment, the compliances λ_{PM} and λ_{PP} have to be evaluated for the initial crack depth ξ_0, otherwise they have to be evaluated for the incremented crack depth $\xi(\overline{M})$ corresponding, through the diagram in fig.2, to the peak moment $\overline{M}$.

From eqs.(13) and (19) the bar reaction takes the form:

$$P_{SC} = \frac{\lambda_{PM}}{\lambda_{PP}}(M_P - \overline{M}) \tag{20}$$

It can be seen that P_{SC} is negative and so it represents an opening force applied to the crack faces. This kind of behavior is called elastic–shake down.

It is interesting to compute the value of the applied peak moment M_{SD} for which the unloading reaction will achieve the reverse yielding limit so generating a process of alternating plasticity. By assuming $P_{SC} = -P_P$ in the eq.(19) one obtains:

$$M_{SD} = 2M_P \tag{21}$$

This kind of behavior is called plastic–shake down.

3 Moment versus Rotation Diagrams

The moment versus localized rotation response of the reinforced beam cross section, can be evaluated by means of the aforementioned arguments. In fact the localized rotation ϕ, due to the crack, takes the form:

$$\phi = \lambda_{MM} M - \lambda_{MP} P \tag{22}$$

and it is valuable when the applied moment $\overline{M}$ and the corresponding reaction P are known.

In order to evaluate the moment versus localized rotation diagram for the r.c. section under a general unidirectional cyclic bending moment, a low cycle fatigue test is herein simulated through the proposed model. The loading process is supposed to be deformation controlled in order to follow the beam behavior even along the unstable branch.

The numerical results of the low cycle fatigue test are illustrated in fig.3, fig.4 and fig.5 by means of a diagram in which the nondimensional moment $\frac{M}{K_{IC} h^{1.5} b}$ versus the nondimensional localized rotation $\frac{\phi E h^{0.5}}{2K_{IC}}$ are illustrated. In particular three beams are examined with the same initial crack depth and position of the bar ($\xi_0 = 0.3$, $\zeta = 0.1$), but with different geometrical and mechanical properties. Beam A is characterized by a brittleness number $N_P = 0.3$, beam B by $N_P = 0.1$ and beam C by $N_P = 0.03$.

The envelope curves in the diagrams are made up by three or four branches which are illustrated below.

Beam A.

(I) Branch 0–F. For $0 \leq M \leq M_F$: the behavior is linear–elastic.

(II) Branch F–P. When the fracture moment is achieved the global behavior is determined by crack propagation which becomes stable after an unstable phase. Unloading and reloading take place with a remarkable stiffness reduction but without residual localized rotations.

(III) Branch P $\rightarrow \infty$. When the plastic moment M_P is achieved subsequent unloadings and reloadings take place with stiffness reduction and residual localized rotations.

Beam B.

(I) Branch 0–P. For $0 \leq M \leq M_P$: the behavior is linear–elastic.

(II) Branch P–F. For $M_P \leq M \leq M_F$: Unloading and reloading in this phase are characterized by a residual localized rotation and by a slight reduction in the initial stiffness.

(III) Branch F $\rightarrow \infty$. When the fracture moment is achieved the global behavior becomes unstable. During unloading remarkable reduction in the stiffness and in the strength can be observed.

Beam C.

(I) and (II) Branch 0–SD. For $0 \leq M \leq M_{SD}$: the behavior is equal to the one characteristic of the beam B.

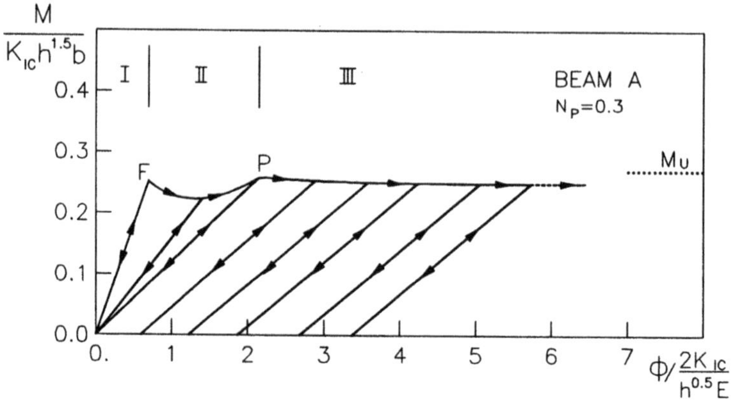

Fig.3. Hardening behavior and elastic shake-down.

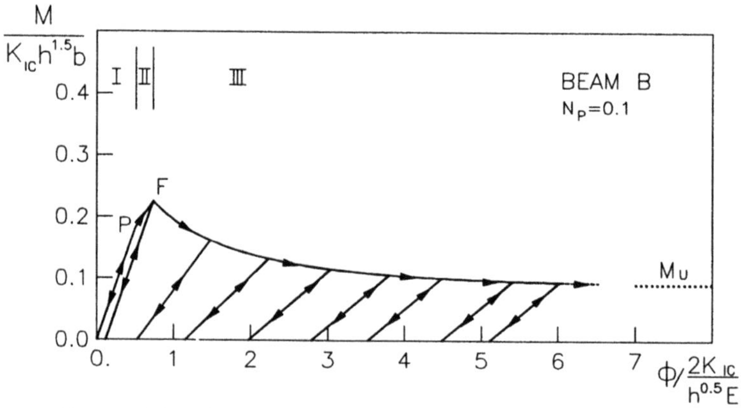

Fig.4. Softening behavior and elastic shake-down.

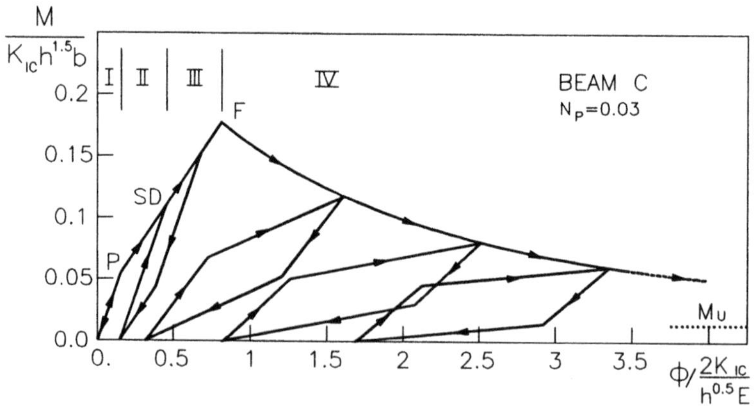

Fig.5. Softening behavior and plastic shake-down.

(III) Branch SD–F. For $M_{SD} \leq M \leq M_F$: the envelope curve stiffness is defined by yielding of the steel bar in tension; during unloading and reloading and due to the yielding of the steel bar even in compression, hystereresis cycles, accompanied with energy dissipation, can be observed.
(IV) Branch F →. When the fracture moment is achieved, the global behavior is represented by unstable crack propagation. If the peak moment during subsequent unloadings in this phase is greater than M_{SD} for the incremented crack, unloadings and reloadings take place with hysteresis loops.

It can be noticed that the envelope curves are identical to the moment versus rotation curves obtained under monotonic loading.

The horizontal dotted lines in the diagrams represent the asymptotes related to the ultimate bending moments for the totally disconnected sections:

$$\frac{M_u}{K_{IC}h^{1.5}b} = N_P(1-\zeta) \tag{23}$$

On the other hand, the collapse due to crushing is not taken into account.

The proposed model is applicable when local damage due to high number of cycles, the so called subcritical crack propagation, may be neglected. Moreover it has been experimentally demonstrated for plain concrete beams that, if the applied peak moment $\overline{M}$ is lower than the moment for which an unstable crack propagation develops, but it is able to cause a stress intensity factor close to the critical one, the subcritical crack propagation rate is very high and the member fails in very few cycles (Baluch and Quereshy 1987). To simulate this phenomenon in reinforced concrete members it is possible to introduce in the model a subcritical crack propagation stage by defining for concrete a fatigue constitutive relationship according to the Paris–Erdogan law (Paris and Erdogan 1963; Baluch et al 1989; Bazant and Xu 1991).

It can eventually be seen that the envelope curves in fig.3, fig.4 and fig.5 show an increasing ductility of the member by increasing the brittleness number N_P, that is to say, by increasing steel percentage or decreasing beam depth. In fact beam A is characterized by an hardening global behavior, while beam B and beam C by a softening global behavior. This has been confirmed by experimental tests on r.c. beams under monotonic loading.

4 Conclusions

A fracture mechanics model suitable for the analysis of reinforced concrete beam section subjected to unidirectional cyclic bending moment has been proposed.

The characteristic moment versus localized rotation relationship for the r.c. section has been obtained when varying geometrical and mechanical properties of the beam. A transitional ductile–brittle behavior has been observed by decreasing steel percentage or increasing beam depth.

Through the analysis of elementary phenomena such as crack propagation, yielding or slippage of steel, opening and closing of cracks, this mechanical model enables to explain the empirical models, usually employed in the damage analysis of structures subjected to earthquake or to low cycle fatigue repeated loadings, such as the Takeda Model.

Future improvements of the model could be obtained by simulating the nonlinear behavior of concrete and the high cycle fatigue crack growth.

Acknowledgements: The Authors wish to thank the Ministry of University Education and Scientific and Technological Research (MURST) and the Italian National Research Council (CNR) for the financial subsidy granted for this research.

5 References

Baluch, M.H. Quereshy, A.B. and Azad, A.K. (1987) Fatigue crack propagation in plain concrete, in **Sem Rilem International Conference on Fracture of Concrete and Rock** (eds S.P.Shah and S.E. Swartz), Houston Texas, pp.80–86.

Bazant, Z.P. and Xu, K. (1991) Size effect in fatigue fracture of concrete, **ACI Material Journal**, 88, pp.390–399.

Bosco, C. and Carpinteri, A. (1992) Softening and snap–through behavior of reinforced elements, **J. of Eng. Mech. ASCE**,118,pp.1564–1577.

Bosco, C. Carpinteri, A. and Debernardi, P.G. (1990), Minimum reinforcement in high–strength concrete, **J. Struct. Eng. ASCE**,116, pp. 427–437.

Carpinteri, A. (1984) Stability of fracturing process in r.c. beams, **J. Struct. Eng. ASCE**, 3, pp.544–558

Carpinteri, A. (1992) **Applications of fracture mechanics to reinforced concrete**, Elservier Applied Science, London and New York.

Carpinteri, Al. and Carpinteri, An. (1984) Hysteretic behavior of r.c. beams, **J. Struct. Eng. ASCE**,110, pp.2173–2084.

Carpinteri, Al. and Massabó, R. (1992) Discontinuitá nella risposta costitutiva di una trave fibrorinforzata inflessa, Proceeding of **AIMETA 92**, Trento, pp.203–208.

Okamura H.,Watanabe K., Takano T. (1973) Application of the compliance concept in fracture mechanics, **ASTM STP 536**, 423–438.

Paris, P. and Erdogan, F. (1963) A critical analysis of crack propagation laws, **Transaction of the ASME**, pp.528–534.

Popov, E.P. (1980), Seismic behavior of structural subassembages, **J. Struct. Eng. ASCE**, pp. 1451–1474.

Rooke, D.P. and Cartwright, D.J. (1976), **Compendium of stress intensity factors** , The Hillingdon Press, Uxbridge, Middx.

Tada H., Paris P. and Irwin G., (1963) **The stress analysis of cracks handbook**, Del Research Corporation, St. Louis, Missouri.

Takeda, T. Sozen, M.A. and Nielsen, N.N. (1970) Reinforced concrete response to simulate earthquakes, **J. Struct. Eng. ASCE**, 96, pp. 2557–2573.

Wang, M.L. and Shah, S.P. (1987) Reinforced concrete hysteresis model based on the damage concept, **Earthquake Engineering and Structural Dynamics**, 15, pp. 993–1003.

22 DAMAGE OF CONCRETE STRUCTURES DUE TO TEMPERATURE CHANGES

A. AJDUKIEWICZ and A. KLISZCZEWICZ
Silesian Technical University, Gliwice, Poland

Abstract
Destructive influences of temperature climatic changes are very often neglected or underestimated by designers of structures as well as by contractors repairing buildings. In large-dimensioned concrete structures these influences may cause serious damages. The behaviour of the large precast building which suffered damages has been presented as an example of temperature destructive action.
Keywords: Concrete Structures, Temperature Influence, Structural Damages, Damage Repair.

1 Introduction

Building structures excepting specific cases, e.g.industrial objects, are usually not analyzed more precisely in view of temperature changes. Designers of buildings with reinforced-concrete bearing structure abridge the consideration to providing of expansion joints in spacings recommended in the codes. In statically indeterminate structures the influence of general temperature changes is taken into account in structural analysis but no local results from extreme changes are usually considered.

Generally, climatic temperature changes are treated as short-term actions without consideration of their cyclic character.

Many examples of structural failures with dominating thermal reasons indicates the necessity of more exact analysis taking into consideration the time factor in destruction process and the extreme conditions of local unequal changes. The influences of temperature together with concrete shrinkage are very often the first destructive factor in concrete structures and as a repeated action with different duration may lead to the failure even as an only reason.

The problem has to be undertaken not only at the original design but also at planning the repair or modernization when a replacement of some elements, e.g.roofing, cladding,

Fracture and Damage of Concrete and Rock – FDCR-2.
Edited by H.P. Rossmanith. © 1993 Taylor & Francis. ISBN 0 419 18470 8.

or insulation layers, may make the thermal action on the structure more dangerous as before.

In the recent publications more attention has been paid to thermal damages in residential buildings, e.g. Schild et al.(1979-1980), than to other kinds of buildings.

2 Influences of climatic temperature changes on concrete structures

Considering separately the thermal action influencing any kind of reinforced-concrete structure, without accompanying actions like moisture, freezing etc, one may indicate three groups of influences:

(i) changes in strength properties of materials due to temperature,

(ii) thermal expansion of materials and entire elements with reference to different expansion coefficient,

(iii)thermal influence on shrinkage and creep of concrete.

In different climatic zones these influences may be of various importance but everywhere the range of temperature changes is significantly greater than the range of mean values recommended in the standards. For the local action of temperature the important factors are the exposure of the element to the sun's rays and the color and/or smoothness of its surface.

The problems are concerned in different manner in the codes or recommendations. The significant international document has been prepared by ISO(1984).

Influences of temperature changes on the strength and deformability of hardened concrete have been precisely described in the CEB-FIP Model Code (1990). It was emphasized that the basic properties of concrete given in the Code are valid in the range from $-20^\circ C$ to $+40^\circ C$. Changes of concrete properties in the range of elevated temperatures $0^\circ C<T<+80^\circ C$ are expressed in the Code as follows:

- compressive mean strength

$$f_{cm}(T) = f_{cm}(1.06 - 0.003T)$$

- tensile flexural strength

$$f_{ct,fl}(T) = f_{ct,fl}(1.1 - 0.005T)$$

- modulus of elasticity

$$E_c(T) = E_c(1.06 - 0.03T)$$

where f_{cm}; $f_{ct,fl}$; E_c are the values at normal conditions.

The similar correction in this range of temperature has been recommended in the Code for the coefficients of shrinkage $\alpha_s(T)$ and creep β_{HT}.

The coefficient of thermal expansion depends mainly on the type of aggregate and on the moisture of concrete; it ranges from $6\cdot10^{-6}$ to $15\cdot10^{-6}$ [$°C^{-1}$]. The average value is commonly recommended equal to $10\cdot10^{-6}$ [$°C^{-1}$] and is good approximation for the concrete made of quarzitic aggregate.

3 Real climatic temperature changes

In typical design work the structural engineer does not consider the position of building in relation to the geographical directions. Similarly, he doesn't take into account the factors which seem to be not important, like the color or smoothness of the surface of roofing or wall cladding. The changes in strength properties of concrete mentioned above are commonly neglected too.

Meanwhile, the differences which can result from the unfavorable combination of thermal influences may cause the significant decrement of the durability of structure.

In many countries the mean external temperatures are given in the standards, usually the same for all the area of the country. For instance, since 1987 in Poland there have been indicated the following values:

- in summer season $t_e = +\ 27°C$,
- in winter season $t_e = -\ 24°C$,
- one day amplitude $a = 8°C$.

Very similar values are recommended in other parts of Central Europe.

The real temperatures measured on the flat roofs or on the south walls range in much wider limits. In summer, on surfaces with full exposure to the sun the temperatures recorded were from ca.70°C (smooth gray surface), to over 90°C (black dull surface), while in winter on the same surfaces the temperature was as low as -35°C. These extreme temperatures may last several hours in summer and half a day or even more in winter.

Exemplary temperature records, obtained by Ajdukiewicz and Kliszczewicz (1991), in the roof structure of concrete building covered with dark-grey roofing (Fig.1a) are present ed for three characteristic points in Fig.1b. The thermal insulation of the building was rather poor as well as internal heating in winter. All the measurements were done near the midday without selecting the days with extreme conditions, because they only accompanied the main

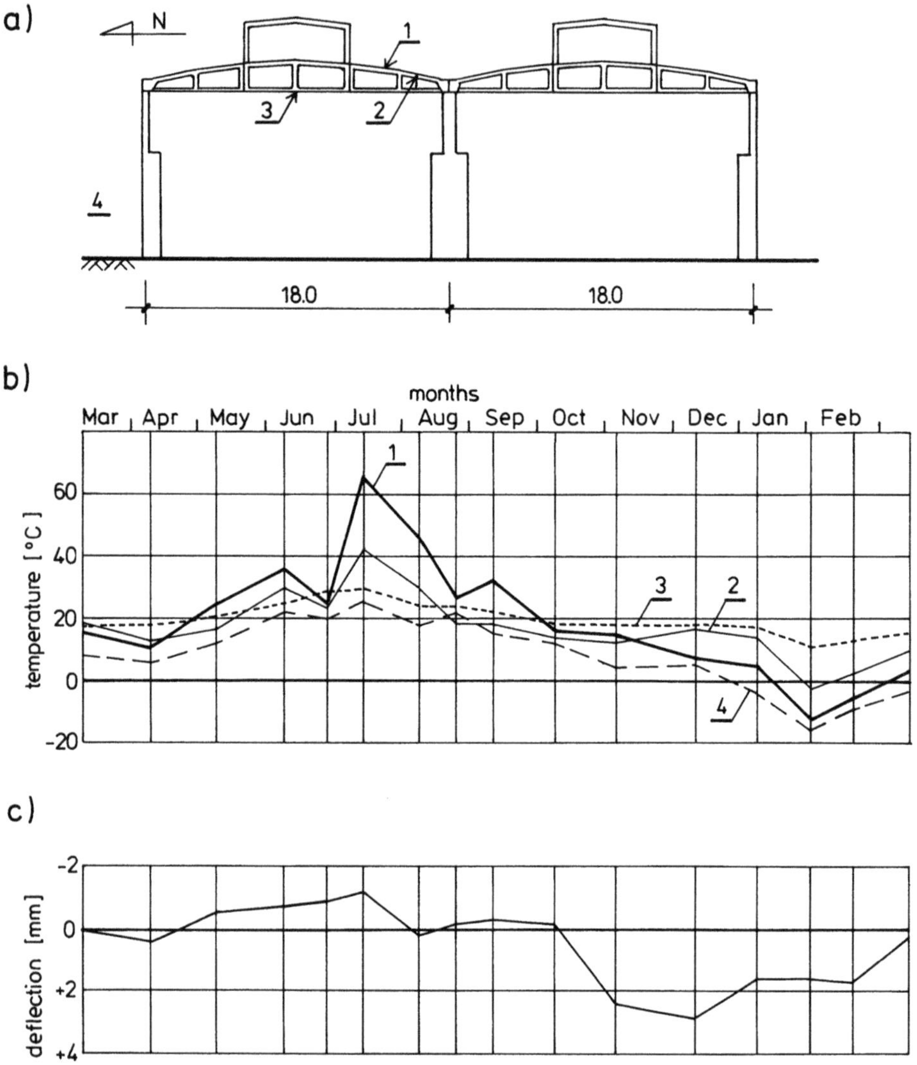

Fig.1. Temperature and deflection measurements in the roof structure of the building with weak thermal insulation and poor heating;
a) location of the points of measurement,
b) comparison of temperatures on the roof surface (1), on the upper part of structure (2), on the lower part of roof structure (3) and the air outside the building (4)
c) changes of deflection in the point (3) during the year

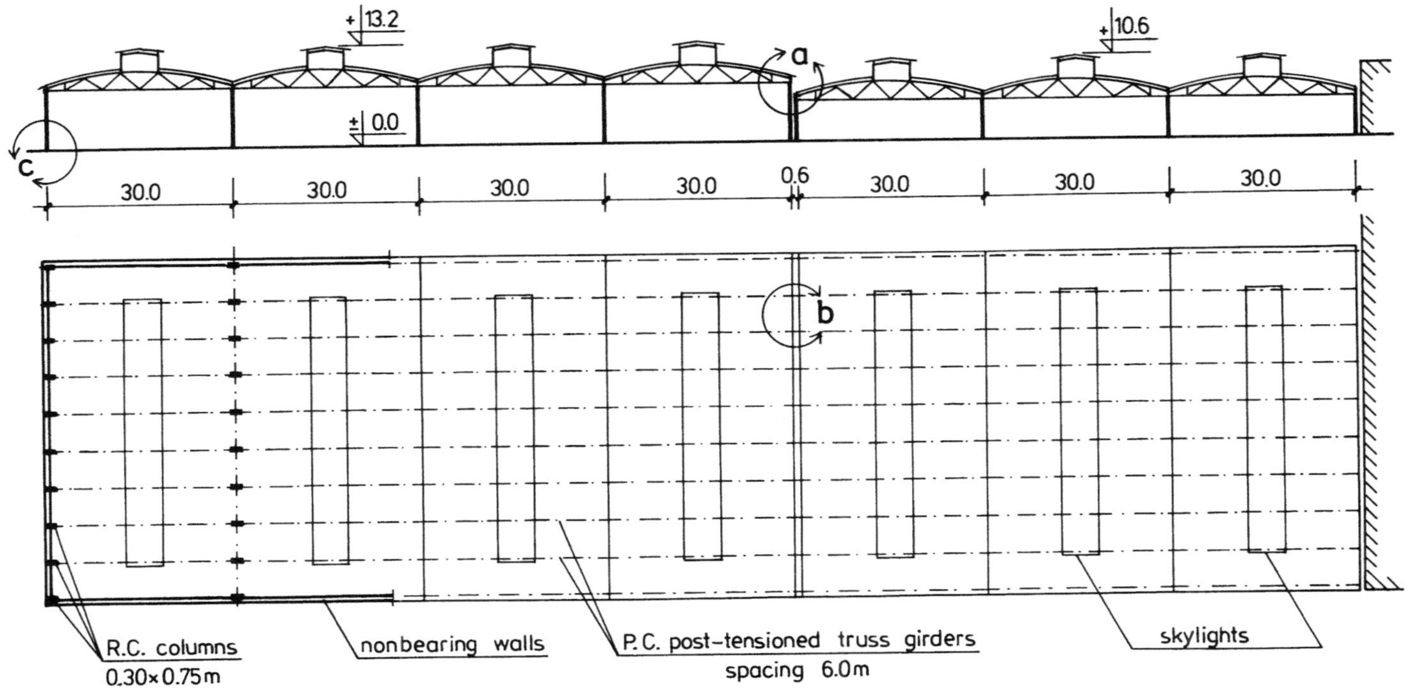

Fig.2. General cross-section and plan of the building

measurements of changes in deflections of the roof girders due to temperature (Fig.1c).

4 Example of damaged structure

The large warehouse is composed of seven bays 30m-span each and 54m long; it has been erected in mid-sixties (Fig.2). The building is divided into two parts: the higher one of 4 bays (over 120m) and the lower one of 3 bays (ca.90m). Along the line of division the expansion joint has been provided in the form of twin columns supported on the mommon foundation pads.

Main transversal bearing structure consists of prestressed-concrete post-tensioned roof girders (truss-shaped) and reinforced-concrete precast columns. The roof has been made from prefabricated slabs 1.50m/5.87m/0.30m supported on the upper chord of P.C. girders and joined together with the supplementary concrete (Fig.3).

For almost 20 years the building had been used without problems and unexpectedly in 1986 the cracks in some parts of walls and roof were noticed. The damages were discovered not long after the general repair of roofing. During the following two years the user was watching the propagation of cracks and then the general inspection was undertaken.

Many reasons which could cause the damages were considered. Finally, it was found that the higher part of building had been additionally supported horizontally by the lower one as a result of unintended filling of expansion joints between the columns. The hot asphalt adhesive flowed down from the roof during repair works and also later due to sun heating of the repaired roof surface. In the shaded space of expansion joint the adhesive hardened and created the additional horizontal support (Fig.4). Moreover, the color of roofing has been changed from grey (sanded surface of the top building paper) to black (asphalt). Both these factors - seeming secondary - significantly changed the character of roof deformability. For the longer part of the building which could originally deform symmetrically (Fig.5a) the horizontal deformation freedom was taken away at one side because the stiff low part of the building supported the columns of the high part. In this new situation the high part deformed asymmetrically (Fig.5b). Due to the change of roofing color the extreme elevated temperature had to increase, therefore $\Delta > 2\delta$.

Owing to relatively slender and flexible high columns the new situation didn't lead to failure of the main bearing structure but it caused damages of secondary structure and finishing.

Direct results of deformation were the serious cracks in nonbearing walls; the cracks exceeded width of 10mm (Fig.6, detail "c").

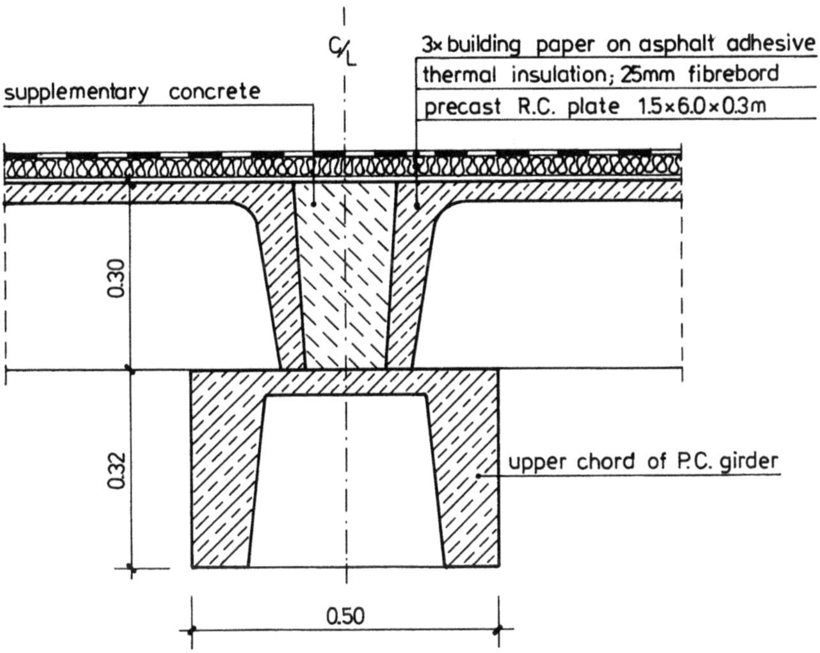

Fig.3. Detail of roof cover and cross-section of the upper chord of P.C. girder joined with roof slabs

More complex damage took place in the roof (Fig.6 - "b"). It resulted from the local horizontal supports of rigid plate of the roof subjected to the thermal action. Solid connections with girders (see Fig.3) held the adjacent parts of roof slabs while the intermediate part of these slabs deformed and cracked diagonally.

The cracks which appeared at the extreme heating of the roof became irreversible in majority at the chilling because scraps and dust "fixed" the cracking state. Every subsequent appearance of significantly elevated temperature could cause the propagation of cracks. Then other factors like water, pollution, freezing would increase the destructive action.

Detailed analysis of the thermal influences in the roof structure, according to recommendations of the Polish Standard (1986) based on the document of ISO (1984), gave the theoretical values of the mean temperature of the roof in the summer season for different roofing surface and color. For the dull, gray surface it was ϑ = 45.0 °C, while for the black surface it was ϑ = 55.8 °C.

The difference over 10°C for the roof long 120m leads to the difference in thermal elongation of concrete structure about 12mm. It could be good presentation of the importance of these secondary factors.

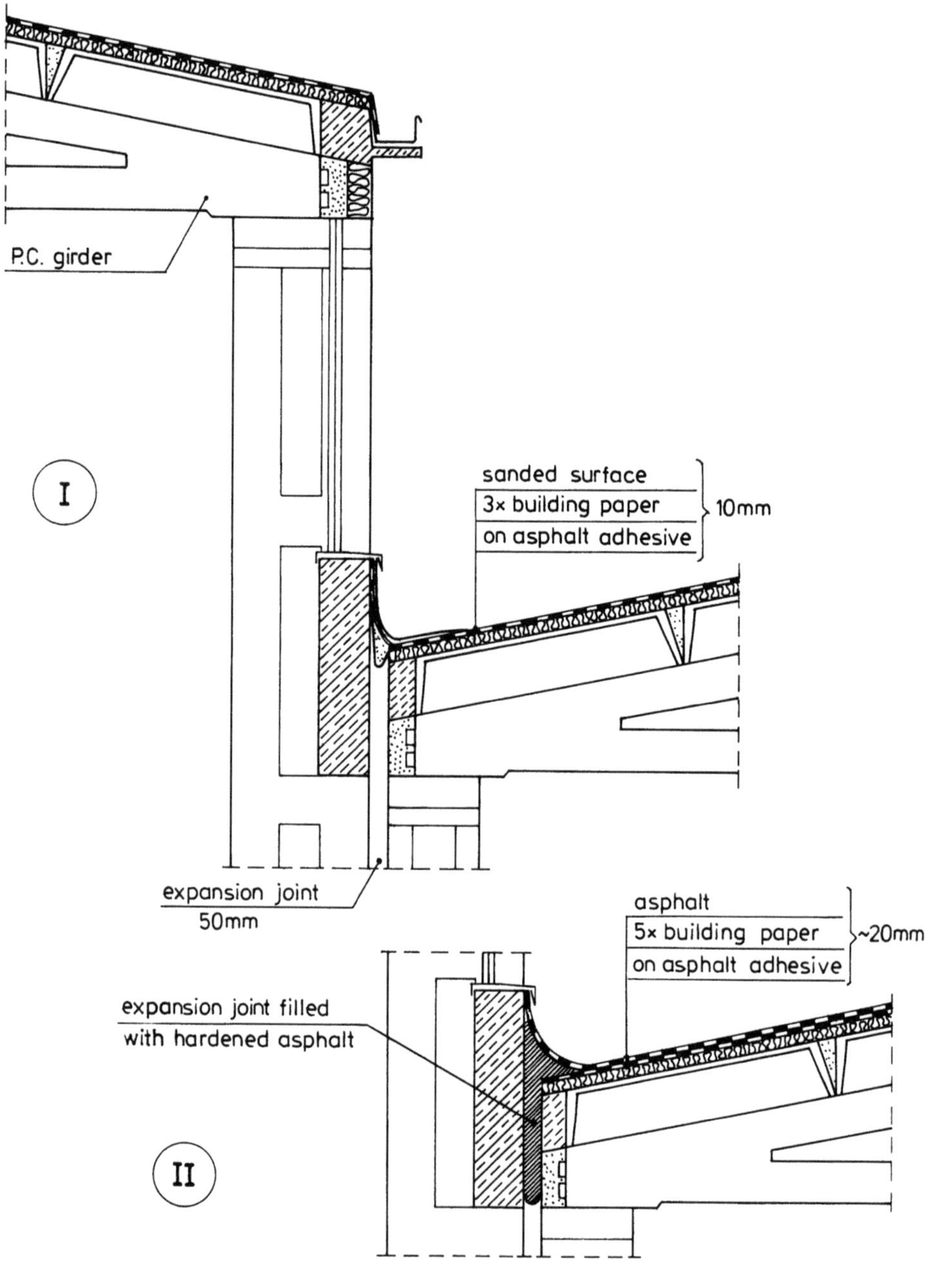

Fig.4. Detail "a" (see Fig.2) executed according to the original design (I) and after roof repair (II)

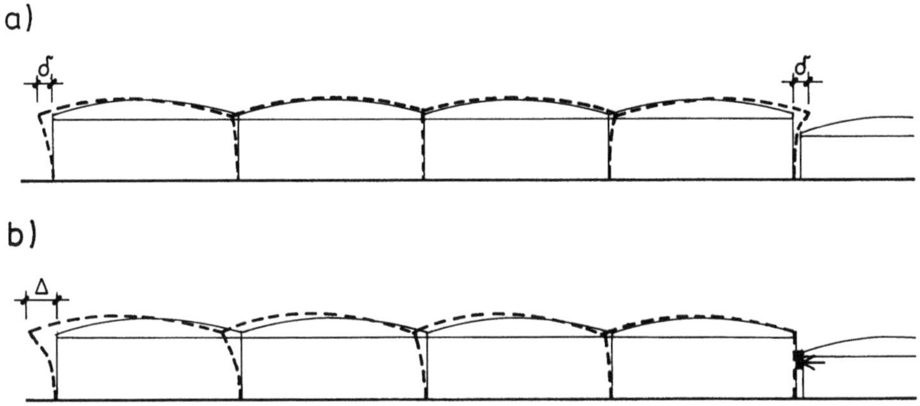

Fig.5. Deformation of the high part of building at elevated temperature (skylights neglected):a) symmetrical deformation in original arrangement; b) asymmetrical deformation due to horizontal unintended support

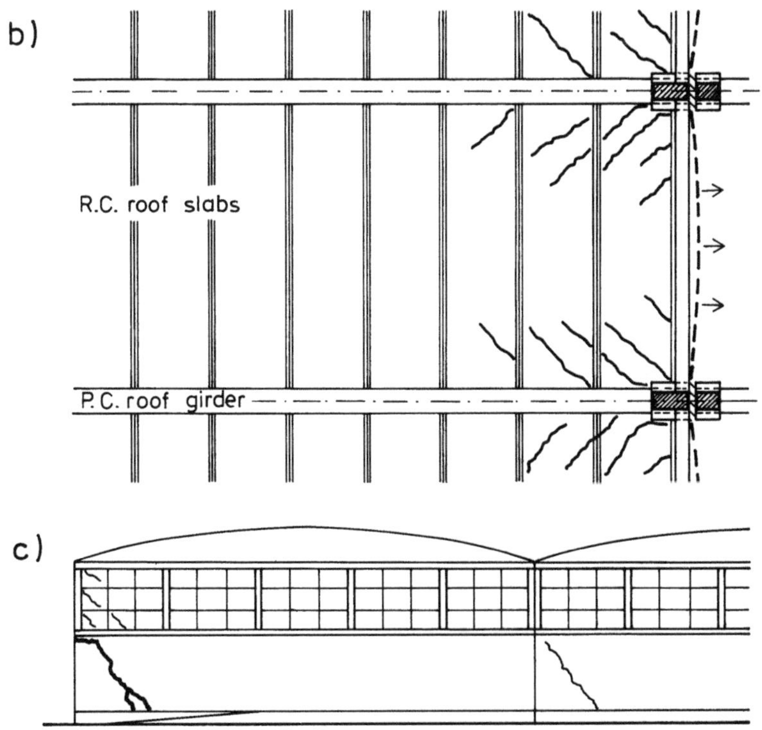

Fig.6. Main damages - details "b" and "c" acc. to Fig.2: view of the cracked roof from the bottom and cracks in the nonbearing walls

5 Conclusions

Thermal climatic influences are more or less ignored at design of bearing structures of buildings. This action is almost always neglected at repairs when the changes in materials' colors or surfaces are very common.

The example of very long building presented in this paper has indicated how repair works - considered usually as of second rank - may cause the serious damages.

6 References

Ajdukiewicz, A. and Kliszczewicz, A. (1991) Post-tensioned roof girders diagnosed by means of continuous measurement of deflection. Proceedings of the **RILEM-IMECO Conference**, Bratislava, 2-6th September'91; pp.258-261.

Comite Euro-International du Beton (1990) **CEB-FIP Model Code MC90.**

International Standard Organization (1984) **ISO TR Draft. Temperature Climatic Actions.**

Polish Standard PN-86/B-02015 (1986) **Obciążenie temperaturą** PKNMiJ, Warszawa.

Schild, E. et al.(1979-1980) **Structural failure in residential buildings.** (Translation from German) Vol.1,2,3. Granada Publishing, London-Toronto-Sydney-New York.

23 DEVELOPMENT OF DAMAGE AND PROGRESSIVE FRACTURE OF AN HETEROGENEOUS ROCK MATERIAL

S. BRICCOLI BATI and G. RANOCCHIAI
Dipartimento di Costruzioni, Universitá di Firenze, Italy

Abstract
The aim of this paper is to provide the qualitative and quantitative study of model prediction for several cyclically varying stress or strain programmes of rock samples. In order to obtain the necessary information on the behaviour of tuff masonry structures under earthquake actions, either the material tuff or simple structural elements made of tuff bricks have been subjected to cycles of static and dynamic actions, in order to allow the identification of the fundamental behaviours with damaging and progressive fracturing, to investigate typical post-peak behaviours and damage rules, and to collect the necessary data for life and failure prediction of such material and structures.
A review and interpretation of representative experimental data is given on the cyclic properties. Emphasis is placed on data concerning the development of damage.
Keywords: Damage, Cyclic loading, Tuff.

1 Introduction

Our interest in tuff material come from our building heritage, since, in many areas of Italy - even liable to earthquakes - the above mentioned material is often used, both in the past and in the present.
As is widely known, the capability of a masonry structure to withstand seismic actions, even if very strong, mainly relies on its mechanical behaviour in post-elastic range. In this field a great deal of work has been done on steel and reinforced concrete structures, but few results are available for masonry construction, especially when they are made of tuff blocks. An important area of practical concern is the response to cyclic disturbances arising from earthquakes and therefore laboratory cyclic loadings are performed over a frequency range of 0.2 to 2 Hz to reflect

Fracture and Damage of Concrete and Rock – FDCR-2.
Edited by H.P. Rossmanith. © 1993 Taylor & Francis. ISBN 0 419 18470 8.

the dominant frequencies associated with earthquake effects.

The tuff material - a sedimentary rock of volcanic origin - looks very heterogeneous because of both non homogeneity of its grains and the simultaneous presence of micro and macroporosity with void diameters varying from a few micron up to a centimeter and therefore it can be classified as an anisotropic porous compressible material. The initial void volume ratio is about 60%; the in-plane mean "nearest-neighbour" porous spacing is approximately 3-4 cm; the mean void areal fraction was about 17% for surfaces parallel to sedimentation layers, and 23% for orthogonal surfaces.

Under the above described circumstances the problem gets more and more difficult to be approached whenever a correct macrodescription of the behaviour of the material is needed.

The main objective of the present investigation is to provide sufficiently comprehensive experimental data on the cyclic behaviour of this material and to attempt a preliminary syntesis of the experimental evidence on various aspects of damage evolution.

2 Compressive behaviour during cyclic loading

In order to understand the load resistance and damage mechanism involved and to expand the experimental database for the improvement of prediction of lifetime of tuff made masonry constructions, both monotonic and cyclic tests were performed on tuff samples and on small walls and panels. Interpretation of test results in terms of stress-strain curve is not attempted because the specimen state is inhomogeneous and then the results are strongly affected by the randomness of material heterogeneity.

In fact the presence of large inhomogeneities causes local fields which vary from one point to another under an external homogeneous field of stress or strain. The inobjective response of a specimen of such material even if loaded in pure tension or compression was clearly related to the presence of inhomogeneities. Some localization process develops within the specimen that prevent relating directly the load displacement curve and the supposed stress-strain curve.

On the basis of the above consideration we cannot deal with characteristic strength etc... but only with average mechanical properties referred to some given size. Such properties, particularly strength and elastic coefficients, are strongly affected by anisotropy of void areal fraction: they show higher values and a better distribution curve in experimental results, when corresponding to an orthotropy

axis which is approximately ortogonal to sedimentation layers.
Therefore emphasis is placed on data concerning the collected phenomenological aspects of the behaviour during cyclic loading and displacement histories. In fact a sufficient amount of experimental evidence is already available concerning mainly the evolution in response to repeated cyclic loading, the shape of the hysteresis loop and the evolution of the form of the unloading and reloading curves.

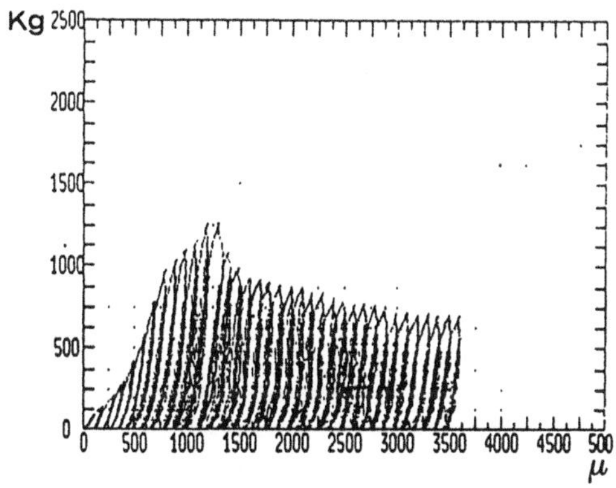

fig.1 Typical cyclic compression response

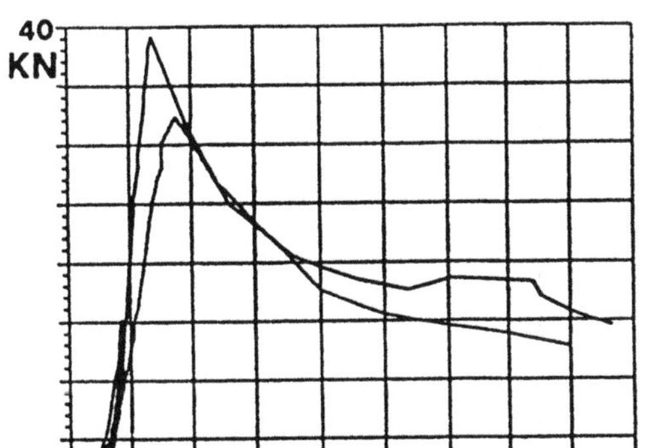

fig.2 Cyclic and monotonic compression response

For lack of space we report just some conclusions drawn from the great deal of experimental work carried out. Over the range of cyclic tests with load control, both on material and on structural elements, some global characteristics may be summarized as follows: two significant features of the mechanical properties of this material is the irreversible compaction and hysteresis during cyclic loading. The permanent volume decrease or progressive compaction results from rearrangement of the granular structure of this material. The difficulty of modelling this feature is due to an impossibility of a macroscopic distinction and numerical decoupling of the two co-existing, different effects induced by loading: porosity reduction and progressive damage.

As the number of cycles increases, the loops approach a closed stable limit loop. A common idealization - confirmed by carried out experiments - is to suppose that the limit loop is reached after one or few cycles. For low load levels, the material is soft, but it becomes stiffer as the load encreases.

An analysis of the response to a single sequence of loading, unloading and reloading indicates that compaction develops depending on the advance in load level.

Further load cycling produces wider and wider hysteresis loops, and the material cannot in general be returned to its initial state, because permanent damage has been created. Loop evolution in repeated unloading and reloading at constant load amplitudes leads to a gradual transformation of a loop; but the evolution rate is remarkably higer at the beginning and at the end of cycling. Also compaction is significant during the first loadings and close to failure. The occurrence of cyclic failure, or fatigue, depends on the ratio of the load amplitude to monotonic ultimate load; for constant load amplitude cyclic failure occures at load level below the monotonic failure value and is reached after a number of cycles which increase with decreasing load amplitude.

When load is repeated within a range of 0.5 Fm (medium monotonic strength), both maximum and residual displacements within the prescribed number of cycles seem to fit an asymptotic diagram. Failure is not usually reached at a number of cycles of technical interest, whereas the peack strength and the ductility of the specimen, in the subsequent monotonic test, seems to be higher then previously (fig.2).

If the maximum value of cyclical load is between 0.67 and 0.8 Fm, we can see both maximum and residual displacements to fit a typical "s" trend before failure occurs within the prescribed number of cycles. The inflection point is approximately at the middle of the load history. For this reason, damage has been related to the dimensionless equation of residual strain, according to the research developed by Hwan Oh. This function (a cubic polinomial) is not,

in our opinion, fully representative of damage growth: until the inflection point is reached, it represents a more compacted material, which causes the strength to increase. In figures 3 and 4 load displacement diagrams are shown. In figures 5 and 6 we can see the characteristic "s" diagrams for damage both of walls and of material.
Herein the damage is assumed to occur at the inflection point of the δ - n curve. This choice is based on the observation that in absence of any fractures, the δ - n curve is flat, otherwise when the curve is concave it is the result of microcracks presence. At this stage the shape of hysteresis loops evolves due to increasing strain accumulation, until it becomes incontrollable and failure occurs. Finally the damage phenomena produced by cyclic and monotonic tests are almost identical and are localized to a region close to the loaded boundary, but the effect of random local inhomogeneities on the localization of damage and failure cannot be captured.
Although the obtained results may be regarded as illustrative only.

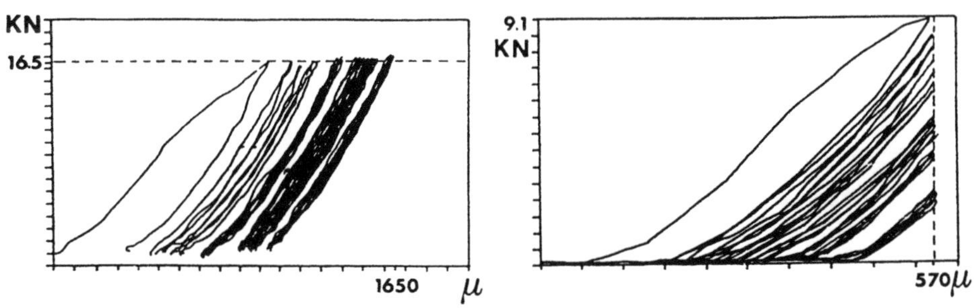

fig.3-4 Load-displacement diagrams of a stress controlled and of a displacement controlled test.

A model taking into account both compaction and damage process can be based on the record of the changing volume of material as a prediction parameter of this phenomenon. Lateral expansion coefficient (the ratio - $\delta a/\delta l$, where δa is the axial strain and δl is the lateral strain) in our opinion can be assumed to represent such variation of volume.
When compaction occurs, volume decreases and often for a short specimen (slenderness ratio about 1) even lateral expansion assumes a negative value. Lateral expansion coefficient has a negative value that becomes smaller and smaller when the loops become stable; if damage takes place, lateral expansion coefficient changes its sign with an asymp-

tote and it approaches to 2 (figure 11) when fracture become visible.

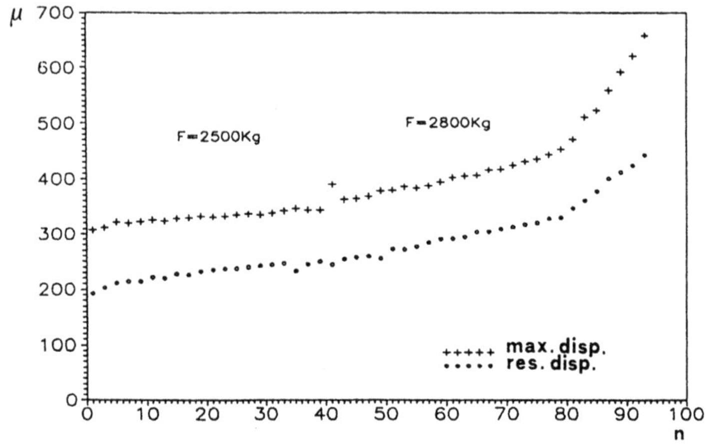

fig.5 Maximum and residual displacement versus cycle number for a tuff specimen

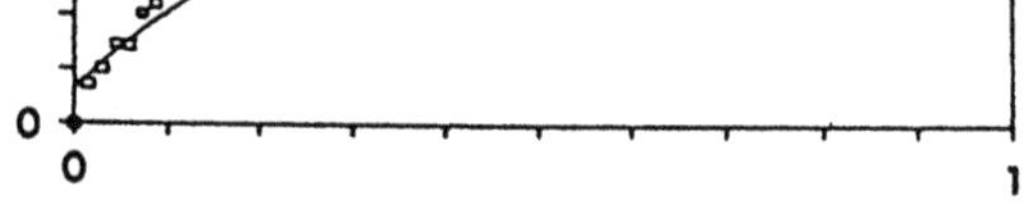

fig.6 Maximum displacement versus cycle number for a panel

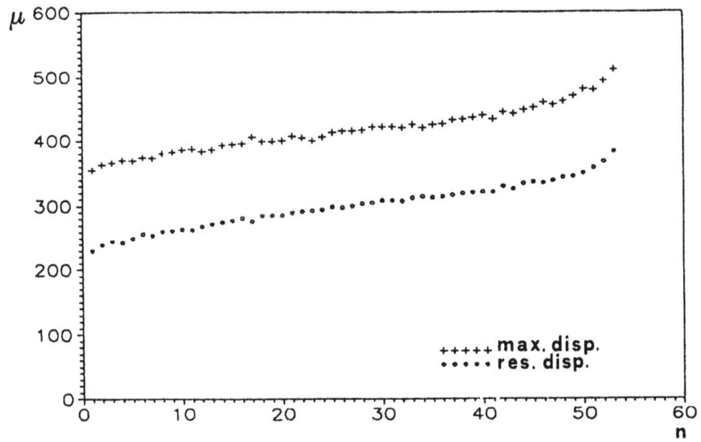

fig.7 Maximum and residual displacement versus cycle number for a tuff specimen

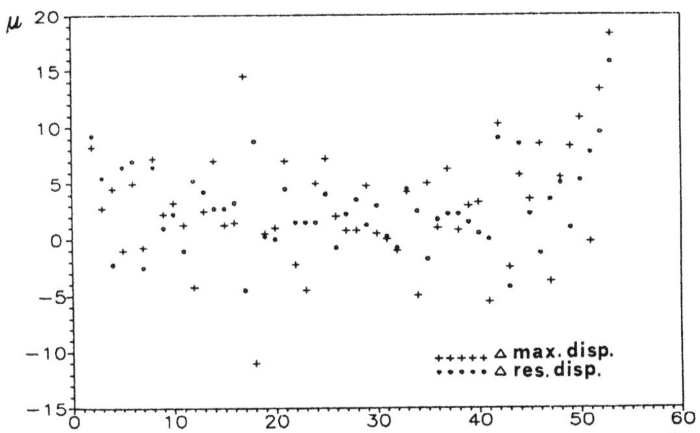

fig.8 Variation of maximum and residual displacement refered to fig.7

3 Conclusions

Laboratory experiments on the uniaxial compressive response of tuff samples show that both damage and permanent deformation occur during the loading process. Therefore it is very difficult to describe analytically the progressive ma-

terial deterioration and growth of individual cracks and their pattern.
The crack initiation is closely related to the localized strain produced by cycling and then depending on the structure of the material.
Evidence indicate that more accurate prediction of the response of such porous material must take into account the effects of large voids presence. It is anticipated that the size, geometry and distribution of the viods can all play a significant role in the elastic and inelastic response of such material.

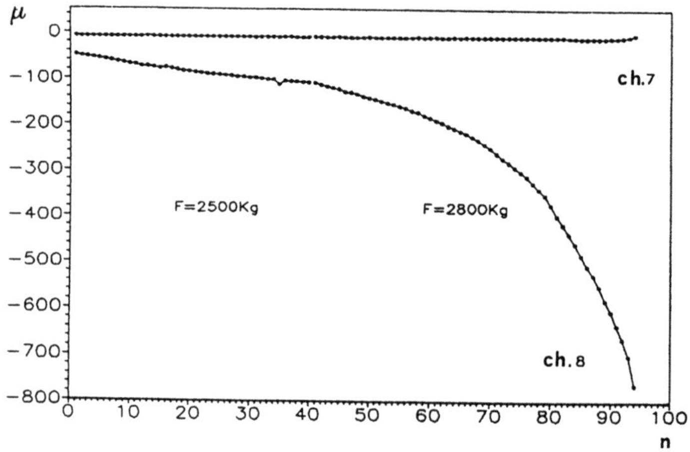

fig.9 Axial displacements for a tuff specimen

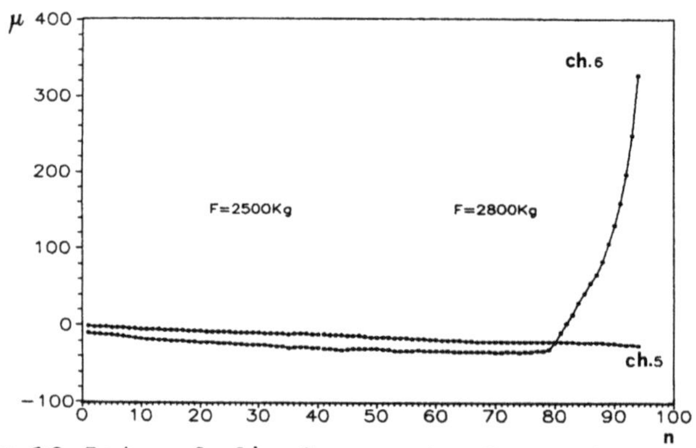

fig.10 Lateral displacements for a tuff specimen

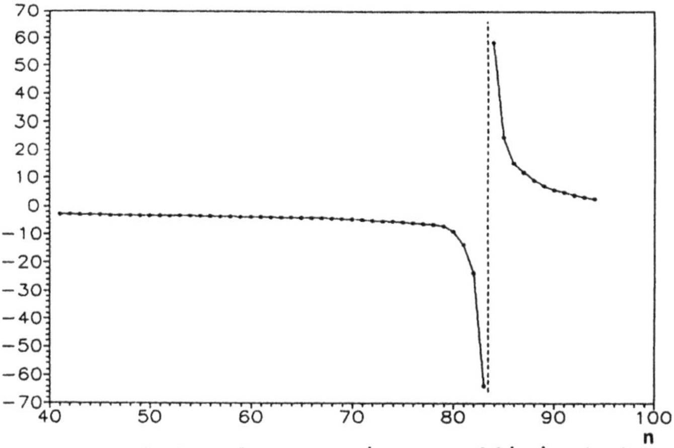

fig.11 Average lateral expansion coefficient for a tuff specimen

6 References

Briccoli Bati, S. and Ranocchiai, G. (1991) Comportamento di elementi strutturali in muratura di tufo sotto azioni cicliche, in **atti V Conv. naz. L'ingegneria sismica in Italia**, Palermo pp. 395-404.

Briccoli Bati, S. and Ranocchiai, G. (1992) Damage and fracture of masonry materials due to cyclic loading, in **proc. 4th Int. Conf. on Structural failure, Product Liability and Technical Insurance**, Vienna.

Fafitis, A. and Shah, S. P. (1984) Rheological model for cyclic loading of concrete. **Jour. Struct. Eng.**, 110, 2085

Horii, H. and Nemat-Nasser, S. (1985) Overall moduli of solids with microcracs; load induced anisotropy. **J. Mech.,Phys. of Solid.**

Hueckel, T. (1991) Damping, cyclic strain buildup and fa tigue of rocks: a generalized Ramberg-Osgood approach. **Computer and Geotechnics**, 12, 235-269.

Hueckel, T. (1991) Dilatancy histeresis in rocks: a gener alized Ramberg-Osgood approach. **Computer and Geotech nics**, 12, 209-234.

Hwan Oh, B. (1992) Cumulative damage theory of concrete un der variable-amplitude fatigue loadings. **ACI Mater. Jour.**, No. 1, 88, 41.

Lemaitre, J. and Chaboche, J. L. (1989) **Mecanique des mate rioux solides**, Dunod, Parigi.

Schatz, J. F. (1976) Models of inelastic volume deformation for porous geologic media, in **The Effects of Voids on Material Deformation** (eds.S. C. Cowin and M. M. Carroll)

ASME, AMD - 16, 141-170.

PART FOUR
DEMOLITION AND RECYCLING

24 GOALS AND BARRIERS TO RECYCLING OF CONCRETE AND MASONRY

E.K. LAURITZEN
DEMEX Consultant Engineers, Denmark

Abstract
Demolition and recycling of building materials are among the important issues of the strategy of cleaner technology within the building and construction industry. Based on the recent work of RILEM TC-121 (Technical Committee on Guidance for Demolition and Reuse of Concrete and Masonry), the current development in the EC and other countries will be presented with a special view to the technical, economical and political backgrounds. The most typical barriers are described and the overcoming of these is discussed.
Keywords: Cleaner technology, Demolition, Recycling, Crushing, Building waste management.

1. Introduction

According to the founder of SONY, Akio Moriata, one of the most important values adopted from older times is contained in the japanese word *moat-tajnaj* [1]. The word expresses that everything in the World is a gift from the Creator. Therefore, the Japanese consider all things as sacred, being entrusted to mankind only for loan and the best possible use. In principle, it is not accepted to waste anything or let things perish.

So far *moat-tajnaj* in Japan. In all communities it has always been common practice to retrieve valuable materials from the arising waste, e.g. metals and building materials. After some decades in this century with an extensive "use-and-throw-away" philosophy it has been realized that we cannot continue this uninhibited use of natural resources and pollution of the world with waste. It is necessary to change our habits and to revise former common practises within the building & construction industry as well as other industries, households etc.

Fracture and Damage of Concrete and Rock – FDCR-2.
Edited by H.P. Rossmanith. © 1993 Taylor & Francis. ISBN 0 419 18470 8.

In the last decade a lot of "green" movements have arisen mostly based on political and idealistic issues, and unfortunately with a rather limited effect on the practical life.

However, within the last few years the World Bank and OECD has emphasised that recycling and the introduction and implementation of environmental friendly technologies or cleaner technologies must be considered as the greatest technological challenges of our time. To encourage the achievement of these objectives, the World Bank [2] has faced the realities and clearly stated that the improvement and protection of the environment is a question of money which should be payed by the developed countries.

2. Cleaner Construction

According to the Danish definition: *The basic strategy of cleaner technology is that the use of resources and load on the environment should be continuously reduced.* Figure 1. shows a model of the life-cycle of buildings/structures, divided into 6 processes and 5 stages.

Generally, European dwellings are supposed to have an average life-time of half a century, and industrial buildings and structures have a considerably shorter life-time. Outside Europe the life-times are often extremely shorter. In the Middle and the Far East a 20 year old building is considered as "old". According to the theory of cleaner technology, the building waste material should be recycled and reused. However, only a very little percentage is at present reused. Most of it is disposed of in landfills or in some cases used as land reclamation.

During each of the mentioned processes there is a need for energy, man-power, water and other resources. Besides the specific purpose of the processes a lot of undesired outputs (waste and emmisions etc.) arise resulting in environmental impacts. Whereas the construction process is carefully considered and planned, the other processes, especially demolition and recycling, are neglected due to a lack of understanding of the importance of integrated resource/environmental management. The achievement of cleaner technology within the building and construction industry (*Cleaner Construction*) requires a detailed knowledge and understanding of all processes, and careful consideration of all sorts of input and output of the single processes with respect to the environment, and brings into focus the related costs.

The following text concentrates on building waste management and the recycling of building materials.

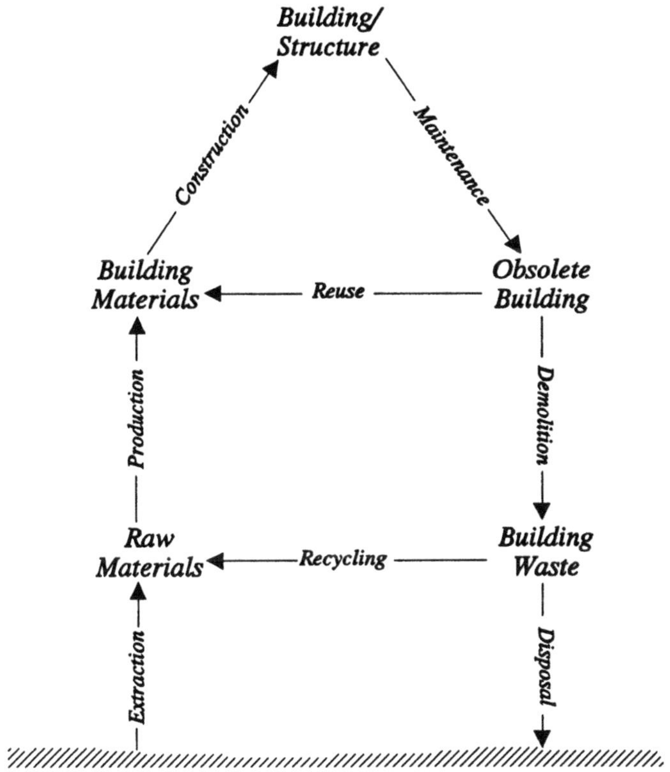

Fig. 1. Model of life-cycle of buildings and structures. Primary (natural) materials (stones, rocks, metal ore, wood) are extracted from the resources of the nature and processed into raw materials (aggregates, steel, timber). Then the materials are used for production of building materials and elements, whereafter being utilized in the construction process for a new building or structure. After having taken the new building/structure into use it will need maintenance and repair, and at the end of use it will be obsolete. Depending on the construction and the purpose of the building the owner decides to rehabilitate the building or to demolish it. After demolition, the building waste materials are crushed to secondary raw materials and the remaining waste is disposed of. All processes require energy, water and man-power, and the processes will be accompanied by an impact on the environment and a risk of health and safety.

3. Building & Construction Waste Material

Building & construction waste materials derive from the different lifecycle processes of buildings and structures:

1. *Extraction and processing of raw materials.*
During the extraction and processing of natural materials a certain amount of low quality and waste material arises. Often, these materials are disposed of in the quarry and have no influence on the general waste streams, and therefore they are not usually considered as waste. However, the mentioned materials need handling and management which require some consideration to environmental impact, costs and planning. Normally, this type of waste material is not taken into account being considered as building waste material.

2. *Production of building materials.*
The industrial production of building materials, cement, construction steel, timber, prefab building elements, windows, doors etc. leads to a certain amount of excess and waste materials. In the prefab concrete element industry and ready mixed concrete production the excess material percentages typically reach figures of 3 - 5 % of the production. Usually, the excess and waste material is treated within the respective industry. However, this quantity of waste needs to be handled and managed as demolition and construction waste. This type of waste has a certain influence on the waste streams and should be taken into consideration in the assessment of building waste material.

3. *Construction waste.*
During the construction process a lot of waste from packaging, concrete forming, excess etc. arises. The packaging waste represent mainly wood, plastic and paper. From the construction process concrete, timber and steel excesses arise. Danish research has proved that in the construction of dwellings an amount of approximately 25 - 50 kg waste and excess is produced per squaremeter of constructed floorage [3]. Failures and changes of design often result in a considerable amount of waste. Scaffolding and form work also produce a lot of waste depending on the construction methods. For instance, reusable formwork systems obviously reduce the excess amounts. In many countries, especially developing countries, scaffold and concrete forms are made of wood, which results in a considerable waste of wood.

4. *Waste from maintenance and repair.*
Maintenance and repair work causes a lot of mixed building waste material. The amount of waste generated depends on the type of repair work. According to Danish investigations, normal renovation work without any change in the building structure itself generates an average of 50 - 100 kg per squaremeter floorage. Considering the

need for repair work and the large number of projects in all countries, the total amount of building waste is considerably high.

5. *Demolition waste.*
Waste material from partial demolition in connection with rehabilitation work and total demolition during urban renewal represents the major component of the building waste materials (see fig. 3.). According to Danish investigations, the output of demolition waste is estimated to 1 - 2 tons per squaremeter floorage.

All together the amount of building and construction waste reaches considerable figures. With reference to recent studies and investigations, e.g. project undertaken for the EC [4], an assessment of the total production of building waste in a number of member countries can be seen in fig. 2. On this basis it can be estimated that in many developed countries an amount of building and construction waste 500 - 1000 t per capita per year is generated, and that the total amount of building waste in the EC lies in the region of 175 million tons p.a. This figure is based on a population of about 350 million and an average production of building waste of 500 kg per inhabitant per annum.

Country:	Inhabitants (Mil.):	Building Waste Prod. (Mil. t):	Average (kg/inh.):
Denmark	5	5	1000
The Netherlands	15	7	470
United Kingdom	57	30 - 50	530 - 890
Germany	62	53	855
Belgium	10	7	700
Spain	39	11	280
Hong Kong	6	6	1000
Kuwait	1.8	1.6	900

Fig. 2. Estimated amounts of building waste, current in 1990 and 1991 for Germany and Hong Kong, excluding soil and asphalt. It should be noted that these amounts are derived from different sources and the amounts should therefore be considered, with reservations, as the best present data.

4. Recycling of Building & Construction Waste

Building waste material consists mainly of concrete and masonry, depending on the construction type. In Europe a typical percentage distribution of the main fractions of demolition waste appears in fig. 3. Naturally, the percentage differs according to the construction/demolition practice and habit of the single region or nation. In building waste a larger percentage of the paper, plastic etc. fraction is expected due to packaging materials, and wasted wood from form work and scaffolding. In road construction waste the concrete fraction will be much larger together with a significant amount of asphalt.

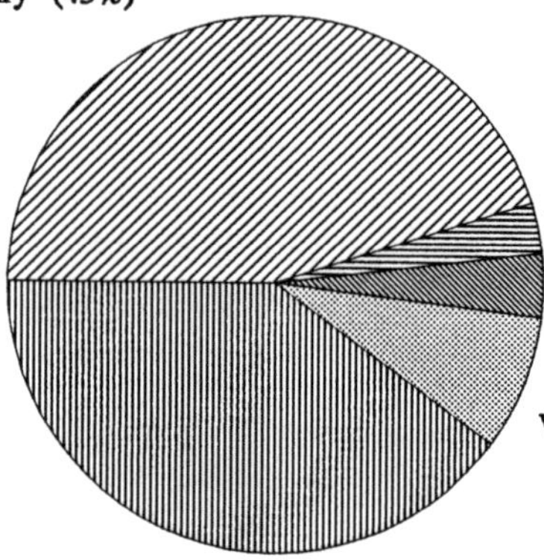

Fig. 3. Percentage distribution of the main fractions of waste in building waste in Europe, excluding earth and asphalt.

In principle, all masonry and concrete waste can be recycled and reused. Fig. 4 shows a table of examples of possibilities for the reuse of brick and concrete rubble. A necessary condition for the recycled products to be marketed as a substitute for natural materials is that the recycled materials satisfy the given technical specifications and are economically competitive. Also cleaning and reuse of undamaged bricks is expected to play an important role in the future reuse industry. In Denmark promising developments in the cleaning processes have taken place in the last two years.

Application	Project example	Waste material
Aggregate in new concrete	Concrete roads	Crushed concrete
	Runways, taxiways and aprons	"
	Concrete pavements in general	"
	Concrete sewage pipes	"
	Concrete culverts	"
	Bridges	"
	Harbour constructions	"
	Environmental plants:	"
	• water treatment plant	"
	• pumping station	"
	• fertilizer tanks	"
	• refuse disposal plant	Crushed concrete/brick
	Buildings (residential, commercial):	Crushed concrete/brick
	• foundations	"
	• floors	"
	• horizontal divisions	"
	• walls	"
	Foundations in general	"
Aggregate in new asphalt	Base course materials in pavements and yards	Crushed concrete
Unbound base course	Bicycle lanes	Crushed concrete/brick
	Pavements	"
	Field roads	"
	Forest roads	"
	Internal building site roads	"
	Primary roads	Crushed concrete/brick/asphalt
	Secondary roads	
	Runways, taxiway and aprons	"
	Parking lots and other yards	"
		"
Fill material	Cable trench	Crushed concrete/brick

Fig. 4. An example of the possibilities for the reuse of crushed brick and concrete rubble.

A number of studies from recent years have been dedicated to the quality and the satisfaction of technical specifications of the recycled materials. These reports were carried out by RILEM TC-37-DRC (Réunion Internationale des Laboratoires d'Essais et de Recherches sur les Matériaux et les Constructions / International Union of testing and research Laboratories for Materials and Structures) on demolition and reuse of concrete and masonry, which was appointed in the period 1981 -

1988. The results of these studies are found in the RILEM Report "Recycling of Demolished Concrete and Masonry" [5].

As a continuation of this work and as a follow up on the documentation of recycled materials, a new work committee was appointed in 1989, RILEM TC-121-DRG on Guidelines for demolition and recycling of concrete and masonry. This committee is expected to conclude the study during 1993.

Research and practical examples with recycled materials have also been undertaken in both the Netherlands and Denmark in the mentioned EC project [4]. Additionally, standards for the use of recycled aggregates for concrete in a passive environmental class have been set in Denmark [6]. These standards form the basis for the RILEM TC-121-DRG's current work on the RILEM Guidelines.

Based on a "Best-Wild-Guess" a global building waste production of 2 - 3 billion tons per year is estimated, and if a 30 to 40 % fraction of this being concrete, there will by the year 2000 arise 1 - 1.2 billion tons concrete waste. This leads to the possibility of replacing up to 5 % of the aggregate materials in new concrete with recycled concrete. According to "Building Waste - Generation and Recycling" [7], there will be a need for approximately 20 - 25 billion tons of aggregates in the year 2000. As a supplement to these aggregates originating from concrete rubble, it is assumed to be possible to produce at least the same amount of high quality aggregates based on recycling of masonry wastes [7].

The wood fraction of the building and construction waste is only partly recyclable. Clean and de-nailed timber and boards can be reused for new constructions, and unpolluted wood can be shredded and used for gardening, farming etc. However, the considerable amount of painted and waterproofed wood must be considered as polluted material and therefore it must be treated as chemical-/hazardous waste due to risk of leachate.

Recycling of metal has taken place for many years by the scrap industry, and at the moment the recycling of metal is performed all over the world as normal routines.

The fourth main fraction covers all other waste materials. Generally, the possibilities for recycling of these materials are rather low compared with the other three fractions. Some recycling of paper, plastic, rubber etc. take place in most developed countries, but it is difficult to organize the recycling within the building and construction industry, and the efforts of recycling of the materials are not considered as economically feasible. The material must be treated as polluted waste and disposed of in controlled landfills or incinerated.

The recycling of building and construction waste requires that the waste fractions are sorted. Mixed building and construction waste is not recyclable and must be handled as polluted waste due to the content of paint and other chemicals and the risk of pollution of the groundwater from leachate.

5. Goals for recycling

With consideration to all fractions of building and construction waste, it should be possible to reuse approximately 90% of the waste material. However, at present only a small part is being reused. Referring to the EC directive of 18th March 1991 [8], Member States shall take appropriate measures to encourage:

- Prevention and reduction of waste production and its harmfulness, particularly by the development of cleaner technologies and reuse of waste.
- Establishment of disposal and recycling facilities.
- Management and planning of waste handling.
- Implementation of *the polluter pays principle*.

In Denmark, the percentage of reuse of building and construction waste in 1986 was assessed to 12 %. After starting an action plan with the goal that before year 2000 at least 50% of all solid waste should be reused the situation today is that 25% of the waste material is reused. A model of the Danish action plan and development programme for the reuse of building and construction waste is shown in fig. 5.

In the revised Danish action plans for reuse of solid waste and cleaner technology the goal for reuse has been strengthened to 60% of all solid waste. In many other countries similar goals are established. The goals for reuse of building and construction waste in Germany is mentioned in fig. 6.

In The Netherlands the reuse of building and construction waste amounts to 40-50% of the production. The goal for the reuse by year 2000 has been raised from 80% to 90% of the total waste production (1986 figures).

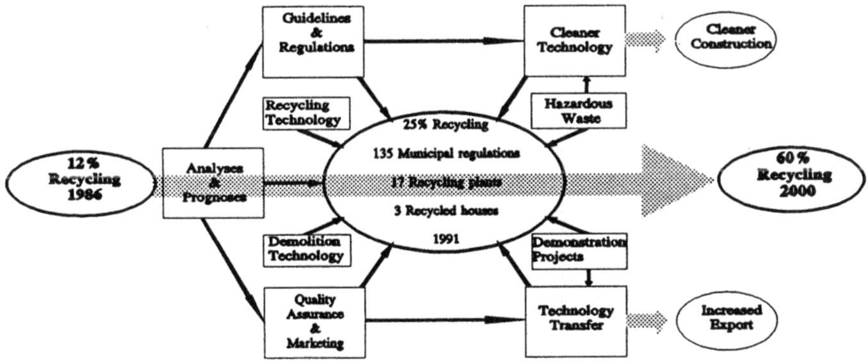

Fig. 5. Model of the Danish development programme for recycling of building and construction waste material in the period from 1986 to 1991. The boxes represent project areas within the research and development programme of the Danish Government.

	Total Waste Amounts	Total Amounts Recycled		1992	1993	1994	1995
	Mio.t. p.a.	Mio.t. p.a.	%	%	%	%	%
Demolition Waste	22,6	3,7	16	30	40	50	60
Construction Waste	10,0	-	-	10	20	30	40
Roadworks Waste	20,4	11,2	55	60	70	80	90

Fig. 6. Draft of official goals for recycling in Germany [9].

6. Barriers

In order to reach these goals, it is necessary that all barriers and obstacles are detected and considered. The overcoming of these barriers must be planned and carried out through a long-term action plan combined with the adequate development and research. The implementation of the necessary legal, economical and technical instruments require that necessary initiatives, involving legislations and regulations, are taken.

Economy

From an economical point of view, recycling of building waste is only attractive when the recycled product is competitive with natural resources in relation to cost and quality. Recycled materials will normally be competitive whenever there is a shortage of raw materials and suitable deposit sites. With the use of recycled materials, economical savings in transportation of building waste and raw materials can be obtained. This is especially apparent in urban development or reconstruction projects where there is a combination of demolition and new construction, and where it is possible to recycle large amounts of building waste at the actual work site or in the vicinity.

With a condition of market economy, the choice between recycled and natural materials depends upon price and quality. The quality of concrete with recycled aggregates is the same as that of concrete with natural aggregates. Hence, recycled concrete materials will be preferred where the price for such aggregates is *considerably* lower than that of the natural materials, provided the recycled aggregates meet the given specifications.

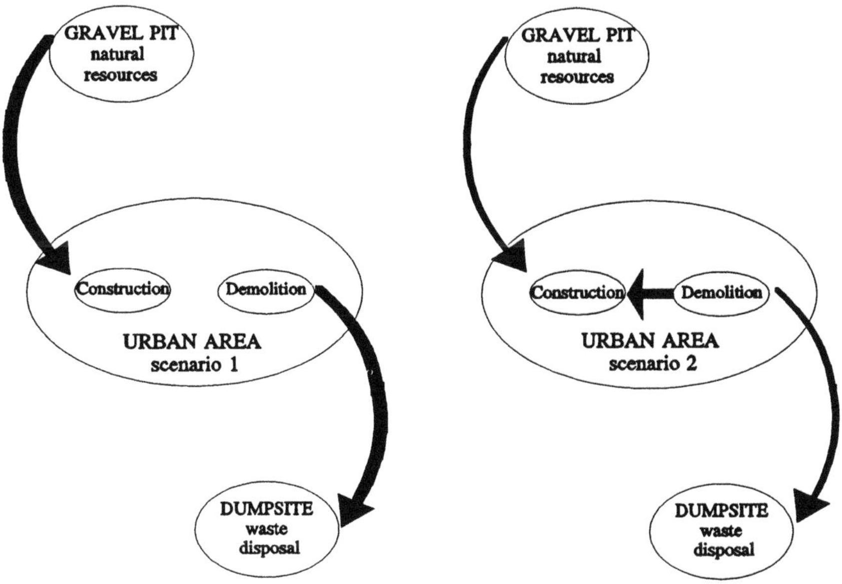

Fig. 7. Macro-economical model showing two urban renovation/rebuilding scenarios. In scenario no. 1, all building waste is removed and disposed of and only new raw materials are supplied to the site. Scenario no. 2 shows how a considerable amount of the demolition waste is reused as raw materials in the new projects. Only refuse waste, including polluted waste, is removed and disposed of, thereby reducing the demand for new raw materials and disposal facilities. The economical savings of scenario no. 2 are primarily related to transport costs and disposal fees, supplemented by a time saving.

With larger recycling projects, such as urban development, renovation of motorways or clearing of war/disaster related damages, the economical model, as shown in fig. 7, will be dominated by the transportation costs, such as the removal of demolition products and supply of new building materials. In these cases, recycling of the waste materials is very attractive.

The economical barriers can be overcome by introducing economic instruments which encourage recycling and the use of recycled materials. As an example it should be mentioned that several countries have introduced special taxes and fees in favour of recycling. Since 01.01.1990, Denmark has, for example, introduced a tax of 130 DKK (approx. 16 ECU) per ton of waste which is not recycled. From 01.01.1993 the Danish governmental fee is expected to be raised with 50% to improve the interest for recycling.

Policies & Strategies

Building and construction waste has to be considered as an individual type of waste connected to the building and construction industry. It is very important that the management and handling of waste is carried out by the industry itself. Generally, the building and construction industry is relatively conservative, and changes in normal procedures often take time and need long time policies and strategies.

One of the most important barriers is the many different interests in building waste. Usually, it is the environmental politicians, departments and offices who prepare the policies and issues concerning waste recycling and reduction, whereas the building and construction industry is controlled by politicians, departments and offices concerning housing, construction and public work. To coordinate the interests of all parties, especially concerning implementation of cleaner technologies in the industry, it is necessary that long term policies and strategies have been prepared and controlled.

According to Danish experiences it is recommended that long-term strategies, eg. for 10 years, with respect to achieving goals for recycling of building and construction waste should be adopted. These must then be continuosly revised in accordance to the political situation, and followed up with sufficient legislation and regulation at all levels: Governmental, regional and local level.

Qualification of recycled materials

The production of recycled materials does not contain technical barriers of greater importance. Demolition and crushing techniques for the production of recycled materials are well known and based on existing technologies. However, some changes in the demolition process, compared with traditional demolition, is required as described below. Presupposed that the recycled materials fulfil current standards for natural material and that the prices are competitive with prices of natural materials, certain barriers still exist.

Due to traditions and psychological barriers, the general attitude towards recycling in the building and construction industry is largely inhibitive to the utilization of recycled materials. Therefore, it is of great importance that recycled materials are qualified and accepted by all parts in the building and construction industry.

It is recommended that considerable emphasis is placed on specifying the areas of utilisation and quality standards for recycled materials. These must be in accordance with the local demand in order to gain confidence in the recycled materials and solve the problem regarding the responsibility of the utilisation of these.

Planning demolition projects
A necessary condition for the recycling of building waste is careful sorting of the waste. Waste from new constructions and rehabilitation is sorted either at production site or at a special treatment site. This separation of material categories is rather simple.

The sorting of the building waste from demolition is, however, a more complicated process. Demolition has until recently been regarded as a low technological process. Rapid demolitions and disposal of the structures were the main aims of the contractor. Special measures to separate the different types of materials were not possible due to the time factor, nor were they desired.

Optimal handling and recycling of the building waste depend on the materials being sorted in-situ and in coordination with the demolition process. It is, therefore, necessary to alter the traditional methods of demolition and introduce *selective demolition*. This requires that before and during the demolition process a concise sorting of the different material categories is carried out, thereby preventing any mix of the materials leading to pollution of the recyclable concrete/masonry rubble with wood, paper, cardboard and plastic etc.

Due to the work involved for in-situ clearing and sorting into the different building material categories before removal from the site, selective demolition is more time demanding and more expensive compared to traditional demolition methods. Economic savings are, however, gained by a better quality of demolition waste and the elimination of the need for further sorting at the recycling plant. Savings in transport costs and disposal fees are also gained. As a whole this results in an economic and environmental profit to the actual project.

It is therefore necessary to plan and conduct demolition work in a completely different way from the traditional methods. Selective demolition is principally carried out in reversal to the construction process and follows the subsequent procedures:

1. Removal of remains and non-fixtures.

2. Stripping, comprising internal clearing, removal of doors, windows, roof components, installations, water, heating and electricity, etc. This leaves only the building bearing structure remaining.

3. Demolition of the building structure.

Since selective demolition needs more time than traditional demolition, detailed planning is considered as mandatory.

In the future it is recommended that demolition projects should be planned and controlled in detail, likewise all other building and construction projects to ensure selective demolition and correct handling of the demolition waste.

Research and development
As mentioned in the used references, a lot of research within technical problems concerning demolition, crushing and use of recycled materials has taken place. In fact no barriers exist due to lack of technical research and development. However, with regard to the treatment of polluted building waste materials, much research and development is still needed.

The establishment of an integrated building waste system in a country or region requires research and development with special focus on management and economical planning.

An effective programme for an increase in reuse of building waste can for example involve the following steps:

- Assessment of amounts and prognoses for production of building waste.

- Implementation and eventual development of appropriate technological methods for demolition, handling and processing of building wastes.

- Establishment of suitable applications for the recycled materials together with setting of standards and quality control systems which can document the applicability of the materials.

- Management and regulations which can ensure the suitability of the recycling process to the given situation or the actual conditions within the building and construction industry.

Education and information
The improvement of cleaner technology and achievement of goals for reuse and waste minimization is inextricably tied to an understanding on the different problems in integrated building waste management and recycling of building and construction waste materials. Therefore, education and information within all parts and on all levels are urgently needed.

As an example it should be mentioned that The Danish Association of Demolition Contractors has planned and conducted education of demolition workers in the member companies with great success. Courses for engineers and planners have also been carried out.

Integrated building waste management
As a concluding recommendation it should be emphasised that an optimal handling of building waste and recycling must be considered in the view of all legal, economical and technical aspects. Isolated considerations in some areas might create inadequate results in other areas and exclude the optimal solutions. It should also be noted that cleaner technology and recycling of building waste materials are not an idealistic issue on its own.

The time of "green-thinking" is over - real environmental improvements and actions must be performed!

References

[1] Akio Moriata: Made in Japan. E.P. Dutton 1986

[2] The World bank: Development and the Environment. World Development report 1992

[3] DEMEX et alt.: Prognose for bygningsaffald (PROBA). Miljøprojekt nr. 150 og 151, Miljøstyrelsen (Danish)

[4] Dansk Vejbeton A/S et alt.: Recyclability in Concrete of demolition Refuses Containing Materials Non-Compatible to the Traditional Cement Matrix. EC Research Action Programme on Materials Secondary Raw Materials. EC-Contract No. MA1D-0022-C, October 1991.

[5] Hansen Torben C.: Recycling of Demolished Concrete and Masonry. E and F N Spon, London 1992

[6] Danish Concrete Association: Recommendation for the use of Recycled Aggregates for Concrete in Passive Environmental Class. Publication no.34 Oct. 1989.

[7] Lauritzen E.K. & Hahn N.J.: Building Waste - Generation and Recycling. ISWA 1991 - 1992 Yearbook

[8] EC Council Directive of 18 March 1991 amending Directive 75/442 EEC on waste

[9] Zielfestlegungen der Bundesregierung zur Vermeidung, verringerung oder Verwertung von Bauschutt, Baustellenafällen, Ersaushub und Strassenaufbruch, Entwurf. Stand 22.02.1992

[10] Lauritzen Erik and Kristensen Nils: KUWAIT: Advanced System for Recycling of Concrete Waste. Rambøll & Hannemann Consulting Engineers and DEMEX Consulting Engineers A/S, August 1990. ISWA Times 1/91.

25 RECYCLED AGGREGATES WITH FLY-ASH: AN ECOLOGICAL AND ENERGY SAVING CONCRETE

R. CAVA, D. GALLO and G. SPADEA
Department of Structural Engineering, University of Calabria, Cosenza, Italy
G. FRIGIONE
Department of Chemistry, University of Calabria, Cosenza, Italy

Abstract
The production of recycled aggregates from demolished concrete and their reuse in manufacturing new concretes is an interesting area of research in the technological field. In addition, in order to give a contribution to the protection of the environment, a wider utilization of industrial waste products is an obligatory strategy for the future.

With this in mind, this experimental investigation deals with the use of recycled aggregates mixed with fly-ash to produce low strength recycled concretes.

Fly-ash is utilized as a cementitious binder alone or by replacing, in various proportions, part of it with Portland cement. In these mixtures the fly-ash is activated by calcium hydroxide which is still present in the cement paste of demolished concrete.

Physical properties of recycled aggregates and compressive strength of hardened concretes are determined.
Keywords: Recycled aggregates, Recycled concretes, Fly-ash, Mix proportions, Carbonation, Compressive strength.

1 Introduction

In recent years, the attention to environmental problems and growing technical and economic difficulties in the preservation of natural resources of raw materials has encouraged a higher utilization of recycled materials.

Within these, demolished concrete has a relevant position since it is estimated that, between now and the year 2000, for the necessity of replacing old constructions with new ones, about 220 million tons of debris will be produced each year in the EEC countries and that about 80% of demolished material will be represented by old concrete, (Hansen, 1986; Sri Ravindrarajah et al., 1987).

At present, very little demolished concrete is currently recycled or reused; the small quantity which is recovered is mainly utilized as unscreened gravel base or sub-base in highway construction, while most of it is dumped or

Fracture and Damage of Concrete and Rock – FDCR-2.
Edited by H.P. Rossmanith. ISBN 0 419 18470 8.

disposed as fill. Moreover, readily accessible disposal sites in the surroundings of larger urban areas are diminishing more and more: distances from sites of demolition become larger and disposal charges and transportation costs increase. In addition, easily reachable supplies of good natural aggregates are also diminishing. For these reasons, it is natural to consider reuse of materials in many useful developments. Recent studies have shown that low strength new concretes can be produced entirely from waste products such as recycled aggregates and fly-ash (Hansen, 1990).

As far as fly-ash is concerned, it is well known that it is a combustion residual product of burned carbon in thermal power plants. It contains silica and alumina in amorphous form and itself doesn't possess cementitious values. However in the presence of moisture, fly-ashes react chemically with calcium hydroxide at room temperature to form compounds with cementitious properties.

The objective of the present work is, then, the study of properties of a recycled concrete manufactured by using coarse and fine recycled aggregates together with low calcium fly-ash, which replaces a part or the whole of Portland cement. Cementitious property of fly-ash is mainly activated by calcium hydroxide still present in the original cement paste of recycled aggregates obtained from demolished concrete.

Generally the reuse of fine recycled aggregates, for many inconveniences which can arise, is avoided in manufacturing recycled concrete (Hansen, 1986; Mulheron and O'Mahony, 1988). Recently, the utilization of fine crushed products from demolished concrete, by mixing them with coarse recycled aggregates and fly-ash, has been considered (Hansen, 1990).

However, when the recycled aggregates are exposed to the air, the calcium hydroxide contained by them is subjected to a carbonation process which reduces the activation capacity of fly-ash. Therefore, in order to evaluate this, one part of recycled aggregates was utilized promptly, by manufacturing test specimens within 24 hours from the crushing of the original concrete, whereas the remainder was employed after two months of exposure to the air. During this period it was stirred several times in order to achieve a higher carbonation of aggregates and to simulate a long-term storage.

2 Experimental program

2.1 Materials

The recycled aggregates were produced by crushing, in a jaw crusher, in one pass, 150 mm laboratory cubes of about one-year old uncontaminated concrete with an average strength of about 37 MPa (in the range 22-56 MPa). The jaw crusher was set at an opening of 65 mm in the closed position.

The crushed products were screened into five size fractions: 25-20 mm, 20-10 mm, 10-3 mm, 3-1 mm, 1-0 mm. Particles greater than 25 mm were rejected.

The recycled coarse aggregate was defined by the size passing through the 25 mm sieve and remaining on the 3 mm sieve, whereas the recycled fine aggregate was defined by the size passing through the 3 mm sieve.

The fine natural aggregate used was river sand with a maximum size of 3 mm.

All the aggregates were then recombined to obtain a grading curve with a good fit to the Fuller's curve.

Type F [ASTM C 618-85] fly-ash (low content of CaO) and type III ASTM cement were used.

2.2 Tests on aggregates

2.2.1 Water absorption

As is well known the most marked difference in physical properties of recycled concrete aggregates compared with conventional aggregates is the higher water absorption. This is due to the presence of old mortar attached to original aggregate particles (Hansen, 1986).

The water absorption of aggregates is determined as the weight of water absorbed referred to the weight of dry aggregates.

Fig. 1 shows the water absorption of recycled aggregates not exposed to the air as a function of elapsed time after their immersion.

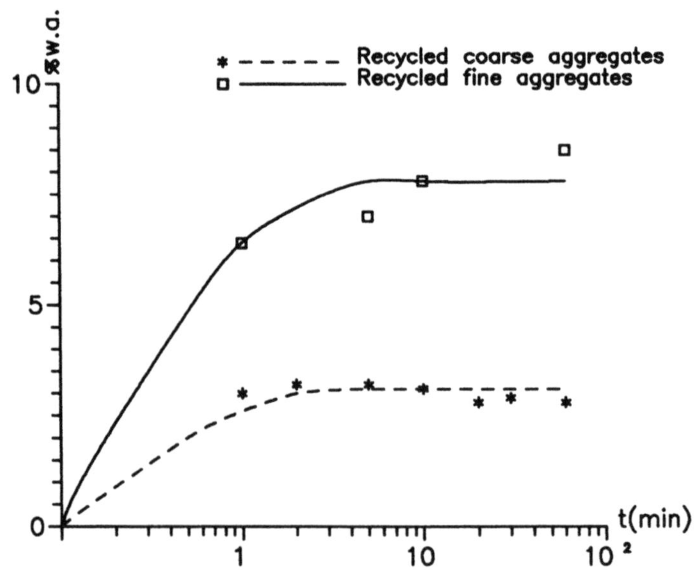

Fig.1. Water absorption as a function of time.

The results show that the water absorption in the aggregate particles increases when the aggregate size decreases.

This higher water absorption was taken into account during the process of concrete mix, particularly from the viewpoint of ensuring a good workability for a sufficient time.

The values of water absorption of recycled aggregates exposed to the air remained, substantially, the same compared to the ones relative to the aggregates not exposed.

2.2.2 Thermogravimetric analysis

Thermogravimetric analyses were carried out on a not exposed to the air recycled aggregate sample of the finest fraction (0-53 μm) and on another corresponding sample exposed to the air. These analyses showed that non-exposed aggregates contain 3.9% by weight of calcium hydroxide whereas exposed aggregates contain 3.5%. The falling of 0.4% of calcium hydroxide corresponds to the increase of percentage of calcium carbonate in the exposed aggregates.

2.3 Mix proportions

In table 1 are shown the compositions of the concrete mixes studied. Two types of mixes were prepared: series RR by using both fine and coarse recycled aggregates and series NR by utilizing natural sand and coarse recycled aggregates. Indices 1, 2 and 3 indicate, recycled concretes made with 300, 250 and 200 kg of fly-ash plus 0, 50 and 100 kg of cement, respectively.

Two batches were prepared for each mix.

Table 1. Mix proportions for 1 cubic meter of concrete

Type of mix	Components				
	Fly-ash (Kg)	Cement (Kg)	Fine recycled aggregate (Kg)	Fine natural aggregate (Kg)	Coarse recycled aggregate (Kg)
RR1	300	0	804	0	1157
NR1	300	0	0	804	1157
RR2	250	50	804	0	1157
NR2	250	50	0	804	1157
RR3	200	100	804	0	1157
NR3	200	100	0	804	1157

From now on, we will indicate the whole content of fly-ash or fly-ash plus cement simply as "binder".

Each mix was prepared within 24 hours from the time of crushing and repeated after about two months of exposure to the air of the recycled aggregates which were, moreover, stirred daily.

The trial mix was made in order to determine the free water content necessary to obtain an adequately workable mix.

The proportions by weight of binder:fine aggregate: coarse aggregate:water were 1:2.7:3.9:0.44.

The water/binder ratio was kept constant and was referred to saturated surface dry condition.

2.4 Batching procedure, casting and curing of specimens

Concrete was mixed in an ordinary laboratory mixer.

Recycled coarse aggregate was placed in the mixer and mixed for 20 seconds. Then 1/3 of the water required to saturate the aggregates was added and mixed for 5 minutes. Afterwards, natural sand and recycled fine aggregates were respectively added to the NR and RR mixes and mixed for 20 seconds. Then, the remaining 2/3 of saturation water was added and mixed for 5 minutes. Finally the binder was introduced, followed by mixing water. All the ingredients were mixed for a further 3 minutes and, then, an adequate number of cubic specimens were prepared.

As will be described in 3.1, before the casting of specimens, tests of workability were done.

Then, cubic specimens were cast in moulds of 150x150x150 mm, compacted on a vibrating table and wet-cured until about 24 hours before compressive tests.

Furthermore, in order to study the influence of the type of curing, specimens made with aggregates exposed to the air were oven-cured at 70°C for 10 hours.

3 Results and discussion

3.1 Tests on fresh concrete

In order to determine the workability of fresh concrete, Walz tests were carried out on each batch. The Walz coefficient was determined in accordance with D.I.N. 1048.

Table 2 shows the values of Walz coefficient of mixes made with aggregates exposed to the air.

Table 2. Walz Coefficient (CW)

Type of mix	RR1	NR1	RR2	NR2	RR3	NR3
CW	6.3	4.7	6.1	3.1	5.8	3.0

The values of Walz coefficient of mixes made with recycled aggregates not exposed to the air are substantially the same.

The results of table 2 show that the use of fine recycled aggregates reduces the workability of mixes. Moreover, the addition of cement, increases the workability especially for mixes made by using natural sand as fine aggregate.

3.2 Tests on hardened concrete

Compressive strength of cubes was determined both at 28 and 90 days.

The oven-cured specimens were tested within 24 hours from the end of treatment.

The results are shown in tables 3, 4 and 5.

Table 3. Compressive strength of specimens made with recycled aggregates not exposed to the air

Type of concrete mixes	RR1	NR1	RR2	NR2	RR3	NR3
28 days compressive strength (N/mm^2)	3.1	0.9	3.7	6.0	11.5	10.0
90 days compressive strength (N/mm^2)	5.0	1.8	6.3	12.1	22.0	19.6

Table 4. Compressive strength of specimens made with recycled aggregates exposed to the air

Type of concrete mixes	RR1	NR1	RR2	NR2	RR3	NR3
28 days compressive strength (N/mm^2)	*	0.1	0.2	7.0	0.5	11.1
90 days compressive strength (N/mm^2)	0.1	0.3	0.2	10.2	0.8	18.1

*Specimens not cemented

Table 5. Compressive strength of oven-cured specimens made with recycled aggregates exposed to the air

Type of concrete mixes	RR1	NR1	RR2	NR2	RR3	NR3
Compressive strength (N/mm^2)	0.1	0.6	0.3	6.6	0.8	10.1

The reported strengths at any age are the average of at least three specimens.

Table 3 shows that compressive strengths, after 28 days, of concretes made with recycled aggregates not exposed to the air are relatively low. However, compressive strengths after 90 days are about twice as high as the ones after 28 days. In addition, compressive strengths of concretes manufactured by using both coarse and fine recycled aggregates are higher compared to those of corresponding concretes prepared by using coarse recycled and natural fine aggregates. This can be attributed to the higher calcium hydroxide content of fine fraction of recycled aggregates which activates cementitious property of fly-ash.

An anomalous behaviour can be observed when part of the fly-ash is replaced by 50 kg of cement (RR2 and NR2 mixes).

Moreover, for mixes produced with coarse and fine recycled aggregates, there are slight differences of strength when the cement content varies from 0 to 50 kg and more notable variations when the cement content changes from 50 to 100 kg.

As far as aggregates exposed to the air are concerned, it can be observed (Table 4) that the increases in compressive strength of concrete after 90 days are relatively lower than of those made with aggregates not exposed to the air.

In addition, the compressive strength of NR concretes is higher compared to that of RR concretes, independently from the type of binder.

This may be due to the carbonation of recycled aggregates and to the consequent decrease in calcium hydroxide content. In this case the cementitious property of fly-ash is less activated.

In particular, exposure to the air of fine recycled aggregates seems to alter the hardening process. This can be seen, for example, by observing that the compressive strength of RR2 mix is the same after 28 as after 90 days (Table 4).

Finally, the effect of 10 hours oven-curing at 70°C is (Table 5) nearly the same as that of 28 days wet-curing at room temperature.

4 Conclusions

From this study the following conclusions can be drawn:

(a) it is possible to produce recycled concretes by using both coarse and fine recycled aggregates mixed with fly-ash and a small amount of cement on condition that the recycled aggregates haven't undergone a carbonation process;
(b) the thermogravimetric analysis showed that the reduction by weight of calcium hydroxide owing to air exposure is percentually small. However, due to the fact that it affects mainly the surface of fine aggregate particles, the fly-ash activation is highly limited;
(c) the compressive strength, after 90 days, especially that of recycled concretes manufactured with non-exposed aggregates, is about twice of that after 28 days. Afterwards it may be conjected that a further increase of strength could be gained with time.

5 References

Hansen, T.C. (1986) Recycled aggregates and recycled aggregate concrete second state-of-the-art report developments 1945-1985. **Materials and Structures**, 19, 201-246.
Hansen, T.C. (1990) Recycled concrete aggregate and fly-ash produce concrete without Portland cement. **Cement and Concrete Research**, 20, 355-356.
Mulheron, M. and O'Mahony, M. (1988) The durability of recycled aggregates and recycled aggregate concrete, in **Demolition and Reuse of Concrete and Masonry** (ed Y. Kasai), Chapman and Hall, London, pp. 633-642.
Sri Ravindrarajah, R. Loo, Y.H. and Tam, C.T. (1987) Recycled concrete as fine and coarse aggregates in concrete. **Magazine of Concrete Research**, 39, 214-220.
Sri Ravindrarajah, R. and Tam, C.T. (1988) Methods of improving the quality of recycled aggregate concrete, in **Demolition and Reuse of Concrete and Masonry** (ed Y. Kasai), Chapman and Hall, London, pp. 575-584.

26 DEMOLITION AND RECYCLING OF A REINFORCED CONCRETE BRIDGE DURING CONSTRUCTION OF THE GREAT BELT LINK IN DENMARK

E.K. LAURITZEN
DEMEX Consultant Engineers, Denmark

Abstract
In the latest 10 years the Danish Government has supported much research and development within the field of demolition and recycling of building waste material. This paper describes a recent demonstration project comprising tests of different demolition techniques with a conclusion on the most appropriate fragmentation of reinforced concrete with respect to recycling of concrete and reducing the loads on the environment and the dynamic loads on the structure itself.
Keywords: Demolition, Concrete-recycling, Blasting, Dynamic Loads, Vibrations, Fragmentation, Shockmeasurements

1. Introduction

In connection with the demolition of a motorway bridge in Nyborg, Denmark, a project on fragmentation and recycling of concrete has been carried out. The bridge was constructed so that the motorway would cross the railway at an angle of 20°. The bridge deck was constructed of a continuous concrete slab over two free spans of 25 m. The deck was heavily reinforced with both post tensed and non-tensed reinforcement bars. The total amount of solid materials gives approx. 11,000 tons, whereof approx. 500 tons are steel.

In construction work, rock blasting etc. it is well known that blast induced vibrations in the ground can cause damage to buildings, structures and installations. Referring to standards of different countries, e.g. the Swedish standard SS 460 4866 and the German norm DIN 4150, it is also known that the effect of blast induced vibrations is normally expressed by the vibration parameter, particle velocity, v (mm/s), and that damage is normally not expected when the level of the particle velocity is lower than 5 - 10 mm/s, depending on the exposed object and natural frequency of the ground etc.

In demolition work vibrations from blasting, jack hammers, wrecking ball and other tools are also a great problem. Compared with the *ground-transmitted external vibrations*, the *structure-transmitted internal vibrations* are of a quite

Fracture and Damage of Concrete and Rock – FDCR-2.
Edited by H.P. Rossmanith. ISBN 0 419 18470 8.

different nature. Due to the transmission in the structure, the dominating frequencies and level of particle velocity are considerably higher in the internal vibrations than the external vibrations. No standards or methods of assessing damage exists, and very little research has been carried out in the field of vibration measurement.

The aim of the research project was:

- Investigation into different demolition methods and their appropriateness for partial demolition (removal of edge beams, hole formation etc.).
- Investigation into the effect of and damage from the demolition methods on the construction adjacent to the demolition areas.
- Cost effective studies of the different demolition methods.
- Analysis of the products resulting from the different demolition methods with regard to use as recyclable aggregate in new concrete.

An illustration of the test project is shown by figure 2. The demolition areas defined were chosen in order to maintain the stability of the bridge during the test project and to avoid the densest areas of reinforcement.

Fig. 1. The motorway bridge at Nyborg, Great Belt Link.

A: Removal of edge beams by blasting.
A: Removal of bridge deck by blasting.
B: Hole formation in bridge deck by rapid blasting.
C: Hole formation in bridge deck by cautious blasting.
D: Removal of edge beams by non-explosive agents.
F: Removal of edge beams by wrecking ball.
H: Removal of edge beam by hydraulic hammer.
H: Removal of bridge deck by hydraulic hammer.
I: Rapid hole formation in bridge deck by hydraulic hammer.
J: Cautious hole formation in bridge deck by hydraulic hammer.

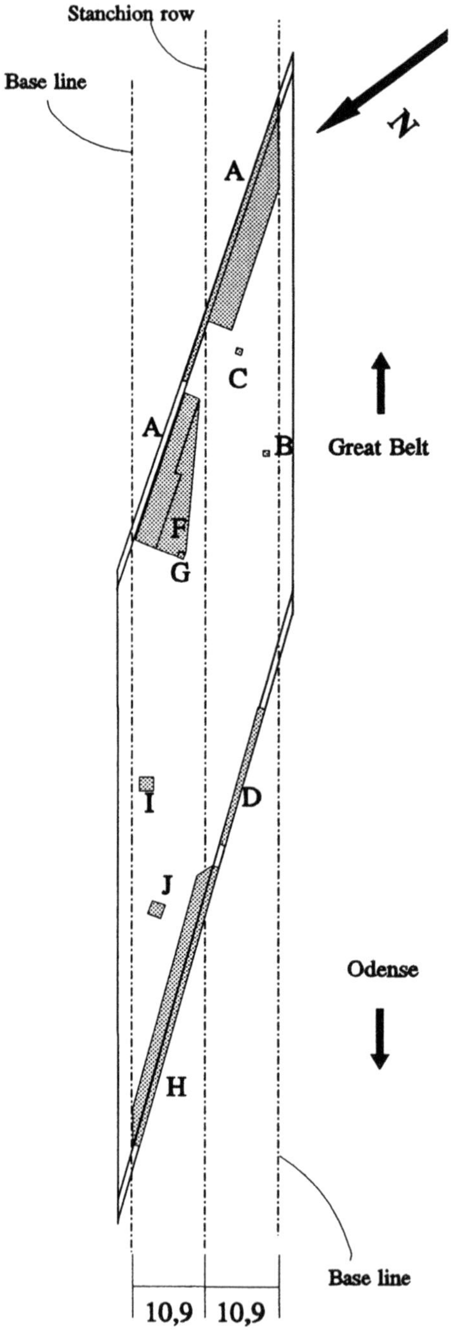

Fig. 2. Plan of bridge deck with identification of demolition areas. In the original planning, area E was for hole formation with diamond cutting, however, this test was later cancelled. The demolition areas defined were chosen in order to maintain the stability of the bridge during the project and avoid the densest areas of reinforcement.

2. Blasting tests

Three different explosives types were used in the tests: Dynamex, Detonating Cord (PETN) and Gurit. The explosives were placed in drill-holes in the edge beam and bridge deck, being detonated by electrical delay detonators.

Determination of the charge sizes and drill-hole spacings was calculated from data gathered through experience of previous blastings [1,2]. From these data it was initially estimated that a specific explosive charge of 0.6 - 1.0 kg/m³ would be appropriate for demolition of the concrete structure.

In order to determine and verify an optimal blasting plan for the removal of the edge beam and bridge deck, pre-tests were carried out with varying parameters, comprising modification of the explosives type, charge size, drill-hole pattern and drill-hole depths.

Drilling of the holes was carried out with pneumatic drills and the drilling was often affected by the heavy reinforcement, leading to a frequent need for re-drilling.

Blasting of Edge Beam
The blasting tests involved (see figure 3):

- Blasting of charges in edge beam (A) and the bridge deck (B) as two separate blastings.
- Blasting carried out as one blast with millisecond delay between (A) and (B).

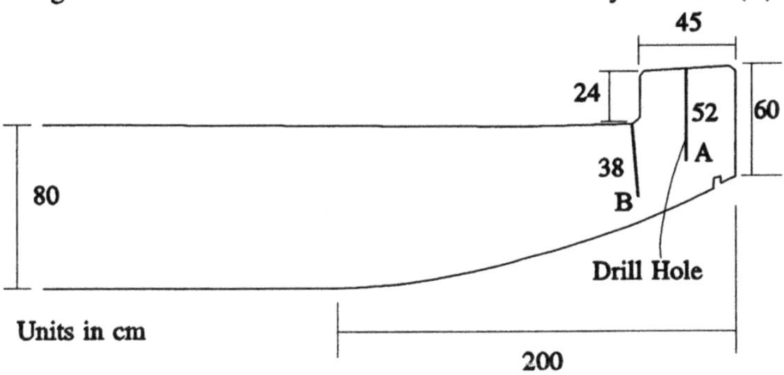

Fig. 3. Cross section of bridge deck and edge beam. The sketch shows the placement of drill holes in the edge beam for the blasting tests.

Blasting tests of the edge beams involved 9 blasting series as illustrated in table figure 4. The blastings were executed with two rows of drill-holes, one row placed in the middle of the beam and one row of shorter holes directly behind the edge beam.

On the basis of the nine test blastings on the edge beam, it was concluded that the optimal charge size for the removal of the beam, as depicted in the bottom row of figure 4., was **approx. 1 kg/m³**.

Series number	Hole Spacing	Row Burden	No. of Holes	Hole Depth	Stemming in Drill Hole	Charge Type	Charge Size	Total Charge	Specific Charge
	cm	cm		cm	cm		g	g	kg/m³
Ak1	30	21	8 A	52	40	Dynamex	62.5	500.00	0.69
Ak2	30	21	4 A	50	25	Gurit	50	200.00	0.56
Ak3	30	21	4 B	35	17	Gurit	40	152.00	0.42
Ak4	30	21	8 A	42	20	Gurit	50	408.00	1.15
	30	21	8 B	35	25	Dynamex	40	336.00	
Ak5	30	21	8 A	52	20	Gurit	70	552.00	1.27
	30	21	7 B	30	15	Gurit	35	238.00	
Ak6	-	21	1 B	35	25	Dynamex	40	40.00	0.44
Ak7	20	21	14 A	52	20	Det. Cord	20	302.40	0.36
Ak8	40	21	7 A	48	25	Dynamex	80	560.00	0.67
Ak9	30	21	6 B	35	20	Det. Cord	10	64.80	0.12
Optimal	30	21	8 A	52	20	Gurit	70	552.00	1.00
	30	21	8 B	35	20	Det. Cord	10	86.40	

Fig. 4. Table with the nine series from the blasting tests for Edge Beam. Shots Ak4, Ak5 and the optimal are one blasting with 30 ms delay between charges in the edge beam (A) and charges in the bridge deck (B). The specific charge is calculated from the actual charge and the amount of concrete removed.

Blasting of Bridge Deck

Figure 5. illustrates the drill-hole pattern used for the removal of the deck. Note that the edge beam has already been removed prior to the deck blastings. Figure 6. presents the 13 different blasting series executed in the test blastings for the removal of the bridge deck. The shots were of varying parameters, being charge size, drill-hole distance and explosives types. As shown in the bottom line in figure 6, the test results in assessment of the most appropriate specific charge: **0,7 kg/m³**.

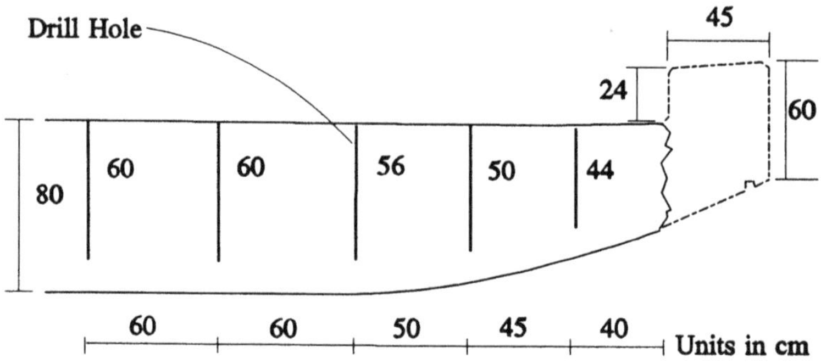

Fig. 5. Cross section of bridge deck with placement of drill-holes used in blasting tests. The edge beam has been removed by blasting/chopping.

Area Number	Row Number	Number of Holes	Hole Spacing	Row Burden	Charge per Hole	Total Charge	Concrete removed depth	length	volume	Specific Charge
			m	m	g	kg	m	m	m^3	kg/m^3
A1	1	8	0.4	0.4	62.5	0.50	0.40	3.20	1.02	0.49
A2a	1	6	0.4	0.4	125	0.75	0.40	2.40	0.77	0.97
A2b	1	3	0.4	0.4	90	0.27	0.40	1.20	0.38	0.71
A2c	1	4	0.4	0.4	90	0.36	0.40	1.60	0.51	0.71
A3	1	3	0.4	0.4	60	10.93	1.60	12.60	16.13	0.68
	2	21	0.6	0.6	250					
	3	22	0.6	0.6	250					
A4	1	20	0.4	0.4	60	9.20	1.60	8.00	10.24	0.90
	2	16	0.6	0.6	250					
	3	16	0.6	0.6	250					
A5	1	36	0.6	0.6	187.5	15.75	1.20	21.60	20.74	0.76
	2	36	0.6	0.6	250					
A6	1	4	0.25	0.5	62.5	0.25	0.50	1.00	0.40	0.63
A7	1	5	0.25	0.5	62.5	0.31	0.50	1.25	0.50	0.62
A8	1	23	0.5	0.5	187.5	10.06	1.00	11.50	9.20	1.09
	2	23	0.5	0.5	250					
A9	1	3	1.0	0.5	250	0.75	0.50	3.00	1.20	0.63
A10	1	3	0.5	0.5	125	0.38	0.50	1.50	0.60	0.63
A11	1	3	0.5	0.5	250	0.75	0.50	1.50	0.60	1.25
A12	1	3	1.0	0.5	125	0.38	0.50	3.00	1.20	0.32
A13	1	27	0.4	0.4	62.5	9.69	1.60	10.80	13.82	0.70
	2	16	0.6	0.6	250					
	3	16	0.6	0.6	250					

Fig. 6. Blasting tests for the removal of bridge deck.

Blasting for Holes

The objective of these tests was to determine the time for penetration through the bridge deck in case of an emergency rescue operation by the different demolition methods. Due to the small quantity of reinforcement in the concrete structure the areas B, C and G were selected as test areas.

In the blasting tests, two methods were used. Cautious blasting, which involves two or more independent shots, and rapid blasting involving one blasting with millisecond delays between the shots. Figure 7 illustrates the drill-hole pattern employed for the rapid blasting of the hole in the bridge deck.

Covering for the blastings comprised of rubber mats, Fibertex and ROBTEX (Kevlar) mats with a 1 - 1.5 m layer of gravel. No covering was used under the bridge due to the vertical fall of the flyrock being taken by the underlying ground. Additionally, the rapid vertical flyrock would not be able to escape the sides of the bridge.

Both the rapid and the cautious blastings resulted in well defined and cleanly cut holes.

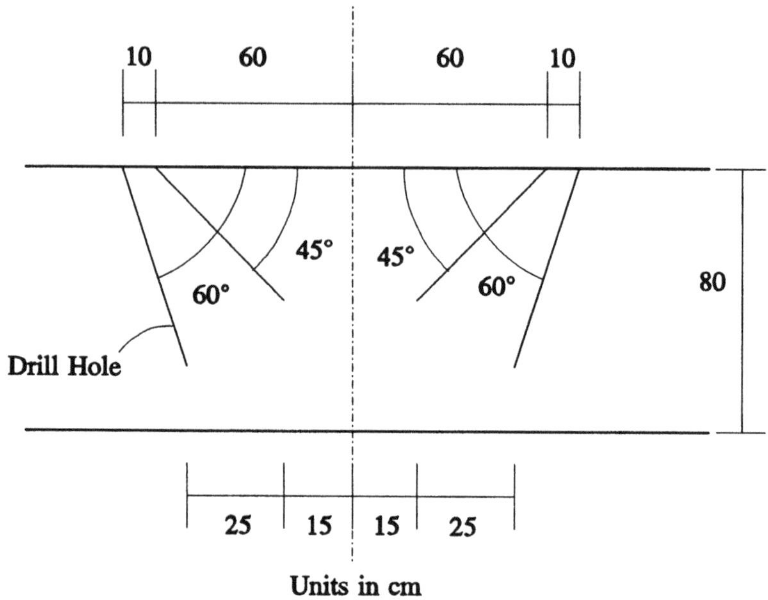

Fig. 7. Diagram showing drill-hole pattern and charge positions for rapid blasting of hole in bridge deck area B and G. The tests showed that with one blasting it was possible to penetrate the bridge deck with a suitable hole. For the blasting the two inner most hole rows were charged with 125 g per hole, and the outer rows with 185 g per hole, thereby a total charge size of 3720 g. Delay between the inner most holes and the outer rows was 25 ms. Square holes of 1.5 x 1.5 m were blasted and the total volume of concrete removed was 1.8 m^3. The specific charge was 2.1 kg/m^3.

3. Other demolition methods

Two hydraulic jack hammers were tested. A big hammer, Montabert 1100, weight 1350 kg, was very competitive and very effective. On the other hand the bridge structure was much affected during the operation of the hammer. A small hammer, Brock 250, weight 250 kg, was not very effective, and did not succeed in penetrating the concrete deck.

A crane and ball, weight 2.2 t, was not as effective as the big hammer. Finally, demolition of some of the edge beam was tested by the use of BRISTAR 100 and BRISTAR SUPER 2000. After a couple of days the agent caused some cracks in the reinforced concrete, but it could not be considered as demolition, only crack damages.

4. Ultrasonic measurement

To control the undesired effects on the remaining concrete structure ultrasonic measurement and recording of vibrations was carried out. Before and after the demolition tests concrete cores were drilled out of the structure and investigated.

Ultrasonic measuring of the concrete was performed in connection with the blasting and chopping tests in areas A, H and I, (see fig. 2). The objective was to register crack propagation in the concrete close to the demolition areas. The measurements were carried out with the use of PUNDIT equipment. One set of measurements were carried out before start of the demolition and compared with measurements taken after completion of the demolition process.

Ultrasonic measurements were performed as indirect, as both the transmitter and receiver were placed on the bridge deck top. With this indirect measuring sequence, the transmission speed will increase with the increase in distance between the transmitter and the receiver. Cracks as a result of the demolition are thereby registered with varying increases in the transmission times. Figure 8 shows the use of the transmitter and the receiver of the PUNDIT equipment.

A considerable increase in the transmission time indicates that there are cracks in the concrete between the two transducers. The influence of this crack will decline with the ensuing increase in distance between the transmitter and receiver since the effect of the crack reduces.

The results from area A show that the greatest percentage increase in "after demolition" times is in the distances of 40 - 50 cm from the demolition area. This applies to both blasting and chopping with hydraulic hammer. These figures confirm those attained from pilot tests performed, and other experiences, eg. according to Molin [3,4].

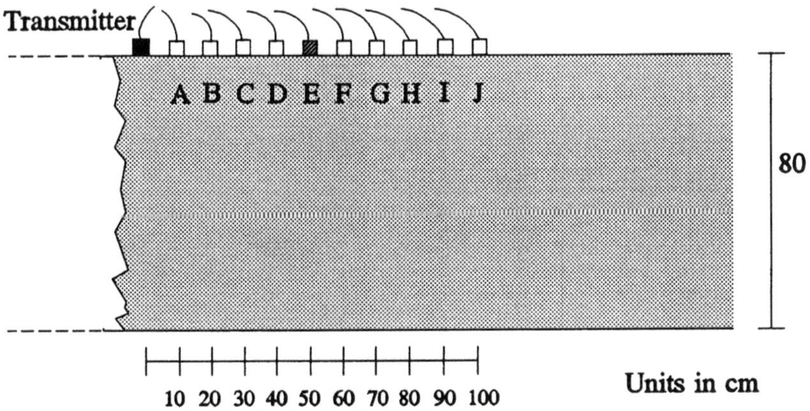

Fig. 8. Diagram showing ultrasonic measuring setup after demolition. The transmitter is stationary and in the diagram, the receiver is at position E (50 cm from transmitter). The measurements register the signal transmission time in the upper portion of the deck. Cracks deep in the deck can not be expected to be registered.

5. Measurement of vibrations

In order to compare the effects of the different demolition methods to the remaining construction, vibration measurements were carried out in connection with the demolition tests as shown at figure 2. Since no standards or specific experience is available for the damaging effects of shock waves in constructions under demolition or neighbour constructions, then it was thought important to examine these effects. The properties and varying levels of the vibrations were studied with regards to acceleration, frequency, displacement and velocity.

The resulting vibrations from all three demolition methods, blasting, wrecking ball and hydraulic hammer, lie between a wide range of frequencies. Depending on the characteristics of the method used, then the different frequencies will be more or less damaging. Therefore it is important to examine the frequencies of the resulting vibrations whereby we can create an impression of the energy levels experienced by the concrete. This energy is greatest with low frequency vibrations which diminish with an increase in distance. Hence, vibrations which are dominated by low frequencies are more damaging to the remaining construction than higher frequency vibrations.

Vibration measurements were performed in lines parallel to the bridge's baselines. The measuring was performed with four accelerometers (Brüel & Kjær 4370) vertically installed in distances 2 m to 32 m from the demolition site as shown in figure 9.

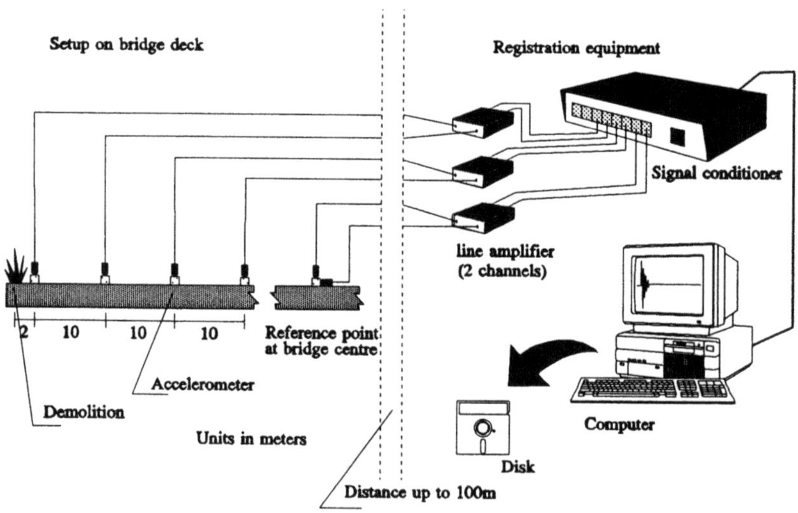

Fig. 9. Diagram with measuring equipment used and setup on bridge for the measuring of vibrations in the construction.

Since the lines were moved for each consecutive demolition area, then it was found necessary to establish a fixed reference point at the centre of the bridge. Here measurements were taken in both the vertical and horizontal planes.

Conditioning of the digital signals was carried out with the software DADiSP 2.0 Beta from DSP Development Corporation. Printouts based on the registered and integrated acceleration signals contained the pattern of acceleration, velocity, displacement, and Energy Spectral Density (ESD). In addition, the following values were calculated:

- Maximum acceleration.
- Maximum velocity.
- Maximum displacement.
- Interval for dominating frequencies of the PSD spectrum.

Blasting

The results from blasting vibration measurements are shown at figures 10, 11 and 12. With regards to the blasting effects to the surrounding concrete, simple arithmetic rules apply, ie. cubic root calculations relating to a spherical blasting medium, and square root calculations to material in a single plane medium. We are thus able to combine the two variables, distance (d) and charge size (L) in a scaled distance ($d_s = d/L^{-2}$) and thereby present the results in a two dimensional graphical display. This gives an impression of the blasting effects to the surrounding material.

It can be seen that the greatest effects originate from blasting in the bridge deck which corresponds to the fact that these blastings involved a considerable larger amount of simultaneously detonating explosive charge (max. 15 kg distributed in two ms-delay steps). The results from the blasting of holes in the deck are included in the figures 10, 11 and 12, where it is obvious that a considerable reduction in the effects to the remaining construction are obtained with cautious blasting, that will say dividing the blasting operation in two separated blasting sequences.

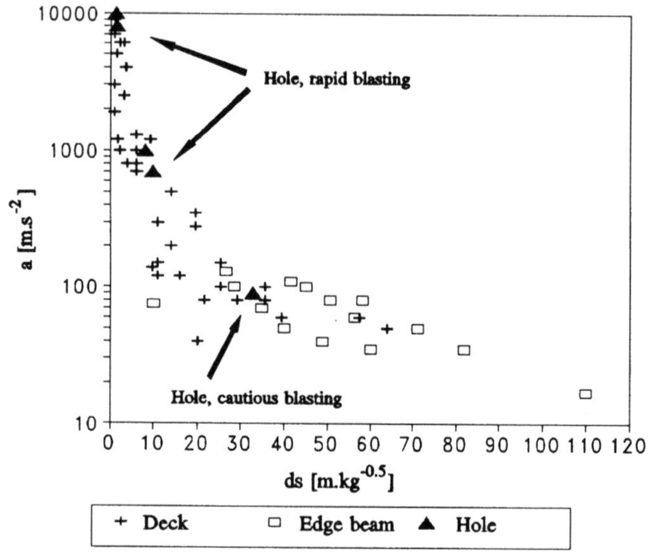

Fig. 10. Maximum registered acceleration with blasting of deck, edge beam and holes.

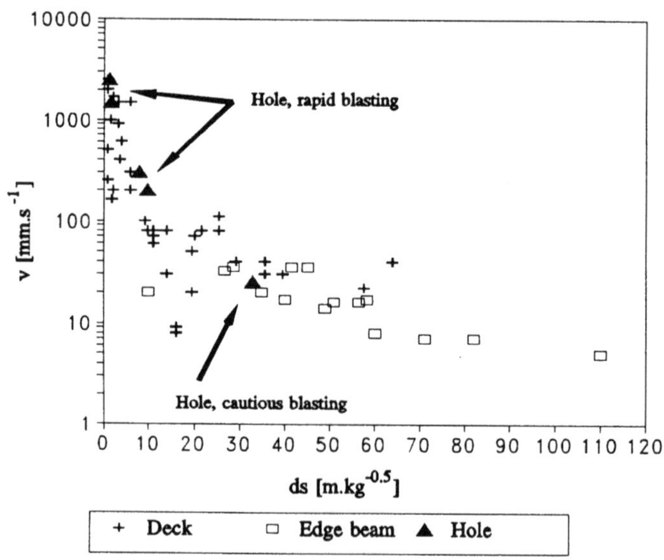

Fig. 11. Maximum registered velocity with blasting of deck, edge beam and holes.

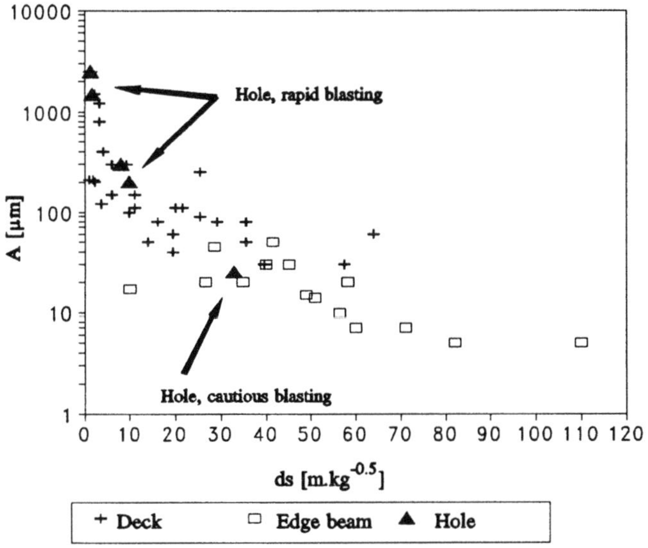

Fig. 12. Maximum registered displacement with blasting of deck, edge beam and holes.

Chopping

Chopping with a Montabert 1100 (weight 1350 kg) hydraulic hammer caused also perceptible dynamic effects on the bridge. The most severe effects were remarked during chopping of the deck, whereas chopping of the edge beam caused less vibrations. The results of the recordings during chopping showed maximum acceleration, approx. 110 m/s^2, velocity 100 mm/s, and displacement 0,1 mm.

6. Evaluation of the energy transmission

The energy which passes through a given point in the concrete deck can be found by summation of the ESD-spectrum of the whole frequency pattern. Below is given a brief account of the ESD results and their interpretation.

ESD spectrums are calculated from the measurement results of blasting tests in the bridge deck and edge beam, as well as chopping by hydraulic hammer. It is also necessary to compare the affectivity of each demolition method when evaluating the energy transmissions. The blastings result in the crushing of varying sized fragments, whilst the energy levels from the hammer give single chops. Therefore it was necessary to carry out multiple impact calculations of the relative energy associated with the demolished concrete volume.

Due to the many uncertainties connected to the formulae used, then it is advised that one does not place too much weight on the results. Under these circumstances then it is proposed that much more research is carried out in this field in order to establish a much broader understanding on the effects of the demolition methods in relation to "cleaner technologies" strategies.

From figures 13, 14 and 15 it can be seen that there is a considerably greater damping effect with blasting than with chopping. Also noticed at the reference point, is that the horizontal effect from blasting is generally larger than the vertical.

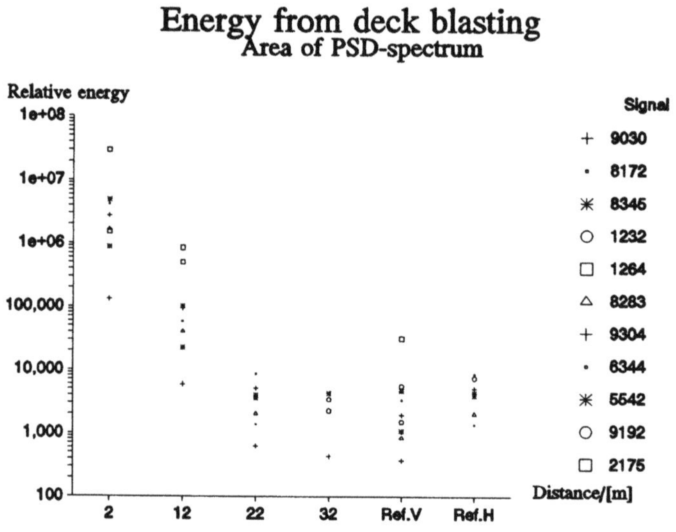

Fig. 13. Relative energy registered by measuring points with bridge deck blasting.

With blasting in the bridge deck, then it is seen that both the larger and smaller charges result in energy of similar magnitude in the furthest of measuring points and at the reference point. This could be a result of greater damping of the larger charges than of the smaller charges.

The energy flow, registered in the four measuring points of the chain, is used as a basis for the comparison of the energy which is distributed in the construction as a result of blasting and chopping. This energy flow must be proportionate to a specified length of edge beam or volume of bridge deck demolished by blasting or chopping. As an example, we can take the case that for the demolition of 1 m of edge beam by hammer necessitates approx. 1000 chops and for 1 m^3 of deck, approx. 3000 chops. In figure 13, blasting no. A 13 resulted in a relative energy level of 1 million, 12 m from the site and demolished 14 m^3 concrete. Demolition of the same amount by chopping requires, according to graph figure 15, an energy level of approx. 14 x 3000 hammer chops x 800 = approx. 34 million.

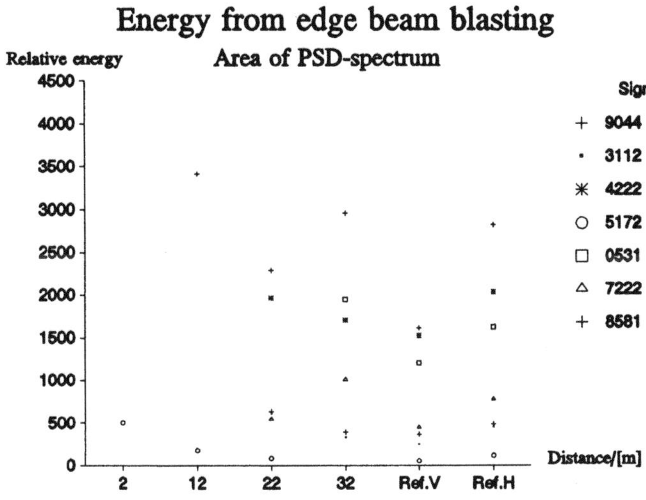

Fig. 14. Relative energy registered by measuring points with edge beam blasting.

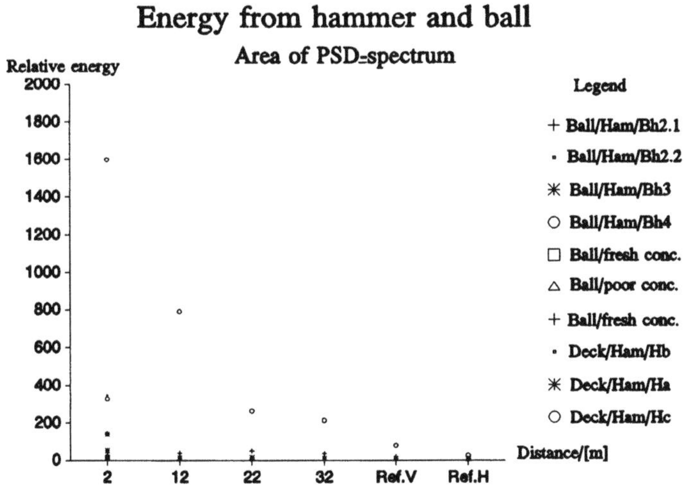

Fig. 15. Relative energy registered by measuring points from single blows with hammer and wrecking ball.

As a result of an evaluation of the relative energy which is involved in the dynamic effects to the construction, it is possible to construct a comparison of the different demolition techniques. This will aid the decision on finding the most appropriate demolition technique.

References

[1] Lauritzen, E.K. (1986) Mini-Blasting for Repair Work, in **Building Research and Practice.**, volume 14, number 5, pp. 274-280.

[2] Lauritzen, E.K. (1992) Blasting of concrete: Localized cutting and partial demolition of concrete structures, in **Recycling of Demolished Concrete and Masonry.** (ed. T.C. Hansen), E & FN Spon, London.

[3] Molin, C. (1983) **Localized cutting in concrete with careful blasting.** Swedish Cement and Concrete, Stockholm, Sweden.

[4] Molin, C. (1984) **A methods development study of localized cutting in concrete with careful blasting.** Swedish Cement and Concrete, Stockholm, Sweden.

27 INVESTIGATION OF ANTI-BACTERIAL ABILITY OF CONCRETE

T. MAEDA and A. NEGISHI
Hazama Corporation, Technical Research Institute, Tsukuba, Japan

Abstract
A series of admixtures which resist the propagation of sulfuric acid producing bacterium are investigated. The sulfuric acid attacks cement, penetrates into concrete, corrodes reinforcement and finally destroys the structures. Various kinds of specimens of portland cement with or without inorganic and organic admixtures were placed in a sludgetank and exposed to sewage. After 2 years of exposure some admixtures were found to have anti-bacterial characteristics.
Keywords: Concrete, Sulfur oxidizing bacteria, Admixture, Inorganic compound, Hydrogen sulfide, Sulfuric acid, Gypsum.

1 Introduction

Recently some concrete structures constructed for sewerage treatment or sewer piping have been damaged by the attack of sulfide oxidizing bacterium (SOB) which have oxidized hydrogen sulfide (H_2S) and produced sulfuric acid (H_2SO_4). The rate of damage has extended over 4mm/year. Thornton (1978) described the deterioration of concrete caused by SOB,in some artificial lakes. Sariciman et.al. (1987) described the concrete in sewage environments in an Arabian gulf country. Amould (1988) reported the bacterial corrosion of calcareous stones caused by acid rain.

Using anti-bacterial admixtures with sulfur cement, Duecker et.al. (1948) reported that the sulfur cement with β -naphthol and selenium was resistible against SOB. Laishley and tyler (1978) reported some inhibitors in sulfur limestone concrete bars. However, no quantitative datum was available in regard to the degree of biological attack on concrete. Emmel et.al.(1988) reported the biocides of SOB adhered to the concrete. Some biocides had a sort of cleaning action on the concrete surface.

With this in mind, we investigated and selected some organic or inorganic admixtures against SOB through exposure tests and laboratory tests.

Fracture and Damage of Concrete and Rock – FDCR-2.
Edited by H.P. Rossmanith. ISBN 0 419 18470 8.

2 Materials and methods

2.1 Admixtures

Admixtures used in the experiments were chosen from organic biocides on the market and inorganic compounds with anti-bacterial ability. Table 1. shows some biocides with anti-bacterial ability used in the tests.

Table 1. List of admixtures used in the tests

Compound	Admixture	Main feature
Organic exposure test	Imide	LD_{50}= 9,000 mg/kg
	Thiocyanate	LD_{50}= 2,900 mg/kg
	Isophthalic	LD_{50}>10,000 mg/kg
laboratory test	Imide	LD_{50}= 9,000 mg/kg
	Thiocyanate	LD_{50}= 2,900 mg/kg
Inorganic exposure test	Copper (Cu)	Powder
	Copper oxide (CuO)	Powder
	Nickel (Ni)	Powder
	Nickel oxide (NiO)	Powder
	Tin (Sn)	Powder
	Tin oxide (SnO)	Powder
	Lead (Pb)	Powder
	Lead oxide (PbO)	Powder
	Cobalt oxide (CoO)	Powder
laboratory test	Stainless steel	Flakes
	Ni	Powder
	CuO	Powder
	Sn	Powder
	Pb	Powder
	Ni + CuO	Powder
	Ni + Sn	Powder
	Ni + Pb	Powder
	CuO + Sn	Powder
	CuO + Pb	Powder
	Sn + Pb	Powder

*Specimens without admixture were also provided for the blank test

2.2 Specimens

In order to raise anti-bacterial ability 2.0,10.0 and 20.0% of admixtures (for the exposure test) and 0.1 ,0.5* and 1.0% of admixtures (for the laboratory test) were added to the specimens by weight of cement. Mortar bars with the size of 4x4x16cm were mixed in the ratios as shown in table 2.

Table 2. Mix ratios of anti-bacterial mortar bars(% w/w)

Test method	Water/Cement	Sand/Cement	Admix./Cement
Exposure test	50.0	200	2.0
	50.0	200	10.0
	50.0	200	20.0
Laboratory test	40.0	200	0.1
	40.0	200	0.5*
	40.0	200	1.0

*Mix ratio of each admixture in the case of specimens with 2 kinds of admixture.

After mixing mortal bars with thiocyanate (admixture/cement = 2.0,10.0,20.0%) and with lead and lead oxide (admixture/ cement=10.0, 20.0%) they failed in hardening. So we did not use all of them. We used 30 specimens with admixtures and 3 specimens without admixtures for exposure test.

2.3 Exposure test

Fig.1. shows specimens exposed to the sludge tank atmosphere. Specimens were hanged from a lid of the sludge tank manhole and were exposed to hydrogen sulphide (H_2S) atmosphere. H_2S gas concentration in the tank was approximately 5-10 ppm. The test was estimated by the observation on the surface of specimens and the measurement of weight loss.

Fig.1. Exposed specimens in the sludge tank

2.4 Laboratory test

We had the laboratory test estimating the limited range of admixture cement ratio and the effect of mixing 2 kinds of admixtures. The specimens after 2 weeks of standard treatment were exposed in a vessel that was kept at the concentration of about 10% of CO_2 and the temperature of 30° C in order to accelerate the carbonation. The same kinds of the specimens were placed in a vessel in which H_2S gas was kept at the concentration of about 150 ppm by means of standard gas producer that evaporates liquid H_2S and changes it into H_2S gas. The vessel was kept in a drying oven at the temperature of 35° C. The specimens were put in the culture medium on which surface SOB were added. We estimated anti-bacterial ability of these specimens by measuring weight loss and elastic coefficient change.

3 Result and discussion

3.1 Exposure test

Fig.2. shows the specimens after 2 years of exposure in the tank. The observation on the specimens shows that ones mixed with Ni or Sn-based admixtures have smooth surfaces, while ones without admixtures have brittle and rough surfaces and showed gypsum formation.

Fig.2. Specimens after 2 years exposure
(upper:with Ni-based admixture lower:blank)

Fig.3. shows weight loss velocity(g/month). The number next to the name of admixture indicates admixture/cement ratio 2.0,10.0 and 20.0%. The specimens with imide-based admixtures increase weight loss velocity as well as admixture/cement ratio. Among inorganic admixtures CuO-based group decreases weight loss velocity, because CuO has reacted H_2S and CuS(copper sulfide) has been produced. The specimens with Pb or PbO-based admixtures show greater weight loss velocity than the blank test. The specimens with Ni or Ni-based admixtures show less weight loss velocity than the blank test.

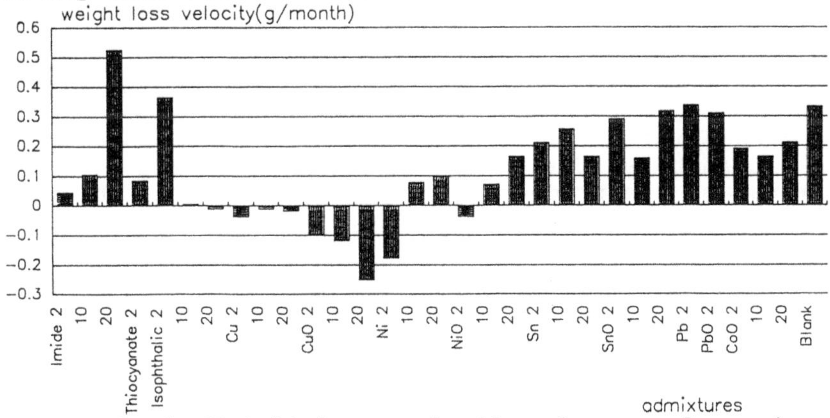

Fig.3. Weight loss velocity of exposed specimens

Thus anti-bacterial ability is to be explained as follows. At the first stage SOB adheres to the surface of the specimens with Ni or NiO-based admixtures and produces H_2SO_4(eq.1). At the second stage H_2SO_4 attacks insoluble Ni or NiO-based admixtures and ionizes the admixtures (eq.2). Finally these products attack SOB and hinder the growth of SOB.

$$SOB + H_2S \rightarrow H_2SO_4 \quad (1)$$

$$Ni \text{ or } NiO + H_2SO_4 \rightarrow Ni^{2+} + SO_4^{2-} + H_2 \quad (2)$$

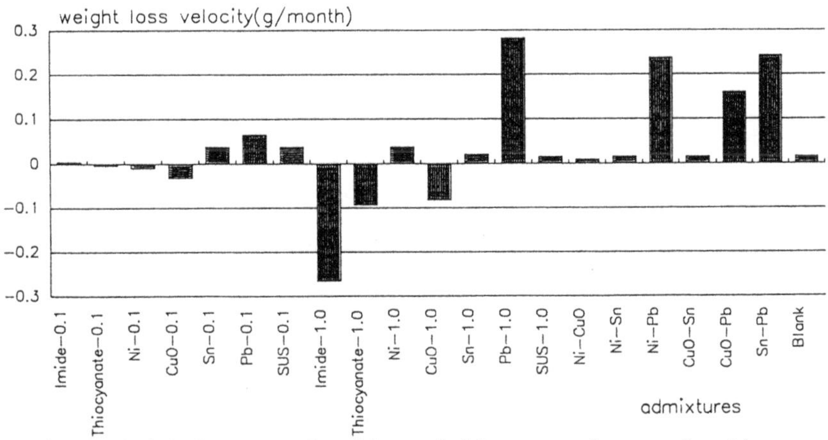

Fig.4. Weight loss velocity of the specimens in the accelerated chamber

3.2 Laboratory test

Fig.4. shows weight loss of specimens in the accelerated chamber after 6 months of exposure. Fig.5. shows elastic coefficient change ratio compared to the blank test without admixtures. These figures show that the specimens mixed with Pb-based admixture indicate less weight loss velocity and less elastic coefficient change ratio than other specimens because Pb-based ones are unsuitable for the cement. The specimens mixed with organic compounds indicate plus weight loss velocity and minus elastic coefficient change ratio. This means that there are some products caused by sulfuric acid on the surface of the specimens and the products are weak in strength, while the specimens mixed with Ni and CuO or Sn have greater elastic coefficient than those with only Ni. The other specimens are in the same level compared to the blank test.

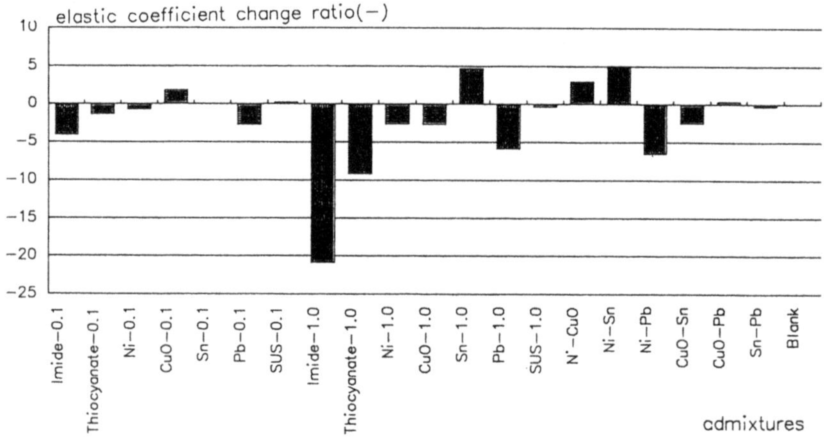

Fig.5. Elastic coefficient change ratio compared to the blank test.

In order to investigate the conditions of metals dissolving out of admixtures, the compositions of the solutions were analyzed. Table 3. shows the compositions of the solutions in which the specimens has been dipped. The solution of the blank test has greater calcium (Ca) and less sulfate (SO_4) than other solutions. The dissolving of metal ions out of specimens becomes greater as quantity of the admixture increases.

3.3 Corroded materials

Corroded materials in both tests consists mainly of gypsum within the range of pH 0-2 and of ettringaite at the much higher pH values. Mori.et.al (1991) reported that pH of concrete in good condition was about 12 and after carbonation SOB began to break out in the neutral pH environment.

Table 3. Composition of the solution in which specimens has been dipped.

Admixtures	Composition(ppm) Ca	SO_4	Ni	Cu	Sn	Pb	Cr
Imide-0.1	52.0	3650					
Thiocyanate-0.1	84.4	3190					
Ni-0.1	54.3	3090	4.4				
CuO-0.1	58.9	3490		6.8			
Sn-0.1	38.8	4000			2.3		
Pb-0.1	35.9	3180				0.5	
SUS-0.1	51.9	3830	2.5				0.2
Imide-1.0	69.8	2930					
Thiocyanate-1.0	52.0	3220					
Ni-1.0	44.9	3770	11.8				
CuO-1.0	50.1	3280		28.8			
Sn-1.0	51.4	3340			2.5		
Pb-1.0	62.2	3130				0.8	
SUS-1.0	76.0	4210	2.4				0.2
Ni + CuO	58.6	2950	7.0	15.1			
Ni + Sn	28.5	4030	6.0		0.5		
Ni + Pb	27.7	4990	7.9			0.5	
CuO + Sn	51.7	3650		16	1.2		
CuO + Pb	43.8	3150		19		0.5	
Sn-Pb	61.1	2610			0.81	0.9	
Blank	70.0	3040					

4. Conclusions

The investigation suggests that inorganic compounds have anti-bacterial ability against SOB and the control materials are inorganic ions dissolved by sulfuric acid which is produced by SOB. In case of constructing a new structure, inorganic admixtures in the concrete such as Ni or Sn are able to prevent the damages caused by SOB. On the other hand admixtures contained in mortal or poly ethylene resin are to coat the surface of deteriorated part of concrete. A set of new tests of concrete piping and fiber reinforcement plastic mortar piping with sewer resisting admixtures is now under preparation and it will give us further information.

5. References

Amould, M. (1988) Bacterial alteration of calcareous stons. **Geologie de l'ingenieur appliquee aux travaux anciens, monuments et sites historiques**, 819-823.

Duecker, W.W. Estep, J.W. Mayberry, M.G. and Schwab, J.W.

(1948) Studies of properties of sulphur jointing compounds. **J. Am. Waterworks Assoc.**, 7, 715-728.

Emmel, T. Brill, H. Sand, W. Bock, E. (1988) Screening for biocides to inhibit biogenic sulphuric acid corrosion in sewage pipelines. **In. Biodeterioration**, 7, 118-122.

Laishly, E.J. and Tyler, M.C. (1978) Biological oxidation of sulphur construction materials. **Proc. Int. Conf. on Sulphur in Const.**, 171-180.

Mori, T. Nonaka, T. Tazaki, K. Koga, M. Hikosaka, Y. and Noda, S. (1991) Interactions of nutrients, moisture and pH on microbial corrosion of concrete sewer pipes. **J. Wat. Res.**, 26, 29-37.

Sariciman, H. Maslehuddin, M. Shamim, M. and Allam, I. (1987) Case study of deterioration of concrete insewage environment in an Arabian gulf country. **Durability of Building Materials**, 5, 145-154.

Thornton, H. T.,JR. (1978) Acid attack of concrete caused by sulfur bacteria action. **ACIJ.**, Nov., 577-584.

PART FIVE
FRACTURE AND DAMAGE OF ROCK

28 MECHANICS OF MICRODEFECTED MATERIALS

V.Z. PARTON and V.A. BURYACHENKO
Moscow Institute of Chemical Engineering, Russia

Abstract
A new general method has been developed in composites mechanics.This is a muliparticle effective field method (EFM) ,which based on replacement of local external fields , containing separate inclusion by the effective field depending on the properties of the inclusion under study and those surrounding it. EFM of averaging the random structures based on the using of Green's function for the respective problem.The effective parameters of strength composites have estimated in the work.
Keywords: Composite materials , effective parameters of elasticity and strength.

1 General relation

The paper discasses a macromodomain w with a characteristic function W containing a set $X=(V_k,x_k,\omega_k),(k=1,\dots,N)$ of ellipsoids v_k with characteristic functions V_k , centers x_k (that forms a Poisson set) , semiaxes a_k^i ($i=1,2,3$, $a_k^1 \geqslant a_k^2 \geqslant a_k^3$) and aggregat of Euler angles ω_k . The local equation for the material state , that connects stress tensors $\sigma(x)$ and strain tensors $\varepsilon(x)$ is given in the form

$$\sigma(x)=C(x)[\varepsilon(x)-\varepsilon^T(x)] \ , \tag{1}$$

where $C(x)$ is a fourth-order tensor for elastisity moduli, $\varepsilon^T(x)$ is a second-order tensor of stress-free strains (also called mismatch strains) . In the matrix $v_o=w\backslash v$ ($v\equiv \cup v_k;k=1,2,\dots$) the tensors $C(x)=C^{(o)}$, $\varepsilon^T(x)=\varepsilon^{T(o)}$ are assumed to be constant in each inclusion $v_k(k=1,2,\dots)$ $C(x)=C^{(o)}+C_1(x)= C^{(o)}+C_1^{(k)}$,$\varepsilon^T(x)=\varepsilon^{T(o)}+\varepsilon^T(x)=\varepsilon_1^{T(o)}+\varepsilon^{T(k)}(x)$; for

Fracture and Damage of Concrete and Rock – FDCR-2.
Edited by H.P. Rossmanith. ISBN 0 419 18470 8.

cavity v_k $C_1(x)=-C^{(o)}$ and for flattened cavity (crack) $a^3 \ll a^2 \leqslant a^1$.

Substituting (1) in the equilibrium equation $\nabla\sigma=0$, we obtain an integral equation (Buryachenko and Lipanov (1986a))

$$\sigma(x)=\sigma^o+\int\Gamma(x-y)\{M_1(y)\sigma(y)+\varepsilon^T(y)-[\langle M_1\sigma\rangle+\langle\varepsilon_1^T\rangle]\}dy \quad , \qquad (2)$$

where $M_1(y)=(C^{(o)}+C_1^{(k)})^{-1}-M^{(o)}$, $M^{(o)}=(C^{(o)})^{-1}$, at $x\in v_k$, $\Gamma(x-y)=-C^{(o)}[I\delta(x-y)+\nabla\nabla G(x-y)C^{(o)}]$, G -the Green tensor of the Lame's equation of a homogeneous medium with an elastic modulus tensor $C^{(o)}$; δ-the delta function, I-the unit forth-order tensor, $\sigma^o\equiv\langle\sigma\rangle$. In (2) and below $\langle . \rangle$, $\langle .|x , \ldots, x_n; x_{n+1}, \ldots x_m\rangle$ stand for the average and the conditional average taken from the ansemble of a statistically homogeneous ergodic field X , on condition that there are inclusions at the points $x_1,\ldots,x_n$ and $x_1,\ldots,x_n \neq x_{n+1}, \ldots, x_m$; $\langle . \rangle_k$ is the volume average involving the ellipsoid v_k.

The effective parameters M^*, ε^* in the macromedium state equation $M^*\langle\sigma\rangle=\langle\varepsilon\rangle-\varepsilon^*$ are defined by relations

$$M^*=M^{(o)}+B_\sigma^* \ , \ \langle M_1\sigma\rangle\equiv B_\sigma^*\langle\sigma\rangle \ , \ \varepsilon^*=\langle\varepsilon^T\rangle+B_\varepsilon^* \ , \ \langle M_1\sigma\rangle\equiv B_\varepsilon^* \qquad (3)$$

where the tensor B_σ^* and B_ε^* are found from the solution of the elastic problem at $\varepsilon^T(x)=0$ and of the thermoelastic problem at $\langle\sigma\rangle=0$, respectivly . Let us introduce $\phi(v_m|x_m;x_1,\ldots,x_n)$, which is a conditional distribution of the m-th inclusion in the domane v_m at fixed inclusions in the domanes $v_1,\ldots,v_n$; $\phi(v_m|x_m;x_1,\ldots,x_n)=0$ at valume of x_m lying inside the domain $v_m^o\supset v_m$ with the characteristic functions V_m^o.

By way of conditional statistical averaging (with the help of various distribution functions $\phi(v_m|x_m;x_1,\ldots,x_n)$), the problem of evaluation of the effective parameters of the medium is reduced to an infinite system of integral equations (n=1,..)

$$\langle\sigma|x_1,\ldots,x_n\rangle- \qquad (4)$$

$$- \sum_{i=1}^{N} \int\Gamma(x-y)V_i(y)\langle M_1(y)\sigma(y)+\varepsilon_1^T(y)|x_1,\ldots,x_n\rangle dy = \sigma^o+$$

$$+ \int\Gamma(x-y)\{\langle M_1(y)\sigma(y)+\varepsilon_1^T(y)|y;x_1,\ldots,x_n\rangle-[\langle M_1\sigma\rangle+\langle\varepsilon_1^T\rangle]\}dy$$

Let us denote the right-hand member of the n-th line of the system by the field $\hat{\sigma}(x)_{1,\ldots,n}$, then each inclusion $v_i(i=1, \ldots,n)$ of the chosen fixed inclusions is in nonhomogeneous field $(x\in v_i)$

$$\overline{\sigma}_i(x)=\sigma(x)_{1,\ldots,n}+\sum_{j\neq i}\int\Gamma(x-y)V_j(y)[M_1(y)\sigma(y)+\varepsilon_1^T(y)]dy \qquad (5)$$

2 The effective field

Let us apply the EFM (Buryachenko(1987) , Buryachenko and Lipanov (1986a,b)) hypotheses to close and subsequetly solve approximately system (4):

H1) Each inclusion v_i has an ellipsoidal form,located in the homogeneous field $\overline{\sigma}(x)$ and $(x\notin v_i, \overline{v}_i=\text{mes}v_i)$

$$\int\Gamma(x-y)V_i(y)[M_1(y)\sigma(y)+\varepsilon_1^T(y)]dy= \qquad (6)$$

$$= \langle\Gamma(x-y)\rangle_i\langle M_1(y)\sigma(y)+\varepsilon_1(y)\rangle_i\overline{v}_i$$

H2) At some sufficiently big n there occurs a closure $\langle\hat{\sigma}(x)_{1,\ldots,j,n+1}\rangle=\langle\hat{\sigma}(x)_{1,\ldots,n}\rangle_i$, where the righthand member of the equality does not contane the index $j\neq i$, $1<j<n$ and $x\in v_j$.

Due to linearity of problem ,there axist constant fourth and second-rank tensors B_i and C_i , such that

$$\langle\sigma(x)\rangle_i=B_i\langle\overline{\sigma}(x)\rangle_i+C_i\ , \qquad (7)$$

$$v_i\langle M_1(x)\sigma(x)+\varepsilon_1^T(x)\rangle=R_i\langle\overline{\sigma}(x)\rangle_i+F_i\ ,$$

where $R_i=\overline{v}_i\langle\Gamma(x)\rangle_i^{-1}(B_i-I), F_i=\overline{v}_i\langle\Gamma(x)\rangle_i^{-1}C_i$.

3 The evaluation of C^*,ε^*

System (4) can be solved by analytical methods , if

$$\langle\hat{\sigma}(x)_{12}\rangle_i=\langle\bar{\sigma}(x)\rangle_i=\text{const.} \quad (i=1,2) \tag{8}$$

Then from (4) , takin account of (7),(8) , we get the expression for effective parameters

$$M^*=M^{(o)}+\sum_{i=1}^{N} R_i D_i n_i \ , \qquad D_i=R_i^{-1}\sum_{i=1}^{N} Y_{ij}R_j \tag{9}$$

$$\varepsilon^* =\varepsilon^{T(o)} + \sum_{i,j=1}^{N} Y_{ij}F_j n_i \ , \tag{10}$$

where n_q is a calculated concentration of ingomogeneous matrix Y^{-1} has elements $(Y^{-1})_{ij}$ (i,j=1,...,N) in the form of submatrixes (6*6)

$$(Y^{-1})_{ij}=\delta_{ij}[I-R_i\sum_{i=1}^{N} T_{iq}(x_i-x_q)Z_{qi}\Phi(v_q|x_q;x_i)dx_q]- \tag{11}$$

$$-R_i\int[T_{ij}(x_i-x_j)Z_{ji}\Phi(v_j|x_j;x_i)-T_i(x_i-x_j)n_j]dx_j$$

$$(Z^{-1})_{ji}=I\delta_{ji} -(1-\delta_{ji})T_{ji}(x_j-x_i)R_i,$$

$$T_{ji}(x_j-x_i)=(\bar{v}_j\bar{v}_i)^{-1}\int\int\Gamma(x-y)V_j(x)V_i(y)dxdy$$

$$T_i(x_i-x_j)=\bar{v}_i^{-1}\int\Gamma(x-x_j)V_i(x)dx$$

4 Strength of composites

Estimated are the average-volume values of elastic stress-fields in components

$$\langle\sigma\rangle_i=B_i D_i\langle\sigma\rangle \ , \ (i=1,2,\ldots) \tag{12}$$

A study of more-compex phenomena ,such as strength and plasticity of composites,however,requires a calculation of the second moment (Parton and Buryachenko(1990)) .For the single-point second-order moments averaged over the valume of the i-th component exact expressions are precise (i=0,1,2,...)

$$\langle\sigma \times \sigma\rangle_i=\partial M^*/\partial M^{(o)}(\langle\sigma\rangle\times\langle\sigma\rangle)/c_i \ , \ c_i\equiv\langle V_i\rangle \ . \qquad (13)$$

Identities (13) were obtatained by varying Hill's conditions with respect to modulus $C^{(i)}\rightarrow C^{(i)}+\delta C^{(i)}$ with $\langle\sigma\rangle$ fixed and $\delta C^{(i)}=const$ assumed .

For a composite under free strains $\varepsilon^T(x)$ and zero boundary conditions in stress we shall have

$$\langle\sigma\rangle=0 \ , \ \langle\varepsilon\rangle=\varepsilon^* \ , \ \langle u^*\rangle=-\langle\sigma(x)\varepsilon^T(x)\rangle/2 \ , \qquad (14)$$

where u^* is the stored elastic energy.An exact estimate was obtained by varyng u^* in compliant tensor $M^{(i)}\rightarrow M^{(i)}+\delta M^{(i)}$ ($\delta M^{(i)}$=const.) with $\varepsilon^T(x)$ fixed and allowance made for Hill's conditional and (14)

$$\langle\sigma\times\sigma\rangle_i=-2c_i^{-1}\ \partial\langle u^*\rangle/\partial M^{(i)}.$$

Table 1 shows calulated values $\alpha_1,\alpha_2$of the normalized average of the second invariant of stress tensor $\alpha=\langle s_{ij}s_{ij}\rangle/\langle\sigma_o\rangle^2$ ($s_{ij}\equiv\sigma_{ij}-\sigma_o\delta_{ij}$,$\sigma_o\equiv\sigma_{ii}/3$) in incompressible matrix under hidrostatic stress of the spherical (α_1;$a^1=a^2=a^3$) and spheroidal (α_2;$a^1=a^2=10a^3$) pores of equally orientătion.

Table 1. Values $\alpha=\alpha(c)$

c	$\alpha_1(c)$	$\alpha_2(\frac{c}{10})$
0.1	0,569	0,202
0.2	1,484	0,409
0.3	3,025	0,605
0.4	5,806	0,837

In the case of composites with the strength of their components decribed by strength criteria

$$\Pi(\sigma)=\Pi^{2(k)}_{ij}\sigma_{ij} + \Pi^{4(k)}_{ijmn}\ \sigma_{ij}\sigma_{mn} =1 \qquad (k=1,\ldots,n) \qquad (15)$$

the estimates (13),(15) justified the assumption of macrostrength criterion

$$\max_k \{\Pi^{2(k)}B_kD_k\langle\sigma\rangle + c_k^{-1}\Pi^{4(k)}\partial M^*/\partial M^{(k)}[\langle\sigma\rangle x\langle\sigma\rangle]\}=1 \ , \qquad (16)$$

where Π^2, Π^4 are the second-and fourth-rank tensors of strength. In particular,for a composite with spherical pores ($a^1=a^2=a^3$) and plane spheroidal cracs ($a^1=a^2\gg a^3$) in incompresible matrix that satisfies strength criterion $s_{ij}s_{ij}=k_p^2$ from (9) , (16) we shall obtain (accordingly) macrostrength criterion

$$I_2+b^*(I_1)^2=(k_p^*)^2 \tag{17}$$

where

$$b^*=\frac{c}{6}(1-35c/24)[(1+5c/24)(1-29c/24)]^{-1}$$

$$(k_p^*)^2=k_p^2(1-c)(1-35c/24)(1+5c/24)^{-1}$$

and

$$b^*=\frac{2}{9}c\,(1-448c/375\pi^2)[(1-848c/375\pi^2)(1-16c/15\pi^2)]^{-1}$$

$$(k_p^*)^2=k_p^2(1-448c/375\pi^2)(1-848c/375\pi^2)^{-1}$$

($c=4\pi(a^1)^3n/3$, $I_1=\langle\sigma_{ii}\rangle/3$, $I_2=\langle s_{ij}\rangle\langle s_{ij}\rangle$) . Relation (17) obtained for additional assumptions

$$Z_{qi}=I\delta_{qi}+(1-\delta_{qi})T_{qi}(x_q-x_i)R_i$$

and

$$T_{ij}(x_i-x_j)=-C^{(o)}\nabla\nabla G(x_i-x_j)C^{(o)}, \tag{18}$$

$\phi(v_j|x_j;x_i)=(1-V_i)n_j$.

The fracture will also occur under hydrostatic compression which contractiots to the conventional strength analyses of the composites based on the assumed homogeneity of the field of stresses in the matrix .

5 Regular structure composites

The analytic calculations of M^* though the use of a fundamental periodical solution (Sangani and Lu (1987)) were employedfor the testing of estimated EFM

$$M^*=M^{(o)}+R_i[I-Q(w)R_i-\sum_j{}'T_{ij}(x_i-x_j)R_j]^{-1}, \tag{19}$$

where x_i - center of the macrodomain w , the summation being made over all centers of ingomogenities $x_j\neq x_i$ and $x_j\in w$. In the case of simple cubic packing of the sperical pores with filling degree of c=0,5 the error of estimating $C^* \equiv (M^*)^{-1}$ (19),(11) is within 5 per cent (comparison with exact estimation SL); it must be pointed out here that c=0,5 corresponds to 93 per cent of the filling maximum degree.

Table 2. Elastic coefficients of composite

c	C^*	SL	(19),(11)	(19),(18)
0,1	C^*_{1111}	2,808	2,808	2,800
	C^*_{1122}	1,116	1,116	1,109
	C^*_{1212}	1,634	1,632	1,634
0,4	C^*_{1111}	1,480	1,463	1,591
	C^*_{1122}	0,420	0,448	0,382
	C^*_{1212}	0,980	0,766	0,422
0,5	C^*_{1111}	1,126	1,138	1,350
	C^*_{1122}	0,261	0,325	0,220
	C^*_{1212}	0,750	0,568	0,014

The EFM was also tested by considering a specific axample of estimating the stress intensity factor (SIF) for normally-loadied regular lattice of colinear cracks over a plane . An accuracy of 6 per cent is ensured at a defect rate as high as 0,95 , which is indicative of legitimacy of gipothesis H_1.

6 References

Buryachenko V.A.(1987) Correlation function of stress field in matrix composites. **Mech. of Solids** , 22, 66-73.

Buryachenko V.A.and Lipanov A.M.(1986a) Stress concentration at ellipsoidal inclusions and effective thermoelastic properties of composite materials. **Soviet Appl.Mech.**,22, 1103-1109.

Buryachenko V.A.and Lipanov A.M.(1986b) Equation of mechanics for gas-saturatied porous Media. **J. Appl. Mech. & Tech Phys.** , 27, 557-581.

Parton V.Z. and Buryachenko V.A.(1990) Stress fluctuation in elastic composites. **Soviet Phys. Dokl.** ,35, 191-193.

Sangani A. and Lu W.(1987) Elastic Coefficients of Composites contaning spherical inclusion in periodic array. **J. Mech. and Phys. of Solids** ,35, 1-21.

29 MACROCRACK PROPAGATION IN MATERIALS WITH MULTI-SCALE MICROCRACKING

A.V. DYSKIN
The University of Western Australia, Perth, Australia
L.N. GERMANOVICH
The University of Oklahoma, Norman, Oklahoma, USA

Abstract
The multiscale approach is introduced to model the interaction between the main crack and surrounding smaller cracks. The smaller cracks are divided into: (a) microcracks, which smaller than the main crack process zone, and (b) mesocracks, i.e., cracks with sizes in the range between the size of the main crack and its process zone size. The different crack sizes are asymptotically considered as different scales. The crack-mesocrack interaction is modeled by introducing an effective medium. The crack-microcrack interaction is modeled at the smallest scale level as a process of fracturing microcracked volume elements at the crack front. A volume element size has been found that minimizes the macroscopic fracture toughness. Fracturing of the volume elements thickens the crack. That makes the macroscopic fracture toughness greater than the microscopic one.
Keywords: Multiscale Approach, Interaction, Mesocrack, Microcrack, Effective Medium, Fracture Toughness, Averaged Stress Intensity Factor.

1 Introduction

Fracture and non-linear deformation of materials with pre-existing cracks such as concrete and rocks are governed by the growth of cracks of a corresponding scale. The crack growth is strongly affected by the interaction with smaller cracks.

The apparent fracture toughness is one of the parameters affected by the interaction. Its value expresses the resistance of the material to crack growth in terms of linear fracture mechanics. In many cases, however, the direct measurement of the apparent fracture toughness is impossible, e.g., in the case of large crack propagation in rock mass. Consequently, modeling of the crack interaction is of great importance.

The crack-microcrack interaction is studied either by considering each microcrack separately (e.g., Hoagland and Embury, 1980; Rossmanith, 1983; Kachanov et al, 1990; Laures and Kachanov, 1991; Chudnovsky and Wu, 1992), or by introducing an effective medium substituting the material with the microcracks (e.g., Evans and Fu, 1985; Dyskin, 1986; Hutchinson, 1987; Ortiz, 1988; Ortiz and Giannakopoulos, 1990).

However the microcrack influence on the crack growth qualitatively depends on the relation between the crack and microcrack sizes. In the present paper, the crack-microcrack interaction is analyzed on the basis of the multiscale approach (Dyskin et al, 1992) by the explicit introduction of different scales and applying the asymptotic methods. One may expect that, similarly to any asymptotic approach, this will give an acceptable accuracy even if the real difference between the sizes is not high.

Fracture and Damage of Concrete and Rock – FDCR-2.
Edited by H.P. Rossmanith. © 1993 Taylor & Francis. ISBN 0 419 18470 8.

2 Multi-scale concept in modeling the crack interaction

The influence of the surrounding cracks on the crack propagation depends on the relation between their sizes and the characteristic sizes of the crack under consideration (the applied load is supposed to be uniform, otherwise the characteristic size of its variation has to be included into the classification). At least three characteristic sizes can be introduced (Fig. 1):

1. **The external size of the main crack** *L*. In the simplest cases it can be the length of a 2-D crack, or the radius of a disk-like 3-D crack;
2. **The length of the process zone** *d*.
3. **The thickness of the main crack** *H*. In principle, real macroscopic cracks cannot be infinitesimally thin. Their thickness, determined by the material microstructure, has to be taken into account in some cases (e.g., Bazant, 1986).

Correspondingly the surrounding cracks fall into three main groups:

1. **Macrocracks**, i.e., cracks equal to or greater than the main crack.
2. **Mesocracks** with sizes less than the external size *L* of the main crack, but greater than the length of its process zone *d* and its thickness *H*.
3. **Microcracks** with sizes less than the process zone length *d* . The crack-microcrack interaction effects both the crack thickness and the process zone size.

A more detailed classification of microcracks depends on the relation between the process zone length and the crack thickness.

We will concentrate only on the crack-mesocrack-microcrack interaction, when the main crack is the largest considered crack.

The multi-scale approach (Dyskin et al, 1992), is based on exaggerating the difference between the sizes. Thus, cracks of different sizes are considered as cracks belonging to different scales. Hence each group of similarly sized cracks can be considered separately at the corresponding scale level.

Each scale level is characterized by its own **minimum length**, which is associated with the corresponding space resolution. The mechanical effect of a considered object is simulated by using the proper asymptotic solution for distances greater than the minimum length. In other words, each scale introduces its own continuum medium with corresponding infinitesimal linear *dl*, area *dA*, and volume *dV* elements. Models using these elements cannot be employed for addressing distances less than the minimum length. On the other hand, physical properties of this continuum medium, i.e., the properties of its volume elements (VE), reflect the situation on the nearest finer scale level. They are effective characteristics representing relationships between the fine-scale fields averaged over the volume elements.

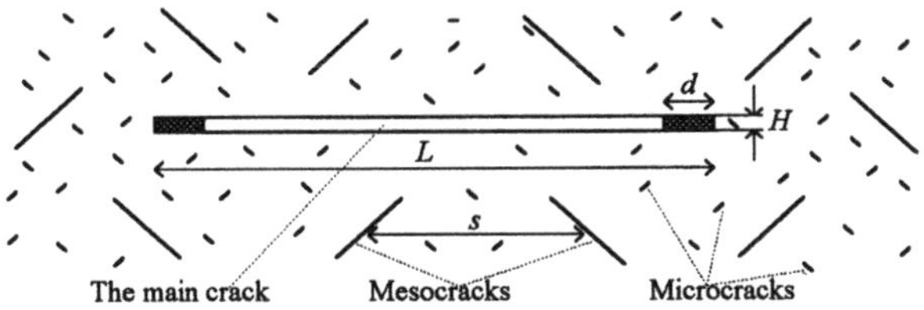

Fig. 1. Relation between the main crack, mesocracks and microcracks.

The crack-mesocrack-microcrack interaction involves the following scales.

1. The **main crack scale** is determined by the minimum length *s* that is greater than sizes of the mesocracks and the distances between them, but smaller than the crack

size L. The crack can be seen clearly at this scale, but not the mesocracks or microcracks. So, the main crack may be considered as immersed in an effective continuum medium that substitutes the material with meso- and microcracks.

2. At the **mesocrack scale** one can distinguish a single mesocrack but the process zone is not visible.

3. The **process zone scales** are determined by the process zone length, d (***d*-scale**), and the crack thickness, H (***H*-scale**). The microcracks are not visible at these scales.

There are three cases concerning relation between these lengths. The first case includes cracks with a developed process zone, i.e., $H << d$. Consideration of such a crack at the d-scale leads to the well-known models of an infinitesimally thin crack with the process zone, like Dugdale's or Barenblatt's models. However, any interpretation of the cohesive forces acting in the zone in terms of the material microstructure requires considering a characteristic thickness of the crack. In the second case, the process zone length is of the order of the crack thickness: $H \cong d$. Then, both d- and H- scales coincide. The hypothetical third case includes cracks without process zone. Similarly to the previous case the situation at the crack front is governed only by the size H.

4. The **microcrack scale** is the scale at which every microcrack can be distinguished. This is the finest scale under consideration. The crack-microcrack interaction at this scale is one of the major reasons of the difference between laboratory and *in situ* criteria of the crack growth.

At the microcrack scale, the process of the main crack growth may be considered as consecutive fracturing of the volume elements (VEs) at the crack front. The fractured VEs form either the process zone or the crack itself. However, regardless of the difference between these cases, the process of the crack propagation starts with onset of the VE fracture. Fracture of the VEs, makes the crack effectively "thick". Therefore, it is the size of the fracturing VEs that determines the effective thickness of the propagated crack.

One has to emphasize that the described multi-scale approach is suited to simplify the calculation of average quantities related to each scale under consideration, as done in the presented paper. Since any step of the propagation requires crashing the majority of the VEs along the crack front (Fig. 2a), the real 3-D crack propagation is naturally described in terms of the average VE strength.

Fig. 2. The VEs located at the crack front.

By introducing the effective characteristics it is easy to calculate the stress intensity factor (SIF) of the main crack averaged over all locations of the mesocracks. However, this may be confusing since the fracture is supposed to be determined by extreme SIF values rather than by averaged ones (see Kachanov, 1991). Nevertheless it is possible to point out several important cases when the "average approach" is relevant.

The first case is fracture caused by the stable growth of cracks. This kind of fracture happens under uniaxial compression and is caused by extensive stable growth of a few cracks (see the analysis and model in Germanovich and Dyskin, 1988). A small increment of stable crack growth does not result in fracture; a number of increments are required to let the crack grow sufficiently to cause a noticeable effect of deformation or fracture. That is why one can consider an **average** increment of the crack growth.

Another similar case is the stable crack growth because the applied load is adjusted, for instance if the crack growth is servo controlled.

The final example is the interpretation of results of strength measurements based on the averaging over many tested specimens. Fracture of a single specimen being caused by starting one of the cracks cannot be modeled by the method. However, the average result of such tests can be described in terms of the average SIF.

3 Crack-mesocrack interaction

3.1 General

Consider the main crack surrounded by mesocracks; two scale levels are involved: the main crack scale, which is the scale greater than both the mesocrack sizes and the distances between the mesocracks (but less than size of the main crack itself) and the mesocrack scale, which is greater than the process zone or the crack thickness. We are considering a quasi-static crack propagation; therefore, the static SIFs can be used.

The mesocrack scale is the finest scale in this consideration. In this situation, the microcracks only change the elastic moduli of the continuum medium in the mesocrack scale. Thus, the microcracks are accounted here by introducing the corresponding continuum medium. This medium plays the role of the original medium for both the crack and the mesocracks considered in this section. Therefore, the microcracks can be left out of consideration in this section.

Consider the influence of the mesocracks on the crack SIFs averaged over all location of the mesocracks. The method of calculating the mesocrack-induced increase in the SIFs consists of two steps (Dyskin, 1986). The first step introduces an effective medium substituting the mesocracked material. This allows calculating the average crack opening. Higher deformability of the effective medium produced by the mesocracks results in corresponding increase of the crack opening. For the isotropic situation this increase is approximately proportional to the ratio E_0/E_*, where E_0 and E_* are the original and the effective Young's moduli respectively (the effective moduli may be calculated by the differential self-consistent method, Salganik, 1973).

The second step of the method is to consider the fine-scale situation in the vicinity of the crack front where the stress state is mainly determined by the SIFs. It can be seen that there are virtually no mesocracks at the crack front. Therefore, the average crack opening calculated at the large scale has to be transferred to the crack tip located in the original material. This leads to the corresponding, E_0/E_* times, increase of the SIF.

The above intuitive consideration was verified by Dyskin (1986) for the elastic 2-D problem of interaction between a crack and parallel mesocracks (the corresponding formulae can also be found in Germanovich and Dyskin, 1988). SIFs obtained by this method were compared to results of the explicit calculation of the crack-mesocrack pair interaction (Dyskin, 1985). The terms responsible for the pair interaction coincided in both methods.

Here, this method will be proved using the method of outer and inner asymptotics. For the sake of simplicity, only the isotropic 2-D situation is considered. Hence, the original material is supposed to be isotropic and the mesocracks are supposed to be randomly oriented.

3.2 Averaged stress intensity factors

The averaged SIFs can be calculated by matching the outer and inner asymptotics (e.g., Gol'dstein and Salganik, 1974). Considering the main crack opening (the displacement discontinuity) at the mesocrack scale is equivalent to treating the crack as being immersed into an effective medium. This implies (a) asymptotic limit transition to infinitesimally small mesocracks, and (b) averaging the stress and strain fields over the representative VEs that are infinitesimally small compared to the crack, but infinitely large compared to the mesocracks. This average is equivalent to average over the mesocrack distribution (e.g., Batchelor, 1974).

The outer asymptotic behaviour of the crack face displacement at the crack tip can be expressed in the form (e.g., Rice, 1968)

$$u_*(x) \sim \frac{2(1-\nu_*^2)}{E_*} K_I^0 \sqrt{\frac{-x}{2\pi}}, \quad x \to 0. \tag{3.2}$$

Here E_*, ν_* are the effective Young's modulus and Poison's ratio for the medium with the mesocracks, K_I^0 is the SIF for the crack in the effective medium, x is a point on the crack line.

The expression (3.1) concerns the mesocrack scale. Therefore, it is valid for $x > s$.

Now consider the tip of the main crack and its vicinity being small compared to the mesocracks, but large compared to the process zone. At this scale both the stress and the crack opening at the tip are asymptotically expressed in terms of the SIFs. For the mode I crack in the plane strain conditions, one can write (e.g., Rice, 1968)

$$u(x) \sim \frac{2(1-\nu_0^2)}{E_0} K_I \sqrt{\frac{-x}{2\pi}}, \quad d/s \to 0. \tag{3.2}$$

Here E_0, ν_0 are the original Young's modulus and Poisson's ratio. After averaging over all mesocrack locations, one has

$$\langle u(x) \rangle \sim \frac{2(1-\nu_0^2)}{E_0} \langle K_I \rangle \sqrt{\frac{-x}{2\pi}}, \quad d/s \to 0. \tag{3.3}$$

This expression regards a crack tip vicinity free of the mesocracks. Hence, it plays the role of the inner asymptotic solution valid for $x < s$. On the other hand, s is the lower limit for the outer asymptotics and the upper limit for the inner expansion. So, it may be used as the matching point for both asymptotics. Reaching this point corresponds to $x \to 0$ for the outer asymptotics and $x \to -\infty$ for the inner one.

Comparing the outer (3.1) and inner (3.3) solutions, one has

$$\frac{\langle K_I \rangle}{K_I^0} = \frac{E_0(1-\nu_*^2)}{E_*(1-\nu_0^2)}. \tag{3.4}$$

3.3 On the energy method of calculating stress intensity factors

Some authors (Evans and Fu, 1985; Hutchinson, 1987; Ortiz, 1988; Ortiz and Giannakopoulos, 1990) suggested another way of calculating the average SIF. The method assumes the path-independent integrals or the energy release due to an elementary step of the crack propagation to be the same in both, the effective and original media. Equating the energy release rates in the effective $G_* = (1-\nu_*^2)E_*^{-1}(K_I^0)^2$ and in the original $G_0 = (1-\nu_0^2)E_0^{-1}(K_I^*)^2$ media, one obtains the expression, that differs from expression (3.4):

$$\frac{K_I^*}{K_I^0} = \sqrt{\frac{E_0(1-\nu_*^2)}{E_*(1-\nu_0^2)}}. \tag{3.5}$$

However this formula implicitly mixes different scales. The consideration of energy release in the effective medium is related to the main crack scale. The corresponding calculations formally operate with infinitesimal steps of the crack growth or infinitesimal elements of the curved path of the J-integral. However at this scale, they cannot be physically smaller than the minimum length. For the mesocrack scale, these lengths are very large: greater than both the mesocracks and the distances between them.

To apply this energy method more consistently, let us introduce the crack elongation Δl. At the coarse scale it is infinitesimal, so the corresponding energy release is $U_* = G_*\Delta l$. At the fine scale the same energy release is

$$\Delta U_0 = (1-\nu_0^2)E_0^{-1}\int_0^{\Delta l} K_I^2(x)dx, \tag{3.6}$$

where $K_I(x)$ is the current fine-scale SIF, x is the coordinate along the crack path. Equating these energy release rates ($\Delta U_* = \Delta U_0$) gives

$$\overline{(K_I)^2}/(K_I^0)^2 = \frac{E_0(1-\nu_*^2)}{E_*(1-\nu_0^2)}, \quad \overline{(K_I)^2} = {1}/{\Delta l}\int_0^{\Delta l} K_I^2(x)dx. \tag{3.7}$$

At the main crack scale, K_I^0 is supposed to be constant due to the infinitesimal character of the crack propagation increment, Δl. Therefore, to ensure direct comparison of the energy release at both scales, one has to keep the mesocrack-scale SIF constant during the whole path, Δl, of the crack propagation. It requires a special loading program to maintain the equality $\overline{(K_I)^2} = \overline{K_I}^2$ along the crack path. This loading program will depend on the particular mesocrack locations. Hence, the averaging over the possible mesocrack arrangements would concern the loading program rather than the SIF itself. This is why this method gives the result different from (3.4).

4 Crack-microcrack interaction

4.1 Mechanism of crack propagation

Suppose, the crack propagation commences with the fracture of VEs at its front. At the moment, it is of no concern whether the VEs are transformed further into the process zone or immediately into the crack. Therefore, we are considering only the onset of the fracture of the VEs that are adjacent to either the crack front or the external contour of the process zone. Let the VEs be fractured by growth of one of the microcracks (Fig. 2*b*).

Since the VEs are not infinitesimal, their fracture makes the crack effectively "thick." Its effective thickness is assumed to equal the VE size *H*.

The strength of the VE depends on its size (the internal scale effect). Inside a greater VE larger microcracks can be found resulting in a smaller strength of the VE. On the other hand, the stress produced by the "thick" crack is not infinite at its contour. Hence, the applied stress decreases with increasing VE size *H*. These observations imply the determination of a size *H* that ensures a minimum critical value of SIF. This minimum value plays the role of the apparent fracture toughness.

4.2 Stress applied to the fracturing volume element

Consider first, the no-process-zone case. At the *H*-scale the crack appears as a semi-infinite notch of thickness *H* (Fig. 2b). The crack generated stresses are non-singular at this scale (singularities at corners should be neglected, since they reflect the situation where details of the real corner geometry are omitted as belonging to a scale finer than the considered scale). This makes it possible to assume that even the VE nearest to the crack front is subjected to the approximately uniform external stress σ. Its value is supposed to equal the stress averaged over the VE (e.g., Neuber, 1946, Novozhilov 1969).

Since in our case the semi-infinite notch has only one dimension *H*, the dimension analysis is immediately followed by the formula (we are considering only mode I cracks)

$$\sigma = D\frac{K_I}{\sqrt{2\pi H}}, \tag{4.1}$$

where the coefficient *D* accounts for a particular shape. It could be found, for instance, from a numerical solution.

Averaging the singular stress field over the VE length, we have, as the first approximation, $D = 2(\sqrt{2}-1)$.

The described approach is also applicable to the more realistic case of a non-developed process zone with the length proportional to *H*.

In this paper, the simplest case is studied when the material is structured in a way that all microcracks are parallel to the main crack trajectory. Hence, the vertical stress σ, normal to the crack trajectory, determines the VE fracture.

4.3 Ultimate tensile strength

Since the fracture of a single VE is not sufficient to result in crack propagation in the 3-D case, the propagation is described as a process of multiple fracturing of VEs located at the crack front (Fig. 2a). Therefore it is possible to consider the strength of the VE as averaged over an ensemble of elements of a given size. Each VE is treated as a sample under uniaxial tension.

Consider the samples containing *n* parallel disk-like microcracks of radius $r_{min} < r < r_{max}$. Let $\Phi(x) = \Pr\{r < x\}$ be the distribution of microcracks by radius. The strength of each sample is defined by the largest microcrack inside it. Thus, if the microcrack distribution by sizes is known, one can determine the sample strength distribution and the average sample strength $\langle\sigma_{min}\rangle$.

Let $F_\sigma(z)=\Pr\{\sigma<z\}$ be the distribution function for the strength σ of a sample with only one microcrack. Consider a sample containing n microcracks. The microcracks are supposed not to interact with each other. Therefore, instead of considering a sample with n microcracks, n similar samples with only one microcrack each can also be considered to choose the sample with the minimum strength. This is similar to the weakest link concept (Freudenthal, 1968). The weakest sample from a group of n samples is characterized by the following distribution:

$$F_n(z)=1-\left[1-F_\sigma(z)\right]^n. \tag{4.2}$$

The strength of the brittle sample with one disk-like microcrack of radius r is given by the Griffith-type formula $\sigma=(1/2)K_{Icm}\sqrt{\pi/r}$. Here K_{Icm} is the **microscopic fracture toughness**. Using it one can express the mean tensile strength of the sample with n microcracks via the microcrack distribution function

$$\langle\sigma_t\rangle=\sigma_{\min}+\int_{\sigma_{\min}}^{\sigma_{\max}}\Phi^n\left(\frac{\pi K_{Icm}^2}{4x^2}\right)dx,\quad \sigma_{\min}=\frac{1}{2}K_{Icm}\sqrt{\frac{\pi}{r_{\max}}},\quad \sigma_{\max}=\frac{1}{2}K_{Icm}\sqrt{\frac{\pi}{r_{\min}}}, \tag{4.3}$$

where $\sigma_{\min}$ and $\sigma_{\max}$ are the minimum and maximum strengths.

Consider the uniform distribution of the microcrack sizes: $\Phi(r)=r/r_{\max}$, if $0<r\le r_{\max}$, $\Phi(r)=0$, if $r\le 0$, $\Phi(r)=1$, if $r>r_{\max}$, $r_{\min}=0$. After substituting it into (4.3) and integrating, one has

$$\langle\sigma_t\rangle=\sigma_{\min}\left(1+\frac{1}{2n-1}\right). \tag{4.4}$$

4.4 Macroscopic fracture toughness for a microcracked medium

Let us compare the stress at the crack front (4.1) with the VE strength (4.4) and substituting $n=NH^3$, where N is the number of microcracks per unit volume. This gives the equation determining the critical value of the crack SIF for the given minimum strength $\sigma_{\min}$ and VE size H.

$$\frac{DK_I}{\sqrt{2\pi H}}=\sigma_{\min}\left[1+\frac{1}{2NH^3-1}\right]. \tag{4.5}$$

The minimum strength is pre-determined by the microcrack distribution and the material micro-properties, whereas the VE size is subject to the crack-microcrack interaction. It is natural to suppose that the crack will choose the VE size that gives the minimum critical SIF value. This value plays the role of apparent (macroscopic) fracture toughness for the microcracked material.

The number of microcracks minimizing the SIF from (4.5) is $n_{\min}=NH^3_{\min}=3.5$. Substituting it into (4.5) and using (4.3) one has

$$K_{Ic}=\frac{3.19K_{Icm}}{D}\left(Nr_{\max}^3\right)^{-1/6}. \tag{4.6}$$

Introducing the dimensionless microcrack concentration $\nu = N\langle r^3 \rangle$ (Salganik, 1973), one finally obtains the apparent K_{Ic} fracture toughness, and the VE size:

$$K_{Ic} = \frac{2.54 K_{Icm}}{D} \nu^{-1/6}, \quad H = 0.96 r_{\max} \nu^{-1/3}. \tag{4.7}$$

The table shows the dependence of the apparent fracture toughness on the microcrack concentration. The apparent fracture toughness is greater than the microscopic one. This corresponds to the increase of the fracture toughness observed from laboratory samples to rock mass. For large microcrack concentrations, $\nu > 1$, the interaction between the microcracks is essential and has to be taken into account.

Table 1. The macroscopic fracture toughness K_{Ic}/K_{Icm} and VE size $H/r_{\max}$ against the dimensionless microcrack concentration ν.

Microcrack concentration ν	VE size $H/r_{\max}$	K_{Ic}/K_{Icm}
1	0.96	3.06
10^{-1}	2.06	4.49
10^{-2}	4.44	6.59
10^{-3}	9.57	9.68
10^{-4}	20.61	14.20

This table shows that the apparent fracture toughness becomes larger with an increase of the crack thickness because a "thick" macrocrack is a weaker stress concentrator than a "thin" microcrack. This increase corresponds to the fracture toughness increase observed in laboratory tests of differently sized samples. The presented simple assessments do not quantitatively explain the increase by one to two orders of magnitude which has been reported by Shlyapobersky et al 1985 with respect to hydraulic fracturing of rock mass However, the results presented in this paper are in general agreement with their observation.

For larger microcrack concentrations, $\nu>1$, the model results in a smaller volume element size compared to the maximum microcrack size, $r_{\max}$. However, for such high microcrack concentrations, the interaction between the microcracks becomes essential and has to be taken into account.

5 Conclusion

The multiscale approach is a convenient frame for modeling the crack-mesocrack-microcrack interaction by asymptotic methods. The explicit introduction of the minimum length at each scale gives ranges of the applicability of the models related to this scale.

The mesocracks increase the average SIFs of the main crack. This results in decreasing the apparent fracture toughness. On the contrary, the microcracks, making the main crack effectively "thick", increase the apparent fracture toughness.

References

Bazant, Z. P. (1986) Mechanics of distributed cracking. **Appl. Mech. Rev.**, 39, 675-705.

Batchelor, G.K. (1974) Transport properties of two-phase materials with random structure. **Annu. Review of Fluid Mechanics.**, 6, 227-255.

Chudnovsky, A. and Wu, S. (1992) Evaluation of energy release rate in the crack-microcrack interaction problem. **Int. J. Sol. Struct.**, 29, 1699-1709.

Dyskin, A.V. (1985) On the calculation of the deformation characteristics of a material with cracks. **Mechanics of Solids**, 20, 126-131.

Dyskin, A.V. (1986) Effective Deformation Characteristics and Fracture Conditions of Solids with a System of Cracks. **PhD thesis**. Moscow. The Institute for Problems in Mechanics of the USSR Academy of Sciences.

Dyskin, A.V., Salganik, R.L. and Ustinov, K.B. (1992) Multi-scale geomechanical modeling. **Western Australian Conference on Mining Geomechanics.** T.Szwedzicki, G.R.Baird and T.N.Little (ed.), 235-246.

Evans, A.G. and Fu, Y. (1985) Some effects of microcracks on the mechanical properties of brittle solids - II. Microcrack toughness. **Acta metall.**, 8, 1525-1531.

Freudenthal, A.D. (1968) Statistical approach to brittle fracture. **Fracture** (H. Liebovitz, ed.), 2. 591-619.

Germanovich, L.N. and Dyskin, A.V. (1988) A model of brittle failure for material with cracks in uniaxial loading. **Mechanics of Solids**, 23, 111-123.

Gol'dstein, R.V. and Salganik, R.L. (1974) Brittle fracture of solids with arbitrary cracks. **Intern. J. of Fracture**, 10, 507-523.

Hutchinson, J.W. (1987) Crack tip shielding by micro-cracking in brittle solids. **Acta metall.**, 35, 1605-1619.

Hoagland, R.G. and Embury, J.D. (1980) A treatment of inelastic deformation around a crack tip due to microcracking. **J. American Ceramic Soc.**, 63, 404-410.

Kachanov, M. (1991). The relationship between fracturing of a microcracking solid and its effective elastic constants. **Toughening Mechanics in Quasi-Brittle Materials**. S.P. Shah (ed.), 373-378.

Kachanov, M., Montagut, E.L.E. and Laures, J.P. (1990) Mechanics of crack-microcrack interaction. **Mechanics of Materials.**, 10, 59-71.

Laures, J.P. and Kachanov, M. (1991) Three-dimensional interactions of a crack front with arrays of penny-shaped microcracks. **Int. J. Fracture.**, 48, 255-279.

Neuber, H. (1946). **Theory of Notch Stresses**. J.W. Edwards, Ann Arbor, Michigan. Page 158.

Novozhilov, V.V. (1969) On a necessary and sufficient criterion for brittle strength. **J. Appl. Math. Mech.**, 33, 201-210. Ortiz, M. (1988) Microcrack coalescence and macroscopic crack growth initiation in brittle solids. **Int. J. Solids Structures**, 24, 231-250.

Ortiz, M. and Giannakopoulos, A.G. (1990) Mixed mode crack-tip fields in monolithic ceramics. **Int. J. Solids Structures**, 26, 705-723.

Rice, J.R. (1968) Mathematical analysis in the mechanics of fracture. **Fracture**, An Advanced Treatise, H. Liebowitz (ed.), 2, 192-311.

Rossmanith, H.P. (1983). Modeling of fracture process zones and singularity dominated zones. **Engineering Fracture Mechanics**, 17, 509-525.

Salganik, R.L. (1973) Mechanics of bodies with many cracks. **Mechanics of Solids**, 8, 135-143.

Shlyapobersky, J. Energy analysis of Hydraulic Fracturing. in Proceedings of the 26th U.S. Symposium on Rock Mechanics. Balkema, Rotterdam, pp. 539-546.

30 A DAMAGE MODEL FOR ROCK WITH INTERACTING MICROCRACKS

TAO ZHEN-YU and LI GUANG-PING
Department of Hydraulics and Electrical Engineering, Wuhan University, China

Abstract
This paper presents a general three-dimensional micromechanical damage model for rock-like materials with interacting microcracks subjected to uniaxial and triaxial compressive loads. In the model, the closing process of preexisting cracks and the interaction of multiple cracks are both considered. The rock-like materials are described as statistically inhomogeneous media with randomly distributed preexisting cracks. The three-dimensional filtered Poisson point field theory is employed to analyze the interaction among cracks. The expected values of stress and strain at any point of materials are also derived on the basis of effective field method. Kinetic equations are developed that describe the microcrack evolution as a function of the applied loads, as it deforms under the action of an effective stress field. In addition, overall nonlinear constitutive relations are derived.
Keywords: damage evolution, frictional sliding, effective field, crack interaction, rock-like materials, microcracks, crack closing process.

1 Introduction

The model of frictional sliding of preexisting cracks which can lead to the formation of tension cracks at their tips has been studied by a number of investigators (Kachanov, 1982; Nemat-Nasser and Horii, 1982; Horii and Nemat-Nasser, 1983, 1985, 1986; Steif, 1984; Nemat-Nasser and Obata, 1988; Ashby and Sammis, 1990; Fanella and Krajcinovic, 1988; Krajcinovic et al., 1991, Ju, 1991; and Lee and Ju, 1991). These damage models, nevertheless, are restricted to Taylor's models or self-consistent models. For Taylor's models, microcrack interaction effects are entirely ignored. Although the self-consistent models are capable of accounting for weak or moderate microcrack interaction, it is very complex for calculating the overall moduli of brittle solids which become nonsymmetric and anisotropic due to frictional sliding of closed microcracks (Ju, 1991; Lee and Ju, 1991). Furthermore, the microcrack evolution analyses (Ju, 1991; Lee and Ju, 1991) are still restricted to Taylor's models.

In the current literature, all investigators assume that preexisting microcracks are all closed under compressive loads. Practically, most of preexisting microcracks in rock-like materials have initial apertures which are not equal to zero before specimens are loaded. Under exter-

Fracture and Damage of Concrete and Rock – FDCR-2.
Edited by H.P. Rossmanith. © 1993 Taylor & Francis. ISBN 0 419 18470 8.

nally compressive loadings, some(or all) microcracks become closed and others are not closed. The closing process of preexisting microcracks depends on the externally compressive loadings (especially, lateral stress q^*) and the normal to the crack surface.

In the present study, a general three-dimensional micromechanical damage model for rock-like materials subjected to uniaxial and triaxial compressive loads is developed within the framework of the effective field method (Kunin, 1983) and damage mechanics (Fanella and Krajcinovic, 1988). The closing process of preexisting microcracks is considered. Both open (not closed) and closed microcracks are studied. The three-dimensional filtered Poisson point field theory is employed to analyze the interaction among cracks Kinetic equations are presented that describe the microcrack evolution as a function of the applied loads, as it deforms under the action of an effective stress field. The mean values of stress and strain at any point of the materials are derived. In addition, overall nonlinear constitutive relations are developed.

2 Framework of effective field

In this section, attention is focussed on the derivation of effective (stress) field and corresponding nonsymmetric and anisotropic compliances for rock-like materials. The derivation is based on Kunin's work (1983).

2.1 Formulation of the problems

The tensors $\sigma(x)$ and $\varepsilon(x)$ at any point of materials take the form (Kunin, 1983):

$$\sigma(x) = \sigma^0 - \sum_K \int_{S_K} S_0(x-x')[n(x')b(x')]_K dS(x') \tag{1}$$

$$\varepsilon(x) = \varepsilon^0 + \sum_K \int_{S_K} K_0(x-x')C^0[n(x')b(x')]_K dS(x') \tag{2}$$

where ε^0 and σ^0 are the external strain and stress fields, respectively, K_0, S_0 are Green's operators(Kunin, 1983), C^0 is the elastic moduli tensor of the homogeneous medium, $b=[u]$ the displacement discontinuity vector across the crack with normal n, S is the microcrack surface. In order to derive the expected values of strain and stress, the three-dimensional filtered Poisson point field theory is employed with the following assumptions:

H1) The placement of all crack centres constitutes a Poisson point field.

H2) The orientation and the diameter of crack are described as the marks of filtered Poisson point field, the marks (for orientation or diameter) are not only independent of each other, but also independent of Poisson counting process $\{N_V\}$, and they have the same distributed function. Where V is the volume of materials.

According to these assumptions, both damage-induced stress and strain due to cracks can be described as filtered Poisson point field:

$$A(x) = \sum_{K=1}^{N_V} A(x - r_K, u_K) \quad (3)$$

where, u_K represents the marks of kth crack, $A(x - r_K, u_K)$ is a weighted function which describes the response to the marks of kth crack at point X. In terms of the properties of filtered Poisson point field, the mean value of A (x) can be expressed as:

$$\langle A(x) \rangle = \int_V \lambda_V \langle A(x - r, u) \rangle dV \quad (4)$$

in which λ_V is intensity function which signifies the number of cracks per unit volume, and $\langle \cdot \rangle$ represents the expected value. In accordance with eq. (4), eqs(1) and (2) can be rephrased as

$$\langle \varepsilon(x) \rangle = \varepsilon^0 + \int_V \lambda_V E[\int_S K_0(x - x') C^0 n(x') b(x') dS(x')] dV \quad (5)$$

$$\langle \sigma(x) \rangle = \sigma^0 - \int_V \lambda_V E[\int_S S_0(x - x') n(x') b(x') dS(x')] dV \quad (6)$$

where E[·] represents the expected value. Eqs(5) and (6) can be applied to the case in which all cracks distribute inhomogeneously in matrix. If λ_V is approximately equal to constant m and external stress field σ^0 is given, eqs(7) and (8) become the simple forms:

$$\langle \varepsilon(x) \rangle = \varepsilon^0 + \varepsilon^* = \varepsilon^0 + ms\langle nb \rangle \quad (7)$$

$$\langle \sigma(x) \rangle = \sigma^0 \quad (8)$$

where ε^* represents the inelastic (damage-induced) strains. If the external field of cracks is assumed approximately equal to the remotely applied stress field σ^0, the mean value of strain at any point of materials becomes:

$$\langle \varepsilon(x) \rangle = \varepsilon^0 + S^* : \sigma^0 \quad (9)$$

where S^* = the damage-induced compliance determined by Taylor's model. If S^* = the damage-induced compliance determined by self-consistent model, eq. (9) becomes the cases of Ju(1991) and Lee and Ju(1991). Practically, the external stress field of cracks is not equal to the applied stress field due to the interaction among cracks, therefore the effective field method is employed to analyze the crack interaction.

2.2 The effective field

As in the previous section, isolated cracks constitute a spatially homogeneous, random field. Let us fix one of its typical realizations and consider an arbitrary crack with number i. If an external field is applied to the medium, this crack is subject to the field $\overline{\sigma_i}(x)$. On the basis of previous consideration $\overline{\sigma_i}(x)$ can be represented in the form:

$$\overline{\sigma_i}(x) = \sigma^0 - \sum_{K \neq i} \int_{S_K} S_0(x - x') [n(x') b(x')]_K dS(x'), \quad x \in S_i \quad (10)$$

The field $\overline{\sigma}_i(x)$ defined by this relation in the region S_i may be interpreted as the external field of the ith crack. In the field $\overline{\sigma}_i(x)$ this crack behaves as an isolated one. The field $\overline{\sigma}_i(x)$ defined in the region $S = US_i$ and equal to the constant $\overline{\sigma}_i$ in S_i will be called the effective field (Kunin, 1983). According to the hypothesis of the effective field method (Kunin, 1983), the effective field is represented as:

$$\langle \overline{\sigma}(x) \rangle = \langle \overline{\sigma}_i(x) \rangle = \sigma^0 - \int_{V-v_0} \lambda_V E\left[\int_S S_0(x - x')\, n(x')\, b(x')\, dS(x')\right] dV \tag{11}$$

where v_0 is the mean volume region occupied by a single crack. If $\lambda_V \approx m$ and σ^0 is given, the effective field takes the form:

$$\overline{\sigma} = \langle \overline{\sigma}(x) \rangle = [I - D: S^*]^{-1}: \sigma^0 \tag{12}$$

where I is the fourth-order identity tensor, and

$$D_{ijkl} = \frac{E}{30} \frac{7 - 5\nu}{1 - \nu^2} (\delta_{ik}\delta_{jl} + \delta_{il}\delta_{jk}) + \frac{E}{15} \frac{5\nu + 1}{1 - \nu^2} \delta_{ij}\delta_{kl} \tag{13}$$

E = Young's modulus of the origin material, ν is Poisson's ratio of the virgin material. Finally, in terms of eq. (7), we get the overall compliance:

$$B = B^0 + B^* = B^0 + S^*: [I - D: S^*]^{-1} \tag{14}$$

where B^0 is the compliance moduli tensor of original materials, and B^* = the damage-induced compliance determined by effective field method.

3 Kinetic equations of the damage evolution

In this section, we derive the kinetic equations of the damage evolution based on the effective field method. Attention is restricted to the following two cases of triaxial compressive loadings (with q denoting the axial compression and q^* denoting the lateral stress, respectively): (1) all preexisting microcracks become closed as a result of the hydrostatical load q^*, (2) some preexisting microcracks are closed and others are not closed. The following kinetic equations are also valid for uniaxial compression by setting $q^* = 0$. The derivation in this section basically follows the work of Fanella and Krajcinovic (1988).

As in the paper of Fanella and Krajcinovic (1988), it is assumed that $D = \rho^2 D_0$, where D_0 is the initial diameter of preexisting microcracks, D is the final size of microcracks, ρ is a scalar between 0 and 1. In addition, the Voigt's notation is adopted for stresses and strains as $\tau_1 = \sigma_{11}$, $\tau_2 = \sigma_{22}$, $\tau_3 = \sigma_{33}$, $\tau_4 = \sigma_{23}$, $\tau_5 = \sigma_{31}$, $\tau_6 = \sigma_{21}$, $e_1 = \varepsilon_{11}$, $e_2 = \varepsilon_{22}$, $e_3 = \varepsilon_{33}$, $e_4 = \varepsilon_{23}$, $e_5 = \varepsilon_{31}$, $e_6 = \varepsilon_{12}$. With uniformely distributed and statistically independent geometric random variables (D, Φ, θ), the tensor S^* in eq. (14) take the form:

$$[S^*] = \begin{bmatrix} S_{11}^* & S_{12}^* & S_{13}^* & 0 & 0 & 0 \\ S_{12}^* & S_{11}^* & S_{13}^* & 0 & 0 & 0 \\ S_{31}^* & S_{31}^* & S_{33}^* & 0 & 0 & 0 \\ 0 & 0 & 0 & S_{44}^* & 0 & 0 \\ 0 & 0 & 0 & 0 & S_{44}^* & 0 \\ 0 & 0 & 0 & 0 & 0 & S_{66}^* \end{bmatrix} \tag{15}$$

It can be proved that the tensor $[I - D: S^*]^{-1}$ has the same structure as the S^*. For simplicity, the tensor $[I - D: S^*]^{-1}$ is replaced by tensor H. Under the action of externally applied loads, the effective (stress) field of crack can be presented in the form according to eq. (12):

$$\bar{\tau}_1 = \bar{\tau}_2 = \tau^* = (H_{11}q^* + H_{12}q^* + H_{13}q) \quad (16)$$

$$\bar{\tau}_3 = \tau = 2H_{31}q^* + H_{33}q \quad (17)$$

In addition, the closing process of cracks is also considered. It is assumed that the external stress field for closing the crack is σ^C. The σ^C can be obtained in terms of the normal deformation of crack ($u_3' = b_3' = 16(1 - \nu^2) a\sigma^C / 3\pi E = t$):

$$\sigma^C = \frac{3\pi E}{16(1 - \nu^2)} \frac{t}{a} = \frac{3\pi E c}{16(1 - \nu^2)} \quad (18)$$

where t is the initial aperture of crack, $c = t/a$. Due to the action of σ^C, the effective normal stress across a closed crack surface decreases σ^C and shear stress increases $\mu\sigma^C$, where μ is the coefficient of friction of microcrack surfaces. If lateral stress q^* is greater than σ^C, all microcracks in materials are closed; if lateral stress q^* is smaller than σ^C, some microcracks are closed and others are not closed (open) under the action of externally applied loads. As the axial stress q is increased, more microcracks become closed. We discuss the two cases of $q^* > \sigma^C$ and $q^* < \sigma^C$.

3.1 Case I : $q > \sigma^C$

In this case, all preexisting microcracks are closed. The normal stress σ^n and shear stress $\tau_{(n,0)}$ on the surface of a typical microcrack at angle θ are

$$\sigma_n = \sigma_3 - \sigma^C = \tau\cos^2\theta + \tau^*\sin^2\theta - \sigma^C \quad (19)$$

$$\tau_{(n,0)} = \tau_4 - \mu\sigma_n = F(\theta)[\tau - (1 + \mu/F(\theta))\tau^*] + \mu\sigma^C \quad (20)$$

or, equivalently,

$$\tau_{(n,0)} = [H_{33}F(\theta) - (\mu + F(\theta))H_{13}]q - [(\mu + F(\theta))(H_{11} + H_{12}) - 2H_{31}F(\theta)]q^* + \mu\sigma^C \quad (21)$$

where $F(\theta) = \sin\theta\cos\theta - \mu\cos^2\theta$. The microcrack surface will slide relative to each other when $\tau_{(n,0)} \geqslant 0$ is met. Hence, eq. (20) can be solved for the upper and lower bounds θ_{u2} and θ_{u1} of microcrack orientations for given values of q and q^*:

$$\tan\theta_{u2,u1} = \frac{1 \pm [1 - 4C_1(C_1 + \mu)]^{1/2}}{2C_1}, \quad q \geqslant \{2\mu[(1 + \mu^2)^{1/2} + \mu] + 1\}q^* \quad (22)$$

where $C_1 = \mu(\tau^* - \sigma^C)/(\tau - \tau^*)$. A microcrack with relative sliding faces will exhibit mode Ⅱ microcrack growth once its crack tip stress intensity factor $k_{\text{Ⅱ}}$ reaches the critical value $k^{if}_{\text{Ⅱ}C}$ along a weak plane. Accordingly, the mode Ⅱ stability criterion can be expressed as

$$\tau_{(n,0)} = \frac{(2 - \nu)k^{if}_{\text{Ⅱ}C}}{4\rho} \left(\frac{2\pi}{D}\right)^{1/2} \quad (23)$$

From eqs (21) and (25), we can solve the q value needed to activate unstable mode Ⅱ microcrack growth from D_0 to D at a specific orientation θ:

$$q = \frac{\frac{(2-\nu)k^{if}_{IIC}}{4\rho}\left(\frac{2\pi}{D}\right)^{1/2} + [(\mu+F(\theta))(H_{11}+H_{12}) - 2H_{31}F(\theta)]q^* - \mu\sigma^C}{H_{33}F(\theta) - (\mu+F(\theta))H_{13}} \tag{24}$$

It is assumed that there exists a higher energy barrier in the matrix so that microcracks become arrested once they reach D. Again, the first cracks to increase in size from D_0 to D will be those for which q is minimum. These cracks have maximum initial size $\rho^2 D_M$, and are oriented at the angle θ_0 given by

$$\theta_0 = \arctan[\mu + (1+\mu^2)^{1/2}] \tag{25}$$

The corresponding threshold value of q_0 is

$$q_0 = \frac{\frac{(2-\nu)k^{if}_{IIC}}{4\rho}\left(\frac{2\pi}{D_M}\right)^{1/2} + [(\mu+F(\theta_0))(H_{11}+H_{12}) - 2H_{31}F(\theta_0)]q^* - \mu\sigma^C}{H_{33}F(\theta_0) - (\mu+F(\theta_0))H_{13}} \tag{26}$$

As the axial stress q increases above q_0, the unstable Mode Ⅱ domain also increases, i. e. more microcracks in the range $(\theta_{u1}, \theta_{u2})$ increase in size from D_0 to D when their particular stability criteria are satisfied. The orientations at which the largest cracks increase in size are obtained from eqs (20) and (23) with $D = D_M$:

$$\tan\theta_{f2,f1} = \frac{1 \pm [1 - 4C_2(C_2+\mu)]^{1/2}}{2C_2}, \quad q > q_0 \tag{27}$$

where $C_2 = \{(2-\nu)k^{if}_{IIC}(2\pi/D_M)^{1/2}/4\rho + \mu\tau^* - \mu\sigma^C\}/(\tau - \tau^*)$. For a given microcrack orientation inside the activation fan $[\theta_{f1}, \theta_{f2}]$, the crack diameters that become activated are

$$D_1(\theta) = 2\pi\left[\frac{(2-\nu)k^{if}_{IIC}}{4\rho G(\theta)}\right]^2 \tag{28}$$

where $G(\theta) = F(\theta)\{\tau - \tau^*[1+\mu/F(\theta)]\}$.

With a further increase in the axial stress q, some of cracks that have increased in size from D_0 to D and that have been arrested by higher energy barrier of matrix material will begin to propagate (kink) into brittle matrix in a stable fashion and eventually align with axial compressive load direction. The kinking threshold stress q_1 for three-dimensional crack can be determined by utilizing the maximum stress criterion (Fanella and Krajcinovic, 1988):

$$q_1 = \frac{\frac{3^{1/2}(2-\nu)k^C_{IC}}{8}\left(\frac{2\pi}{D_M}\right)^{1/2} + [(\mu+F(\theta_0))(H_{11}+H_{12}) - 2H_{31}F(\theta_0)]q^* - \mu\sigma^C}{H_{33}F(\theta_0) - (\mu+F(\theta_0))H_{13}} \tag{29}$$

With an increase of q above q_1, more cracks of diameter $D_0 = D$ kink into the brittle matrix . The orientations for which the largest cracks ($D=D_M$) kink are determined from

$$\tan\theta_{k2,\,k1}=\frac{1\pm[1-4C_3(C_3+\mu)]^{1/2}}{2C_3},\quad q>q_1 \tag{30}$$

with $C_3=\{[3^{1/2}(2-\nu)k_{IC}^C(2\pi/D_M)^{1/2}/8]+\mu\tau^*-\mu\sigma^C\}/(\tau-\tau^*)$. For the fan $\theta_{k2}>\theta>\theta_{k1}$, the cracks which kink into the matrix have the diameters given by

$$D_2(\theta)=2\pi\left[\frac{3^{1/2}(2-\nu)k_{IC}^C}{8G(\theta)}\right]^2 \tag{31}$$

As the axial stress q is increased to a certain level, the last cracks (with minimum initial diameter $\rho^2 D_m$) to increase in size are those for which q is maximum. The corresponding threshold value of q_2 can be computed from eq. (26), with D_M replaced by D_m. As the axial stress q increases above q_2, the orientations at which the smallest cracks increase in size are determined from eqs(21) and (23) with $D=D_m$.

$$\tan\theta_{m2,\,m1}=\frac{1\pm[1-4C_4(C_4+\mu)]^{1/2}}{2C_4},\quad q>q_2 \tag{32}$$

where $C_4=\{[(2-\nu)k_{IIC}^{if}(2\pi/D_m)^{1/2}/4\rho]+\mu\tau^*-\mu\sigma^C\}/(\tau-\tau^*)$.

Similarly, the angle fan domain $\{\theta_{l2},\theta_{l1}\}$, within which all microcracks become kinking, is evaluated by using the maximum stress criterion with $D=D_m$:

$$\tan\theta_{l2,\,l1}=\frac{1\pm[1-4C_5(C_5+\mu)]^{1/2}}{2C_5},\quad q>q_3 \tag{33}$$

where $C_5=\{[3^{1/2}(2-\nu)k_{IC}^C(2\pi/D_m)^{1/2}/8]+\mu\tau^*-\mu\sigma^C]/(\tau-\tau^*)$, and q_3 can be calculated from eq. (29), with D_M replaced by D_m.

3.2 Case Ⅱ : $q^*<\sigma^C$

In this case, if $q<(\sigma^C-2H_{31}q^*)/H_{33}$, all microcracks in materials are not closed. Once the normal stress $\overline{\tau}_3'$ on a crack face is equal to σ^C, the crack becomes closed. The closed domain is $[0,\theta_b]$, the open(not closed) domain $[\theta_b,\pi/2]$. The angle boundary separating the domains of open and closed microcracks can be found by setting $\overline{\tau}_3'=\sigma^C$. Hence, we obtain

$$\cos\theta_b=[(\sigma^C-\tau^*)/(\tau-\tau^*)]^{1/2} \tag{34}$$

The process of damage evolution for closed domain $[0,\theta_b]$ is similar to the case of 3.1. It is emphasized that if $(\sigma^C-2H_{31}q^*)/H_{33}<q<(\sigma^C+2H_{31}q^*)/H_{33}$, then set $\theta_{u1}=0$, and if $C_1\sim C_5<0$ or θ_{u2}, θ_{f2}, θ_{k2}, θ_{m2}, $\theta_{l2}>\theta_b$, then the corresponding upper limits θ_{u2}, θ_{f2}, θ_{k2}, θ_{m2}, θ_{l2} are setted equal to θ_b. In addition , the angle θ_0 for the calculations of q_0, q_1, q_2, q_3 must take the min $\{\theta_0,\theta_b\}$. The kinetic equations of damage evolution for open domains are analyzed as follows . Where we ignore the effect of normal compressive stress on the stress intensity factor . In the open domain $[\theta_b,\pi/2]$, the normal stress and shear stress on a crack face at angle θ become:

$$\sigma_n = \bar{\tau}_3' = \tau\cos^2\theta + \tau^*\sin^2\theta \ , \ \tau_{(n,0)} = \bar{\tau}_4' = \sin\theta\cos\theta(\tau - \tau^*) \qquad (35)$$

Once $\tau_{(n,0)}$ meets the condition of eq. (23), the crack grows along a weak plane, and increases in size from D_0 to D. The domain for crack growth is $[\theta_b, \theta_f^m]$, θ_f^m determined by

$$\theta_f^m = \frac{\pi}{2} - \frac{1}{2}\arcsin\left[\frac{(2-\nu)k^{if}_{IIC}}{2\rho}\left(\frac{2\pi}{D_M}\right)^{1/2} \Big/ (\tau - \tau^*)\right], \ q > q_0^m \qquad (36)$$

$$q_0^m = \left[\frac{(2-\nu)k^{if}_{IIC}}{2\rho}\left(\frac{2\pi}{D_M}\right)^{1/2} \Big/ \sin2\theta_b + (H_{11} + H_{12} - 2H_{31})q^*\right] \Big/ (H_{33} - H_{13}) \qquad (37)$$

In the derivation of eq. (36), we assume that θ_b is greater than 45°; if $\theta_b \leqslant 45$°, the case can be treated in a similar way. For a given crack orientation inside the activation fan (θ_b, θ_f^m), the crack diameters that become activated are

$$D_3(\theta) = 2\pi\left[\frac{(2-\nu)k^{if}_{IIC}}{4\rho\sin\theta\cos\theta(\tau - \tau^*)}\right]^2 \qquad (38)$$

As the axial stress q increases to a certain level, some arrested cracks will begin to propagate (kink) into brittle matrix, the kinked domain is $[\theta_b, \theta_k^m]$, the θ_k^m given by

$$\theta_k^m = \frac{\pi}{2} - \frac{1}{2}\arcsin\left[\frac{3^{1/2}(2-\nu)k^C_{IC}}{4(\tau - \tau^*)}(2\pi/D_M)^{1/2}\right], \ q > q_I^m \qquad (39)$$

and q_I^m can be computed from eq. (37), with k^{if}_{IIC}/ρ replaced by $3^{1/2}k^C_{IC}/2$. For the fan $\theta_b < \theta < \theta_k^m$, the cracks which kink into the matrix have the diameters $D_4(\theta)$ given by eq. (38) with k^{if}_{IIC}/ρ replaced by $3^{1/2}k^C_{IC}/2$. If q increases further, the case can be analyzed in the same way.

4 The effective field and overall compliance/strain tensor

4.1 Case I: $q^* > \sigma^C$

In this case, once q becomes greater than q*, the tensor S* in eq. (21) can be computed in the same way as Fanella and krajcinovic (1988). It is noted that the angle θ, function $G(\theta)$ and diameter $D(\theta)$ in our case are different from the case of Fanella and Krajcinovic (1988). The kink length l is determined by

$$l = \frac{1}{\pi}\left\{\frac{k^C_{IC} - [(k^C_{IC})^2 + 4\tau^* D\tau_{(n,0)}\cos\theta]^{1/2}}{2\tau^*}\right\}^2 \qquad (40)$$

The effective field analysis for closed cracks requires the use of the yet-unknown compliance S* to calculate the activated domains of cracks. Therefore, the proposed damage model naturally requires iterative schemes to obtain S* corresponding to a specified axial stress q. Once the angles $(\theta_u, \theta_f, \theta_k, \theta_m, \theta_l)$ converge, the effective field and overall compliance tensors can be computed by eqs (12) and (14), respectively. The overall strain tensors can be determined by

$$\varepsilon_{ij} = \varepsilon_{ij}^{C} + \varepsilon_{ij}^{t} + \varepsilon_{ij}^{0} + \varepsilon_{ij}{}^{*} = \varepsilon_{ij}^{C} + \varepsilon_{ij}^{t} + B_{ijmn}: \sigma_{mn}^{0} \tag{41}$$

where ε_{ij}^{t} are the inelastic strains induced by the normal closure of all microcracks, and given by (with the probability density functions $p(D)\,p(\Phi)\,p(\theta) = [\pi^2(D_M - D_m)]^{-1}$)

$$\varepsilon_{ij}^{t} = \frac{N_V \rho^6 c}{32\pi V}\,(D_M^2 + D_m)^2 (D_M + D_m) \int_0^{2\pi} d\Phi \int_0^{\pi/2} n_i n_j \sin\theta \, d\theta \tag{42}$$

where the second order transformation tensor [g] with $g_{3i} = n_i$ is determined from eq. (29) in Fanella and Krajcinovic (1988). ε_{ij}^{C} are the inelastic strains induced by $\mu\sigma^C$ (eq. (20)). In this paper, the inelastic strains due to initial cracks relative sliding are ignored (Fanella and Krajcinovic (1988)). Hence, for $q > q_3$, the ε_{ij}^{C} are given by (other cases can be treated in the same way)

$$\varepsilon_{ij}^{C} = \frac{\eta_{f1} N}{V} \int_0^{2\pi} d\Phi \int_{\theta_{f1}}^{\theta_{m1}} d\theta \int_{D_1(\theta)}^{D_M} f_{ij}\, dD + \frac{\eta_{f2} N}{V} \int_0^{2\pi} d\Phi \int_{\theta_{m2}}^{\theta_{f2}} d\theta \int_{D_1(\theta)}^{D_M} f_{ij}\, dD$$

$$+ \frac{\eta_{f3} N}{V} \int_0^{2\pi} \int_{\theta_{m1}}^{\theta_{m2}} d\theta \int_{D_m}^{D_M} f_{ij}\, dD \tag{43}$$

where

$$f_{ij} = \frac{16(1-\nu^2)}{3E(2-\nu)} \left(\frac{D}{2}\right)^3 (g_{3j} g_{2i} + g_{3i} g_{2j})\, \mu\sigma^C p(D)\, p(\Phi)\, p(\theta) \sin\theta \tag{44}$$

$$\eta_{f1} = \frac{2}{\pi(1-\gamma)} \left\{ (\theta_{m1} - \theta_{f1}) - \int_{\theta_{f1}}^{\theta_{m1}} \frac{D_1(\theta)}{D_M}\, d\theta \right\},\quad \gamma = D_m / D_M \tag{45}$$

$$\eta_{f2} = \frac{2}{\pi(1-\gamma)} \left\{ (\theta_{f2} - \theta_{m2}) - \int_{\theta_{m2}}^{\theta_{f2}} \frac{D_1(\theta)}{D_M}\, d\theta \right\} \tag{46}$$

$$\eta_{f3} = (\theta_{m2} - \theta_{m1}) / \pi/2 \tag{47}$$

4.2 Case Ⅱ : $q^* < \sigma^C$

In the closed domain $[0, \theta_b]$, the tensor S* can be computed similar to the case of 4.1. In the open domain $[\theta_b, \pi/2]$, the tensor S* due to the cracks which have increased in diameter from D_0 to D can be computed in the same way as Krajcinovic and Fanella (1986), and the tensor S* due to kinked cracks can be calculated similar to the case of 4.1. The overall S* for materials is equal to the sum of S* of closed domain and S* of open domain. The effective field and overall compliance and overall strain tensors can be computed in the same way as the case of 4.1. But, due to the complexity of crack closing, the ε_{ij}^{t} are ignored in this case.

5 Conclusions

Due to limited space, the results of numerical simulations are not given in this paper . But , our numerical simulations indicate that the proposed damage model is simpler than the self-consistent damage model (Ju , 1991 ; Lee and Ju , 1991) and patterns available experimental results quite well . The proposed model can be applied to the case of $N_V \langle a \rangle^3/V=2$. However , the maximum allowable density $N_V\langle a \rangle^3/V$ for self-consistent model is only 0.5625.

6 References

Ashby, M. F. and Sammis, C. G. (1990) The damage mechanics of brittle solids in compression. Pure Appl. Geophys. , 133, 489 - 521.

Fanella, D. and Krajcinovic, D. (1988) A micromechanical model for concrete in compression. Eng. Fract. Mech. , 29, 49 - 66.

Ju, J. W. (1991) On two-dimensional self-consistent micromechanical damage models for brittle solids. J. Eng. Mech. , 117, 227 - 258.

Horii, H. and Nemat-Nasser, S. (1983) Overall moduli of solids with microcracks: load-induced anisotropy. J. Mech. Phys. Solids, 31, 155 - 171.

Horii, H. and Nemat-Nasser, S. (1985) Compression-induced micro crack growth in brittle solids: axial splitting and shear failure. J. Geophys. Res. , 90, 3105 - 3225.

Horii, H. and Nemat-Nasser, S. (1986) Brittle failure in compression : splitting, faulting, and brittle-ductile transition. Phil. Trans. R. Soc. , 319, 337 - 374.

Kachanov, M. (1982) A microcrack model of rock inelasticity, Part Ⅱ : propagation of microcracks. Mech. Mater. , 1, 29 - 41.

Krajcinovic, D. and Fanella, D. (1986) A micromechanical damage model for concrete . Eng. Fract. Mech. , 25, 585 - 596.

Krajcinovic, D. Basista, M. and Sumarac, D. (1991) Micromechanically inspired phenomenological damage model. J. Appl. Mech. , 58, 305 - 310.

Kunin, I. A. (1983) Elastic media with microstructure Ⅱ . Springer, Berlin.

Lee, X. and Ju, J. W. (1991) Micromechanical damage models for brittle solids. Ⅱ : compressive loadings. J. Eng. Mech. , 117, 1515 - 1536.

Nemat-Nasser, S. and Horii, H. (1982) Compression-induced nonplanar crack extension with application to splitting, exfoliation, and rock-burst. J. Geophys. Res. , 87, 6805 - 6821.

Nemat-Nasser, S. and Obata, M (1988) A microcrack model of dilatancy in brittle materils . J. Appl. Mech. , 155, 24 - 35.

Steif, P. S (1984) Crack extension under compressive loading . Eng. Fract. Mech. , 20, 463 - 473.

31 DAMAGE PROPAGATION OF JOINTED ROCK WITH CONSIDERATION OF JOINT ROUGHNESS

G. SWOBODA
University of Innsbruck, Austria
F. ITO
Taisei Corporation, Japan

Abstract
Rock masses have originally many discontinuities. To design tunnels, dams, slopes, we need mechanical parameters of discontinuities considering these dimensions of structures due to the change of stress condition. Often these discontinuities are oriented more or less regularly but, these mecahnical behaviour is often dependent on the character of jointing. In this paper, to understand these behaviour considering joint roughness, we have applied a second-order tensorial representation of damage and treat the cracks as a continuum phenomenon on a suitable macroscopic scale. Our model accounts for the progressive failure of damaging materials due to the growth of these defects. We shall call this failure damage propagation. Assuming the existence of a Helmholz free energy function and a dissipation potential in the space of affinities we have derived a constitutive equation and a damage propagation law by means of the thermodynamic principles of irreversible processes.
Keywards: Damage, Crack, Propagation, Joints, Roughness, Thermodynamics, Constitutive Equation

1 Introduction

Experimental observations reveal that the deformation of jointed rock is accompanied by irreversible changes in their internal structure. The presented work considers the inelastic strains and the gradual degradation of the elastic stiffness, caused by the complex interaction of two dissipative mechanisms: plasticity and the nucleation and growth of microcracks.

The short-term deformation and failure of rock-like materials is governed by the propagation of cracks on microscopic and macroscopic level. It is evident that the classical analysis that is applied in all fields of rock engineering cannot predict the behaviour of such materials beyond their elastic range. Numerical simulation based on a perfect plasticity model with unlimited ductility does not provide any information on magnitudes of strains and displacements occuring before failure. Therefore there is a need for new models which consider the specific properties of these materials. Damage mechanics models the response of cracked bodies by treating the cracks as a continuum phenomenon on a suitable macroscopic scale.[6] In the present case damage is represented by a second-order symmetric tensor.

As mentioned above, the deformation of rocks is governed by the propagation of cracks. When an initial damage field exists in a direction other than the applied principal strain, the stresses in the plane of the cracks are predominantly shear in character. In this case the initial defects will grow in a plane different from the original one. The proposed model can account for cracks which propagate in mixed-mode and, thus, it is not restricted to proportional loading.

Fracture and Damage of Concrete and Rock – FDCR-2.
Edited by H.P. Rossmanith. © 1993 Taylor & Francis. ISBN 0 419 18470 8.

2 Damage variable

2.1 Damage tensor

Since bodies with regularly oriented cracks show a certain degree of anisotropy in their mechanical response, we must require that this variable can provide information about the orientation of the cracks. Tensors of second and higher ranks constitute a natural tool for the description of internal damage and the resulting influence on the macromechanical material properties such as the gradual degradation of the elastic stiffness.

A second-order damage tensor can be introduced as[7,11]

$$\Omega_{ij} = \omega^{(k)} n_i^{(k)} n_j^{(k)} \tag{1}$$

where $\omega^{(k)}$ is the share of voids (cracks) per surface unit of the joint plane orientated normal to the unit vector $\mathbf{n}^{(k)}$. Thus, $\omega = 0$ indicates virgin material whereas $\omega = 1$ would correspond to complete seperation.[4] Compared with scalar or vectorial damage variables used in parallel approaches, the present representation has the advantage that it can account for cracks which deviate from their original plane.

$[\Omega]$ defines the extent of microstructural rearrangements within a volume-element. $[\Omega]$ will neither enter equilibrium equations nor boundary conditions but appears only in constitutive equations and in expressions for the thermodynamic potentials.

2.2 Effective damage concept

A theory which intends to predict the behaviour of rock-like materials under compression must consider the directional nature of the cracks. When cracks close because the normal strain perpendicular to the damage plane is compressive, they have no influence on the mechanical properties of the system. We shall discuss briefly how this strain dependent behaviour can be considered.[10]

In the local coordinate system defined by the principal directions of $[\Omega]$, the damage tensor takes on a diagonal form. We call this cordinate the *damage cordinate.* We shall use $[\Omega]$ can be transformed from the global system to the local one by means of its general transformation matrix $[\mathbf{T}]$:

$$\Omega_{IJ} = T_{Ik}\Omega_{kl}T_{Jl} \tag{2}$$

The strain tensor obeys the same transformation rule. In general however, ε_{IJ} will not be diagonal.

$$\varepsilon_{IJ} = T_{Ik}\varepsilon_{kl}T_{Jl} \tag{3}$$

We shall separate ε_{IJ} into the diagonal part containing the normal strains ε^n_{IJ} and the shear part ε^s_{IJ}. For the plane problem, we can express as follows.

$$\begin{bmatrix} \varepsilon_{11} & \varepsilon_{12} \\ \varepsilon_{21} & \varepsilon_{22} \end{bmatrix} = \begin{bmatrix} \varepsilon_{11} & 0 \\ 0 & \varepsilon_{22} \end{bmatrix} + \begin{bmatrix} 0 & \varepsilon_{12} \\ \varepsilon_{21} & 0 \end{bmatrix} \tag{4}$$

The diagonal elements of ε_{IJ} represent the normal strains perpendicular to the corresponding sets of cracks whereas the non-diagonal components denote the shear strains acting on the damage planes.

Let us introduce two further second-order tensors

$$\begin{aligned} C^n_{ij} &= c^n_k n^k_i n^k_j \\ C^s_{ij} &= c^s_k n^k_i n^k_j \end{aligned} \qquad \text{(sum over all } k) \tag{5}$$

$\mathbf{n}^k$ is the normalized eigenvector of $[\Omega]$ which corresponds to the principal value ω^k and c_k depends on the normal strain ε_{kk} in the direction of $\mathbf{n}^k$:

$$\begin{array}{rl} c_k^n = c_k^s = 1 & \text{if } \varepsilon_{kk} \geq 0 \\ 0 = c_k^n \leq c_k^s \leq 1 & \text{if } \varepsilon_{kk} < 0 \end{array} \quad \text{(do not sum)} \tag{6}$$

Then we can define two *effective damage tensors* $[\Omega^n]$ and $[\Omega^s]$

$$\Omega_{ij}^n = \begin{bmatrix} c_1^n\omega_1^n & 0 \\ 0 & c_2^n\omega_2^n \end{bmatrix} \quad \Omega_{ij}^s = \begin{bmatrix} 0 & c_1^s\omega_1^s \\ c_2^s\omega_2^s & 0 \end{bmatrix} \tag{7}$$

that correspond to $[\varepsilon^n]$ and $[\varepsilon^s]$, respectively. The idea for the *effective damage concept* is following; If the two adjacent faces of a crack touch each other, some of the applied shear stresses and all normal stresses (in this case they are compressive) can be transmitted through friction and contact forces. We consider this in our numerical model by reducing the area of the crack, separately with respect to the normal stresses and to the shear stresses.

3 Structure of constitutive relation

Mathmatically thermodynamics formulation has already been discussed in detail.[1,2]

3.1 Helmholz free energy

To follow the *effective damage concept*, free energy function could be defined by the normal strains and shear strains.

Defining the elastic strain tensor $[\varepsilon^e]$

$$\varepsilon_{ij}^e = \varepsilon_{ij} - \varepsilon_{ij}^p \tag{8}$$

and splitting it up as shown in the foregoing section,we presume as the *Helmholz free energy function*

$$\rho\psi = \tfrac{1}{2}\varepsilon_{ij}^{e,n} D_{ijkl}^n \varepsilon_{kl}^{e,n} + \tfrac{1}{2}\varepsilon_{ij}^{e,s} D_{ijkl}^s \varepsilon_{kl}^{e,s} \tag{9}$$

Here $[\varepsilon^{e,n}]$ is the diagonal part of the elastic strain tensor, containing the normal strains, and $[\varepsilon^{e,s}]$ is the shear part. Equation (9) implies that in the absence of dissipative mechanisms the material of consideration behaves linear elastic.

Making use of the theory of higher algebra we assume that $[D]$ is given by

$$\begin{aligned} D_{ijkl} = {} & \lambda\, \delta_{ij}\delta_{kl} + \mu\,(\delta_{ik}\delta_{jl} + \delta_{il}\delta_{jk}) \\ & + C_1\,(\delta_{ij}\Omega_{kl} + \Omega_{ij}\delta_{kl}) \\ & + C_2\,(\delta_{ik}\Omega_{jl} + \delta_{il}\Omega_{jk} + \delta_{jk}\,\Omega_{il} + \delta_{jl}\Omega_{ik}) \\ & + C_3\tfrac{1}{3\bar{\omega}}\Omega_{ij}\Omega_{kl} \\ & + C_4\tfrac{1}{3\bar{\omega}}(\delta_{ij}\Omega_{km}\Omega_{ml} + \Omega_{im}\Omega_{mj}\delta_{kl}) \\ & + C_5\;(\delta_{ik}\Omega_{jm}\Omega_{ml} + \delta_{il}\Omega_{jm}\Omega_{mk} \\ & \quad + \delta_{jk}\Omega_{im}\Omega_{ml} + \delta_{jl}\Omega_{im}\Omega_{mk}) \qquad (10a) \\ & -\delta_{KI}\delta_{LJ}\langle\delta_{IJ}\rangle(1 - \delta_{MK}\delta_{NL})\cdot C_6\Omega_{mn}(\text{IJKLMN do not sum}) \qquad (10b) \end{aligned} \tag{10}$$

The symbol δ denotes the *Kronecker delta.* λ and μ are *Lame's* constants, and hence, the first two terms on the right hand side of equation (12) correspond to the stiffness of an isotropic solid ($[\Omega] \equiv 0$). The remaining parameters define the influence of the cracks on the free energy.[2]

$\bar{\omega}$ is an average measure for the principal values ω_i, ω_j and ω_k of the damage tensor, calculated as

$$\bar{\omega} = \frac{I_1 - 2I_2 + 3I_3}{3 - 2I_1 + I_2} \tag{11}$$

where I_1, I_2 and I_3 are three independent invariants of $[\Omega]$:

$$\begin{aligned} I_1 &= \Omega_{kk} \\ I_2 &= \tfrac{1}{2}\Omega_{ii}\Omega_{jj} - \tfrac{1}{2}\Omega_{ij}\Omega_{ji} \\ I_3 &= \tfrac{1}{6}\Omega_{ii}\Omega_{jj}\Omega_{kk} + \tfrac{1}{3}\Omega_{ik}\Omega_{kl}\Omega_{li} - \tfrac{1}{2}\Omega_{ij}\Omega_{ji}\Omega_{kk} \end{aligned} \tag{12}$$

The term containing (10b) accounts for the strain energy release observed in the case of uniaxial compression when cracks propagate parallel to the loading axis. It vanishes when either the lateral strain is compressive or the axial strain is tensile. This conforms with the fact that such cracks do not change the mechanical properties of a specimen subjected to uniaxial tension. This strain dependent behaviour is taken into consideration by a function $\langle\ \rangle$ defined as

$$\langle \delta_{IJ} \rangle = \begin{cases} \delta_{IJ} & \text{if } \varepsilon_{IJ} < 0 \\ 0 & \text{if } \varepsilon_{IJ} \geq 0 \end{cases} \tag{13}$$

3.2 Conjugate force

From the thermodynamics restriction, the stress tensor $[\sigma]$ is given by the *stress relation*

$$\sigma_{ij} = \rho \frac{\partial \psi}{\partial \varepsilon_{ij}} = D_{ijkl} \varepsilon^e_{kl} \tag{14}$$

and the damage rate $[\dot{\Omega}]$ obeys the *internal dissipation inequality*

$$R_{ij}\dot{\Omega}_{ij} - \sigma_{ij}\dot{\varepsilon}^p_{ij} \leq 0 \tag{15}$$

Derived from equation (8),(14),(15), $[\mathbf{R}]$ is the generalized thermodynamic force *conjugate* to the state variable $[\Omega]$, defined by

$$\begin{aligned} R_{ij} &= \rho \frac{\partial \psi}{\partial \Omega_{ij}} \\ &= C_1 \varepsilon^e_{ij} \delta_{kl} \varepsilon^e_{kl} + 2C_2 \varepsilon^e_{ik} \varepsilon^e_{jk} \\ &+ C_3 \left(\tfrac{1}{3\bar{\omega}} \varepsilon^e_{ij} \Omega_{kl} \varepsilon^e_{kl} + \tfrac{1}{2} G_{ij} \varepsilon^e_{mn} \Omega_{mn} \Omega_{kl} \varepsilon^e_{kl}\right) \\ &+ C_4 \left(\tfrac{1}{3\bar{\omega}} [\delta_{km} \varepsilon^e_{km} \Omega_{jl} \varepsilon^e_{il} + \delta_{lm} \varepsilon^e_{lm} \Omega_{ki} \varepsilon^e_{kj}] \right. \\ &\qquad \left. + \ G_{ij} \varepsilon^e_{op} \Omega_{om} \Omega_{mp} \delta_{kl} \varepsilon^e_{kl}\right) \\ &+ C_5 \left(2\,\delta_{km} \varepsilon^e_{mi} \Omega_{jl} \varepsilon^e_{kl} + 2\,\delta_{km} \varepsilon^e_{ml} \Omega_{li} \varepsilon^e_{jk}\right) \\ &- \tfrac{1}{2} \varepsilon^e_{mn} \varepsilon^e_{kl} \delta_{KI} \delta_{LJ} \langle \delta_{IJ} \rangle (1 - \delta_{MK} \delta_{NL}) \cdot C_6 (\text{IJKLMN} \quad \text{do not sum}) \end{aligned} \tag{16}$$

For simplicity this relation has been derived for $c_k^n = c_k^s = 1$ $(D^n = D^s = D)$. We note that in general the tensors [**R**] and [Ω] are not coaxial.

The constant [**G**] in $(16)_2$ has been introduced as

$$G_{ij} = \frac{\partial}{\partial \Omega_{ij}}\left(\frac{1}{3\bar{\omega}}\right) = \frac{\delta_{ij}[I_1 - 2] - \Omega_{ji}}{3I_1 - 6I_2 + 9I_3} - \frac{(3 - 2I_1 + I_2)(3\delta_{ij}[1 - 2I_1 + 3I_2] - 3\Omega_{ji}[3I_1 - 2] + 9\Omega_{jk}\Omega_{ki})}{(3I_1 - 6I_2 + 9I_3)^2} \tag{17}$$

By the *conjugate force*, damage propagation could be produced.

Taking into consideration plane strain condtion($\varepsilon_{zz} = 0$) and assumption of $\omega_{33} = 0$, we can get stiffness matrix of damage elasticity from equation (14).

$$\begin{bmatrix} \sigma_1 \\ \sigma_2 \\ \tau_{12} \end{bmatrix} = \begin{bmatrix} D_{11} & D_{12} & 0 \\ D_{21} & D_{22} & 0 \\ 0 & 0 & D_{33} \end{bmatrix} \begin{bmatrix} \varepsilon_1 \\ \varepsilon_2 \\ \gamma_{12} \end{bmatrix} \tag{18}$$

Where,

$$\begin{aligned} D_{11} &= \lambda + 2\mu + [2C_1 + 4C_2]c_1^n\omega_1 + [C_3 + 2C_4]\frac{1}{3\bar{\omega}^n}c_1^{n2}\omega_1^2 + 4C_5c_1^{n2}\omega_1^2 \\ D_{12} &= \lambda + C_1[c_1^n\omega_1 + c_2^n\omega_2] + C_3\frac{1}{3\bar{\omega}^n}c_1^n\omega_1c_2^n\omega_2 + C_4\frac{1}{3\bar{\omega}^n}[c_1^{n2}\omega_1^2 + c_2^{n2}\omega_2^2] \\ D_{21} &= \lambda + C_1[c_2^n\omega_2 + c_1^n\omega_1] + C_3\frac{1}{3\bar{\omega}^n}c_2^n\omega_2c_1^n\omega_1 + C_4\frac{1}{3\bar{\omega}^n}[c_2^{n2}\omega_2^2 + c_1^{n2}\omega_1^2] \\ D_{22} &= \lambda + 2\mu + [2C_1 + 4C_2]c_2^n\omega_2 + [C_3 + 2C_4]\frac{1}{3\bar{\omega}^n}c_2^{n2}\omega_2^2 + 4C_5c_2^{n2}\omega_2^2 \\ D_{33} &= \mu + C_2[c_1^s\omega_1 + c_2^s\omega_2] + C_5[c_1^{s2}\omega_1^2 + c_2^{s2}\omega_2^2] \end{aligned} \tag{19}$$

4 Damage evolution law

Damage propagation is denoted by the increment of damage which derived from a potential Φ . This potential provided that Φ depends on the strains only through the conjugate thermodynamic force to [Ω].[5] In this case the damage increments are directed along the outward normal to the surface $\Phi = 0$, i.e.

$$\dot{\Omega}_{mn} = \Gamma\frac{\partial \Phi}{\partial R_{mn}} \tag{20}$$

This surface separates the elastic domain ($[\dot{\Omega}] \equiv 0$) from the inelastic domain and shall be called *damage surface*. In general the dissipation potential will be function of [**R**], the state of damage and loading history. To take account of nonproportional loading, we must express the damage criterion in a form depending on a tensorial hardening parameter [**H**]. The mechanical response of the damaged body should not be affected by the orientation of the employed coordinate system. Therefore we shall express the potential Φ in terms of the invariants of its two tensor arguments, namely

$$\Phi = H_{ij}^{-1}H_{jk}^{-1}R_{kl}R_{li} - 1 \tag{21}$$

To complete our model we must specify an evolution law for [**H**]. Any damage increment is accompanied by a variation of [**H**]. We can therefore assume that

$$H_{ij} = \lambda q\left(\frac{\kappa}{\lambda}\right)^{r} \Omega_{ij} + \lambda v_o^2 \, \delta_{ij} \qquad \kappa = \int_0^t -R_{ij}\dot{\Omega}_{ij} \, dt \tag{22}$$

The constant v_o defines the initial threshold against brittle cracking and thus, it can be obtained from the condition that the onset of microcracking corresponds to the stress level at which *virgin* material starts exhibiting nonlinearity.

In the damage cordinate, H_{ij} should be diagonal matrix.

$$H_{11} = \lambda q\left(\frac{\kappa}{\lambda}\right)^{r} \Omega_{11} + \lambda v_o^2 \qquad H_{22} = \lambda q\left(\frac{\kappa}{\lambda}\right)^{r} \Omega_{22} + \lambda v_o^2 \tag{23}$$

Since every single increment of a propagating prismatic crack must lie entirely in one single surface the damage rate tensor $[\dot{\Omega}]$ may have only one non-vanishing principal value. However, a tensor that is derived from a potential Φ according to equation (20) has as many principal values as $[\mathbf{R}]$, namely two if we use its upper left partition.

If we use our considerations to calculate the unit normal $\dot{\mathbf{n}}$ of the damage increments instead of the damage rate itself this condition can be met quite easily. We need both the direction and magnitude of damage increment. In this case we must take the unit vectors pointing in the principal directions of $[\Omega]$ as state variables and define the conjugate thermodynamic forces as

$$p_i^k = \frac{1}{\omega^k} \cdot \rho \frac{\partial \psi}{\partial n_i^k} \tag{24}$$

The incremental change of $\mathbf{n}^k$ is then assumed to be proportional to the gradient of a potential $\tilde{\Phi}$, i.e.

$$\dot{n}_i^k = \Gamma \frac{\partial \tilde{\Phi}}{\partial p_i^k} \tag{25}$$

where $\tilde{\Phi}$ is defined in the space of the conjugate forces $\mathbf{p}^k$. They are calculated as

$$p_i^k = \frac{\rho}{\omega^k} \cdot \frac{\partial \psi}{\partial \Omega_{mn}} \frac{\partial \Omega_{mn}}{\partial n_i^k} \qquad = R_{mn}(\delta_{im} n_n^k + \delta_{in} n_m^k) \tag{26}$$

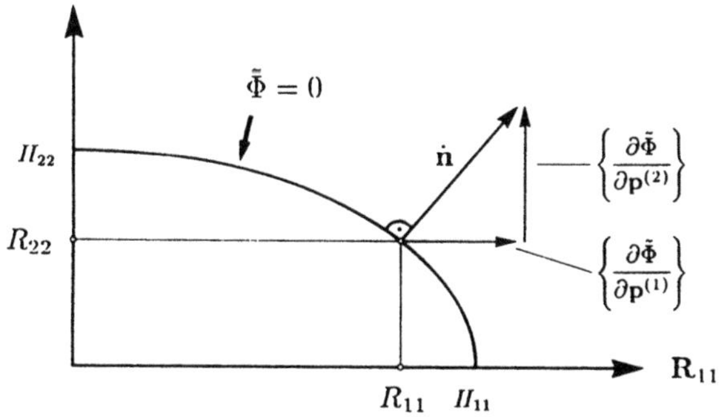

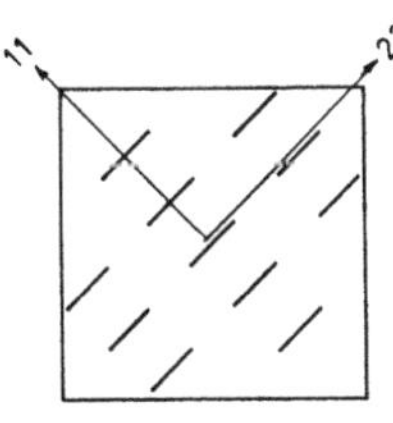

Fig. 1 Direction of the damage increment

To get element of $\partial\tilde{\Phi}/\partial p_i^k$, we shall express the potential $\tilde{\Phi}$ analogous to Φ in equation (21) as

$$\tilde{\Phi} = \frac{1}{4} p_l^k H_{lm}^{-1} H_{mn}^{-1} p_n^k - 1 \tag{27}$$

We shall note that $R_{12}, R_{21} = 0$ because of plane strain condition. Taking into consideration these conditions, we can get the element from (26),(27) as follows.

$$\frac{\partial\tilde{\Phi}}{\partial p_i^k} = \frac{1}{2} H_{im}^{-1} H_{mn}^{-1} p_n^k = \frac{1}{2} \left\{ \frac{p_1^1}{H_{11}^2}, \frac{p_2^2}{H_{22}^2} \right\}^T \tag{28}$$

$$\mathbf{p}^1 = \pm \begin{Bmatrix} 2R_{11} \\ R_{12} + R_{21} \end{Bmatrix} = \pm \begin{Bmatrix} 2R_{11} \\ 0 \end{Bmatrix} \quad \mathbf{p}^2 = \pm \begin{Bmatrix} R_{12} + R_{21} \\ 2R_{22} \end{Bmatrix} = \pm \begin{Bmatrix} 0 \\ 2R_{22} \end{Bmatrix} \tag{29}$$

Since the increments of a propagating prismatic crack can have only one unit normal we shall assume

$$\dot{\mathbf{n}}_{\mathbf{i}} = \Gamma \sum_k \dot{n}_i^k = \Gamma \sum_k \frac{\partial\tilde{\Phi}}{\partial p_i^k} = \Gamma \left\{ \pm\frac{R_{11}}{H_{11}^2}, \pm\frac{R_{22}}{H_{22}^2} \right\}^T \quad \frac{\mathbf{1}}{\mathbf{\Gamma}} = \sqrt{\frac{R_{11}^2}{H_{11}^4} + \frac{R_{22}^2}{H_{22}^4}} \tag{30}$$

Once $\dot{\mathbf{n}}$ is given we are able to calculate the damage rate as

$$\dot{\Omega}_{kl} = \tilde{\omega} \dot{n}_k \dot{n}_l \tag{31}$$

The scalar valued quantity $\tilde{\omega}$ is determined by the consistency condition of the potential $\tilde{\Phi}$.

$$\dot{\Phi} = \frac{\partial\Phi}{\partial R_{op}} \dot{R}_{op} + \frac{\partial\Phi}{\partial H_{ij}} \dot{H}_{ij} \doteq 0 \tag{32}$$

In the damage cordinate,

$$\frac{\partial\Phi}{\partial R_{op}} = \begin{Bmatrix} \frac{\partial\Phi}{\partial R_{11}} \\ \frac{\partial\Phi}{\partial R_{22}} \end{Bmatrix} = \begin{bmatrix} \frac{2R_{11}}{H_{11}^2} & 0 \\ 0 & \frac{2R_{22}}{H_{22}^2} \end{bmatrix} \tag{33}$$

and

$$\frac{\partial\Phi}{\partial H_{ij}} = \begin{Bmatrix} \frac{\partial\Phi}{\partial H_{11}} \\ \frac{\partial\Phi}{\partial H_{22}} \end{Bmatrix} = \begin{bmatrix} -\frac{2R_{11}R_{11}}{(H_{11}^3)} & 0 \\ 0 & -\frac{2R_{22}R_{22}}{(H_{22}^3)} \end{bmatrix} \tag{34}$$

In view of (22) the evolution law for the hardening parameter $[\mathbf{H}]$ is given as

$$\dot{H}_{ij} = \frac{\partial H_{ij}}{\partial \Omega_{kl}} \dot{\Omega}_{kl} + \frac{\partial H_{ij}}{\partial \kappa} \dot{\kappa} \; r_{ijkl} \dot{\Omega}_{kl} = [\delta_{ik}\delta_{jl}\lambda q \left(\frac{\kappa}{\lambda}\right)^r - rq \left(\frac{\kappa}{\lambda}\right)^{r-1} \Omega_{ij} R_{kl}] \dot{\Omega}_{kl} \tag{35}$$

Combining these equations shown in this section, we can obtain an expression for $\tilde{\omega}$

$$\tilde{\omega} = -\frac{(\partial\Phi/\partial R_{op})}{(\partial\Phi/\partial H_{ij})r_{ijkl}\,\dot{n}_k\dot{n}_l} \cdot \dot{R}_{op} \tag{36}$$

and our damage evolution law between damage increment and conjugate force increment reads

$$\dot{\Omega}_{mn} = a_{mnop}\dot{R}_{op} \quad a_{mnop} = -\frac{(\partial\Phi/\partial R_{op})\,\dot{n}_m\dot{n}_n}{(\partial\Phi/\partial H_{ij})r_{ijkl}\,\dot{n}_k\dot{n}_l} \tag{37}$$

$(\partial\Phi/\partial H_{ij})r_{ijkl}\dot{n}_k\dot{n}_l$ should be constant. We shall note that a_{13}, a_{23}, a_{33} should be zero in the plane strain condtion because of the damage cordinate.

Since $[\mathbf{R}]$ is a function of $[\varepsilon]$ and $[\Omega]$ the incremental relation is given as follows

$$\dot{R}_{ij} = \frac{\partial R_{ij}}{\partial \varepsilon_{mn}}\dot{\varepsilon}_{mn} + \frac{\partial R_{ij}}{\partial \Omega_{kl}}\dot{\Omega}_{kl} \quad = b_{ijmn}\dot{\varepsilon}_{mn} + c_{ijkl}\dot{\Omega}_{kl} \tag{38}$$

The two partial derivatives

$$b_{ijmn} = \frac{\partial R_{ij}}{\partial \varepsilon_{mn}} \quad c_{ijkl} = \frac{\partial R_{ij}}{\partial \Omega_{kl}} \tag{39}$$

will be calculated preferably in the damage coordinate system. There it is comparatively easy to consider the directional nature of cracks. The result can be rotated immediately by the well known transformation laws for higher-order Cartesian tensors.

We may use $(38)_2$ to substitute $[\dot{\mathbf{R}}]$ in (37) which leads to

$$d_{ijkl}\dot{\Omega}_{kl} = a_{ijop}b_{opmn}\dot{\varepsilon}_{mn} \tag{40}$$

where

$$d_{ijkl} = \frac{1}{2}(\delta_{ik}\delta_{jl} + \delta_{il}\delta_{jk}) - a_{ijmn}c_{mnkl} \tag{41}$$

By inverting $[d]$ we obtain a relation between $[\dot{\Omega}]$ and the strain rate:

$$\dot{\Omega}_{kl} = d^{-1}_{ijkl}a_{ijop}b_{opmn}\dot{\varepsilon}_{mn} \tag{42}$$

This relation is the most important. Because we can get damage increment from strain increment.

$$\dot{\Omega}_{kl} = d^{-1}_{ijkl}a_{ijop}b_{opmn}\dot{\varepsilon}_{mn} \quad = h_{klmn}\dot{\varepsilon}_{mn} \tag{43}$$

Using matrix formulation, we can express as follows.

$$\begin{bmatrix} \dot{\Omega}_1 \\ \dot{\Omega}_2 \\ \dot{\Omega}_{12} \end{bmatrix} = \begin{bmatrix} d_{11} & d_{12} & 0 \\ d_{21} & d_{22} & 0 \\ d_{31} & d_{32} & 1 \end{bmatrix}^{-1} \begin{bmatrix} a_{11} & a_{12} & 0 \\ a_{21} & a_{22} & 0 \\ a_{31} & a_{32} & 0 \end{bmatrix} \begin{bmatrix} b_{11} & b_{12} & b_{13} \\ b_{21} & b_{22} & b_{23} \\ 0 & 0 & 0 \end{bmatrix} \begin{bmatrix} \dot{\varepsilon}_1 \\ \dot{\varepsilon}_2 \\ \dot{\gamma}_{12} \end{bmatrix} \tag{44}$$

5 Constitutive equation

In view of the extremely nonlinear character of material response during damage propagation the stress – strain relation must be cast into an incremental form. Since the elements of $[\sigma]$ depend on $[\Omega]$, differentiation leads to:

$$\dot{\sigma}_{ij} = \frac{\partial \sigma_{ij}}{\partial \varepsilon_{mn}} \dot{\varepsilon}_{mn} + \frac{\partial \sigma_{ij}}{\partial \Omega_{kl}} \dot{\Omega}_{kl} \tag{45}$$

We can calculate the two partial derivatives immediately:

$$\frac{\partial \sigma_{ij}}{\partial \varepsilon_{mn}} = D_{ijmn} \qquad \frac{\partial \sigma_{ij}}{\partial \Omega_{kl}} = e_{ijkl} \tag{46}$$

We shall note that e_{13},e_{23}, e_{33} should be zero because of the damage'cordinate. After inserting the above relations in (45), the equation reads

$$\dot{\sigma}_{ij} = D_{ijmn}\dot{\varepsilon}_{mn} + e_{ijkl}\dot{\Omega}_{kl} \ = D_{ijmn}\dot{\varepsilon}_{mn} + e_{ijmn}h_{mnop}\dot{\varepsilon}_{op} \tag{47}$$

where we can define the damage propagation stiffness as F_{ijop}.

$$F_{ijop} = e_{ijgh}h_{ghop} \ = e_{ijgh}d^{-1}_{klgh}a_{klmn}b_{mnop} \tag{48}$$

If we define the effective "stiffness" of the deteriorating material in the damage cordinate as

$$G'_{ijmn} = D_{ijmn} + F_{ijmn} \tag{49}$$

we may express the incremental stress – strain relationship in its usual form:

$$\dot{\sigma}_{ij} = G'_{ijmn}\dot{\varepsilon}_{mn} \tag{50}$$

So, we can get matrix formula in the plane strain condition as follows.

$$\begin{bmatrix} \dot{\sigma}_1 \\ \dot{\sigma}_2 \\ \dot{\sigma}_{12} \end{bmatrix} = \begin{bmatrix} D_{11}+F_{11} & D_{12}+F_{12} & F_{13} \\ D_{21}+F_{21} & D_{22}+F_{22} & F_{23} \\ F_{31} & F_{32} & D_{33}+F_{33} \end{bmatrix} \begin{bmatrix} \dot{\varepsilon}_1 \\ \dot{\varepsilon}_2 \\ \dot{\gamma}_{12} \end{bmatrix} \tag{51}$$

So far, all equation can be calculated preferable in the damage cordinate system. Finally, we can transform these equations to global cordinate system. Global stiffness matrix [G] should be transformed by [T] for the plane strain problem.

$$[\mathbf{G}] = [\mathbf{T}][\mathbf{G}'][\mathbf{T}^T] \quad [\mathbf{T}] = \begin{bmatrix} \cos^2\theta & \sin^2\theta & -2\sin\theta\cos\theta \\ \sin^2\theta & \cos^2\theta & 2\sin\theta\cos\theta \\ \sin\theta\cos\theta & -\sin\theta\cos\theta & \cos^2\theta - \sin^2\theta \end{bmatrix} \tag{52}$$

6 Analytical results

As an illustration of the proposed model, a numerical uniaxial compression experiment on an initially cracked specimen was performed (Fig. 2,Tab. 1).

The angle of the initial crack has been chosen to be 15°, 30° and 45°, respectively. Its normalized length is 0.1 in all three cases. As the damage propagation grows, we can get the rotation of the damage tensor.

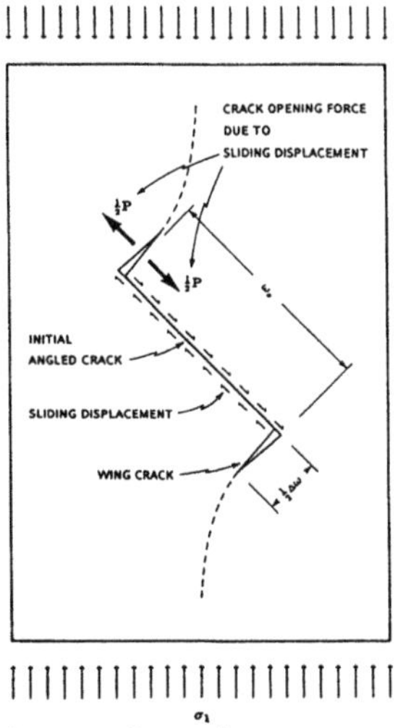

Fig. 2 Crack growth under a compressive stress σ_1

λ	40000	kN/m^2
μ	40000	kN/m^2
C_1	−29000	kN/m^2
C_2	−20000	kN/m^2
C_3	120000	kN/m^2
C_4	−33000	kN/m^2
C_5	0	kN/m^2
C_6	60000	kN/m^2
v_o	0.14	%
q	0.08	−
r	0.55	−
c^s	1.0	−

Tab. 1 Material parameters used in computation

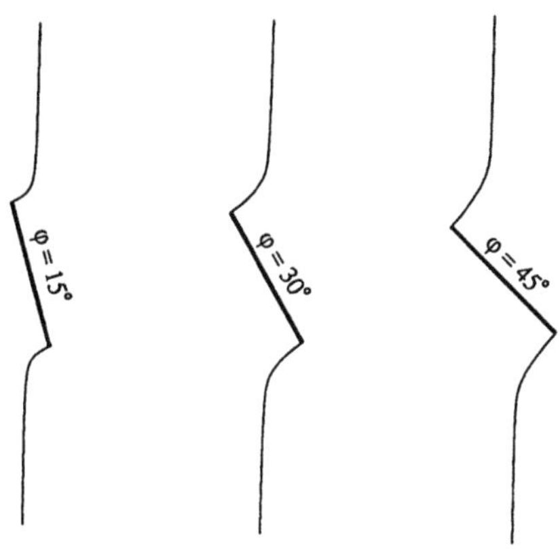

Fig. 3 Shape of the wing cracks

The calculated stress – strain curves are similar to those that can be found in relevant literature.[8,9] Fig.3 however is new. It shows the way the wing cracks extend from the sliding crack as the compressive load is increased. The calculated shape of the wing cracks corresponds well with experimentally obtained real cracks.[13] This proves the validity of the two potentials (9) and (27). Joint deformation can take the normal closure,openning, shear and dilation. Evidence that rock joint properties are dependent on surface characterictic such as roughness and wall strength can be deducted in part from the early autuors. Direct physical evidence for influence of surface joint properties were obtainted by Barton and Bandis as JRC(joint rouness coefficient and JCS(joint wall compression sterngth)[14]. Under the general shear-mode II, the behaviour of rock is evidently dependent on joint roughness, wall strength. Our model has the possibility to show the shear behaviour considering joint roughness as a continuum phenomenon on a suitable macroscopic scale. We shall discuss how the influence of joint character can be considered in our model. The constant c^s can denote the shear friction acting on the damage plane. Our choice of the constant c^s makes a conribution to this phenomenon. c^s on the joint has physical meaning discussed by many reserchers. Popularly, we can introduce Mohr-Coulomb Law in shear, using friction angle ϕ, effective shear cohesion c_s, which act on the joint.

$$\tau_f = \sigma_n \tan \phi + c_s \tag{53}$$

Moreover, about the friction angle, Barton and Bandis have utilized the joint roughness coeffcient JRC, joint wall compression strength JCS, residual friction angle ϕ_r, to predict post-peak behaviour for joints,

$$\tau_f = \sigma_n \tan \left(JRC \cdot \log \left(JCS/\sigma_n' + \phi_r \right) \right) \tag{54}$$

Taking into account of each equation, we can define the suitable constant c^s between 0 and 1 in our model.

$$c^s = \begin{cases} 0 & \text{if } \tau_f > \tau_{max} \\ 1 - \dfrac{\tau_f}{\tau_{max}} & \text{if } \tau_f \leq \tau_{max} \end{cases} \tag{55}$$

Hence, In our model, we can consider the influence of joint roughness as the joint friction. Fig.3 dipicted the way of the wing crack extend from the sliding crack, which means, during damage propagation friction is not acting on the damage plane. These calculated results showed the crack propagation gradually become parallel to the loading axis. On the other hand, Fig.4 illustrates the results of crack propagation after changing the constant c^s. Compared with Fig.3, the direction of crack propagation gradually become inclined toward initial angle because of friction effect as the change of constant c^s. From these results of calculation, the change of this constant can give us suggestion of understanding the mechanical behaviour of damaged rock as a continuum phenomenon.

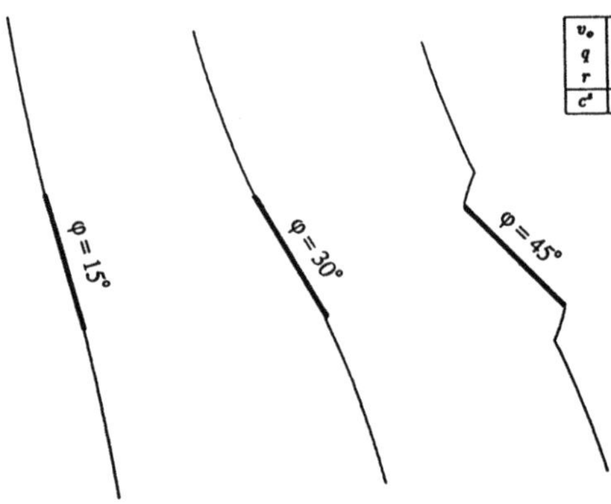

v_o	0.14	%
q	0.001	-
r	0.55	-
c^s	0.1	-

Fig. 4 Change of the damage orientation

7 References

[1]Stumvoll, M. Kyoya, T. Ichikawa, Y. & Swoboda, G. (1991) *A Damage Model for discontinuous Rock Mass based on Thermodynamic Principles,* **Proc. Comp. Meth. and Adv. in Geomech.**, Beer et al., eds., Balkema, pp. 427–434

[2]Stumvoll, M.(1991) *Numerical Model for Discontinuous Rock Mass considering Damage Propagation under Non-proportional and Compressive Loading,* **Ph.D. Thesis**, University of Innsbruck, Austria

[3]Coleman, B.D. & Gurtin, M.E. (1967) *Thermodynamics with Internal State Variables,* **J. Chem. Phys.**, 47, pp. 597–613

[4]Rabotnov, Y.N. (1969) *Creep Problems in Structural Members,* **North Holland**, Amsterdam

[5]Rice, J.R. (1971) *Inelastic Constitutive Relations for Solids: An Internal-Variable Theory and its Application to Metal Plasticity,* **J. Mech. Phys. Solids**, 19, pp. 433–455

[6]Lemaitre, J.(1971) *Evaluation of Dissipation and Damage in Metals, Submitted to Dynamic Loading,* **Proc. ICM1**, Kyoto

[7]Murakami, S. & Ohno, N. (1981) *A Continuum Theory of Creep and Creep Damage,* **Creep in Structures**, Springer, Berlin, pp. 422–444

[8]Krajcinovic, D. (1983) *Constitutive Equations for Damaging Materials,* **J. Appl. Mech.**, 50, pp. 355–360

[9]Suaris, W. (1987) *A Damage Theory for Concrete Incorporating Crack Growth Characteristics,* **Constitutive Laws for Engineering Materials**, C.S. Desai et al., eds., Elsevier, pp. 931–938

[10]Kawamoto, T. Ichikawa Y. & Kyoya, T. (1988) *Deformation and Fracturing Behaviour of Discontinuous Rock Mass and Damage Mechanics Theory,* **Int. J. Num. Anal. Meth. Geom.**, 12, pp. 1–30

[11]Murakami, S. (1988) *Mechanical Modeling of Material Damage,* **J. Appl. Mech** 55, pp. 280–286

[12]Ashby ,M.F. & Hallam, S.D. (1986) *The Failure of Brittle Solids containing Small Cracks under Compressive Stress States,***Acta metall.**, 34, pp. 497–510

[13]Barton, N. Bandis, S.& Bakhtar, K. (1985) *Strength, Deformation and Conductivity Coupling of Rock Joints,* **Int. J. Rock Mech. Min. Sci. and Geomech**, vol. 28, No.1, pp.121–140

32 PROGRESSIVE DAMAGE OF INTERFACES IN LAYERED ROCK DUE TO CYCLIC IMPACT

H.P. ROSSMANITH, R.E. KNASMILLNER, C.F. XU
Institute of Mechanics, Technical University, Vienna, Austria
J. WEGNER
Department of Mechanical Engineering, University of Victoria, Canada

Abstract

The interaction of elastic SH-waves with imperfectly bonded thin interfaces between two elastic media is examined. Small strains and displacements prevail. The interface conditions are defined as linear relations between interface displacement discontinuities and related traction. Progressive damage of the interface is taken into account by an evolution law in continuum damage mechanics whereby successive incident wave/pulse interaction events promote the development of damage at the interface. The system of coupled equations consisting of one coupled partial differential equation for the propagating wave and one nonlinear equation for the damage can be solved iteratively for any given initial damage. Lifetimes, measured in numbers of wave pulses to failure, have been calculated for various dissimilar material parameter combinations.

Keywords: damage, joint, interface, layered rock, impact, wave propagation, fractured rock, percussion drilling, blasting.

1 Introduction

In recent years the response of joints and faults in fractured rock to seismic or mining induced wave interaction has become of great importance in geotechnical engineering, mining and rock mechanics. When waves of seismic or other origin are propagated through fractured rock mass not only will the reflection and transmission coefficients depend on the acoustic impedance mismatch and the nature of the interface layer but the physicomechanical behaviour of the interface will change as a function of time. This change can partly be attributed to degradation and ageing processes of the interface material which may quantitatively be formulated by means of continuum damage mechanics principles.

Earlier work on fracture related reduction of stiffness of the rock-interface system and attenuation of elastic waves including frictional sliding and scattering may be found in Walsh (1966), Mavko (1979) and Hudson (1981).

Elastic wave behaviour across imperfectly bonded linear slip interfaces separating two elastic media is investigated in Schoenberg (1980), where the displacement

Fracture and Damage of Concrete and Rock – FDCR-2.
Edited by H.P. Rossmanith. © 1993 Taylor & Francis. ISBN 0 419 18470 8.

discontinuity, or slip, is taken to be linearly related to the stress traction which is continuous across the interface for both, the normal and tangential direction. Applying the theoretical results of Schoenberg 1980 to seismic wave propagation model experiments in a jointed columnar basaltic rock formation acceptable agreement between observed and predicted results is achieved in a recent contribution by Myer et al.(1990).

For many practical applications in geotechnical engineering analysis the deformation behaviour of rock joints has to be taken into account. Stress strain relations for both normal and shearing stresses are strongly nonlinear in loading and unloading and often interdependent, see Goodman (1976), Bandis et al.(1983) and Ming and Horii (1992). A proper constitutive model for rock joints must be able to predict the deformation behaviour involving ageing effects on a long time scale as well as deterioration effects due to repeated loading and unloading. In percussion drilling work the interface is repeatedly dynamically loaded at high intensity and the interface characteristics will change with each consecutive blow of the percussion/drill tool. Improved knowledge about the impact-induced time-variation of the mechanical behaviour of the rock joints appears to be instrumental and vital for the improvement of drill efficiently and the development of *smart* mining machinery.

This paper discusses the dynamic behaviour of single rock joints and a system of two adjacent rock joints subjected to medium frequency ($f \sim 40Hz$) cyclic impact loading. The theory of elastic wave propagation across a displacement discontinuity at the boundary between two half-spaces is combined with continuum damage mechanics as represented by the application of a Kachanov-type damage evolution law, Kachanov (1986), Lemaitre (1985). This theory predicts changes of the reflection and transmission coefficients on the basis of damage-controlled time-varying rock joint characteristics. In the following the evolution laws for normal and shear damage were allowed to be dissimilar. However, the inevitable coupling effects have been neglected. In addition, plane interfaces with linear normal and slip conditions have been employed up to complete degradation, i.e. complete separation, of the joint surfaces.

2 Linear interface conditions with damage

Consider a nominally plane interface between two dissimilar linear elastic solid half-spaces. Imperfect bonding is assumed to prevail. Tractions are assumed to be continuous across the interface but the small displacement field is not. The condition of continuous displacement is replaced by a so-called *linear slip condition*, Schoenberg (1980), where the dependence may be complex and frequency dependent simulating viscoelastic spring conditions.

Introduce a cartesian coordinate system at a point on the surface such that x_2 and x_3 are directions tangential to the interface and x_1 is normal to the surface. Let $\vec{u}$ be the displacement vector and $\Delta\vec{u} = \vec{u}(0^+, 0, 0) - \vec{u}(0^-, 0, 0)$ be the displacement discontinuity vector at the point. The traction vector, $\vec{t}$, at that point on the interface, has components σ_{11}, τ_{12}, τ_{23}, and represents the force per unit area of the material transmitted from the x_1^+-side to the x_2^--side of the joint.

A functional relationship between the jump in the displacement across the joint/-interface and the external tractions may be expressed as a power series in $\Delta\vec{u}$,

$$\vec{t} = F(\Delta\vec{u}, D) = k(\Delta\vec{u}; D) + O(\Delta u_i \Delta u_j; D) \tag{1}$$

where the dependence on the time-varying damage, i.e. changing material characteristics of the interface, has been explicitly included by the symbol D.

A positive definite displacement discontinuity energy density function, U, associated with the slip interface requires the form

$$U = L_j \Delta u_j(D) + \frac{1}{2} K_{ij} \Delta u_i(D) \Delta u_j(D) + \cdots \tag{2}$$

It is more convenient to characterize the interface slip behaviour in terms of compliance where the compliance matrix is the inverse of the stiffness matrix. In addition, by generalization via the harmonic elastic-viscoelastic analogy, slip boundary compliances may be generalized to allow the modeling of linear viscoelastic slip interfaces.

A linear damage-controlled viscoelastic slip interface constitutive law of the form

$$P(\Delta\vec{u}) = Q(\underset{\sim}{D} \bullet \vec{t}) \tag{3}$$

will be employed where P and Q are polynomial operators with respect to time and $\underset{\sim}{D}$ represents the damage tensor. For mutually independent damage behaviour in the normal and shear directions, the damage tensor diagonalizes and the traction-displacement discontinuity relations decouple.

For SH-waves, a constitutive law of the form

$$P(\Delta u_3) = Q\left(\frac{\tau_{13}}{1 - D}\right) \tag{4}$$

with

$$\begin{aligned} P() &= \alpha_0 + \alpha_1 \frac{\partial}{\partial t} + \cdots + \alpha_m \frac{\partial^m}{\partial t^m} \\ Q() &= \gamma_0 + \gamma_1 \frac{\partial}{\partial t} + \cdots + \gamma_l \frac{\partial^l}{\partial t^l} \end{aligned} \tag{5}$$

will be employed. With respect to the numerical simulations discussed in later sections of this paper it will be anticipated that the damage variable D will be kept constant per cycle, which considerably simplifies the analytical computations.

Damage development at the interface will be assumed to occur according to a Kachanov-type evolution law,

$$\frac{d\Psi}{dt} = -A\left(\frac{\tau}{\Psi}\right)^m \tag{6}$$

where $\Psi = 1 - D$, A and m are constitutive parameters and τ is the stress applied which is also responsible for the damage. In static creep tests τ is the stress acting

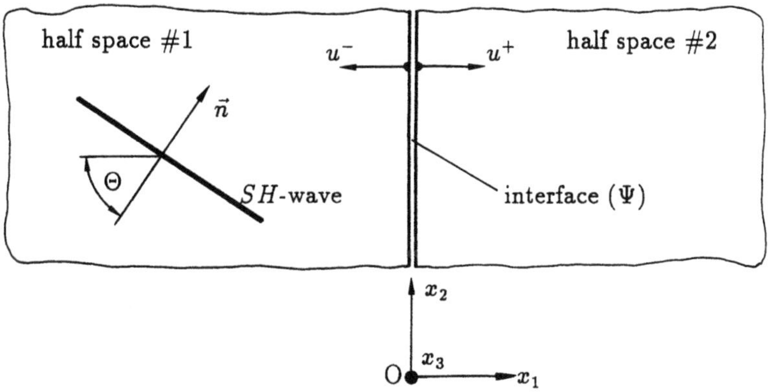

Figure 1: Interaction of incident SH-waves with a damaged interface between two elastic halfspaces

on the volume element. In a dynamic situation the proper definition of σ is more difficult. An upper limit to the damage would follow from identification of τ with the maximum shear load intensity during one load cycle. This approach has been followed here.

3 Plane SH-wave reflection and refraction

Consider two isotropic, homogeneous, linear elastic half spaces in contact along a plane (possibly predamaged) interface along $x_1 = 0$. Predamage is considered to be uniformly distributed. Let the elastic media posses densities ρ_i, compressional wave speeds $c_{Pi} = (\lambda_i + 2\mu_i/\rho_i)^{1/2}$ and shear wave speeds $c_{Si} = (\mu_i/\rho_i)^{1/2}$ and occupy the regions $x_1 < 0$,$x_1 > 0$.

In this analysis, any realistic repetitive cyclic impact loading such as the action exerted by the percussion drilling tool on the rock material will be Fourier-decomposed and the harmonic content will be treated in the following. Thus, consider an incident harmonic plane SH-wave of frequency ω and unit amplitude whose propagation vector lies in the x_1, x_2-plane which impinges on the interface from medium 1 ($x_1 < 0$), see Fig. 1.

For an incident SH-wave only one reflection and one transmission coefficient are present in the general wave solution. The only non-zero component of the displacement, $u_3 = (x_1, x_2)$ assumes the form, Schoenberg (1980),

$$\begin{aligned} u_3 &= \exp i\omega\left[(x_1 \cos\Phi_1 + x_2 \sin\Phi_1)/c_{S1} - t\right] + \\ &\quad + R \exp i\omega\left[(-x_1 \cos\Phi_1 + x_2 \sin\Phi_1)/c_{S1} - t\right] \qquad \text{for} \quad x_1 < 0 \\ u_3 &= T \exp i\omega\left[(x_1 \cos\Phi_2 + x_2 \sin\Phi_2)/c_{S2} - t\right] \qquad \text{for} \quad x_1 > 0 \end{aligned} \tag{7}$$

with $\omega \sin \Phi_1/c_{S1} = \omega \sin \Phi_2/c_{S2} = k_1$ and where R and T denote the reflection and transmission coefficients, respectively. In the sequel the $\exp i(k_2x_2 - \omega t)$ dependence of the wave field will be suppressed.

The non-trivial interface conditions are:

$$\tau_{13}^{+} = \tau_{13}^{-} \quad \text{and} \quad \Delta u_3 = \eta_T' \tau_{13}^{\pm} \tag{8}$$

where $\eta_T' = \eta_T/\Psi$ and η_T is the transverse compliance.

The combination of Hooke's law (valid for the elastic halfspaces)

$$\tau_{13i} = \rho_i c_{Si}^2 \frac{\partial u_{3i}}{\partial x_1}, \tag{9}$$

the interface constitutive equations, eqs (4) and (5) and the derivatives

$$\frac{\partial^n}{\partial t^n}\tau_{13} = \rho c_S^2 \frac{\partial}{\partial x_1}\left(\frac{\partial^n u_3}{\partial t^n}\right) = (-i\omega)^n \rho c_S^2 \frac{\partial u_3}{\partial x_1} \tag{10}$$

yields two equations,

$$\begin{aligned} \left(\frac{\partial u_3}{\partial x_1}\right)^{+} &= \left(\frac{\partial u_3}{\partial x_1}\right)^{-} \\ u_3^{+} - u_3^{-} &= C' \tau_{13}^{\pm} \end{aligned} \tag{11}$$

where the coefficient C' follows from the frequency domain representation of the interface constitutive law and is given by

$$C' = \left\{\Psi^{-1}\sum_{l=0}^{L}\gamma_l(-i\omega)^l + \dot{\Psi}^{-1}\sum_{l=1}^{L} l\gamma_l(-i\omega)^l + \ddot{\Psi}^{-1}() + \cdots\right\}\left\{\sum_{m=0}^{M}\alpha_m(-i\omega)\right\} \tag{12}$$

For slowly increasing damage, terms containing time derivatives of Ψ may be neglected.

Equations (11) allow the determination of the, now damage controlled, reflection and transmission functions

$$\begin{aligned} R(\Psi) &= \frac{Z_1 - Z_2 - i\omega C' Z_1 Z_2}{Z_1 + Z_2 - i\omega C' Z_1 Z_2} \\ T(\Psi) &= \frac{2Z_1}{Z_1 + Z_2 - i\omega C' Z_1 Z_2} \end{aligned} \tag{13}$$

Figs 2 and 3 show the variations of R and T as a function of progressive damage for an obliquely impinging SH-wave of various frequencies. In Fig 2 the frequency is assumed to be $40Hz$ ($\omega = 2\pi.40$) and the angle of wave incidence is varied, whereas in Fig 3 the effect of frequency on damage evolution is studied for a normal and an oblique (45°) incident wave. The frequency of $40Hz$ corresponds to the basic frequency at which percussion drilling tools operate and, therefore, corresponds to the impact frequency at the tool-rock interface.

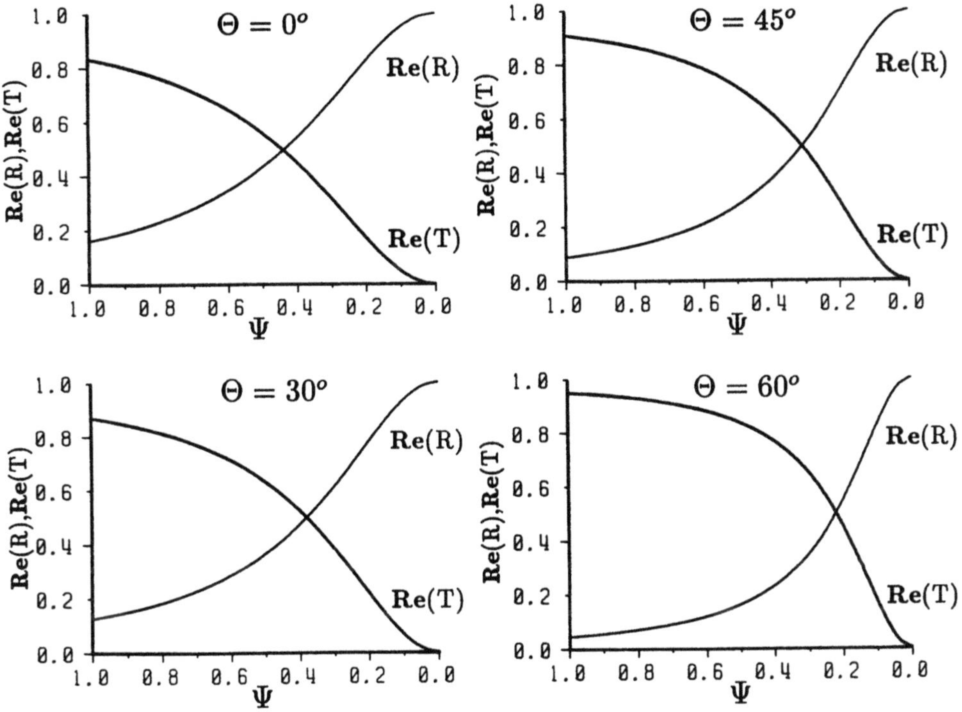

Figure 2: Variation of $R(\Psi)$ and $T(\Psi)$ as a function of damage Ψ for obliquely impinging SH-waves of frequency $f = 40Hz$

4 Damage evolution during repeated impact loading

In percussion drilling work a series of pulse trains is applied from the tool to the rock. Physically, each pulse will interact with the interface and dissipate some amount of energy at the interface which will contribute to the ageing of the bond material and finally lead to rupture and separation of the interface. Hence, the level of damage, Ψ, after each impact will progressively different such that Ψ will be approximated by a monotonically decreasing step function. In the limit, a smooth decay curve for the interface integrity (Ψ) will be obtained. The damage is assumed to occur during the time of wave interaction only. Thus, environmentally controlled ageing and degradation processes will not be taken into account in this study. Also, interaction effects between normal and shearing damage will be excluded.

Consider an initially undamaged ($\Psi(0) = \Psi_0 = 1$) or predamaged ($\Psi(0) = \Psi_0 < 1$) state to prevail at the interface, which is characterized by the initial reflection coefficient $R(\Psi_0) \leq 1$ and transmission coefficient, $T(\Psi_0) \leq 1$. The development of damage as a function of time, and hence also the time-variation of $R(\Psi)$ and $T(\Psi)$

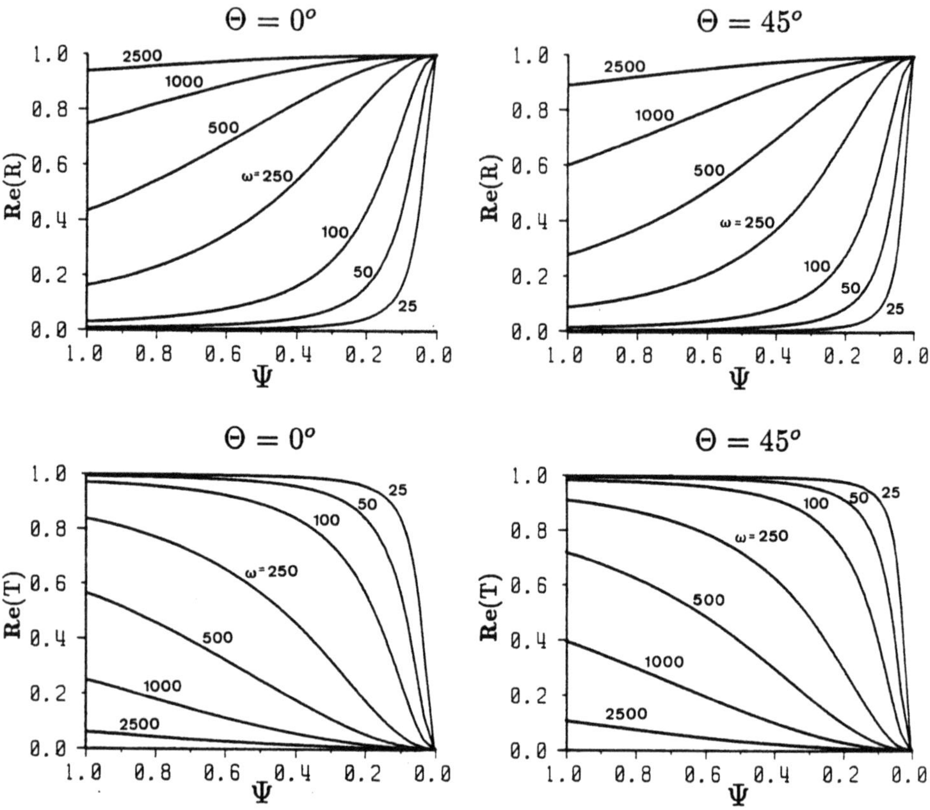

Figure 3: Frequency dependent variation of $R(\Psi)$ and $T(\Psi)$ as a function of damage Ψ for normal and obliquely impinging SH-waves ($\omega = 2\pi f$)

strongly depends on the related load characteristics such as periodicity, frequency, pulse shape, amplitude etc. In percussion drilling the loading exerted by the tool on the rock consists of periodic wave trains as shown in Fig 4. These wave trains (labelled I, II, III in Fig 4) are associated to an input frequency spectrum characteristic of the tool.

The wave trains can be decomposed into a static compressive pre-load exerted from the tool to the rock to ensure positioning of the tool. This preload of possibly varying intensity is superimposed by the sequences of impact blows during the working periods I, II, III etc. The impact blows (percussive blows) are periodic within the wave trains with a repetition time interval of T seconds. Their duration is typically in the order of $\Delta t = 200\mu s$ and their load profile is shown in the blow-up of Fig 4 together with the frequency spectrum. For a typical percussion drill tool operating

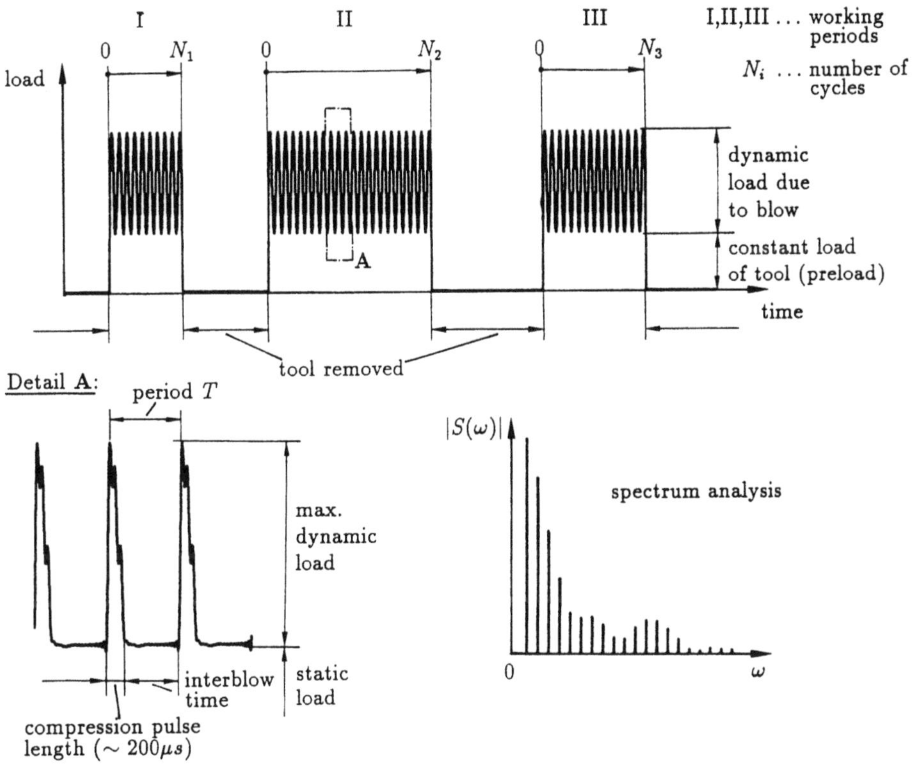

Figure 4: Wave train/pulse and spectrum analysis for a percussion tool in a typical work period

at a repetition rate of $40Hz$ and a load application time of $200\mu s$ the load-time trace would give rise to a sequence of spike-shaped wave pulses as shown in Fig 5. Each one of these spike-shaped pulses carries a SH-pulse component acting on the interface. In this case all the impact-mechanical damage will be done during the spike interaction and, thus, the evolution in time of the damage will occur stepwise.

The development of interface damage can thus be characterized by a cycle description of the form

$$\frac{d\Psi}{dn} = -A\left(\frac{\tau_{inc}}{\Psi}\right)^m \tag{14}$$

where τ_{inc} denotes the peak value of the shear stress spike. This point of view corresponds to cycle-by-cycle damage accumulation where interaction effects between two consecutive load pulses will not be taken into account. This situation obviously holds if the tool hits the rock only once per hammer blow and the pulse length of the

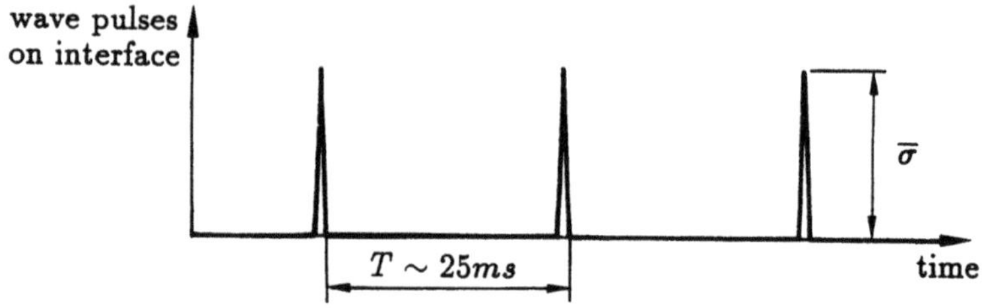

Figure 5: Periodic sequence of pike-shaped wave pulses impinging on an interface

shear stress spike is short as compared to the wave length of the percussion cycle.

It is interesting to note that in a situation where interface damage is also environmentally controlled, the total damage may be represented in a form similar to creep-fatigue,

$$\Psi_{total} = \Psi_{impact} \oplus \Psi_{environment} \tag{15}$$

where the $\oplus$-sign characterizes the interaction of the forms of damage: accumulative, competitive or a hybrid form thereof. A possible form for this type of damage would be given by equ (14) where the coefficients A and possibly also m depend on the environmentally controlled damage, Ψ_{env}:

$$\frac{d\Psi}{dn} = -A(\Psi_{env}) \left\{ \frac{\tau_{inc}}{\Psi} \right\}^{m} \quad \text{and} \quad \frac{d\Psi_{env}}{dt} = -B(\Psi) \left\{ \Psi^{-1} \right\}^{m_e} \tag{16}$$

where the time rate amplitude B may depend on the mechanical degree of degradation of the interface.

On the basis of equ (14) an iterative scheme can be developed for the actual local damage.

5 Conclusion

Progressive evolution of damage at interfaces between rock layers is modeled by combining the fundamental law of continuum damage evolution and constitutive equations for a viscoelastic interface material.

Reflection and transmission coefficients associated with obliquely incident horizontally polarized shear waves (SH) depend on the degree of connectivity of the layers and therefore become time-dependent functions. Obviously, for loose joints (damage $\Psi = 0$) no energy will be transmitted but all will be reflected. There is a strong dependence on the frequency of the incident SH-wave.

Percussion drilling tools are customarily operated at a frequency $f = 40Hz$ and according to our analysis this corresponds to smooth operating conditions with gradual decrease of the interconnectivity of the interface. Generalizations to include environmental effects (e.g. due to joint hydraulics) will lead to more complex systems.

6 Acknowledgement

The first three authors would like to acknowledge financial support of this work by the *Austrian National Science Foundation FWF* under project number *P*6989.

7 References

Bandis, S.C. Lumsolm, A.C. and Barton, N.R. (1983) Fundamentals of rock joint deformation. **Int.J.Rock Mech. Mining Sci. Geomech.Abstr.,** 20(6), 249-268.

Goodman, R.E. (1976) **Methods of Geological Engineering in Discontinuous Rocks.** West, St.Paul, MN, USA.

Hudson (1981) Wave speeds and attenuation of elastic waves in material containing cracks. **Geophys.J.Royal Astron.Soc.,** 64(1), 133-150.

Kachanov, L.M. (1986) **Introduction to continuum damage mechanics.** Martinus Nijhoff Publishers.

Lemaitre, J. and J.L. Chaboche (1985) **Mécanique des matériaux solides.**Dunod, Paris.

Mavko, G.M. and Nur, A. (1979) Wave attenuation in partially saturated rocks. **Geophysics,** 44(2), 161-178.

Ming Cai and Horii, H. (1992) A constitutive model of highly jointed rock masses. **Mechanics of Materials,** 13, 217-246.

Myer, L.R. and Pyrak-Nolte, L.J. and Cook, N.G.W. (1990) Effects of single fractures on seismic wave propagation. **Proc. ISRM Symp.on Rock Joints,** Loen, Norway (eds Barton & Stephansson), Balkema, Rotterdam, 467-473.

Schoenberg (1980) Elastic wave behavior across linear slip interfaces. **J.Acoust. Soc.Am.,** 68(5), 1516-1521.

Walsh, J.B. (1966) Seismic wave attenuation in rock due to friction. **J.Geophys. Res.,** 71(10), 2591-2599.

33 EVALUATION OF BRANCHING BY MARMOR UNDER UNIAXIAL COMPRESSIVE LOADING

K. BAGI, I. BOJTÁR and M. GÁLOS
Technical University of Budapest, Hungary

Abstract
In the analysis of fracture processes of rocks the traditional fracture theories assume the continuity of the material. This involves that the whole internal structure is assumed to be destroyed at the fracture. The reality seems to be different: in uniaxial compression tests fracture is localised in planes bounded by cracks and there are intact parts between them. A physically and geometrically non-linear numerical model was applied to analyse the fracture processin the microstructure which simulates the behavior of two-dimensional assemblies ofperfectly circular grains. Displacements of each individual grainare followed during statical loading processes. Since the model consideres large displacements of the grains, moving of broken parts of the material can also be followed. Laboratory experiments were performed on marmor with crystallic-granular internal structure. Uniaxial stress was applied and increased until the rock was totally destroyed. Both types of experiments showed that load-bearing strips developed during the process and fracture planes arose along them.
Keywords: Rock mechanics, Fracture process, Crushing, Numerical modelling, Uniaxial compression test.

1 Introduction

Uniaxial compression test is a standard method for the qualification of rocks in several countries. In spite of its long history and simple mechanical conditions, the process of fracture and the way of crack opening is still an open problem for the scientists.

The aim of our research is to analyse the fracture mechanism of rocks during uniaxial loading process. The traditional models assume that the material can be represented by an infinitesimal volume element which is homogeneous and usually isotropic. In the experiments the sample is assumed to correspond to the infinitesimal element. It is loaded by uniaxial stress which is continuously increased until the material is totally destroyed; a stress-strain curve is obtained this way which can be used for the qualification and categorisation of the material. Special analysis is needed in case of rough heterogeneous or layered rocks.

The most important result of the uniaxial test is the ultimate compression stress of the material (magnitude of force destroying the

Fracture and Damage of Concrete and Rock – FDCR-2.
Edited by H.P. Rossmanith. ISBN 0 419 18470 8.

sample devided by the orthogonal area; or, according to the energetical approach, the flux of internal energy change caused by the external work which already cannot be dissipated by the structure of the material but has to be dispersed).

Fracture is caused by the internal stress-state of the sample. Two ways of destroying is observed: if the friction at the loaded bottom and top faces is high, shearing planes develope; if the friction is low, tensional cracks will open because of crosswise strains - Everling(1961). Both mechanisms show that the material cannot be assumed homogeneous at fracture: merging craks form fracture planes with intact parts between them.

We considered the material as a three-phase structure - its basis is assured by a petrophysical model, outlined by Gálos and Kertész (1983) - in which minerals and other solid parts (like fossils) give the solid phase, and liquid and gas in the pores are the two other phases.

Behavior of the rock is determined by the texture of the components and (joints) bonds between them. So the texture is considered as the fundamental characteristic in our description. Stresses and strains in the elementary volume were analysed by Wittke (1984) who compared the behavior of homogeneous and discrete models.

In our analysis we modeled the internal structure of the material as a 2D assembly of touching discs (their size corresponded to the size of solid components); the bonds between the particles were replaced by forces and moments between the discs (Figure 1.).

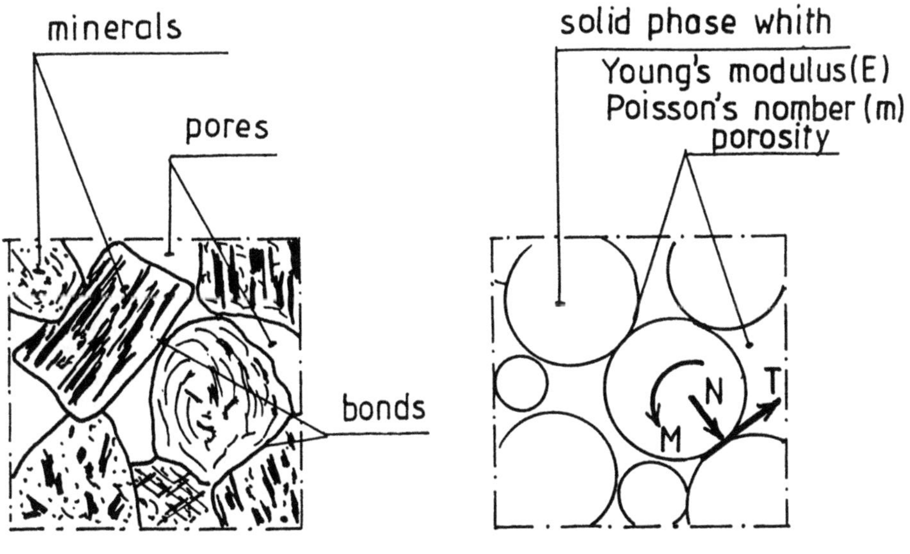

Fig. 1. Structure of rocks in the petrophysical model and in the numerical modelling

Greek marble was used in the laboratory experiments. Marble consists of calcit cristalls as solid components and pores between them so the geometry of the numerical model was defined by the sieve curve of marble components determined under microscope.

The results proved that the continuum assumption is very rough and the global description of the material has to be based on the above introduced fracture mechanism.

2 The numerical model

Numerical modelling is a powerful tool in the research of micromechanics of materials: computer simulations give detailed information on the state of the micro-structure (forces transmitted between the particles; displacements of each individual particle etc.). The models can be built on two different principles: dynamic (Cundall, 1988) or quasi-static (Zubelewicz - Bažant 1987).

Quasi-static processes have a basic role in the engineering practice; but the recently used dynamic models have serious disadvantages when they are applied for finding the equilibrium state of an assembly. Unfortunately the quasi-static models are not very developed yet and the existing ones also have serious drawbacks.

A quasi-static model based on the equilibrium equations of each individual particle will be shortly introduced here (the detailed description can be found elsewhere: Bagi, 1991; Bagi, 1992). This model can exactly find the equilibrated and compatible state of the material; besides, it follows the large displacements, rearrangements and the instabilities of the structure in a natural way.

In contrast with the usual dynamical models this algorithm does not consider any time effects: it searches for the equilibrium state of the material that corresponds to the external loads, regardless of how this state is reached in time.

The basic idea of the algorithm originates from the method generally used in 3rd order calculations of engineering structures. This method was strongly modified and developed to be able to consider topological changes (new and lost contacts) and instabilities (like particles with no contacts, or free groups of particles). The basic unknowns are the displacements (translations and rotations) of the particles caused by the given external loads. First the global equilibrium equations are compiled and solved to get a linear estimation of the displacements. This estimation is then gradually modified according to the instabilities, large displacements, and topological changes. (The exact description of the algorithm can be found in Bagi (1991) and Bagi (1992).)

The recently existing version works in 2D only. Since the effect of change of shape and size of individual particles is negligibly small, the material is modelled as a random assembly of perfectly circular discs with deformable contacts between them.

Figure 2. illustrates how the initial unloaded assembly is created: particles are generated with random size (following any arbitrarily givens sieve curve) at random places on the 'top' of the given rectangular area; then the area is filled up by 'dropping down' the particles. A rather dense assembly of touching grains is generated this way.

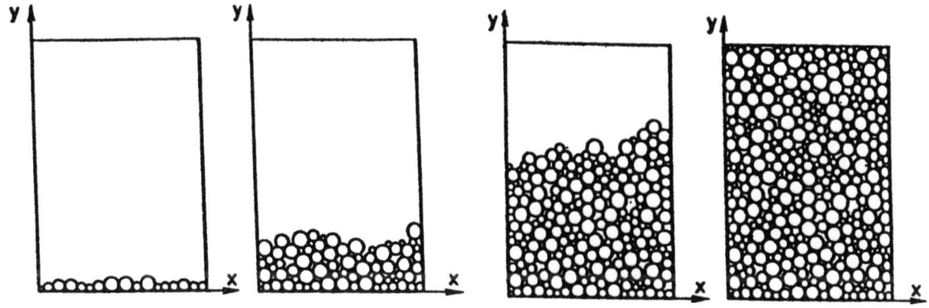

Fig. 2. The creation of the assembly

Four rigid walls bound the assembly; they are also able to move. Any particle or wall can be supported rigidly or by elastic springs. Loads (external forces or prescribed displacements) can act on the walls or directly on the grains. The meaning of the unknowns is shown in Figure 3. (the walls have also the same three degrees of freedom).

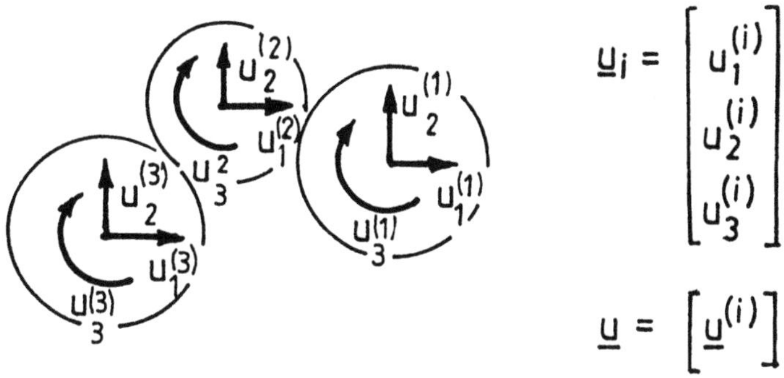

Fig. 3. The displacement variables

The material properties of the bonds are expressed by the micro-material-law of the contacts. The following two types of contacts can be used in the model.

Figure 4. illustrates the frictional contacts: they can transmit negative normal force (compression), and frictional tangential force (the normal stiffness is assumed to be constant and the tangential stiffness depends on the magnitude of normal force). The Coulomb-law limits the magnitude of tangential force. In case of very small ($|N|<N_o$) compression the contacts are still assumed to have resistance to shear; the effect of not perfectly smooth surface is considered this way.

If the magnitude of compression force exceeds the fracture limit of the material of the grains, the smaller grain is 'destroyed', and deleted from the assembly.

Relative displacements at contact 'c':

v_n^c : normal displacements

v_t^c : tangential displacement

v_φ^c : relative rotation

Transmitted forces:

N^c : normal force

T^c : shear force

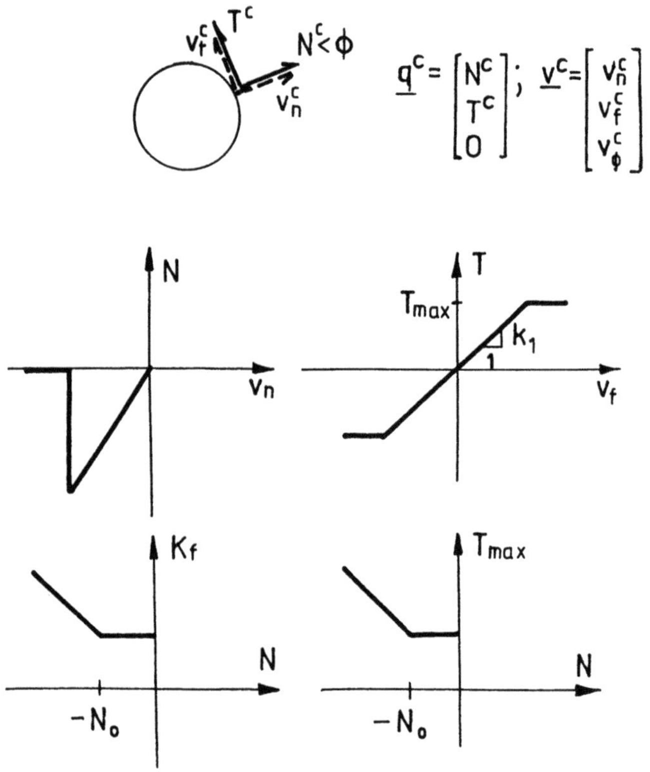

Fig. 4. Frictional contacts

Figure 5. illustrates the micro-material-law of the cemented contacts: they resist compression and tension, tangential force, and bending moment. The tangential and bending stiffness are independent of the normal force. If a contact is broken for tangential force or bending moment, and there is compression (negative normal force) between the two grains, the contact becomes frictional and it will behave as shown in Figure 4.

Relative displacements at contact 'c':

v_n^c : normal displacement

v_t^c : tangential displacement

v_φ^c : relative rotation

Transmitted forces:

N^c : normal force

T^c : shear force

M^c : bending moment

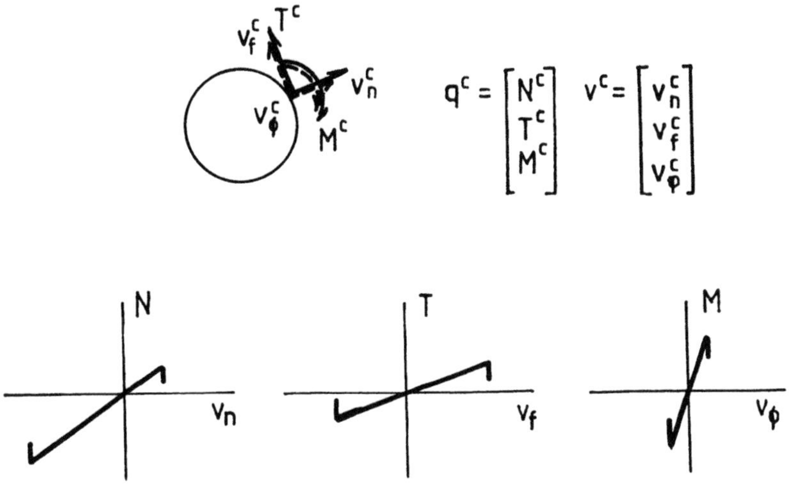

Fig. 5. Cemented contacts

3 Fracture simulation of cemented materials

The usual numerical models that exist for the simulation of concrete -like materials (see for instance - Zubelewicz-Bažant (1987) - follow only the range of small displacements and the contact deletion: they are not able to treat large displacements; instabilities (like small parts of the material broken down and falling out); slip planes; creation of new contacts etc. The main advantage of the presented model is that these phenomena can easily be followed.

In the initial state all the contacts are assumed to be cemented (the micro-material-law is explained in Figure 5.). Contacts are deleted if the fracture criteria are satisfied; new contacts with frictional material law are created if any two grains come into contact.

A simulation test on greek marble is shown in Figures 6.-7. In Figure 6. the initial geometry of the assembly can be seen . The internal structure is a geometrical model of greek marmor. The contact properties were defined in such a way that the global behavior of the model would correspond to the real behavior.

Both walls in the numerical test are frictionless; one of the walls is loaded by prescribed displacements towards the other wall.

$R_{min} : R_{max} = 1:5$,
gravity is zero in the plane of the discs

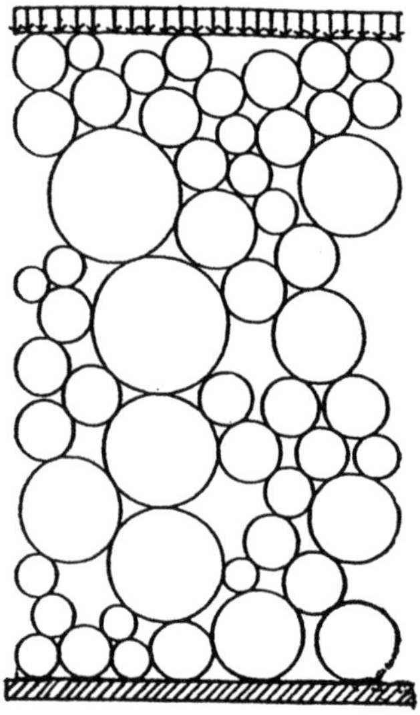

Fig. 6. The initial assembly

Figure 7. illustrates how the internal skeleton changed during the compression. (The internal skeleton is modelled by the replaced graph of the assembly: nodes correspond to the centers of the grains, branches mean the contacts between them: solid and dotted lines represent cemented frictional contacts respectively; F is the load parameter). The sample was destroyed at F = 0.565.

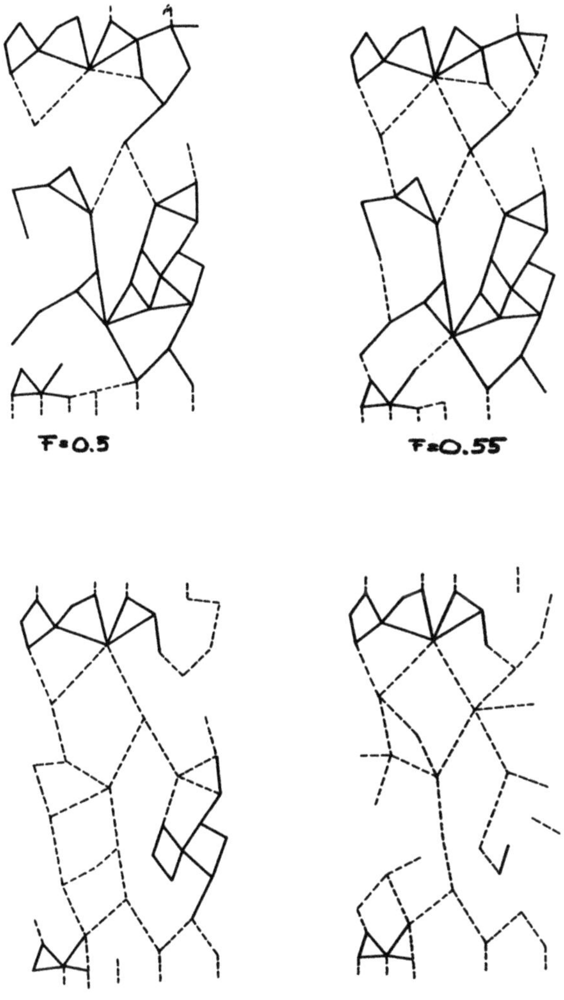

Fig. 7. Changing of the internal structure

4 Laboratory experiments

Samples of greek marble with size 7,8x3,4x6,4 cm were tested. Their front plane was covered with a plastic layer so the developing of stresses during the lodaing process could be followed with the help of polarised light. One example will be introduced here. Figures 8.,9. and 10. show the stress versus strain, the magnitude of load versus time, and he axial strain versus time, respectively.

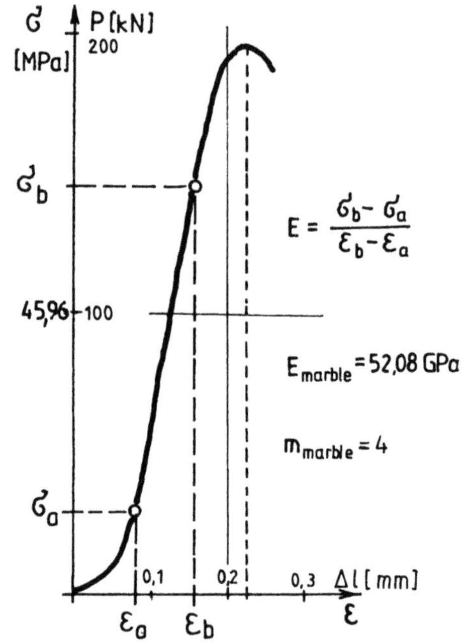

Fig. 8. Stress-strain (load-deformation) of marble (Greek)

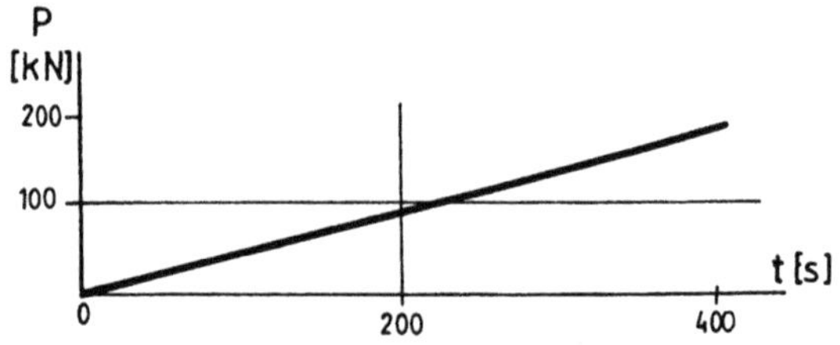

Fig. 9. Load-time diagram

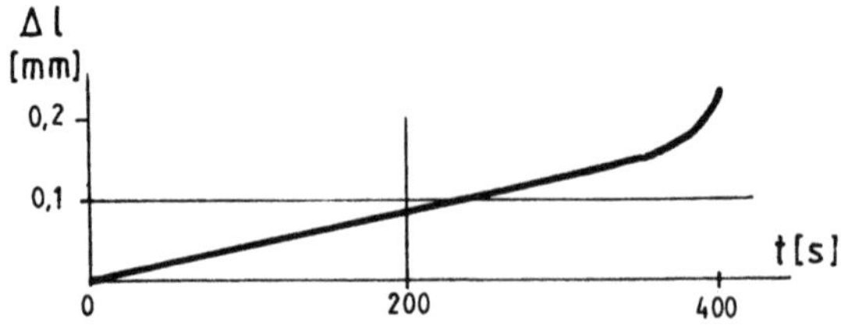

Fig. 10. Deformation-time diagram

Stress distribution in the sample was determined through analysis of photoelastic coating performed by L. Borbás (Technical University of Budapest). Figure 11. shows the results.

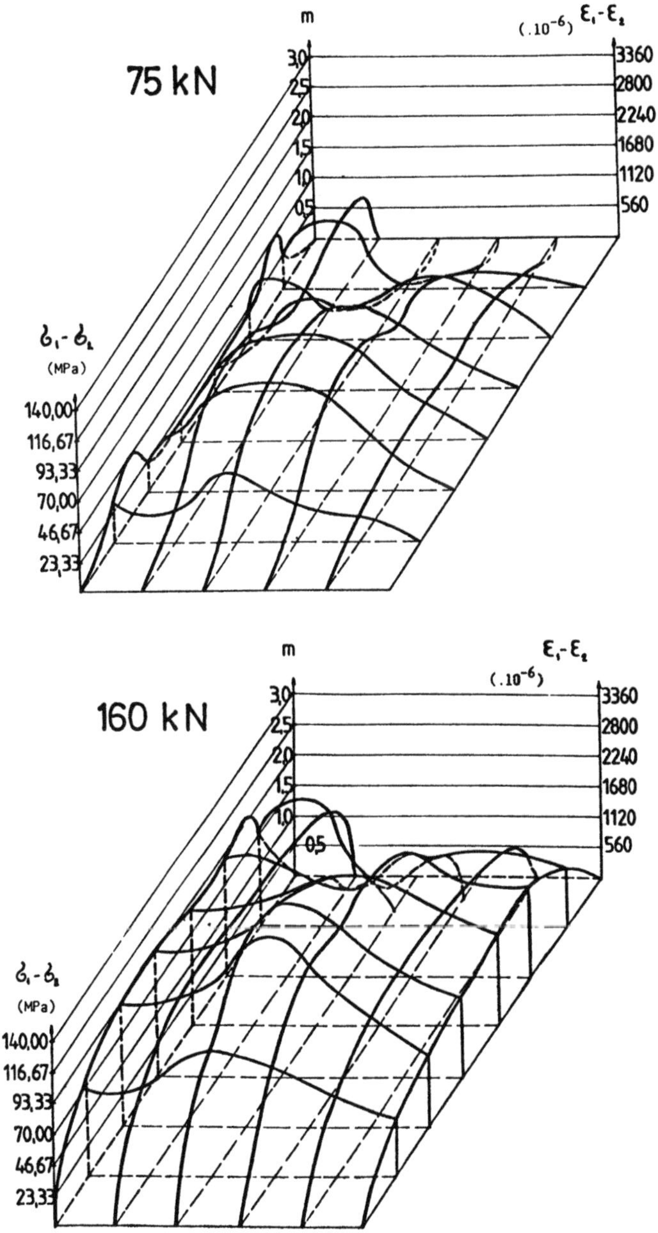

Fig. 11. Stress distribution in the sample

5 Conclusions

The introduced numerical model is a good basis for the micro-mechanical analysis of fracture process in rocks. The internal structure of the material is in volved in the model so the simulated behavior correspondes to the reality. Our results proved that in cont vast with the theoretical assumption of isotropic behaviour, discrete fracture planes develope from the merging cracks and this process leads to the distortion.

The effect of the internal structure can be evaluated through numerical fracture analysis of samples with different textures. Effects of the properties of solid components, pores and bonds can also be determined. The method could serve as a basis for energetic approach as well.

By changing the boundary conditions the method can be used, of course, not only for uniaxial tests but also for any other usual experiments (triaxial, shearing etc.) where the fracture process can be analysed.

6 References

Bagi, K. (1991) Numerical analysis of granular assemblies. Doctorial thesis, Technical University of Budapest (in Hungarian)

Bagi, K. (1992) A quasi-statistic numerical model for the micro-level analysis of granular assemblies. Mechanics of Materials (in press)

Cundall, P.A. (1988) Computer simulation of dense sphere assemblies. Micro- mechanics of Granular Materials, ed. M. Stake and J.T. Jenkins, pp 113-123, Elsevier

Everling, E. (1961) Das Endflächenproblem beim Druckversuch. Bergbau--Archiv 22. Heft 1. pp 27-37

Gálos, M. - Kertész P. (1983) Results of rock mechanics in Hungary - An engineering geological model of Rock Mechanics. Melbourne, pp F 259-F 263

Wittke, W. (1984) Felsmechanik. Grundlagen für Wirtschaftliches Bauen im Fels. Springer-Verlag, p 1050

Zubelewicz, A. and Bažant, Z. P.(1987) Interface Element Modeling of Fracture in Aggregate Composites. Journal of Engineering Mechanics, Vol. 113. No 11. pp 1619-1630

34 PHYSICAL AND NUMERICAL MODELLING OF SLIDING MODE CRACKS IN ROCK

W. STEGER and R. POISEL
Institute of Geology, Technical University of Vienna, Austria

Abstract
Rocks of medium strength form shear fractures when their compressive strength has been exceeded. Such Mohr-Coulomb type, discrete shear fractures often determine the stability of rock engineering structures (e.g., foundations, tunnels). Plasticity theory, however, cannot model these phenomena and particularly discontinuities of the displacement field properly. Discontinuum and fracture mechanical approaches become necessary. Numerical models of discrete fractures are compared with the results of eccentric compression tests of rock. Thus the influence of shear fractures on the stability of rock can be caught much closer to reality.
Keywords: Shear Fractures, Numerical Modelling

1 Introduction

Shear fractures are the most important features when the strength of rock is exceeded. They dominate the stability of rock slopes, dam foundations and tunnels. They can be observed in compression test of rock of medium strength and be described mathematically by Mohr-Coulomb's theory. Their origin and set-up can be studied by plasticity theory. E.g., Cundall (1990) showed that plasticity theory can model their propagation and arrest (Fig. 1a).

On the other hand shear fractures may be regarded as Mode-II fractures in the sense of fracture mechanics. It can be shown that the direction of a plastic fracture and of a Mode-II fracture with max. K_{II} and Coulomb friction inside the crack coincide (Poisel, 1979; Melin, 1986).

Observing shear fractures in rock under inhomogenous stress it becomes evident that shear fractures may propagate as described by plasticity theory (as K_{II} fractures, Fig. 1b) as well as tensile fracture may be formed, which is in common called kinking (Fig. 1c).

The decision whether a crack grows by extending into shear direction or by kinking is very important, when engineering measures have to be designed. This shall be out-

Fracture and Damage of Concrete and Rock – FDCR-2.
Edited by H.P. Rossmanith. © 1993 Taylor & Francis. ISBN 0 419 18470 8.

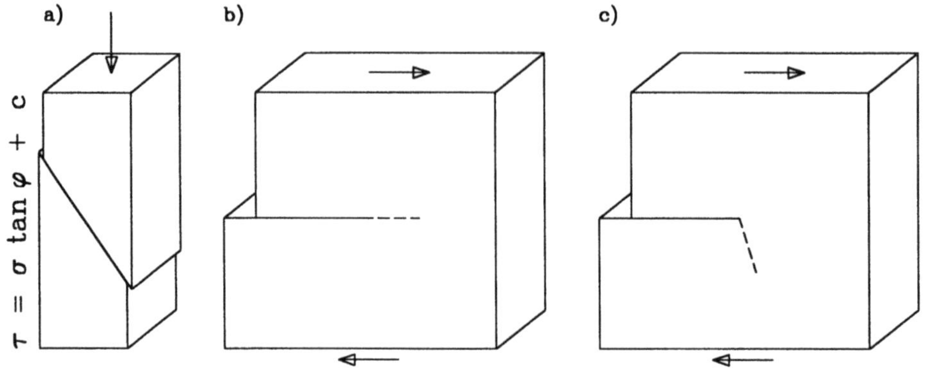

Fig 1: Shear fractures in rock specimen
a) Mohr Coulomb fracture under uniaxial loading
b) Crack propagating in shear direction
c) Mode-II crack

lined for the example of shear fractures occurring around tunnels.

2 Shear and tensile fractures around tunnels

The mechanisms of failure occurring typically around tunnels are often described by plastic continuum theory. How-

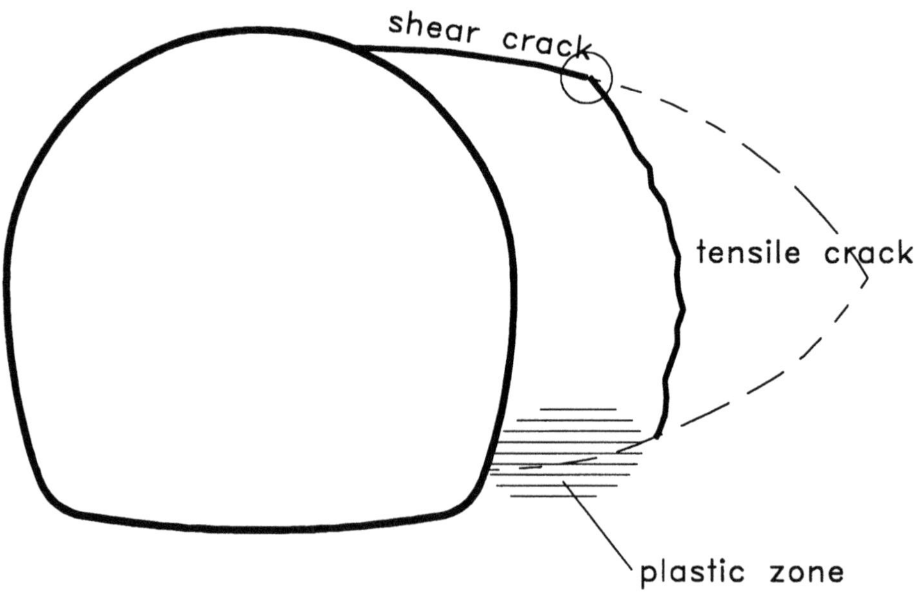

Fig. 2: Failure Mode around a deep tunnel.

ever, at least in rather brittle rock it becomes evident that continuum mechanics does not reproduce correct behaviour because of discrete shear failures , which are beyond the limits of continuum mechanics. This fact is pointed out clearly by the ISRM-report (1989).

A typical failure around a deep tunnel in rock is depicted in Fig 2. This kind of failure was observed in Austrian tunnels like the Arlbergtunnel and the Tauerntunnel. A shear failure extends from the tunnel boundary inside the rock, in some place turns into a tensile crack delimiting the rock body at the back. At the foot of the rock body probably plastic deformations occur.

The accurate investigation of this failure mechanism should give better insight into the working of supporting measures. Especially the influence of rock bolts crossing the tensile cracks, which should work as tensile reinforcement, could be much better understood.

To investigate the problem of this kind of tunnel failure a series of laboratory experiments were carried out. The fracture set-up observed in these experiments was modelled in Finite-Element calculations.

3 Model experiments

The model experiments were designed as eccentrically loaded uniaxial compression tests to reproduce the most important features of the stress distribution around a

Fig. 3: Typical specimen after the experiment. Harder stressed edge left
1) shear crack
2) starting kinks

tunnel: the concentration of tangential stress and the absence of radial stresses. The experiments were carried out on phyllite specimen with the dimensions of (ca.) 11.4x10x5cm (hxbxd). The load platens were designed carefully to avoid frictional effects. This was achieved by using a material with the same E/ν ratio. Thus in the initially intact specimen a linear stress distribution can be assumed. Fig. 3 shows a typical specimen after the experiment: The experiment was stopped after kinks had formed.

All specimen showed the same behaviour of macroscopic cracks:

- At first, the eccentric loaded prisms show a shear fracture originated at the harder stressed edge.
- After propagating as a shear fracture for a certain length a tension crack more or less parallel to the loading direction is formed.
- This kink grows until it hits another shear fracture and a trapezoid body is formed.

A closer look at the kinks shows, that they start out into a direction of about 70° from the shear crack and then turn into a direction parallel to the unloaded edge. This is in good agreement with fracture mechanical considerations.

4 Finite Element Calculations

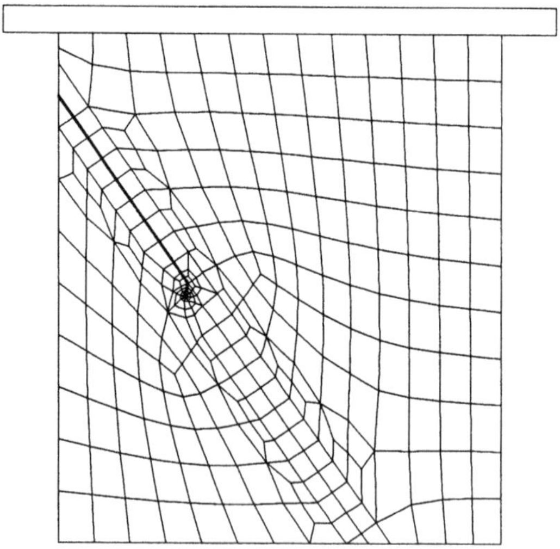

Fig. 4: Finite-Element mesh
Set-up with shear crack and kink
— crack and kink

4.1 Prerequisites

In order to interpret the experimental data and to gain more insight into the stress distribution inside the model samples a series of Finite-Element calculations was carried out. The calculations were carried out on a HP 9000/835 mini computer using the general purpose Finite-Element package MARC and the pre- and postprocessor MENTAT. Fig. 4 shows one of the Finite Element meshes used in the calculations. It consists of 300 8-node plane stress elements and 1018 nodal points.

The load was applied by a rigid load platen at the top. The shear crack was simulated using Gap-Friction Elements which allow accurate simulation of opening and shear friction. Kinks were introduced by simple discontinuities in the mesh, as they always opened immediately.

The well-known quarter-point method was used to introduce the crack-tip singularity. The following formulae were used to evaluate K_I and K_{II} from crack-tip elements (From Linsbauer, 1985; See Fig. 5 for node numbers):

$$K_I = \left(\frac{2\pi}{L}\right)^{1/2} \frac{G}{\kappa+1} \left[4(v_4 - v_2) - (v_3 - v_5)\right] \quad (1)$$

$$K_{II} = \left(\frac{2\pi}{L}\right)^{1/2} \frac{G}{\kappa+1} \left[4(u_4 - u_2) - (u_3 - u_5)\right] \quad (2)$$

Where G is the shear modulus of the material and

$$\kappa = \frac{3-\nu}{1+\nu} \quad \text{for plane stress} \quad (3)$$

$$\kappa = 1-4\nu \quad \text{for plane strain} \quad (4)$$

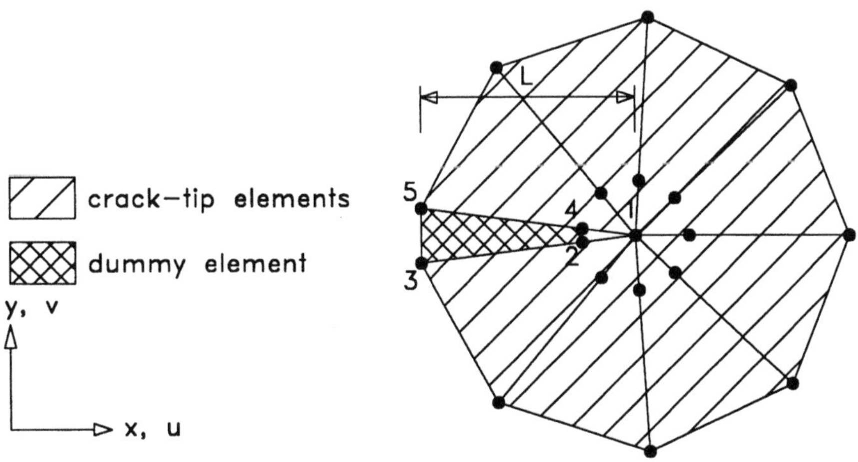

Fig. 5: Crack-tip and dummy elements.

In order to automatically collect the data needed for the above formulae, a user-defined dummy element was used. This element had no stiffness and loads, so it didn't disturb the calculations. The use of this element gave quick access to nodal coordinates and displacements, so K_I and K_{II} could be calculated during analysis and did not need further evaluations.

4.2 Calculation set-up 1: only shear crack present

Of course Finite-Element calculations of perfect (uncracked) eccentrically loaded prisms give a stress distribution which varies linearly with the width of the prism and which is constant over the height. So the initial crack position was selected according to the experiments. The crack direction was chosen according to Mohr-Coulomb's theory as 45-φ/2 from the direction of maximal compression stress (this is also the direction of maximal K_{II}). A friction angle of 20° and a cohesion of 0,4 kN/cm² were assumed for the joints.

The calculations for different lengths always gave pure Mode-II conditions for the crack tip. This is shown clearly in Fig. 6 (a typical Mode-II fringe pattern,

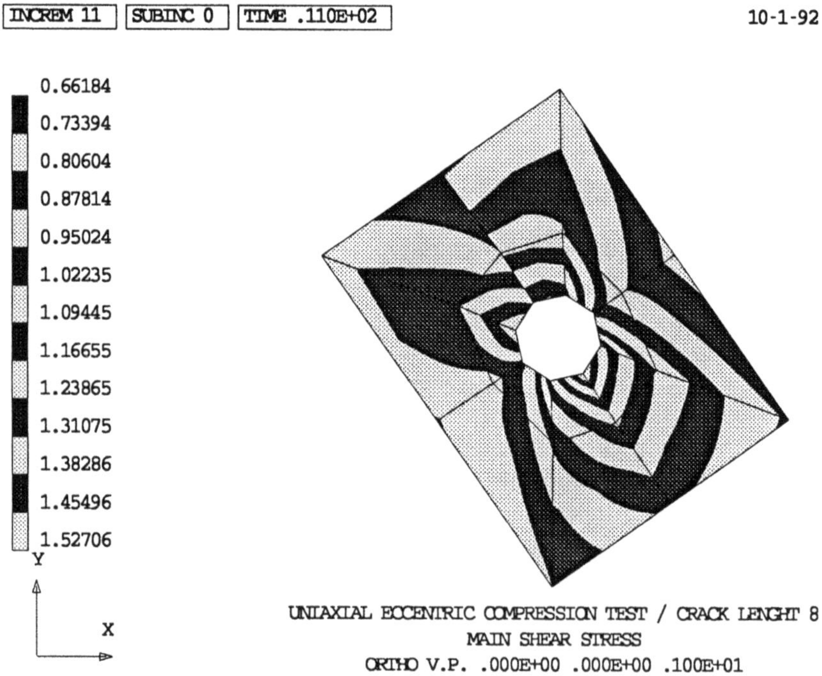

Fig. 6: Contours of main shear stress near crack-tip (fringe pattern) showing Mode-II conditions with an overlying far field.

comp. Rossmanith, 1979).

Fig. 7 shows the variation of K_{II}, Mohr-Coulomb flow function F and max. tangential stress σ_θ (the latter two calculated at about 1 mm distance from the crack tip) with the crack length. While K_{II} and F are rather proportional and both have a maximum at a certain crack length, σ_θ increases further and has it's maximum at a longer crack length.

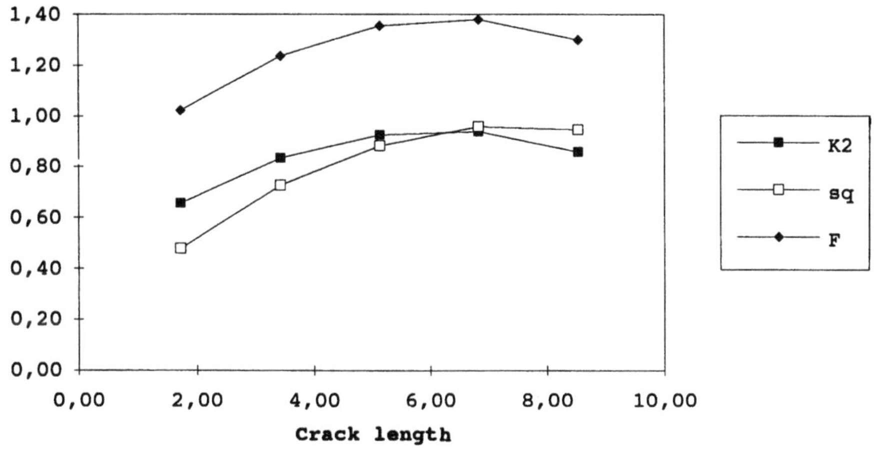

Fig. 7: Values of K_{II}, Mohr-Coulomb flow-function F and max. tangential stress σ_θ for different shear crack lengths

4.3 Calculation set-up 2: shear crack and kink present. As a next step the behaviour of specimen with a shear fracture and a kink was investigated where the kink angle was varied (comp. Fig. 4).

Fig. 9 shows a typical fringe pattern for such a calculation set-up. Fig. 8 hows K_I, K_{II} and max. tangential stress σ_θ for different kink angles. The kinks show mixed Mode behaviour, where K_I has a maximum at about 70-80° kink angle. Looking at the distribution of σ_θ around the kink tip (Fig. 10) it becomes obvious that for longer lengths of the kink it will eventually turn parallel to the unstressed edge. This is in good agreement with the experiments and fracture mechanical considerations.

5 Conclusion

The main purpose of these investigations was to find out if there are criteria for the decision whether a given

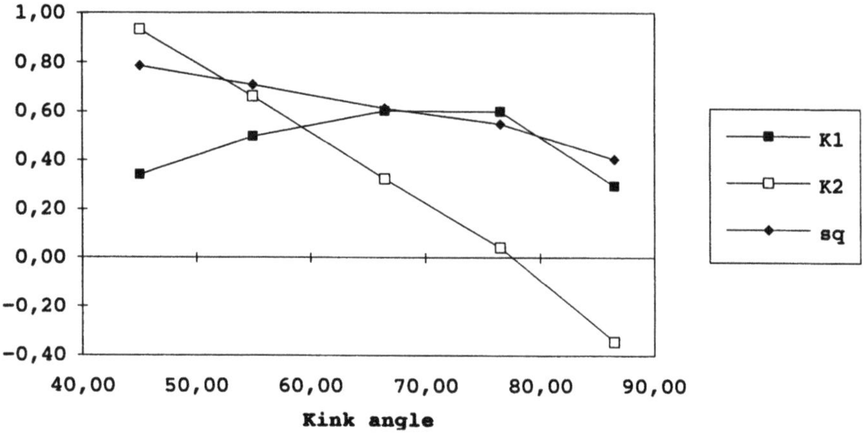

Fig. 8: Values of K_{II}, Mohr-Coulomb flow-function F and max. tangential stress σ_θ for different kink angles.

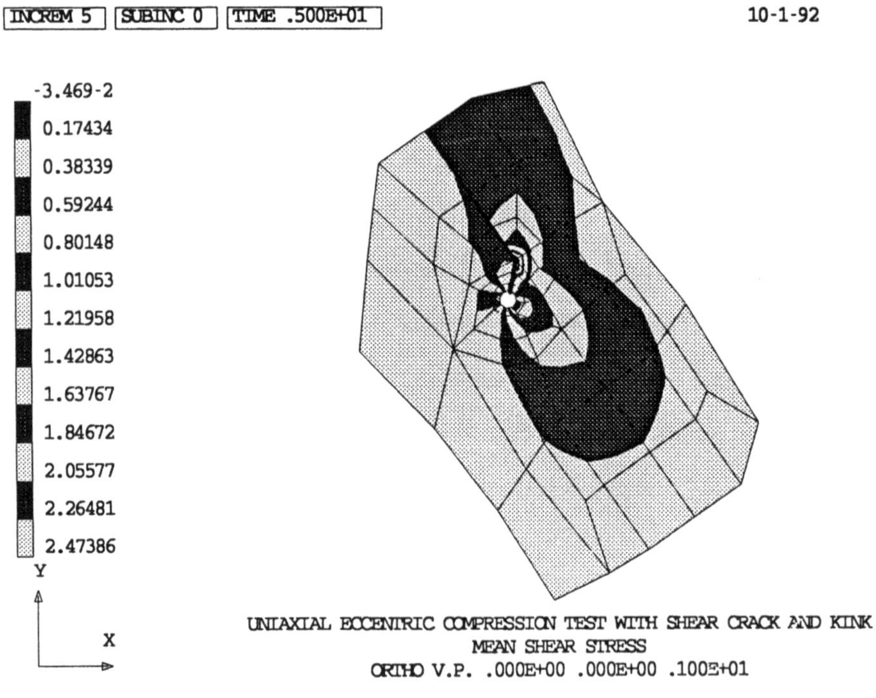

Fig. 9: Contours of main shear stress near crack-tip (fringe pattern) for sample with kink

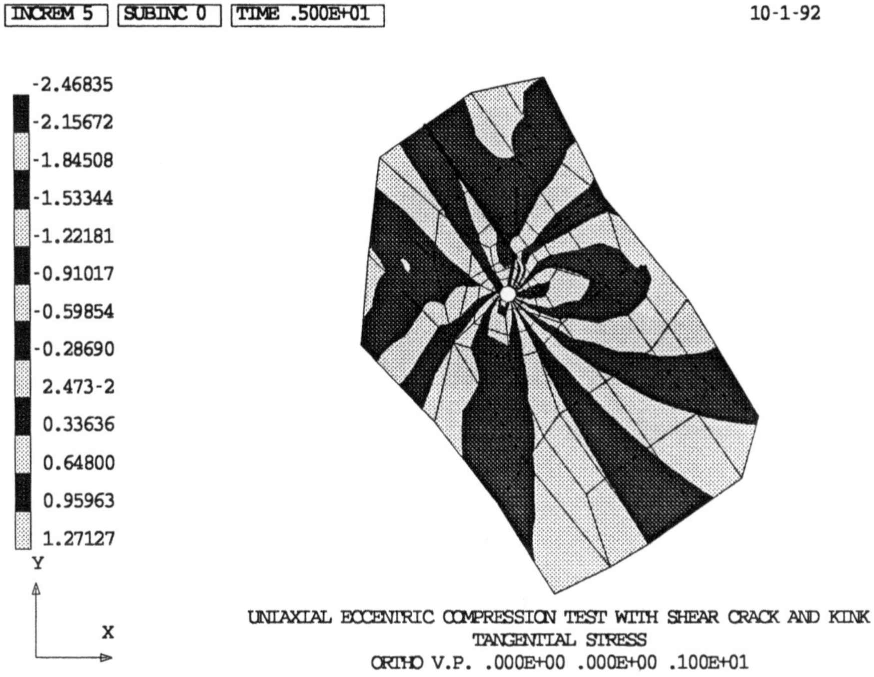

Fig. 10: Contours of tangential stress near crack-tip for sample with kink

crack will continue to grow in shear direction or it will form a kink under the experimental conditions and further under the conditions near a tunnel like that in Fig. 2. In the laboratory test the crack growth into shear direction could be observed.

The conditions for crack growth in shear direction are not very clearly described in the literature. However, a positive ratio of σ_1/σ_3 seems to be necessary to inhibit the growth of a kink (e.g., Petit and Barquin, 1988).

Melin (1986) assumed that K_{II} for the straight crack is decisive for shear crack growth , while K_I for an 'infinitesimal' kink is decisive for kink growth, both in relation to critical values K_{IIc} and K_{Ic}.

He gave a criterion by calculating $\kappa = K_{II}/K_I$ and showed that this ratio decreases with the value of p/τ, where $p = (\sigma_1+\sigma_3)/2$. As this criterion is valid only for the "far-field" stress, it cannot be used in the situation of the eccentrically compressed, cracked rock prism. However, comparing the stress fields around the crack tips of the Finite-Element calculations without a kink, it seems that the overall normal stress level decreases with crack length, while the shear stress increases. This is a hint for a correspondence of Melin's criterion and the test observations on principle. But the decrease in the normal stress level is low, and this agrees with the observation that the point, where the crack stops growing into shear direction, cannot be reproduced exactly in the model tests.

As the results of model tests and FE-calculations coincide, it is shown that fracture mechanics approaches are suitable and necessary for the exact investigation of rock failure.

6 Acknowledgement

The authors would like to thank the Austrian Funds for Research Promotion (Forschungsförderungsfonds, FFF) and the Allg. Baugesellschaft PORR A.G. for their invaluable support.

7 References

Cundall, P.A. (1990) Numerical modelling of jointed and faulted rock, in **Mechanics of jointed and faulted rock,** (ed. H.P. Rossmanith), A.A. Balkema, Rotterdam, pp. 11-18.

ISRM Commission on failure mechanisms around underground excavations (1989) First report: observations, researches and recent results about failure mechanisms around single galleries.

Linsbauer, H. (1987) **Das Tragverhalten von Betonbauwerken des konstruktiven Wasserbaues - Einfluß von Rißbildungen**. Inst. f. konstruktiven Wasserbau, Technische Univ. Wien, Bericht Nr. 21.

Melin, S. (1986) When does a crack grow under Mode-II conditions ? **Int. Jour. of Fracture**, 30, 103-114.

Petit, J.-P. and Barquins, M. (1988) Can a natural fault propagate under Mode-II conditions ? **Tectonics**, Vol. 7, No. 6, 1243-1256.

Poisel, R. (1979) Ein Beitrag zur Mechanik des Gebirges um einen tiefliegenden Hohlraum im Post-Failure-Bereich. **Rock Mechanics**, 12, 61-76

Rossmanith, H.P. (1979) Analysis of mixed-mode crack-tip fringe patterns. **Acta Mechanica**, 34, 1-38.

35 A NEW PRINCIPLE OF EFFECTIVE DESTRUCTION OF ROCK

N.V. NOVIKOV, I.A. SVESHNIKOV
and A.L. MAISTRENKO
Institute for Superhard Materials, Ukrainian Academy of Sciences, Kiev, Ukraine

Abstract
Results are given of the solving of typical problems on interaction between cutters of intricated shape and rocks which gives a system of cracks and on the control of their paths due to the variation of loading conditions. This formulation of the structural strength of a cutter itself based on the criteria of limiting state of cemented carbides with due account of defects stored and variation of a working profile becouse of wear, and needed crack paths in rocks as well. Based on the models developed the computer-aided realization of a numerical experiment was possible using a variety of limiting conditions. The results are shown graphically. The adequacy of the fracture model is verified experimentally.
Keywords: Rock Block, Bench, Cutter, Crack, Path, Limiting Conditions, Crack Detour.

1 Introduction

Recent years have seen the development of highly efficient methods to destruct rocks with the aim both to mine mineral resources and tunneling. The methods range from the classical mechanical one to hydrodynamic, shock- wave, thermal methods and their modification. Let's analyze one of the most-used methods, i.e. mechanical destructuion, which depending on its application undergoes substantial changes, the main feature - local contact interaction between an operating element of a tool and rock - being preserved. Theoretical models were based, in some measure, on hertzian concepts and then were refined in a number of solutions of contact problems of the deformable-solid mechanics in a variety of formulations both for a singgle indentor and a system of indentors of various shapes and straining conditions. A number of transformations and the introduction of the present high-strength and wear resistant tool materials and a powerful drive have made it possible to reduce the mechanical method of rock destruction to one of widely used schemes of a

Fracture and Damage of Concrete and Rock – FDCR-2.
Edited by H.P. Rossmanith. © 1993 Taylor & Francis. ISBN 0 419 18470 8.

step-by-step cutting of a rock block using a system of cutters (Fig.1a). Locally-applied stresses produce in a tool-rock contact zone (Fig.1b) a zone of a real removal of rock (1) and a zone of rock damaging to form a crack system (2). At the same time, due to a specific feature of the stressed state being realized, which stems because of the influence of the side wall of a bench, extension of brittle failure cracks in observed (3). The volume of the split off rock is much higher than that of the damaged rock directly at the contact zone. Just this excess of the split off rock volume the damaged rock volume is the essence of the efficiency of the use of the energy being consumed, which goes not only into irreversible dispersion processes at the damage zone and initiation of a crack system, but also both into the use of the effect of spontaneous crack propagation, which attain a critical size for a given stress state, and into the subsequent crack propagation at the expense of elastic energy stored in the tool-rock block system. However, it is evident that even with the simplicity of the realization of the method by the use of production-type tools it has fundamental limitations, which depend on the ratio of the distance between the load-application point and the bench edge in a rock block to the rock block height. As the zone size does not vary in proportion to the applied load, the wanted effect of spontaneous crack propogation along the shearing path will cease to work at some values of the ratio. Altiero (et. al., 1975) have shown that the ratio can attain 2.5.

The further rise of efficiency of the mechanical method under condition is possible by the use of an untraditional scheme of tool action on rock to form directly the stress state at which larger volumes of rock can be split off.

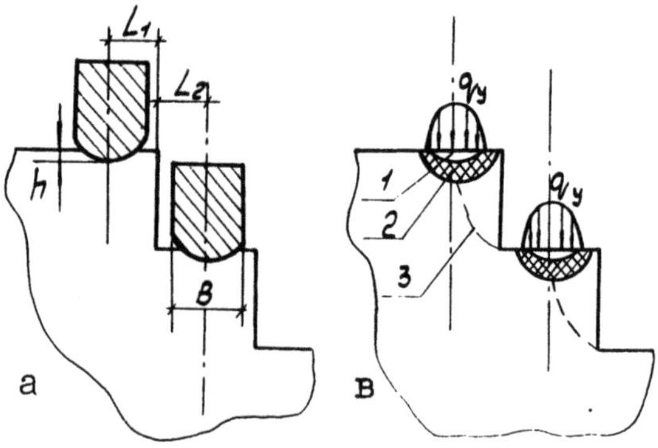

Fig.1. Schematic diagram showing a conventional way of bench-type destraction of rock
a-distribution of cutters on benches;
b-boundary conditions: 1-dispersed rock zone,1-damaged rock zone, 3-path of rock spalling.

A scheme of loading with a tool not at the upper bench edge but under it (Fig.2a) has been proposed at the Institute for Superhard Materials. In this case the tool itself made of cemented carbide is an unsymmetrical wedge, which indentation into rock allows the formation of distributed normal (qx) and shearing (qy) components of the applied load(Fig.2b). The tool cutting-in is accompanied by the formation of a median crack and a hertzian crack, the latter propogates by the mechanism of mixed fracture , i.e. by the combined mechanism of normal fracture (K_{Ic}) and antiplane shear load (K_{IIc}). Once formed, an initial crack does not propagate. The hertzian crack is directed at an angle Θ_0=20...25° with respect to the surface, and with a specific Px/Py ratio and the bench block height one should expect the crack turning with respect to its initial direction as well as its spontaneous extension to come to the surface (Fig.2b and 3).

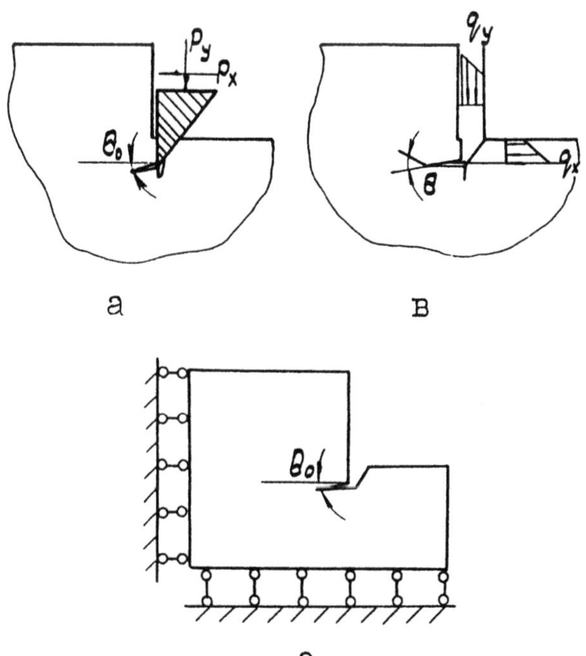

Fig.2. Schematic diagram showing the realization of a new way of rock bench breaking
a-the catting-in of an asymmetrical wedge;
b-boundary conditions; c- calculation model.

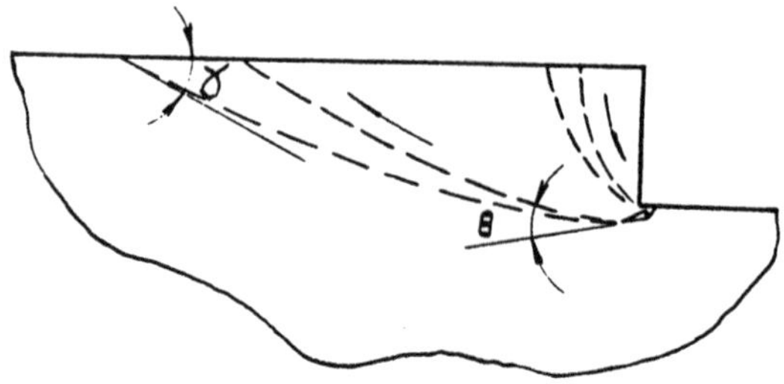

Fig.3. A comparison of bench spalling paths constructed for the two ways of rock breaking.

Thus, a definite Px/Py ratio value has to be fitted for each specific case (rock properties and a bench height), which allows the crack extension at an angle $\Theta > 0$ with respect to the initiated crack direction (Fig.2c) An angle at which the crack will turn is derived from a known relation by Cherepanov (1974)

$$\Theta = 2 \operatorname{arctg} [(1 - \sqrt{1 + 18 n^2}) / 4n] \quad (1)$$

where $n = K_{II} / K_I$, and the limiting equilibrium of the rock block having a growing crack is derived based on a critical stress intensity factor from the condition $K_{ef} > K_{1c}$, where

$$K_{ef} = \cos \frac{\theta}{2} [K_I \cos \frac{\theta}{2} - 3 K_{II} \sin \frac{\theta}{2}] \quad (2)$$

here, Kef is the value of the effective stress intensity factor in case of the mixed fracture mechanism.

2 Results and Discussion

A singular problem under consideration was computer-aided solved by finite element method. A two-dimensional model for calculation of a bench and a scheme for boundary-condition stipulation are given in Fig.2. For the purpose of illustration. Alghet litholimestone having the properties stated below, was taken: elastic modulus E = 30 GPa; Poisson's ratio $\nu = 0.35$; fracture toughness in normal tear $K_{Ic} = 0.55$ MPa$\sqrt{m}$, that in shear $K_{IIc} = 0.47$ MPa$\sqrt{m}$; an initial crack length is 4 mm; an angle of a crack inclination $\Theta_o = 20^\circ$; Py = 6300 N; Px= 700 N. The problem was solved using an elastic two-dimensional formulation. The model region was devided into 700 triangular elements. Thus, we have consider only a section of the fracture front, which propagates in a rock block, as the analysis of a complete picture has to be done in a volume formulation, because of the fact, that a real picture of fracture is more complicated and the modelling of the picture may be of the interest as a separate study. The calculation procedure was as follows. At the initial crack tip K_I and K_{II} values were calculated based on which an angle Θ of the crack reorientation with respect to an initial surface was found from Eq.(1). Then using a triangulation grid of fine elements a the breaking of a bond was simulated over the length of 1 mm at a corresponding Θ angle with respect to an initial crack and the calculations were recommenced. This resulted in a construction of a piecewise linear path of a crack front up to its coming to the surface. In spite of important assumptions which simplify an accepted model for calculations, this approach to the construction of a possible shear path has a more real logic basis than the approaches to the construction of calculated shear paths, which are predicted from fracture function lines which corresponds to one of the continuous medium strength theories. From our point of view, the approach which is based on solution of fracture mechanics problems with account of singularity is proved to be more correct as it accounts for the stress redistribution in a rock block as a crack propagates. Moreover, on a calculated crack path the coordinates were defined when the condition $K_{ef} < K_{Ic}$ was fulfilled, this corresponded to the attainment of a critical size, followed by a period of the crack spontaneous extention.

The calculations permitted the construction of crack propagation paths in a rock block in the bench heigth range 3 - 30 mm. It is found that an angle of a crack coming to the surface $\Theta = 30...40^\circ$ (Fig.3), which is in a good agree- ment with experimental data. The optimal height of the rock bench in our case is 20 mm, as its further increase does not contribute to an essential increase in a volume of the rock spalled even under high loads. The optimal P_y and P_x values at which a crack tend to the surface of a rock block were found to be 3000 and 350 N, respectively. Below these values the crack will extent deep into the rock block, i.e. $\Theta = 0$, and the spalling of the rock block will not occur in the case.

And, finally, the section area of the rock bench spalled and, consequently, its volume, produced by the procedure proposed, will be no less than 1.3 times that produced by a known procedure (spalling from the upper shelf of a rock bench) and this elevation may attain 3.7 times (Fig.3). This potentials do not leave practical wor- kers indifferent, and therefore, a special carbide-tipped asymmetric cutting tool was developed at ISM, which allows to apply the system of distributed normal (qx) and shearing (qy) loads at the contact zone under the bench basis. The application of the tool has permitted the implemention of the idea proposed. The application of this tool in combination with a common-used one lowers the power - intensity of the bench bottom hole destruction by a factor of 7 - 12. This results provides the basis for the de- velopment of new highly efficient rock-destruction tools.

3 References

Altiero, N.J. and Sikarskie, D.L. (1975) Fracture initiation in elastic brittle materials having non-linear fracture envelopes. Inter. J. Fracture, 11, 431-440.

Cherepanov, G.P.(1974) Brittle Fracture Mechanics, Moscow, Nauka.

36 INFORMATIONAL MODEL OF ROCK DESTRUCTION AND THE PRINCIPLE OF MINING TOOL IMPROVEMENT

L.L. MISHNAEVSKY Jn
Institute for Superhard Materials, Ukrainian Academy of Sciences, Kiev, Ukraine

Abstract
The mathematical model of hard rock destruction at drilling which is based on the information, percolation and probability theories methods was developped. The model allows to take into account the influence of the crack systems and tool shape on the destruction process and can be used as a theoretical basis for mining tool improvement.The general principle of drilling tool design was formulated.
Keywords:Rock Destruction,Drilling Tool Design,Cracking, Informational Model.

1 Introduction

The purpose of the investigation is a development of a mathematical model of rock disruption under drilling, which can be used as a theoretical basis for mining tool improvement. The rocks contain a great number of cracks,pores,and other stress concentrators.Because of interference of stress fields of all stress concentrators the stress in a point becomes a random value. Hense,the stress state in hard rocks may be described by probability methods. The tool form influences on the rock destruction process through the unevenness of contact stress distribution on the contact surface.To characterize the contact stress distribution unevenness we suggest to use the informational entropy of contact stress distribution.
Based on the conception about cracking and crack growth as the joining of defects we may use the percolation theory approach to describe a cracking. Using the methods of mathematical modelling of rock destruction we can analyze the possibilities to increase the drilling efficiency and to improve mining tools.

2 Information theory approach for description of tool shape

2.1 Informational entropy of contact stress distribution

Owing to the roughness of the tool and rock contact surface the contact stress in point contains a great random component,and so the contact stress distribution law may be presented as a probability function:

$$p(\sigma k) = \frac{d}{S_K \, d\sigma_K} \iint_{(S_0)} dS_K \qquad (1)$$

Fracture and Damage of Concrete and Rock – FDCR-2.
Edited by H.P. Rossmanith. © 1993 Taylor & Francis. ISBN 0 419 18470 8.

where $p(\sigma k)$ is the probability density of the value σk, $S k$ is contact area, σk is contact stress, S_o is the part of contact area,in which contact stress is bigger than given value σk.

The unevenness of contact stress distribution can be characterized by the informational entropy of distribution (1):

$$Hk = - \int_{\sigma_K} p(\sigma k) \ln p(\sigma k) d\sigma k \qquad (2)$$

2.2 Comparison of indentors shape on the basis of contact stress entropy

The contact stress entropies for indentation with different indentors were determined.The formulas for the contact stress distribution laws for loading with the spherical indentor,cylindrical and conical indentors are given in the book of Zhlobinsky (1990).The formulas were substituted in (1),and then in (2).The value σk was made discrete. The integral in (2) was transformed into the sum.In the table 1 the contact stress entropy values for the cases are given.

Table 1. Informational entropies of contact stresses under loading with diverse indentors

Indentor's form	Contact stress entropy	Volume of rock chipped out at pressing into limestone (cmm) (from book of Zhlobinsky,(1990))
Cylindrical	0.47	69.0
Conical	0.58	86.8
Spherical	0.32	43.3

Also the experimental data of volume of rock chipped out at the pressing by diverse form indentors into limestone (from the book of Zhlobinsky (1990)) are given. On can conclude from the table that the contact stress entropy characterizes the indentor form and it's performance.

3 Percolation theory approach to cracking modelling

The cracking can be considered as the result of defects joining. If the microcracks density is sufficient,the cracking takes place under the low load.So the cracking can be considered as a percolation process. Rock massif is assumed to be medium with many defects evenly distributed throughout the volume or as a great number of some cells.Here we understand "defects" as the smallest microcracks which merge under the most possible stresses at drilling (Sveshnikov and Mishnaevsky-Junior (1991a)).

The"cell"is understood as a rock volume between two defects and when the defects merge the cell is considered to pass into a new state. The present size crack formation is equivalent to infinite cluster formation in volume which diametral section size is equal to that of the crack. Then the probability of present size crack formation is equal to the probability that there are xn destroyed cells among all cells, where x is the critical density of destroyed cells, n is cells amount. Thus,crack sizes are distributed by binomial law.

When r is small and n is large the binomial distribution transforms into Poisson's one:

$$p(S) = (nr)^{nx} [(xn)!]^{-1} \exp(-nr) \tag{3}$$

where p(S) is the probability that a given crack area is equal to S,r is the cell destruction probability,S is the crack area,$n=0.09\ S^{3/2} l^{-3}$,l is the size of small area in which the correlations between the stress in diverse point are stronger than random factors influence. Sveshnikov and Mishnaevsky-Junior (1991b) determine the value l as size of particles,which are formed at crushing.

The crack growth process depending on the cell destruction probability can be described by formylae from Stanley (1972)

$$S \sim (1 - r/x)^{-2\nu} \tag{4}$$

where ν is a index of critical phenomena theory. The critical crack size depends on the value r by formyla

$$S_{cr} \sim (1 - r/x)^{2(\beta - 3\nu)}$$

where β is also critical phenomena theory index. For the present instance, $\beta \approx 0.4$ and $\nu \approx 0.88$ (Stanley (1972)). Hense the model allows to describe the cracking by simple mathematical method, taking into account the connection between macro-and microcracks formation.

4 Probabilistic description of stress state

The stress state in cracked rock mass is very complete and stress in the point depends on many of random factors(crack sizes and placing,inclusions etc).Because of that and stress fields interference,the stress becomes a random value and can be described by probabilistic methods.

The contact surface is discrete.For all contact point the Boussinesque solution is supposed to be carried out. In near points of contact surface the contact stresses are close.Hense the stress in point of rock mass is inversely proportional to distance between the point and nearest point of contact surface and proportional to contact stress in the contact point.Taking into account the solution of problem about stress distribution around the crack (Timoshenko and Goodier (1970)),we obtain:

$$\sigma_i = c\,\sigma_k\, S^{1/4} \tag{5}$$

$$c = \pi^{-3/2} r_1^{-1/2} l\ r_2^{-1} \cos\alpha\ f(\varphi) \tag{6}$$

where σ_k is the contact stress in the point nearest to present point,σ_i is the stress intensity in the present point, S is nearest crack area, c is a coefficient including the cracks placing and orientation characteristics, r_1 is a distance between the point and the nearest crack tip, r_2 is a distance between the point and the nearest contact point,φ is a angle between the plane in which the crack is,and the vector from the crack tip to the point,α is the cylindrical coordinate of the point (the angle between the vertical line and a vector from the point to nearest contact point), $f(\ )$ is a product of trigonometrical functions (Timoshenko and Goodier, (1970)).

The formula (6) was deduced for plane deformation and can be generalized easy for any other case.

If man knows the distributions of values $\sigma k,c$ and S,the probability function of stress distribution can be determined as the convolution of distributions of random value logarithmes. Mishnaevsky-Jr.(1992) has shown, that the distribution of value c is approximmately the exponente. Taking into account the redistribution of contact stress between rock and tool owing to rock crushing,which is directed to contact stress entropy maximization, we can use the entropy maximization method (Wilson (1970)) to find the contact stress distribution function.Maximising the corresponding Lagrange polynomial we obtain:

$$p(\sigma k) = \lambda k \exp(-\lambda k \sigma k) \tag{7}$$

where $1/\lambda k$ is the mean contact stress. Taking into account the definition of probability function of random value logarithm and presenting the exponential functions in the Poisson distribution form we have:

$$p(\sigma i) = (1/\sigma i) \sum_{Y} \psi(1;Y)\,\psi(nx; \lambda_c c + \lambda k \sigma k + nr - Y) \tag{8}$$

where $p(\sigma i)$ is the probability that stress σi is equal to given value, ψ is Poisson distribution, Y is variable. Taking into account the stability of Poisson distribution we have:

$$p(\sigma i) \sim p(C)\, p(S)\, p(\sigma i) \tag{9}$$

The function (9) describe the stress state in cracked rock mass under loading by tool with complex form.

To calculate the cell destruction probability the following condition was used:cell is destroyed if $\sigma i > \sigma cr$ where σcr is rock tensile strength.Basing on that we have:

$$r = \int_{\sigma_{cr}}^{\infty} p(\sigma i)\, d\sigma i \tag{10}$$

5 Information transformation at rock disruption and principle of tool improvement

5.1 Destruction as information transformation process

The unevenness of stress field in rock mass can be characterized by informational entropy of distribution (9).Determining the entropy by formula of Boltzmann and taking into account the additiveness of entropy we have:

$$Hi = Hk + Hs + Hc + Ho \tag{11}$$

where Hi, Hk, Hs, Hc are informational entropies of stress distribution in rock mass,contact stress,crack areas,parameter of cracks placing into massif accordingly, Ho is the nonlinearities including member.Since the entropy is equivalent to non-information one can conclude that the information which is contained in tool form (Hk) and the one,which is contained in rock massif structure (Hs and Hc) are transformed into stressed state information,and then into information of being formed cracks distribution. There is the feedback: new cracks distribution determines the rock massif structure.When the stress distribution can be described by exponential law the connection between stress state information Hi and cell destruction probability looks as follows:

$$r = \exp\left[-\frac{\sigma_{cr}}{\sigma_m(H_i+1)}\right] \quad (12)$$

where σm is mean of stress.

5.2 Principle of maximization of entropies of tool distributed parameters

Analysing the formulas (11),(12) we came to the following conclusions.To improve the tool form the contact stress unevenness must be increased.The mutual placing the cutters must ensure the various orientation of forming cracks,what increase Hc.The loading conditions must ensure large cracks forming and reduction of the number of small cracks what increase Hs.The more Hk and Hi the more intensive the rock disruption goes on.

Generalizing the conclusions one can formulate the following principle: the tool construction becomes optimal when the informational entropies of distributed parameters of tool are increased.The distributed parameters can mean both contact stresses,cutting edges orientations and wear resistances in diverce points of tool working surface,etc.

5.3 Results of patents analysis

To test the theoretical generalization the patents analysis in the area of drilling tool design was conducted. Approximately 250 patents were considered,and as a result the patents were divided into the seven following groups: the unflat and uneven shape of tool working surface,the asymmetry of contact surface about cutting speed vector, the diversity of teeth shapes and orientations on the same auger,irregularity of teeth arrangement on the auger,the interaction of tool with diverse mechanisms of loading, the variety of wear resistances in diverse points of tool working surface,self-sharpening and self-organization of tool working area.

The first group includes such versions of techical solutions:convexity or concavity of cutting face,prismatic or cylindrical lugs on cutting face ,stepped contact surface,bevels,slopes or cavities on tool working surface.For example,the patents number 1044765A,1023062A, 1323706A1,623958 (USSR),1284539 (England), 57-35357 (Japan) etc were included in the group.In the second group the diverse versions of cutting face or it's part inclines to the cutting speed vector were included (for example,the patents of USSR number 723123 and 1046465A).The third group contains the diverse combinations of radial and tangential cutters on the same auger as well as the change of cutting and wedge angles,and insert strength from auger periphery to that axis.The patents number 395559,1536880A1,1366627A1 etc relate to the group.The fourth group includes the teeth arrangement in paars or in groups as well as various distance between teeth.In the group the American patents number 3726350 and 3158216, patent of USSR number 1472623A1 etc. were included.The fifth group contains such technical solutions as the combinations of mobile and fixed tools or progressive and rotatory moving that (for example,the Japanese patent 52-48082,and Soviet patent 697711).The sixth group includes the availability of various strength layers in cutters,as well as the diamant coatings and cavities of reguired shape in cutters (for example, Soviet patents 714003,281349,145496, 693000,609884 etc. The seventh group contains the patents number 717327,719192,and number

4230193 (USA). Generalising the ideas one can draw the following conclusion:to improve the drilling tool construction the such tool parameters as contact stress,cutting edges and faces inclines,loading mechanisms,insert forms and strength etc,are to be distributed unevenly.That confirms the principle of tool distributed parameters entropies maximizing formulated above.

6 Conclusions

The informational entropy of contact stress distribution can be used as a general characteristic of tool shape. The cracking in rock can be described by the percolation theory model.The model allows to take into account the connection between macro- and microcracks formation,to describe the crack growth process depending on the microcracking and the being formed crack sizes distribution.

The stressed state in cracked rock massif can be described by probability function.The function is proportional to product of the probability function of contact stress multiplied by the one of crack sizes and crack location distributions.

The rock failure process can be considered as a information transformation process.The informational characteristic of tool shape and the one of rock massif structure transform into the stress distribution information and then into sizes distribution information of being formed cracks.

The general principle of drilling tool improvement is formulated.To improve the tool shape the informational entropies of distributed parameters of tool are to get maximal.To test the principle,the patents analysis was made,and the analysis results confirms the principle.

7 References

Mishnaevsky,Junior,L.L.(1992) Informational model of rock destruction and its application, in New elaborations in the area of superhard materials and coatings, Institute for Superhard Materials Press,Kiev,Ukraine, pp.42-48 (in Russian).

Stanley H.E. (1972) Introduction to Phase Transitions and Critical Phenomena. Oxford University Press, London.

Sveshnikov I.A.and Mishnaevsky,Jr.L.L.(1991a)Mathematical Simulation of Failure of Brittle Nonuniform Rocks. Strength of Materials,Vol.23,12,pp.71-76.

Sveshnikov I.A.and Mishnaevsky,Jr.L.L.(1991b) Theoretical investigations of hard rock destruction at drilling. Institute for Superhard Materials Press, Kiev,Ukraine, (in Russian).

Timoshenko S.P.and Goodier J.N. (1970) Theory of Elasticity,Mc.Graw-Hill, N.Y.

Wilson A.G.(1970)Entropy in urban and regional modelling, Pion Ltd, London.

Zhlobinsky B.A.(1970) Impact failure of rock under indentation, Nedra,Moscow, (in Russian).

37 LOCAL ROCK FRACTURE UNDER SHOCK INTERACTION WITH HARD IMPACTOR

A.L. MAISTRENKO, I.A. SVESHNIKOV
and A.A. SIDORKO
Institute for Superhard Materials, Ukrainian Academy of Sciences, Kiev, Ukraine

Abstract
The present paper discusses the results of experimental study of local fracture of granite rock under shock loading with a free-falling spherical 9-mm-radius tungsten carbide impactor. The process of step-by-step rock fracture is analyzed along with the effects of energetic impact parameters on the impactor penetration depth, mechanism and direction of crack propagation in the rock. It is shown that by varying collision energy and velocity one can control both the mechanism and direction of fracture propagation. It is found that there are certain threashold values of the impact parameters above which the damaged volume of rock increases insignificantly.
Keywords: Rock, Local Fracture, Hard Impactor, Plastic Zone, Energy Dissipation, Crack

1 Introduction

A typical fracture mode of rock when interacting dynamically with working elements of drilling bits is the surface crushing. The mechanism of chipping and the process of material dispersion can be analyzed from the standpoint of mechanics of contact interaction of solids using various physical techniques for in-depth study of fracture propagation.

Analysis of the literature on the subject shows that the results of investigation into fracture kinetics of rocks during their contact interaction with solids are rather limited and often discrepant. This can be attributed to dificulties encountered in recording individual stages of fracture which propagates at high speeds. However, the test results for model transparent materials, such as crystaline and amorphous glasses, allow to assert unabiquously that shock fracture of quasi-brittle media features a complex array of cracks which can be arbitrarily divided into median, conic, radial and lateral cracks. The mechanism of rock fracture during its dynamic interaction with an impactor was found (Kirchner et al.,1983) to differ from

Fracture and Damage of Concrete and Rock – FDCR-2.
Edited by H.P. Rossmanith. © 1993 Taylor & Francis. ISBN 0 419 18470 8.

that during static impression in succession and contribution of fracture phases. The fracture phases under static and dynamic loading are of the same type.

2 Experimental

One of the main procedure-related problems is to ensure experimental conditions which should simulate actual rock loading as well as the phusical similarity of parameters of a model and an actual material. As the similarity criteria we have chosen:

-geometry of the drilling-bit impactor,

-dynamic interaction parameters wich governed by equivalent mass of impactor and impact velosity, and

-physical and mechanical properties of the specimen.

The similarity criteria requirements were satisfied completely: the impactor geometry, dimensions and material corresponded to actual ones, the tests were carried out on granite conventionally drilled with the bit type under study.

Dynamic interaction parameters were contrrolled by varying impactor drop height within the range encountered in real drilling bit application.

The tests were performed using a gravitational impact tester with freely falling load (Novikov et al.,1979). A steel rod of 1085.6g with a cylindrical impactor attached to its end was used as an indenter. The impactor was made of BK-6 cemented carbide grade, its spherical working surface had a radius of curvature of 9mm.

The tests were carried out on cubic (70mm on edge) granite specimens which had been surface-ground and polished with a diamond wheel of 80/63 grain size.

Durig the tests the impactor dropping height was varied so that the impact velocity could be changed from 1.14 to 5.2m/s.

The interaction force between the specimen and impactor was determined from P(t) plots, which were recorded for the cases of kinetic impactor energy ranging from 1.2 to 150.0J. The tested specimens were studied by fractographic and microstructure analyses. To study internal local zones of rock deformation and fracture the specimens were examined using a scanning electron microscope.

3 Results and Discussion

The P(t) diagrams constitute the primary information about parameters of interaction between the impactor and the rock. They can be used to determine the rate of increase in P at different stages of the impactor penetration into the rock (or the mean rate), maximum force on the impactor, time of the impactor penetration to a maximum depth, and their derivatives such as impulse of force, work during the impactor penetration into the rock and during unloading, and thus, work of fracture.

Furthermore, the P(t) diagrams reflect, at least qualitatively, the kinetics of accumulation of local damage and microfracture. Much information can be also extracted from the results given in Fig.1 which suggests a nonlinear dependence of maximum interaction force on impact energy the load grows less intensively. The maximum force vs. impact energy diagram is well approximated by power function.

The results of fractographic and structure analyses of fracture zones are of particular interest within the scope of the present study. Fig.2 demonstrates profiles of chip craters produced at impact velocities of 1.14 and 2.80m/s. The crater profiles were found to have the shape of a saucer irrespective of the impact velocity. However, in the case of lower impact velocity less damaged material was in the crater centre. At higher impact velocities one could observe an increase in density of cracks near the impact site with simultaneous decrease in depth of the damaged zone propagarion. At lowest impact velocities the cracking zone below the crater has almost conic shape but coincides with Hertz stress isolines under quasi-static loading. An interesting feature of the rock fracture was observed at high impact velosities. In these cases the crater has a distinct contour beyond which there is no visible damage and macrocracks at the crater-bottom level are oriented parallel to the spacimen working surface.

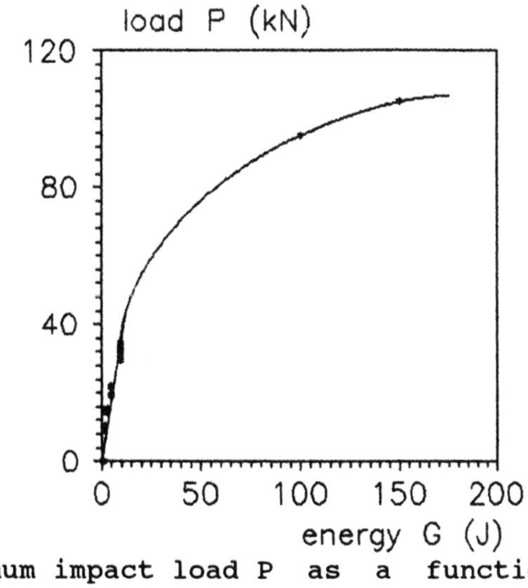

Fig.1.Maximum impact load P as a function of impactor potential energy

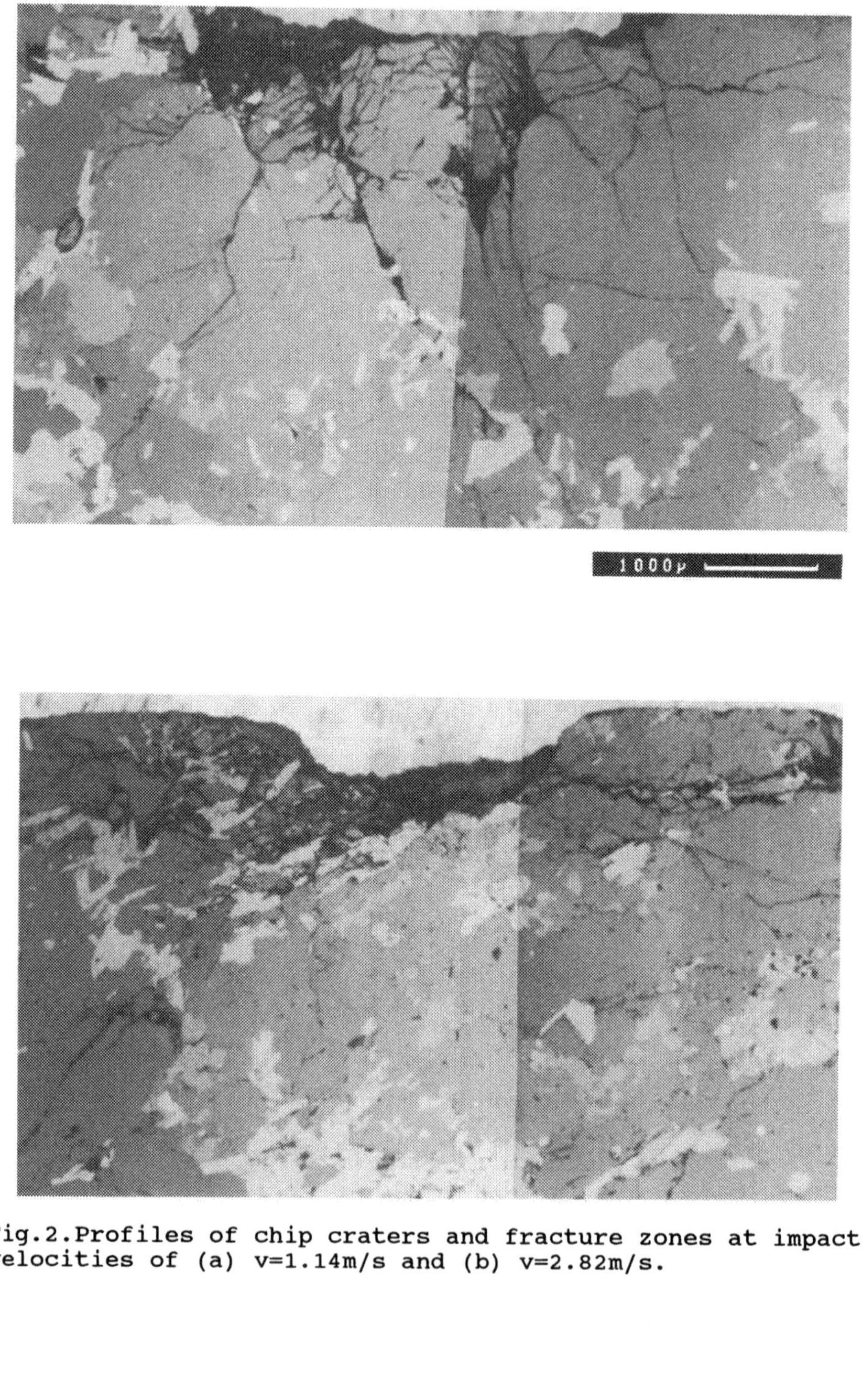

Fig.2.Profiles of chip craters and fracture zones at impact velocities of (a) v=1.14m/s and (b) v=2.82m/s.

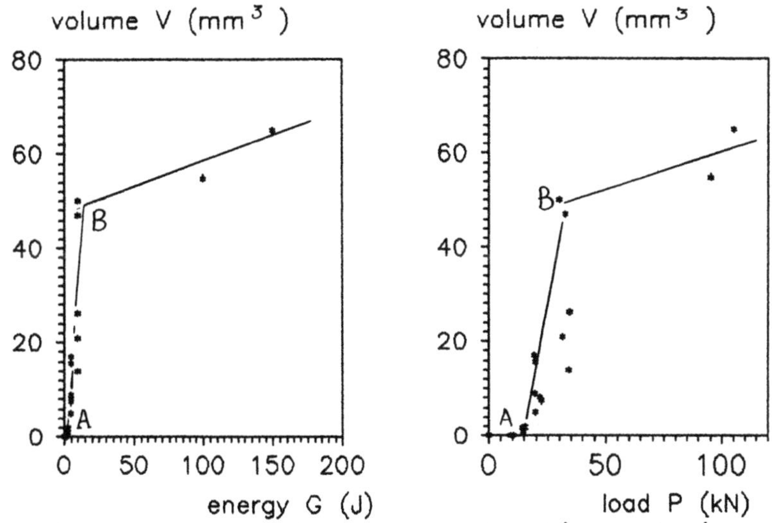

Fig.3.Fracture zone volume as a function of impactor potential energy and maximum impact force.

Almost for all specimens the fracture zone which includes material volume with distorted initial structure has the shape of spherical segment. Some asymmetry of the fracture zone profile may be attributed to fluctuation of the material properties.

Geometrical interpretation of the data obtained is given in Fig.3. It is evident that the plot of fracture zone volume as a function of force and energy parameters of impact can be divided into two nearly linear sections: the section AB exhibiting active growth of the impactor penetraton depth with increasing interaction force and energy, and the section of less active penetration. This indicates that there is a critical value of the impact energy parameters (which does depend on the rock structure and properties) above which the material resistance to impactor penetration grows considerably. This may be associated with an abrupt rise in material density under the impactor due to closing ("healing") of initial defects of the rock structure in zones of high compression stresses.

4 Conclusions

The process of rock fracture during the interaction with an impactor is shown to be of stage-by-stage nature. At the first in dynamic loading mode resulting in its complete fracture to form a "powdered" layer. The properties of the layer can be simulated by those of a loose medium with low Coulomb friction between particles. The latter enables the impactor action on the rock at the second stage to be simulated by a uniformly distributed pressure over the crater surface formed at the first stage of the impactor action. At the second stage the impactor penetrates much slower than at the first one, i.e. essentially in the static loading mode. It has been experimentally established that for a given impactor geometry there are optimum values of the impact energy parameters above which an increase in depth of the tool penetration into rock becomes less active. The impact energy parameters effect not only the tool penetration depth but also the mechanism of rock fracture. Furthermore, as the impact velocity increases starting form a certain level, the material disintegrates with no cracks propagating in the impact direction. By varying impact energy one can control both the mechanism and the direction of fracture propagation: from deep fracture at lower impact energies to near-surface one at high energies. The present study of kinetics of the rock fracture under impact loading with a tool gives experimentally substantiated grounds for calculating stress-strain state of the rock upon tool action to optimize bit design and its loading conditions.

5 References

Kirchner, H.P. and Ragosta, J.A. (1983) Relation of load to radial crack length for spherical indentations in hot-pressed ZnS. J. Amer. Ceram. Soc., 66, 293-296.

Novikov, N.V. Devin, L.N. and Maistrenko, A.L. (1979) Determination of fracture toughness of superhard materials and tunsten carbide under dynamic loading. Sverkhtverdye Mat. (in Russian), 3, 15-18.

38 NONLINEAR THEORY OF VIBRATIONS IN FRAGMENTED ROCKS AND SOILS

V.N. NIKOLAEVSKIJ
Institute of Earth Physics, Russian Academy of Sciences, Moscow, Russia

Abstract
Nonlinear mathematical models of geomaterials with microstructure are discussed. The choice of generalized rheological law is made for account of effects of dominant frequency oscillations and of ultrasound generation. The three-dimensional grade-consistent micropolar theory is considered in more details. Applications for enhanced oil recovery are mentioned.
Keywords: rocks, soils, fragments, grains, vibrations, dominant frequencies, ultrasound, creep, acoustical emission.

1 Introduction

The experiments with seismic wave have shown that seismic waves possesses some dominant frequency which can depend not only on stratification of rock masses but also on microstructure if rocks are in fragmented state due to explosions or to natural tectonic events. Such phenomenon is used for monitoring when the choice of observation frequency is made for each seismic station with account of its distance from test site and of its local conditions.

Another application is enhancement of oil recovery from water-flooded reservoirs. This effect was mentioned as a response of oil production wells to earthquakes and to heavy vibrator action at the free surface. The effect is very sensitive to frequencies of waves generated by the vibrator (Asan-Djalalov et al, 1990).

The problem of mathematical description is mainly connected with the proper formulation of rheological constitutive laws accounting granulated internal structure of geomaterials.

2 Generalized elastoviscous model

The first possibility was found by study of traditional rheological schemes after adding of internal concentrated masses to elastic springs. Analytically it means the

Fracture and Damage of Concrete and Rock – FDCR-2.
Edited by H.P. Rossmanith. © 1993 Taylor & Francis. ISBN 0 419 18470 8.

introduction of higher time derivatives into the constitutive law. In general it has the following form (Nikolaevskij, 1990)

$$\sigma + \sum_{p=1}^{m} a_p \frac{d^p \sigma}{dt^p} = Ee + \sum_{p=1}^{n} b_p \frac{d^p e}{dt^p} \tag{2.1}$$

where the derivatives has to be the Oldroyd nonlinear ones. Here σ is the stress, e is the strain, E is the Young modulus.

The choice of values m, n and of rheological coefficients a_p, b_p is such that some oscillations have to be unstable in linear approximation. Of course, this step is unusual for mathematical analysis but my idea is to bound the corresponding amplitude growth by nonlinear terms.

One can see that the rheological coefficients have to include the internal length scale $\varkappa$. For example,

$$\varkappa^2 = b_2/\rho \tag{2.2}$$

where ρ is the density and b_2 is playing the role of spring mass M indeed.

If one adds balances of mass, impulse and the connection between strains and the velocity field to the constitute laws and uses conventional method of small parameter expansion in the moving coordinate system ξ = x-ct, then the following evolution equation can be found (Nikolaevskij, 1988):

$$\frac{\partial v}{\partial \tau} + Nv\frac{\partial v}{\partial \xi} + \sum_{p=1}^{n} \alpha_{2p+1} \frac{\partial^{2p+1} v}{\partial \xi^{2p+1}} = \sum_{p=1}^{n} \beta_{2p} \frac{\partial^{2p} v}{\partial \xi^{2p}} \tag{2.3}$$

Here v is the displacement velocity. The terms with α -coefficients correspond to dispersion and are connected with soliton wave phenomenon. The β -coefficients are connected with wave dissipation. Here c is the sound velocity.

According to linear approximation the wave attenuation coefficient is presented by the formula

$$B = -k^2 (\beta_2 - \beta_4 k^2 + \beta_6 k^4) \quad (2.4)$$

where k is the wave number.One can see now that the attenuation could be negative for interval of k^2-values between two roots k_+^2, k_-^2 of the equation $B=0$, $k \neq 0$ if

$$\beta_2 - \beta_4 k_*^2 + \beta_6 k_*^4 < 0 \quad (2.5)$$

where $k_- < k_* < k_+$.

The numerical calculations (Beresnev, Mitlin and Nikolaevskij, 1991) has shown that white-noise seismic signal is transforming to the wave number k_* corresponding to minimum value of the attenuation coefficient B . The bigger is the coefficient N of nonlinearity, the quicker is such a transformation.

Analytical treatment (Malomed, Mitlin and Nikolaevskij, 1991) was used to prove the stability of such dominant oscillations.

The experiments with marine sands partly saturated with water had shown (Vilchinskaya and Nikolaevskij, 1984) that the seismic wave was transformed to running oscillation with the dominant frequency equal to 25 Hz.

The possibility of nonequality (2.5) means that the fragmented material has to possess two Young's modulus one of which has to be much more than the another. That is, the matrix of fragmented rock has to be much weaker than the grains themselves. The same conditions have to be formulated for viscosities at contacts and of bulk material, etc.

The conditions for effective monitoring of nuclear explosions at 1-3 Hz mean that the internal length or fragmentation of the Earth crust along the seismic path from intense events has the order of 100 meters (Nikolaevskij, 1989).

The effect of dominant frequencies was mentioned at field test of vibroaction at water-flooded oil reservoirs. In one of experiments such a frequency was equal to 12 Hz, in another - to 6Hz (Asan-Djalalov et al, 1990).

3 Model of ultrasound generation

The physical explanation of enhance oil recovery is the clasterization of residual oil-drops which were separated earlier by water due to flooding of pore space. It means drastical changes of oil relative permeability under vibration action (Nikolaevskij, 1988).

However, the size of oil residual drops is so small that only ultrasound waves can disturb them and change their positions. It is known due to field experiments that ultrasound logging of oil wells can recreate their oil productivity but under such a treatment the ultrasound waves penetrate into the stratum not farther than the first meters.

So, the field experiments have shown that the surface vibrator can create an ultrasound at finite stage of its mechanical action inside deep reservoirs. For account of transformation of seismic energy to ultrasound one it is necessary to use the micropolar elastic model developed earlier (Parfitt and Eringen, 1969). N.G.Mazur and the author have constructed the solution of corresponding equations:

$$\frac{\partial^2 u}{\partial t^2} = (c_2^2 + c_3^2)\frac{\partial^2 u}{\partial x^2} + c_3^2 \frac{\partial \varphi}{\partial x}$$

$$\frac{\partial^2 \varphi}{\partial t^2} = c_4^2 \frac{\partial^2 \varphi}{\partial x^2} - 2\omega_0^2 \varphi - \omega_0^2 \frac{\partial^2 u}{\partial x^2} \tag{3.1}$$

which describes shear wave stationary radiation. Here u is the displacement and φ is the microrotation, c_i are wave velocity, ω_0 is the characteristic frequency, $\omega_0 \sim c/\ae$. The boundary conditions

$$u(0,t) = e^{i\omega t}, \qquad \varphi(0,t) = 0$$

and zero initial conditions are satisfied by

$$u(x,t) = \exp\left[i\omega t - \frac{x}{c_2'} + \delta t\right] - i\varepsilon e^{i\omega t} \exp\left[\left(\frac{1}{2}\varepsilon\frac{\omega_0^2}{r_0} - 1\right)\frac{x}{c_4}\sqrt{2\omega_0^2 - \omega^2}\right], \qquad \varepsilon = c_3^2 \ll 1$$

$$\varphi(x,t) = i\frac{\omega\omega_0^2}{r}\left\{1 + \varepsilon\left[\frac{\omega_0^2}{r_0}\left\{1 - \varepsilon\left[\frac{\omega_0^2}{r} - \frac{1}{c_2^2}\right]\frac{b_0}{2r_0}\right\}\left\{\exp\left[i\omega\left(t - \frac{x}{c_2'}\right) + \delta t\right] + \exp\left[i\omega\left(t + \delta t\right)\right]\exp\left[\left(\frac{1}{2}\varepsilon\frac{\omega_0^2}{r_0} - 1\right)\frac{x}{c_4}\sqrt{2\omega_0^2 - \omega^2}\right]\right\}\right.\right. \tag{3.2}$$

where $r_0 = 2\omega_0^2 c_2^2 - \omega^2 (c_2^2 - c_4^2)$,

$$\delta t = \frac{\varepsilon h}{\omega}, \quad c_2' = c_2\left[1 - \left(\frac{\omega_0^2}{r_0} - \frac{1}{c_2^2}\right)\frac{\varepsilon}{2}\right], \quad h = \frac{\omega\omega_0 c_2 c_4}{r_0^2}\sqrt{2\omega_0^2 - \omega^2}$$

One can see that in linear approximation the existence of micromotion does not generate high frequency signal (of the order ω_0). So, the only possibility can be connected with nonlinear long-short wave resonance between seismic ultrasound modes.

Let us use the grade-consistent nonlinear model (Brulin and Hjalmars, 1981) in which the free energy had depended on higher space derivatives of u than of microrotation φ. A.L.Krylov, N.G.Mazur, G.A.El' and the author have transformed the corresponding system of equation to the following two ones:

$$\rho \ddot{u}_k = a \varepsilon_{k\alpha\beta} \varphi_{\beta,\alpha} + b_1 u_{k,\alpha\alpha} + b_2 u_{\alpha,\alpha k} + b_3 u_{k,\alpha\alpha\beta\beta} + b_4 u_{k,\alpha\beta\beta k} + b_5 \varepsilon_{k\alpha\beta} \varphi_{\alpha,\beta\gamma\gamma} + Q^u_k , \tag{3.3}$$

$$\rho j \ddot{\varphi}_k = a \varepsilon_{k\alpha\beta} u_{\beta,\alpha} - 2a\varphi_k + c_k \varphi_{k,\alpha\alpha} + c_2 \varphi_{\alpha,\alpha k} + c_3 \varepsilon_{k\alpha\beta} u_{\alpha,\beta\gamma\gamma} + Q^{\varphi}_k \tag{3.4}$$

where Q^u_k, Q^{φ}_k are the symbols of sums of nonlinear terms, $\varepsilon_{k\alpha\beta}$ is the Levi-Chevita alternating tensor, j is the moment of inertia. The waves are combinations of longitudinal and shear motions and it is properly to name those waves as longitudinal or shear which are such waves in linear approximation.

It is possible to show that the equation system for longitudinal waves is following one

$$\ddot{u} - c_1^2 u'' - \delta u'''' - \nu u' u'' + 2\chi \varphi' \varphi + Q^u_* = 0 \tag{3.5}$$

$$\ddot{\varphi} - c_2^2 \varphi'' + \omega_0^2 \varphi - u' (c_2^2 \varphi'' - \mu \varphi) + Q^{\varphi}_* = 0 \tag{3.6}$$

Here Q^u_* and Q^{φ}_* are nonlinear terms which are not essential for dynamic analysis. The analysis with $Q^u_* = Q^\varphi_* = 0$ was given earlier (Krylov, Nikolaevskij and El', 1991) and had shown that under condition

$$c'_g(k_{us}) = c_1 \tag{3.7}$$

of equality of group ultrasound velocity c'_g and phase velocity c_1 of seismic wave at $k \to 0$. The wave length of generated ultrasound can be evaluated by following expression

$$\lambda_* = 2\pi \frac{c_2}{\omega_0}\left(\frac{c_2^2}{c_1^2} - 1\right)^{1/2} \tag{3.8}$$

that is, the ultrasound velocity c_2 has be bigger than seismic wave velocity c_1 .Physically, the ultrasound generation can be explained by motion of microelements (grains) in contact.

4 Acoustic emission under creep of rocks

Let us the same micropolar model under conditions when displacements are governed by viscous constitutive law but microrotational motion is elastic one. Then the simplest variant of the dynamic theory is described by the following equation

$$j\ddot{\varphi} + 4\beta\dot{\varphi} - \Lambda\varphi'' + w_{,\varphi} = F \tag{4.1}$$

where Λ is the elastic coefficient, β is the coefficient of interaction of displacement and rotation modes, w is the free energy, F is source of energy. If $w_{,\varphi\varphi} < 0$,then the unstable creation of acoustical pulse is possible (Dinariev and Nikolaevskij, 1992).

5 References

Asan-Djalalov A.G., Kusnetzov V.V., Kissin I.G., Nikolaev A.V.,Nikolaevskij V.N. and Urdukhanov P.I. (1990). The method of oil recovery from water-flooded reservoir. USSR Author Certificate N 1596081, cl.E21 B 43/00. Bull. of USSR Inventions, N 36.

Beresnev I.A., Mitlin V.S.and Nikolaevskij V.N. (1991). The role of coefficient of nonlinearity at generation of dominant seismic frequencies. Proc. USSR Acad. Sci. (DAN), v.317, N 5, pp.1103-1107 (in Russian).

Brulin O. and Hjalmars S.(1981). Linear grade-consistent micropolar theory. Int. J. Engng. Sci., v.19, N 12.

pp.1258-1269.
Dinariev O.Yu. and Nikolaevskij V.N. (1992). Mechanism of acoustical emission under creep of rocks with microstructure (in press).
Krylov A.L., Nikolaevskij V.N. and El' G.A.(1991). Mathematical model of nonlinear generation of ultrasound by seismic waves. Proc. USSR Acad. Sci. (DAN), v.318, N 6, pp.1340-1345 (in Russian).
Malomed B.A., Mitlin V.S. and Nikolaevskij V.N. (1991). Short wave bifurcation in the model seismoactive medium and dominant frequencies. Sov. Appl. Math. Mech. (PMM), v.55, N 5, pp. 795-802 (in Russian).
Nikolaevskij V.N. (1988). Nonlinear waves in soils and fractured rocks. Phys.-Tech. Probl. of Recovery of Mineral Resources, N 6, pp.31-38 (in Russian).
Nikolaevskij V.N. (1989). Dynamics of viscoelastic media with internal oscillations. In: Recent Advances in Engineering Sciences, Lecture Notes in Engineering, N 39, Berlin: Springer, pp.210-211.
Nikolaevskij V.N. (1990). Mechanics of Porous and Fractured Media. Singapore: World Scientific.
Parfitt V.R. and Eringen A.C. (1969). Reflection of plane wave from the flat boundary of micropolar elastic halfspace. J.Acoust. Soc. America, v.45, N 5, pp. 1258-1269.
Vilchinskaya N.A. and Nikolaevskij V.N. (1984). Acoustical emission and seismic signal spectrum. Solid Earth Physics (Izvestia), N 5, pp.91-100.

39 MECHANICS OF DILATANT FRACTURING OF GEOMATERIALS

V.N. NIKOLAEVSKIJ
Institute of Earth Physics, Russian Academy of Sciences, Moscow, Russia

Abstract
Theory of rock dilatancy based on concepts of plasticity is presented. The motion of failure surface in the stress space corresponds to damage of geomaterials due to a microcrack growth. The parameters of model for natural rocks are given. The examples of numerical calculations are discussed. They are calculations of intense underground explosions, rock damage in vicinity of superdeep wells. Dynamic strength is interpreted as result of the limit crack velocity and of overloading of material under explosion action.
Keywords: rocks, fracture, dilatancy, shear, strains, strength, explosions, faults, deep wells.

1 Introduction

There are two well-known alternatives for study of fracturing of rocks and concrete. One is connected with the Griffith concept of failure by individual cracks. The policrystalline or granular microstructure of real geomaterials opens another possibility of fracturing connected with microcrack growth. The latter one is described in terms and concepts of usual plasticity theory but with one exception. The dilatancy proportionality (Nikolaevskij, 1967) of shear and volume strain increments is more general restriction than conventional plastic incompressibility (of metals).

It is essential that the plasticity laws do not account internal time of geomaterials and this possibility is proved also by scale analysis of damage zones after intense underground explosions (Rodionov et al, 1972). So, small characteristical time connected with finite limit crack velocity can not influence fracturing even under fast dynamical process as explosion motion although it explains the dynamical strength phenomenon. At longer time interval the rock creep can be essential and it change to some extent the plasticity calculation data.

Not so long ago the problem of proper choice of mathematical model seemed to be unsolved (Nemat-Nasser et al, 1984) but nowadays one can see that the theory of dilatant

Fracture and Damage of Concrete and Rock – FDCR-2.
Edited by H.P. Rossmanith. © 1993 Taylor & Francis. ISBN 0 419 18470 8.

plasticity is quite sufficient for many practical applications.

2 The constitutive dilatancy and plasticity laws

Growth of micricracks in geomaterials happens if the stress state corresponds to the Mohr failure condition

$$\sqrt{I_2} = \alpha p + Y \tag{2.1}$$

where I_2 is the second invariant of stress-deviator tensor, p is the pressure, α is the friction coefficient, Y is the cohesion.

The dilatancy kinematical condition is formulated as follows

$$de_{ij}^{p} \delta_{ij} = \Lambda \left| d\gamma^{p} \right| \tag{2.2}$$

where e_{ij} is plastic deformation, $d\gamma^{p}$ is the intensivity of plastic shear increments, Λ is the dilatancy rate, δ_{ij} is the unit tensor.

Even in isotropic case the constitutive flow rule has to include two unknown scalar coefficients $d\lambda$ and $d\xi$

$$de_{ij}^{p} = \sigma_{ke} \delta_{ke} \delta_{ij} \, d\lambda + 2\sigma_{ke} \delta_{kj} \delta_{ei} \, d\xi \tag{2.3}$$

Therefore two limit conditions (2.1) and (2.2) are necessary for determination of $d\lambda$ and $d\xi$. However, the latter can be found (or be excluded) only during solution process of the complete system of equations, that is, of (2.1) - (2.3) altogether with balances of mass and impulse

$$\partial \rho / \partial t + \partial \rho v_j / \partial x_j = 0 \tag{2.4}$$

$$\partial \rho v_i / \partial t + \partial \rho v_i v_j / \partial x_j = \partial \sigma_{ij} / \partial x_j + F_i \tag{2.5}$$

Here ρ is the density, v_i is the displacement velocity,

σ_{ij} is the stress and $\mathcal{F}_i$ is the body force. The stresses are supposed to be symmetrical ($\sigma_{ij} = \sigma_{ji}$). Elastic part of strains are determined by the Hooke law.

Every model is complete only if the conditions at strong discontinuities are determined. Of course, they have to be well-known balances of mass, impulse (and energy if it is necessary), but there are some percularities.

Firstly, nonholonomic plastic flow rule does not permit to get the Hugoniot adiabate for normal shock independent of motion outside of the discontinuities. Secondly, tangential discontinuities (known also as slip-lines or strain-localization bands) are determined by condition of coincidence of strong discontinuities of stress field and of characteristics of velocity field. As the result (Nikolaevskij et al., 1991) the effective Coulomb angle of friction along the discontinuity has to be less than internal Mohr friction angle φ , that is , $\chi < \varphi$:

$$tg\,\chi = \frac{\sin\varphi\cos\nu}{1-\sin\varphi\sin\nu} \tag{2.6}$$

where ν is the dilatancy angle conventional for plane problems.

So, the important conclusions follow: localization process has to diminish the bearing capacity of the body under consideration and has to be modelled in a full scale by continuum theory which permits description of micro-deformation (and change of the friction angle in a slip-band).

3 Underground explosions with account of dilatancy

In the vicinity of the explosion cavity the irreversible deformation is so essential that elastic part of strains can be neglected. Then the dilatancy condition (2.2) has the form

$$\frac{\partial v}{\partial t} + 2\frac{v}{r} = \Lambda\left(\frac{v}{r} - \frac{\partial v}{\partial r}\right) \tag{3.1}$$

where v is the velocity radial component and the spherical symmetry is taken into account. If Λ = const, the first kinematical integral follows

$$v(r,t) = \frac{C(t)}{r^n}, \qquad n = \frac{2\sqrt{3}+2\Lambda}{\sqrt{3}-2\Lambda} \tag{3.2}$$

One can see that incompressibility of geomaterials (n=2) is happened only in the case of zero dilatancy (= 0). The density distribution is governed by the second kinematical integral which is sequence of (3.2) and the mass balances (2.4)

$$\rho(r,t) = \frac{1}{r^{2-n}} f\left(r^{n+1} - \int \frac{\rho - \rho_o}{\rho_o} dR^{n+1}\right) \tag{3.3}$$

where ρ_o is the initial value of ρ ,$R(t)$ is the coordinate of the shock wave front and f is the function which has to be selected from shock compression rule (Nikolaevskij, 1990).

If one uses both integrals in the impulse balance equation, the dilatant modification of the Taylor-Penny dynamical equation (Chadwick et al., 1964) follows. The latter can be integrated. Many variants of such a theory were studied analytically although they needed simple numerical calculations.

The numerical methods were developed also on the base of artificial viscosity concepts. They did include shock front as zone of high gradients of variables and did not need suggestion of friction, cohesion and dilatancy constancy that permitted to account real properties of geomaterials.

One of essential results was the size of zones of explosions fracturing in rocks. The scaled radius was equal to 1 $m/kg^{1/3}$ in granite and 0,47 $m/kg^{1/3}$ in weak porous rocks where weights of TNT equivalent charges were used. Moreover, there is no long cracks in porous rocks. The main effect was a pore collapse if the initial porosity was more than 0.15 (Nikolaevskij, 1990a).

There was the opinion that the physical viscosity has to be introduce to get the explosion wave length corresponding to field observation. However, the factor of a limit crack velocity which was not introduced in the mentioned above calculation can influence upon the wave form.

4 Dynamical strength of rocks as overloading effect.

Some experiments (Schock and Heard, 1974) had shown that under intense impact rocks appeared to be essentially more hard than in static conditions. Therefore the concept of dynamical strength was developed with introduction of special relaxation time. I had alternative idea that the fracture front velocity was restricted by limit crack velocity close to one half of shear wave velocity. The latter value can be found experimentally for each material.

Under high velocity impact the boundary at which the stress are reaching the static strength condition can move faster than the cracks themselves. Between them the zone of overloading appears where stresses are higher than admissible by static laws (Nikolaevskij, 1990a). The observer can measure and announce the effect of dynamical strength of material.So, the sapphire would have such a strength equal to 400 kilobars. Even plexiglass would have tremendous strength under short action of electronic beams (with time interval of nanoseconds). In the zone of overloading the small defects begin to develop and the specific prefracture damage is appearing. But if the impact is short enough the construction or the body will stand it without macrofailure.

5 Damage in the vicinity of superdeep wells

There is quite unusual behavior of granite masses at depths of some kilometers during drilling. It is found (Juhlin, 1990) that intact rocks show breakouts but not the naturally crushed geomaterials which are inside the fault bodies crossed by the borehole.

The possible explanation is connected with dilatancy effect. After borehole drilling the rock is removed to its axis and the particle velocity field is described by the same expression (3.2) but in this axisymmetrical case

$$n = \frac{3+2\Lambda^2}{3-4\Lambda^2}\left(1 + \frac{2\Lambda^2 - 2\Lambda\sqrt{1-(\Lambda^2/3)}}{1-(4\Lambda^2/3)}\right) \tag{5.1}$$

In the outer zone the damage is characterized by condition of compaction ($\Lambda < 0$) and the porosity of fault geomaterial is diminishing,its cohesion and friction resistance are growing. The numerical calculations (Nikolaevskij, 1990b) have shown that the hard arch is created around the well and stresses are essentially higher than in elastic case (Obert and Duvall, 1967). The internal zone is characterized by loosening ($\Lambda > 0$) and can be identified with potential breakout masses. In case of intact rocks the elastic zone is in contact with loosening zone directly and the induced dilatant strengthening is absent.

Recently the elastoplastic calculations of dilatant states around wells with full description of rock properties (Kapustyansky and Nikolaevskij,1985) were performed.

6 References

Chadwick P., Cox A.D. and Hopkins H.G. (1964). Mechanics of deep underground explosions. Phil. Trans. Roy. Soc. London, A, N 1070, v.256.

Juhlin Ch. (1990). Seismic attention, shear wave anisotropy

and some aspects of fracturing in the crystalline rock of the Siljan Ring Area, Central Sweden. Acta Universitatis Upsaliensis, 30, Uppsala, 167 p.

Kapustyanskii S.M. and Nikolaevskii V.N. (1985) Parameters of elastoplastic model for geomaterials. J. Appl. Mech. Tech. Phys., N 6, p.145-150 (in Russian).

Nemat-Nasser S., Asaro R.J. and Hegemeir G.A. (eds.), (1984). Theoretical foundation for large-scale computation of nonlinear material behavior. Nijhoff Publ.

Nikolaevskii V.N. (1967). On relation of volume and shear plastic deformation and on shock waves in soft soils. Proc. USSR Acad. Sci.(DAN), v.177, N 3, p.542-545 (in Russian).

Nikolaevskii V.N. (1990a). Mechanics of porous and fractured media. Singapore: World Scientific, 472 p.

Nikolaevskii V.N. (1990b). Dilatant ring compaction around borehole (skin-effect). In: Proc. Int. Conf. on Gascondensate recovery. Krasnodar, May 29-June 2, 1990, sect.2, p.43-47.

Nikolaevskii V.N., Kuznetsov A.S. and Belendir E.N. (1991). Mathematical dilatancy theory and conditions at strong discontinuities. Int. J. Engng. Sci., v.29, N 11, p.1375-1389.

Obert L. and Duvall W.I. (1967). Rock mechanics and the design of structures in rock. New York and London: Wiley, 650p.

Rodionov V.N., Adushkin V.V., Kostyuchenko V.N., Nikolaevskii V.N., Romashov A.N. and Tsvetkov V.M. (1971). Mechanical effect of an underground explosion, Moscow: Nedra (English Transl.- 1972 US Atomic Energy Commission, Tech. Inf. Center, UCRL - Trans - 10676).

Schock R.N. and Heard H.C. (1974). Static mechanical properties and shock loading response of granite. J. Geophys. Res., v.79, N 11, p.1662-1666.

40 INFLUENCES OF LOADING RATES ON THE FRACTURE TOUGHNESS OF ROCK

ZHANG ZONGXIAN, YU YONG and ZHAO QING
Mining Research Institute, University of Science and Technology, Beijing, China

Abstract
An experimental method is described for determining the fracture toughness of rock over a wide range of loading rate. The fracture tests at both low and high loading rates are performed with the same type of specimens, the short rod (SR,for short) specimen suggested by ISRM. The experimental results show that the fracture toughnesses of marble and gabbro is almost constant at lower loading rates ($\dot{K} < 10^1 MPam^{1/2}s^{-1}$), but increase with loading rates in excess of $10^5 MPam^{1/2}s^{-1}$. Finally, the fractured surfaces of some tested specimens are analysed by SEM observations, and the experimental results are disscussed in the paper.
Keywords: Rock Fracture Toughness, Loading Rate, Gabbro, Marble, Dynamic Fracture Mechanism, Experimental Technique

1 Introduction

In many rock engineerings, such as drilling, blasting, earthquake, landslide, rockburst, oil–well fracturing, protection engineering of nuclear explosion etc. the rock materials are fractured under dynamic loading with widely varied loading rates. Therefore, it is very important to understand the relation between fracture toughness and loading rates. Some works have been done in this field. For examples, Costin [1] measured the dynamic fracture toughness of oil shale, but he did not get the relation between the fracture toughness and loading rates. Wu [2] and Liu [3] obtained the fracture toughness of granite, marble, tuff, and red sandstone, which showed an increasing tendency with loading rates. However, due to the limitation of their test equipments, the maximum loading rate in their tests was only $10^3 MPam^{1/2}s^{-1}$. Employing SR specimen suggested by ISRM for determining the fracture toughness of rock, Zhang et al [4,5] obtained the relation between loading rates and fracture toughness of marble. The maximum loading rate is about $1 \times 10^7 MPam^{1/2}s^{-1}$. This paper is a continuous and further study of [4,5], and the tested rock is gabbro at suburb of Beijing which is a typical igneous rock.

2 Fracture tests

The geometry and dimensions of SR specimen are illustrated in Fig.1 and the parameters are listed in Table 1. Observing the requirement given in reference [6] and the SHPB method, we use D = 30mm and $t \leqslant 0.5$mm in the present study. Fracture testing needs a tensile load to be directly applied to the specimen. Therefore, a special

Fracture and Damage of Concrete and Rock – FDCR-2.
Edited by H.P. Rossmanith. ISBN 0 419 18470 8.

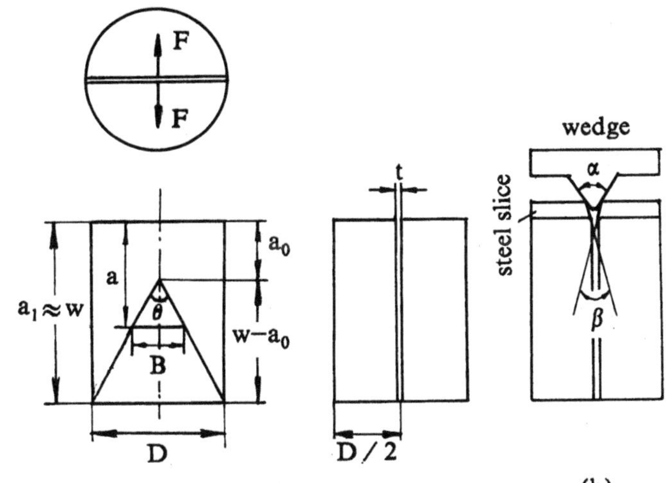

Fig.1. The SR specimen and the wedge as well as steel slices

(a) the SR specimen of ISRM suggested methods

(b) the SR specimen and wedge in this study

a_1 = maximum depth of chevron flanks

B = crack front length

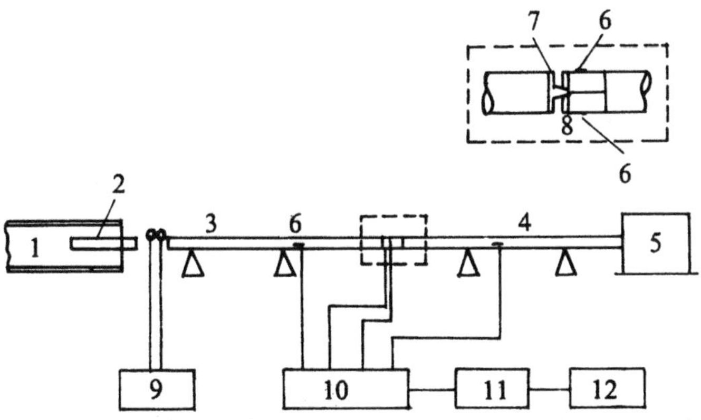

Fig.2. The test equipment of SHPB and measuring system

1. gas gun; 2. striker bar ($\varphi = 30$mm); 3. incident bar ($\varphi = 30$mm)
4. transmitted bar ($\varphi = 30$mm); 5. shock absorbor; 6. strain gauges;
7. wedge; 8. specimen with steel slices; 9. device of measuring velocity;
10. strain amplifier; 11. transient memory; 12. computer

loading wedge and two semi–cirle steel slices are designed to load a SR specimen so that the specimen can be used in both of the quasi–static and the impact fracture tests (Fig.1b). Two steel slices are firmly bonded on the specimen by an adhesive numbered 914 on the market. A V–shaped notch with an angle $\beta=42°$ is formed in the middle of the slices so that the wedge acts on the notch. Application of the wedge with an angle $\alpha=46°$ enables compressive force P to be applied, whereas the crack tip remains loaded in tension.

Table 1. SR specimen dimensions of the ISRM suggested methods [3]

Geometry Parameter	Value	Tolerance
Specimen diameter	D	> 10 × grain size
Specimen length, W	1.45D	± 0.02D
Subtended chevron angle, θ	54.6 °	± 1.0 °
Chevron V tip position, a_0	0.48D	± 0.02D
Chevron length, a_1-a_0	0.97D	± 0.02D
Notch width, t	<0.03D or 1mm *	

* Whichever is greater.

The fracture toughness K_{SR} may be calculated in terms of the following formulas [6,7]

$$K_{SR}=24.0C_K F_C / D^{1.5} \tag{1}$$

and

$$F_c=\frac{P_c}{2tg(\alpha/2+tg^{-1}\mu)} \tag{2}$$

where P_c is the failure compressive force acting on the wedge, F_c the failure load on the specimen, and θ the angle of the wedge, μ denotes the coefficient of friction between the contact surfaces of the wedge and the slices (sliding friction coefficient between the steel wedge and slices is 0.15), K_{SR} is the fracture toughness of rock and C_K is a correct factor. According to the reference [6]

$$C_K=1-0.6\Delta W/D+1.4\Delta a_0/D-0.01\Delta\theta \tag{3}$$

The symbols Δw, Δa_0, and $\Delta\theta$ denote the differences between practical values and the standard ones, for example, $\Delta w/D=w/D-1.45$ etc. With the failure load F_c in KN and the specimen diameter D in cm, the fracture toughness will be in $MPam^{1/2}$.

In quasi–static tests the wedge is loaded by compressive force P, which is recorded during fracture testing. The test is performed according to level I in the suggested method [6] with a testing machine MTS 810 and COD (the crack mouth opening displacement) is also recorded in order to judge the failure load.

The dynamic testing equipment is a gas gun, which together with the measuring system is shown in Fig.2. Once the gun is loaded to a prescribed pressure, the striker bar in the bore is launched. The speed of striker bar is measured by device 9, so that the results of strain gauges on the bars are caliberated. The impact of the striker bar develops a longitudinal compressive wave. The wave propagates down into the bar–specimen system. It is transmitted and reflected many times on the interfaces between the specimen and the wave conduct bars.

Since the striker bar is about eight times as long as the total of the wedge and the specimen, the loading process of the specimen can be approximately treated as quasi–staic. This is the basic assumption of the method.

Strain gauges are used to pick up the wave signals on the bars and the specimen. The signals are then stored in a transient memory and processed by computer as shown in Fig.2.

To describe quantitatively the dynamics of specimen fracturing, a long elastic wave was employed. In this case a force equilibrium in the specimen is very quickly attained. When the crack reaches the critical state the critical force P_c can be calculated according to Eq.(4):

$$P_c(t_c) = EA\varepsilon_T(t_c) \tag{4}$$

where t_c is the time when the crack just starts to propagate, ε_T the strain measured on the transmitted bar, E and A are Young's modulus and the cross section area of the bar, respectively.

It is rather important to determine the critical time t_c of the crack instability. Two strain gauges are attached on the appropriate places of specimen surface, see Fig.2. The principles of the technique are as follows: stress conceneration exists at near the crack tip when the stress acts on the specimen, the strain signal starts to grow up, while the stress rises to certain value the crack becomes instable and starts to propagate. Meanwhile, the strain signal goes down. Consequently, the critical time can be thus determined appoximately. Fig.3 is the strain wave profile.

Thus, fracture toughness of rock under impact loading can be calculated by equations (4), (1), and (2). Typical strain profiles recorded in this study are shown in Fig.4, where ε_I, ε_R, and ε_T denote the incident, reflected, and transmitted strains, respectively.

3 Experimental results

The tested rock is black–grey gabbro taken from the suburbs of Beijing. All of the tested specimens are manufactured from one block of gabbro. The mineralogical composition and maximum grain–size of the gabbro material are shown in Table 2. The obtained relation between the fracture of gabbro and loading rates is shown in Fig.5. Experimental results indicate that the gabbro fracture toughness increases fast when loading rate is over $10^5 MPam^{1/2}s^{-1}$. This result is much same as that of marble [5] (see Fig.5). Unfortunately, there is an experimental data gap within the loading rates $10^1 MPam^{1/2}s^{-1} \leqslant \dot{K} \leqslant 10^5 MPam^{1/2}s^{-1}$ due to the limitation of the testing equipment. But according to the results got by WU [2] and Liu [3], the fracture toughness of marble, granite, tuff and red sandstone increased linearly with increasing loading rates within $10^1 MPam^{1/2}s^{-1} \leqslant \dot{K} \leqslant 10^4 MPam^{1/2}s^{-1}$. Let it be supposed that the fracture toughness of gabbro increase linearly with loading rates within $10^1 MPam^{1/2}s^{-1} \leqslant \dot{K} \leqslant 10^5 MPam^{1/2}s^{-1}$, so the data gap in Fig.5 can be supplemented by a straight line.

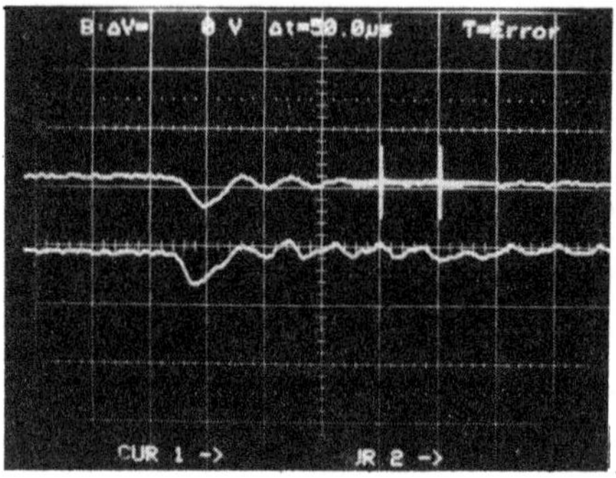

Fig.3. Osillograms of T_c

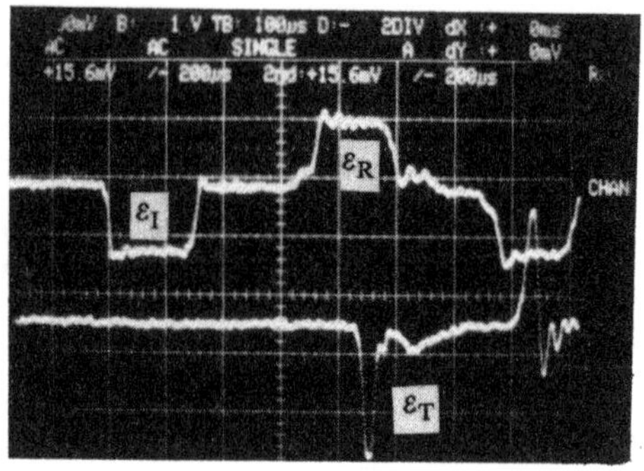

Fig.4. Osillograms of ε_I,$\varepsilon_R$and ε_T

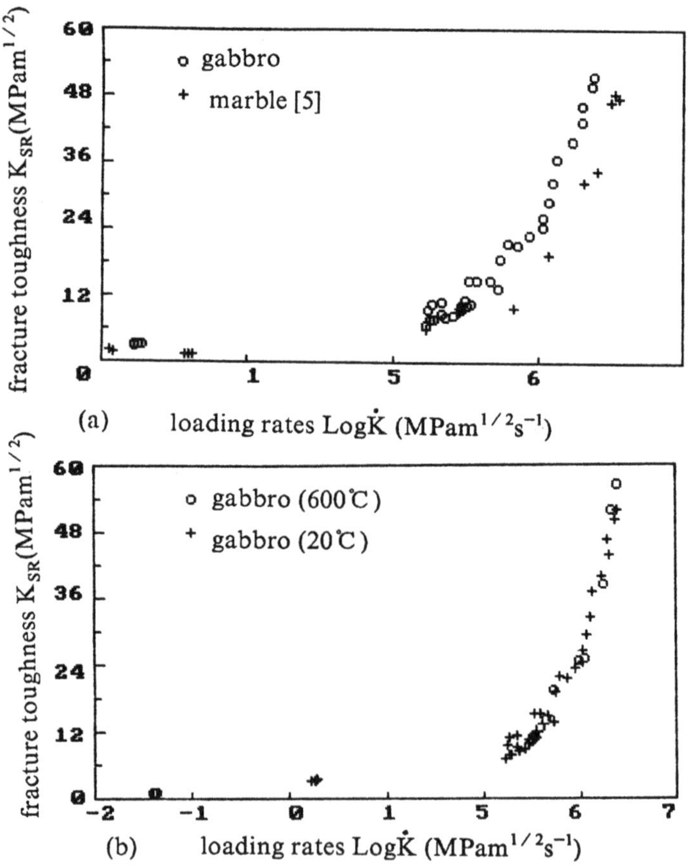

Fig.5. Loading rate — fracture toughness curves

(a) The result of non–treated rock

(b) The comparison results of treated and non–treated gabbro

Table 2. Mineralogical Composition (M.C.) and Maximum Grain–Size (M.G.S.) of Gabbro

rock type	M.C.	(%)	M.G.S. (mm)
gabbro	plagioclase	48–52	1.9
	pyroxene	35–38	1.4
	olivine	5–7	0.6
	hornblende	5	1.1
	biotite	3	0.4
	magnetite	1	0.4

4 Electron microscopy observations

The scanning electron microscope model S–250MK3 is used to observe the fracture surfaces of some specimens fractured at different loading rates.

Considering the mechanisms of final seperation, the cracking observed in various tested specimens is essentially brittle or no plasticity involved. The most common crack modes induced by static and dynamic loading are shown in Fig.6. They occur as cleavage planes, boundary cracks and microcavities. These crack modes are also the characteristic damage patterns created by cyclic loading [8]. Consequently, crack modes in rock can not be used to differentiate between static and dynamic loading or monotonic and cyclic loading induced damage.

However, some other damage characteristics are found on the dynamic fracture surfaces. They are: (1) dynamic fracture surface is much rougher than static one; (2) there are obvious widened boundary microcracks which are perpendicular to fracture surface.

The microcracks is similar to those on the static fracture surface of heated gabbro specimen [9]; (3) secondary cracks on the fracture surface are seen in Fig.6; (4) there are many microcavities and some loosening which may be produced by shearing.

5 Discussion

Rock is a complex solid consisting of heterogeneous aggregates of polycrystalline mineral constituents. These constituents are characteristically anisotropic and have different sizes. Under the condition of impact fracturing, the region near the crack–tip is practically a adiabatic one since rock has lower thermal conductivety than common metals. In the region, stress, energy and temperature are suddenly increased due to fast loading, which leads to formation of microcracks (We can see such microcracks from Fig.6). These microcracks have various oritations and sizes, and they help form more stress concentration locations.

The arbitrary distribution of the locations decreases the stress and energy concentration of the tip of the main crack, so the crack developing resistence increases, i.e. the fracture toughness of rock increases. According to above analyses, it can be induced that the higher the loading rate, the greater the fracture toughness of rock.

To testify our analyses, an experimental result is provided. Under the condition of stress wave loading, the fracture toughness of gabbro heat–treated at 600℃ is not less

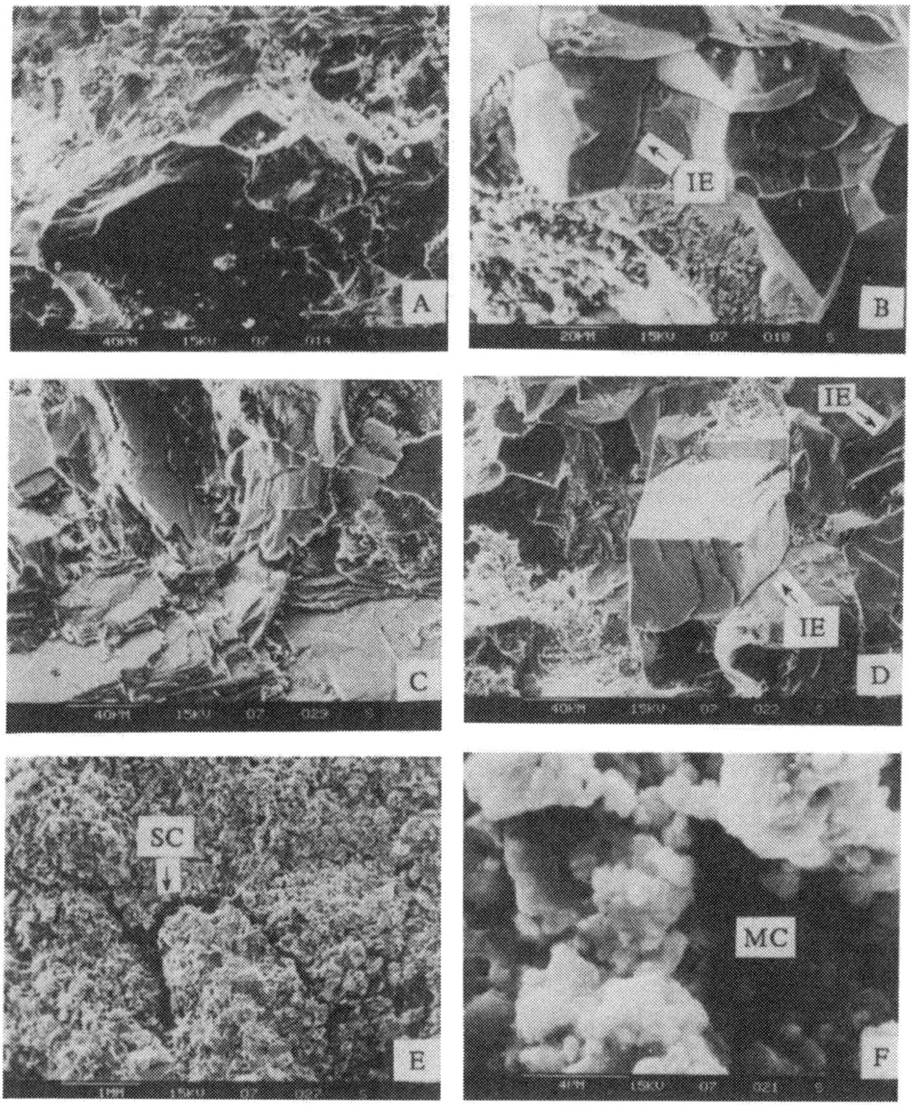

Fig.6. SEM micrographs

A — static loading, B ~ F — stress wave loading ($\dot{K} = 10^5 - 10^6 \mathrm{MPam}^{1/2}\mathrm{s}^{-1}$)

IE — intergranular crack, SC — secondary cracks, MC — microcaveties than that of non–treated one under the condition of the same loading rate [9] see Fig.5b. But in the crack–tip region as well as other region there are more microcracks of the former than that of the latter. That is to say, when the microcracks near the crack–tip increase, the fracture toughness also shows increasing.

In order to explain why the fracture toughness of rock increases with loading rates, various further experimental and theoretical studies are needed.

6 Conclusions

(a) Fracture toughnesses of gabbro and marble rapidly increase with loading rates rising over $10^5 MPam^{1/2}s^{-1}$.

(b) Critical time tc decreases with increasing loading rates, and the higher loading rate, the greater the difference between $t_c$and t_{max}(reaching maximum of transmitted wave).

(c) The experimental technique described in this paper can be used to measure the rock fracture toughness over a wide loading rates, especially to study the effects of loading rates on rock fracture toughness.

Acknowledgements

This work belongs to the Joint Research Project A17 with Professor Dr. H. P. Rossmanith Technical University Vienna, and is sponsored by Committee of Science and Technology China and Department of Mines, Ministry of Metallurgical Industry China. The authors express their sincere gratitude to Professor Wang Chunkui and Yang Jinhui, Mr. Zheng Rong, Dr. Zheng Jian and other people in Mechanics Center of UST and SHPB Lab of Institute of Mechanics, Chinese Academy of Sciences for their enthusiastic supports in testing experiments, and to Professor Duan Zhuping for his enthusiastic support and fruitful suggestions to our impact fracture tests.

References

[1] Costin L. S., Fracture Mechanics for Ceramics, Rock and Concrete, **ASTM SPT745** (1981), 169–184

[2] Wu Mianba, **Mechanics and Practice,** Vol. 10, No. 2 (1986) 21–23

[3] Liu Daan, **Advanced Computer Aid Measuring Techneque and Study on Rock Fracture Mechanics,** Doctor Thesis, Central South University of Technology, 1991

[4] Zhang Zongxian, Zhao Qing, Kou Shaoquan and Yu Jie, **Journal of University of Science and Technology Beijing,** Vol. 14, No. 2 (1992) 210–214

[5] Zhang Zongxian, Zhao Qing, Kou Shaoquan and Yu Jie, **Proc. of the 2nd Int. Symp. on Intense Dynamic Loading and Its Effects,** June 9–12, 1992, Chengdu, China, 396–400

[6] ISRM, **Int. J. Rock Mech. Min. and Geomech. Abstr.,** Vol. 25, No. 2 (1988) 72–96

[7] Klepaciko J., **Inst. Phys. Conf. Ser.,**No. 47, chapter 2 (1979) 201–214

[9] Moelle K. H. R., Li G. and Lewis J. A., **Eng. Frac. Mech.,**Vol. 35, No. 1/2/3 (1990), 367–376

[10] Zhang Zongxian and Yu Yong, in this proceedings.

41 EFFECT OF TEMPERATURE ON FRACTURE TOUGHNESS OF ROCK

ZHANG ZONGXIAN and YU YONG
Mining Research Institute, University of Science and Technology, Beijing, China

Abstract
Gabbro and marble samples were treated at temperatures ranging from −200 to 600℃ and then subjected to static and impact fracture testing. The fracture testing was performed on testing machine MTS 810 and SHPB (Split Hopkinson Pressure Bar), and a short rod (SR, for short) specimen suggested by ISRM was employed. The testing results showed that the static fracture toughnesses of gabbro and marble obviously decreased with increasing heat–treatment temperatures, and that the dynamic fracture toughnesses of heat–treated gabbro and marble were not less than that of non–treated gabbro and marble under the condition of the same loading rate over $10^5 MPam^{1/2}s^{-1}$. Finally a scanning electron microscope (SEM) was used to analyze the thermal cracking of the samples, and the experimental results were disscussed.
Keywords:Temperature, Rock Fracture Toughness, Loading Rate, Experiment Study, Gabbro, Marble.

1 Introduction

In many energy recovery and nuclear waste disposal studies, a fundamental understanding of thermal cracking mechanism of rock is imperative. Therefore, rock mechanics has recently considered the factor of temperature as important [1].Moreover, in rock drilling, fragmentation and crushing, it is very important to study the effects of temperatures on the mechanical properties of rock so as to find out a more efficient and more economical rock fragmentation and crushing method than current single mechanical one.

The effects of temperatures on the mechanical properties of rock obviously change with types of rock, heating ways and heating rates. But many experimental studies indicated that the compressive and tensile strengths of all tested rocks (granite, sandstone, andesite limestone etc [2]), which were first heated, then rapidly cooled and finally tested,basically decreased with increasing heat–treatment temperatures . These results showed that the effects of temperatures on the strengths of all fast cooled rocks have a similar result, i. e. fast cooling reduced the rock strengths.

In order to further study the influence of fast heating and cooling on the mechanical properties of rock, investigating the relation between the temperature and the fracture toughness of rock is rather imperative. Atkinson [3] et al and Kou [4] investigated the effect of temperature on the fracture toughness of granite. In this paper, gabbro and marble are choosed for testing, and the relation between heat or cool treatment temperature and the fracture toughnesses of the rocks has been obtained by experiment. Moreover, the effects of loading rates on the fracture toughness of the heat–treated gabbro and marble have been studied too.

Fracture and Damage of Concrete and Rock – FDCR-2.
Edited by H.P. Rossmanith. © 1993 Taylor & Francis. ISBN 0 419 18470 8.

2 Experimental methods

2.1 Rock types

The gabbro and the marble at suburbs of Beijing were selected for testing (Table 1). They belong to igneous and metamorphic rock respectively.

The gabbro has an equigranular texture, a compact block structure and a grey–white to grey–black colour. It has a good shape of mineral crystals. Plagioclase occurs as self–shaped crystals. The cleavage of pyroxene and olivine develops well.

The marble is white. Its mineralogical composition includs calcite and a small amount of magnesia olivine and magnetite. The cleavages of the calcite amd magnesia olivine grow well. The magnetite is a homogeneous metamorphic product. It has a equigranular and crystalcharged texture, compact block structure.

Table 1 Mineralogical Composition (M.C.) and Maximum Grain–Size (M.G.S) of Gabbro and Marble

rock type	M. C.	(%)	M. G. S. (mm)
Gabbro	Plagioclase	48–52	1.9
	Pyroxene	35–38	1.4
	Olivine	5–7	0.6
	Hornblende	5	1.1
	Biotite	3	0.4
	Magnetite	1	0.4
Marble	Calcite	93–97	1.4
	Magnesia olivine	3	0.4
	Magnetite	1	0.4

2.2 Sample preparation and thermal or cold treatment

Firstly cylindrical cores of gabbro and marble, 30mm in diameter and 100–120mm long, were taken from the same block of each rock.Homogeneous samples were selected with the naked eye. Then,the selected samples were treated at temperature −200 to 600℃ . Finally, the samples were manufactured for fracture toughness and the Brizilian tests.

Heat–treatment at temperature 100–600℃ : The samples were divided into six (gabbro) and five (marble) groups, each group were shock heated to the temperatures mentioned above. The samples were put in a hot oven and left there for 2h at a constant predetermined temperature. They were then taken out of the oven and simply left on a table to cool down.

Cold–treatment at temperature −200℃ : The samples were put in a liquid nitrogen container and left there for 7 days at the temperature −200℃ . They were then taken out of the container and simply left on a table to cool down, and then put into a desicator until they were tested.

2.3 Experimental methods

All tests were carried out at room temperature which may fluctuate a couple of degrees from one day to another.

2.3.1 Brazilian tests

These tests were conducted in accordance with the ISRM's Suggested Methods [5] (Brown, 1981) apart from the fact that all tests were carried out at a constant displacement rate (0.01mm / s) rather than at a constant loading rate. The 100KN MTS–810 press was utilized in these tests.

2.3.2 Static fracture toughness tests

The static fracture toughness tests were carried out according to the procedure of the Level I testing of the ISRM's Suggested Method for Determining the fracture toughness of rock using SR Specimens [6]. The length of the tested specimen is about 44mm and the notch width is less than or equal to 0.5mm. All of the specimen geometry parameters are within the tolerances of the ISRM's suggested method. The 100KN MTS–810 was utilized too.

The testing was carried out at a constant tensile displacement rate of 0.001–0.01mm / s

2.3.3 Dynamic fracture toughness tests

These tests were carried on the SHPB test bed. The testing method was described in reference [7] in this proceedings.

3 Experimental results

All data points in the diagrams presented here are mean values from three to six tests except the data points of dynamic fracture toughness.Each of the points is one test value.

3.1 Brazilian tests

The Brazilian test is a rather insensitive test for studing variations in the tensile strength of rocks due to microstructural changes in the starting material, but it is widely used in Rock Mechanics. Equation 1 gives the formula used for calculating this strength.

$$\sigma_t = \frac{2P}{\pi D t} \tag{1}$$

where P is the maximum failure load, D the diameter of the specimen, and t the thickness of the specimen, which in this case was equal to half the diameter. Our results are summarized in Fig.1. We see that the tensile strengths of gabbro and marble decrease with increasing heat treatment temperatures, and that the decrease of tensile strength in marble is faster than in gabbro tensile strength. In addition, the tensile strength of marble treated at −200℃ is less than that of marble non–treated, but the tensile strength of gabbro treated at −200℃ is greater than that of non–treated gabbro.

3.2 Static fracture toughness tests

Fracture toughness of rock is an important material property in the modelling of rock fragmentation like rock cutting, hydraulic fracturing, gas driven fracturing, explosive stimulation etc. The formula used for calculating this parameter is

$$K_{SR} = 24 C_K F_{MAX} / D^{1.5} \tag{2}$$

K_{SR} is the fracture toughness of Short Rod Specimen, F_{MAX} the failure load, D the

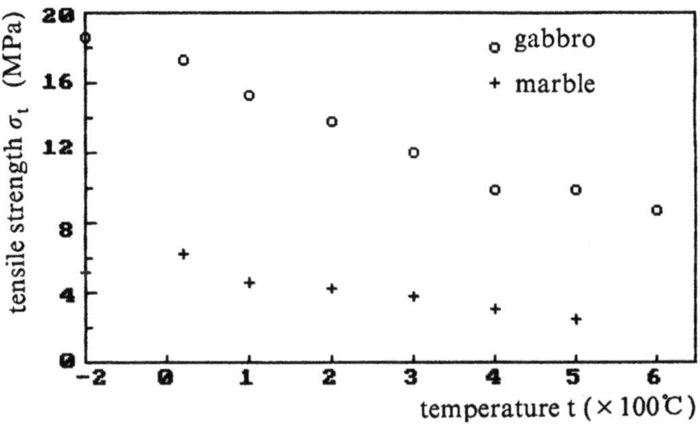

Fig.1. Temperature — tensile strength curves

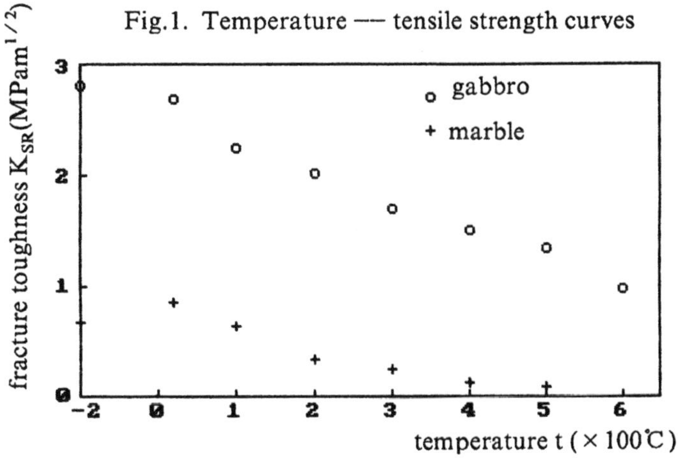

Fig.2. Temperature — fracture toughness curves

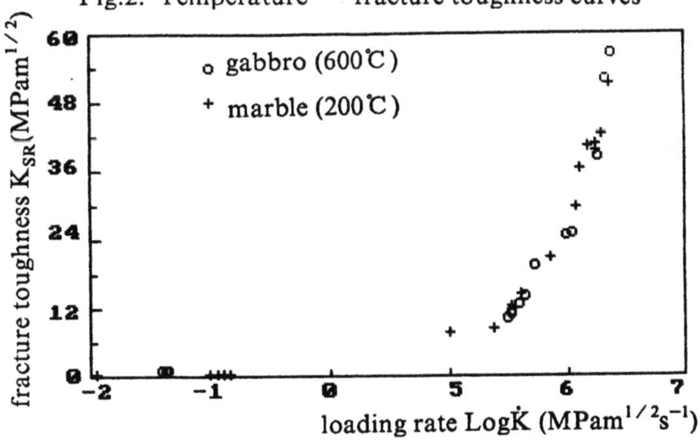

Fig.3. Loading rate — fracture toughness curves

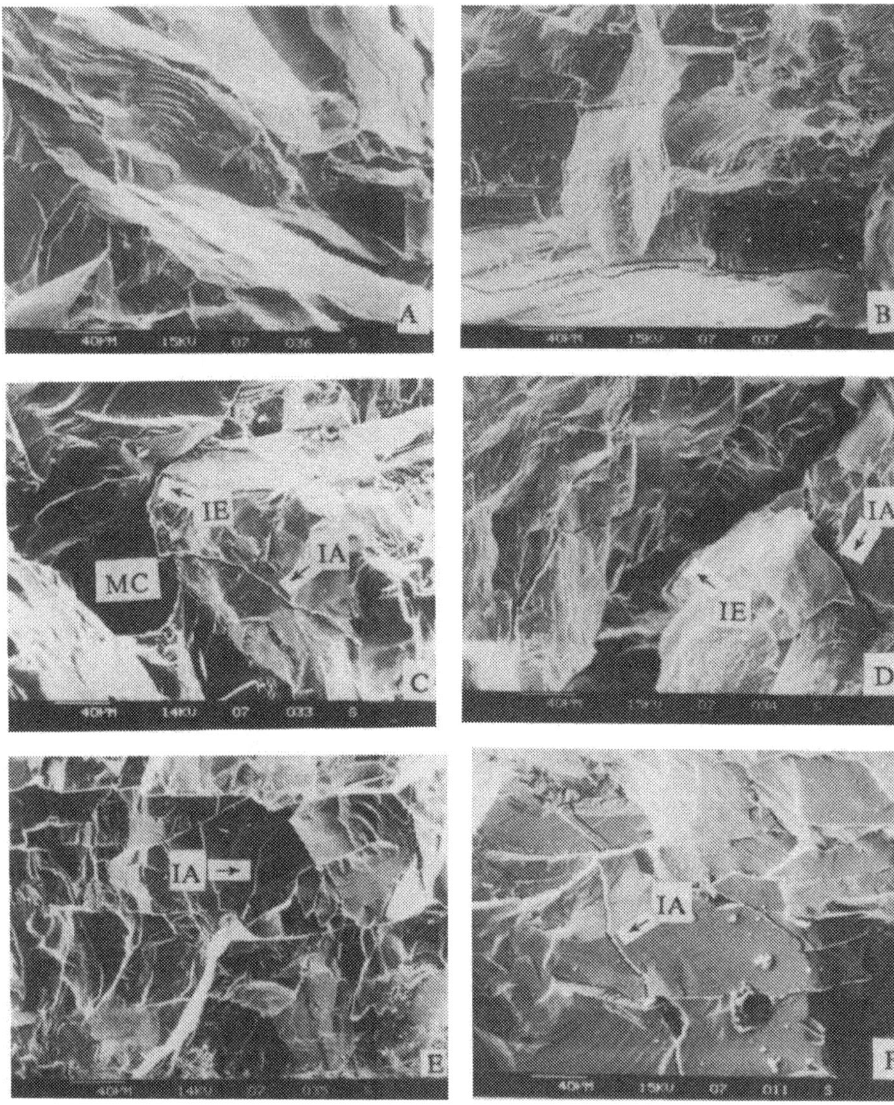

Fig.4. SEM micrographs

A ~ B — 100℃, C ~ E — 300℃, F — 500℃

IA — intragranular cracks, IE — intergranular cracks, MC — microcavities diameter of specimen, and C_K the correction factor. It shall then be calculated:

$$C_K = (1 - 0.6\Delta w / D + 1.4\Delta a_0 / D - 0.01\Delta\theta) \tag{3}$$

The symbols Δw, Δa_0, and $\Delta\theta$ denote the differences between measured values and the nominal ones, i. e. $\Delta w / D = w / D - 1.45$ etc. Our results are shown in Fig.2. It can be seen that the fracture toughnesses of gabbro and marble decrease with increasing heat–treatment temperatures, and that the fracture toughness of marble decreases more rapidly than that of gabbro. This result is similar to that of the tensile strengths of gabbro and marble.

3.3 Dynamic fracture toughness tests

The tested gabbro and marble were heat–treated by previous described method at 600℃ and 200℃ respectively. The testing results are summerized in Fig.3. The results show that the dynamic fracture toughnesses of heat–treated gabbro and marble increase with loading rates increasing. This change is similar to non–treated gabbro, i.e. heat–treatment does not practically affect the changing law between fracture toughness of the rock and loading rate.

3.4 Microscope

The microcracks on some thin granite sections representing all heat– treatment temperatures had been directly observed in an optical and a scanning electron microscope by Alm et al [8] and Homand–etienne et al [1]. Their investigations indicated that the mean cracking density increased with heat treatment temperatures rising. In our study, we only observed the fractured surfaces of heat–treated rock by SEM (S–250 MK3). The SEM micrographs are shown in Fig.4.

From Fig.4 we find that there are hardly any intergranular and intragranular cracking in the samples heat–treated at 100℃, but there are several intergranular and intragranular cracks in the 300℃ and 500℃ samples. In addition,an obvious microcavity is seen in the 300℃ sample (Fig.4), and an intragranular and a few intergranular cracks seem to develop from the microcavity. Similarly, in Fig.4 a small intragranular crack seems to start from a large intergranular one.

According to observed results in SEM, it can be concluded that intergranular cracking is by far the most important cracking mechanism in thermally induced microcracking, that intragranular cracking increases with heat–treatment temperatures rising and that the microcavities and big intergranular microcracks are possibly the damage sources of thermal cracking.

4 Discussions

Rock is a complex natural material consisting of heterogeneous aggregates of polycystalline mineral constituents which are characteristically anisotropic and have different sizes. Moreover, there are great differences between tensile, shearing and compressive strengths of rock, and in general, the tensile strength is much less than the compressive one. Heating a block of rock usually induces a triaxial state of tensile stresses within the rock [9]. The tensile stresses induced by heating often results in formation of intergranular and intragranular cracking in the rock. Thermal fragmentation and crushing, therefore, makes full use of the intrinsic attributes of rock, such as a great number of innate microcracks, microcavities, and other defects within rock, great differences between tensile and compressive strength of rock etc. Thirumalai [9] studied rock crushing by thermal shock about twenty years ago, however, this thermal crushing method has not been used in engineerings so far. Zhang [10] considered that combining thermal and mechanical fragmentation and crushing was possibly a more effective and economical method than a single thermal or mechanical one. But this need to be proved by further experimental and theoretical investigations.

5 Conclusions

(a) The static fracture toughnesses of gabbro and marble decrease with thermal treatment temperatures increasing. This change consists with that of the tensile strength of the rocks.

(b) With heat–treatment temperatures increasing, the fracture toughness and tensile strength of marble decrease more rapidly than those of gabbro. This result indicates that rock type deeply affects the extent of thermal cracking damage.

(c) After gabbro is heat–treated, the relation between its fracture toughness and loading rate is much same as that of non– treated one.

(d) To fracture rock efficiently, treating it by shock–heating is a possibly better method for some rock, especialy at lower loading rates.

Acknowledgements

This work is sponsored by Department of Mines, Ministry of Metallurgical Industry China and supported by Committee of Science and Technology China. The authors express their sincere gratitude to Prof. Wang Chunkui, Yang Jinhui and Duan Zhuping, Mr. Zheng Rong, Dr. Zheng Jian, and other people in Mechanics Center of UST and SHPB Lab in Institute of Mechanics, Chinese Academy of Sciences for their supports in testing experiment.

References

[1] Homand–etienne,F. and Houpert, R. **Int. J. Rock Mech. Min. Sci. & Geomech. Abstr.**, vol.26, no.2 (1989) 125–134

[2] Lin Muzeng, **Thermal physics of rock and Its Engeineering Applications**, Chongqing University Press, China (1991)

[3] Atkinsion, B. K. et al, **Proc. 3rd. Int. Conf. on AE / Microseismic Activity in Geologic Structures and Materials**,Ed. by H. R. Hardy & F. W. Leighton (1984)

[5] Brown E. T. (editor), **Rock Characterization testing and Monitoring: ISRM Suggested Methods**, Pergamon Press, Oxford, pp.113–121 (1981)

[6] ISRM, **Int. J. Rock Mech. Min. Sci. and Geomech. Abstr.**, vol.25, no.2 (1988) 73–96

[7] Zhang Zongxian, Yu Yong and Zhao Qing, in this proceedings

[8] Alm, O. Jakeland, L. L. and Kou Shaoquan,**Physics of the Earth and Planetary Interiors**,40 (1985) 161–179

[9] Thirumalai, K. and Cheung J. B., A study on a New Concept of Thermal Hard Rock Crushing, **ASCE Proc. of the 14th Symp. on Rock Mech.**(1973) pp. 527–554

[10] Zhang Zongxian, Effect of heat–treatment on rock damage, **Nonferrous Metal**(quarlerly) (to be published)

PART SIX
APPLICATIONS

42 HYDRAULIC FRACTURING OF CONCRETE AND ROCK

J.H.M. VISSER and J.G.M. van MIER
Stevin Laboratory, Delft University of Technology,
The Netherlands

Abstract
Hydraulic fracturing of porous media is shown to be a coupled problem between the poroelastic deformation, the fluid flow in the porous medium and the fluid flow in the fracture. Seven material parameters are needed to describe the fractured porous medium. These parameters strongly influence the fracture initiation and propagation. To determine this influence, a new pressure cell has been built with which fluid-driven fractures can be obtained in laboratory scale specimens. Pilot test with this cell have been performed on concrete cylinders subjected to a radial fluid pressure. The preliminary results are very promising with regard to the ability to perform stable, deformation controlled tests.
Keywords: Hydraulic Fracture, Poroelasticity, Fluid Flow, Fracture initiation, Fracture Propagation, Pressure Cell, Concrete, Experiments.

1. Introduction

Hydraulic fractures often are observed in concrete structures in a water environment, such as dams and off-shore structures. The fractures can lead to extensive damage, and even to a breakdown of the concrete structures. However, the hydraulic fracturing process is still one of the least understood problems in materials science.

Hydraulic fracturing can be defined, in general terms, as the fracturing of a solid due to a fluid pressure. It is a highly coupled problem between fracture mechanics and fluid mechanics. Fracture mechanics predict the fracture geometry once the pressure distribution in the crack is known. On the other hand, fluid mechanics predict the pressure distribution in the fracture once the fracture geometry is known. The problem becomes even more complicated when the solid is porous and also the flow in the porous rock has to be taken into account. The mutual dependency between the fracture geometry and the fluid pressure distribution in the fracture and rock pores is defined largely by the constitution of the fracturing fluid and the solid. However, the influence of the material parameters on the hydraulic fracturing process is still very obscure. In order to determine these influences, a new pressure cell has been built with which fluid driven fractures can be obtained in laboratory scale specimens. With this cell, the hydraulic fracturing process can be monitored for different materials and for different loading conditions.

This paper will give a review of the hydraulic fracturing of porous media and states the parameters involved. To complete the hydraulic fracturing problem, a fracture initiation and propagation criterion is set. The hydraulic fracturing process is monitor using a new pressure cell, of which the design is given in this paper together with some preliminary results.

Fracture and Damage of Concrete and Rock – FDCR-2.
Edited by H.P. Rossmanith. © 1993 Taylor & Francis. ISBN 0 419 18470 8.

2. Theory

In a porous medium containing a hydraulic fracture, three distinct but coupled physical phenomena can be distinguished. These phenomena are (1) the deformation of the poroelastic medium, (2) the fluid flow in the medium and (3) the fluid flow in the fracture. In this section, these three physical phenomena will be described, under the assumption that it is a fully saturated medium, i.e. both the pores and the fracture are filled by a homogeneous fluid.

2.1 Deformation of the poroelastic medium

The assumptions made for the porous medium is that it is isotropic and elastic. The deformation of the porous medium is then described by the linear poroelastic theory of Biot (1941,1956). However, we will use the formulation of Rice and Cleary (1976). The first constitutive equation for the state of equilibrium when the pore pressure is constant throughout is:

$$\sigma_{ij} = 2G\varepsilon_{ij} + \frac{2G\nu}{(1-2\nu)}\varepsilon_{kk}\delta_{ij} - \frac{3(\nu_u-\nu)}{B(1-2\nu)(1+\nu_u)}p_p\delta_{ij} \tag{1}$$

This constitutive equation gives the stress σ_{ij} in terms of the strain ε_{ij} and the confining strain ε_{kk}, following Hooke's law, plus the static increment in fluid pressure p_p. The effective stress $\tilde{\sigma}_{ij}$ that governs the elastic deformation is thus given by:

$$\tilde{\sigma}_{ij} = \sigma_{ij} + \alpha p_p\delta_{ij} \quad , \quad \alpha = \frac{3(\nu_u-\nu)}{B(1-2\nu)(1+\nu_u)} = \text{Biot's poroelastic constant} \tag{2}$$

The second constitutive equation is:

$$p_p = \frac{-2GB(1+\nu_u)}{3(1-2\nu_u)}\varepsilon_{kk} + \frac{2GB^2(1-2\nu)(1+\nu_u)^2}{9(\nu_u-\nu)(1-2\nu_u)}\zeta \tag{3}$$

This equation gives the pore pressure p_p is terms of a hydrostatic strain ε_{kk} plus a variation in water content ζ.

The poroelastic medium is described by four poroelastic parameters, namely the drained rock shear modulus G; Skempton's coefficient or the pore pressure coefficient B, defined as the ratio of variation of the pore pressure over the variation of the confining pressure under undrained conditions; the drained Poison's ratio υ, refering to the asymptotic value obtained when any alternation of the pore pressure field induced by loading has disappeared; and the undrained Poison's ratio υ_u, describing the short term response of the porous solid when no fluid movement has occurred (Detournay et al., 1986).

The linear poroelastic theory reduces to the linear elastic theory in the following cases: (1) ζ=0, this is for the undrained case, i.e. the fluid in the pores has yet had no time to escape. The pore pressure then is proportional to the volumetric strain (see eq.(3)) and constitutive eq.(1) reduces to Hooke's law with an undrained Poison's ratio.
(2) p_p=0, this for the drained case, i.e. the permeability is so large that, compared with the loading time, the pore fluid has ample time to dissipate. Eq.(1) reduces now to Hooke's law with a drained Poison's ratio.

The field equations which describe the deformation of the poroelastic medium (by neglecting body forces) are the equation of equilibrium:

$$\frac{\partial \sigma_{ij}}{\partial x_j} = 0 \quad , \quad \sigma_{ij}=\sigma_{ji} \tag{4}$$

and the equation of compatibility (from eq.(1) and eq.(4)):

$$\nabla^2[(1+\nu)\sigma_{ij}-\nu\sigma_{kk}\delta_{ij}]+\frac{\partial^2\sigma_{kk}}{\partial x_i\partial x_j}+\frac{3(\nu_u-\nu)}{B(1+\nu_u)}[\ \nabla^2 p_p\delta_{ij}+\frac{\partial^2 p_p}{\partial x_i\partial x_j}] \tag{5}$$

2.2 Flow in the poroelastic medium

Next, the transient phenomena in the poroelastic medium have to be described, i.e. those equations governing the distribution of the stress, water content and settlement as a function of time under given loads. It is assumed that the constitutive law governing pore fluid diffusion is given by Darcy's law, for the isotropic case:

$$q_{p_i} = -\kappa\frac{\partial p_p}{\partial x_i} \tag{6}$$

in which q_p, the fluid flow rate in the pores, is proportional to the pore pressure gradient. In this constitutive equation for the pore fluid flow, only one additional material parameter is needed. This is the permeability κ, defined as the ratio of the intrinsic permeability k over the dynamic viscosity of the pore fluid μ_p.

The field equation is obtained from the mass conservation law:

$$\frac{\partial q_{p_i}}{\partial x_i}+\frac{\partial V_p}{\partial t}+q_l=0 \tag{7}$$

with V_p the volume of pore fluid in the porous medium and q_l the fluid loss (or gain) rate. Substitution of eq.(3) and eq.(6) in eq.(7) then gives;

$$\frac{2GB^2\kappa(1-\nu)(1+\nu_u)^2}{9(\nu_u-\nu)(1-\nu_u)}\nabla^2(\sigma_{kk}+\frac{3}{B}p_p)=\partial_t(\sigma_{kk}+\frac{3}{B}p_p)+q_l \tag{8}$$

2.3 Fluid flow in the fracture.

The physics of the transport of flow in the fracture is significantly different from that in the porous medium. Also, the fluid in the fracture need not to be similar to that in the pores. Therefore the fluid flow in the fracture is considered separately.

Assume that the fracture fluid is homogeneous (single-phased) and incompressible. Then the constitutive equation for the Newtonian fluid is given by:

$$\sigma_{ij}=-p_f\delta_{ij}+2\mu_f v_{ij} \tag{9}$$

where the subscript f now denotes fracture fluid. The only material parameter in this equation is the viscosity of the fracturing fluid μ_f. The fluid movement is governed by the boundary conditions at the fracture walls and the equation of transport (e.g. Navier-Stokes). From mass balance, the following field equation is found:

$$\frac{\partial q_{f_i}}{\partial x_i}+\frac{\partial V_f}{\partial t}+q_l=0 \tag{10}$$

in which the first two terms give the increase in fluid mass via the fracture opening for an incompressible fluid and the third term q_l gives the fluid loss or gain via the boundaries with the porous medium.

2.4 Fluid-loss

The poroelastic medium containing a fracture has now been described. Its formulation is complete except for one additional constitutive equation. In the equations for the mass conservation for the fluid in the poroelastic medium (eq.(8)) and that for the fluid in the fracture (eq.(10)), a fluid loss (or gain) term q_l is added. It is assumed that the permeable porous medium is sealed except on the boundaries of the fracture, then the term in both equations is similar except for the sign. The rate of exchange depends on the pressures in the systems and the permeability of the porous medium. Because the fracture is in general pressurized from the fracture entrance, the pressure in the fracture is normally higher than the pressure in the porous medium. Fluid will then leak off from the fracture in the porous medium. Assuming that the fluid movement is governed by pressure propagation only, the constitutive equation for the fluid flow is $q_l=\beta(q_p-q_f)$ in which β is the coefficient describing the intensity of the fluid exchange (Shapiro, 1987). Mostly, β is taken proportional to the product of the porous matrix permeability and the specific surface area of the fracture. In the oil-well recovery branch, another more or less empirical relation is adopted (Howard and Fast, 1970). Under assumption of a piston-like Darcy flow and infinite media, they assume that an instant steady-state flow develops in the porous medium such that:

$$v=C_T/\sqrt{t-\tau} \tag{11}$$

in which v is the instantaneously steady-state flow through the region, C_T is the fluid-loss coefficient and $t-\tau$ is the time elapse since the fluid was first exposed to the surface. Following Williams (1970), the fluid-loss coefficient is given by:

$$C_T=2C_cC_v/(C_v+\sqrt{C_v^2+4C_c^2}) \tag{12}$$

with C_c the pore fluid compression control coefficient and C_v the viscous fracture fluid coefficient given by (Settari, 1985):

$$C_c=\sqrt{\frac{\kappa c_p\phi\Delta p^2}{2\mu_p}}\ ,\ C_v=\sqrt{\frac{\kappa\phi\Delta p}{\mu_f}} \tag{13}$$

The fluid loss coefficient C_T is thus a function of (1) pressure differential between fracture and porous medium $\Delta p=p_f-p_p$, (2) the permeability κ and porosity ϕ of the solid, (3) the pore fluid viscosity μ_p, (4) the fracturing fluid viscosity μ_f and (5) c_p the compressibility of the pore fluid. It is assumed that the liquids in the pores are incompressible. Consequently the compressibility of the pore fluid $c_p=0$. The coefficient C_c is zero as is then C_T and there will be no fluid leak-off. This is of course a very rude approximation and will in general hold only when the pore fluid and the fracturing fluid are similar. The assumption of incompressibility will however not hold for gases. In case the pores are filled by air, it is assumed that the air in infinitely compressible. The medium is then called unsaturated. In this case C_c becomes infinite and the fluid- loss rate C_T depends now only on the resistivity of the porous medium to the invading fluid, given by C_c.
As Settari (1985) noted, this linear theory has several severe restrictions, such as not taking into account partial saturation of the porous medium due to progressive invasion of the fracture fluid and by taking a constant pressure difference between fracture pressure and pore pressure. But in section 3.2, it will be shown that the above approximation will hold in quite some cases.

3. Fracturing

In the preceding section, we have set the equations governing the stress distribution in a poroelastic medium containing a fracture. In this section, we will add a fracture initiation and propagation criterion, to come to a full description of the hydraulic fracturing problem.

3.1 Fracture initiation

A fracture will be initiated when the imposed pressure load is larger than the failure strength of the material. Failure in the presence of pore fluid pressures is assumed to be governed by Terzaghi effective stress such that:

$$\sigma'_{ij} = \sigma_{ij} + p_p \delta_{ij} \tag{14}$$

where σ'_{ij} is the total stress and p_p is the pore pressure (Boone and Ingraffea, 1988). The effective stress governing the deformation is however given by Biot's effective stress (eq.(2)). The direction of the initiated fracture will be normal to the least principal stress, or largest tensile stress, given by the field equations in section 2.

There are two parameters which might alter the apparent failure strength of a material: the hydrostatic stress or confining pressure and the pore pressure. The effects of the pore and confining pressure depend largely on the constitution of the solid (section 2.1). Depending on the constitution of the solid, they may give an increase or a decrease in failure strength. As seen in section 2.1, the volumetric deformation of fluid saturated solids exhibits a sensitivity to the rate of isotropic loading, i.e. the solid appears to be stiffer under fast than under slow loading when the fluid has ample time to diffuse through the solid (Detournay and Cheng, 1988).

3.2 Fracture propagation

Once a crack has been initiated, the fracturing fluid will enter the crack. Because the fluid is under pressure, it will apply an extra force on the walls of the fracture. Fracture propagation will occur when the resultant of the internal and external stresses together will be large enough to overcome the failure strength of the material.

To determine the influence of the material properties and variables on the internal loading, we will distinguish to cases of fracture propagation, namely static and dynamic fracture propagation.

3.2.1 Static propagation.

It is assumed that the loading rate is so low that the crack propagation can be considered static, i.e. there's no fluid flow in the crack. The diffusivity transport which takes place to reduce the excess pore pressure is still a much more slower process, so undrained conditions will be assumed throughout. Then the following cases can be distinguish:

(1) no porosity: because no fluid can leak off, the pressure in the fracture p_f is constant. The pressurized fluid in the crack will impose an extra body force normal to the fracture walls. The contribution of the internal loading to the hydraulic fracturing process thus depends solely on the geometry of the fracture.

(2) high porous, but impermeable solid: no fluid can leak-off from the crack into the surrounding solid, and the pressure in the crack is maintained. However, for a saturated solid, the pore pressures have to be taken into account, which might act as a closing force, reducing the pressure in the crack.

(3) porous, permeable solid: the pressurized fluid in the crack may now leak off in the

surrounding material. This reduces the pressure in the crack and gives rise to a pore-pressure build-up in the surrounding material. The fluid loss is said to be static because there is no flow across the fracture surface at which fluid leak-off occurs (Williams, 1970). For the saturated specimen, it was argued in section 2.4 that there's no leak-off current in the concrete, at least no pressure difference flow. Therefore, pressure can be built up faster, giving rise to a higher crack propagation speed. In other words, less pressure is needed for a similar crack speed as for the unsaturated specimen.

3.2.2 Dynamic propagation
The main implication of a high crack speed is that the fluid actually flows in the crack. The flow pattern e.g. laminar or turbulent, depends heavily on the fracture geometry. Due to the flow in the crack, the pressure will drop with distance from the crack opening, which will change the stress situation in the solid considerably. This pressure drop depends not only on the type of flow, but on the fluid properties as well: for high viscous fluids, the pressure drop will be largest, due to large friction losses along the boundaries of the crack. Also now there exist a trade-off between fracture fluid velocity and leak-off. The higher the flow velocity, the lower the leak-off will be. This trade-off is also dependent on the type of flow. For the present, fluid flow in the crack will be discarded.

To check the influence of the material parameters on the fracture initiation and propagation, they have to be varied one by one. This can be done by performing experiments on concrete. The properties of concrete can be varied very easily by changing the mixtures.

4. Experiments

In order to determine the influence of the hydraulic fracturing parameters, a simple pressure cell has been developed. In this cell, fluid driven fractures can be obtained in laboratory scale specimens. Concrete or rock cores of 100 mm diameter and 300 mm length are subjected to axial tension or compression and a radial fluid pressure of up to 100 bar.

In this chapter, the construction of the test cell will be described and the first preliminary results will be given. But first the back ground principle on which the design of the pressure cell is based will be given.

4.1 Background idea of the pressure cell

The idea behind the design of the pressure cell is as follows: consider the cross-sectional view of a cylindrical solid subjected to a radial fluid pressure in Fig.1.
Following the superposition rule, the loading on the cylinder can be thought to consist of a hydrostatic pressure and an axial tensile stress. The hydrostatic pressure causes a volume change only but gives no net force. Therefore, the response of the cylinder subjected to the radial fluid pressure will be largely similar to the response of the same cylinder under uniaxial tension (Clayton and Grimer, 1978). Fracture initiation is assumed to occur due to external loads only. In response to the radial fluid pressure, the solid will thus fracture as if it was submitted to an uniaxial force (see Fig. 2). Thus crack initiation will occur in mode I, perpendicular to the axis of the cylindrical specimen. This was found in several pilot tests, see section 4.4. On the other hand, fracture propagation does not only depend on the external forces, but on the internal forces due to the pressurized fluid in the crack

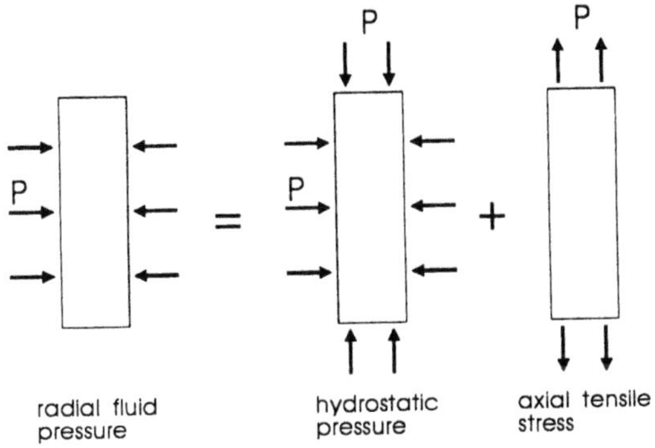

Fig. 1. Superposition principle.

as well (see Fig.3), see the discussion in section 3.2.

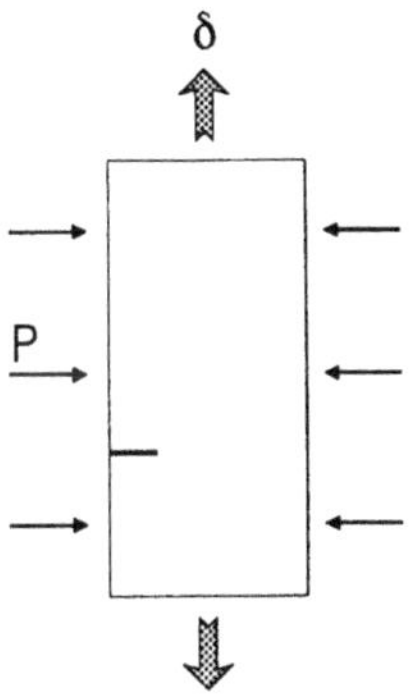

Fig. 2. Fracture initiation.

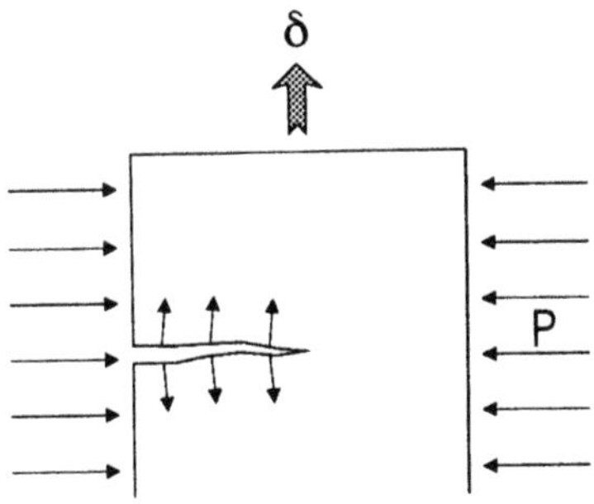

Fig. 3. Fracture propagation.

4.2 Materials and specimen

The specimens used in the pressure cell are mortar (d_{max} = 2 mm) cylinders of 300 mm length and 100 mm diameter. The composition of the concrete is given in Schlangen and van Mier (1992). The manufacturing of the specimen can be found in Visser and van Mier (1992). Preparation of the specimens consists of notching them to define the region of fracturing and sealing the ends of the cylinder. A circular notch of 5 mm depth and 5 mm width is sawn 100 mm from the top of the specimen. The specimens are completely sealed except for the notch so no fluid can enter the specimen from the water reservoir. Sealing of the cylinder ends is done with epoxy resin. In Fig.4, the cylindrical specimen is shown.

4.3 Experimental set-up

The pressure cell has been developed primarily for triaxial tension- compression-compression states of stress. Therefore, the pressure cell, of which a cross sectional view is given in Fig.5, consists of three parts: an upper plate, a circular shaft and a lower plate which are held together by safety bolts. The upper and lower plate of the cell are not solid, but contain a circular hole of 100 mm diameter in which the cylindrical specimen is contained. In this way, the specimen can move freely in longitudinal direction. Also, an axial force (either compressive or tensile) can be applied on the specimen ends. Radial compressive forces on the specimen can be applied by pressurizing the water in the cell. As can be seen in Fig.5, not the total length of the circumference of the specimen is pressurized, but only 200 mm of the specimen is surrounded by water. Rod seals between the upper c.q. lower plate and the specimen prevent water leakage from the cell.

In the upper plate of the cell, a filler cap and a valve are inserted so the cell can be filled with water before pressurization. On top of the upper plate, a safety plate is attached to prevent the specimen from shooting out of the cell in case of uncontrollable fracture.

In the lower plate of the cell, a small channel has been drilled to the water reservoir of the cell through which water can be transported either in or out the cell in order to adjust the pressure. Four additional channels are leading the LVDT-wires out of the cell. A second plate is attached under the lower plate. The specimen can be attached to this so-called base-plate, and an axial force can be applied to the specimen.

Only two parameters can be measured in the pressure cell. The first parameter is the pressure in the cell, which is measured with a pressure transducer. The second variable is the axial elongation of the specimen which is measured with two high pressure LVDT's (type W1S/6S, linear stroke +/- 1 mm, measuring length 80 mm) 180 degrees apart. The LVDT's can be used up to pressures of 200 bar. For a full description of regulation system see Visser and van Mier (1992).

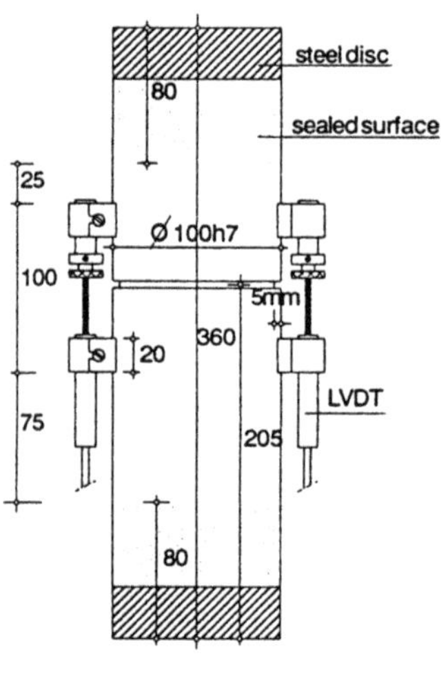

Fig. 4 Specimen.

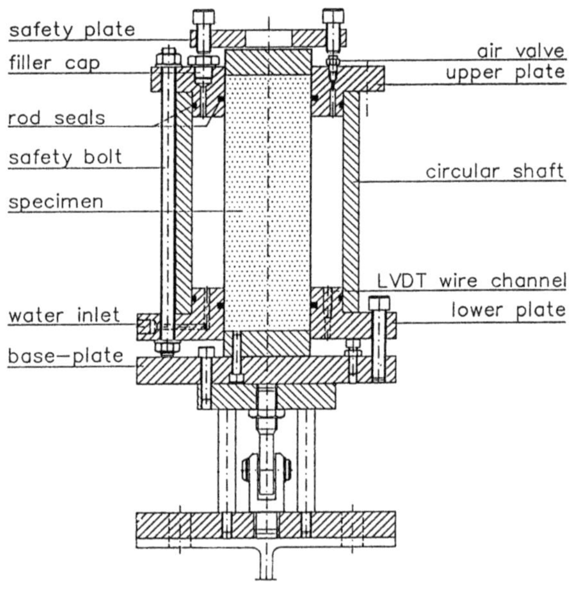

Fig. 5 Pressure cell.

4.4 Preliminary results

The experiments performed so far are all pilot tests to adjust the regulation system of the pressure cell. Water is used as a pressure fluid and the specimens are loaded by a radial fluid pressure only. The tests done with the hydraulic cell are all deformation controlled, so the control variable is the average axial elongation of the specimen measured with two LVDT's. The prescribed deformation rate is kept constant at 0.12 μm/s. The specimen are unsaturated, but with a very low porosity.

An example of a pressure-displacement curve obtained in a hydraulic test is shown in Fig.6. The pressure-displacement curve has a remarkably similar shape with the load-displacement curves from the uniaxial tensile tests of Schlangen and van Mier (1992). Although the test have been performed on specimen with different geometry and material, the Young's moduli and tensile strengths are almost equal. Moreover, a descending branch is seen beyond peak, followed by a stable tail until complete separation occurs.

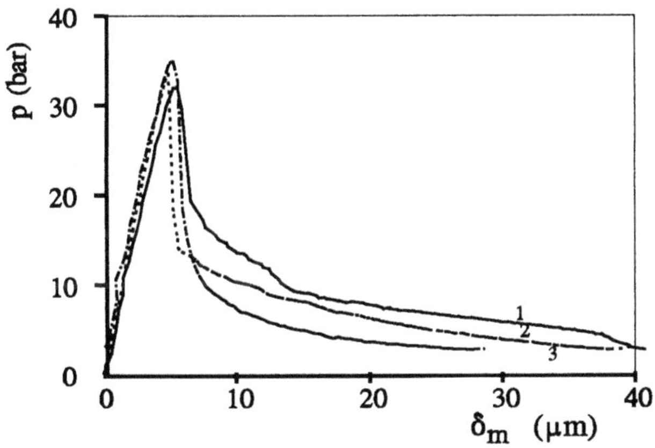

Fig. 6 Pressure-displacements curves for three hydraulic tests.

It thus seems that the fracturing mode of the specimens is purely tensile, or mode I. This conclusion is confirmed by the fact that the fracturing plane is indeed normal to the axis of the cylinder. A third confirmation can be found from fact that the peak strengths of the curves ($\bar{f}_t$=3.3±0.2 MPa)are almost equal to the tensile strength of the mortar, as derived from the cubic tensile splitting tests ($\bar{f}_{spl}$=3.4±0.1 MPa).

No further conclusions can be drawn at present. The tests are only preliminary and are all performed on the same material and under the same loading conditions. The test results are seen to be very promising. As far as we know, it is the first time that stable, deformation controlled test have been carried out in a water environment.

5. Conclusions

In the theory of hydraulic fracturing of porous media, seven material parameters are of importance. Four parameters describe the poroelastic medium, namely the shear modulus, the Skempton pore pressure coefficient and the drained and undrained Poison's ratio. One additional parameter, the permeability, describes the fluid flow in the poroelastic media. The pore fluid and the fracture fluid are described by their viscosity. It is shown that the fracture initiation and propagation strongly depends on the mentioned parameters. To determine the effect of the material parameters on the hydraulic fracturing process, a new test cell for laboratory scale specimen is developed. Only some pilot tests have been performed in which the concrete cylinders were loaded along the circumference by water pressure only. The tests were all deformation controlled and the deformation rate was kept constant. Similarity between the load-displacement curves of the hydraulic tests and uniaxial tensile tests indicate a pure tensile fracturing mode. This is supported by the peak load of the specimens which is equal to the tensile strength of the material. The pressure cell seems to be a powerful experimental set-up for gaining a better insight in hydraulic fracturing process and parameters involved.

6. Acknowledgement.

The authors are greatly indebted to the valuable assistance of Mr. G. Timmers with the experiments. The research was made possible through a grant of Delft University (Committee Beek) and the Commission of the European Community.

7. References

Biot, M.A. (1941) General theory of three-dimensional consolidation. *J.Appl.Phys.*,12,155-164.

Biot, M.A. (1956) General solution of the equation of elasticity and consolidation for a porous material. *J.Appl.Mech.*, 23, 91-96.

Boone, T.J. and Ingraffea, A.R. (1988) Simulation of fracture propagation in poroelastic materials with application to the measurement of fracture parameters, in *Fracture Thoughness and Fracture Energy* (eds H. Mihashi et al.), Balkema, Rotterdam, pp. 264-284.

Clayton, N. and Grimer, F.J. (1978) The diphase concept, with particular reference to concrete, in *Developments in Concrete Technology-1* (ed F.D. Lydon), Applied Science Publishers Ltd., London, pp. 283-318.

Detournay , E. and Cheng, A.H.D. (1988) Poroelastic response of a borehole in a non-hydrostatic stress field. *Int.J.Rock Mech.Min.Sci. & Geomech.Abstr.*,A25(3), 171-182

Detournay, E., McLennan, J., and Roegiers, J.C. (1986) Poroelastic concepts explain some of the hydraulic fracture mechanics. SPE 15262, *SPE Unconventional Gas Technology Symposium.* Louisville (KY).

Howard, G.C. and Fast C.R. (1970) *Hydraulic Fracturing.* Monographic Series Vol II, Soc.Petr.Eng., Dallas, pp 11-31.

Rice, J.R. and Cleary, M.P. (1976) Some basic stress diffusion solutions for fluid-saturated porous media with compressible constitutes. *Rev.Geophys. and Space Phys.*, 14, 227-241.

Schlangen, E. and van Mier, J.G.M. (1992) Experimental and Numerical Analysis of Micromechanisms of Fracture of Cementbased Composites. *Cement and Concrete Composites*, Vol.14, no. 2.

Settari, A. (1985) A new general model of fluid loss in hydraulic fracturing. *S.P.E.J.*, Aug, 491-501.

Shapiro, A.M. (1987) Transport equations for fractured porous media, in *Advances in Transport Phenomena in Porous Media* (eds J. Bear and M. Yavuz Corapcioglu), Martinus Nijhoff Publishers, Dordrecht, pp 405-472.

Visser J.H.M. and van Mier, J.G.M. (1992) Hydraulically driven fractures of concrete and rock: a new test cell, in *Fracture Mechanics of Concrete Structures* (ed Z.P. Bazant), Elsevier Applied Science, London, pp 512-517.

Williams, B.B. (1970) Fluid loss from hydraulically induced fractures. *J.Pet.Tech.*, 22, 882-888.

43 CAVITATION DAMAGE AND DETERMINATION OF TENSION AND FATIGUE RESISTANCE IN CONCRETE SURFACES

G. ECHÁVEZ
Engineering Faculty, National University of Mexico, Mexico City, Mexico

Abstract

In this paper the evolution of damage due to cavitation in a concrete probe is shown and a method to estimate the tension and fatigue resistance of the upper layer of concrete surfaces is presented.

The method consists of cementing small vertical cylinders of $7mm$ in diameter and $12mm$ in height of an adhesive to a concrete surface and to register the tension force necessary to separate the cylinder from the surface. To find the fatigue resistance of the surface similar small cylinders were subjected to cyclic loads and both the amplitude and the number of loads applied when they failed were registered.

The experimental results were plotted in a S-N-P diagram in which:

S amplitude of the cyclic load, given as a percentage of the ultimate tension load,
N number of cycles applied until failure occurred,
P probability of failure for each test,

and an analytical expression of the form

$$P = 1 - 10^{-aS^b(\log N)^c}$$

was adjusted to the experimental data with the method of minimum squares, obtaining the values: $a = 0.000049$, $b \doteq c = 6.7$ for $10^2 < N < 10^6$.

1 Introduction

Damage due to cavitation in concrete spillways of high dams is a problem that has come to the attention of specialists during the last decades. As reported by the ACI Committee 210 (1984), the progression of cavitation erosion in concrete is not as well documented as it is in metals, so it is of interest to observe the evolution of damage in concrete specimens tested in high velocity flow installations and to see the causes of erosion.

In this work the evolution of this kind of damage and a method to estimate the tension and fatigue resistance of the upper layer of concrete surfaces is presented.

Fracture and Damage of Concrete and Rock – FDCR-2.
Edited by H.P. Rossmanith. © 1993 Taylor & Francis. ISBN 0 419 18470 8.

Figure 1: Side view of a concrete surface with rounded protuberance

2 Cavitation erosion

Damage due to cavitation in concrete surfaces starts with the detachment of small portions of the cement paste. This occurs during the first seconds when the surface becomes subjected to cavitation. A few minutes later, the upper cement layer is gone and the finer aggregate is eroded from the surface exposing the gravel.

Apparently the gravel is not damaged, but eventually gets loosened with the erosion of the conglomerate until it is carried away by the flow.

In Figure 1 a concrete surface with a rounded protuberance is shown and in Figures 2 and 3 the same surface, after 5 and 10 minutes under the influence of cavitation, is shown. There, it can be seen the typical pattern of successive eroded zones, each one bigger than the one immediately upstream.

When this occurs in a tunnel spillway, it can lead to steadily increasing damage zones, until they are equal to the width of the tunnel. This can be seen in Figure 4 where the erosions at the end of the vertical curve of a $13m$ in diameter concrete lined tunnel are shown.

Although the pressure fluctuations associated with cavitation are of high enough intensity to damage even steel surfaces, Arreguin and Ruiz (1990) found that the compressive strength of the concrete has an influence on the rate of erosion and, in some cases, there is doubt whether the water jet striking the eroded zone is enough to continue the damage; a decision depends on the knowledge of the resistance of the surface. In the next sections, an experimental procedure to estimate the resistance to both tension and fatigue of the top layer of a concrete surface is presented.

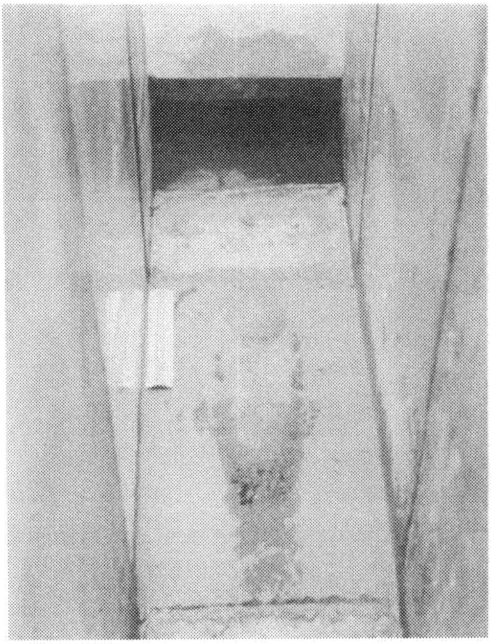

Figure 2: Cavitation erosion after 5 minutes, flow from left to right

Figure 3: Cavitation erosion after 15 minutes, flow from left to right

Figure 4: Cavitation damages in a 13m diameter tunnel

3 Experimental procedure

Static and fatigue test experiments were performed. The static tests were done on 40x40x10*cm* blocks and in standard cylinders, which have been cured in a moisture-controlled room.

On the surface of the blocks tested, small paper cylinders, 7*mm* in diameter and 12*mm* in height, were located and filled with an adhesive, Araldit-SW418, as shown in Figure 5.

After hardening of the adhesive a slowly increasing tensile load was applied until the glued-on probes separated from the surface. A small portion of concrete glued to the adhesive slightly larger than the face of the cylinder came off . those tests that were not representative – about 5% of the total – were discarded. The ultimate resistance was defined as the quotient between the final load and the concrete area. In order to find the maximum compressive strength three standard cylinders were tested on the same day.

Fatigue tests were conducted employing the cyclic load apparatus shown in Figures 6 and 7. Cyclic loads of 60 cycles per second were applied by a calibrated spring connected to an eccentric wheel. The load amplitude was changed by varying the length of the arm attached to the eccentric wheel. The loads were applied and maintained without interruption until failure. The ultimate resistance was defined and evaluated similarly to the static tests.

Figure 5: Concrete block with the adhesive cylinders

Figure 6: Side view of the vibrator

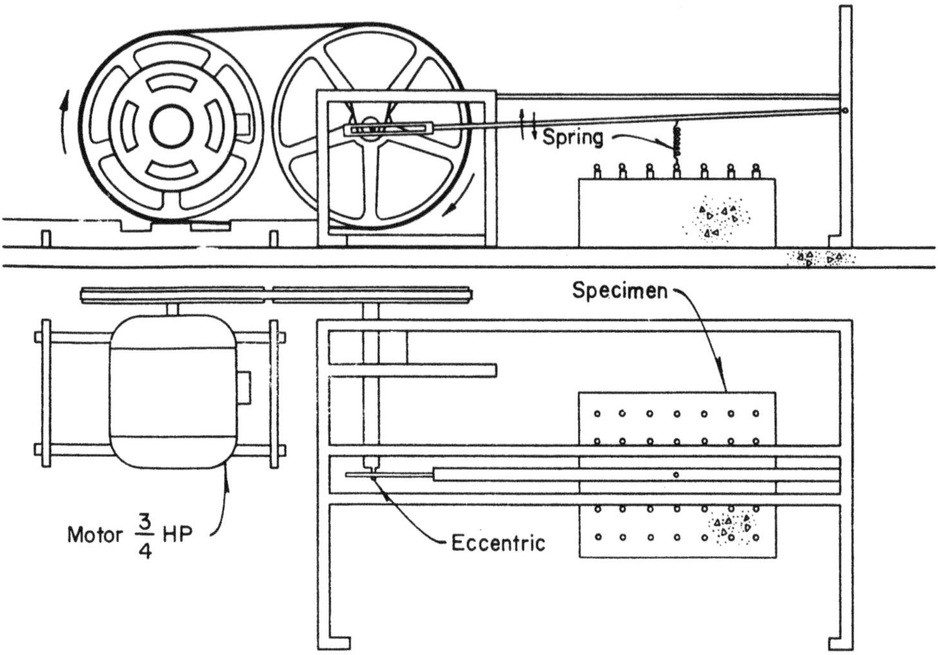

Figure 7: Vibrator to find the fatigue resistance of concrete surfaces

4 Presentation of results

4.1 Results of static tests

A set of 310 specimens grouped in three different ages (28, 40, and 54 days) were tested in tension. Simultaneously three standard cylinders were tested in compression. The results are shown in Table I and in Figure 8. In Figure 8 the standard compression is correlated with the ultimate load on the surface.

The experimental data points associated with concrete of age 54 days can be curve-fitted to yield the empirical relationship

$$f'_T = 0.15 f'_C - 10 \qquad \text{for} \quad 150 < f'_C < 270\, kp/cm^2$$

where f'_C = compressive strength and f'_C = tensile strength.

4.2 Results of fatigue tests

To estimate the fatigue resistance of the top surface layer of concrete 36 specimens were tested. The number of cycles, N, at which the failure occured, and the stress level, S, defined as the ratio between the cyclic load and the ultimate load where recorded. When plotting the results in the form of S versus $\log N$, common with this kind of experiments a large amount of dispersion has been noted. In order to improve the presentation of results a diagram S-N-P was employed. Here curves of constant

Table I: Compressive and Tensile Strengths

Test	Age days	Number of Probes	Tensile Strength kp/cm^2	Compressive Strength kp/cm^2
1	40	14	21.1	200.5
2	40	22	11.1	144.0
3	48	31	17.0	163.0
4	49	14	25.7	207.0
5	30	15	23.4	251.0
6	56	10	30.0	263.3
7	56	17	34.6	231.0
8	28	38	25.9	248.2
9	28	54	11.2	166.5
10	28	51	17.0	165.5
11	61	44	21.1	186.3

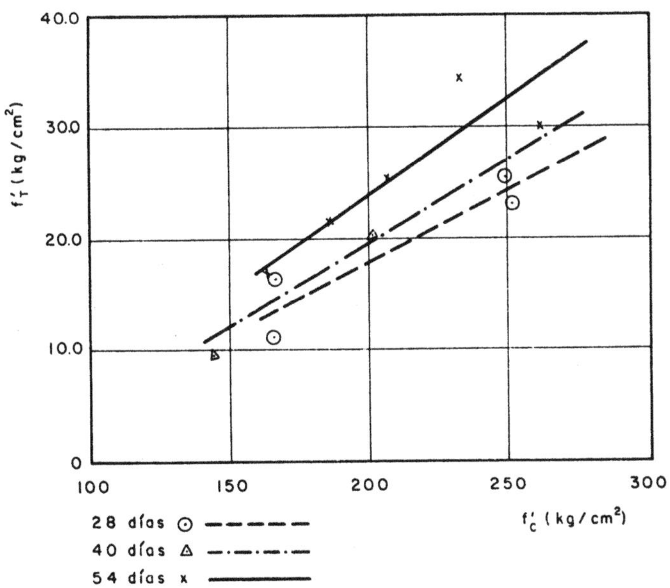

Figure 8: Compression maximum stress versus tension of the surface

Table II: Experimental Values (S-N-P) of the tension fatigue test

Test	N	log N	S	P
1	361	2.5508	0.834	0.50
2	499	2.6980	0.589	0.09
3	939	2.9729	0.728	0.33
4	1 420	3.1527	0.713	0.33
5	15 790	4.1987	0.656	0.60
6	456	2.6590	0.597	0.06
7	1 080	3.0342	0.728	0.32
8	12 610	4.1002	0.494	0.08
9	12 350	4.0910	0.737	0.75
10	85 800	4.9320	0.817	0.98
11	4 920	3.6330	0.684	0.50
12	356	2.5514	0.900	0.80
13	2 140	3.3290	0.900	0.96
14	8 700	3.9390	0.835	0.96
15	8 390	3.9243	0.660	0.51
16	1 211	3.0828	0.785	0.63
17	414	2.6170	0.830	0.60
18	204 000	5.3096	0.660	0.83
19	36 100	4.5597	0.693	0.83
20	2 360	3.3738	0.611	0.15
21	284	2.4548	0.764	0.25
22	150 000	5.1760	0.650	0.80
23	54 900	4.7393	0.651	0.75
24	28 900	4.4602	0.812	0.95
25	65 900	4.8189	0.823	0.93
26	6 850	3.8351	0.483	0.05
27	12 080	4.0838	0.486	0.09
28	36 000	4.5552	0.501	0.20
29	36 900	4.5679	0.441	0.08
30	48 700	4.6866	0.576	0.40
31	1 582 000	6.2000	0.320	0.05
32	1 567 000	5.1960	0.350	0.10
33	1 587 000	6.2030	0.390	0.20
34	792 000	5.9000	0.360	0.10
35	126 000	5.1010	0.675	0.85
36	77 600	4.8900	0.810	0.98

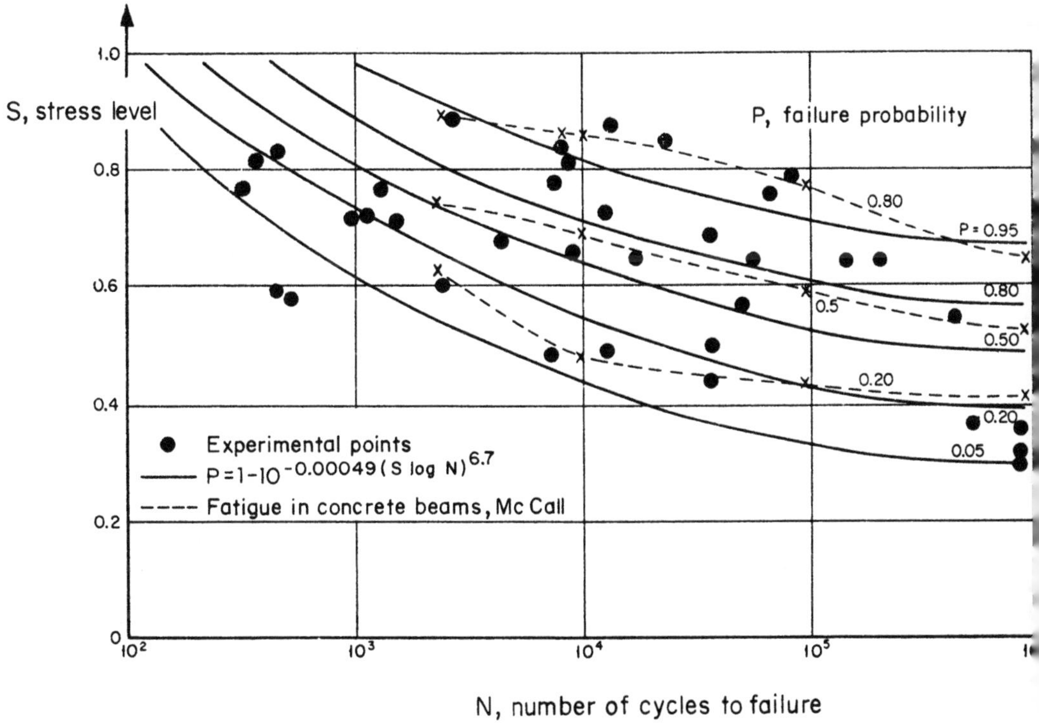

Figure 9: S-N-P tension diagram for concrete surfaces

probability are plotted assuming a normal density of probability distribution in the vertical direction. In this way each point is characterized by three values (S_i-N_i-P_i) in which P_i is the probability of failure of that test. The values obtained in this way are shown in Table II.

McCall (1958) suggested the use of a function of the type

$$P = 1 - 10^{-aS^b(\log N)^c}$$

where a, b and c are constants to be evaluated by the method of least squares from the experimental data points. In this study, the values

$$\begin{aligned} a &= 0.00049 \\ b &= c = 6.7 \end{aligned}$$

were obtained and the equation takes the form

$$P = 1 - 10^{-0.00049(S\log N)^{6.7}} \quad \text{for} \quad 10^2 < N < 10^6.$$

The multiple correlation coefficient of $r = 0.98$ suggests that the function is acceptable.

Figure 9 shows the experimental data points and five curves pertaining to different failure probabilities. For comparison three curves for concrete beams without reinforcement (McCall (1958)) are also shown.

From this figure, it can be seen that for $N > 10^5$ both families of curves are similar and depend less on the number of cycles to failure, so they can safely be extrapolated to $N > 10^6$. With this hypothesis and for a failure probability of 5% the stress level of the concrete surfaces tested is 0.30. This value is larger than any pressure fluctuation associated with turbulence or jet action.

It can be concluded that the resistance of the concrete surface to both static and fatigue excitations is large enough to withstand the pressure fluctuations associated with the turbulence of the flow, but it is not large enough to resist the action of cavitation which can induce pressures concentrated in small areas in the order of $1000 MPa$.

5 Conclusions

Some aspects of the evolution of erosion of concrete surfaces due to cavitation were presented.

Static and fatigue tests were performed and the value of the coefficients for both cases are reported.

The resistance of the concrete surface to both static and fatigue excitations is large enough to withstand the pressure fluctuations associated to the turbulence of the flow but it is not large enough to resist the action of cavitation.

6 Acknowledgments

This work was sponsored by *The Federal Commission of Electricity* and was performed at *The Engineering Institute* of *The National University of Mexico.* The cooperation of Mr. Rafael Rosales during the test program is kindly acknowledged.

7 References

ACI Committee 210 (1984) Erosion of Concrete in Hydraulic Structures. **ACI Materials Journal.** Title 84-M17, 84, No.2, Detroit.

Arreguin, F.I., and Ruiz, R.M. (1990) Cavitación en obras con flujos de alta velocidad. **XI Congreso Nacional de Hidráulica Asociación Mexicana de Hidráulica,** Zacatecas, Zac., México.

McCall, J.T. (1958) Probability of Fatigue Failure of Plain Concrete. **Journal ACI,** Volume 30, 2 Detroit.

44 STUDY ON CRACK PROPAGATION TESTS USING LARGE-SCALE CONCRETE SPECIMENS

H. ISHIDA and T. ENDO
Central Research Institute of Electric Power Industry,
Abiko, Japan

Abstract
The loading tests using concrete dam specimens were performed and the fracture mechanics discrete cracks approach was applied to explain the experimental results. In the tests, large trapezoidal fracture specimens (1.8m × 1.0m × 0.4m) were used and statically loaded to failure. The variation of initial notch length and location was intended in order to investigate the difference of crack propagation process. Load-deflection curves and concrete strains were measured during crack propagation. The process of crack propagation was simulated using the finite element fracture analysis with personal computer. The behavior observed in experiment was able to successfully simulated using this analytical method.
Keywords: Crack propagation, Discrete crack model, Fracture toughness

1 Introduction

The analysis procedure based on classical material strength approach has been adopted in order to evaluate the safety of concrete dams. However, this approach can not be adequately applied to concrete dams because dams are huge concrete structures and concrete strengths are dependent on sizes. Recently, fracture mechanic approaches, which are expected to explain properly the size effect on strength and load-bearing capacity, is increasingly being studied in order to assess dam safety. So, the loading tests using large-scale concrete specimens were performed to investigate crack propagation behavior and finite element fracture calculations using discrete cracks were carried out to explain the experimental results.

2 Test program

The diagrammatic sketch of the specimen used is shown in Fig.1. There was a notch in the mid height of the specimen, at which crack propagation started. Test specimens are listed in Table 1. Notch length and its location were different between specimens. Total number of the specimens was five. The mix proportion for all specimens was identical. The physical properties of concrete at test are shown in Table 2. The

Fracture and Damage of Concrete and Rock – FDCR-2.
Edited by H.P. Rossmanith. © 1993 Taylor & Francis. ISBN 0 419 18470 8.

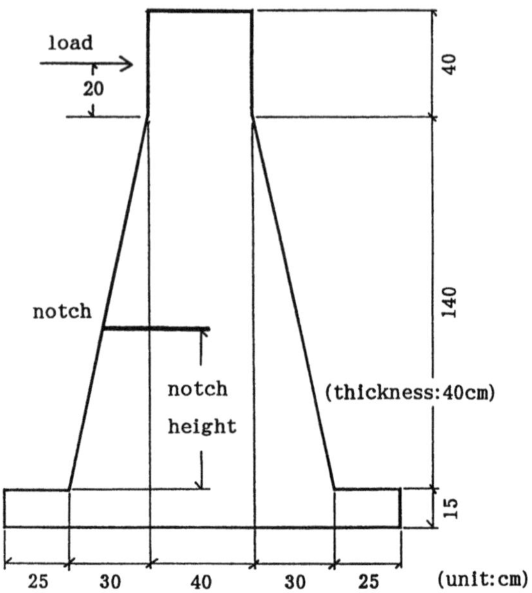

Fig.1. Configuration of specimen

specimens were loaded very slowly to failure by displacement control to avoid occurring sudden unstable crack propagation. Displacements of the specimen were measured by the displacement transducers that were fixed to the bolts embedded in the specimen. Concrete strains normal to expected crack propagation path were also measured by strain gauges. The crack propagation process was able to be monitored by checking the strain values. Wedge splitting tests were also performed to obtain fracture energy of concrete at test from which we calculated the fracture toughness value from classical linear elastic fracture mechanics for plane stress condition. Fracture energy G_F and tensile strength f_t obtained by the test were 0.1061 kgf/cm and 21.4 kgf/cm^2, respectively.

3 Experimental results

3.1 Load-displacement curve

Figure 2 shows the relationship between load and top displacement (displacement at the opposite side of the loading point). The specimen (A-1) with 20 cm length notch fractured suddenly when the maximum load reached 8.27 tonf. Therefore, falling branch was not able to be measured for this specimen. But for the other specimens, falling branch was able to be obtained because they fractured stably. Test results (maximum load, top displacement at the maximum load, crack opening displacement at the maximum load) for each specimen are listed in table 3.

Table 1. Test parameters

specimen	notch height	notch length
A-1	40cm	20cm
B-1 B-2	80cm	40cm
C-1 C-2	60cm	40cm

Table 2. Concrete properties

maximum size aggregate, mm	20
cement content, kg/m^3	381
water cement ratio	0.47
compressive cylinder strength, kgf/cm^2	410
splitting tensile strength, kgf/cm^2	28.2
Young's modulus, $\times 10^5$kgf/cm^2	3.1

3.2 Crack propagation process
Measured concrete strains normal to the crack propagation path (C-2) are shown in Fig.3. When the load was 1.00 tonf (30% of the maximum load), the tensile strain nearest to the notch tip was 26×10^{-6} and this value was almost same as the maximum concrete strain at the compressive zone. The neutral axis where measured concrete strain equaled to zero existed about 10 cm far from the notch tip. The strain distribution at the notched section was proportional to the distance from the neutral axis at this loading stage. However, at 2.02 tonf (70% of the maximum load), the tensile strain nearest to the notch tip was 119×10^{-6} and strain distribution became nonlinear. The neutral axis did not move from 1.00 tonf to 2.02 tonf. When the maximum load (2.96 tonf) attained, the tensile strains within 10 cm far from the notch tip became over 200×10^{-6}. This meant that the crack propagated as far as this position. The neutral axis also shifted to the position of fifth or sixth concrete gauge from the notch tip. After the maximum load, crack propagated slowly and the neutral axis also shifted towards the

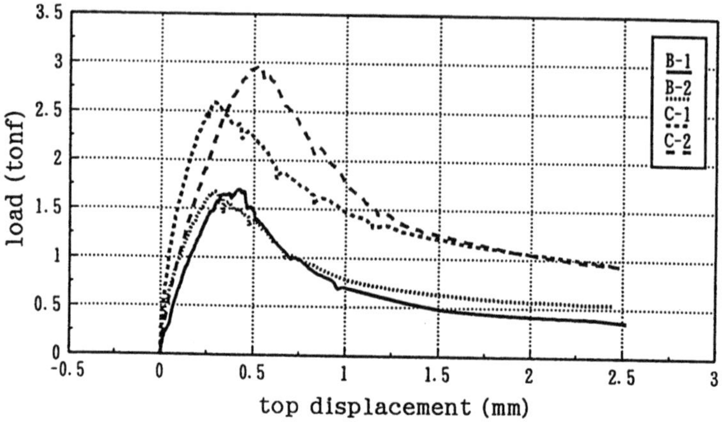

Fig.2. Load-displacement curve

Table 3. Maximum load and displacements

specimen	maximum load (tonf)	top disp. (mm)	crack opening disp. (mm)
A-1	8.27	2.973	not measured
B-1	1.71	0.417	0.022
B-2	1.68	0.288	0.031
average	1.70	0.353	0.027
C-1	2.59	0.285	0.035
C-2	2.96	0.536	0.061
average	2.78	0.411	0.048

compressive zone. Thus for the specimens that had 40 cm length notch, crack propagated stably. The final trajectory of the crack (C-2) is shown in Fig.4.

4 Finite element analysis

The process of crack propagation was simulated using the finite element fracture analysis with discrete crack model developed for personal computer. Used personal computer had a transputer that was a device for calculating stiffness matrix quickly. The program used in this analysis had some typical features; automatic regeneration of the mesh, zoomed details of meshes, selection of mixed-mode fracture initiation theory. The program adopted three mixed-mode fracture initiation theories. That is, maximum circumferential tensile stress

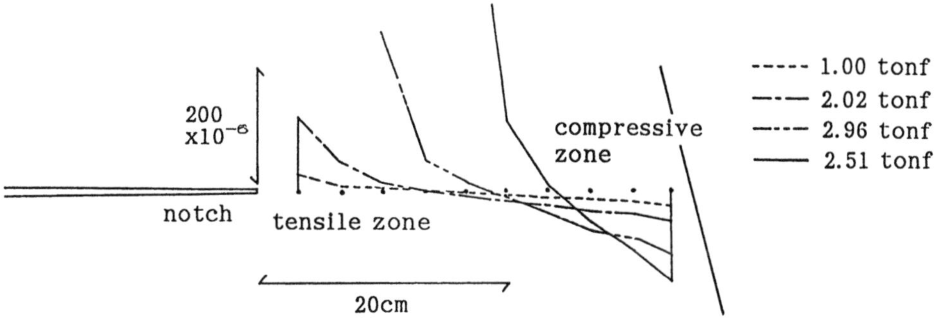

Fig.3. Concrete strain distribution (C-2)

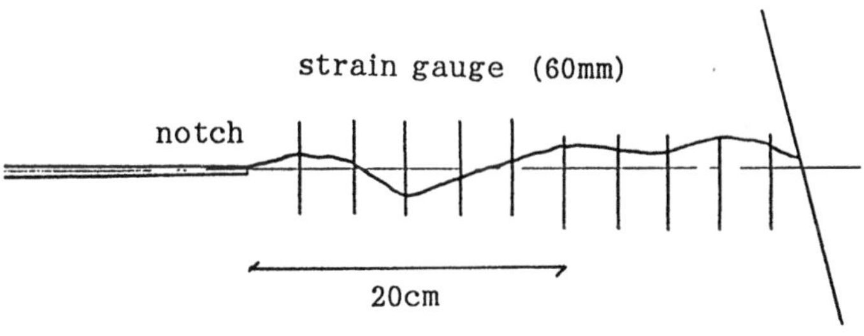

Fig.4. The trajectory of crack (C-2)

theory, maximum energy release rate theory and minimum strain energy density theory. We were able to choose mixed-mode fracture initiation criteria from these three theories. Determination of a crack increment length was based on the concepts of energy release rate and the materials resistance to a unit crack extension (R-curve method).

4.1 Load-displacement curve

Figure 5 shows the effect of fracture toughness value (K_{IC}) on the load-displacement curve for the specimens (C-1,C-2) using maximum circumferential tensile stress theory. The maximum load increased as K_{IC} value increased. The K_{IC} value was calculated to be 181.4 kgf/cm$^{3/2}$ using the relation between K_{IC} and G_f ($K_{IC}=\sqrt{G_f \cdot E}$) from classical linear elastic fracture mechanics for plane stress condition. But, in the analysis with this value, the calculated maximum load was 4.34 tonf. The average value of the maximum loads (C-1,C-2) was about 2.8 tonf. Therefore, K_{IC} value for the specimen (C-1,C-2) was estimated to be 120.0 kgf/cm$^{3/2}$. In the analysis with K_{IC}=120 kgf/cm$^{3/2}$, calculated top displacement (0.38mm) corresponded to the experimental one (0.41mm) fairly well. But, after the maximum load, calculated load decreased linearly with displacement, whereas experimental load decreased bi-linearly (See Fig.6). It seemed that this discrepancy was attributed to assumptions adopted in the analysis.

The one may be R-curve method adopted in the analysis in which R

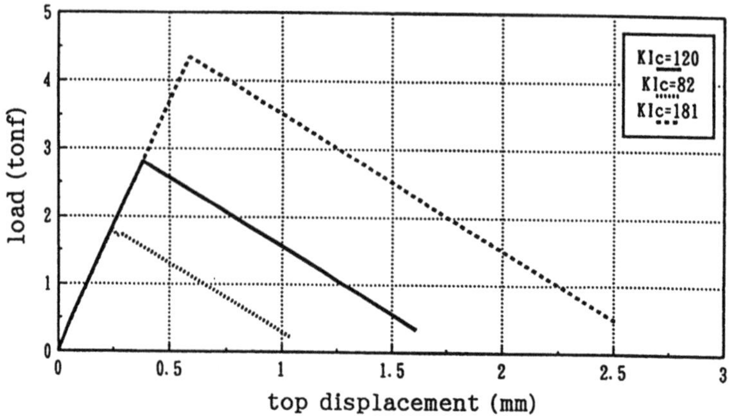

Fig.5. The effect of K_{IC} on load-displacement curve

value was assumed as a constant. For quasi-brittle material such as concrete, it has been said that R-curve increases with crack extension. In further research program, it is more desirable to carry out the analysis with rising R-curve.

The other may be the difference of loading conditions between the test and analysis. That is, in the test, the specimen was loaded by the actuator head (loading area; 40 × 40 cm) which was fixed to the specimen by bolts. In the analysis, load was given to the specimen horizontally at nodal point. While the displacement of the specimen was small, the loading condition in the analysis was not so different from that in the test. So, calculated load-displacement curve coincided with experimental one fairly well up to about 1mm displacement. But when the displacement became large, the difference of loading conditions also became large enough to affect the

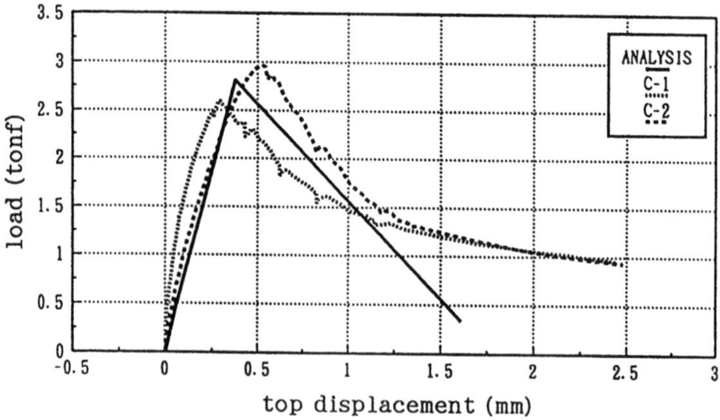

Fig.6. Load-displacement curve (C-1,C-2)

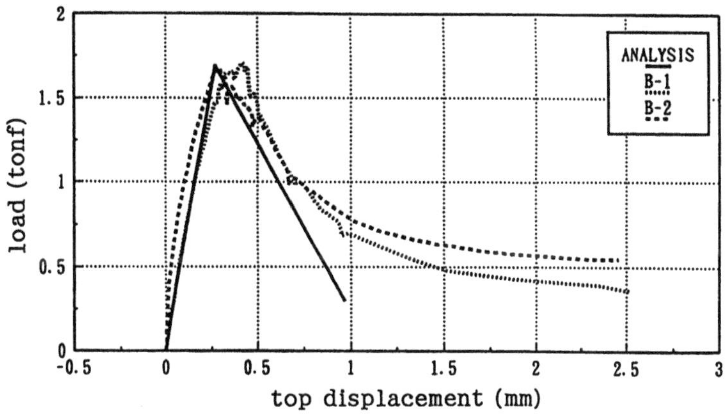

Fig.7. Load-displacement curve (B-1,B-2)

load-displacement curve.

For the specimens (B-1,B-2), $K_{I}c$ value was estimated to be 90.0 kgf/cm$^{3/2}$(See Fig.7). Therefore, $K_{I}c$ values were different for type-B and type-C specimens. The reason $K_{I}c$ values were different is not yet clarified.

Figure 8 shows the effect of mixed-mode fracture initiation theory on the load-displacement curve of specimens (C-1,C-2) with $K_{I}c$=120.0 kgf/cm$^{3/2}$. Judging from the figure, the load-deflection curves were almost same irrespective of the using mixed-mode fracture initiation theory The trajectories of cracks were also almost same irrespective of the using mixed-mode fracture initiation theory. Therefore, in this study, mixed-mode fracture initiation theory did not affect the load-displacement curve so much. It seemed that crack

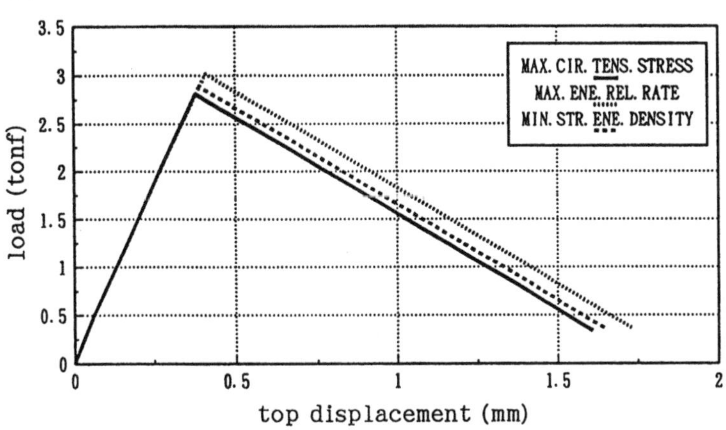

Fig.8. The effect of mixed-mode fracture initation theory on load-displacement curve

propagation took place almost under mode I.

Figure 9 shows the mesh size effect on the load-deflection curves. The figure may indicate that mesh size scarcely affected the load-displacement curve.

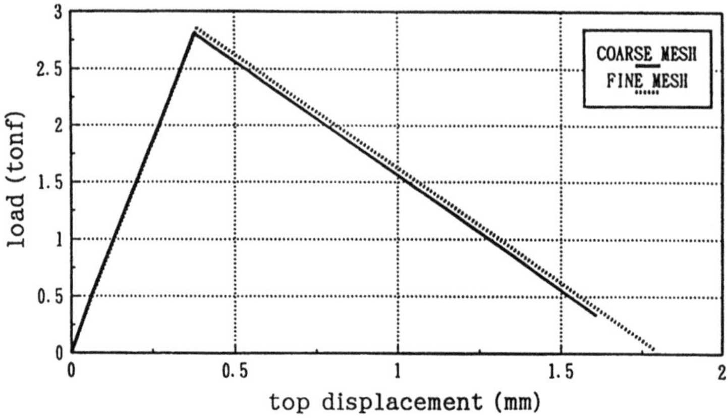

Fig.9. The effect of mesh size on load-displacement curve

4.2 Crack propagation process

The trajectory of crack shown in Fig.10 was obtained for type-C specimen using maximum circumferential tensile stress theory. The crack propagated from the notch tip towards the compressive zone in zigzag way. The trajectories were almost same irrespective of using mixed-mode fracture initiation theory. The real trajectory (C-2) was already shown in Fig.4. In the figure, crack propagated more zigzag than analysis. Figure 11 shows the trajectory obtained from the analysis using fine mesh. Compared with Fig.10, the trajectory was a little different especially for the final part. But, as mentioned above, load-displacement curves were almost same for coarse and fine mesh.

5 Conclusions

The conclusions obtained from the present study are as follows.

1)The behavior of specimen observed in the experiment was able to successfully simulated using the finite element fracture analysis with discrete crack model.

2)In the analysis, calculated load-displacement curve was greatly influenced by fracture toughness value ($K_{I}c$). Consequently, it was very important to determine this value properly.

3)On the other hand, calculated load-displacement curve were not so much influenced by using mixed-mode fracture initiation theory and mesh size in this study.

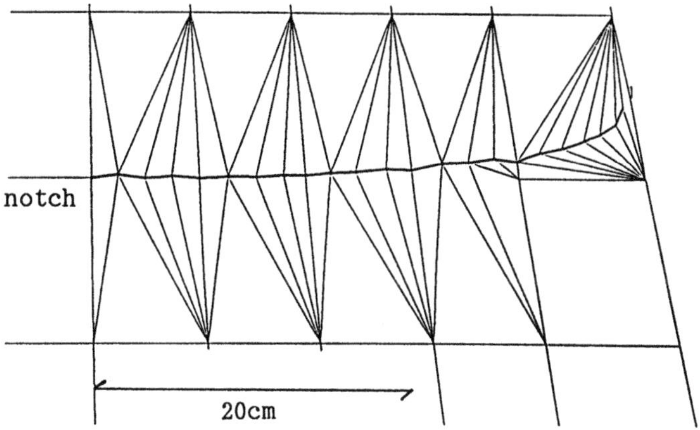

Fig.10. The trajectory of crack (analysis,type-C,coarse mesh)

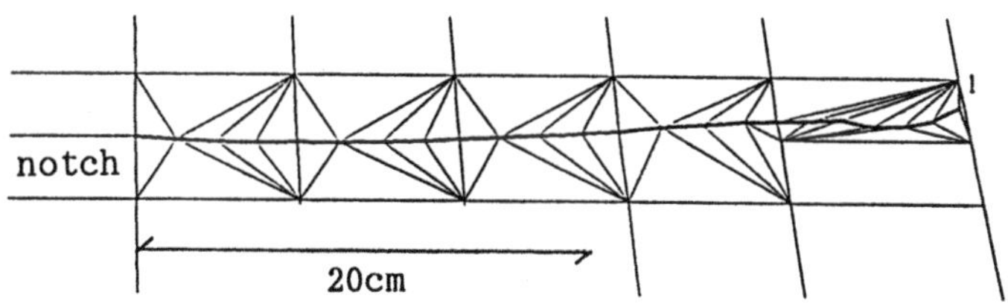

Fig.11. The trajectory of crack (analysis,type-C,fine mesh)

Further research works will be carried out to improve this analytical program by adopting the concepts of nonlinear fracture mechanics model.

6 References

Ingraffea, A.R. and Saouma, V. (1985)
Numerical modeling of discrete crack propagation in reinforced and plain concrete,Fracture mechanics of concrete: Structural application and numerical calculation, Edited by Sih, G.C. and DiTommaso, A.,MARTINUS NIJHOFF PUBLISHERS, DORDRECHT, pp.171-225.

Ouyang, C. Mobasher, B. and Shah, S.P. (1990)
Prediction of Fracture of Concrete and Fiber Reinforced Concrete by the R-Curve Approach,APPLICATIONS OF FRACTURE MECHANICS TO REINFORCED CONCRETE,Edited by Carpinteri, A., ELSEVIER APPLIED SCIENCE, LONDON and NEW YORK, pp.137-168.

Saouma, V.E. (1981)
INTERACTIVE FINITE ELEMENT ANALYSIS OF REINFORCED CONCRETE: A FRACTURE MECHANICS APPROACH, Report 81-5, Cornell Univ.

45 A CONTRIBUTION TO THE PHYSICALLY AND GEOMETRICALLY NONLINEAR ANALYSIS OF REINFORCED CONCRETE SHELLS: APPLICATION TO BUCKLING OF COOLING TOWERS

I. CHAHROUR, J.M. REYNOUARD, M. DJERROUD and O. MERABET
Laboratoire Bétons et Structures, INSA, Lyon, France

Abstract

The development of a simple and efficient numerical model describing correctly the behavior of reinforced concrete structures is of great interest, especially when dealing with curved structures.

The aim of this study is to contribute to the physically and geometrically nonlinear computer analysis of general reinforced concrete shells subjected to time invariant loads up to failure.

The shell element results from the superposition of a plate element based on the discretization of the MINDLIN theory and the CST element. The initial curvature is incorporated using the MARGUERRE shallow shell theory. The effect of coupling membrane and bending creates the membrane locking phenomenon which is treated by the mode decomposition technique.

The constitutive model of concrete is based on the elastoplastic theory and describes correctly the dissymetry of its behavior in tension and compression. This model is incorporated in a layered approach. The description of motion is made in corotational Lagrangian description and the nonlinear structural response is computed by incremental-iterative path-tracing algorithms.

The numerical part of the paper contains a detailed study of a built hyperbolic cooling tower. It is shown that buckling loads resulting from linear and geometrically nonlinear prebuckling analysis are considerably larger than the ultimate load.

Key words : reinforced concrete - Finite Elements - nonlinear analysis - cooling tower - buckling

1 Introduction

Reinforced concrete shells are of special interest since they possess a high load bearing capacity while using an extremely low amount of material. Such optimized structures, however, often exhibit a strongly nonlinear behavior. Due to its clear mathematical describability, the questions of geometrically nonlinear behavior can be viewed as mostly answered. In the case of physically nonlinear responses, however, there are still many open questions.

The present work therefore aims at developing a finite element numerical model for general layered shells, taking into account geometrical and physical nonlinearities. This model is used to study a typical cooling tower submitted to its dead load and to the pressure of wind load.

Fracture and Damage of Concrete and Rock – FDCR-2.
Edited by H.P. Rossmanith. © 1993 Taylor & Francis. ISBN 0 419 18470 8.

2 Theoretical concept

2.1 - Finite Element model

The element used results from the superposition of the Constant Strain Triangle (CST) and the Discrete Shear Triangle (DST) proposed by BATOZ (1989). The introduction of the MARGUERRE shallow shell theory to incorporate the effects of the initial curvature gives a curved shell element (DSTM). The effect of coupling membrane and bending creates the membrane locking phenomenon which is treated by the mode decomposition technique (STOLARSKI & al 1983).

The formulation of the DST element is based on a generalization of the discrete KIRCHHOFF technique to include the transverse shear effects. DST has a proper rank and is free to shear locking.

The DSTM element now possesses five degrees of freedom at each node. The sixth degree associated with the shell normal rotation is the drilling rotation proposed by BERGAN and used in MARGUERRE theory by FREY. Fig. 1 shows the stages of the construction of the DSTM element

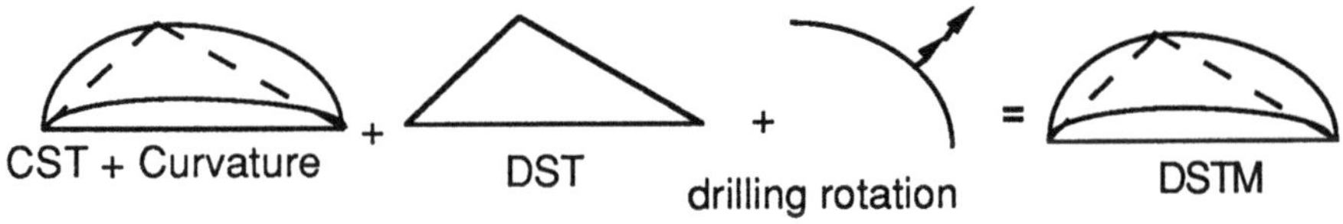

Fig. 1 - Construction of the DSTM element

The two materials (concrete and steel) are individually represented : the concrete by the DSTM elment and the steel by the CST element. A layered approach is realized by the computation of the integral expressions at GAUSS points according to the thickness of the element (Fig. 2).

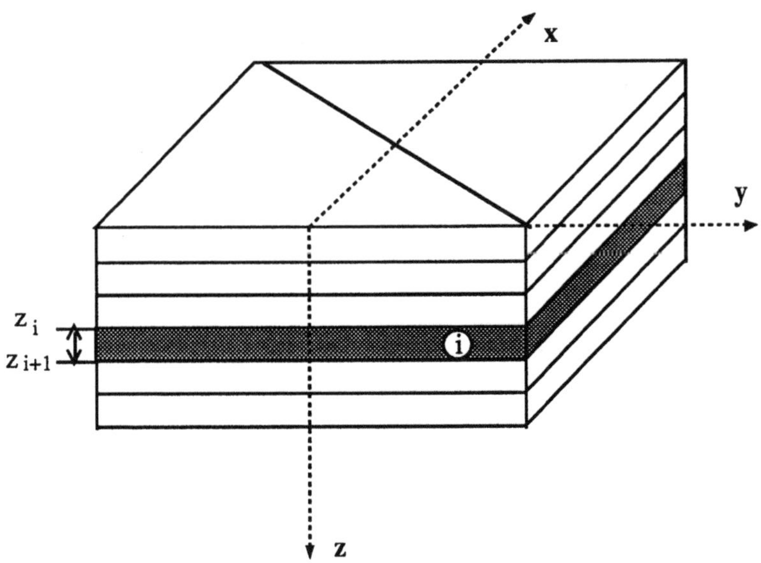

Fig. 2 - Layered approach

2.2 - Geometric relationships

Nonlinear strain-displacement equations restricted to the consideration of large rotations and small strains are applied within the framework of a corotational Lagrangian description.

On the basis of MINDLIN's normal hypothesis, the strains are obtained as :

$$e_{ij} = e_{ij}^{(0)} - z K_{ij}$$

where $e_{ij}^{(0)}$ are the middle surface strains and K_{ij} are the changes of curvature of the middle surface. It is assumed that also for the cracked subregions of the shell MINDLIN's normal hypothesis is an acceptable approximation.

The middle surface strains can be written as :

$$e_{ij}^{(0)} = \varepsilon_{ij} + \eta_{ij}$$

where ε_{ij} is the linear and η_{ij} is the nonlinear part of the strain tensor. ε_{ij} can be decomposed as :

$$\varepsilon_{ij} = \varepsilon_{ij}' + \varepsilon_{ij}^{*}$$

ε_{ij}^{*} is the effect of the initial curvature in MARGUERRE theory.

Details concerning ε_{ij}^{*} and η_{ij} are given elsewhere (CHAHROUR, 1991).

2.3 Constitutive equations

Concrete

The assumption of a membrane state of stress in the concrete layers allows biaxial constitutive models to be used. The concrete is considered as elastic-plastic material. The NADAI criterion is used to approximate the yield surface. It can be expressed as follows (compression) (MERABET, 1990 - DJERROUD, 1992) :

$$F(\sigma, K) = \frac{\sqrt{\frac{2}{3} J_2 + A_1 I_1}}{B_1 f_c(K)} - 1 \leq 0$$

K : hardening parameter

I_1 : first invariant of the stress deviator

J_2 : second invariant of the stress deviator

f_c (K) : maximum compressive strength

A_1, B_1 : constants deduced form uniaxial and biaxial compressive strengh

To determine the beginning of cracking, a principal tensile stress criterion is applied.

Steel

Reinforced steel is considered as an elastoplastic material. The stress-strain diagram is assumed to be identical in tension and compression.

2.4 Linear stability criterion

The stability of a structure is investigated using the method of adjacent equilibrium which is also called Euler's method. If λ is the critical load level leading to instability and if the initial displacements can be considered as very small, the stability criterion is :

$$\det (K_0 + \lambda K_\sigma) = 0 \qquad \text{(e)}$$

K_0 is the infinitesimal matrix and K_σ is the initial stress matrix.

3 Numerical investigation

This section contrains a study of the cooling tower erected in Port Gibson, Miss., USA. Its purpose is to study the behaviour of the structure from the uncracked state until final collapse and to compare the linear buckling load with the ultimate load.

Figure 3 shows the characteristic dimensions of the shell. The equations of the two parts of the generatrix (hyperbola and ellipse) are obtained by substituting the coefficients A, B ...F in table 1 into the equation for a general curve of second order, given as :

$$Az^2 + 2Brz + Cr^2 + 2Dz + 2Er + F = 0$$

Table 2 contains the positions and the thicknesses of the two orthogonal reinforcement nets (smeared-steel layers). The wind pressure function is given as :

$$p(z, \alpha) = p_0 H(z) G(\alpha)$$

$$H(z) = (az + b)^K$$

$$G(\alpha) = \sum_{n=0}^{12} A_n \cos(n\alpha)$$

p_0 = 413,031 Pa   K = 2/7
a = 3,281
b = 393,696

The coefficients A_n are given in Table 3. Because of symmetry of the wind load, only one half of the cooling tower needs to be analysed. Figure 4 shows the FE mesh for one half of the shell. It consists of 12 x 12 x 12 = 288 curved elements. The simplifying assumption of a rigid, hinged base was made.

As shown in Figure 5, the load-displacement path consists of two main parts, characterized by intact and fractured concrete. The latter can be divised into three sections. Section AB is termed the "crack plateau". Section BC is characterized by hardening of the partially fractured structure. The last section beginning at point C, is termed the "yield plateau". Because of hardening of the steel, the yield load is smaller than the ultimate load.

Figure 6 refers to a study of the influence of geometric nonlinearity. It is seen that disregard of geometric nonlinearity results in a "stiffer" solution.

Figure 7 contain results form a study of the influence of the reinforcement percentage. It is seen that the ratio of the crack loads for the three considered reinforcement percenages (no reinforcement, regular, doubled) is equal to the ratio of the corresponding membrane stiffnesses. The length of the crack plateau decreas with increasing reinforcement percentage.

Finally, the critical load intensity corresponding to linear buckling using equation (e) is 4.60 x wind load. It is considerably larger than the ultimate load. Consequently, analysis concepts which are based on buckling are inadequate for even an approximate assessment of failure laods of reinforced-concrete coolong tower. Because failure of concrete in tension would initiate the collapse of cooling towers, the tensile strength of concrete is of crucial importance for the safety of these shells.

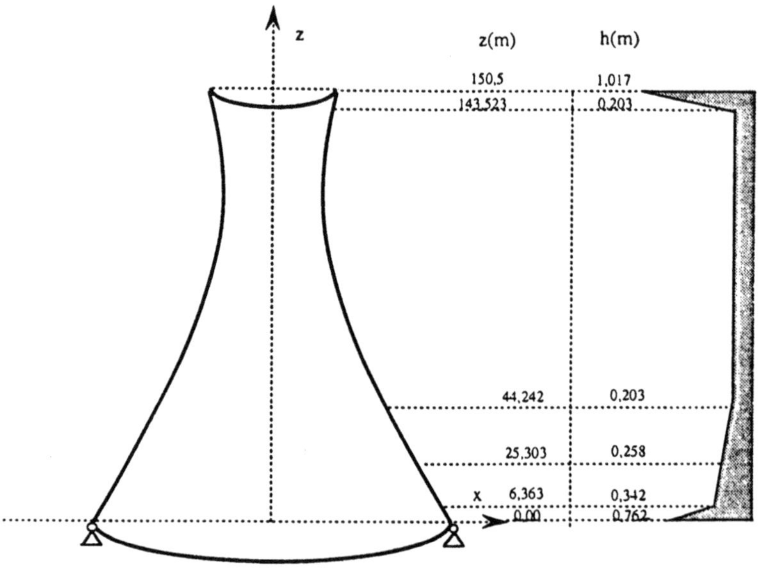

Figure 3 : Dimensions of the shell

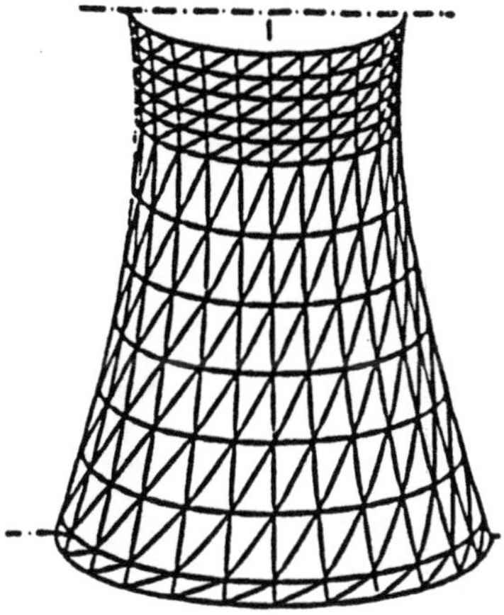

Figure 4 : FE mesh

Table 2

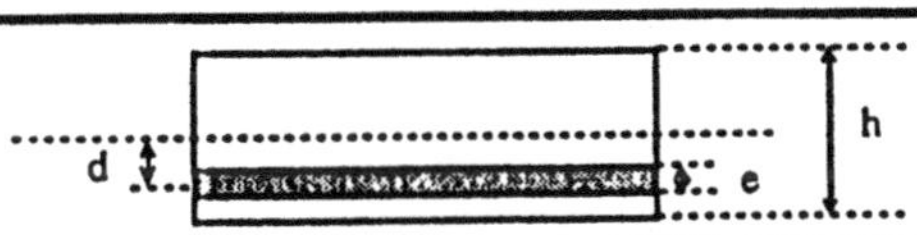

	$\left\lvert \frac{d}{h} \right\rvert$	e_c	e_m
1	0,406	0,142	0,1083
2	0,339	0,0573	0,116
3	0,312	0,0504	0,1143
4	0,278	0,0499	0,1247
5	0,251	0,0499	0,117
6	0,25	0,0356	0,0986
7	0,266	0,0356	0,0513
de 8 à 11	0,266	0,0356	0,0369
12	0,41	0,2031	0,063

n	An	n	An
0	-0,1167	7	-0,04474
1	-0,27918	8	-0,00833
2	-0,61978	9	-0,00972
3	-0,50927	10	-0,01356
4	-0,09167	11	-0,00597
5	0,11794	12	-0,01667
6	0,03333		

Table 3

	Hyperbole	Ellipse
A	-0,002601	0,8602
B	-0,038937	0,7104
C	1	1
D	1,4174	-25,8598
E	-35,7642	-247,2319
F	1278,6678	166674,1904

Table 1

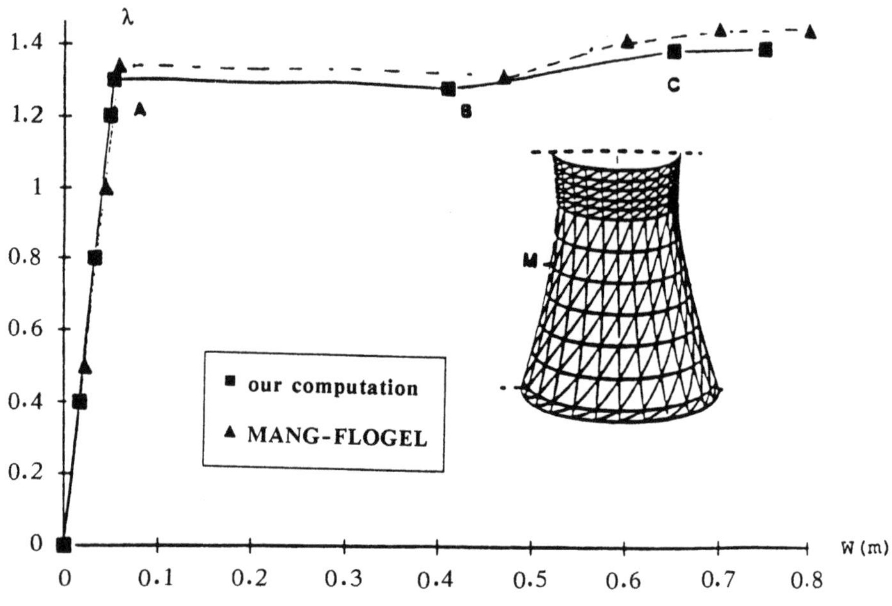

Figure 5 : Load displacement path

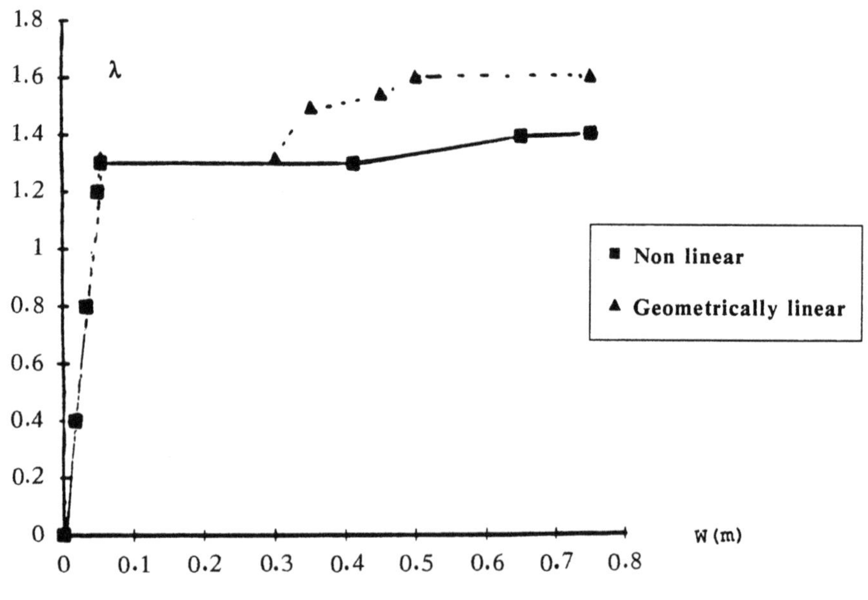

Figure 6 : Load displacement path
(influence of geometric nonlinearity)

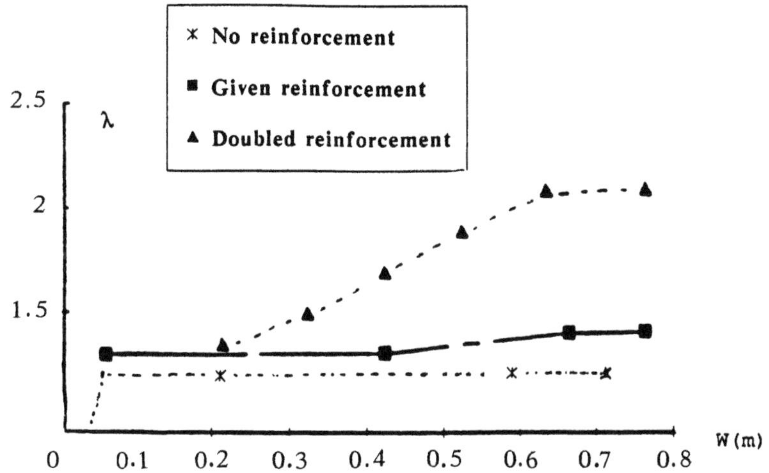

Figure 7 : Load displacement path (influence of the reinforcement percentage)

REFERENCES

BATOZ J.L., LARDEUR P.(1989)
"A discrete shear triangular nine dof element for the analysis of thick to very thin plates", IJNME, Vol. 28, p. 533-560

CHAHROUR I., 1991
"Contribution à l'analyse non linéaire matérielle et géométrique des coques en béton armé", Thèse de Doctorat, INSA LYON

DJERROUD M., 1992
"Contribution à l'analyse des pièces fléchies en béton armé sous chargements monotone et cyclique : développement et validation des modèles élastoplastiques", Thèse de Doctorat, INSA LYON

MERABET O., 1990
"Modélisation des structures planes en béton armé sous chargements monotone et cyclique. Construction et validation d'un modèle numériqué", Thèse de Doctorat, INSA LYON

STOLARSKI H., BELYTSHKO T., CARPENTER N., 1983
"Bending and shear mode decomposition in C° structural elements", J. Struct. Mech. Vol. 11 n° 2, p. 153-176

46 NONLINEAR ANALYSIS OF DAMAGE IN REINFORCED CONCRETE STRUCTURES SUBJECTED TO SEISMIC LOADING

F. FLEURY, O. MERABET and J.M. REYNOUARD
Laboratoire Bétons et Structures, INSA, Lyon, France

Abstract
In order to analyse damaged reinforced concrete structures submitted to dynamic or seismic loads, we propose a numerical model capable of describing the behaviour of reinforced concrete beams and columns under alternate cyclic loading, which can be efficiently used also in dynamic analysis: with the assumption of a locally uniaxial stress state, we are able to obtain the rapidity needed for the integration in time of the dynamic equilibrium equations for large structures, within an important time interval.

The model developed is presented: path-dependent constitutive law for uniaxial reinforced concrete and finite element formulation. The results given by this model can be local and global : displacements, maximum load, stress and strain at any point in the structure. Damage of the members is described directly by the local state of the materials: crack distribution, state of compressed concrete, permanent strains.

A number of applications are presented, serving to evaluate the performance of the model for different kinds of structures.
Keywords: Finite Elements, Reinforced Concrete, Nonlinear Dynamics, Damage, Seismic Loads

1 Introduction

There exists many methods for the analysis of reinforced concrete structures, which can be classified mainly according to the procedure in calculating the internal restoring force, and to the nature of the kinematic conditions imposed. The global modeling usually reduces the displacements of the structure to those at the ends of the structural elements of which it is composed (a column, beam, or a storey of shear wall), and enforces the static experimentaly observed deformed shapes (point of zero curvature at midlength, for example). The internal force is directly obtained by the experimental construction of the force-displacement relations for typical structural elements, subjected to simple loading and limit conditions, thus reducing the domain of validity (Song & al.(1991), Fardis (1991), Meyer & al. (1991)). The results obtained from this kind of analysis do not permit the description of the local damage mechanisms responsible for the behaviour.

In contrast, the local approach, which needs only the stress-strain relations at each point for constitutive equations, permits the search for a solution in the greatest space, eventualy reduced by kinematic considerations imposed through the finite-element formulation. Even though this method is capable of producing a maximum of information for

Fracture and Damage of Concrete and Rock – FDCR-2.
Edited by H.P. Rossmanith. ISBN 0 419 18470 8.

any kind of geometry and loading conditions, it suffers from major drawbacks when used to model real structures subjected to long lasting dynamic loadings :

- its computational and memory requirements are very high;
- the behaviour of concrete under general biaxial and triaxial cyclic loading conditions is still not well understood, and few efficient models of such behaviour are presently available.

The model presented herein has been constructed in order to overcome these drawbacks, provided two extra assumptions, which are justified for an important class of structures: those constructed with lineic members. The choice of a classic two node Kirchhoff beam element, with three degrees of freedom per node yields the first assumption, which is kinematic. This hypothesis permits an important reduction of the overall number of degrees of freedom compared to the use of surface elements, and therefore also of the memory and computational costs. The second assumption consists in degenerating the stress tensor to one direction, and letting only a uniaxial stress field develop within the element. This restriction, which yields to neglecting the effect of shear, permits a finer description of the cyclic aspect of the behaviour of concrete.

2 The material behaviour models

Among the various local approaches possible for the modeling of concrete behaviour, our laboratory has up to now successfully experimented strain-hardening elasto-plastic laws in compression, allied to a correct description of the opening and closing of cracks throughout the loading cycles (Merabet (1990), Charhour (1991), Djerroud (1992)). The uniaxial version, (Merabet & al. (1991)) based on experimental observations, has two extra features, which refine its cyclic behaviour:

- the elastic modulus of concrete is progressively damaged when the inelastic deformations increase;
- as soon as a crack starts to close, the concrete develops some compression, according to the imperfect imbrication of the crack lips.

The problem is thus to define at any point of an element, the criteria for opening and closing of cracks, regarding all possible loading paths. Indeed, when a beam is subjected to a cyclic flexure, for example, a crack initiated at the beginning in the tension area will close with the inversion of the load, as a new crack will be created in the initially compressed zone, and so on. Let us examine both the behaviour of the concrete initially cracked, and that of the material initially in compression.

Fig. 1 represents the behaviour of concrete initially in tension: when the tensile stress reaches f_t, (path 1&2), a crack is created perpendicular to the direction of traction, and the stress falls to zero until the material is subjected to compression (path 3). As the crack closes progressively, the stress then becomes negative up to $-f_t$, the level at which the rigidity is completely restored (path 4). The modeling of path 3 is founded on the tests of Reinhardt & al. (1986), and Ramtani (1990). Along path 5, the compressed concrete follows the unloading line ($E_2 \neq E_0$) passing through a focal point (f_c, ε_0) as suggested by Mander & al. (1988), and Park (1990). Indeed, the experiments (Ramtani (1990), Sinha (1964)) show that the elastic modulus at unloading is different from the initial modulus E_0 due to the damage of concrete through compression. When the stress becomes inferior (in absolute value) to $|-f_t|$ (path 6), modulus E_1 corresponding to the closing, is restored. Paths 7, 8 and 9 obey the same criteria as paths 2, 3 and 4.

1- Elastic tension
2- Crack opening
3- Crack closing
4- Non-linear compression
5- Damaged unloading (modulus E_2)
6- Damaged unloading (modulus E_1)
7- Reopening of crack (f_t=0)
8- Closing of crack
9- Linear compression (modulus E_2)

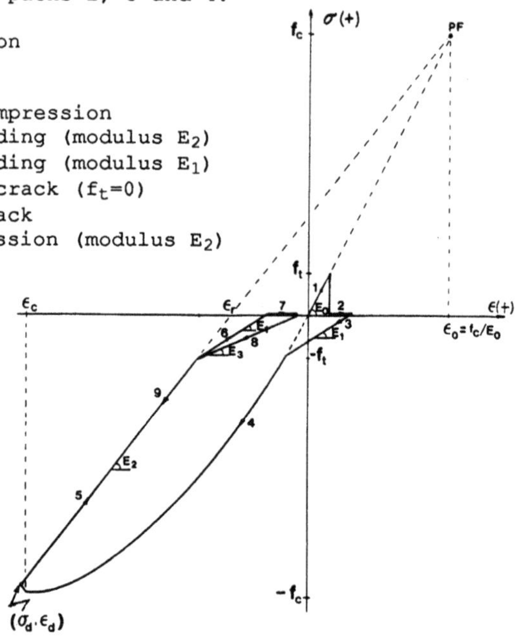

Fig.1. Point initially in tension

Let us now follow in Fig. 2 the history of an initially compressed concrete: the virgin concrete is subjected to a compressive stress followed by unloading with a damaged modulus E_2 (paths 1&2). The stress then changes sign and increases up to the new tensile resistance (path 3), modified as suggested by Morita and Kaku (1975). We are able in this way to take into account the degradation of the mechanical properties of concrete in tension when the material has been previously damaged in compression. Once the limit compressive stress is reached (path 4), the descending branch is simply obtained from the ultimate deformation in compression and the deformation at onset of collapse. Paths 5 and 6 are identical to paths 2 and 3, and paths 7 and 8 proceed with the same treatment of the opening-closing-reopening presented for a point initially in tension.

This law enables us to calculate from the deformation field, the corresponding stress state, described at a number of points, distributed in the volume of the element. Thus the state of the material is not the same everywhere in the element. By integration of the stress at these integration points, the corresponding internal restoring forces are obtained.

The modeling of steel can be elastic-perfectly plastic, multiple line strain hardening, or with the real stress-strain curve described

1- Non-linear compression
2- Damaged unloading (modulus E_2)
3- Elastic tension with resistance f'_t
4- Softening behaviour in compression
5- Damaged unloading (modulus E'_2)
6- Damaged tension (modulus E'_2)
7- Crack opening
8- Crack closing

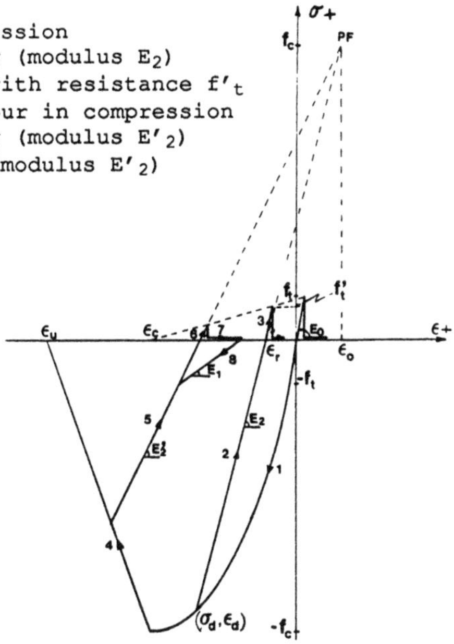

Fig.2. Point initially in compression

point by point. The hardening has up to now always been taken as isotropic, and we assume perfect bond between steel and concrete.

3 Application

This model has been validated with simple structures, composed of slender elements, subjected to a reduced number of quasi-static alternate cycles (Merabet & al. (1991), Pinto-Barbosa (1992)). The calculations presented here serve to evaluate the applicability of the model to real structures, subjected to seismic loads. Indeed, when addressing dynamic problems, one has to face:

- larger time and memory consumption ;
- usually more than five alternate cycles ;
- more complex loads, the production and the effect of which have to be modeled correctly ;
- damping, which is taken into account implicitely, only through material hysteretic dissipation .

The behaviour model itself (calculation of the internal restoring force) is not modified, and thus, the effects of viscosity, fatigue, shear, sliding between steel and concrete, are neglected.

3.1 Simple column subjected to earthquake

The first example is the study of a reinforced concrete column tested by the C.E.A.(Comission for Atomic Energy) on a shaking table (Gauvin (1978)), described on Fig. 3. It is subjected to a base acceleration which has a maximum value of 1.9 $m.s.^{-2}$, and lasts about 4s.

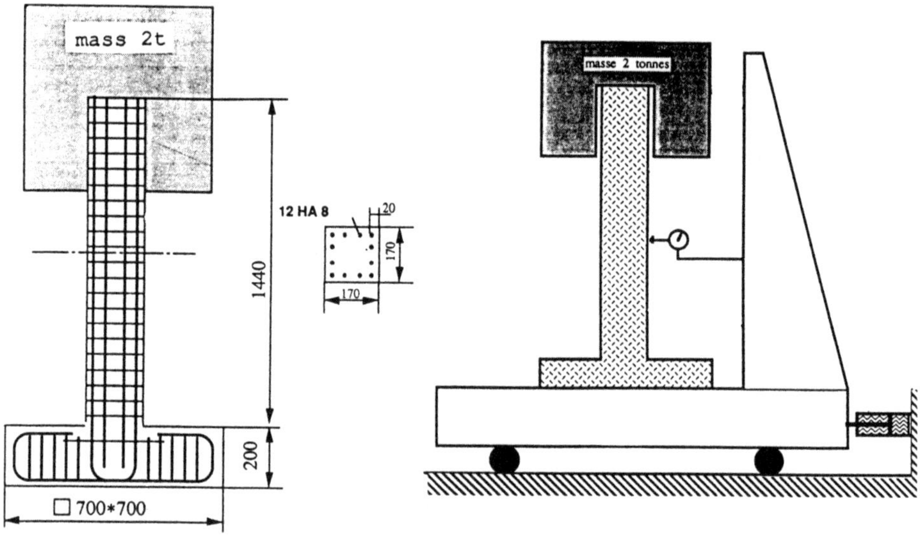

Fig.3. Specimen geometry (dimensions in mm.)

The experimental results available concern the horizontal displacement versus time for a point in the middle of the column which is compared to the analytical, on Fig. 4. The measured and calculated maximum amplitudes and frequencies are summed in Table 1.

Even though the corresponding extrema do not occur at the same times, the results are considered to be satisfactory. Indeed, in reality the column tested had previously been subjected to smaller earthquakes, and is thus initially damaged when undergoing this accelerogram. Since the initial conditions of displacement and material state are different from the ones imposed in the numerical study, there is no meaning in trying to superpose the curves. Furthermore we observe indeed that the numerical global results are closest to the experimental ones after the column has been sufficiently damaged.

Table 1. Global results

	experimental	calculated
amplitude max	1.64 mm.	1.55 mm.
frequency	3.33 hz.	3.21 hz.

One can also note that the absence of the damping matrix [C], justified by the intrisic modeling of the hysteretic dissipation, through the material nonlinear cyclic law, is regretted only once the earthquake has stopped ($t \simeq 3.5s.$), and the column is left to its free vibrations: in contrast with the experimental results, the column continues to vibrate with a constant amplitude. This is due to the fact that the materials have entered a new nonlinear but elastic domain, reducing the hysteretic dissipation to zero. The case of elastic materials is indeed the only case for wich the usual artificial damping

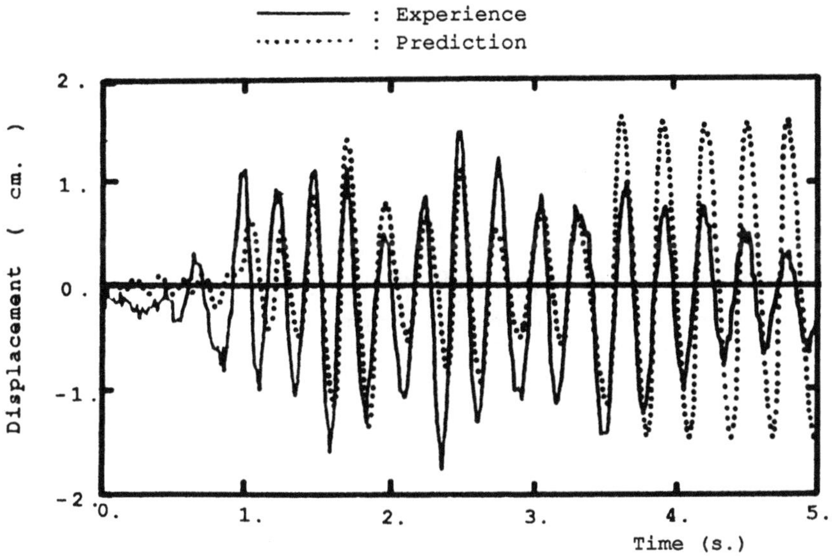

Fig.4. d = f(t) at mid-height of the column

matrix is felt to be justified.

This first study shows that through the local treatment of the damage, the present model is able to reproduce quantitatively its influence at the global level: increase of the period of vibration and tracing of the amplitude. The crack pattern, strain field, and deformation are also available anywhere in the structure, at every time step.

3.2 Limit case: study of a wall

One of the major drawbacks of this model is the restriction of its validity domain to structures composed of lineic elements. Therefore, it is necessary to evaluate the performances of the model in limit

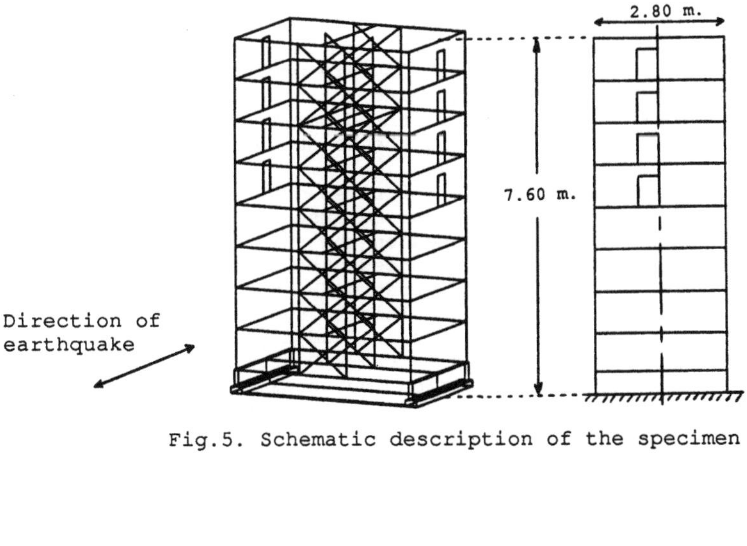

Fig.5. Schematic description of the specimen

cases, here in the case of a shear wall building, eight storeys high, with a lenght-to-width ratio of 2.7, described in Fig. 5. This structure is studied by a number of french teams, within the C.A.S.S.B.A. project, which is supported by the french department of research and technology. A 1/3 scale model of this building (weighing 90 t.) will be tested on the Azalée Shaking table of the C.E.A. (Comission for Atomic Energy). The direction of the earthquake being parallel to the walls, these are the only structural components modeled.

The object is thus to evaluate the effect of degenerating the stress field to one dimension in the calculated behaviour of a structure of such slenderness, in order to eventually use the present model for the prediction of its dynamic behaviour. In order to do this, the results of the one-dimensional model are compared to those obtained with the bi-dimensional approach, for a static-equivalent load, up to failure. Figs. 6 & 7 compare the load-displacement curves, and the crack patterns for a load level close to the maximum.

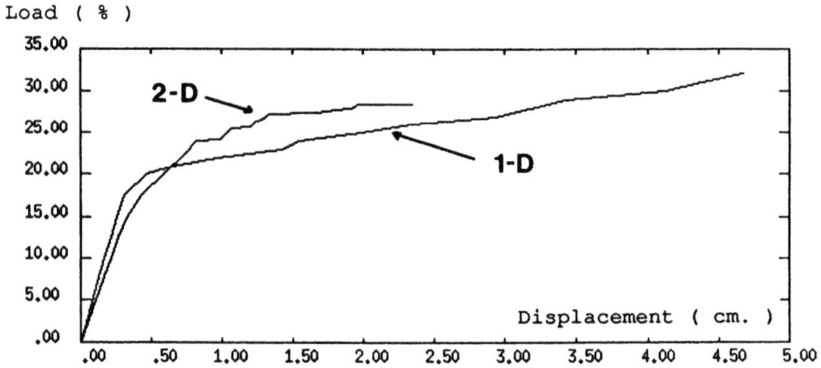

Fig.6. Comparison of global 1-D and 2-D results

It can be seen that:

* The initial stiffness for the 1-D model is 1.25 times that obtained with the 2-D model. This comes from the fact that the distortion and the warping of the section are prevented.

* Cracking is similar at the beginning for both calculations: it initiates in the same area, and is oriented horizontally. Nevertheless, it progresses much more rapidly along the height of the wall with the 1-D model, bringing more flexibility.

* The estimated ultimate load is 1.2 times greater for the uniaxial approach than for the biaxial calculus. Because in this case the loads and not the displacements were imposed on the structure, one cannot compare the ultimate displacements. Indeed, the calculations were interupted by a limiting number of iterations, and not through divergence of the solution procedure. This is typical of a plateau behaviour, for which the displacements are undefined.

* The mode of collapse described by the 2-D analysis is primarily a flexural mode, as the crack orientation indicates. Since yielding of the steel appears only for the non-converged points in both cases, it can be concluded that both analysis do describe the same failure mode: yielding of the steel where a flexural crack becomes too important.

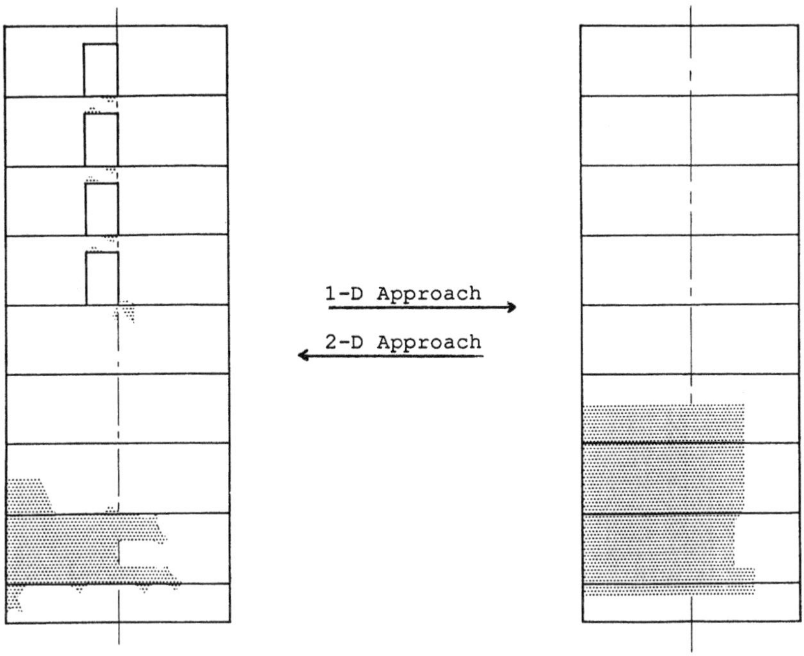

Fig.7. Crack patterns for P=27 %

It can be drawn from this example that the use of the present uniaxial model for static monotonous analysis of structures having a slenderness ratio inferior to 3 is delicate, but gives in the present case still acceptable results. Assuming that the similarity between the behaviours obtained still hold for a cyclic load on one hand and for a dynamic load on the other, (which is still to be justified,) the present model will be able to yield the seismic levels corresponding to the onset of cracking, yielding of steel, values of a failure criteria, the qualitative effect of non-linearities on the dynamic behaviour.

3.3 Two-bay/four-storey frame subjected to earthquake

The object of this last calculation is to try the proposed model for the seismic analysis of a more complex structure, and evaluate its performance through comparison with experience and other models: the structure considered, (shown Fig.8) is to be tested on the Ispra Joint Research Centre reaction wall, and a number of research teams are to perform predictive simulations with each their own numerical model.

At this stage, a first comparison with the global model of the C.E.A. (Comission for Atomic Energy) is available. This model (Bairrao, 1986), developed and implemented in the general purpose finite-element program Castem 2000, works in two successive stages: first the moment/curvature skeleton curves are constructed analytically for each particular section, through the Samson program, using the local material properties of concrete and steel, and the geometry of the structural element. The nonlinear dynamic analysis of the structure can

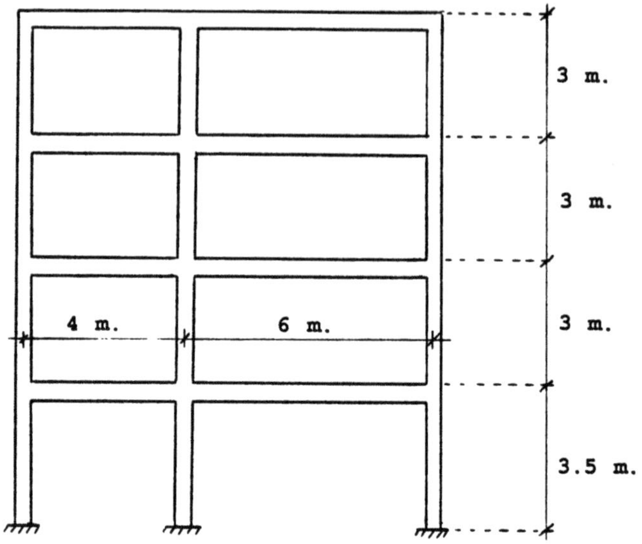

Fig. 8. The Ispra J.R.C. frame

then be carried out in a classical fashion.

The (top-horizontal displacement/time) plots are compared on Fig. 9. Though we cannot claim one to be closer to reality than the other, we can observe very close results for the intervall during which the structure is increasingly damaged, then a stronger damping for the global model once the major acceleration input peaks are past, and the structure is already quite damaged.

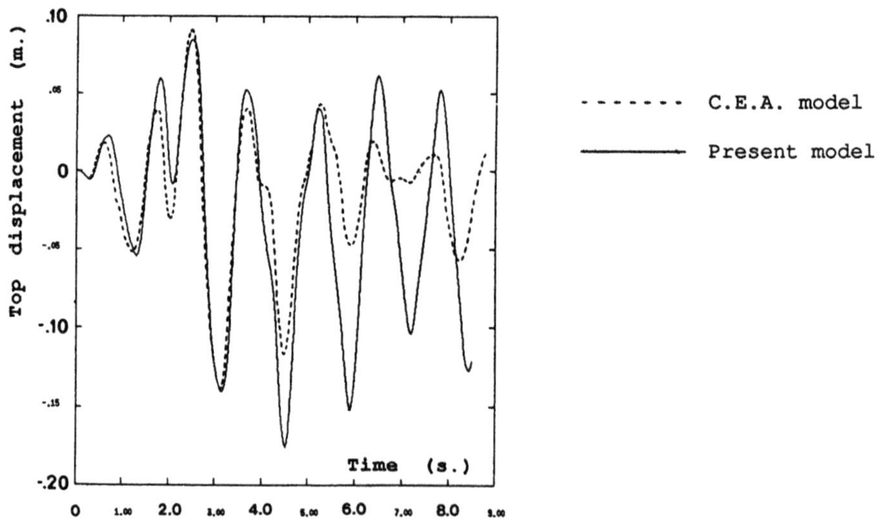

Fig. 9. Comparison C.E.A. model/Present model

While the global approach :

* needs in this particular case the construction of 41 different moment-curvature curves, which is engineer-time consuming;
* is quicker in the time integration of the dynamic equilibrium equations;
* assumes constant axial loads throughout the time history;
* does not yield directly information on the distribution of damage along the length of the members;

The proposed model :
* works directly with the material characteristics;
* is more computer-time and memory consuming for the dynamic analysis itself;
* takes into account the variation of axial load naturally;
* permits a detailed description of the state of damage of the materials.

Because the two calculations were performed on different computers, by different persons and with a different number of other simultaneous calculations, it is practically impossible to compare quantitatively the times involved for each of them. Nevertheless, after comparing the quantitative factors of efficiency, it can be concluded that the proposed model is an interesting tool for the analysis of damage in frames subjected to long lasting dynamic loads.

4 Conclusion

These three applications demonstrate the performance of the model in diverse situations.

Firstly, it is shown that the local description of damage at the material level produces the correct global effects. From this local description, one can calculate classic member damage indicators, as those used with global models.

Since bibimensional dynamic calculations are still unaffordable, this model can be a good approximation even for relatively low shear span ratios, under the condition that its reliability be assessed for each particular problem: the effect of neglecting shear is measured through static comparison with the 2-D model.

Finally we were able to check that the computer time consumption stays reasonable, even for a real structure.

The present model is therefore an efficient tool when it is used within the range of application corresponding to the main following assumptions:

* the effect of shear is negligeable;
* Kirchhoff hypothesis;
* the effect of the steel-concrete slip is negligeable;
* the effect of fatigue is negligeabe.

Nevertheless, it will be interesting to extend its application with relatively simple developments, in order to take into account shear and slippage between steel and concrete. This can be done without reducing its efficiency.

5 References

Berrao, R. (1986) "Modèle de comportement à la ruine des structures constituées de poutres en béton armé. Prise en compte des phénomènes d'endommagement et de plasticité.", **Doctoral dissertation,** LMT/ENS Cachan , Paris 6, 1986.

Chahrour, I. (1991) "Contribution a l'Analyse Non-linéaire Matérielle et Géométrique des Coques en Béton Armé", **Doctoral dissertation,** INSA of Lyon, France.

Djerroud, M. (1992) "Modélisation des Structures en Béton Armé sous Chargement Monotones et Cycliques dans le Cadre Elastoplastique", **Doctoral dissertation,** INSA of Lyon, France.

Fardis, M.N. (1991) "Member-Type Models for the Nonlinear Seismic Response Analysis of Reinforced Concrete Structures", in **Experimental and Numerical Methods in Earthquake Engineering** (eds J. Donea and P.M. Jones), Kluwer Academic Publishers, Dordrecht, Netherlands, pp. 247-280.

Gauvin, J.; Jeandidier, C.; Gaubert, J.C.; Queval, J.C. and Vaghi, H. (1978) "Essais Sismiques de Poteaux en Béton Armé." **Note Technique C.E.A.- E.M.T.**/78/177.

Mander, J.B.; Priestley, M.J.N. and Park, R. (1988) Theoretical Stress-Strain Model For Confined Concrete. **J. of Struct. Eng.**, ASCE, Vol. 114, No. 8, PP. 1804-1825.

Merabet, O. (1990) "Modélisation des Structures Planes en Béton Armé sous Chargements Monotone et Cyclique. Construction et Validation d'un modèle numérique", **Doctoral dissertation,** INSA of Lyon, France.

Merabet, O.; Djerroud, M.; Chahrour, I. and Reynouard J.M. (1991) Développement d'un Modèle Semi-Global pour le Calcul des Systèmes de Poutres en Béton Armé Sous Chargements Alternés Cyclés, **Rapport Scientifique du GRECO,** (ed. Reynouard, J.M.), INSA Press, Lyon, France, pp 417-422.

Meyer, C.; de Borst, R.; Bicanic, N. and Filippou, F. (1991) Dynamic Loading, in **Intl. Workshop on Finite Element Analysis of Reinforced Concrete,** Columbia University, New York.

Morita, S. and Kaku, T. (1975) Cracking and Deformation of Reinforced Concrete Beams Subjected to Tension, in **Liege Colloquium Inter-Association.**

Park, Y.J. (1990) Fatigue of Concrete Under Random Loadings. **J. of Struct. Eng.,** ASCE, Vol. 116, No. 11, PP. 3228-3235.

Pinto-Barbosa, M. (1992) Sémi-Rigidité des Noeuds d'ossatures en Béton Armé: Simulations numériques, **Doctoral dissertation,** INSA of Lyon, France.

Ramtani, S. (1990) Contribution à la Modélisation du Comportement Multiaxial du Béton Endommagé avec Description Du Caractère Unilatéral, **Doctoral dissertation,** University of Paris 6, Paris.

Reinhardt, H.W.; Cornelissen, H.A.W. and Hordijk, D.A. (1986) Tensile Tests and Failure Analysis of Concrete. **J. of Struct. Eng.,** ASCE, Vol. 112, No. 11, PP. 2462-2477.

Sinha, B.P.; Gerstle, K.H. and Tulin, L.G. (1964) Response of Singly Reinforced Beams to Cyclic Loading. **J. of Amer. Concr. Inst.,** pp. 1021-1037.

Song, S. and Maekawa, K. (1991) Dynamic Nonlinear Finite Element Analysis of Reinforced Concrete. **J.of the Faculty of Eng., University of Tokyo,** Vol. XLI, No. 1, 73-158.

47 3D-NUMERICAL ANALYSIS OF THE SLAB-COLUMN CONNECTION

J. OŽBOLT
Stuttgart University, Germany
S. BLAŽIĆ
University of Civil Engineering, Rijeka, Croatia

Abstract
In the present paper the results of the three-dimensional nonlinear finite element analysis of the slab-column connections are reported. The slab is first loaded by the service level vertical (shear) load, introduced through the normal column force, and then followed by ten cycles of the service bending moment that is applied through the shear column force. After damage caused by cycling loading, the slab is further loaded by vertical load up to the certain level followed by bending moment up to failure.
The analysis is performed by the use of the general nonlocal microplane model for concrete. In this fully 3D cyclic material model, including the rate effect, concrete behavior is simulated through a number of planes called microplanes. The fracture mechanics is introduced by the use of the nonlocal strain concept.
Numerical results obtained by the finite element analysis are compared with previously published test results. Similarly as in the experiment, the numerical results indicate diagonal shear failure mode and a significant influence of the damage introduced through the cycling loading on bearing capacity of the plate.
Keywords: Microplane model, Cycling, Rate effect, Viscosity, Fracture

1 Introduction

In the current engineering practice the slab-column connection is often used. Gravity, wind or earthquake forces produce shear and bending moment in flat-plate floors. Bending moment is often referred to as the "unbalanced" moment. Flat-plate buildings usually have structural systems to provide adequate stiffness to resist lateral loads. However,despite of the existence of such a system some shear-moment transfer between the slab and the column must take place.

The shear force and the unbalanced moment produce shear, bending moment and torsional moment in the slab. Brittle

Fracture and Damage of Concrete and Rock – FDCR-2.
Edited by H.P. Rossmanith. ISBN 0 419 18470 8.

failure in the column vicinity can occur due to the high shear stresses that are produced by such a transfer. A slab-column connection reinforced by well-anchored shear reinforcement and with flexural reinforcement may fail by punching shear or by flexure. Thus, this type of structure must satisfy two basic requirements:
(1) To avoid brittle failure due to the diagonal shear
(2) The ductility of the connections must be assured, especially in seismic active regions.

In the present paper the results of the three-dimensional nonlinear finite element analysis of the slab-column connection are reported with attention focused on the punching shear failure.

2 Finite element analysis

2.1 Geometry of the plate and finite element mesh

The flat plate slab placed over the interior square column, distant enough from the boundaries, is analyzed (see Fig.1). The structure of the same geometry has been previously tested [4]. The FE model used in the analysis represents a full-scale interior column connected to a slab part bound by the lines of contraflexure around the column. In the model the slab was simply supported near the edges along the perimeter of a square having the side length of 1800 mm. Due to the symmetry only one-half of the plate is considered.The geometry, the FE mesh of the plate and the reinforcement details for the adopted numerical model are shown in Fig.1 and Fig.2.

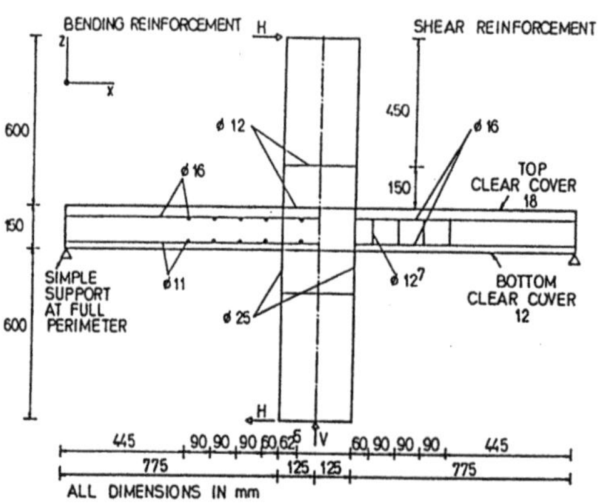

Fig.1. Geometry of the plate

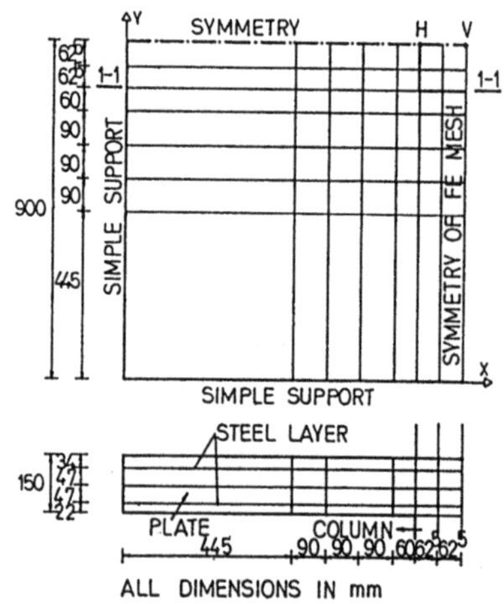

Fig.2. FE mesh

In the study two types of the finite elements are

employed. Eight nodes 3D linear displacement field finite elements for concrete and 1D truss elements for the bending and shear reinforcement. It is assumed that the steel bar elements crossing the concrete elements in the nodes and no bond-slip relation is used. Boundary conditions are shown in Fig.2. Cross section area and position of reinforcement are approximately the same as in the experimental study [4].

The bending reinforcement of the plate was designed to prevent flexural type of the failure, i.e. to assure failure of the plate in diagonal shear.

2.2 Loads

The present plate is analyzed for the action of the vertical (punching) load and bending moment. In order to demonstrate the effect of damage introduced in the plate through the cycling loading on the failure capacity, two different loading histories are applied.

In the first, the slab is loaded by the service level vertical (shear) load, introduced trough the normal column force (V=75.kN), and kept constant, followed by ten cycles of the service level bending moment (M=65.kNm), introduced by the shear column force, to simulate service load repetition. After damage caused by the cycling loading the slab is further loaded by vertical force up to the constant value (V=150.kN), followed by the increasing bending moment up to failure (M_f=130.kNm).

In the second loading history the slab is first loaded by the test level vertical (shear) load, introduced by normal column force (V=150.kN), and kept constant, followed by bending moment introduced by shear column force, increased from zero until failure (M_f=162.kNm).

In Figs.1 and 2 the directions of application for vertical force V and bending moment M are plotted.

2.3 Material models used for concrete and steel

The numerical analysis is performed by the use of the general 3D cyclic and rate sensitive nonlocal microplane model for concrete. In the microplane material model [1,2,5], the material properties are characterized separately on planes of various orientations within the material, called microplanes, on which there are only a few uniaxial stress and strain components and no tensiorial invariance requirements need to be observed. The state of each microplane is described by normal stains (deviatoric and volumetric) and shear strain. In order to avoid spurious localization instabilities due to the strain softening and the consequent mesh-sensitivity problems in finite element analysis the concept of nonlocal continuum with local stains is adopted. Basic concrete parameters used in the analysis are: Young modules E=32000 MPa Poisson's ratio ν=0.18 Characteristic length of the nonlocal continuum l=2.50 cm and Viscosity ρ=0.01 s^{-1}.

The microplane material parameters are chosen such to fit the standard tension and compression strength of concrete that is used in the experiment [4].

Fig.3 shows microplane material model stress-strains response in the case of tension and compression obtained using the local single finite element analysis (3D eight node finite element) in monotonic and cyclic loading. The rate of deformations applied in the present case ware $d\varepsilon/dt=0.1$ 1/s in tension and $d\varepsilon/dt=1.0$ 1/s in compression.

Behavior of steel in the analysis is assumed to be ideal elasto-plastic with yielding limit $\sigma_y=460$ MPa.

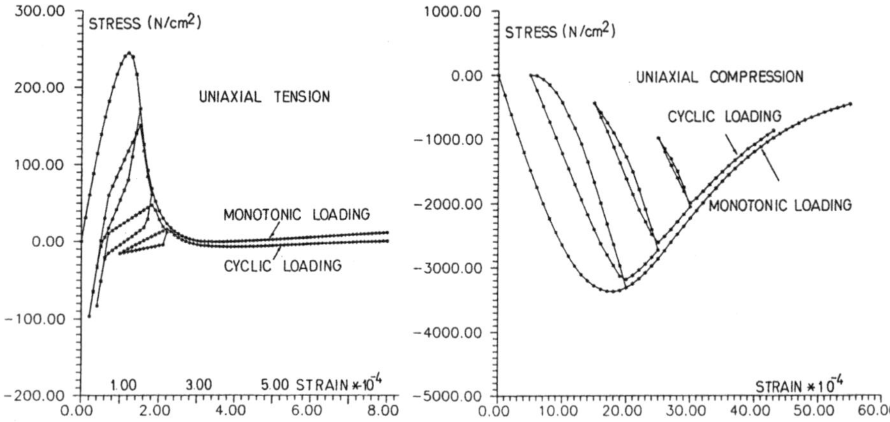

Fig.3. Strain-stress response in tension and compression

2.4 Differences between numerical model and test specimen

Since the results of the analysis are compared with test results it is worthwhile to mention the main differences between both studies. Namely, there are a few minor differences such as the fact that in the analysis the self weight is not taken into account and that the load was not locally introduced into the structure exactly as in the experiment. These differences have practically no influence between test and numerical results.

However, in the test the anchorage of the vertical shear reinforcement is made with top anchor plate and the bottom anchor strips while in the numerical model the anchorage of the studs at the top and the bottom are modeled such that that the horizontal bending reinforcement does not contribute to anchorage of shear reinforcement since horizontal steel bar has no stiffness in shear reinforcement direction (truss element). It is evident that the anchorage in the model is not equivalent to the anchorages in test specimen. This matter will be studied in separate study.

3 Results of the analysis and comparison with test data

The results obtained by the finite element analysis are reported and compared with test results. In order to show the failure mode obtained in the analysis, the principal strains for typical cases at different load stage are plotted in vertical section 1-1 (see Fig.2). Also, the load-displacement curves and stress distribution in reinforcement are plotted and compared with measured data.

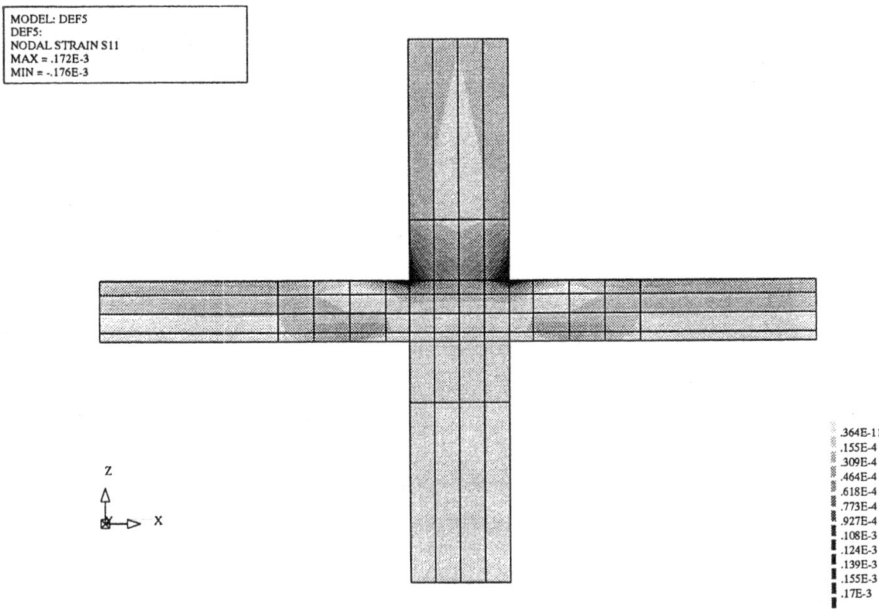

Fig.4. Principal strains ε_{11} at service level vertical load

In Figs.4,5 and 6 the principle strains at different load stages for the plate without shear reinforcement, loaded by cycling loading are shown. Fig.4 shows the principal strains ε_{11} at service level vertical load. Fig.5 shows the principal strains ε_{11} after introducing ten cycles of the service level bending moment. Comparison between Figs.4 and 5 indicate significant increase of residual strains and damage due to cycling loading.

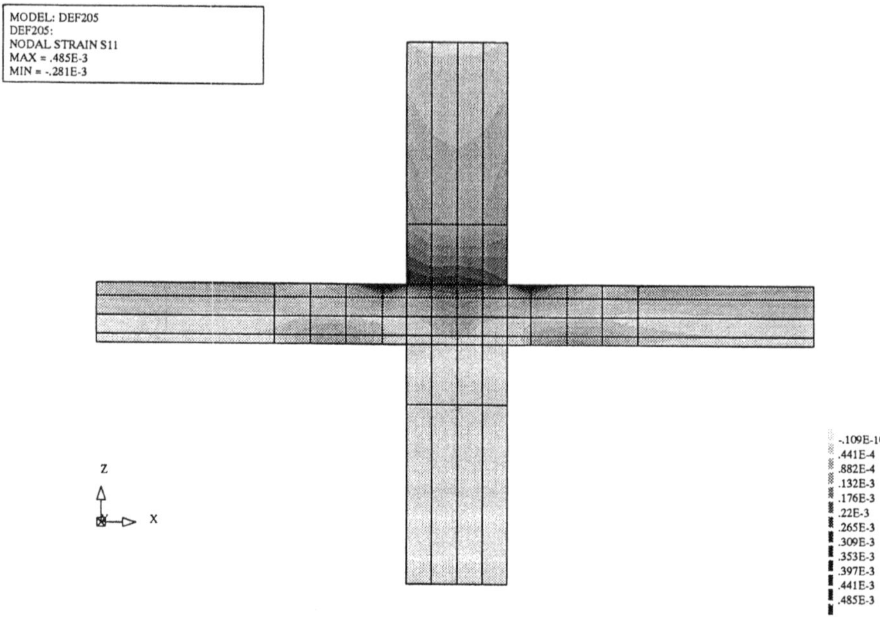

Fig.5. Principle strains after service level bending moment

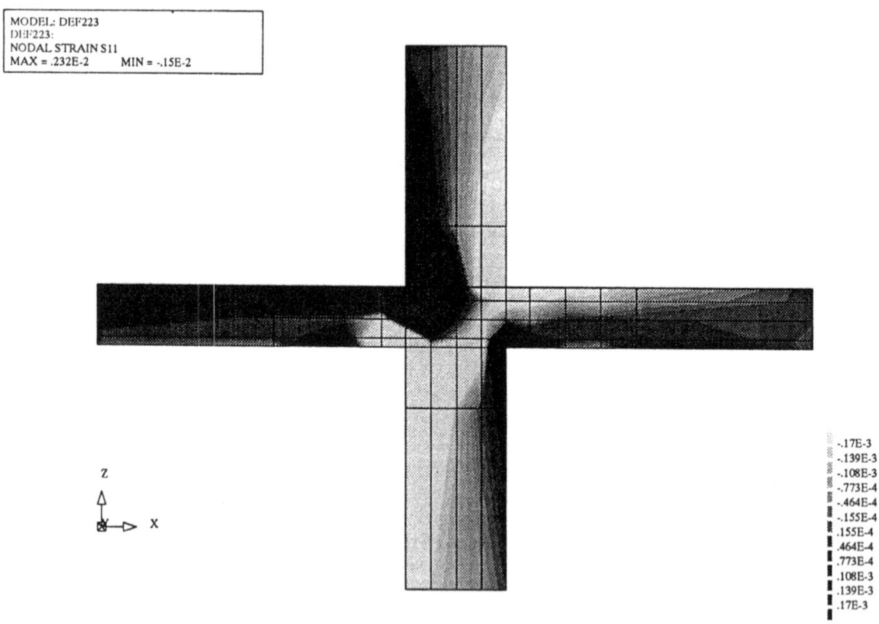

Fig.6. Principle strains at failure

Fig.6 shows the principle strains ε_{11} at failure. It can be seen that the failure is caused by shear in an inclined plane near the column face since in this plane concentration of the principle strains ε_{11} can be observed (see Fig.6). The shear brittle failure type is also confirmed by the fact that the bending reinforcement is still in the elastic range as well by the fact that the divergency of the solution was very strong.

In the experiment the shear failure was followed by punching of the column through the slab and splitting of the slab at the top flexural reinforcement layer.

Results of the analysis indicate 30% higher failure load than measured in test. Due to the complexity of the problem it is difficult to say what is the reason for this. However, overall behavior as well as failure mode is similar as observed in the experiment.

As expected, results of the analysis for the specimen that was not damaged by the cycling loading indicate increase of the loading capacity for approximately 25% in comparison to the previously damaged specimen. Failure mode observed in this analysis is similar as shown in Fig.6. However, difference in failure load between measured and calculated results are slightly larger and reaches ≈ 60% .

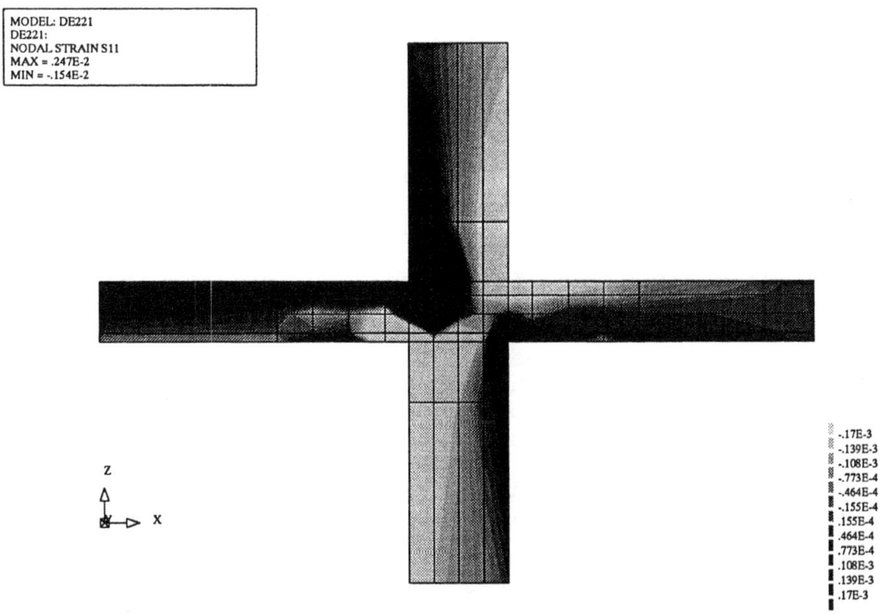

Fig.7. Principle strains at failure with shear reinforcement

In Fig.7 the same results of the analysis as shown in Fig.6 are present except that the plate was analyzed with shear reinforcement. Basically the same failure mode is observed as in the case of the plate without shear reinforcement and the effect of cycling load was similar as well. As can be expected, deformations were large and failure load higher than in the case of the plate without shear reinforcement. Present results indicate that the cycling load causes large decrease of failure load in both cases - with and without shear reinforcement. Namely, the relative calculated decrease of the failure load due to cycling loading was for the unreinforced plate 20% and for the reinforced plate 15% , compared with the corresponding plate loaded without cycling loading.

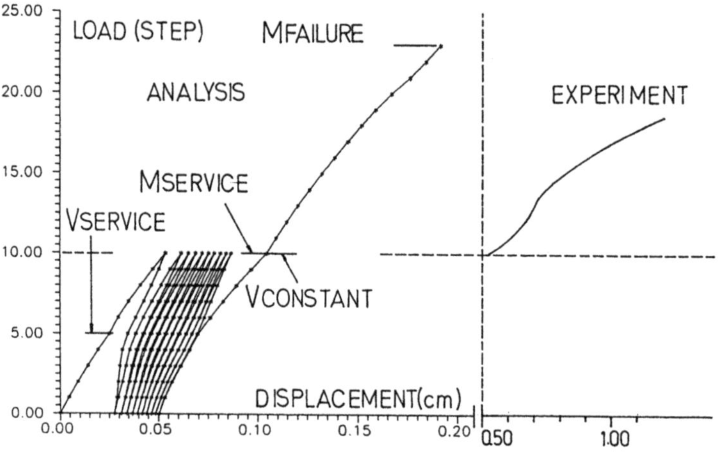

Fig.8. Load-displacement curves

In Fig.8 load-displacement curves for cases without shear reinforcement measured in experiments and obtained in the analysis are shown. It can be seen that the shape of the calculated and measured load-displacement curves are similar. However, calculated displacements are much smaller than measured in experiments. This is partly caused by relatively coarse FE mesh used in the analysis. However, the main reason is probably due to the fact that in experiments the displacements have been measured at the position where the local damage was high. Again, comparison between calculated load-displacement curves obtained in the case of cycling loading and without cycling loading indicate that damage due to cycling loading significantly contributes to the increase of the displacement and decrease of the failure load.

In Fig.9 stresses in vertical shear and bending reinforcement in section 1-1 at failure are shown. In all cases analyzed, the stresses in the bending reinforcement were in elastic range. This is in agreement with experimental observations. Calculated stresses in vertical shear reinforcement are also in elastic range. However, in the test it has been observed that the stresses in shear reinforcement, close to column, are almost at yield limit. As already mentioned, the reason for this is probably due to the fact that the anchorage of the shear reinforcement in the FE model was to weak.

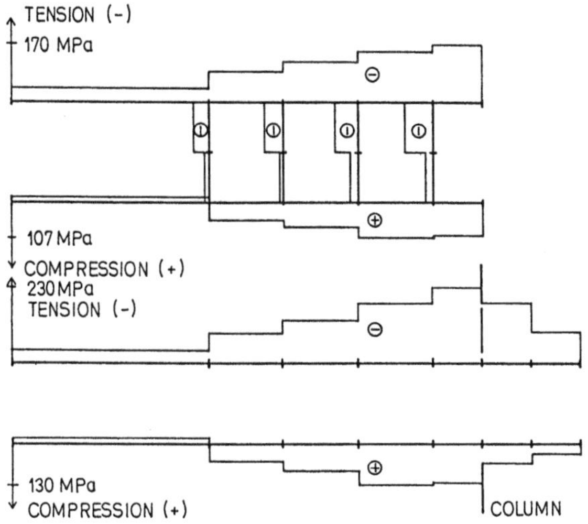

Fig.9. Stresses in shear and bending reinforcement

4 Conclusions

It is demonstrated that the numerical analysis based on the general nonlocal model for concrete is capable to predict the failure mode of the plate-column structure rather realistically. Similarly as in the experimental study, it is found that the damage that has been introduced through cyclic loading decreases the bearing capacity of the structure and significant increase of displacement failure. If the shear reinforcement is not present decrees of failure load is more pronounced. Numerical studies as well as the test results indicate diagonal shear failure mode in all cases studied. As can be expected, plates with shear reinforcement are more ductile and deformations at failure are smaller.

Further studies are needed in order to investigate the influence of the anchorage of the shear reinforcement as well as the size effect study for plates with and without shear reinforcements.

5 References

1 Bažant, Z.P., and Prat, P.C., Microplane model for brittle plastic material - parts I. and II. J. of Eng. Mech., ASCE, 114(10), pp.1672-1702, 1988.

2 Bažant, Z.P., and Ožbolt J., Nonlocal microplane model for fracture , damage and size effect in structures. J. of Eng. Mech., ASCE, 116, (11), 1990, pp. 2485-2505.

3 Bažant, Z.P., and Chern, J.-C., Strain softening with creep and exponential algorithm. J. of Eng. Mech., ASCE, 111(3), pp. 381-390, 1985.

4 Elgabry, A.A., and Ghali, A., Tests on concrete slab-column connections with stud-shear reinforcement subjected to shear- moment transfer. ACI Structural Journal, September -October1987, pp. 433-442.

5 Ožbolt, J., and Bažant, Z.P., General microplane model for cyclic triaxial behavior for concrete. J. of Eng. Mech., ASCE, 1992, in press.

6 Pijaudier-Cabot, G., and Bažant, Z.P., Nonlocal damage theory. J.of Eng. Mech., ASCE, 113, pp. 1512-1533, 1987.

7 Sinha, B.P., Gerstle, K.H. and Tulin, L.G., Stress-Strain relations for concrete under cycling loading, J. Am. Concr. Inst., 61(2), pp. 195-211, 1964.

48 STRENGTH AND DEFORMATION CAPACITY OF SLENDER RC BEAMS

A.R. MEGAHID, H. SOGHAIR, M.A.A. MAGEED
and A.M.A.A. HAFEZ
Assiut University, Assiut, Egypt

Abstract
Slender beams loaded in the usual manner tends to fail laterally of stress often far below the yield point of the material. This fact is provided for in design specifications by reducing the basic allowable stress in the extreme compression fiber for beams in which the ratio of length to width is greater than a prescribed maximum limit. The specified reduction is based more or less on semiemperical considerations rather than on a strict analytical study. The behaviour of slender reinforced concrete beams is of high degree of complexity because of, the instability failure which takes the form of lateral buckling accompanied by twist, moreover the complex character of the material (inhomogeneity, partial cracking, limited elasticity,...etc). The analytical treatment of the problem becomes complex if an attempt is made to asses all features of reinforced concrete behaviour realistically. Because of the insufficient experimental evidence against which a theoretically derived critical load could be tested convicingly. The best design approaches therefore employ both theoretical and experimental studies, together with judgements of their limits.In addition , an appreciation is needed for the requirements of the designer and the standards of construction that may reasonably be expected. The objective, is therefore, to produce an approach that combines technical accuracy with case of application and realistic fabrication demands. This study deals mainly with the problem of lateral instability of slender reinferced concrete beams, which have no lateral supports, with special refference to rectangular sections.
Keywords: Strength and Deformation Capacity, Lateral Buckling and Displacement of Selender beam, Capacity reduction factors.

1 Introduction

For the lateral stability of reinforced concrete beams, the stability criteria, reduced to simplified formulae involving the ratios of (L/b) and (d/b), and based on conservative assumptions which suggested for three types of loading commonly met in practice. Both the ratios(d/b) and (L/b) are concurrent criteria for lateral stability of rectangular beams. Stability provisions based on (L/b) ratio alone is insufficient to completely define the required physical dimensions.

Fracture and Damage of Concrete and Rock – FDCR-2.
Edited by H.P. Rossmanith. © 1993 Taylor & Francis. ISBN 0 419 18470 8.

The results show that any regulation which bases instability solely on the (L/b) ratio is unsound and that the (d/b) ratio must also be taken into account. They also show that these seems to be justification for the clause in the draft code which prohibits a(L/b)ratio greater then a limited value.

It is very probable that the requirements of the (ACI-Code)are too restricitve for rectangular beams with low (d/b) ratios. The (L/b) limitations specified in the ACI-Building Code are extremely conservative for ordinary strength steel and relatively small(d/b) ratios.

2 Lateral bucking of slender beams

A beam which is bend in the plane of greatest flexural rigidity may buckle laterally at a certain critical value of the load. This lateral bucking is of importance in the design of beams without lateral support (or slender beams), providing the flexural rigidity of the beam in the plane of bending is large in comparison with the lateral bending rigidity. As long as the load on such a beam is below the critical value, the beam will be stable. As the load is increased, however, a condition is reached at which a slightly deflected (and twisted) form of equilibrium becomes possible. The plane configuration of the beam is now unstable, and the lowest load at which this critical condition occurs represents the critical load for the beam.

The beam with two planes of symmetry shown in Fig(1) is assumed to be subjected to arbitrary loads acting in the yz plane which is the plane of maximum rigidity. It is assumed that a small lateral deflection occurs under the action of these loads.

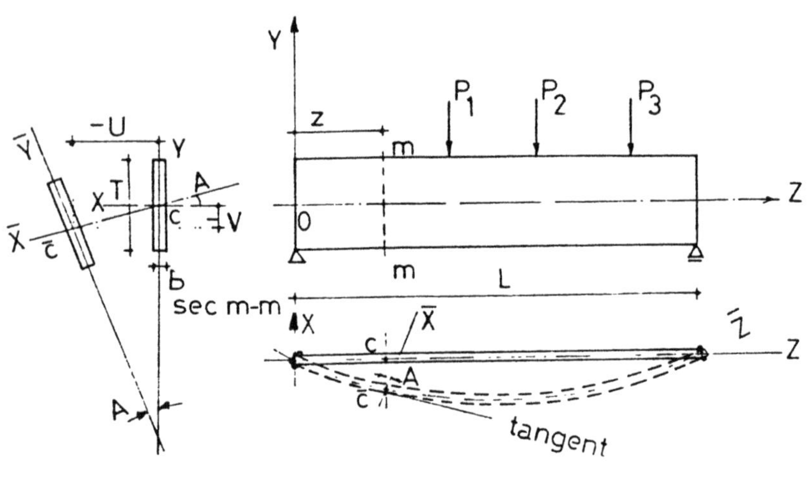

Fig.1.

3 Basis of analysis

In this study, the exact displacement functions in the horizontal, vertical and torsional directions have been used to get the critical buckling or flexural load for simply supported reinforced concrete slender beams of rectangular cross section and subjected to contentrated load at mid span length. The equilibrium equations can be written as follows:

$$d^2V/dz^2 = M/FR_x \,, \tag{1}$$

$$d^2u/dz^2 = (M/FR_y).\ A, \text{ and} \tag{2}$$

$$dA/dZ = (M/TR)\ .\ (du/dz) \tag{3}$$

where u is the horizontal displacement, V is the vertical displacement and A is the angle of rotation, M is the bending moment, FR_x is the flexural rigidity about the major axis (x axis), FR_y is the flexural rigidity about the minor axis (y axis), and TR is the torsional rigity of the section.

From equations (1), (2), and (3)

$$d^2A/dz^2 + (M^2/FR_y.\ TR)\ .\ A-(dM/dz)\ .\ (1/M).\ (dA/dZ) = 0 \tag{4}$$

Equation(4) is the basic differential equation of slender beams which can be solved numerically by using a numerical technique to get the angle of rotation at each segment of the divided beam length.

3.1 Basic assumptions

The main basic assumptions are :

The strain distribution is linear at ultimate strength.

The maximum strain at the extreme fiber of the concrete compression zone is assumed equal to 0.003.cm/cm.

The tensile strength of concrete is neglected.

The concrete stress distribution in compression at ultimate strength is rectangular with an intensity of fc' over a depth of 0.85 z from the extreme edge of the concrete compression zone.

The stree-strain relationship of steel in tension and compression is elastic-perfectly-plastic.

The modulus of elasticity of steel is equal to 20×10^6 kg/cm^2.

The stress in reinforcement and concerte are derived from the stress-strain curve shown in Appendix A.

The member is free to deflect in any direction.

Effect of impact and repetition of loads is not considered, assuming that the loads are static and do not repeatedly vary.

3.2 Evaluation of various structural parameters

The basic variables governing the critical load which cause lateral buckling of reinforced concrete slender beams are, the grade of concrete, the type of steel, the ratio of reinforcement , the width of the beam, the depth of the beam, and the slenderness ratio (L/b). In designing the selected beams (4320 beams), the following assumptions were considered :

The cross section is rectangular in shape.
The beam is simply supported and loaded one concentrated load concentrically at the mid-span section.
Calculated ultimate load is to be governed by ACI-Building code requirements.
The grade of concrete (varying between 180-300 Kg/cm^2).
The Types of steel were to be of grades St. 37, St. 44, and St.52.
The ratios of reinforcement (varying between 1% up to balanced ratio), and the section was taken to be singly under-reinforced.
The width of the beams were chosen to be 12,25, and 40 cms.
The depth of the beams were chosen to be 40,60,80, and 100 cms.
The slenderness ratios defined as (L/b) (Variable from 30 up to 100).
The shear span to depth ratio should be greater than 2.5 to avoid failure due to shear.
The formula for the torsional rigidity at any stage of loading is given as :

$$TR = Tb^3/3\ (1.0 - 0.63\ b/T)\ .\ (E/2(1 + \nu)) \tag{5}$$

Where E is the modulus of elasticity and ν is Poisson's ratio.

The Poisson's ratio was to be 0.15.
The formula for the flexural rigidity of reinforced concrete slender beam with rectanqular cross-section about y axis at different stages of loading is as follows:
a) Stage I (uncracked state):

$$FR_y = Tb^3E\ /12.0 \tag{6}$$

b) Stage II (cracked elastic state) :

$$FR_y = (M.n.E/c.a)\ (\ b_1^2\ /4(d-n) + b^2/6n) \tag{7}$$

Stage III (ultimate or plastic state):

$$FR_y = (M.b^2/12.\lambda.\ \varepsilon.a) \tag{8}$$

where the parameters of these formulae are as shown in Fig.(2).

A computer program has been prepared to facilitate the calculation of the different coefficients concerning the problem of lateral instability of concrete beams. From the program, the value of capacity reduction factor (ϕ) can be determined based on the three values of the loads, which are ultimate load(P_u), critical load(P_{cr}), and serviceability load (P_{ser}). See Hafez (1989).

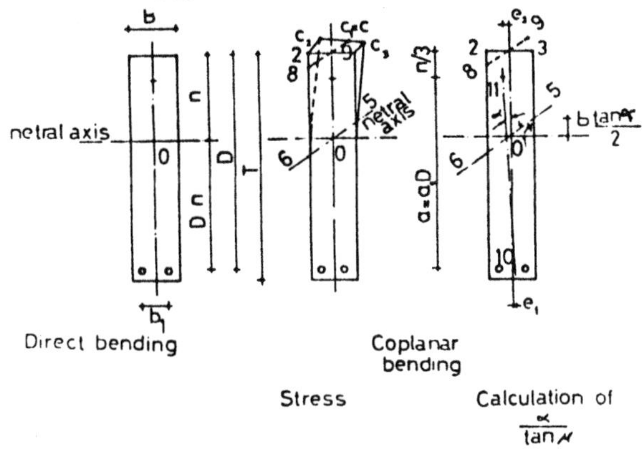

Fig.2. Stresses in narrow rectangular cross-section.

4 Results and discussion

The grade of concrete has a slight effect in increasing capacity reduction factor. This effect can be noticed at the high values of slenderness ratio (L/b) Corresponding to high values of the depth(d). By using high grade of reinforcement, the increase of the grade of concrete may be taken with no effect on the capacity reduction factor for all depthes used. This effect can be noticed as shown in Fig.(3)

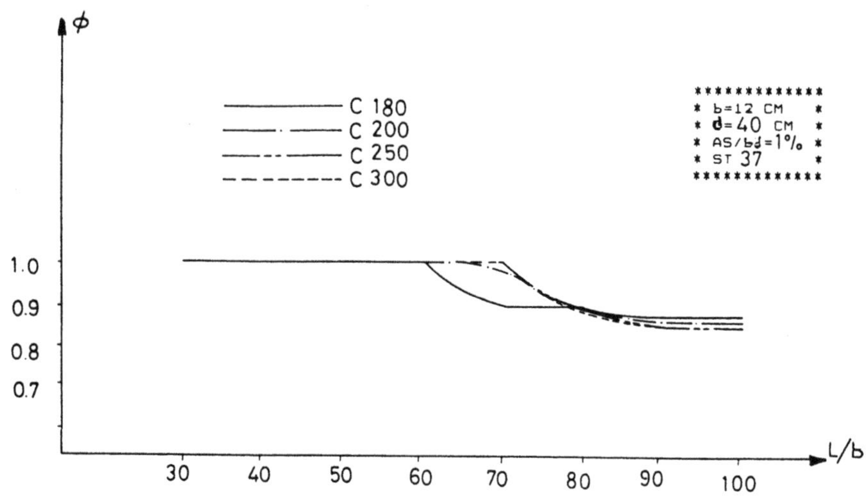

Fig.3. Effect of grade of concrete on the capacity reduction factor

High grade of steel, with low ratios of reinforcement decreases the capacity reduction factor for big slenderness ratio specially at large depthes, while with high ratios of reinforcement, this effect was noticed at small depthes as shown in Fig.(4).

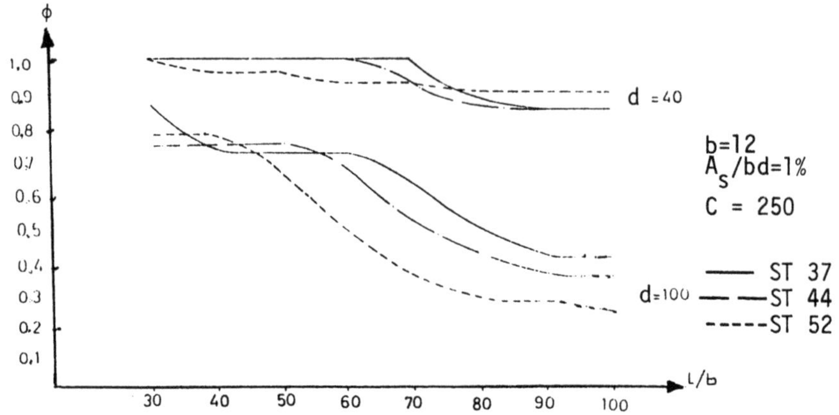

Fig.4. Effect of grade of steel on the capacity reduction factor

By increasing the amount of reinforcement, the capacity reduction factor decreases, and this effect is of high evidence at high depthes. The rate of variation of capacity reduction factor(ϕ) depends on the values of both grade of concrete and depth of cross-section, and its value is high at high grades and high values of depthes as shown in Fig.(5).

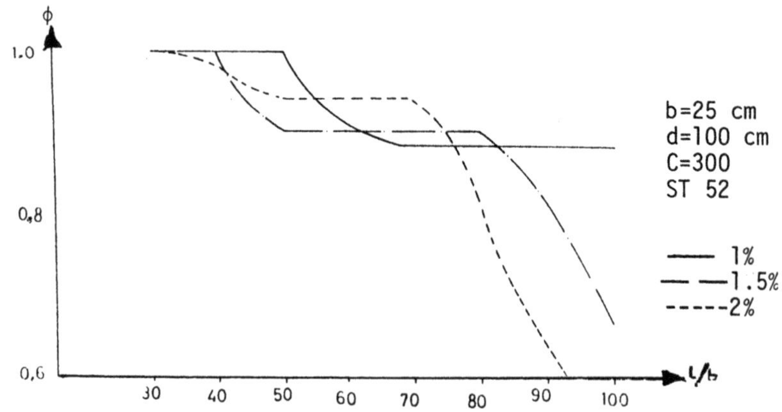

Fig.5. Effect of amount of reinforcement on the capacity reduction factor.

Increasing the width of the beam perimts the use of high slenderness ratio with high capacity of the section. The section with high width is safe against lateral instability and excessive deflection. It is found that high values of width (b) give safety against lateral instability problems.
Fig.(6) shows the effect of width of the beam on the capacity reduction factor.

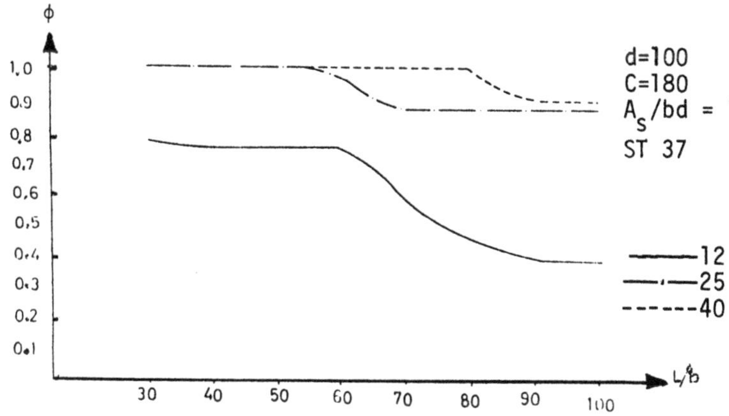

Fig.6. Effect of width of beam on the capacity reduction factor

The variation of the depth of the beam is of intercurrent dependent effect. The increase of depth increases the slenderness factor (d/b), hence decreases the capacity reduction factor(ϕ), while it provides degree of goodness against excessive vertical delection. Generally, by increasing the depth of the cross-section the capacity reduction factor decreases. This effect is of great evidence with small values of the width of the cross-section. It is desirable to use high values of depth with high values of width to avoid lateral instability failure. This effect can be noticed as shown in Fig.(7).

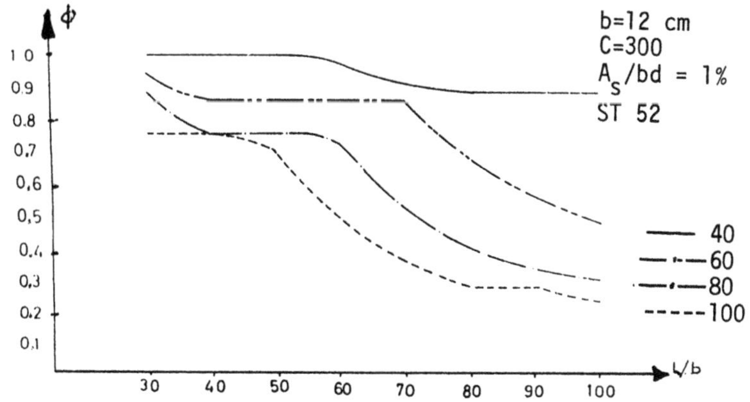

Fig.7. Effect of depth of beam on the capacity reduction factor

As the slenderness ratio (L/b) is progressively increased, the value of capacity reduction factor (ϕ) decreases significantly. That means the value of (ϕ) not only depends on the grade of concrete, type, and amount of reinforcement, width, and depth of the beam, but also depends mainly on the slenderness ratio (L/b) for all beams assumed.

Based on numerical technique and using the multiple regression method, a general formula taken into consideration most of the important provisions for slender reinforced concrete beams (lateral buckling and vertical deflection), has been derived for computing the capacity reduction factor of under reinforced simply supported concrete beams. This formula has been expressed as :

$$\phi = 1.0 - |0.001\ (Ld/b^2) - 0.022\ (As\%/bd)(fy/C)|\ , \not> 1.0 \qquad (9)$$

A comparison between the obtained results and proposed equation for some flexural reinforced concrete slender beams, as shown in Fig.(9), shows a variation as result of neglegting all other parameters, however the proposed equation conform with the results obtained from available experimental studies.

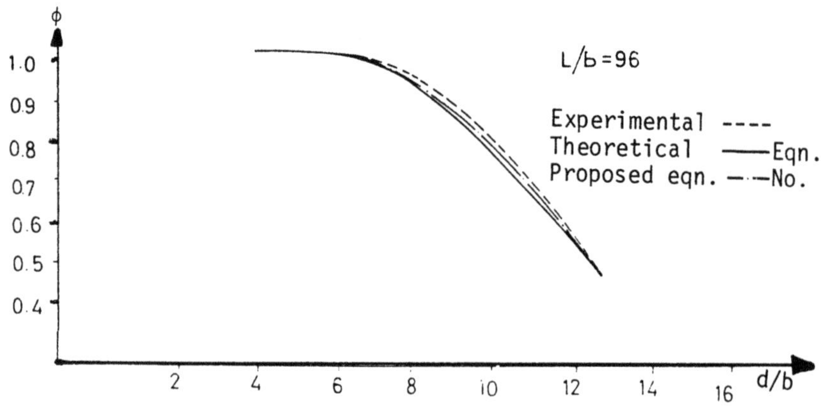

Fig.9. Comparison between theoretical and experimental results by Sant and Bletzacker (1961) and proposed eqn. (9).

5 Conclusions and recommendations

The limited scope of the investigation reported heren substaintially restricts the conclusions which can be rigorously supported by the data. Nevertheless, several important points have been demonstrated and are summarized below:

The lateral stability of reinforced concrete beams depends on both lateral rigidity, and the torsional rigidity which depends in-turn on the steel, concrete dimensions, and on the state of stress at the critical load.

A comprehensive study for declaring the different parameters

affecting the capacity reduction factor of slender reinforced concrete beams, has been considered.
Results proved that not only the length and width of flexural beam affect its capacity reduction factor(ϕ) but also the latter is considerably affected by other parameters including the grade of concrete (C), the grade of steel (f_y), the amount of reinferencement (μ = As/bd), and the depth of the beam (d).
It has been found that the parameters b,d and L/b play a great and significant role in the value of ϕ, while C, f_y and As/bd% have a very little contribution on it.
In general, the effect of varaition of a certain parameter,holding the other ones constant, on the value of capacity reduction factor is found to be as follows:

ϕ increases with the increase of b
ϕ increases as d decreases.
ϕ decreases rapidly beyond high limiting value of L/b depending on other parameters.
ϕ does not significantly affected with the variation of each C,f_y and As/db% .

Based on numerical technique a general formula has been derived for computing the capacity reduction factor as:

$$\phi = 1.0 - |0.001 \ (Ld/b^2) - 0.022(As\%/bd)(fy/C)|, \not> 1.0$$

6 References

ACI Committee 318 (Detroit 1983) Building Code Requirements for Reinforced Concrete"(ACI 318M-83), American Concrete Institute .
Bleich, F.(1962) Buckling Strength of Metal Structures, McGraw-Hill 300K co., Inc., New York, pp. 116-119.
Marshall, W.T. (1969) A survey to the problem of lateral instability in reinforced concrete beams., Journal, Institute of Civil Enginees (London), V.42, pp. 397-406.
Salvadori, M.G. (1951) Numerical computation of buckling loads by finite differences., Trans. ASCE, Vol. 116 .
Sant, Jagadish K. and Bletzacker, Richard W. (1961) Experimental study of lateral stability of reinforced concrete beams,ACI Journal Proceedings, V. 58, No. 6, pp. 713-736.
Siev, A. (1960) The lateral buckling of slender reinforced oncrete beams ., Magazine of Concrete Research (London), V. 12, No.36, pp. 155-164.
Timoshenko, S. and J.M. Gere, (1961) Theory of Elastic Stability ., 2 nd edition, MaCGraw Hill Book, Co., New York.

Hafez, A.M. (1989) Static load and deformation capacity of slender reinforced concrete beams. M.Sc. Thesis, Assiut University, Egypt.

7 Appendix A. Assumed stress-strain diagrams for concrete and steel.

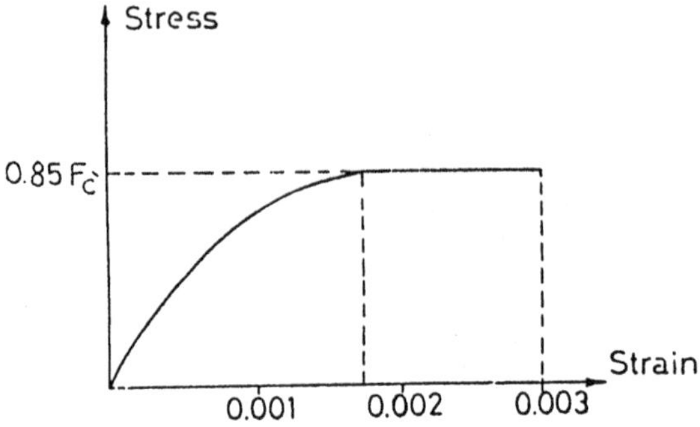

Assumed Stress-Strain Diagram For Concrete.

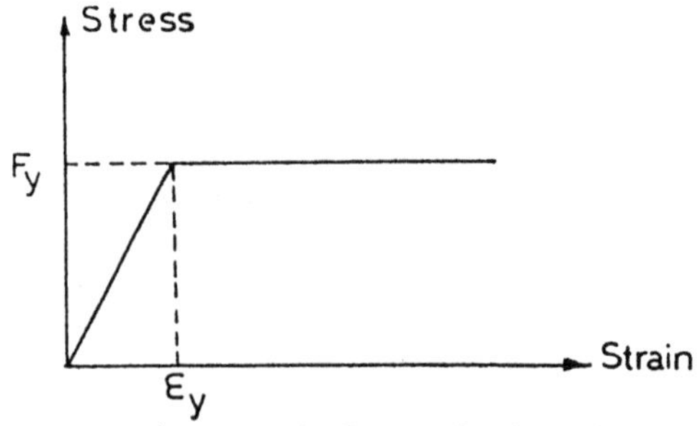

Assumed Stress Strain Diagram For Steel.

8 Notation

A_s = area of the steel tension reinforcement
b = width of cross section
C = grade of concrete
d = effective depth of cross section
E = modulus of elasticity
f_y = yield stress of steel
L = span of beam
M = bending monant
P = applied load
P_{cr} = critical load (caused lateral buckling)
P_{ser} = serviceability load (caused excessive deflection)
P_u = ultimate load (according to ACI-Code)
P_f = failure load (the lowest of P_{cr}, P_{ser} & P_u)
ν = Poisson's ratio
λ = block coefficient of the compression zone
ϕ = capacity reduction factor ($\phi = P_f / P_u$).

49 PROBABILITY ANALYSIS OF THERMAL CRACKS IN MASS CONCRETE

LI-HONG CHEN, MING-RONG MEI and ZUO-XIN FU
Department of Mechanics, Hohai University, Nanjing, China

Abstract
An index, called the overall crack probability, is proposed to analyze the crack probability of concrete which includes not only the effect of the value of tensile stress, but also that of the size of tensile zone. Both mean value and standard deviation of tensile strength have been taken into account. In combination with three-dimensional FEM, this index can be used to solve some practical problems of mass concrete in stage construction, such as the selection of placement plans, the determination of quality control standard and the design of construction joints. As the index does not depend on the size of FEM mesh, the method proposed here is effective and reasonable.
Keywords: Probability Analysis, Thermal Crack, Mass Concrete, Construction.

1 Introduction

For the two same concrete beams loaded as shown in Fig. 1, there are two questions: (1) Which one is easier to crack when M_1 is equal to M_2? (2) If M_1 is a little bit more than M_2, which beam is easier to crack?

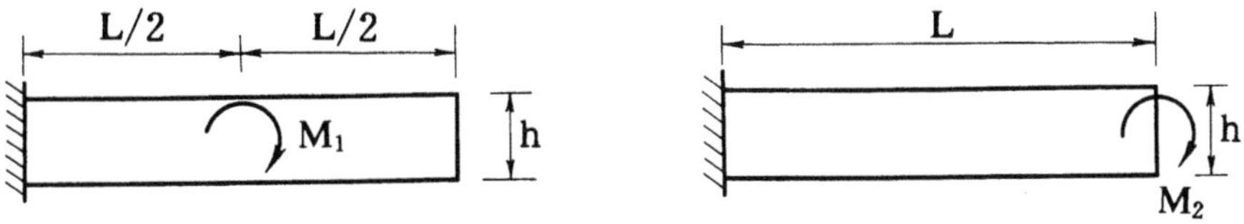

(A) (B)

Fig. 1. Two beams under the loading of a concentrated moment

Let us start with question (1), according to the maximum tensile stress criterion, the crack probabilities of both beams are the same; actually

Fracture and Damage of Concrete and Rock – FDCR-2.
Edited by H.P. Rossmanith. © 1993 Taylor & Francis. ISBN 0 419 18470 8.

beam B is easier to crack. The reason is that the tensile strength of concrete is a random variable due to the non-homogeneity of the composition of concrete and the quality of construction. Although the value of maximum stress of beam B is as same as beam A, the size of tensile zone of beam B is larger than that of beam A. In the light of probability theory, the crack probability of beam B is greater than that of beam A. This illustrates that the crack probability of a concrete structure not only depends on the value of tensile stress, but also relates to the size of tensile zone. However, there is no such crack criterion that can take those two factors into consideration. For example, we can not answer above question (2) up to now. An index is proposed in this paper to solve this problem.

2 Overall crack probability and its solution

2.1 Overall crack probability P_f

The maximum principal stress σ_1 of any structure D under some given loads is generally a function of coordinates, i. e.,

$$\sigma_1 = \sigma_1(x, y, z) \qquad (\forall (x, y, z) \in D) \tag{1}$$

Suppose that the maximum value of σ_1 in domain D is σ_{max} and $\sigma_{max} > 0$, The region $(-\infty, \sigma_{max}]$ can be divided into n+1 parts: $(-\infty, 0)$, $(\sigma^{i-1}, \sigma^i]$ $(i=1, 2\cdots, n)$, where $\sigma^i = \sigma_{max} * i/n$. In the light of the value of σ_1 at any point in domain D, the domain D can also be divided into n+1 subdomain $D_i (i=0, 1, 2, \cdots, n)$:

$$D_i = \{(x, y, z)/\ (i-1)/n * \sigma_{max} < \sigma_1(x, y, z) \leqslant i/n * \sigma_{max}\} \quad (i = 1, 2, \cdots, n) \tag{2}$$

$$D_0 = \{(x, y, z)/ -\infty < \sigma_1(x, y, z) < 0\} \tag{3}$$

$$D = \bigcup_{i=0}^{n} D_i \tag{4}$$

Let Ai be the event that any unit volume taken from the structure D belongs to subdomain D_i and let B be the event that any unit volume from the structure D is cracked. The A_i's form a partition of the sample space. The probability of event A_i is $P(A_i) = r_i = D_i/D$ $(i=0, 1, 2, \cdots, n)$. Assume that the tensile strength of the material has a normal distribution with mean $\bar{R}$ and standard deviation σ_R. According to the maximum tensile stress criterion, the conditional probability of event B within the subset of the sample space defined by A_i is

$$P(B/A_i) = p_i = \int_{-\infty}^{\sigma'} \frac{1}{\sqrt{2\pi}\sigma_R} exp[-\frac{1}{2}(\frac{R-\overline{R}}{\sigma_R})^2]dR \qquad (i = 1, 2, \cdots, n) \tag{5}$$

$$P(B/A_0) = p_0 = 0 \tag{6}$$

Then

$$P(B) = \sum_{i=1}^{n} P(B/A_i)P(A_i) = \sum_{i=1}^{n} p_i * r_i = \sum_{i=1}^{n} p_i * D_i/D \tag{7}$$

P(B) is the crack probability of any unit volume taken from the structure D. We call it overall crack probability, denoted by P_f. When $n \to \infty$, Eq. (5) (7) give

$$P_f = \frac{1}{D}\int_{D_T} dD \int_{-\infty}^{\sigma_1(x,y,z)} \frac{1}{\sqrt{2\pi}\sigma_R} exp[-\frac{1}{2}(\frac{R-\overline{R}}{\sigma_R})^2]dR \tag{8}$$

where D_T is the tensile zone and $D_T = \lim_{n\to\infty} \bigcup_{i=1}^{n} D_i \subset D$. Eq. (8) shows that the overall crack probability P_f has nothing to do with the number of subdomains. When the mean and standard deviation of tensile strength of the material are determined, the value of P_f depends on the value of maximum principal stress $\sigma_1(x, y, z)$ and the size of tensile zone. Therefor, P_f is an overall index representing the crack probability of whole structure. Its value indicates what percentage of the entire structure may crack.

For the two beams shown in Fig. 1, the overall crack probability P_f of each beam can be calculated directly from Eq. (8):

$$P_f^A = \frac{1}{8} + \frac{h^2\sigma_R}{24\sqrt{2\pi}M_1}\sum_{k=0}^{\infty}\frac{2^k}{(2k+1)!!}\{J_k[(\frac{\overline{R}}{\sqrt{2}\,\sigma_R})^2] - J_k[(\frac{6M_1}{\sqrt{2}\,h^2\sigma_R} - \frac{\overline{R}}{\sqrt{2}\,\sigma_R})^2]\} \tag{9}$$

$$P_f^B = \frac{1}{4} + \frac{h^2\sigma_R}{12\sqrt{2\pi}M_2}\sum_{k=0}^{\infty}\frac{2^k}{(2k+1)!!}\{J_k[(\frac{\overline{R}}{\sqrt{2}\,\sigma_R})^2] - J_k[(\frac{6M_2}{\sqrt{2}\,h^2\sigma_R} - \frac{\overline{R}}{\sqrt{2}\,\sigma_R})^2]\} \tag{10}$$

where

$$J_k(x) = e^{-x}[x^k + kx^{k-1} + k(k-1)x^{k-2} + \cdots + k!]$$

By the formulas above, we can definitely answer which beam in Fig. 1 is easier to crack when M_1 is different from M_2.

On the analogy of the most probable speed in molecular probability analysis, we call the mathematical expectation of tensile stress in the tensile zone the most probable tensile stress, denoted by σ_j. It is another index for overall structure which states the level of tensile stress in a structure. It can be calculated by

$$\sigma_j = \sum_{i=1}^{n} \sigma^i \cdot D_i / D_T \tag{11}$$

when $n \to \infty$, Eq. (11) becomes

$$\sigma_j = \frac{1}{D_T} \int_{D_T} \sigma_1(x, y, z) dD \tag{12}$$

2. 2 FEM solution of P_f and the effect of mesh

For a general problem, it is difficult to calculate P_f directly by Eq. (8). FEM is a suitable method to solve this problem. The problem domain D is divided into m finite elements. The maximum principal stress $\sigma^k (k=1,2,\cdots,m)$ on the center of each element should be calculated first. Let m1 be the number of the elements of which the maximum principal stresses are tensile stresses. The volume of each element is D_k and the tensile zone $D_T = \bigcup_{k=1}^{m1} D_k$. From Eq. (8) and Eq(12), we have

$$P_f = \frac{1}{D} \sum_{k=1}^{m1} D_k p_k \tag{13}$$

$$\sigma_j = \frac{1}{D_T} \sum_{k=1}^{m1} D_k \sigma^k \tag{14}$$

The changes in the size of finite elements influence the values of maximum or minimum stresses, but have little effect upon the overall crack probability P_f and the most probable tensile stress σ_j. The reason can be explained by Eq. (7) and Eq. (11). When the elements become smaller, the value of maximum stress may increase and their corresponding volumes D_i will decrease. As a result, P_f and σ_j show little dependence on the size of elements. It also can be illustrated by the results of numerical analysis of a intake tower shown in Fig. 2. We use three different meshes to compute the P_f and σ_j. The results are shown in Table 1. Because of their independence on meshes, P_f and σ_j are reliable and objective criteria for numerical analysis.

Table 1. P_f and σ_j of the intake tower by three different meshes

σ_1 (MPa)	p_i	$r_i = D_i/D$ Mesh A	Mesh B	Mesh C
0.0—1.4	0.0003	0.7100	0.7078	0.6988
1.4—1.8	0.0004	0.1401	0.1498	0.1586
1.8—2.2	0.0202	0.0745	0.0691	0.0664
2.2—2.6	0.2206	0.0641	0.0648	0.0671
2.6—3.0	0.6954	0.0075	0.0000	0.0090
3.0—3.4	0.9292	0.0000	0.0085	0.0000
3.4—3.6	0.9633	0.0038	0.0000	0.0000
P_f		0.0248	0.0239	0.0227
σ_j (MPa)		1.0583	1.0399	1.0620

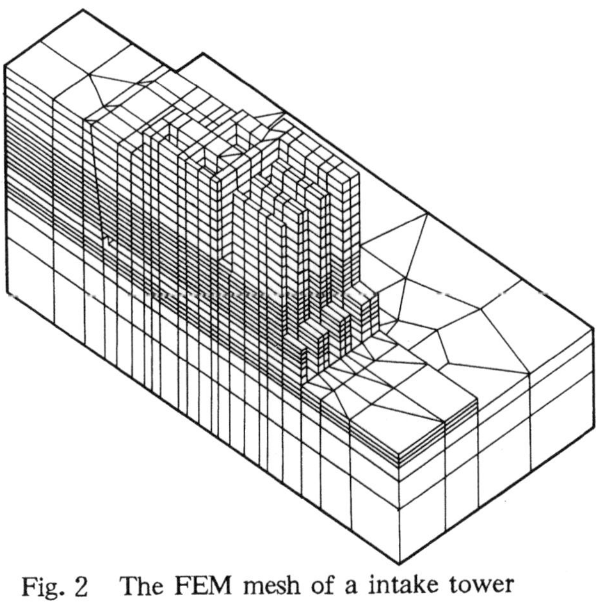

Fig. 2 The FEM mesh of a intake tower

3 The overall crack probability of mass concrete

3.1 $P_f(t)$ of mass concrete

Mass concrete structures are hyperstatic structures. Some cracks in local parts do not cause the failure of whole structure. Thus, it is reasonable to reduce the crack probability of entire structure, not parts of the structure, to the lowest degree in the design of mass concrete. In addition, because the values of tensile stresses are usually less than the mean of tensile strength, and the tensile zones are considerably large in mass concrete, the nonuniformity of tensile strength is one of main causes of cracking in mass concrete. Therefor, the overall crack probability P_f is an effective index for mass concrete.

Thermal crack of mass concrete depends on the history of temperature and creep in the construction and operation periods. The overall crack probability P_f is then the function of time. The history of temperature, creep and stress of mass concrete, from stage construction period to operation period, has been simulated by FEM time history analysis method [1][2]. Combined with FEM, $P_f(t)$ is obtained from Eq. (13). $P_f(t)$ can be used to solve some practical problems of mass concrete during construction. The following are examples of its applications.

3.2 The determination of placement plan

In addition to the properties of material and structure, the placement plan is another main factor affecting the crack probability of mass concrete which includes the height of each lift, the interval time between two lifts, the season of placement, the methods of surface protection and the required placement temperature and so on. With the $P_f(t)$ in the construction period, it is easier for us to determine a best way of placement. For example, there were many possible plans for the base slab of intake tower shown in Fig. 2. According to the conditions of construction, there were 3 possible heights of each lift, 3 possible interval times, 2 dates of placement, 2 surface protection methods and 2 placement temperatures. Using FEM and above formulas, the different curves $P_f(t)$ of whole construction period under different placing ways have obtained. By the maximum values of $P_f(t)$ of every possible plan, the best placing plan has been determined effectively. Table 2. shows some results of these analyses.

3.3 The standard of placement quality

After the determination of placing plan, the crack probability of mass concrete depends on the quality of placement. Table 3. shows the relation between the coefficient of variation Cov. of tensile strength and the overall crack probability P_f of the intake tower in Fig. 2. It is obvious that the crack probability of whole structure rises with the increase of Cov. of tensile strength. Thus, a standard for quality control on construction site can be established in the light of the grades of structures.

Table 2. P_f^{max} of the intake tower under different placing plans

Height of lift (m)	Date of placement	interval time (day)	P_f^{max}
4	Oct. 1st		0. 0984
	Mar. 1st		0. 2610
2	Oct. 1st	5	0. 0251
		7	0. 0205
		10	0. 0175
	Mar. 1st	5	0. 2040
		7	0. 1900
		10	0. 0643
1. 3	Oct. 1st	5	0. 0154
		7	0. 0143
		10	0. 0132
	Mar. 1st	5	0. 0822
		7	0. 0688
		10	0. 0232

Table 3. The relation between Cov. of tensile strength and P_f

Cov.	0. 10	0. 12	0. 14	0. 18	0. 20	0. 22
P_f	0. 0094	0. 0100	0. 0110	0. 0160	0. 0194	0. 0240

3. 4 The design of construction joints

Contraction joints are commonly used to prevent mass concrete from cracking. Since the overall crack probability P_f has a clear relation with the distance between two joints, it can help us to design suitable construction joints. Table 4. shows the relation between the P_f and the number of vertical joints within the intake tower in Fig. 2. It illustrates that the overall crack probability of the whole tower placed without any vertical joint is more than three times as much as that of the structure in which two vertical joints are made.

Table 4. The relation between the number of joints and P_f of the intake tower

Number of vertical joints	0	1	2
P_f	0.0345	0.0233	0.0112

4 Conclusion

The overall crack probability P_f, presented in this paper, offers an effective index which takes both the effects of the value of tensile stress and the size of tensile zone into consideration. Combined with FEM, it is suitable to analyze the probability of thermal crack of mass concrete during construction. Because of the independence of P_f on FEM mesh, the method proposed here is effective and reliable.

5 References

Chen Li-hong (1989) Analysis of the creep and fracture of concrete dams during construction with the consideration of softening. Ph. D. Dissertation, Department of Mechanics, Hohai University, Nanjing, P. R. China.

Chen Li-hong and Fu Zuo-xin (1992) Simulation of thermal crack of mass concrete in stage construction, in **Fracture Mechanics of Concrete Structures** (ed Z. P. Bazant), Elsevier Science Publishers Ltd., England, pp. 983—986.

PART SEVEN

SPECIAL SEMINAR ON ANCHORAGE ENGINEERING

50 BONDED ANCHORS SUBJECTED TO LONG TIME AND CYCLIC LOADS

LENNART ELFGREN
Division of Structural Engineering, Luleå University of Technology, Sweden
GUNNAR SÖDERLIND
Swedish National Testing and Research Institute, Borås, Sweden

Abstract
Results are presented from long time and cyclic tests with bonded anchors of capsule type. The long time tests have been running for seven years. The cyclic tests have in most cases been running during 2 million cycles. The tests have been carried out at SP, the Swedish National Testing and Research Institute in Borås, and at Luleå University of Technology.
Keywords: Bond, Anchor Bolts, Creep, Cyclic loads, Fatigue, Polymeric resin

1 Introduction

Bonded anchors can be used as a term for all mortar-filled embedments including cementitious as well as polymeric-based systems. In this paper mainly anchors of capsule type will be treated with resins consisting of unsaturated polyesters.

Bonded anchors with polymeric resins were developed for rock strengthening in Germany around 1955 and were further developed for use in concrete structures during the sixties, Klöker (1984), Schuermann et al (1970). The first approval for general use was given by "Institut für Bautechnik" in Berlin in 1975.

The quick and easy installment and the moderate price of bonded anchors have made them very popular. They are nowadays used :

- to anchor columns and other structural elements to slabs and walls
- to anchor machines and other equipment
- to fasten pipes and other installations

General reviews of fastenings to reinforced concrete and masonry structures have recently been published by Eligehausen et al (1991) and Rehm et al (1992).

In this paper results are given from long time test series that were started in 1984. The first results from the tests have earlier been presented by Elfgren et al (1988).

2 General about bonded anchors

The resin used in the tests is an unsaturated polyester with the name of Leguval K27, Bayer(1981). It is made of diols and dicarbolic acids. The acids are unsaturated, that is they have double bonds between some of their coal atoms which make them very

Fracture and Damage of Concrete and Rock – FDCR-2.
Edited by H.P. Rossmanith. © 1993 Taylor & Francis. ISBN 0 419 18470 8.

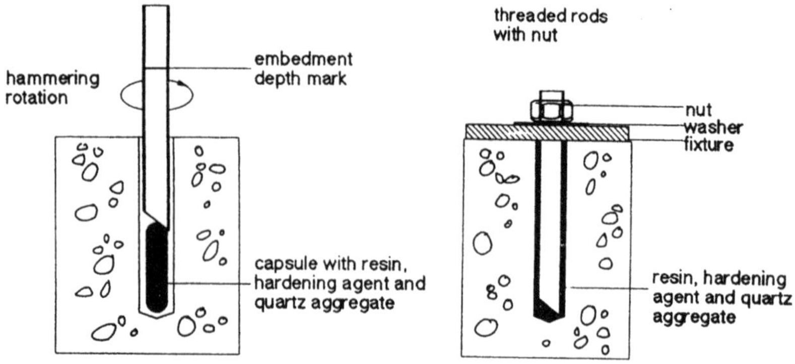

Fig. 2.1. Bonded anchor of capsule-type, Eligehausen et al (1991)

reactive. This is a prerequisite for the curing (cross-linking) later in the process. The reaction between the diols and the dicarboxylic acids is a polycondensation in which a polyester and water is formed. The polyester contains long chains of molecules and it is initially very viscous, that is it can be hard and brittle.

In order to adopt the polyester to further processing, it is dissolved in a liquid as monomeric styrene, which is also unsaturated and has reactive cross-linking agents. During curing, the unsaturated, reactive groups of the cross-linking agent (the styrene) react with those of the polyester. In this reaction the molecules of the cross-linking agents are incorporated as connecting links between the polyester chains resulting in steric crosslinking.

To start the curing process the formation of radicals is necessarry. Radicals are highly reactive molecule fragments which stimulate the crosslinking. Organic peroxides are used as donors and their mode of action is set off by adding an accelerator (initiator). Tertiary amines are particulary effective as accelerators with benzoyl peroxide.

During curing the unsaturated polyester resin shrinks 6-8% by volume (chemical shrinkage). It is brought about by the chemical linkage which takes place between the individual molecules during curing. The linkage process causes the chains to draw closer together. By adding a filler of ballast, the shrinkage can be greatly reduced as in the case with adhesive resins where quartz fragments and glass are added. The proportion of resin to quartz may usually vary from 1:1.5 to 1:4.

The cured Leguval K27 resin without filler or ballast has the following mechanical properties according to the producer:

Tensile strength	f_{rt}	50 MPA
Elongation at rupture	ε_{ru}	2 %
Compression yield strength	f_{rc}	160 MPa
Modulus of elasticity	E_r	4 GPa
Shear modulus	G	1.5 GPa
Poissons ratio	ν_r	0.33
Thermal expansion	α	$1.15 \cdot 10^{-4}$

Fig. 3.1 Test arrangement for long time outdoor tests in Luleå. The loads are applied by means of lever arms (approximately 1 to 10) loaded by concrete blocks.

For most bonded anchors the resin is delivered in glass cartridges (glass phials). There are two systems. In one system the radical donor (benzoyl peroxide) is stored as a powder in an inner glass phial. The phial is placed in an outer phial togetheher with the resin, quartz sand and an initiator (tertiary amines). We have tested four products with this system (Anchors of type B, C, D, and E). In the other system the quartz sand and the radical donor are stored in the inner glass phial whereas the resin and the initiator are stored in the outer phial. We have tested one product with this system (Anchor type A).

To start the curing, the glass cartridge is placed in a hole drilled into the concrete, see Fig. 2.1. An anchor rod is then mounted to a rotary drill hammer. The rod is driven into the hole. The cartridge is then crushed, the components of the resin are mixed with each other by the rotating rod and the curing starts. Curing time varies from 10 to 20 minutes at room temperature and up to several hours at - 5°C.

In order to adopt the polyester to further processing, it is dissolved in a liquid as monomeric styrene, which is also unsaturated and has reactive cross-linking agents.

Recently, resins with improved properties have been developed, e.g. epoxy acrylats and vinyl esters, Ammann (1991), Eligehausen et al (1991), Zavliaris & Speare (1992).

3 Test program

3.1 Materials

The concrete used in the tests had mix and strength properties according to Table 3.1.The compression strength was determined on 150 mm cubes. The tensile strength was determined by splitting tests on the same size of cubes according to Swedish Concrete Standard (1978). The steel in the bolts had a yield stress of 240 or 400 MPa.

Table 3.1 Concrete mix and properties

	Long-time tests Luleå	Borås	Fatigue tests Borås
Cement, kg	210	255	235
Gravel 0 - 8 mm, kg	1050	1245	1245
Stone 8 -16 mm, kg	950	620	620
Silica, kg	10	-	-
Water reducer, kg	0.8	-	-
Water-cement ratio	0.81	0.82	0.89
Compression/tensile strength f_{cc}/f_{ct}, MPa/MPa			
7 days			13.8/-
13 days			17.4/-
29 days			19.0/-
100 days	37/3.0		
150 days	40/3.5		
163 days			22.9/-
269 days		32/2.9	
345 days			21.8/-

3.2 Long time tests

The concrete was cast in foundations of the dimensions 0.40 x 0.75 x 0.75 m provided with transport reinforcement of 12 mm Ks 400 ($f_{yk} \geq 400$ MPa). The top 200 mm of the foundations were reinforced only along the edges.

Bonded anchors of dimension 16 and 20 mm were used of Type A, B and E. The anchors were tested under the following environmental conditions:

I 14 anchors were placed indoors with approximately constant temperature and humidity (20 °C, RH 30 - 40%)
O 4 anchors were placed outdoors in Luleå in order to check the influence of varying temperatures and humidities
S 6 anchors were placed indoors but had an additive (salt) of 2 % Pozzolith 122 He to the cemnent in the foundation blocks.
W 2 indoor anchors had a 10 mm layer of watrer on the top of the foundation around the anchor.

The resin had an age of 24 hours or more when tests were started.

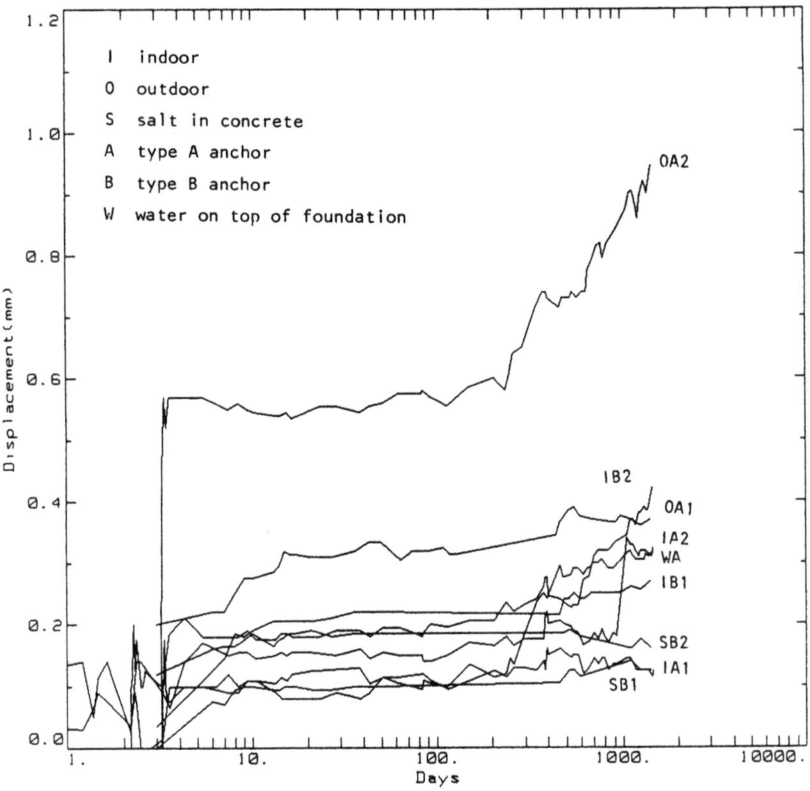

Fig. 4.1. Creep tests with M16 anchors of types A and B loaded with 15 kN under different environmental conditions

Table 3.2 Test program for cyclic loading

Mean Load Level (kN)	±5	±10	±15	±20	±25	±17.5	±40	±55
M16								
20			1B					
25				3B				
30	1A		1B	1C	1B			
35	1A							
40	2A	1B	1C					
45	2A							
50	1A							
M24								
22.5						1A		
60							1A	1A
							1B	
							1C	
							1D	

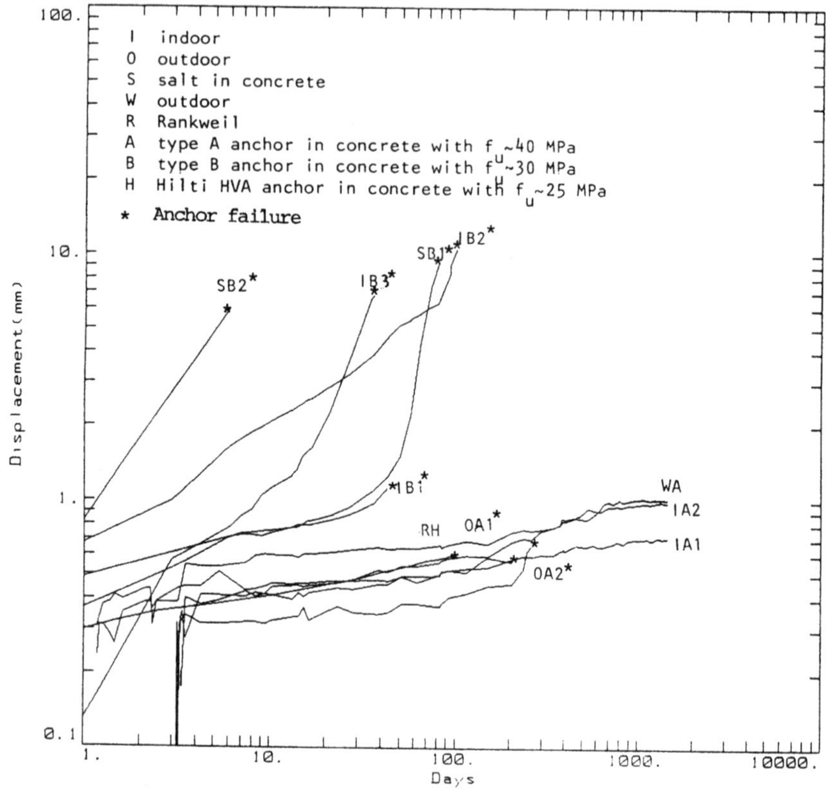

Fig. 4.2. Creep tests with M16 anchors loaded with 45 kN. Results for anchors of type H are quoted from Rankweil (1980).

3.3 Fatigue tests

Two bolt dimensions were tested, M 16 and M24. They were tested in thin slabs 0.2 x 0.9 x 2.0 m and thick slabs 0.5 x 0.9 x 2.0 m. The slabs were reinforced in the top and bottom with welded mesh reinforcement made of 4 mm Ps 500 ($f_{yk} \geq 500$ MPa) in the thin slabs and of 8 mm Ks 400 ($f_{yk} \geq 400$ MPa) in the thick slabs. The spacing of the bars was 200 mm and the concrete cover was 20 mm. 17 M16 bolts and 6 M24 bolts were tested with different load levels according to Table 3.2.

4 Test results

4.1 Long time tests

Some of the test results are summarized in Table 4.1. and in Figs 4.1 and 4.2

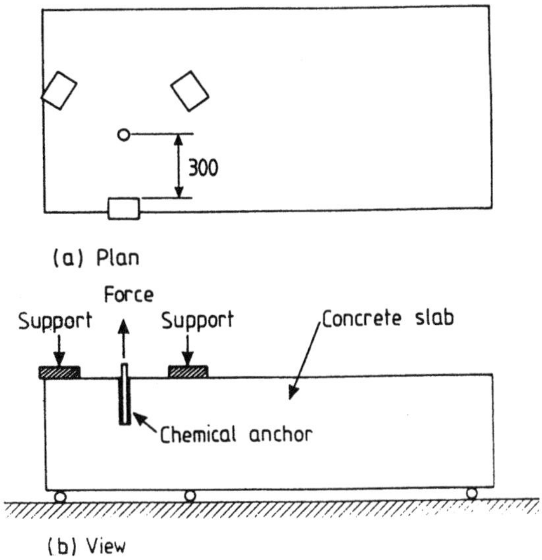

Fig. 4.3. Test set-up for cyclic loading of M24 bolts.

Table 4.1. Summary of test results for long time loading

Test no	Load F (kN)	Mean stress τ (MPa)	Creep deformation after 1 year (mm)	Creep deformation after 3 years (mm)	Creep deformation after 7 years (mm)	Age at failure (days)
I 15 A1	15.9 M16	2.25	0.14	0.15	0.13	-
I 15 A2	15.6 M16	2.21	0.33	0.35	0.34	-
I 15 B1	15.0 M16	2.12	0.22	0.26	0.26	-
I 15 B2	15.0 M16	2.12	0.35	0.37	0.38	-
O 15 A1	16.6 M16	2.35	0.24	0.32	0.46	-
O 15 A2	17.5 M16	2.48	0.72	0.80	1.00	-
W 15 A	15.5 M16	2.19	0.25	0.30	0.33	-
S 15 B1	15.0 M16	2.12	-	0.10	0.10	-
S 15 B2	15.0 M16	2.12	-	0.12	0.15	-
I 30 B1	30.0 M16	4.24	1.71	2.10	2.15	-
I 30 B2	30.0 M16	4.24	0.52	0.54	0.54	-
I 30 B3	30.0 M16	4.24	0.83	0.96	1.02	-
I 45 E 1	45.0 M20	3.37	0.99	-	(1.06)*	-
I 45 E2	45.0 M20	3.37	0.58	-	(0.62)*	-
S 45 E1	45.0 M20	3.37	0.43	-	(0.47)*	-
S 45 E2	45.0 M20	3.37	0.47	-	(0.54)*	-
W 45 A	39.9 M16	5.64	0.80	1.00	1.02	-
I 45 A1	43.7 M16	6.18	0.62	0.70	0.75	-
I 45 A2	41.4 M16	5.86	0.79	0.95	1.04	-
I 45 B1	45.0 M16	6.37	-	-	-	42
I 45 B2	45.0 M16	6.37	-	-	-	96
I 45 B3	45.0 M16	6.37	-	-	-	35
S 45 B1	45.0 M16	6.37	-	-	-	79
S 45 B2	45.0 M 16	6.37	-	-	-	2
O 45 A1	46.8 M16	6.62	-	-	-	235
O 45 A2	43.4 M16	6.14	-	-	-	284

* After 5.75 years

Fig 4.4 Some bolts after cyclic loading tests

4.2 Cyclic tests

Test results are summarized in Tables 4.2. and 4.3 and in Figs. 4.3 to 4.5. In the Tables the number of load cycles and the type of failure are given for each individual bolt. For some of the bolts tested with cyclic loading the tests were stopped after two million load cycles. A loading to failure with a monotonically increasing load was then performed

The cyclic tests are illustrated in a Wöhler diagram in Fig. 4.6 and in a Haigh diagram in Fig. 4.7. The Wöhler diagram shows the influence of the stress range σ_r on the fatigue capacity. Different mean stresses σ_m have been used for different bolts which must be born in mind when the diagram is interpreted. The Haigh diagram shows the influence of the mean stress σ_m and the stress amplitudeσ_a. Stress levels σ in the steel are calculated by dividing the applied load with the stress area A_s of the bolt (A_s = 157 mm^2 for M16 and A_s = 353 mm^2 for M24).

As a comparison the mean shear stresses τ_b (bond) between the resin and the concrete can also be calculated by dividing the applied load with the mantle area A_m of the resin cylinder. We get $A_m = \pi \cdot 18 \cdot 125 = 7069$ mm^2 for M16 bolts and A_m = 18473 mm^2 for M24 bolts of Type A and C and A_m = 21112 mm^2 for M24 bolts of Type B and D.

Ultimate and allowable steel stresses for bolts with roll-formed threads are given in the figures according to Swedish Steel Standard (1976) for both monotonic and cyclic loading. Allowable bond stress for monotonic loading are given according to German Anchor Bolt Standard (1987). Although the number of tests is very limited the tendency is quite clear. Bonded anchors seem to have a good capacity to withstand cyclic loading. No bond failure occurred for fatigue stresses $\sigma_m \pm \sigma_a$ giving bond stresses under the inclined line A in Fig. 4.7. With a safety factor of three this leads to proposed allowable bond stresses as indicated with a shadowed area in the lower left corner of Fig. 4.7.

Table 4.2 Results from cyclic tests with M 16 bolts. Hole depth 125 mm. Hole diameter 18 mm

Test No	Type	Load level $F \pm \Delta F$ (kN)	R= F_{min} / F_{max}	Number of cycles (kilocycles)	Ultimate load F_u (kN)	Type of failure
F11	A	30±5	0.71	>2000	64.5	
F12	A	40±5	0.78	280	-	Slip
F13	A	35±5	0.75	>2000	70.5	
F14	A	40±5	0.78	>2000	-	
F15	A	50±5	0.82	6.6	-	(a)
F16	A	45±5	0.80	>2000	71.5	Slip
F17	A	45±5	0.80	>8000	77.0	
F18	A	Monotonic	-	-	63.5	
			-			
F21	B	Monotonic	-	-	60.0	
F22	B	Monotonic	-	-	82.5	
F23	A	Monotonic	-	-	74.4	
F24	A	Monotonic	-	-	60.1	
F25	B	30±25	0.09	1.8	-	Slip
F26	B	25±20	0.11	0.4	-	Slip
F27	B	25±20	0.11	79.0	-	Slip + Bolt
F28	B	20±15	0.14	161.2	-	Slip
F31	B	20±15	0.14	356.7	-	Bolt
F32	B	25±20	0.11	60.6	-	Slip(b)
F33	B	30±15	0.33	231,7	-	Bolt
F34	B	10±10	0.0	1166.0	-	Bolt
F35	C	30±20	0.20	119.4	-	Bolt
F36	C	Monotonic	-	-	77.8	
F37	C	40±15	0.45	290.2	-	Slip
F38	C	Monotonic	-	-	51.5	

(a) Interruption in power supply
(b) The bolt had been hit by a sudden peak load with a maximum value of 80 kN.

Table 4.3 Results from cylic tests with M24 bolts. Hole depth 210 mm. Hole diameter 28 mm (A,C), 32 mm (B, D)

Test No	Type	Load level $F \pm \Delta F$ (kN)	R= F_{min} / F_{max}	Number of cycles (kilocycles)	Ultimate load F_u (kN)	Type of failure
F61	A	60±55	0.04	16.2	-	Slip
F62	A	Monotonic	-	-	154	
F63	A	60±40	0.20	30.0	167	
F71	B	Monotonic	-	-	123	
F72	B	Monotonic	-	-	153	
F73	C	Monotonic	-	-	156	
F81	D	Monotonic	-	-	144	
F82	B	Monotonic	-	-	>156	Old capsule
F83	A	22.5±17.5	0.13	>2950	162	
F91	B	60±40	0.20	981.0	-	Bolt failure
F92	C	60±40	0.20	288.2	-	Concrete failure
F93	D	60±40	0.20	>1000	-	Bolt failure

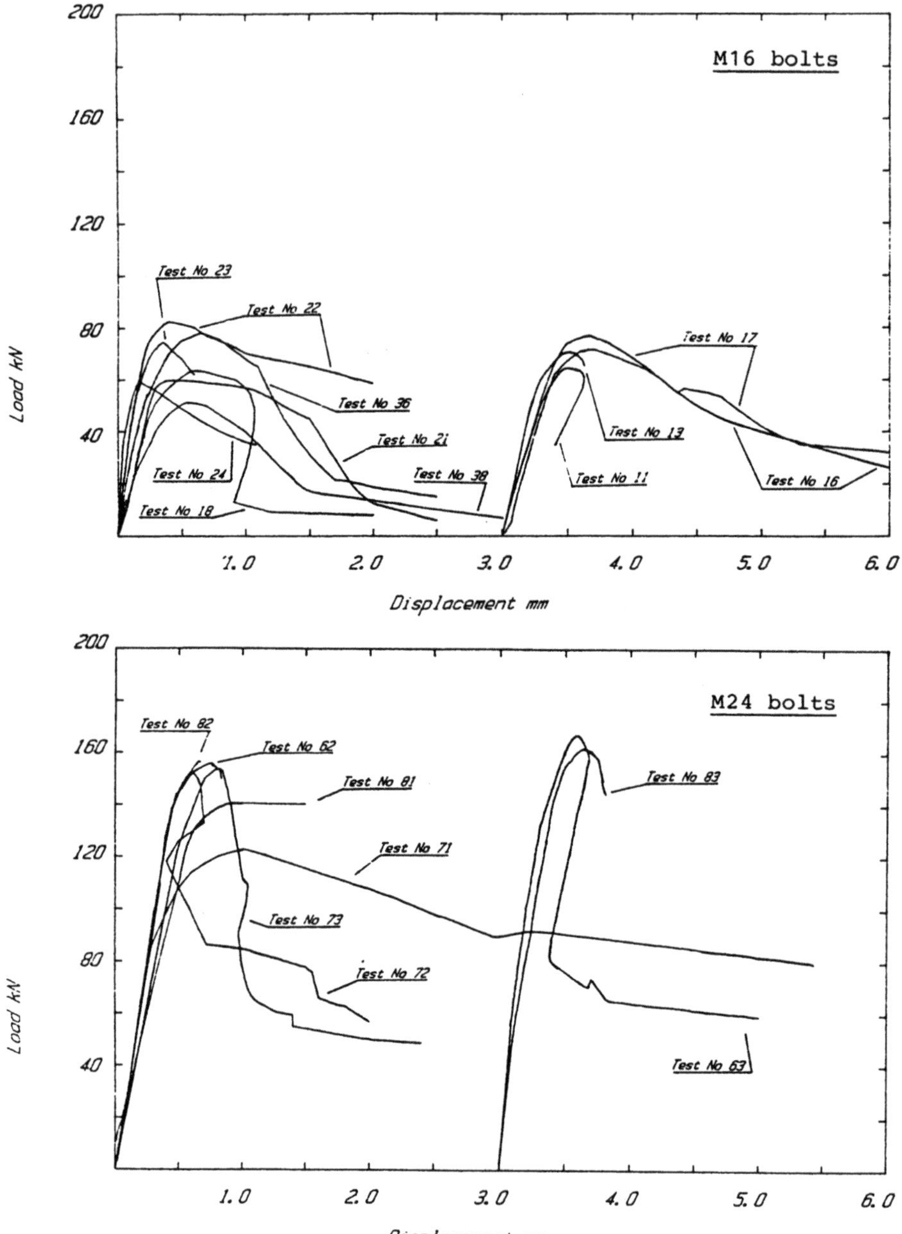

Fig. 4.5 Load displacement curves for monotonic loading

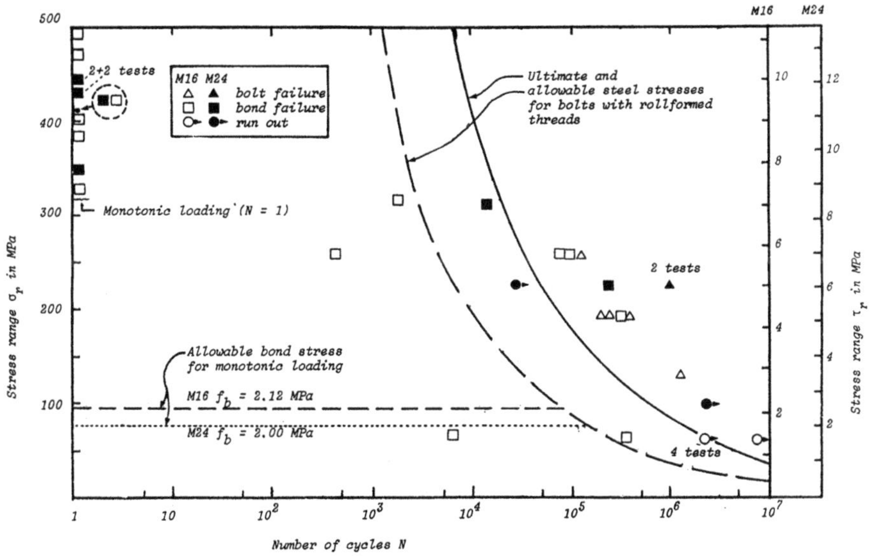

Fig. 4.6 Wöhler diagram for M16 and M25 bolts. The stress areas used for calculating the tensile stresses are 157 mm^2 and 353 mm^2 and the stress areas used for calculating the bond streses are $\pi \cdot 18 \cdot 125 = 7070$ mm^2 and $\pi \cdot 28 \cdot 210 = 18472$ mm^2 (Type A and C)

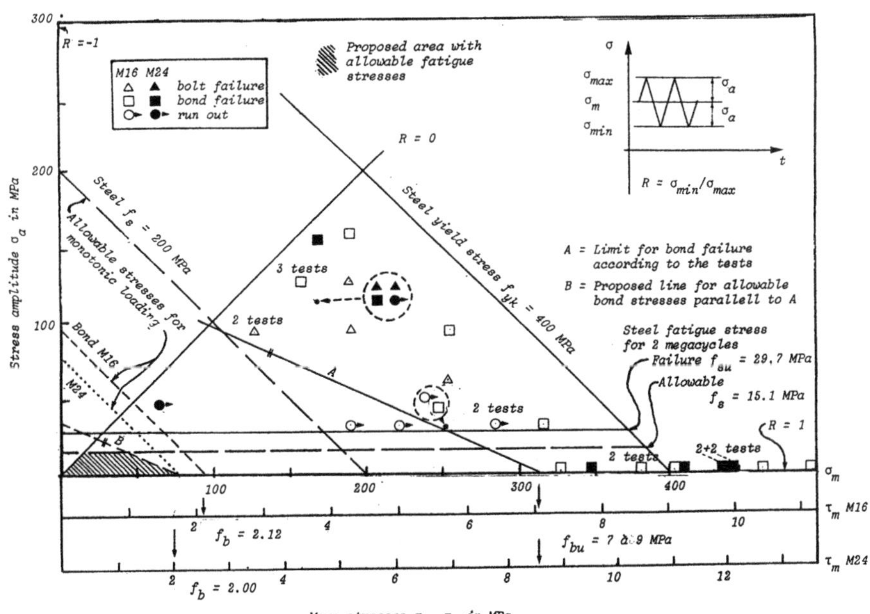

Fig. 4.7. Haigh diagram. The shaded area in the lower left corner of the diagram indicates a proposed area with allowable fatigue stresses for a concrete compressive strength f_{cc} of 22 MPa.

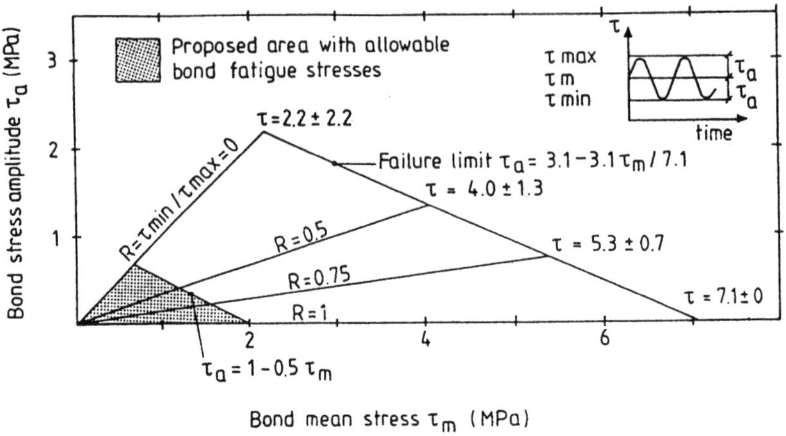

Fig. 4.8. Haigh diagram giving failure and proposed allowable bond stresses for 2 million load cycles. M16 and M24 bolts in concrete with a compressive strength f_{cc} of 22 MPa.

5 Analysis and conclusions

5.1 Long time tests

Normal design loads for static conditions (15 kN for M16 bolts) have in general not caused any notable time-dependent displacements.

The anchors that were stored indoors have not shown any deformation increase during the last four years. The two anchors that were stored outdoors show on the other hand a continous increase in deformation (about 0,05 mm/year). However, the concrete that the anchors are installed in is not freeze-thaw resistant as it has a high water-cement ratio and contains no air-entrainment. The daily and annual temperature changes with freeze-thaw cycles can here lead to a gradual break down of the concrete and a slow pull-out of the anchor. The load level is about one fourth of the bearing capacity in short time loading and the friction can alone, under favorable conditions, stand up to this load, even if the adhesion has completely broken down.

Higher load levels (30 kN for M16 anchors) have caused considerable deformations but no failures.

Still higher load levels, equal to three times the normal static design load (45 kN for M16 anchors), have caused failure in 7 out of 10 anchors after 2 - 284 days.

Tests reported by Ammann (1991) and Eligehausen et al (1991) indicate that newly developed epoxy accrylate resins have better resistance to water saturation and freeze-thaw cycles than the unsaturated polyesters discussed here.

Tests reported by Håkansson et al (1981) indicate that so called non-shrinkage grouts, used for anchors grouted in holes and recesses, usually have higher shrinkage and creep than ordinary concrete made of portland cement.

To sum up, it can be said that the bonded anchors tested here have shown good long time properties. However, it must be emphasized that the concrete properties and the environmental conditions are of vital importance. Exposure of bonded anchors to water and to outdoor temperature variations and freeze-thaw cycles may increase deformations consdiderably. For that reason caution should be exercised in the design of anchors and high quality freeze-thaw resistant concrete and a regular (e g at five years intervals) checking of the remaining load-carrying capacity might be prescribed for vital anchors subjected to outdoor climate.

5.2 Cyclic loading

The cyclic tests indicate that stress levels that are higher than stresses given by the equation $\tau_a = 3.1 - 3.1\ \tau_m / 7.1$ lead to bond fatigue for a concrete wtih a compressive strength $f_{cc} = 22$ MPa, see Figs 4.7 and 4.8. With a safety factor of 3 a simple limit line for allowable stresses could be

$$\tau_a = 1.0 - 0.5\ \tau_m$$

As the number of tests is very limited, the stress level given above shall only be seen as a preliminary indication of the level of the stresses that govern the bond failure. However, it should be pointed out that the steel fatigue stresses are lower than the bond fatigue stresses given above for load cases with high amplitudes and low mean stresses. This means that steel fatigue is decisive rather than bond failure for low values of $R = \tau_{min} / \tau_{max} = \sigma_{min} / \sigma_{max}$. For an M16 rolled bolt with an allowable stress range $\sigma_r = 2\ \sigma_a = 15.1$ MPa this gives a bond stress range of $\tau_r = 2\ \tau_a = \sigma_r\ A_s/A_m = 15.1 \cdot 157/7069 = 0.34$ MPa. This corresponds to the point $\tau_a = 0.34/2 = 0.17$ MPa and $\tau_m = 2(1\text{-}0.17) = 1{,}66$ MPa and it gives an R-value of $R = \tau_{min} / \tau_{max.} = (1.66 - 0.17)/(1.66 + 0.17) = 0.81$. This means, that for this case, steel fatigue is governing for $R = \sigma_{min} / \sigma_{max.} < 0.8$ and that bond is governing for $R > 0.8$.

No tests has been running with negative R-values, that is tests where σ_{min} and τ_{min} have orher sign than σ_{max} and τ_{max}

One way to reduce the stress ranges σ_r and τ_r is to prestress the anchor, see e.g. Eligehausen et al (1991).

Analysis according to fracture mechanics methods are given in Elfgren (1989), and Elfgren and Shah (1991).

6 Acknowledgement

The research program has been sponsored by the Swedish Council for Building Research; Luleå University of Technolog; SP, the Swedish National Testing and Research Institute in Borås; and by producers of bonded anchors. In the preparation of the program guidance has been given by Krister Cederwall, Kent Gylltoft and Lennart Ågårdh. The program was planned and directed by Lennart Elfgren and Anders Eriksson. The following persons have made substantial contributions to various phases of the program during the10 years it has been running: Roger Anneling, Stig-Ola Granlund, Ingvar Holm and Gunnar Söderlind.

7 References

Ammann, Walter J. (1991) **Static and dynamic long-term behavior of anchors.** Paper SP 130-8 in "Anchors in Concrete - Design and Behavior "(Edited by George A Senkiw and Harry B Lancelot III), Special Publication SP-130, American Concrete Institute, Detroit 1991, pp 205 - 220.

Bayer (1981) **Leguval - Unsaturated polyester resins.** Bayer AG, KL Division, D-

5090 Leverkusen. Order No KL 43016e, Edition 6.81, 40 pp and Leguval K27. Unsaturated polyester resin (UP Resin), Order No KL 43123e, Ed. 1.12.1975, 10 pp.

Elfgren, Lennart; Anneling, Roger; Eriksson, Anders and Granlund, Stig-Ola (1988) **Adhesive anchors. Tests with cyclic and long-time loads**. Swedish National Testing Institute, Technical Report SP-RAPP 1987:39, Borås 1988, 87 + 25 pp (ISBN 91-7848-080-9

Elfgren, Lennart, Editor (1989) **Fracture Mechanics of Concrete Structures. From theory to applications.** A RILEM Report. Chapman and Hall, London 1989, 407 pp (ISBN 0 412-30680-8)

Elfgren, Lennart and Shah, Surendra P., Editors (1991) **Analysis of Concrete Structures by Fracture Mechanics.** Proceedings of the International RILEM Workshop dedicated to Professor Arne Hillerborg. Chapman and Hall, London 1991, 305 pp (ISBN 0-412-369870-x)

Eligehausen, Rolf; Bergmeister, Konrad and Silva, John F., Editors (1991) **Fastenings to Reinforced Concrete and Masonry Systems. State-of-art report**. CEB - Comite Euro-International du Beton. Bulletin d'Information No 206 + 207. Lausanne 1991, 492 pp (ISBN 2-88394-011-8 and 012-6).

German Anchor Bolt Standard (1987) **Verzeichnis der allgemain bauaufsichtlich zugelassenen Dübel und Ankerschienen**, Gruppe Verbundanker. (Table of approved anchor systems. In German) Institut für Bautechnik in Berlin. Stand 2.1.1987.

Håkanson, Mats; Johansson, Håkan E.; Broms, Carl Erik and Elfgren, Lennart (1981) **Ingjutningsbruk. Tidsberoende egenskaper** (Time dependent properties of grouts for anchor bolts. In Swedish. Summary in English). Division of Structural Engrg, Luleå University of Technology, Technical Report 1981:47T, Luleå 1981, 41pp.

Klöker, Werner (1984) 30 Jahre Reaktionsharzmörtel, -beton und -kunststein auf Basis ungestättiger Polyesterharze (Reaction resin mortar, reaction resin concrete and artificial stone based on unsaturated polyester resins - 30 years' experience. In German) **Fourth International Congress on Polymers in Concrete**, 19-21 September 1984, Proceedings ICPIC '84 (Edited by Herbert Schultz), Technische Hochschule Darmstadt 1984, pp 11 -19.

Rankweil (1980) **Long-time performance of HILTI HVA adhesive anchors.** Test reports issued by Höhere Technische Bundes-, Lehr- und Versuchsanstalt, Rankweil, Austria. Size M12 No 186/78, Nov 1978, 15+15 pp; Size M20, No 170/79, July 1979, 16+24 pp; Size M16, No 52/80, April 1980, 5+26 pp (In German); Size M8, No 143/80, June 1980, 15+20 pp (In German).

Rehm, Gallus; Eligehausen, Rolf and Mallée, Rainer (1992) **Befestigungstechnik** (Fastening technique. In German). Reprint from Betonkalender 1992, Ernst & Sohn, Berlin 1991, 119 pp.

Schuerman, Fritz; Jankowski, Alfons and Novotny, Rudolf (1970) Die Weiterentwicklung des Klebeankers (Further development of adhesive anchors. In German). **Glückauf** (Essen), 26 Nov 1970, pp 1145 - 1151.

Swedish Concrete Standard (1978) Concrete testing - Hardened concrete - Cube strentgh. SS 13 72 10, 3 pp and SS 13 72 13, 3 pp

Swedish Steel Standard (1976) Skruvförbandsnorm St BK-N3 (Code for bolted connections. In Swedish). Statens Stålbyggnadskommité, Svensk Byggtjänst, Stockholm 1976, 80 pp.

Zavliaris, K D and Speare, P R S (1992) **The behaviour of adhesive anchorages installed in concrete**. In "Proceedings from the International Conference **Bond in Concrete** - From research to practise", Riga, Latvia, October 15-17,1992 (Edited by A Skudra and R Tepfers), Technical Univ, Riga LV-1658, Latvia 1992, pp 11-1 to 10.

51 FASTENING ELEMENTS IN CONCRETE STRUCTURES – NUMERICAL SIMULATIONS

J. OŽBOLT and R. ELIGEHAUSEN
Institut für Werkstoffe im Bauwesen, Stuttgart University, Germany

Abstract
Anchoring elements such as headed and expansion studs and grouted or undercut anchors, are often used for local transfer of loads into concrete members. In order to better understand the failure mechanism, a large number of experiments have been carried out in the past. However, due to the complicated three-dimensional load transfer a very few or no numerical studies have been performed for a number of different fastening situations i.e. influence of the embedment depth, crack-width influence (fastening in cracked concrete), influence of the edge distance etc. Therefore, in the present study some results of the axisymmetric and three-dimensional numerical analysis of the headed studs embedded in plane concrete block are presented. Influence of different geometrical and material parameters have been studied employing finite element method and nonlocal microplane model. Comparison between experimental and numerical results indicate reasonable good agreement. Generally it has been observed that the failure mechanism is governed by fracture energy rather than by tensile strength of concrete. As a consequence, the size effect is strong and close to linear elastic fracture mechanics.
Keywords: Concrete, Fastening, Finite element analysis, Nonlocal microplane model, Fracture, Size effect.

1 Introduction

In current engineering practice, headed anchors are used to transfer load into reinforced concrete members. Experience, a large number of experiments and numerical studies confirm that fastenings are capable to introduce the force into the reinforced concrete member without using a reinforcement. As shown by Eligehausen and Sawade (1989) and Eligehausen and Ožbolt (1991), the main reason for this is a stable crack propagation up to peak load, what is a consequence of a three-dimensional stress-strain state in the load transfer zone. However, since no reinforcement is present, the fracture process zone is relatively small and failure is rather brittle i.e. very close to a linear elastic fracture mechanics type of failure. Therefore, it is im-

Fracture and Damage of Concrete and Rock – FDCR-2.
Edited by H.P. Rossmanith. © 1993 Taylor & Francis. ISBN 0 419 18470 8.

portant to clearly understand the failure mechanism and the influence of different geometrical and material parameters on the load capacity. This is of course not simple because in the practice many types of fastenings exist as well as a number of different geometrical situations.

The simplest fastening case is a single headed stud anchor transferring a force into a large concrete block. Provided the steel strength of the stud is high enough, headed studs embedded in a concrete block subjected to tensile loading fail by pulling a concrete cone out of it. The failure is due to the failure of concrete in tension by forming a circumferential crack growing in so-called mixed mode (Eligehausen and Sawade, 1989). In recent years several attempts have been made to understand this growth and to predict the pull-out load of headed studs (Ottosen, 1981; Eligehausen and Sawade, 1985; Krenchel and Shah, 1985; Ballarini, Shah and Keer, 1986). Summarizing this activities it can be said that material models based on plasticity and on macroscopic stress-strain relationships together with stress criteria are not capable to predict behavior of anchors as observed in experiments (Eligehausen and Sawade, 1989). Furthermore, the predicted failure load depends on the finite element size and load step size (Cedolin and Bažant, 1980). A better explanation of anchorage failure mechanism can be expected using more general material models based on fracture mechanics. Generally, the numerical analysis has to be carried out using three-dimensional finite elements together with a realistic three-dimensional material model for concrete. Not only material nonlinearity, but also fracture process in concrete must be correctly simulated.

Due to above mentioned complexity, for geometrically complicated types of fastenings, presently only experimental results are available. Based on these test results design rules were developed that are used in the engineering practice. However, recently rather sophisticated nonlocal microplane model for concrete is developed and implemented into a three-dimensional finite element code (Bažant and Ožbolt, 1990; Ožbolt and Bažant, 1992). This model has been successfully used in three-dimensional numerical studies of different problems (Bažant, Ožbolt and Eligehausen; 1992) and therefore it is also employed here.

In the present study axisymmetric finite element analysis of the single fastening element embedded in uncracked concrete block is carried out in order to investigate the influence of macroscopical concrete properties (tensile and compression strength, fracture energy) and the geometry of the structure (embedment depth, size of the headed stud, edge distance) on the concrete cone failure load. Beside this, in order to study the influence of the distance of the anchor from the edge as well as the influence of the crack width on the pull-out failure load, three-dimensional nonlinear analysis is also carried out. Results of the analysis are compared with the experimental ones.

2 Axisymmetric analysis - material parameters and size effect

2.1 Headed anchor with embedment depth h_v= 130 mm

A single headed anchor embedded in a large uncracked concrete block is analyzed using axisymmetric four-node finite elements. The typical finite element mesh is shown in Fig. 1. Pulling of the anchor is performed by prescribing displacements at the bottom of the stud. Contact between steel stud and concrete exists only under the head of the stud. To account for the restraining effect of the embedded anchor, the displacements of the concrete surface along the steel stud in the vicinity of the head are fixed in direction perpendicular to the load direction. The supports are supposed to be fixed in both directions.

Summary of geometrical and material parameters that are used and varied in the analysis are shown in Table 1.

In order to demonstrate that the present numerical approach is able to correctly prescribe failure load the numerical results are first compared with experimental results. Therefore, basic (reference) material parameters and geometry of the specimen used in the analysis are similar as in the experiment (See. Fig. 1 and Table 1 - example 1).

Table 1. Parameters varied in the analysis

Example	head size d (mm)	tension strgth. f_t (MPa)	compr.strgth. , f_c (MPa)	fract. energie G_f (N/mm)	Young.modul. E_c (MPa)	char. length l_c (mm)	Influence
1	35	1.72	17.6	0.064	23500	6	test
2	35	1.72	17.6	0.064	23500	6	G_f
3		1.70		0.117			G_f
4		1.96		0.165			G_f
5	52	1.72	17.6	0.064	23500	6	f_c
6			20.7				f_c
7	35	1.72	17.6	0.064	23500	6	d_h
8	52						d_h
9	35	1.41	17.6	0.070	23500	6	f_t
10		1.70					f_t
11		1.90					f_t
12	35	1.90	17.6	0.070	23500	6	E_c
13			25.1		33500		E_c
14			32.6		43500		E_c
15	35	1.90	25.1		33500	6	l_c
16						12	l_c
17						24	l_c

After showing that the used approach is effective, the influence of the major macroscopical concrete properties has been investigated i.e. compression and tension strength, concrete fracture energy and size of the head are considered as parameters. Influence of the characteristic length of the nonlocal continuum (l_c) is also studied. In all examples the embedment depth (h_v= 130 mm) and the diameter of the steel bar (ϕ= 22 mm) are keept constant. The finite element mesh for the larger headed stud case (d_h= 52 mm) is slightly modified. Basic value of the concrete Young's modulus (E_c= 23500 MPa) and Poisson's ratio (ν= 0.18) are taken the same as in the above mentioned experiment. Used fracture energy values (G_F) were calculated employing a unit area finite element with length l= 360 mm (similar as in experiment). To demonstrate the influence of the initial Young's modulus of elasticity and of the characteristic length, parametric studies with three different initial Young's modulus of elasticity i.e. three different characteristic length of nonlocal continuum, keeping all other parameters constant, are also carried out.

2.2 Headed anchor with variable embedment depth

In order to investigate size effect in fastening of headed stud in concrete block, the finite element analysis as proposed by RILEM TC 90-FMA have been carried out. The specimen shape and the typical finite element mesh are shown in Fig. 2. The size effect is analyzed using three geometrically similar specimens with the size increase by a factor of three. The geometry of the specimens is related to the embedment depth h_v. Three embedment depths are considered: h_v= 50, 150 and 450 mm. The distance between support and anchor is $2h_v$, so that the unrestricted formation of the failure cone is possible. In all cases analyzed pulling out of the anchor is performed by prescribing displacement at the bottom of the stud. Material parameters are taken as follows: uniaxial tensile and compression strength f_t= 3 MPa and f_c= 40 MPa, concrete fracture energy G_F= 0.1 N/mm (assuming a tension specimen length of 360 mm), Initial Young's E_c= 30000 MPa and Poisson's ratio ν= 0.20. Characteristic length of the nonlocal continuum is taken l_c= 12 mm.

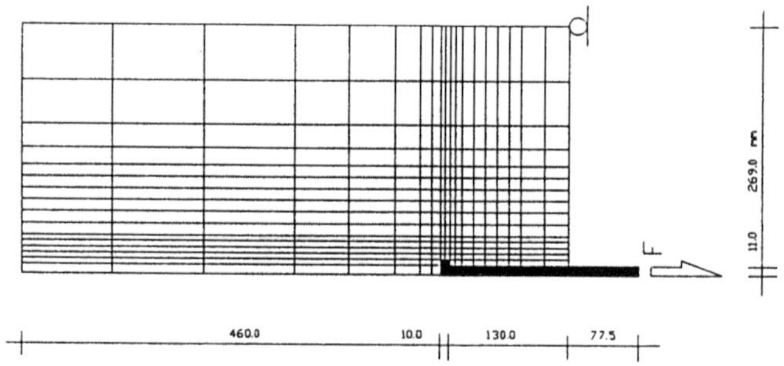

Fig. 1 Geometry and FEM used in parameter study

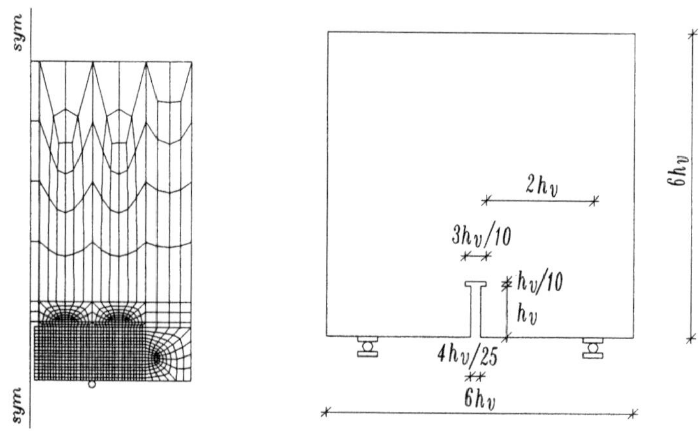

Fig. 2 Geometry and FEM used in the size-effect study

2.3 Results of the axisymmetric analysis

2.3.1 Comparison of Numerical and Test Results for Headed Anchor with h_v= 130 mm

In Fig. 3, the numerically obtained load-displacement curve is compared with the measured one. As can be seen, the calculated and measured failure load agree rather well. This could be expected, because the characteristic length was fixed according to this experiment (l_c= 6 mm).

While the calculated initial stiffness agrees well with the measured value, the displacements at failure are smaller than observed in the experiment. It should be noted that the area of the anchor head is relatively small (d_h = 35 mm), and due to this the average compression stress under the head at failure load is about 10 times larger than the uniaxial concrete compression strength f_c (the maximum calculated value in one integration point is $20f_c$). Therefore, most of the displacements in the load displacement diagram (see Fig. 3) is due to concrete compression softening under the anchor head. The concrete pre- and post-peak behavior under very high compression in 3D stress-strain state have to be correctly modeled to predict load-displacement curve. As can be seen from Fig. 3, assumed peak stress in the analysis is too high in comparison to the behavior found in the test i.e. calculated displacement are too small. According to Eligehausen and Sawade (1989), most of the total displacement is due to concrete compaction under the anchor head. However, in general the calculated displacement field is in good agreement with the experimental observations.

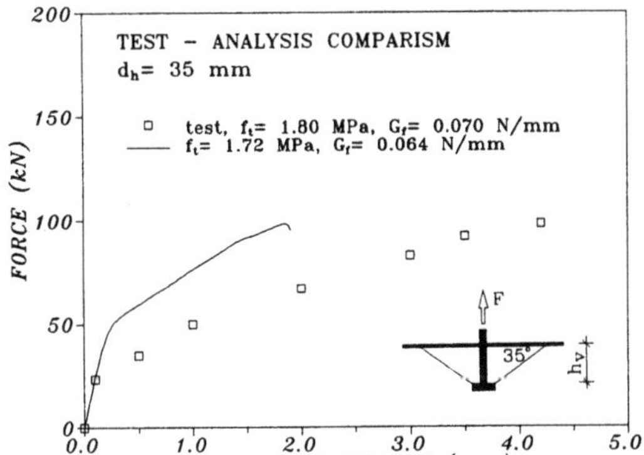

Fig. 3 Test - analysis load displacement curves

It is observed that the circumferential cracking starts relatively early at the anchor head and progresses in a stable manner towards the concrete surface, radial cracks are initiated at much higher loads at the surface. Failure is caused by circumferential cracking. The calculated slope of the failure cone surface is in good agreement with the experimentally measured results (Eligehausen and Sawade,

1989). The average angle between the pulled cone surface and the concrete surface amounts to $\alpha = 35^0$.

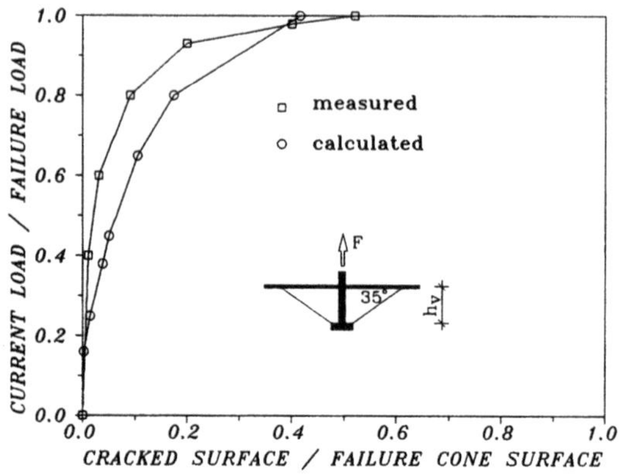

Fig. 4 Calculated and measured crack surface area

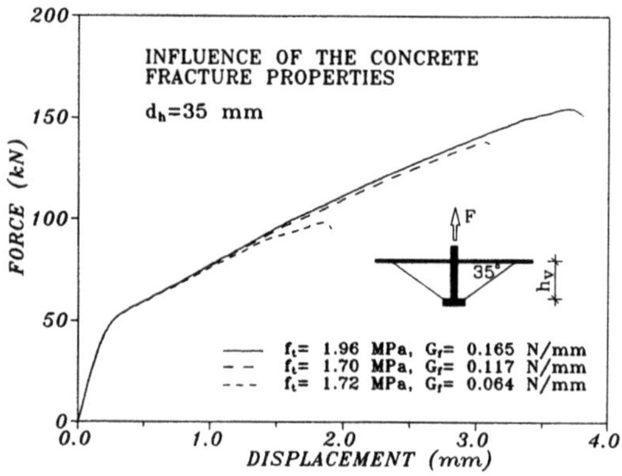

Fig. 5 Load-displacement curves for different G_F values

In Fig. 4, the ratio between the crack cone surface area and the total crack cone surface area (calculated for $\alpha = 35^0$) is plotted. Stable crack growth starts at about 30 % of the failure load and further crack growth is very stable up to the peak load, when unstable crack development starts. These results are again in good agreement with experimental observations based on measurements of concrete strains (Eligehausen and Sawade, 1989).

Summarizing, the obtained results are in good agreement with experimental findings. This is valid for the fracture process, displacement field and failure load.

2.3.2 Influence of the material parameters on the anchor behavior

In Fig. 5 the influence of the fracture energy (G_F) on the load displacement behavior is shown. Three different G_F values are considered (see Tab. 1). The concrete compression strength (f_c = 17.6 MPa) is fixed, and average tension strength used was f_t = 1.79 MPa. Fig. 5 clearly indicates a strong dependency between fracture properties of concrete and failure load. Roughly, the failure load is proportional to the square root of the fracture energy (Fig. 6). This finding is confirmed by analytical and experimental results obtained by Sawade (1992).

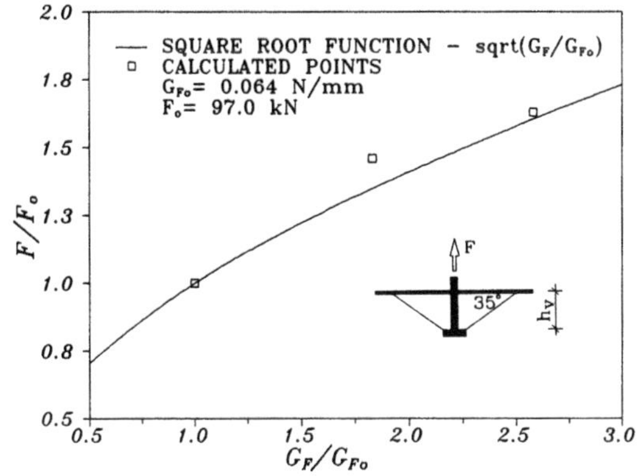

Fig. 6 Relation between failure load and G_F

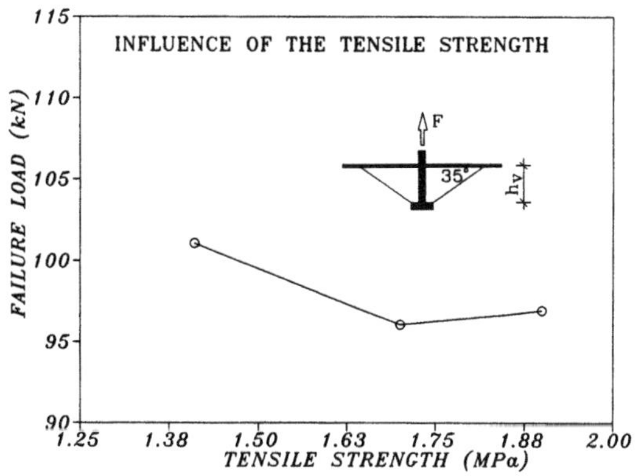

Fig. 7 Tension strength influence on the failure load

In Fig. 7 the influence of the tension strength on the failure load is shown. Namely, concrete with three different values of tension strength (see Tab. 1) and approximately constant fracture energy is analyzed. Concrete compression strength and head diameter were constant. The results indicate no significant influence of the tension strength on the failure load. According to this, the fracture energy seems to be the governing parameter in the analysis of headed anchors.

The influence of the concrete compression strength is demonstrated in Fig. 8, where load-displacement curves for two specimens with head diameter $d_h = 52$ mm are plotted. Tension strength and the fracture properties of the concrete are constant, while two different concrete compression strengths are considered (see Tab. 1). Results clearly indicate insensitivity of the failure load on the concrete compression strength, but the displacement at failure is smaller if the concrete compression strength is higher.

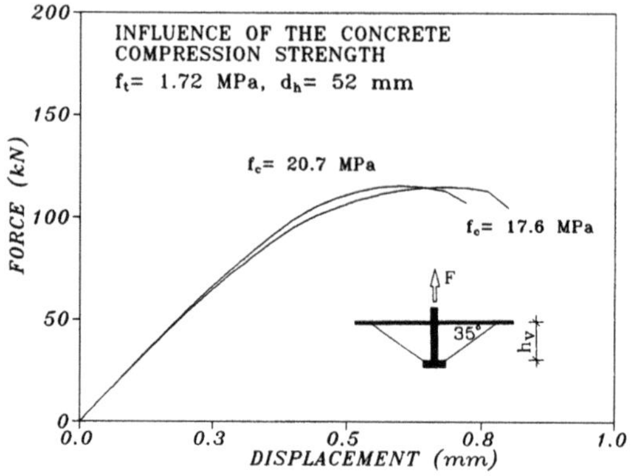

Fig. 8 Compression strength influence on the failure load

In Fig. 9 the load-displacement curves of two specimens with the same concrete properties but different head diameters ($d_h = 35$ and 52 mm) are compared. It can be seen that, within the range of the assumed head diameters, the failure load does not depend significantly on the head diameter, that means of the concrete compression stress-strain state under the head. When increasing the effective head area by a factor of 3, the increase of the calculated failure load is only 17 %. For a head diameter $d_h = 52$ mm the average compression stress under the load at failure is only about 3.75 f_c compared to 10 f_c for $d_h = 35$ mm. Due to this reduction of the compression stress the displacement of the anchor head is much smaller. Therefore, with decreasing head diameter concrete compression softening under the head contributes larger displacement at failure load. This numerical results are in agreement with test evidence (Elfgren et. all., 1988; Furche, 1988). For anchor head $d_h \ll 35$ mm, the displacement of the anchor head will increase even more, reducing the actual embedment depth and thus the concrete cone failure load (Furche, 1988).

From the numerical point of view, analysis of the specimen with smaller head size, converges very slowly and requires considerably more computer time than analysis of specimen with larger head size. This is a consequence of compression softening in 3D stress-strain state under the anchor head.

In Fig. 10 the influence of the initial modulus of elasticity on the failure load is shown. Geometry and all concrete material parameters were fixed (see Tab. 1). According to the numerical results, the increase of the failure load is proportional to the $E_0^{2/3}$. This increase is slightly larger than is obtained by Sawade (1992), using linear elastic fracture mechanics (LEFM), where the increase was found to be to the power of 1/2. The differences are probably due to the fact that the numerical analysis is based on the nonlinear fracture mechanics approach. Namely, in contrast to LEFM where the modulus of elasticity is kept constant up to reaching the concrete tensile strength, in the present analysis this modulus is continuously changing from $E = E_c$ (at origin) to E = 0 at the peak stress.

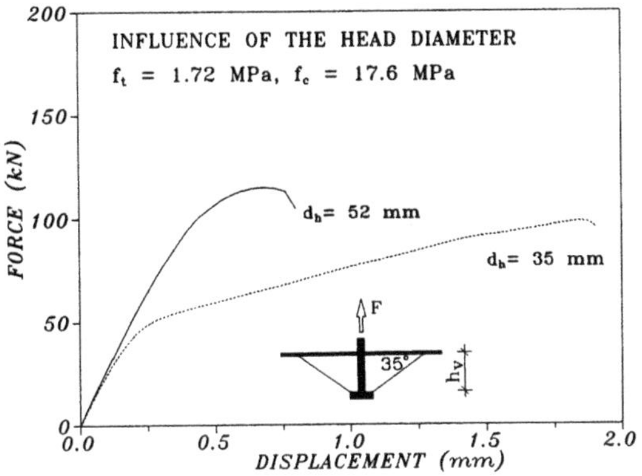

Fig. 9 Influence of the head stud size on the load-displacement curve

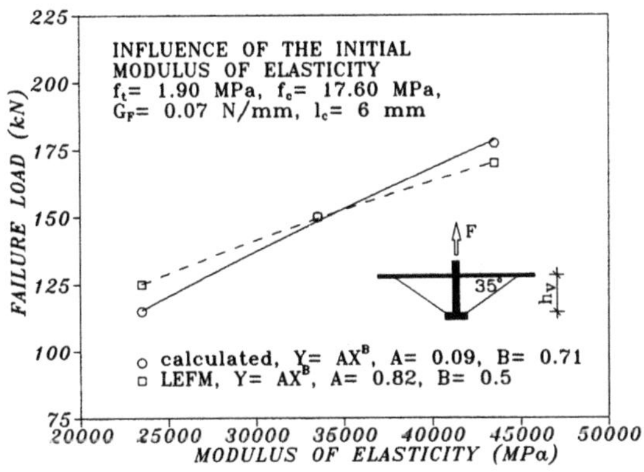

Fig. 10 Influence of the Young's modulus on the failure load

Finally, to demonstrate that the characteristic length of the nonlocal continuum (l_c) has a considerable influence on the failure load, a parameter study is performed (see Tab. 1) in which the concrete properties and geometry were kept constant and only l_c is varied. The results of the analysis, shown in Fig. 11, exhibit significant influence of the characteristic length on the concrete cone failure load in the analysis of the headed anchors. This increase of the failure load is approx. proportional to $l_c^{1/3}$. Consequently, the question arises: how this characteristic length can be correctly determined and what is the relation between the fracture concrete properties and l_c?. According to the proposal by Bažant and Pijaudier-Cabot (1989), which is based on the smeared crack approach, l_c can be correlated with the maximum aggregate size. However, numerical analysis indicate that l_c is not a material parameter but depend also on stress-strain state (Bažant, Ožbolt and Eligehausen, 1992). Therefore, recently new nonlocal concept has been proposed (Bažant, 1992) which is based on the crack interactions. In this new approach the characteristic length defines crack-interaction volume that is not constant but also depend on the stress-strain state in the material. Further work is required in order to verify this new approach.

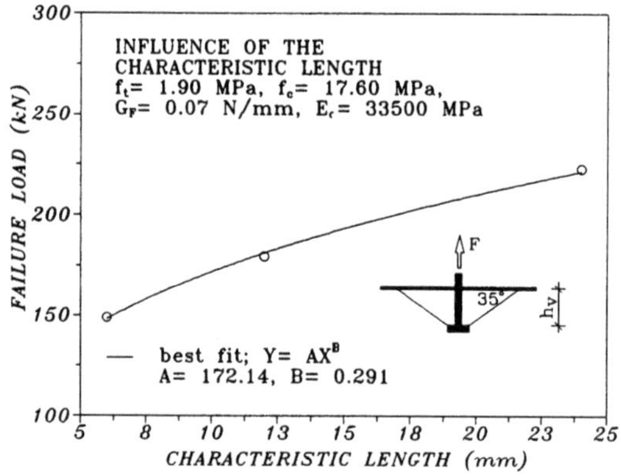

Fig. 11 Influence of the characteristic length on the failure load

Summarizing, the failure of the analyzed headed stud embedded in concrete is due to circumferential cracking of concrete in tension. The compression behavior of concrete under the head is mainly responsible for the displacement at failure. The concrete fracture energy plays a dominant role.

2.3.3 Influence of the Embedment Depth on the Failure Load - Size Effect
Load displacement curves for three different embedment depths are shown in Fig. 12. The displacement is monitored under the head. It can be seen that the displacement at peak load increase with increasing embedment depth. Assuming no size effect, the peak load should increase in proportion to h_v^2, that means by a factor of 9 when tripling the embedment depth. The results of the analysis show that the increase in failure load is much less (approximately by a factor 5.7) which is a consequence of the size effect. Using the size effect law (Bažant, 1984), the concrete cone failure load can be calculated as:

$$F_N = F_U B(1 + h_v/d_0)^{-1/2} \tag{1}$$

where F_N represents load at failure including size effect, F_U the would be failure load without size effect, h_v the embedment depth while B and d_0 represents constants which are calculated from the linear regression analysis of the obtained failure loads. No size effect ultimate load F_U is calculated using the formula:

$$F_U = a(\sqrt{f_c})h_v^2 \tag{2}$$

where f_c represents the concrete compression strength and a is factor to calibrate calculated failure loads with the measured values and to assure dimensional correctness of Eq. (2). Eq. (2) is proposed by ACI 349, appendix B for the prediction of the concrete cone failure load.

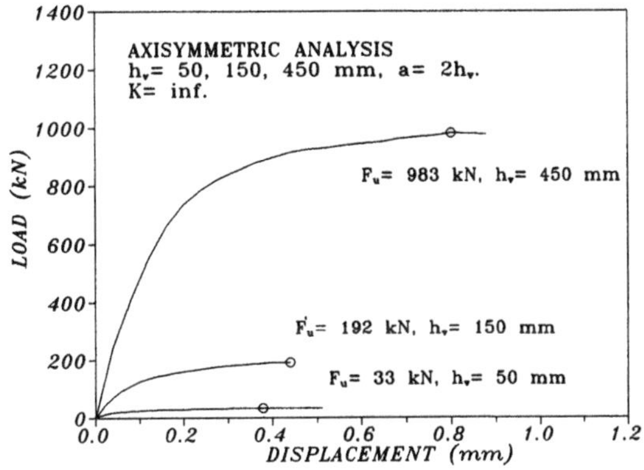

Fig. 12 Load-displacement curves for three different embedment depths

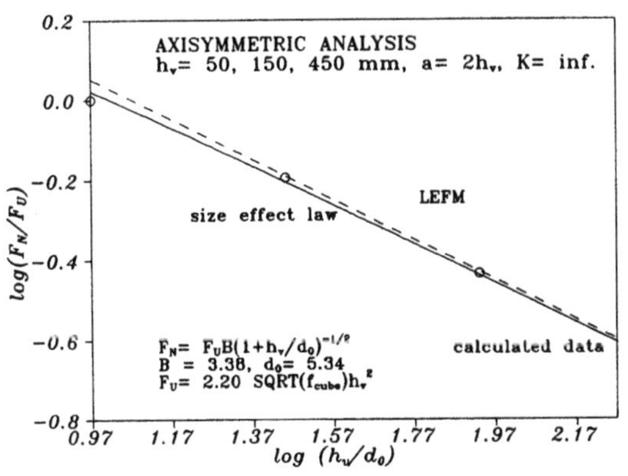

Fig. 13 Size-effect observed in the analysis

In Fig. 13 the results of the analysis are plotted and compared with the size effect law (Eq. (1)). The coefficient a in Eq. (2) is fixed such that the numerically obtained failure load for anchors with an embedment depth $h_v = 50$ mm is predicted correctly. As can be seen from Fig. 13, the concrete cone failure load exhibit a strong size effect, since the numerical results are close to the linear elastic fracture mechanic

(LEFM) solution.

In Fig. 14 the results of the analysis are compared with the different failure equations. The relative failure loads as a function of the embedment depth are shown. The failure load for an embedment depth $h_v = 150$ mm is taken as reference value. Plotted are the relative failure load according to the size effect law (Eq. (1)), a formula that neglects the size effect (Eq. (2)) and a formula derived on the basis of linear elastic fracture mechanics (Eq. (3)) (Eligehausen and Sawade, 1989):

$$F_N = a_1(\sqrt{EG_F})h_v^{3/2} \tag{3}$$

In Eq. (3), a_1 is a constant. The failure loads predicted by Eq. (3) agree rather well with the test results (Eligehausen and Sawade, 1989). Assuming no size effect, the failure loads should increase in proportion to h_v^2, that means by a factor of nine when tripling the embedment depth. The results of the analysis show that the increase of the failure load is much less (approximately by a factor of 5.7). Therefore the size effect should be taken into account in the design of anchorages, otherwise the failure loads are underestimated for small embedment depths and are overestimated for large embedment depths (Fig. 14). The agreement between the size effect formula and the formula based on the linear fracture mechanics is good in the entire embedment range. The size effect has also been observed in tests by Bode and Hanenkamp (1985) and by Eligehausen et al. (1989), (1992). According to these authors, the failure load increases in proportion to $h_v^{1.5}$.

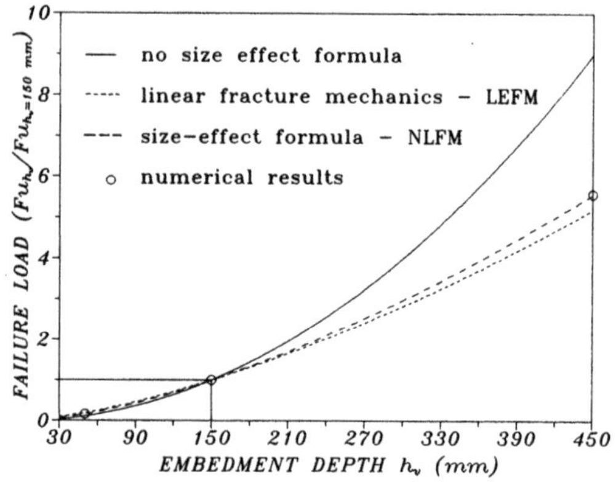

Fig. 14 Different failure equations

The relative shapes of the failure cone for three different embedment depths at peak load, estimated from the numerical analysis, are plotted in Fig. 15. In Fig. 16 the distribution of the tensile stresses perpendicular to the failure cone surface are shown as a function of the ratio $l_h/l_{h_{max}}$ where l_h represents the distance from the anchor and $l_{h_{max}}$ is the failure cone radius taken from Fig. 15. These distributions are estimated from the results of the numerical analysis. From Fig. 15 and 16 the size effect can be explained as follows: With increasing embedment depth the ratio diameter of the failure cone to embedment depth decreases, that means the effective relative cone surface area decreases as well. Furthermore, the average stress over the failure surface decreases with increasing embedment depth because the stress distribution is more triangular like, in the case of a large embedment depth, and more parabola like in the case of smaller embedments.

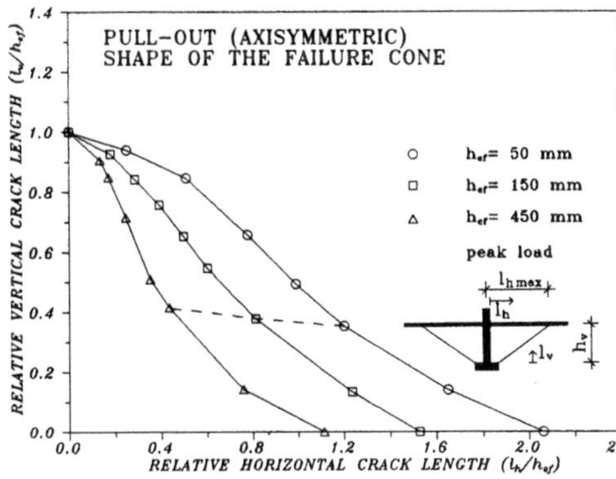

Fig. 15 Shapes of the failure cone surface area

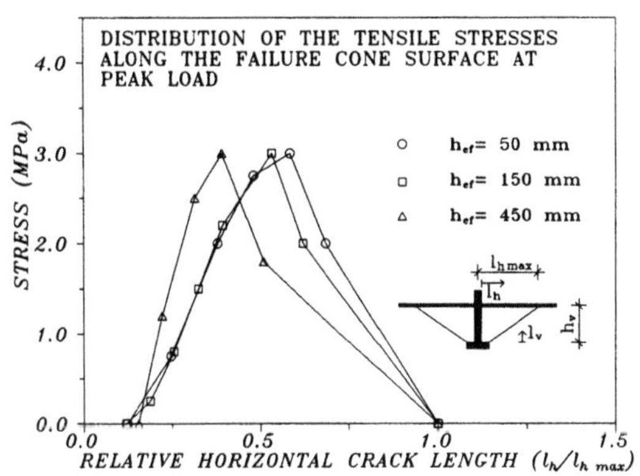

Fig. 16 Tensile stress distribution along the cone surface at peak load

Summarizing, the analysis indicate significant size effect that should be taken into account in design practice (Eligehausen and Ožbolt, 1990; Bažant, Ožbolt and Eligehausen, 1992).

3 Three-dimensional analysis - the edge and crack width influence

3.1 Influence of the edge distance on the failure Load

To investigate influence of the distance between the headed stud and the free edge on the pull-out failure load, nonlinear finite element analysis is performed on the concrete cube specimen (Fig. 17) (Ožbolt, 1991). In the analysis specimen edges are free except the bottom three edges which are fixed in vertical direction. However, the bottom edge of the specimen at the side where the stud is placed is taken as a free. Fixing embedment depth as h_v= 90 mm, the distance of the stud from the free edge, a_r, is varied as $a_r = 40, 80, 120, 160$ and 200 mm,

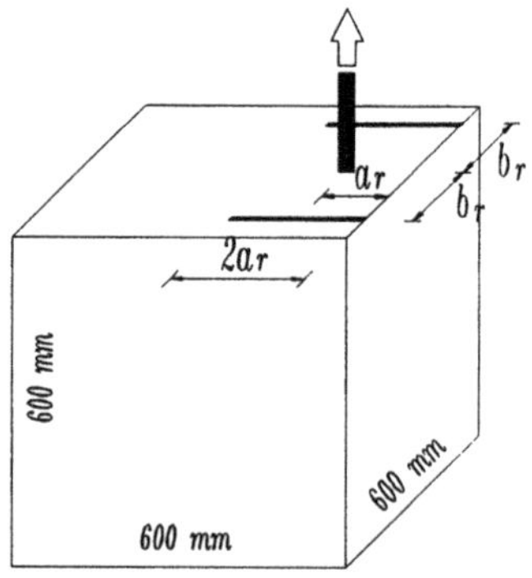

Fig. 17 Edge influence - geometry of the specimen

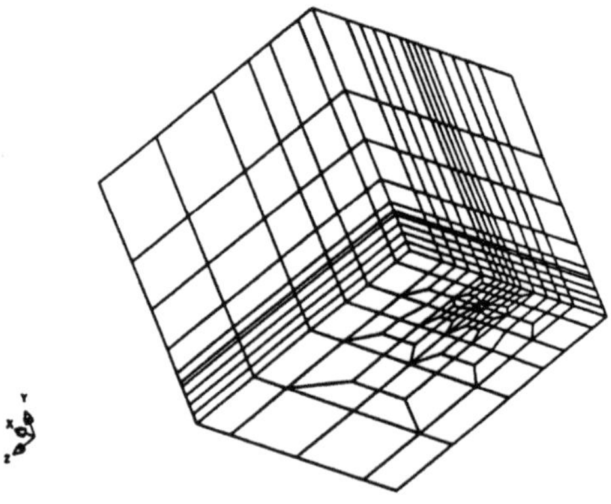

Fig. 18 Edge influence - finite element mesh

In the finite element analysis three-dimensional finite elements with eight nodes and eight integration points are used. A typical finite element mesh is shown in Fig. 18. The spatial discretization is made such that the mesh is relatively fine in the fracture process zone and coarse elsewhere.

Analysis is performed assuming basic material properties, Young modulus and Poisson's ratio as: E_c= 30000 MPa and ν= 0.18. Microplane model parameters are set such that the calculated uniaxial concrete compression strength yields to f_c= 34 MPa and the uniaxial concrete tensile strength $f_t \cong 1.8$ MPa.

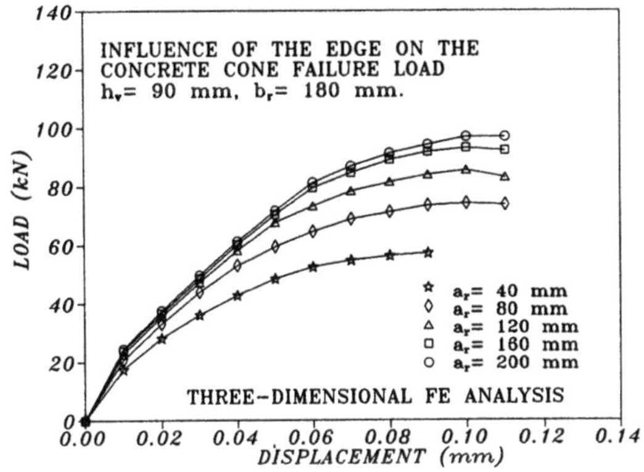

Fig. 19 Load-displacement curves for different edge distances

In order to correctly represent the concrete fracture energy (G_F), it is necessary to properly set the characteristic length of the nonlocal continuum (l_c). However, since no explicit relation between G_F and l_c exists, the characteristic length is here chosen by fitting the concrete cone failure load for the case a_r= 200 mm (no influence of the edge) using empirical formula (Eligehausen et al., 1988).

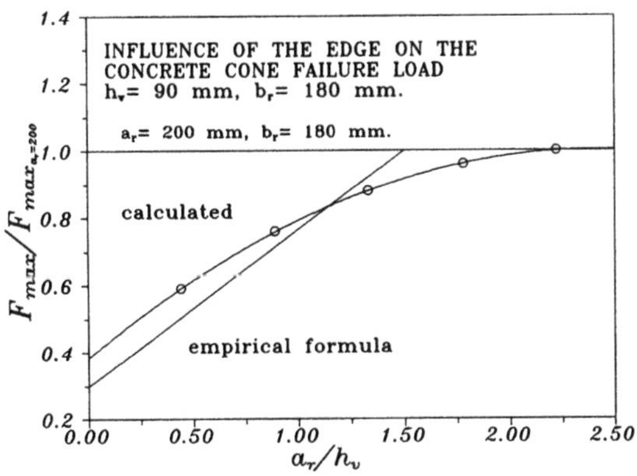

Fig. 20 Relation between failure load and edge distance

Fig. 19 shows calculated load-displacement curves for different a_r values. These curves exhibit an increase of the peak load with increasing distance of the stud from the free edge. Displacement at the peak load, for all a_r values, is approximately constant and does not change significantly by increasing the a_r value. Initial stiffness of the stud is practically constant and independent of a_r.

In Fig. 20 calculated relative values of the peak loads for different values of a_r and the best fit curve are plotted as a function of the distance of the stud from the free edge. Results clearly indicate an increase of the calculated failure load with the increase of a_r up to approximately $a_r = 2h_v$. Further increase of a_r, however, does not significantly increase the peak load. For comparison in the same figure empirical relation between a_r and relative peak load is also plotted (Rehm, Eligehausen and Mallée, 1988). As can be seen from Fig. 20, comparison between results of the numerical analysis and prediction formula based on the test results show rather good agreement.

3.2 Influence of the crack width on the failure load

The influence of the crack width on the concrete cone pull-out is studied (the same as in the experiments) on the reinforced concrete thick plate dimensions 1000 x 1040 x 300 mm (Fig. 21) (Ožbolt, 1991). In the center of the specimen single anchor is embedded with embedment depth h_v= 120 mm. In order to initiate crack propagation in the anchor plane, a notch of a length 100 mm is introduced on the both vertical sides of the specimen (see Fig. 21). The crack width in the specimen is controlled through 12 reinforcing bars. The specimen is vertically supported around the bottom edges as well as around the anchor at the distance $2h_v$.

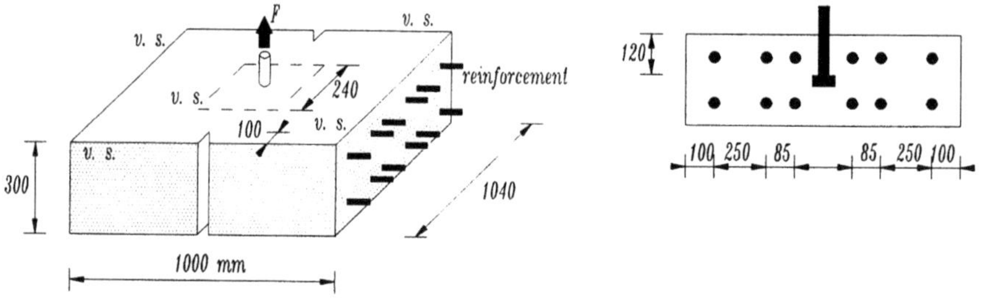

Fig. 21 Influence of the crack width - geometry of the specimen

The loading procedure used in the analysis is the same as it is common in the experiments. First, tensile stresses are introduced through the reinforcement bars (controlling displacement) producing the cracks of a certain width. In the next step the reinforcement extension is keept fixed and pulling out of the anchor is performed (controlling force). This is repeated for different tension stresses in the reinforcing bars i.e. for different crack widths.

In the finite element analysis only 1/4 of the specimen is modeled i.e. symmetry is used. The material properties as well as the finite element type used are the same as in previous example. The finite element mesh is shown in Fig. 22.

Since the analysis is based on smeared crack approach the crack width that is introduced through the reinforcement can not be obtained explicitly from the analysis. Therefore, empirical formula is used in order to calculate the crack width from the known reinforcement strains (Rehm and Martin, 1968).

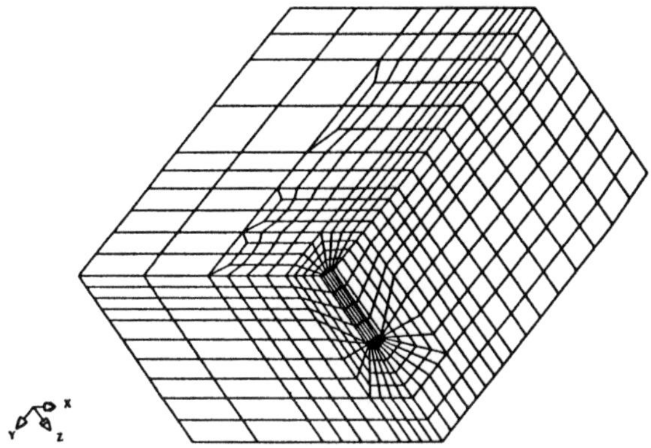

Fig. 22 Influence of the crack width - finite element mesh

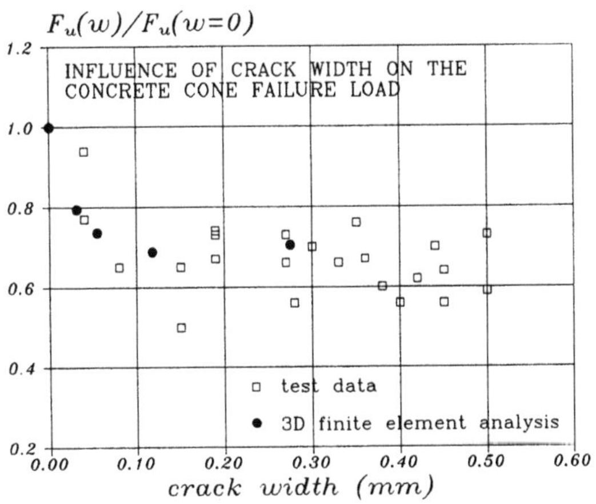

Fig. 23 Influence of the crack width on the failure load (test, calculations)

The relation between the crack width and the failure load, measured relative to the uncracked concrete cone failure load, is shown and compared with test results (Rehm, Eligehausen and Mallée, 1988) in Fig. 23. It can be seen that the failure load decreasing with the increase of the crack width rather steep up to the crack width Δw= 0.2 mm. For further increase of the crack width the failure load is constant and reaching approximately 70 % of the failure load obtained for the uncracked concrete specimen (Δw= 0). Calculated relations between failure load and crack width are in good agreement with experimental results (see Fig. 23).

More detailed description of the failure mechanism at different crack widths is given by Ožbolt (1991).

4 Conclusions

1. Comparison between axisymmetrical finite element analysis of the single anchor that is embedded in a plane concrete block and test results indicate good agreement. In all analyzed examples failure is due to failure of concrete in tension (circumferential cracking) rather than in compression. Circumferential crack growth starts at about 30 % of failure load and it is very stable up to the peak load. After the peak, unstable crack develops and the final failure cone is formed.

2. The fracture load is mainly influenced by the concrete fracture energy G_F and is approximately a square root function of G_F. The size of the anchor head as well as the concrete compression properties have a rather small influence on the failure load, however they influence the load-displacement relationship significantly. The influence of the concrete tension strength, assuming constant G_F value, has small influence on the failure load. On the other side, initial Young's modulus and the characteristic length of the nonlocal continuum significantly influence the failure load.

3. The numerical study of the embedment depth influence on the failure load indicate a significant size effect that is close to linear elastic fracture mechanics. The failure loads increases approximately in proportion to $h_v^{1.5}$. This is in good agreement with experimental evidence. Prediction formulas which do not take into account the size effect overestimate the failure load for larger embedment depths. The size effect is due to the fact that the elastic energy at peak load is released from the entire structure and as a consequence, the effective relative failure load surface as well as the area under the tensile stress distribution decrease with increasing embedment depth. The size effect should be take into account in design practice.

4. Results of the 3D finite element analysis indicate decrease of the failure load if the edge distance is decreasing. This influence practically disappear when a_r is greater than $2h_v$. Experiments confirm that calculated results shown in the present analysis are reliable.

5. Three-dimensional finite element study of the crack width influence on the concrete pull-out load, similar as in the experiments, confirms that the concrete cone failure load decrease with the crack width increase. This decrease is relatively fast up to crack width $\Delta w = 0.2$ mm and reaching approximately 70% of the failure load for uncracked concrete. However, further increase of the crack width does not decrease failure load since reaching critical crack width

(approx. 0.2 mm) final load transfer mechanism is formed and further increase of crack width does not influence failure load. Three-dimensional numerical analysis of the headed anchor embedded in a cracked concrete indicate that the calculated relation between concrete cone failure load and crack width exhibit qualitatively and quantitatively good agreement with experimental observations.

6. Present numerical approach based on the nonlocal microplane model seems to be powerful tool in the analysis of the complicated 3D problems of the concrete structures. However, open question is the calibration of the material model and characteristic length of the nonlocal continuum l_c which seems to be stress-strain dependant rather than in advanced fixed concrete property. Therefore, more promising results can be expected by the use of the new nonlocal concept that is based on the microcrack interaction approach. Indeed, first results obtained using this approach seems very promising (Ožbolt, 1992).

5 References

1. ACI 349-80 (1980) **Code Requirements for Nuclear Safety Structures**, Appendix B - Steel Embedments, American Concrete Institute, Detroit.

2. Ballarini, R.; Shah, S. and Keer, L. (1986) Failure Characteristics of Short Anchor Bolts Embedded in a Brittle Material, **Proc. R. Soc. London**, A 404, pp. 35-54.

3. Bažant, Z.P. (1984) Size effect in Blunt Fracture: Concrete, Rock, Metal. **Journal of Engineering Mechanics**, ASCE, 110(4), 518-535.

4. Bažant, Z.P., and Ožbolt, J., (1990) Nonlocal microplane model for fracture, damage, and size effect in structures. **J. of Engrg. Mech.**, ASCE, 116(11), 2485-2504.

5. Bažant, Z.P., Ožbolt, J. and Eligehausen, R., (1992) Fracture Size Effect: I. Review of Evidence for Concrete Structures. Submitted to **Journal of Str. Eng.**, ASCE.

6. Bažant, Z.P., Ožbolt, J. and Eligehausen, R., (1992) Fracture Size Effect: II. Should it go Into Concrete Design Code. Submitted to **Journal of Str. Eng.**, ASCE.

7. Bažant, Z.P. (1992) New Nonlocal Damage Concept Based on Micromechanics of Crack Interactions, Report No. 92-t/C457n, Department of Civil Engineering, Northwestern University, USA.

8. Bode, H. and Hanenkamp, W., (1985) Zur Tragfahigkeit von Kopfbolzen bei Zugbeanspruchung. **Bauingenier**, 1985, 361-367.

9. Cedolin, L. and Bažant, Z. P. (1980) Effect of Finite Element Choice in Blunt Crack Band Analysis. **Comp. Meth. in Appl. Mec. and Eng.**, 24, 305-316.

10. Elfgren, L., Broms, C. E. and Johansson, H. (1980) Anchor Bolts in Reinforced Concrete Foundations, Report No. 1980:36, University Lulea, Sweden.

11. Eligehausen, R. and Sawade, G. (1985) Behavior of Concrete in Tension. **Betonwerk + Fertigteil Technik**, No. 5 and 6.

12. Eligehausen, R., Fuchs, W. and Mayer, B. (1988). Loadbearing Behaviour of Anchor Fastenings in Tension. **Betonwerk + Fertigteil Technik**, 12/1987(1), 826-832, and 1/1988(2), 29-35.

13. Eligehausen, R. and Sawade, G. (1989) Analysis of Anchoring Behavior (Literature Review), **Fracture Mechanics of Concrete Structures** - RILEM Report, (ed L.Elfgren), Chapman and Hall, pp.263-280.

14. Eligehausen, R. and Sawade, G. (1989) A fracture mechanics based description of the pull-out behavior of head studs embedded in concrete, **Fracture Mechanics of Concrete Structures** - RILEM Report, (ed L. Elfgren), Chapman and Hall, pp. 263-280.

15. Eligehausen, R. and Ožbolt, J., (1990) Size effect in Concrete Structures, **Application of Fracture Mechanics to Reinforced Concrete**, (ed A. Carpinteri), Elsvier Applied Science, Torino, pp. 17-44.

16. Eligehausen, R., and Ožbolt, J., (1991) Use of the Tensile Strength in Anchorage to Concrete, **Structural Concrete**, IABSE Colloquium, IABSE-AIPC-IVBH, Stuttgart.

17. Eligehausen, R., Bouška, P., Červenka, V. and Pukl, R. (1992) Size Effect of the Concrete Cone Failure Load of Anchor Bolts, **FramCoS 1**, (ed Bažant, Z.P.), Elsevier applied science, Breckenridge, USA, pp. 517-525.

18. Furche, J. (1988) Ausziehverzuche an Kopfbolcen mit kleinen Unterkopfflachen und Flachen Hinterschnittwinkeln. Report No. 9/4-88/1, Institut ftr Werkstoffe im Bauwesen, Stuttgart University.

19. Krenchel, H. and Shah, S. (1985) Fracture Analysis of the Pull-Out Tests. **Material and Structures**, 18(108), 439-446.

20. Ottosen, N. S. (1981) Nonlinear Finite Element Analysis of Pull-Out Tests. **Journal of Structural Division**, 107(4) 591-603.

21. Ožbolt, J., and Eligehausen, R., (1990) Numerical Analysis of Headed Studs Embedded in Large Plain Concrete Blocks, Internal Report Nr. 4/10-90/9, Institut für Werkstoffe im Bauwesen, Stuttgart University, Germany.

22. Ožbolt, J., (1991) The Edge Influence on the Pull-out Failure Load of the Headed Stud Anchors, Internal Report Nr. 4/12-91/11, Institut für Werkstoffe im Bauwesen, Stuttgart University, Germany.

23. Ožbolt, J., (1991) The Influence of the Crack Width on the Concrete Cone Failure Load, Internal Report Nr. 4/13-91/12, Institut für Werkstoffe im Bauwesen, Stuttgart University, Germany.

24. Ožbolt, J. and Bažant, Z.P., (1992) Microplane Model for Cyclic Triaxial Behavior of Concrete. **Journal of Eng. Mech.**, ASCE, 118(7), 1365-1386.

25. Ožbolt, J., (1992) Smeard Crack Analysis - New Nonlocal Microcrack Interactions Approach, Internal Report Nr. 4/14-92/19, Institut für Werkstoffe im Bauwesen, Stuttgart University, Germany.

26. Rehm, G., and Martin, H. (1968). Zur Frage der Rissbegrencung im Stahlbetonbau. **Beton - und Stahlbetonbau**, (8), Berlin, 1-8.

27. Rehm, G., Eligehausen, R., and Mallie, R. (1988) Befestigungstechnik. **Betonkalender 1988**, 2, Wiehelm Ernst and Sohn, Berlin, Germany.

28. Sawade, G. (1992) A contribution to the behavior of concrete in tension. Thesis in the preparation, Stuttgart University.

52 COMPARISON OF FINITE ELEMENT CONCRETE MODELS SIMULATING PULL-OUT TESTS

A. di TOMMASO and O. MANFRONI
Faculty of Engineering, University of Bologna, Italy
G. VALENTE
Faculty of Engineering, University of L'Aquila, Italy

Abstract

A comparison of different numerical approaches simulating Pull-Out tests for a prototype of Anchor-Bolt embedded in a concrete plate is presented. On the basis of previous Round-Robin Pull-Out results (Elfgren, 1991) some characteristic parameters are supposed to influence the scatter in the results themselves, and besides they are now investigated using two different F.E. programs, hereafter referenced to as **MODEL-1** and **MODEL-2** as described below.

Thus the present study covers the main topics which seem to the Authors influencing the numerical results in such a formulation, spanning from mesh geometry for F.E. discretizzation to constitutive laws for concrete behaviour idealisation and numerical solution approaches for solving the non-linear equations governing the model behaviour of Pull-Out tests.

From the present report some helpful considerations can be achieved for further investigations regarding concrete mechanics modelling.

Keywords: Band Width, Energy Criterion, Fracture Energy, Fracture Mechanics, Pull-Out, Softening Shape, Strain Softening, Strength Criterion.

1 Introduction

Recently a lot of interest is growing up about Pull-Out numerical and experimental tests. Namely it seems to be attractive for engineers and researchers to investigate such a phenomenon either for its practical use in a wide range of engineering applications, either for the main aspects regarding fracture mechanics which arise during the tests. From a numerical point of view, a recent Round-Robin (Elfgren, 1991) has put in evidence a great scatter in the results achieved with different models. Therefore it seemed to the Authors of great interest to investigate the main reasons which give rise to such differences in order to establish some helpful criteria for further improved numerical investigations. In the present work the Authors have focused their attention to the results of the previous Round-Robin establishing some gap-source parameters which seem slightly to infirm the numerical simulations. Afterwards these parameters were checked by use of two different

Fracture and Damage of Concrete and Rock – FDCR-2.
Edited by H.P. Rossmanith. © 1993 Taylor & Francis. ISBN 0 419 18470 8.

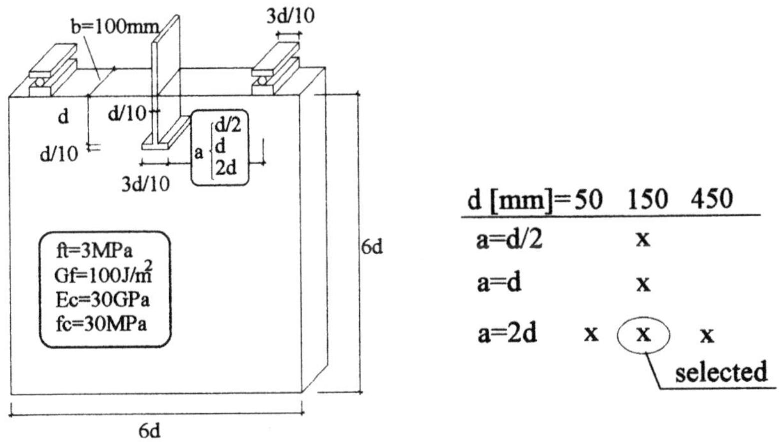

Fig.1 Pull-Out model set-up

F.E. codes (Bathe, Hibbit), in order to establish their influence on the final results. So firstly it was checked where the programs differ each-other, and secondly a wide range of tests was carried out separately by use of both programs. Since each of the above codes presents some meaningful aspects in treating the concrete behaviour, it was so possible to investigate the influence of each of them on the final results.

For the purpose of the present work the attention was focused toward predefined geometry, boundary conditions and material properties for the Pull-Out model, as stated by the above mentioned Round-Robin, and depicted in fig.1.

2 Recent Round-Robin on Pull-Out tests

2.1 Main results

An international Round-Robin has been organised by RILEM TC-90 FMA (Fracture Mechanics of Concrete-Applications). During a meeting held in Delft (The Netherlands) on June 24th, 1991, a preliminary report was discussed and the main results were shown to the participants. It's the scope of the present work to analyse and discuss the scatter which influence those results.

As you can see from fig.1, the model set-up is arranged in such a way that a parametric definition of the geometry and boundary conditions can be chosen, meanwhile for the material all the parameters are constant. Namely a lot of contributions deal with a particular set-up, which is the most recommended for the test and which is also the only one the Authors have chosen for compiling the present work.

The requests for the contributions were focused toward an output macroscopic overall behaviour of the model, such as the peak load and the load-displacement curve, which can be thought as the main results which have an engineering value for technical purposes. Since a quite scatter is found among all the results (fig.2), it's

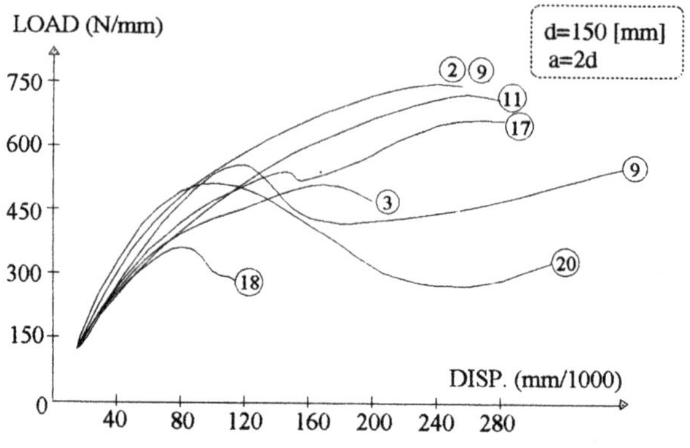

Fig.2 Load-displacement Round-Robin results

unrealistic that a "mean result" can be chosen to be representative of the "real" behaviour of the model. Furthermore the call for paper did not set any restriction nor any classification on the use of the F.E. programs; only a summary report on their performances was requested. By inspection of the Round-Robin preliminary report you can see a large range of different hypothesis assumed by the participants, which reflects the state of the art on the numerical approach to the fracture mechanics. These hypotheses range from crack model assumptions (smeared or discrete) to the softening shape (linear or non linear) and finally to the numerical solution techniques (tangent or secant stiffness). By a more detailed analysis of that report a list of influencing parameters can be chosen as shown in the next section.

2.2 Estimation of Gap-Source parameters influencing the main results

The Authors of the present work have tried to investigate the reasons which give rise to such a scatter in the numerical results of the above mentioned Round-Robin. They are listed in fig.3, and they can be summarised in the following three main categories:

a) Mesh geometry; this includes F.E. type (triangular, quadrilateral, three-four or more Gauss points), mesh orientation and F.E. size. Generally for some kinds of analysis (elasticity, plasticity ...) these geometric aspects don't notable affect the numerical solution; instead for non linear fracture mechanics analysis it's well demonstrated that the final results are quite sensitive to them.

b) Constitutive law; it seems to be the core of the problem, since almost every research group has focused it's attention to well defining this law. It deals with the idealisation of fracture initiation and its propagation (mode I, mode II or mixed mode). Besides it concerns primary the way by which cracks are treated inside the model (smeared, discrete), and the possibility to take into account for crushing in compressive stress state and

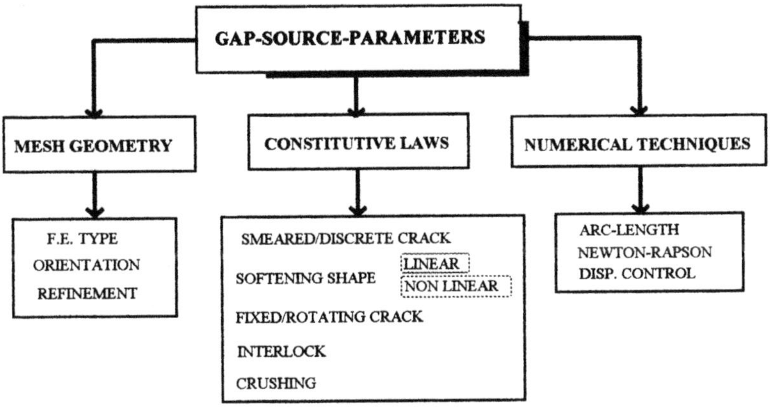

Fig.3 Gap-Source parameters list

interlocking in mixed mode.

c) Numerical solution; all those computational procedures which let to perform the numerical run for the solution of the non-linear equations which arise from the model idealisation are collected in this category. Although most of them seem to be almost equivalent, when dealing with fracture mechanics the choice of one procedure instead of the others can slightly affect the results.

In the subsequent it will be shown the influence each of the above mentioned categories applies to the Pull-Out test, by reproducing it's behaviour separately by two different F.E. programs as depicted in the next section. Besides it's not intended to assess the validity of the theoretical hypotheses which lie under each numerical idealisation, but all the above mentioned topics are assumed only for purpose of comparison.

3 Description of the models used for further comparisons

3.1 General

Two different F.E. programs (hereafter named MODEL-1 and MODEL-2) are used for further comparison of the above specified parameters. Both programs assume the smeared crack approach in simulating fracture process in concrete materials; in particular MODEL-2 was opportunely modified for taking into account for fracture mechanics.

The programs are not perfectly equivalent, as it will be shown, so different results can be achieved by them.

3.2 Stress-strain laws in uniaxial and multiaxial states

Both programs treat fracture process via stress-strain law, that, in uniaxial behaviour, is represented by a polyline which can assume any non-linear configuration according to the user's choice. In bi-

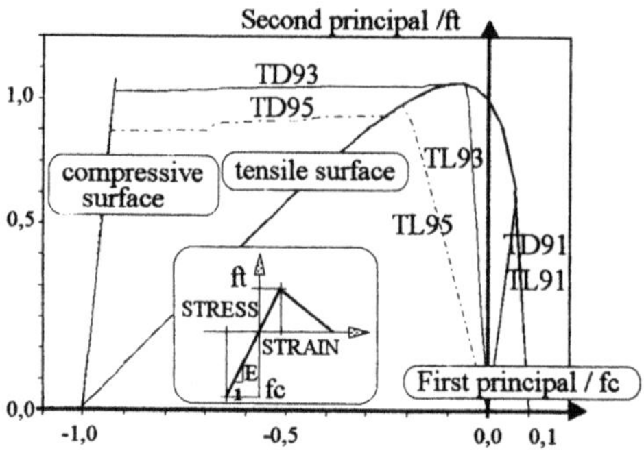

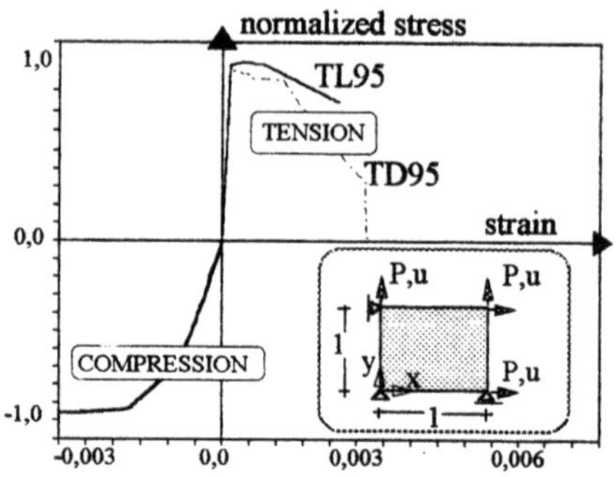

(a) MODEL-1 (b)

Fig.4 MODEL-1 and MODEL-2 comparison
TL=Test with Load-control
TD=Test with Displacement-control
(a) failure surfaces in stress domain for MODEL-1
(b) uniaxial stress-strain behaviour for MODEL-1
(c) failure surfaces in stress domain for MODEL-2
(d) uniaxial stress-strain behaviour for MODEL-2

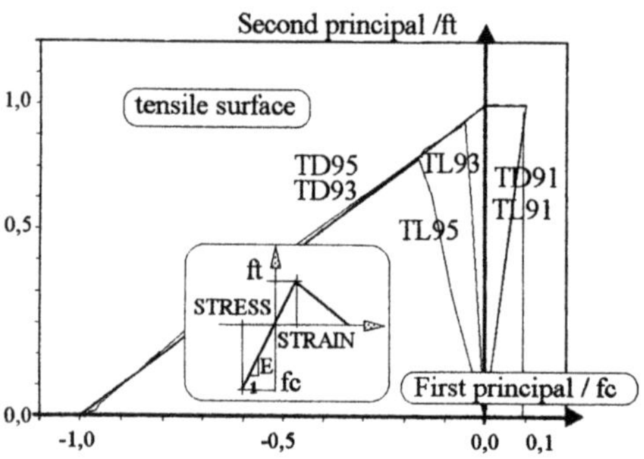

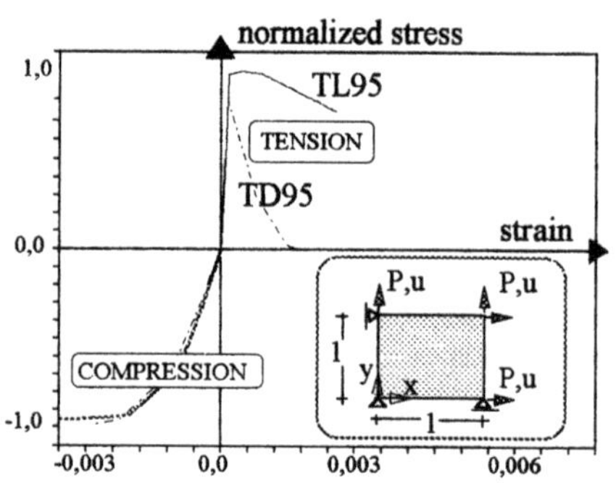

(c) MODEL-2 (d)

axial state the failure envelope is an approximation of the well known experimental domain (Kupfer). MODEL-1 assumes a closer approximation to the above stress domain, defining a set of two continuous curves as a limit state for the stress extent. Instead MODEL-2 adopts a polyline (i.e. made up of straight lines) as a representation of that domain.

Since in Pull-Out test the most representative stress state is the mixed one (i.e. tensile in one principal direction and compression along with the other), it happens that a different representation of the failure envelope in the stress domain can perform a sensible influence on the computational results. Besides in tensile-tensile region the tension cut-off (adopted only by MODEL-1) is also very meaningful for the final results (fig.4).

3.3 Mode I fracture simulation (stress-crack opening)

As well known in smeared crack approach the crack distribution over the material volume generally is approximated by crack localisation in sampling points, over which numerical integration is performed in order to realise the stiffness (or compliance) matrix for the finite element. Therefore in the constitutive formulation of the material it needs to define the stress-strain law, which characterises both the elastic and the damaged state. In doing so, a particular attention is generally paid in defining the shape of the strain-softening branch, which generally reduces to the uniaxial tensile stress limit definition and to the fracture energy consumption, which corresponds to the amount of area under the constitutive law.

The latter argument is quite non-objective and generally user's dependent, since any softening shape can perform the same amount of fracture energy and then scatter the computational results, as described below. Moreover the softening-shape is generally related to the mesh size and to its orientation with respect to the crack path, as the request is for calibrating the band-width on which macro-cracks are supposed to be spread-out (Bazant 1983, Rots 1988), so the bridging from an experimental stress-displacement result to an hypothetical stress-strain constitutive law is not unique.

With respect to these topics, both MODEL-1 and MODEL-2 assume the same hypotheses and therefore they are almost identical.

3.4 Mode II fracture simulation (shear stress-crack sliding)

Even if both models used in the present work can perform the MODE II fracture simulation (or via shear retention factor or with a more sophisticated shear-displacement formulation), for the purpose of simplicity the crack-sliding effect here is not taken into account, either for it complicates the numerical tests and subsequent comparisons, either for the correspondence with experimental tests for calibrating the MODE II fracture parameters yet is not well known.

4 Check of parameters' influence in the simulation of Pull-Out tests

4.1 Mesh Geometry

In F.E. analysis mesh choice is generally problem dependent and user's skill influenced, so it's not possible to define a unique procedure for all the problems. However, in fracture mechanics analysis, in order to overcome the general problem in defining the process zone for smeared models or the crack path for discrete ones, it seems the best choice be dictated mainly by numerical difficulties rather then by physical aspects. In particular in the smeared crack approach based on fracture energy one main difficulty concerns the choice of the crack band width on which macro-cracks are supposed to be spread out in to micro-cracks in order to get a contin

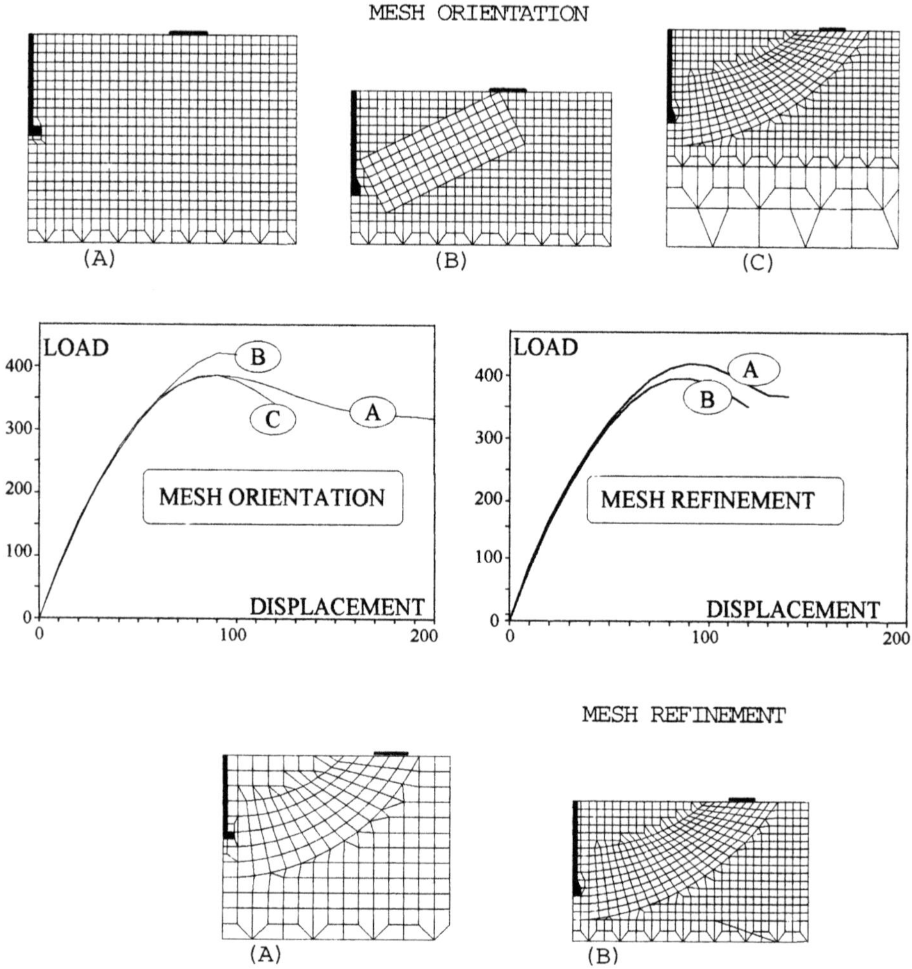

Fig.5 Mesh geometry influence

uum material. So a regular mesh of quadrilateral elements generally fits the above requirements, because it permits the fracture energy to be spread out homogeneously on the volume of the band width. Also F.E. orientation can produce a difference in computational results because generally cracks "prefer" to follow mesh sides, which represent a weaker path than the diagonal crossing one.

For this reason many Authors have chosen their mesh as a "radius-like" type connecting the upper-bolt head to the outer support of the model, in order to follow almost the same way the real cracks will assume in the test. By this assumption, however, it's not possible to define an identical geometry for all the elements; consequently local differences are generated mainly by the transition from the stress-displacement to the stress-strain law for uniaxial behaviour, which performs a non homogeneity in the constitutive relationship of the model from element to element.

By inspection of fig.5 it's possible to observe the influence that mesh size and mesh orientation perform on the final result of the Pull-Out test. It is not only on the peak load but also (and mainly) on the post-peak behaviour that we recognise a manifest scatter in the results.

To overcome this "mesh-orientation sensibility" some Authors have proposed the use of triangular elements which minimise crack-path disturb. Instead the use of refined interpolating functions for the field variables of the model seems not to be as important as the above topics.

4.2 Constitutive models

The parameters grouped under this category are demonstrated to play a fundamental role in the successful of the overall behaviour of the numerical models.

Historically fracture mechanics has been developed starting from a **strength criterion** assumption as a crack initiation and its propagation inside the materials. Later on it was found that only incorporating also an **energy criterion** it was possible to overcome some problems that arose in simulating crack propagation, while a strength criterion is still utilised to describe the crack initiation (Bazant, 1983). All these assumptions were checked by simulating simple tests for which a lot of experimental data was available, such as the direct tensile test and the three points notched beams bending test. Generally in all the above tests fracture propagates mainly in mode I along with a well known path, letting so easier a numerical simulation to predict the experimental results. Besides many discrepancies based on the assumption of a smeared or discrete crack approach seemed to disappear when also an energy criterion was introduced in to the model, and only a straight line was assumed for the softening shape.

By looking at fig.2 it seems that for the Pull-Out test the previous results do not match at all, then some other reasons seem to influence the results themselves. Then the question is: **which are these reasons that distinguish the behaviour of Pull-Out test from the behaviour of classical model tests?** To answer this question we have to remind that any numerical simulation must always respect two conditions; first, the model must be intrinsically unconditioned (e.g. objectivity on mesh size and its orientation for F.E.); second, the model must incorporate some testing parameters which must be respected during the run (e.g. Gf consumption rate).

In Pull-Out test combined stresses produce crack initiation and mixed mode is active during crack propagation. For this reason the numerical results are affected by the way the models take into account for failure envelope and crack growth mechanics (crushing, tension cut-off, interlocking). Moreover in Pull-Out test the crack path is not known in advance, for which conditioned models give rise to a scatter in the results for different meshes (fig.5)

A third fracture parameter (after tensile strength and fracture energy) which play a significant role in F.E. simulation, is

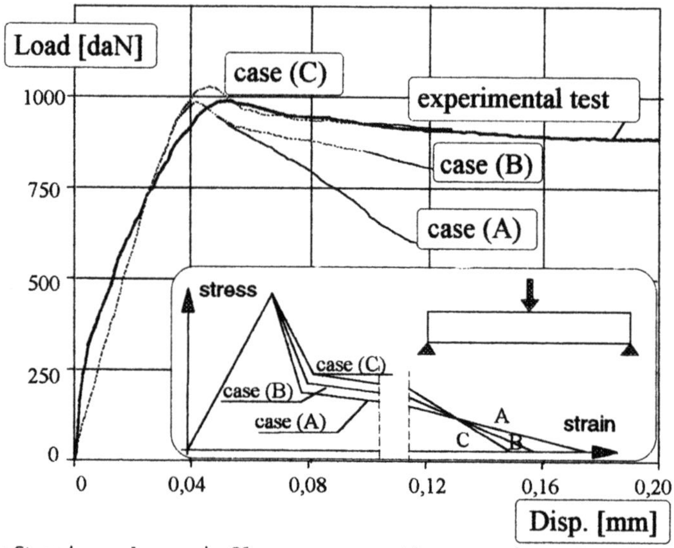

Fig.6 Softening-shape influence on a three-points bending test

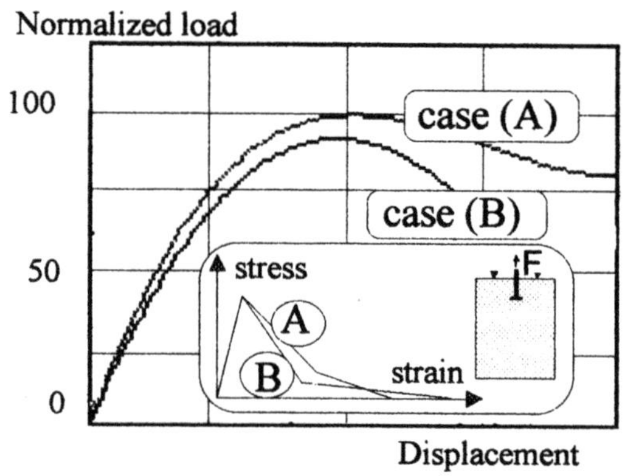

Fig.7 Softening-shape influence on a Pull-Out test

demonstrated to be the **softening shape**, that, as well known, also for the classical three points bending test can deeply influence the results (Petersson, 1981-Rots, 1988).

In a displacement method based model the stepping evaluation of the tangent stiffness leads the non-linear path for numerical runs; hereafter it's well demonstrated, as a consequence of the displacement compatibility, a spurious cracking may arise from the runs, leading to a loss of strain localisation and consequently to an unstable crack path orientation (Rots, 1988). This unstable behaviour is more pronounced in a Pull-Out test then in a classical model test for the above mentioned reasons, and therefore it can be thought (and demonstrated) to play a great role in scattering the results (figg.6-7).

The conclusion is that not only the fracture energy should be issued for the purpose of analysing Pull-Out test, but also (and mainly) the softening shape ought to be defined in order to predict more realistic results.

Furthermore it seems that the less stiff the model is the best the numerical results fit the experimental data. So a rotating crack is able to predict a more realistic path than a fixed one since it mitigates the effect of "spurious MODE-I interlocking" due to the non correct crack orientation. Also the accounting for interlock itself in Mode II can be in conflict with the spurious interlocking arising in Mode I fixed crack path propagation.

Finally the crushing phenomenon, which physically seems not to influence the test except for the limited volume near the bolt head in which high pressures give rise to it, can deeply modify the numerical performance for the shape assumed by the biaxial failure envelope, which in tensile/compressive stress state reduces the peak tensile stress modifying locally the material weakness.

4.3 Numerical solutions

From a strictly numerical point of view numerical solution is the core of any non-linear computational run. Generally well tested procedures are employed in a wide class of structural analyses. For non-linear problems Newton-type procedures are at present the most implemented in F.E. programs. Among these we remind the Newton-Rapson method with displacement or load control and the arc-length method with a constraint between loads and displacements. All these classes are based on the evaluation of the tangent stiffness-matrix and therefore they all need the definition of a tangent constitutive relationship for the material.

Another class of procedures, instead, is based on the stepping evaluation of the secant stiffness of the model, which assumes the definition of a secant constitutive relationship for the material.

Till now it is not well demonstrated the influence each procedure acts on the final result. However our test performs a comparison of the behaviour of a single element and subsequently of the whole Pull-Out model adopting in both cases the arc-length method and the displacement control method. As you can see from fig.4 the test performed on a single element shows a great difference on the computational performance. It can be explained by the fact that stretching element nodes by imposing a displacement field can run behind the peak load and follow the softening shape of the material; instead a load control is not able to do so, and after reaching the peak load it falls down by unloading the element.

On the other hand the test performed on the Pull-Out model doesn't show any sensible difference on the final result. This can be explained by the fact that imposing a load or a displacement at the anchorage doesn't matter for the rest of the model, since its effect is mitigated by the presence of many elements which absorb the load (or the displacement) and also because globally the model doesn't exhibit any abrupt snap-back phenomenon in its load-displacement curve.

5 Concluding remarks

In the present work a full inspection of previous Round-Robin results for a Pull-Out test of anchor-bolt embedded in concrete specimens is performed, with the aim to clarify the scatter shown by those results. Three main categories of parameters which influence the computational runs are identified, and consequently they all are investigated by use of two F.E. models, with the goal to identify the influence each of them acts on the final results.

It is demonstrated that all the categories influence the results, but the softening shape is surely the most significant parameter.

It is also shown that it's quite difficult in a computational procedure to distinguish the influence each parameter performs on the final results, because all of them influences each-other.

Acknowledgement

For the present study the Authors are grateful to the Italian Ministry of Research (M.U.R.S.T.) for the financial support, and also to Mr. M. Arduini for his fruitful co-operation.

6 References

Bathe, K.J., (1975) **ADINA - A Finite Element Program for Automatic Dynamic Incremental Non-linear Analysis**, Report 82448-1, Acoustic and Vibration Laboratory, Dep. of Mech. Eng., M.I.T., Cambridge.

Bazant, Z.P., (1986) Mechanics of distributed cracking. **Appl. Mech. Rev.**, 5, 675-705.

Bazant, Z.P. and Oh, B., (1983) Crack band theory for fracture of concrete. **RILEM Materials and Structures**, 16, 155-177.

Elfgren, L., (1991) **Round-Robin analysis of anchor bolts.** RILEM TC-90 FMA, Fracture Mechanics of Concrete-Applications, Luleå, Sweden.

Eligenhausen, U and Sawade, G., (1989) Anchorage to concrete. **Fracture Mechanics of Concrete Structures**, L. Elfgren editor, Chapman and Hall, London.

Hibbit, Karlsson & Sorensen, Inc., **ABAQUS user's manual, rel. 4.9**, Medway street, Providence, RI 02906.

Kupfer, H., Hilsdorf, H.K. and Rusch, H., (1969) Behaviour of concrete under biaxial stresses, **J. Am. Concrete Inst.**, 656-666.

Petersson, P.E., (1981) **Crack growth and development of fracture zones in plain concrete and similar materials.** Rep. TVB-M-1006, Div. Build. Mat., Lund Inst. of Techn., Lund, Sweden.

Rots, J.G., (1988) **Computational modelling of concrete fracture.** Dissertation, Delft University of Technology, Delft, The Netherlands.

53 FINITE ELEMENT ANALYSIS OF ANCHOR BOLT PULL-OUT BASED ON FRACTURE MECHANICS

J. WANG, P. NAVI and C. HUET
Laboratoire de Matériaux de Construction, Département des Matériaux, École Polytechnique Fédérale de Lausanne, Switzerland

Abstract
Analysis of anchor bolts problem in concrete structures is made through the application of finite element method and a granular microcracked model. This analysis is based on the principles of linear elastic fracture mechanics (LEFM) applied at the level of the matrix and takes both material's heterogeneity and pre-existing microcracks into consideration. Numerical results concerning crack trajectory, load-displacement curve and K-a curve are presented. Comparison is made between the results for the specimen composed of a granular microcracked material and those obtained through the application of LEFM to the same specimen composed of a homogeneous material with one crack. The numerical results are also compared with experimental ones. The influences of the granular structure of concrete and pre-existing microcracks on the crack trajectories and stability of crack propagation are discussed.
Keywords: Analysis of anchor bolts, Pull-out, Material's heterogeneity, Microcracks, Linear elastic fracture mechanics, Numerical simulation.

1. Introduction

It is generally recognized that crack propagation in concrete structures is accompanied with the development of a relatively large damage zone (fracture process zone) where microcracks accumulate. The development of the fracture process zone may be attributed to the heterogeneity of concrete material and the existence of distributed pre-existing microcracks. For concrete considered as a homogeneous material, the existence of the fracture process zone makes the fracture of concrete structures deviate from linear elastic fracture mechanics (LEFM), which deals with the problem of one crack with a point singularity at crack tip. This is the reason for which the application of fracture mechanics to concrete structure needs nonlinear approaches at the macroscopic level. Nevertheless, in the framework of RILEM Round-Robin analysis of anchor bolts (Elfgren [1991]), it was found that the numerical results based on LEFM for the case of concrete considered as homogeneous material given by Wang et al. [1991] gives features similar to those observed in experiments performed in other laboratories, see Elfgren [1991,1992], Helbling et al. [1991] and Ohlsson et al. [1991]. This is true both for the value of the peak load and, at least partially, for the main features of the crack trajectory. In fact, some of the features were obtained by none of other simulations performed on the same problem in this Round-Robin analysis. Thus it is worthwhile to examine how this can be explained by taking into account more realistically the heterogeneous character of concrete. This is done in this paper through a new numerical tool. The latter involves a granular microcracked concrete generated by computer and

Fracture and Damage of Concrete and Rock – FDCR-2.
Edited by H.P. Rossmanith. © 1993 Taylor & Francis. ISBN 0 419 18470 8.

the finite element analysis. It is described in another paper to this conference, Wang et al. [1992c], to which the reader may refer for details.

In the present investigation concrete structure is taken to be composed of hard inclusions and cement paste matrix. Microcracks are introduced around the inclusions to simulate the interface layer between the inclusions and hardened cement paste. It is assumed that crack propagation in the matrix can be studied by LEFM. Application of LEFM at the micro-mechanics level has been already done in the past, see, for instance Zaitsev [1983]. The results presented here includes the crack trajectory, the load-displacement curve and the relation between the stress intensity factor and the crack length (K-a curve). Those results are compared with experimental ones. Comparison is also made between the results for the specimen made of the granular microcracked material and those obtained through the application of LEFM to the same specimen composed of the homogeneous material with only one crack. The influences of material's heterogeneity and pre-existing microcracks on the crack trajectory and on the stability of crack propagation are discussed.

2. Analysis procedures

The geometry of the specimen used in the Round-Robin analysis of anchor bolts was defined by RILEM TC-90-FMA Committee and is redrawn in Fig. 1(a). All the specimen dimensions are proportional to the embedded depth of the steel bar (d). Three values are given to d, namely, 50, 150 and 450 mm for the case of plane stress, which leads to the square specimen size of 300, 900 and 2700 mm respectively. Only the case of d equal to 50 mm is analyzed in this paper. The overall dimensions of the specimen are thus: 300x300x100 mm. The distance between the supports at the upper edge of the specimen and the central line is 2d=100 mm. External force is applied at the upper end of the steel bar and two supports are situated symmetrically. Except two supports, free boundary conditions are imposed on the external surface of the specimen. The assumption of symmetry is made for this specimen. Therefore only half of the specimen is taken in the calculation. Inclusions and microcracks embedded around them are

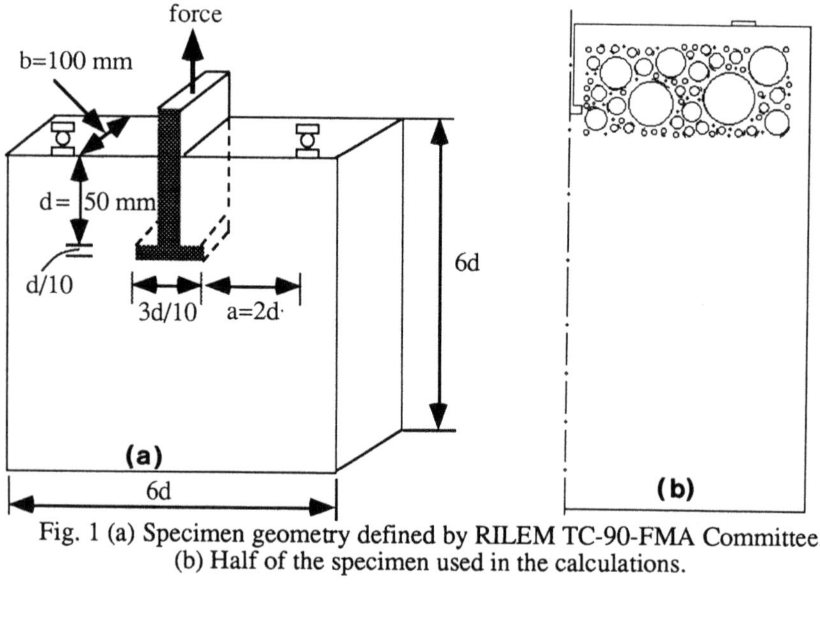

Fig. 1 (a) Specimen geometry defined by RILEM TC-90-FMA Committee.
(b) Half of the specimen used in the calculations.

introduced within the upper part of the specimen, whereas the other part of the specimen is taken to be undamaged homogeneous material as shown in Fig. 1(b). The diameters of the inclusion are calculated from Fuller curve of the particle size distribution. The positions of the inclusions and microcracks and the lengths of the microcrack are computed randomly.

The details of the numerical procedure is described in Wang at al [1992c]. This numerical procedure of determination of displacement and force provide us with the equilibrium curve of Berry [1960], which is valid for all types of loading. The analysis of the stability of crack propagation under different types of loading can be made from this equilibrium curve directly. Once the crack trajectory is determined, it is also possible to study the stability in terms of crack propagation. This is important since, in civil engineering structures, sudden fractures due to the unstable crack propagation are always considered as unacceptable while stable propagation of limited distance is very often tolerable. The analysis of stability must be performed after the type of loading condition for the practical problem under consideration has been chosen.

The stability analysis in this paper is performed from the concepts developed by Huet [1971a, 1971b, 1973, 1974, 1978, 1982]. The latter, being based on general thermodynamics, can be used for both linear and nonlinear fracture mechanics. They make use of energetic parameters. Here a transposition to the use of the stress intensity factor can be performed since LEFM is applied. Nevertheless, this prevents us to study the possible extensions of the crack beyond the position predicted by the equilibrium condition. This may have an influence on the comparison of the computed crack patterns with the experimental ones.

3. Numerical results

In the calculation the displacement is imposed at the upper end of the steel bar. Young's modulus and Poisson's ratio are taken to be 20 GPa and 0.2 for the matrix, 60 GPa and 0.2 for the inclusion, 210 GPa and 0.3 for the steel. The critical stress intensity factor K_{IC} of the matrix is given as 20.0 N/mm$^{-3/2}$. Plane stress condition is assumed in the calculations.

Two numerical examples are studied corresponding to two random realizations of the same heterogeneous structure. The maximum and minimum diameters of the inclusions

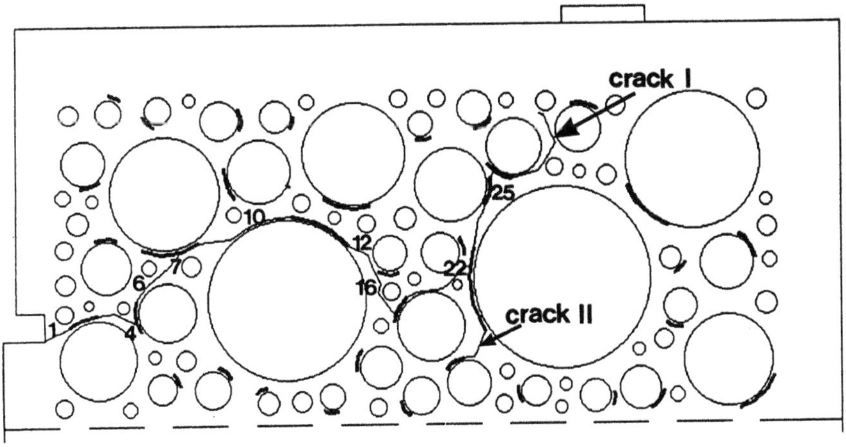

Fig. 2 Crack trajectory for example 1.

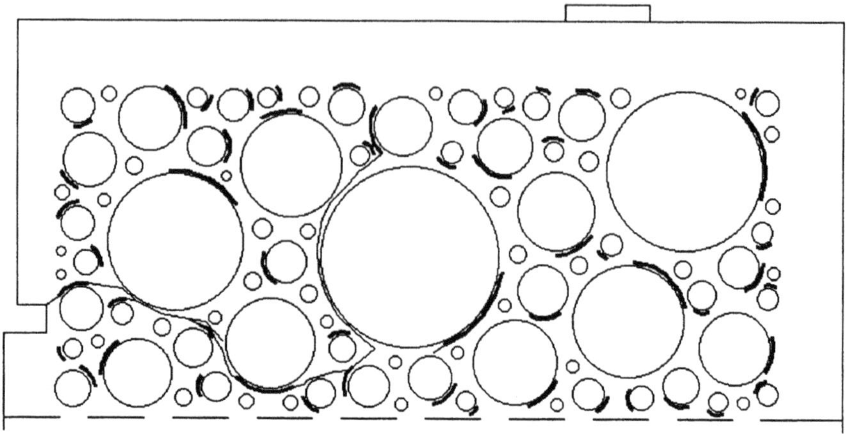

Fig. 3 Crack trajectory for example 2.

are 32 mm and 2 mm for both specimens respectively. Microcracks are embedded around the inclusions with diameter larger than 4.0 mm only. Figs. 2 and 3 give the crack trajectories at the end of analysis for example 1 and example 2 respectively. Bold lines in these figures represent the original position of microcracks. The load-displacement curves of the two specimens are plotted in Fig. 4., where the displacement corresponds to that at the upper end of the steel bar. Fig. 5 gives the variation of the relative stress intensity factor $\bar{K} / K_{IC}$ in terms of crack length *a* for crack branch I shown in Fig. 2 for the first specimen. The results obtained by applying LEFM to the same specimen composed of a homogeneous material with only one crack (Wang et al. [1992b]) are also plotted in Fig. 5 (K-a curve), Fig. 6(a) (load-displacement curve) and Fig. 6(b) (crack trajectory) for the sake of comparison. The point corresponding to the region of the instability in the crack trajectory for the homogeneous body with one crack is represented by a dot shown in Fig. 6(b). The small rectangles located at the upper edge of the specimen shown in Fig. 2, 3 and 6(b) represent the position of the support.

4. Discussion of the numerical results

From Figs 2 and 3 it appears that for the two granular specimens studied, the crack propagates generally towards to the support at the upper edge of specimen. Apparently the crack will terminates at the left side of the support, although the calculation was not performed until that point. It is observed that at the end of the calculation, the crack branch II in Fig. 2 might propagate further. Contrary to this is the homogeneous material with only one crack (Fig. 6(b)), for which the crack runs slightly upward before entering the region below the support and then downwards.

Numbers in Fig. 2 are the successive steps of crack propagation in the simulation. The corresponding values of force and displacements are shown with the same numbers in Fig. 4. Since the load-displacement curve is the equilibrium one [Berry, 1960], the stability domain in Fig. 4 is the area encompassed by the loading line before the peak load, the equilibrium curve and the displacement axis for all the possible loading programs. A loading program is globally stable if the representing point starting from

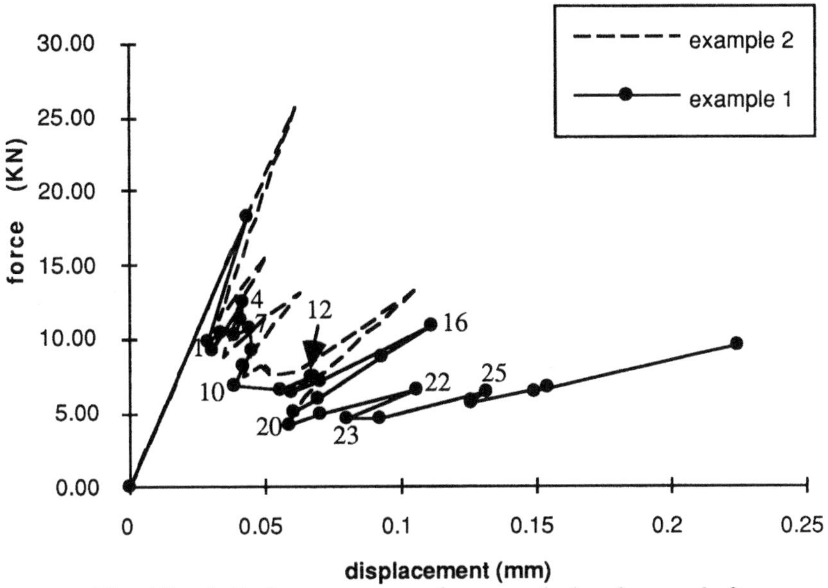

Fig. 4 Load-displacement curve for example 1 and example 2.

the equilibrium curve enters into the stability domain and remains inside it to the end. The local stability means that the representing point having entered the stability domain leaves it by crossing the equilibrium curve. Going out of the stability domain instead of running along the equilibrium curve may occur in the experiments and may be caused by dynamic effects in the material or inertia effects of the loading system. This should not be considered as a material's property.

The load-displacement curve shown in Fig. 6(a) for the case of homogeneous body with one crack provides a region just after peak load, within which no snap-back is observed. This corresponds to the fact that, in this region, the propagation of the crack is stable under fixed displacement and unstable under fixed force.

The documented crack trajectory presented in Fig. 2 for the heterogeneous microcracked body shows that the crack propagation process starts from the pre-existing internal microcrack nearest to the bottom corner of the bolt where the strongest stress concentration is. This microcrack propagates first backward through the matrix until it reaches the bottom corner of the bolt. This backward propagation is associated with a large decrease of the reaction force, which is reduced by almost one half of the peak load. The equilibrium curve exhibits snap-back, showing that this propagation is unstable both under constant displacement and constant force condition. For fixed displacement conditions, this is confirmed by Fig. 5, which shows that in this snap-back region, $\overline{K}_{max} / K_{IC}$ increases with the crack length a. This can be related to the fact that, for a homogeneous body under the condition of pure tension, an internal crack with two tips exhibits always unstable propagation once the critical load is reached.

From step 2 onwards, the equilibrium curves (Fig. 4) shows a series of indentations around a general trend. These indentations are similar to the ones observed in the pure tension case, Wang et al. [1992c]. From beginning to the step 20, this general trend is the decrease of the force with increasing displacement. From step 20 onwards, it is the increases for both force and displacement. The latter is consistent with the increase

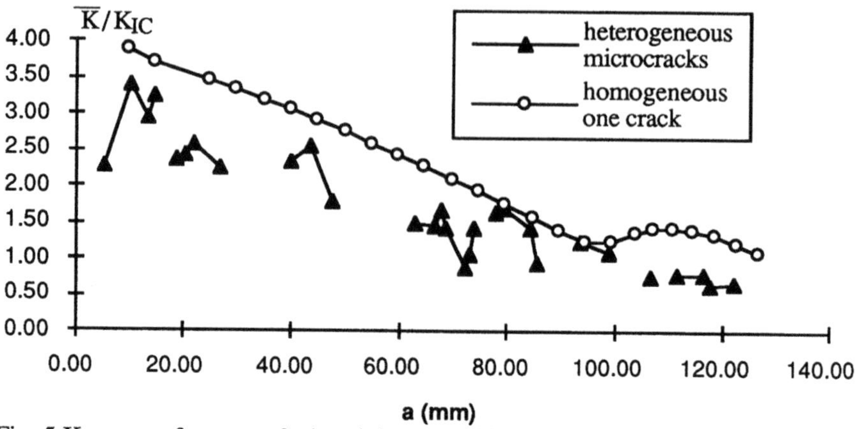

Fig. 5 K-a curve for example 1 and the case of homogeneous body with one crack

observed in Fig. 6(a) for the homogeneous body with one crack before the region under the support is reached and with the fact that compression parallel to the crack trajectory may be expected in this region.

It is appeared in Fig. 4 that an indentation has two positive slopes, indicating the existence of local instabilities even at fixed displacement. The first slope of the indentation is obtained when the crack tip approaches an inclusion where there is no pre-existing microcracks around that inclusion. A change in the direction of crack propagation occurs which is accompanied with a decrease of both the force and displacement. When the crack reaches a pre-existing microcrack around the inclusion, crack branching may occur, as in the steps 4-5 or 22-23.

As shown in Fig. 5 giving $\overline{K}_{max} / K_{IC}$ in terms of the crack length a, the general trend of negative slope observed both for the homogeneous body with one crack and for granular microcracked body. Nevertheless the indentations illustrated in

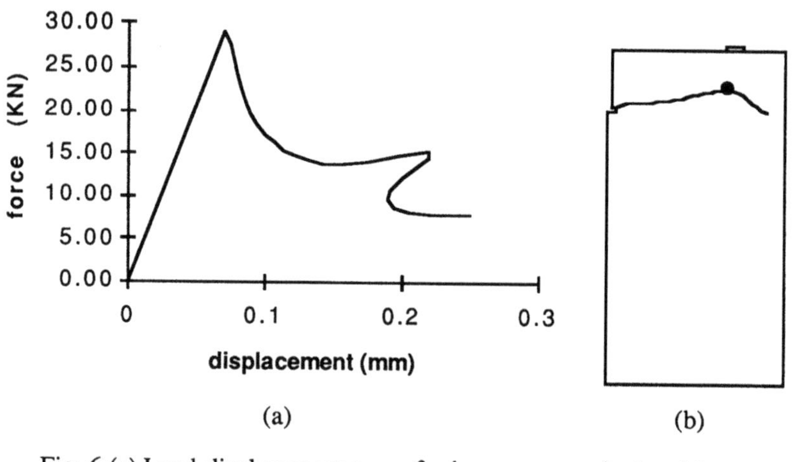

(a) (b)

Fig. 6 (a) Load-displacement curve for homogeneous body with one crack.
(b) Crack trajectory for homogeneous body with one crack.

Fig. 4 appear as a series of non-monotonic segments for heterogeneous case. A single non-monotonic segments is also found on the K-a curve for the homogeneous body with one crack. This corresponds to the part of the crack trajectory that goes down beyond the region under the support shown in Fig. 6(b), and to the snap-back region in the load-displacement curve of Fig. 6(a). In this region the crack propagation is unstable at both fixed displacement and fixed load. It becomes stable again when approaching the lateral edge of the specimen where free boundary conditions are imposed.

The unstable crack branch shown in Fig. 6(b) for the homogeneous body with one crack has disappeared for the granular microcracked body. This may be attributed to the fact that large inclusions are present in the region under the support and prevents this unstable branch of the crack to develop. In place of this unstable branch, there is a branching of the crack in two opposite directions that are almost perpendicular to the general direction of the crack trajectory before the branching. The upper branch corresponds still to a stable propagation, as appears in Fig. 5.

5. Comparison with experiments

It is interesting to compare out numerical results with the experimental ones obtained, in the framework of the same Round-Robin, by Helbling et al.[1991]. Fig. 7 presents, redrawn from the paper quoted above, the experimental crack trajectories obtained for 3 concrete specimens with the dimension 900 x 900 x 100 mm and with the position of the supports located at the upper edge of the specimen a=2d. The maximum size of aggregates is 16 mm.

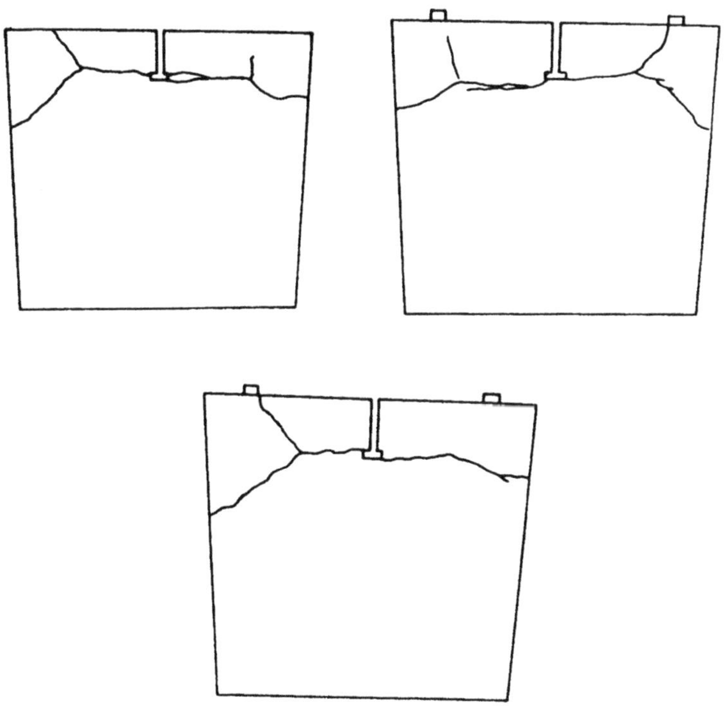

Fig. 7 Crack trajectories observed in experiments (from Helbling et al. [1991])

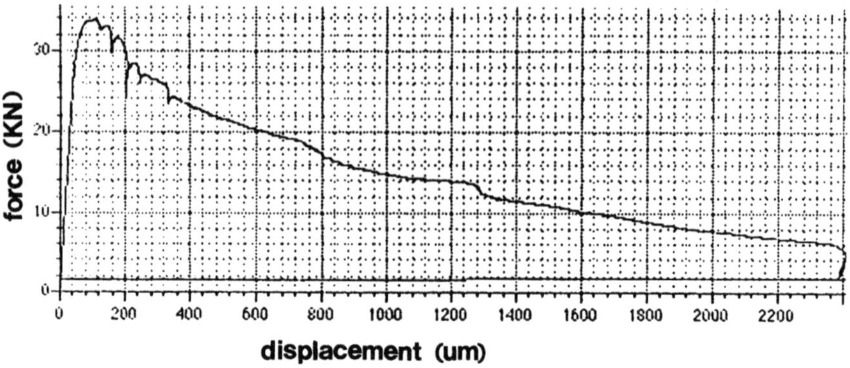

Fig. 8 One example of load-displacement curve obtained from experiments (from Helbling et al. [1991])

The numerical results discussed above have several features in common with the experimental ones: the crack propagates first in the horizontal direction; it branches when its tip enters the region below the support. A descending branch beyond the support exists in 5 cases over 6 presented in Fig. 7 (each case corresponding to one half of specimen). The upward crack branch corresponds to our numerical results for the heterogeneous material with microcracks. The other slightly downward branch is represented by the numerical results of the homogeneous material with only one crack. It has to be pointed out that the maximum aggregate size used in experiments is 16 mm and dimension parameter of the specimen d=150 mm, whereas in our numerical calculation the maximum diameter of the inclusion is 32 mm and d=50 mm. Thus the ratio between the maximum diameter of inclusion and specimen size is six times greater in our calculations than that in the experiments. Since the inclusions with larger size may be considered as more efficient crack arrestors than smaller ones, it is plausible that the numerical calculation for the heterogeneous material with microcracks might possibly provides two branches of crack trajectories observed in experiments if the specimen used in simulation were constructed with the inclusions of relatively smaller sizes.

It must be pointed out that, for the distance ratio a=d/2, the experimental crack trajectories form a triangular pattern, which is similar to the cone pattern obtained in the case of 3D. For the ratio a=d, intermediate features are obtained.

Some randomness is apparent, but the general features are quite stable for the case with supports far enough from the steel bar. This randomness is also observed in the numerical calculations shown by the comparison of example 1 with example 2 in the Fig. 3 and 4.

The experimental force-displacement curve obtained in the case of a=2d are presented on Fig. 8. They all exhibit snap-back and reloading regions in the beginning of the post-peak part. In these regions the sudden decrease of the load at almost constant displacement occurs. Then a new increase of the load is necessary in order to continue the crack propagation. These sudden decreases followed by increases of the load seemes to be the local instabilities and crack arrests corresponding to the indentations obtained in the numerical simulation, Fig. 4 and 5. For large enough displacements, the instabilities almost disappear and the experimental curve becomes very smooth and always descending.

But ascending trends of the load-displacement curve very similar to our numerical results have been found in the experiments performed by Ohlsson et al. [1991]. The corresponding crack patterns are very similar to that predicted by our homogeneous LEFM computation, shown in Fig. 6(b).

6. Conclusions

In this paper the numerical analysis of anchor bolts is performed with the finite element method. This numerical analysis is based on the application of LEFM at the micro-mechanics level and taking account of the granular structure of concrete and of the pre-existing microcracks. The results obtained gives interesting informations about the development of cracks in the concrete structures corresponding to the anchor bolts problem both from qualitative and quantitative viewpoints.

In particular it turns out that the influence of hard inclusions are not always on the same side, since the stabilizing effects of the inclusion are partly compensated by the local snap-backs. The same holds for the pre-existing microcracks. When there are no pre-existing edge cracks, the first stage of crack propagation is unstable, but of limited extent and followed by a temporary crack arrest that needs new increases of the load to continue crack propagation. Thus the behavior in this beginning of the post-peak region is general softening behavior with snap-back tempered by local hardening. This seems to be confirmed by the corresponding load-displacement curves obtained experimentally on several specimens.

From our study it is shown that such a numerical model, despite its limitation, is capable of giving realistic crack trajectories for 2D problems. It seems that this was not the case with other numerical models used in Round-Robin analysis of anchor bolts, see Elfgren [1991, 1992].

In particular, no numerical model based on macroscopic nonlinear fracture mechanics was able to provide the various crack branches obtained in the experiments. Further, no one was able to predict the unstable crack propagation occurring beyond the support.

The experimental results quoted here are thus clarified by our numerical calculations, since a clear distinction can be made between that parts of the trajectory that correspond to stable propagation and that to unstable one.

It is also confirmed that, despite the stabilizing effect provided by the inclusions in normal concrete, far running unstable propagation may occur in this material even at fixed displacement, when the geometry of the structure is appropriate. In addition, it turns out that LEFM applied to the homogeneous case is able to predict crack trajectories that occur in real structures and may not be predicted by the macroscopic nonlinear fracture mechanics.

Since the prediction of LEFM from this viewpoint are related to the structure itself and not to the material, it seems worthwhile to recommend that every structure analysis involving fracture mechanics in concrete structures begins with a LEFM calculation.

Of course, many improvements have still to be made to the numerical tool applied here. Some of them are presently under study. Nevertheless, the information that it can bring in its present state appear to be meaningful and useful. In particular, it is expected that it will help in clarifying some questions that remain open in the elaboration of realistic nonlinear models.

7. Acknowledgment

Financial support from Swiss National Science Foundation is gratefully acknowledged.

8. References

Z. P. Bazant (ed.) (1992), Fracture Mechanics of Concrete Structures, Elsevier.

L. Elfgren (ed.) (1991), Round-Robin Analysis of Anchor Bolts, Preliminary Report of RILEM TC-90 FMA Committee, Second Edition, Division of Structural Engineering, Lulea University, May.

L. Elfgren (1992), Round Robin Analysis and Tests of Anchor Bolts, in **Fracture Mechanics of Concrete Structure** (Ed. Z. P. Bazant), pp. 865-869.

F. Erdogan, G.C. Sih (1963), On the crack Extension in Plates under Plane Loading and Transverse Shear, **J. Basic Engng.**, 85, pp. 519 - 527.

C.E. Freese, D.M. Tracey (1976), The Natural Isoparametric Triangle versus Collapsed Quadrilateral for Elastic Crack Analysis, **Int. J. Fracture**, 12, pp. 767 - 770.

A. Helbling, A.M. Alvaredo and F.H. Wittmann (1991), Round Robin Test of Anchor Bolts, in **Preliminary Report of RILEM TC-90 FMA Committee** (Ed. L. Elfgren), Second Edition, Division of Structural Engineering, Lulea University.

C. Huet (1971a), Détermination expérimentale de l'énergie spécifique de rupture et du facteur intensité de contrainte critique de céramiques de construction et de divers roches naturelles, **C.R. Acad. Sc.**, Paris, t. 272, p. 1474-1477, juin.

C. Huet (1971b), Note sur la rupture fragile des maçonneries porteuses, **5e Congrès du CIB, Versailles**, II, p. 139-142.

C. Huet (1973), Méthode de détermination de l'énergie spécifique de rupture et application aux céramiques et à divers matériaux, I: Bases théoriques, **Cahiers du Groupe Français de Rhéologie**, novembre, t. 3, n° 3. pp.128-142.

C. Huet, (1974) Méthode de détermination de l'énergie spécifique de rupture et application aux céramiques et à divers matériaux, II: Conséquences pratiques et expérimentation , **Cahiers du Groupe Français de Rhéologie**, septembre, t. 3, n° 4, pp. 164-174.

C. Huet (1978), Concepts de la Mécanique de la Rupture, in **La Fragmentation**, Collège International des Sciences de la Construction, SEBTP, Paris, 12-14, décembre.

C. Huet (1982), Rupture fragile des matériaux, in **Mécanique de la rupture**, Collège International des Sciences de la Construction, SEBTP, Paris, 8-11 juin, pp.3-34.

A.R. Ingraffe, V. Saouma (1985), Numerical Modeling of Discrete Crack Propagation in Reinforced and Plain Concrete, in **Fracture Mechanics of Concrete** (Editors, G.C. Sih, A. Ditommaso), Martinus Nijhoff Publishers, 171-225.

U. Ohlsson, L. Elfgren (1991), Anchor Bolts in Concrete Structures: Two Dimensional Modelling, in **Analysis of Concrete Structure by Fracture Mechanics** (Editors, L. Elfgren, S.P. Shah), Chapman and Hall, p. 281-301.

J. Wang, P. Navi, C. Huet (1992a), Analysis of Anchor Bolts Based on Linear Elastic Fracture Mechanics And Finite Elements Method, will be published in **Round Robin Analysis and Tests of Anchor Bolts** (Eds., Elfgren et al).

J. Wang, P. Navi, C. Huet (1992b), Numerical Study of Granule Influences on the Crack Propagation in Concrete, in **Fracture Mechanics of Concrete Structure** (Ed. Z. P. Bazant), pp.373-378.

J. Wang, P. Navi, C. Huet (1992c), Application of Fracture Mechanics to the Study of Crack Propagation in Concrete Structures Using a Granular Microcracked Model, accepted for **International Conference on Fracture and Damage of Concrete and Rock**, Vienna, Austria, november 9-13, 1992.

Y. Zaitsev (1983), Crack propagation in a composite material, In Facture Mechanics of Concrete (Ed. F. H. Wittmann), Elsevier, pp. 251-300.

54 PULL-OUT OF STEEL ANCHORS EMBEDDED IN CONCRETE

A.H.J.M. VERVUURT, and J.G.M. van MIER
Stevin Laboratory, Delft University of Technology,
The Netherlands
E. SCHLANGEN
KFA, HRLZ, Jülich, Germany

Abstract
In this paper some preliminary results are presented of a numerical and experimental analysis of pull-out tests of anchor bolts. In the numerical model the material is schematized as a network of brittle breaking beams. The heterogeneity of the material can be implemented in three different manners, i.e. (1) by assigning random strength values to the various beam elements, (2) by generating an aggregate structure of the material and assigning according strengths to the beam elements or (3) by assigning a constant strength to the beam elements in a lattice with random beam lengths. Comparison with recent tests of anchor bolts embedded in rock by Fathy et al. (1992) shows that the lattice model is very capable in predicting crack growth in anchor plates with different boundary conditions.
Keywords: Fracture Mechanics, Anchor Bolts, Lattice Model, Experiments.

1 Introduction

In order to demonstrate the application of fracture mechanics in design problems, RILEM committee TC 90-FMA proposed a round-robin analysis on the pull-out behaviour of anchor bolts. It was decided to contribute to the round robin, both by carrying out experiments, as well as by performing analyses with the recently developed lattice model. The geometry used for both the simulations and experiments is based on the round robin proposal as it was published in Material and Structures, volume 23, January 1990 and shown in figure 1. In this study only the plane stress problem is simulated and tested. In table 1 the variation of the parameters is given. In stead of the proposed embedded depth of d=450 mm and a support span of a=2d, experiments with d=100 mm (a=2d) have been performed, because of the limitations of the experimental set-up.

As mentioned, for the simulations the lattice model has been used. The model has proven to give realistic predictions of crack growth in concrete specimens under various loading conditions, see for example Schlangen and Van Mier (1992a).

When writing this paper, the numerical simulations were completed, the experiments were just started and only a limited number of results are included in this paper. Additional results will be presented at the conference.

Fracture and Damage of Concrete and Rock – FDCR-2.
Edited by H.P. Rossmanith. © 1993 Taylor & Francis. ISBN 0 419 18470 8.

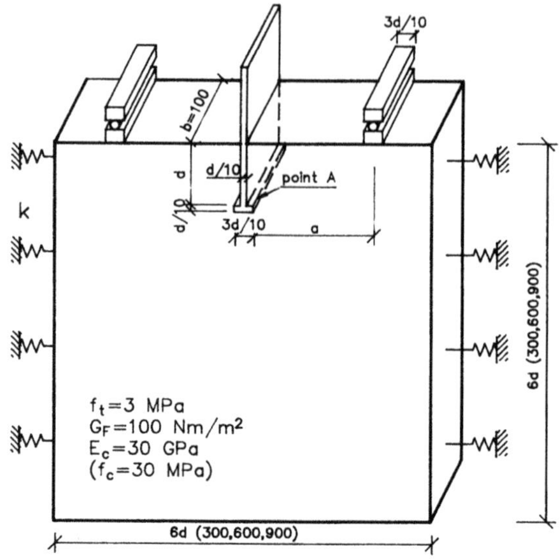

Fig.1. Geometry.

Table 1 Variation of parameters in the numerical simulations (o) and in both numerical simulations and experiments (+).

b=100 mm	k=0* d=50 mm	d=100 mm	d=150 mm	k=∞* d=150 mm
a=d/2	o		+	
a=d			+	+
a=2d	+	+	+	+

* k = lateral stiffness, see Fig.1.
all dimensions are explained in Fig.1.

2 Lattice Model

2.1 Basics

Lattice models, as described for example by Herrmann et al. (1989) seem to be a promising tool to predict crack patterns in heterogeneous materials like concrete. In the model the material is schematized as a lattice of brittle breaking beam elements. Only the area where cracks are expected to appear is modelled as a lattice. The remainder of the specimen is modelled using linear, plane stress elements available in DIANA. Fracture of the material is obtained by removing a beam out of the mesh when the stress σ reaches the tensile strength of the beam (σ_t). The stress σ is derived from:

$$\sigma = \beta * (F/A + \alpha * (|M_i|,|M_j|)_{max} / W)$$

where F is the normal force in the beam element, M_i and M_j are the bending moments in the nodes i and j in the beam, A=b*h is the cross sectional area of the beam and $W=b*h^2/6$. The coefficient α is introduced in order to select a failure mode where bending plays either a dominant or a restricted role. The coefficient ß is a scaling factor for the global stress which must be obtained from a simulation of a uniaxial tensile tèst. The forces F and M are obtained from a lineair elastic analysis with a prescribed external load of unity.

The heterogeneity of the material can be introduced by either varying the beam strength and stiffness (regular lattice) or through a variation of the beam length (random lattice).

2.2 Disorder: regular lattice

The first approach for introducing heterogeneity is to choose a regular lattice with different beam strengths σ_t and/or stiffnesses. The value of σ_t can be obtained by either choosing at random strength values, or by projecting the lattice on top of a generated grain structure and assigning different strength and stiffness to the beams at different locations in the grain structure (i.e. aggregate (A), bond (B) or matrix (M), see figure 2). For the distribution of the grain structure in the concrete a Fuller curve has been chosen. For more specific information see Schlangen and Van Mier (1991).

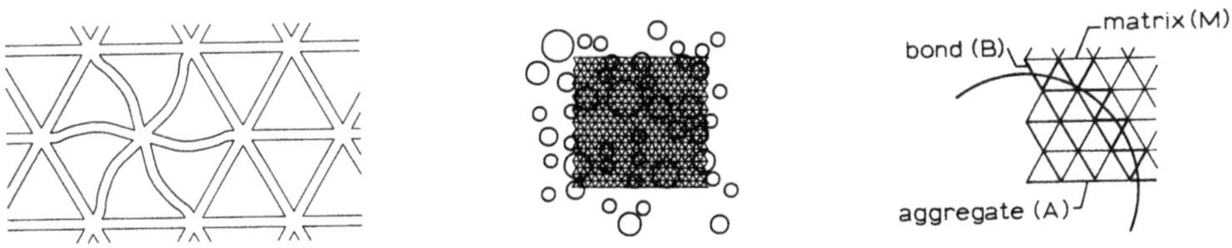

Fig.2. Regular lattice projected on a generated grain structure.

Simulations of displacement controlled uniaxial tensile tests on double-edge-notched (DEN) specimens ($50*50*150$ mm^3) are used to tune the input parameters (Schlangen and Van Mier (1992b)). The parameters shown in table 2 on the following page gave the best fit.

2.3 Disorder: random lattice

A disadvantage of using the regular lattice is the enormous amount of elements that have to be used for obtaining a realistic discription of the material. Since the size of the aggregate particles is defined, also the maximum length of the beam elements is fixed. To overcome this problem a random lattice (randomly choosen beam lengths) with a larger average beam length can be used. On the other hand, due to the larger elements in the random lattice, the amount of details in the crack patterns are considerably less in comparison with the regular lattice with the projected aggregate particles. In this paper a random lattice is chosen which has a regular square lattice (size of cells s*s mm^2) as starting point. In each cell a point is selected at random with uniform distribution. Subsequently the random lattice is defined by connecting always the three points which are closest to each other. The advantage of making the lattice in such a way is that it is vectorizable (see Mouzarkel and Herrmann (1992)). The parameters which are used for the random lattice are shown in table 2. Analyses with two different square cell sizes have been performed.

Table 2 Input parameters for simulations

Regular lattice	σ_A = 10 MPa E_A = 70 GPa α = 0.005 b = specimen thickness l = 5/3 mm	σ_M = 5 MPa E_M = 25 GPa ß = 2.0 h=0.68*l	σ_B = 1.25 MPa E_B = 25 GPa
Random lattice (s=5 mm)	σ_t = 1 MPa α = 0.005 size of cells in square lattice 5*5 mm^2 b = specimen thickness h = 4.0 mm	E = 30 GPa ß = 4.2	
Random lattice (s=10 mm)	σ_t = 1 MPa α = 0.005 size of cells in square lattice 10*10 mm^2 b = specimen thickness h = 7.3 mm	E = 30 GPa ß = 3.0	

3 Numerical Results

When modelling a specimen with an embedded depth of d=50 mm with a regular lattice already a large amount of beam elements is needed (i.e. over 20,000). Therefore the specimens have all been simulated with the random lattice. To compare the two models (regular and random) only the smallest specimen (d=50) has been modelled with both the regular and the random lattices (s=5 mm and s=10 mm). Figure 3 shows the applied load versus the displacement of the anchorhead (point A in figure 1).

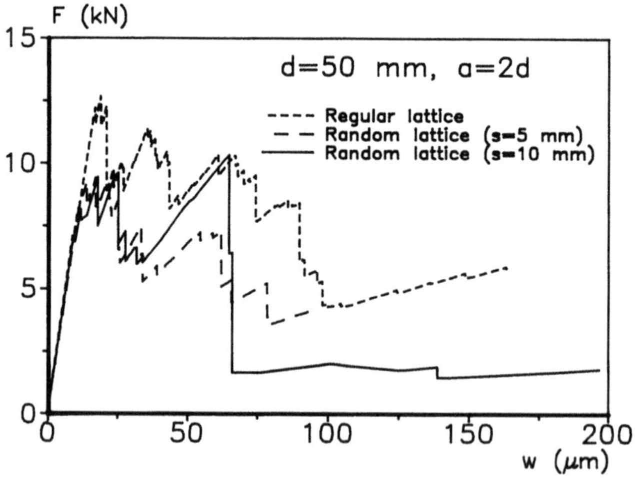

Fig.3. Load-displacement curves for three lattice models.

The fact that the random lattices with s=5 mm and s=10 mm are very much alike near the peak is a result of tuning ß for s=10 mm on basis of the specimen d=50 mm (a=2d) with s=5 mm because modelling of a tensile specimen (50*50*150 mm^3) with the random lattice with s=10 mm would be much to rough. This tuning showed also to be correct for a=d/2 (d=50 mm, see table 3 further on in this paper) which confirms the choösen value for ß. For simulating the specimen with an embedded depth of the anchorhead of 50 mm, a random lattice with s=5 mm seems to be satisfactoring size. The larger specimens (d=100 mm and d=150 mm) have only been simulated with s=10 mm.

The results of the analyses with total confinement (k=∞), showed that after an initial plateau in the load-displacement diagram the curve increased again. Such behaviour has never been observed experimentally. However as mentioned by Schlangen and Van Mier (1992a) this too stiff behaviour can likely be explained from the fact that in experiments infinite confinement cannot be reached. In addition to the two extreme confinements (k=0 and k=∞), the situation of the experiments is simulated. In the experiments confinement has been realized by connecting the two sides of the specimen with steel bars (6φ20, Fe360) fixed to stiff beams. In the analysis this has been simulated by connecting the two sides with beam elements corresponding to the bars in the tests. The load-displacement curves (F versus the displacement w of point A, see Figure 1) which are given in figure 4 show that there is a considerable influence of reducing the confinement (k=500 N/mm^2).

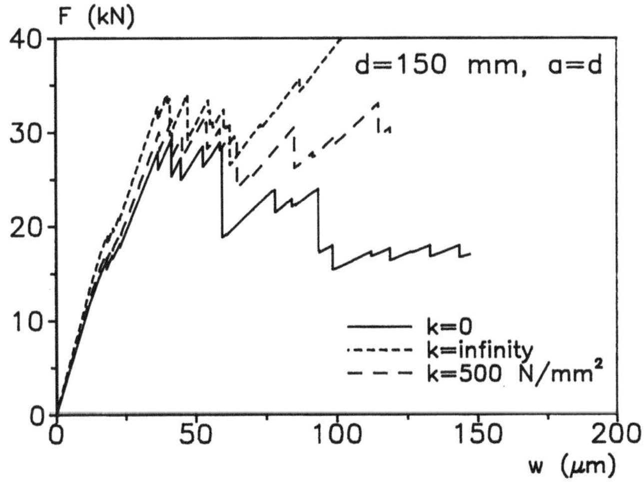

Fig.4. Load-deformation curves for free edges (k=0), total confinement (k=∞) and partial confinement (k=500 N/mm^2).

In figure 5 the final crack patterns with corresponding and the load-displacement curves are shown for three specimens of different size but with the same boundary conditions. Experiments on anchor bolts embedded in rock by Fathy et al. (1992) and simulations by Rossi and Wu (in RILEM TC 90-FMA (1991)) show comparable crackgrowth. This crack growth confirms that modelling only half of the specimen on basis of symmetry is absolutely not correct which of course must be expected as a consequence of the heterogeneity of the material.

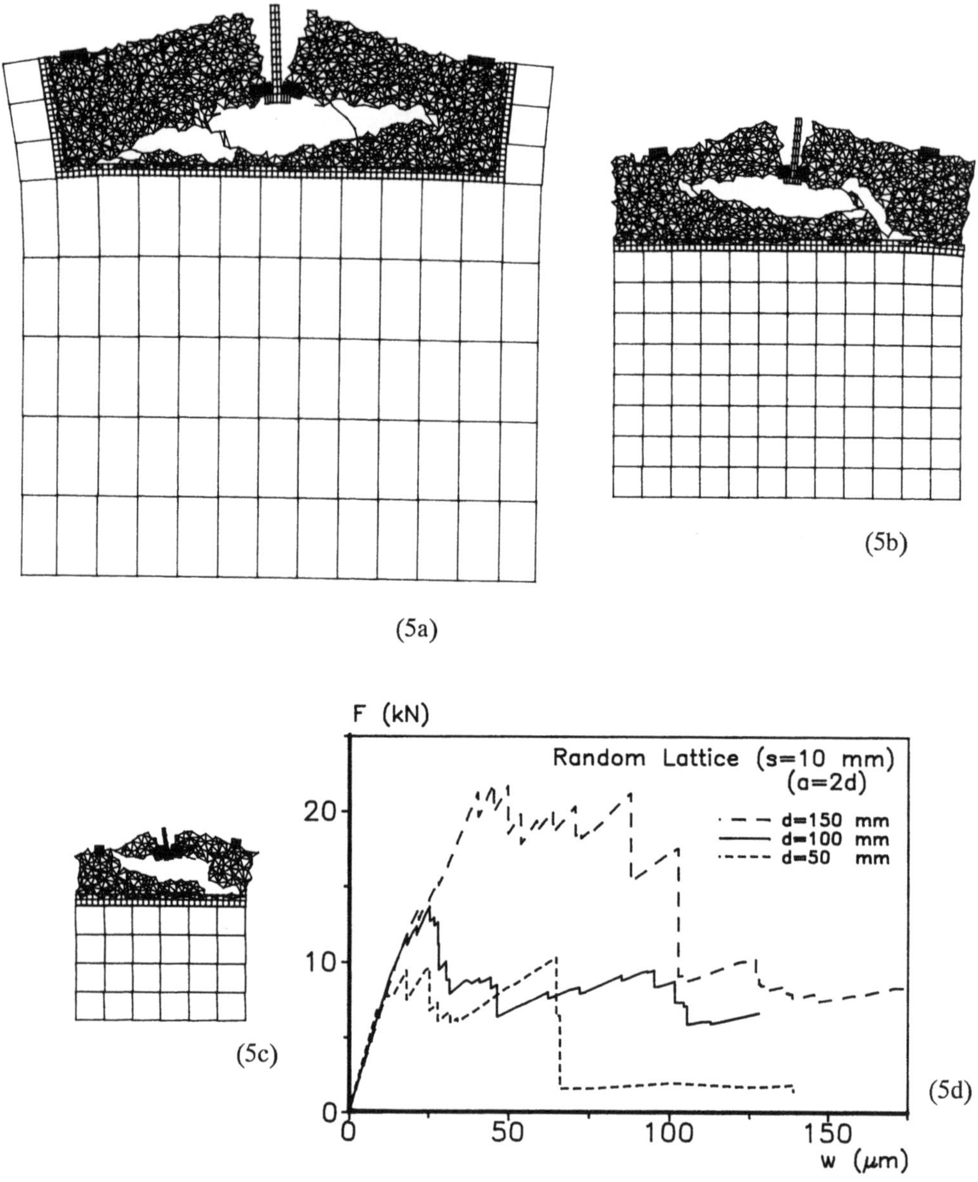

Fig.5. Final crack patterns for three different specimens (d=150 mm (5a), d=100 mm (5b) and d=50 mm (5c)) and corresponding load-displacement curves (5d) with equal boundary conditions (a=2d).

Table 3 contains a summary of the peak loads obtained in the numerical simulations.

Table 3 Summary of peak-loads of the numerical analyses

	k=0			k=∞	k=500 N/mm^2
d (mm)	50	100	150	150	150
	Regular lattice				
a=2d	12 kN				
	Random lattice s=5 mm				
a=d/2	29 kN				
a=2d	10 kN				
	Random lattice s=10 mm				
a=d/2	29 kN		50 kN		
a=d			29 kN	33 kN	31 kN
a=2d	10 kN	13 kN	21 kN	28 kN	24 kN

4 Experiments

4.1 Test set-up

The dimensions of the plates that are being tested are given in table 1 and an illustration of the test set-up is shown in figure 6. To minimize the influence of friction in the supports which occurs when they are loaded in compression, four double hinged tensile bars are used in order to support the specimen. Confinement is obtained by 6 tensile bars ϕ20 fixed to stiff steel beams. The plates are loaded under displacement control. The average signal of four LVDTs measuring the displacement w of the anchorhead (point A, left, right, front and rear in figure 1) is used as feed back signal in the servo control.

Details of the concrete mix that has been used for all tests are given in the appendix. In order to obtain the different plates 4 moulds were designed: two moulds measuring each 900*900*100 mm^3 (d=150 mm), one mould 600*600*100 mm^3 (d=100 mm) and one 300*300*100 mm^3 (d=50 mm). After two days of casting the specimens were demoulded, covered by plastic and cured in the laboratory until the time of testing. In order to obtain the compressive and splitting tensile strength 6 cubes (150 mm) were made and cured in the wet-room (100% RH) and tested after 28 days. Also six 150 mm cubes were made and cured under the same conditions as the specimens. These "dry" cubes were tested at the same day the first specimen of that particular batch was tested. All cubes were tested conform the Dutch codes. The compressive and splitting tensile strength of each batch for both the wet and dry cured cubes are included in the appendix.

4.2 Results

At the moment of preparing this paper, the tests were not completed, and only limited results can be presented. In order to give the most recent information table 4 is submitted with the results of a few tests. Figure 7 contains the crack patterns of these two experiments.

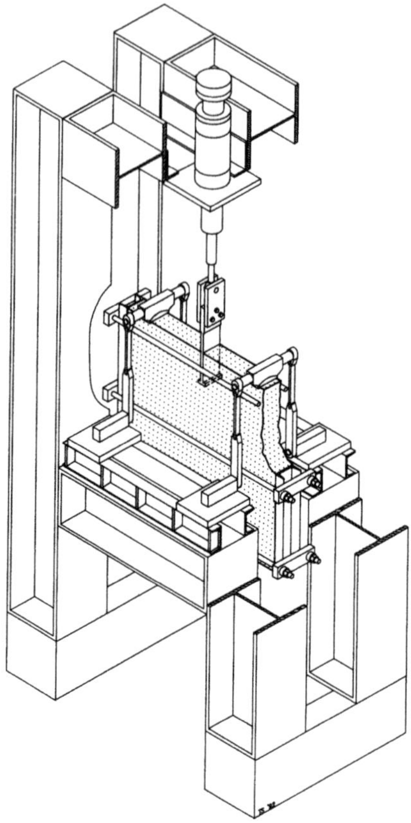

Fig.6. Test set-up.

Table 4 Maximum loads and corresponding displacements

Batch	Specimen No.	Geometry d [mm]	a	k [N/mm^2]	Age [days]	F_{max} [kN]	w^* at F_{max} [μm]
2	02k002	50	2d	0	28	14	37
2	02G003	150	2d	0	36	31	82

* w is the average displacement of point A, shown in figure 1.

Comparing the maximum loads according the simulations (table 3) and the experiments (table 4) it can be concluded that the experiments show a rather high peak-load. However the crack patterns simulated by the lattice model show a very good agreement to the experimental results.

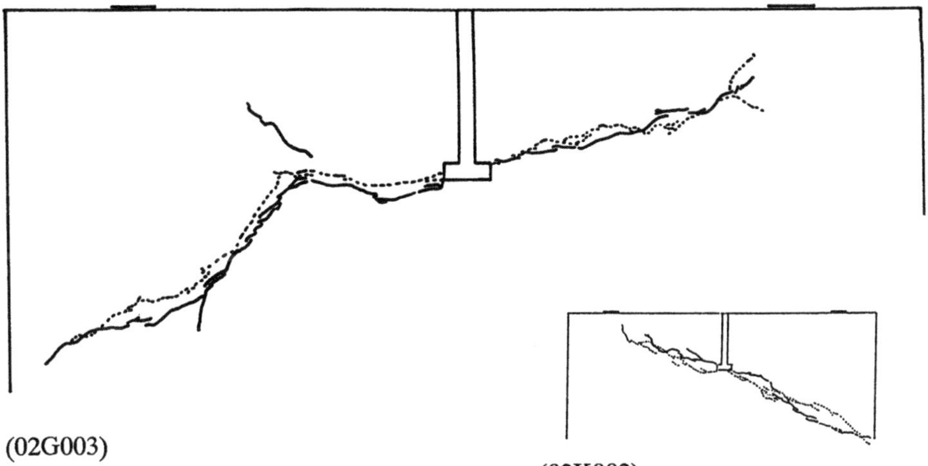

(02G003) (02K002)

Fig.7. Crack patterns of two experiments with support span of a=2d for d=50 (02K002) and d=150 (02G003). The solid and dashed lines represent the cracks at the front and rear face of the specimen respectively.

5 Conclusions

Lattice models are considered to be an excellent tool for studying crack growth in heterogeneous materials like concrete. A small number of parameters is needed, which easily can be obtained from standard tests. The linear elastic approach of the model makes it rather simple in comparison with non-linear models. With the lattice model fracture in different plate geometries can be simulated where as crack interface bridging, which causes a ductile load displacement response (Schlangen and Van Mier (1991)).

A disadvanage of the lattice model is the great amount of computer time, especially when the aggregative structure of the concrete is modelled. Partly this can be compensated by using a random lattice model which easily can be blown up by enlargening the square lattice size. However in that case the amount of detail in the crack geometry is greatly reduced.

The influence of the boundary conditions on the load-displacement curve is considerable for the confined and unconfined simulations. Failure tends to be through a single straight crack. At first this crack grows at both sides of the anchor bolt perpendicular to the loading axes. After the peak-load however one of these cracks bends towards the supports while the other one bends in the opposite direction. The failure mechanism is never conical which is in agreement with the present tests and the experiments by Fathy et al. (1992). Therefore extrapolation to 3D axi-symmetric pull-out tests is useless.

Furthermore it should be mentioned that tests show a less brittle behaviour than the simulations. This brittle behaviour can be reduced by modelling more small particles in the used grain structure or probably also by using a three dimensional mesh. Due to the computional limitation this is not considered possible yet.

6 Acknowledgement

Financial support from the Dutch Technology Foundation (STW) is gratefully acknowledged. The authors are greatly indebted to Mr. A.S. Elgersma for his assistance in carrying out the experiments.

7 References

Ballarini, R., Shah, S.P. and Keer, L.M. (1986), Failure characteristics of short anchor bolts embedded in a brittle material, *Proc. R. Soc. Lond.* A 404, pp 35-54.

Fathy, A., Planas, J., Elices y, M. and Guinea, G.V. (1992), Fracture of anchors in rock, *Anales de Mechanica de la Fractura*, vol. 9, pp 69-74.

Herrmann, H.J., Hansen, A. and Roux, S. (1989), Fracture of disordered, elastic lattices in two dimensions, *Physical Review*, vol. 39, no. 1, January 1, pp 637-648.

Moukarzel, C. and Herrmann, H.J. (1992), A vectorizable lattice. *Report no. HLRZ 1/92* (preprint).

RILEM TC 90-FMA (1991), Round Robin analysis of anchor bolts in fracture mechanics of concrete applications (ed Elfgren), Division of Structural Engineering, Luleå University of Technology, Sweden.

Schlangen, E. and Van Mier, J.G.M. (1991) Lattice model for numerical simulation of concrete fracture, *in Proceedings Int'l conference on DAM fracture* (eds. Saouma, Dungar and Morris), Boulder, Colorado, sept 11-13, pp 511-527.

Schlangen, E. and Van Mier, J.G.M. (1992a), Fracture modelling of granular materials, in Proceeding Material Research Society Spring Meeting, Symposium 'W' Computional Methods in Material science, MRS Pittsburg (in press).

Schlangen, E. and Van Mier, J.G.M. (1992b), A new lattice model for predicting localized fracture in concrete, *in Localized Damage II, Vol I: Fatigue and Fracture Mechanics* (eds. Aliabadi, Cartwright and Nisitani), Computional Mechanics Publications Elseviers Applied Science, London/New York, pp 425-438.

8 Appendix

CONCRETE MIX PROPORTIONS:
Portland cement Type B, 375 kg/m^3; w/c-ratio=0.5; 0.125-0.25 mm, 127 kg/m^3; 0.25-0.5 mm, 234 kg/m^3; 0.5-1 mm, 272 kg/m^3; 1-2 mm, 272 kg/m^3; 2-4 mm, 363 kg/m^3; 4-8 mm, 540 kg/m^3.
STRENGTH RESULTS:
The average 28 days (RH=100%) compressive strength f_{cc}=48.30 N/mm^2 (st.dev. 1.18 N/mm^2) and splitting tensile strength $f_{c,spl}$=3.42 N/mm^2 (st.dev. 0.11 N/mm^2). The other cubes (RH=70%) were also tested after 28 days and had a compressive strength of f_{cc}=47.67 N/mm^2 (st.dev. 1.18 N/mm^2) and a splitting tensile strength of $f_{c,spl}$=3.34 N/mm^2 (st.dev. 0.11 N/mm^2).

55 TESTS OF THE LOAD-BEARING CAPACITY OF ANCHOR BOLTS

U. OHLSSON and L. ELFGREN
Department of Civil Engineering, Luleå University of Technology, Sweden

Abstract
In this paper results are presented from tests of the load-bearing capacity of anchor bolts. The tests were conducted according to the RILEM Round Robin Analysis of Anchor Bolts.

1. Introduction

Anchor bolts are important structural details. Design loads are mostly determined by testing or by simplified empirical formulas. It is however possible to analyse the failure with fracture mechanics methods.

Discussions in RILEM Technical Committee TC-90 FMA (Fracture Mechanics of Concrete-Applications) led to an invitation to a Round Robin Analysis of Anchor Bolts in 1989/90, RILEM (1990). Sixteen contributions were submitted with quite a large scatter. It was then decided to give more precise rules for what to be calculated and what to be presented. It was also decided to invite laboratory tests of the proposed loading cases. A second invitation was issued, RILEM (1991), figure 1.

Several new contributions were now submitted. The results are very interesting and give a possibility to compare different methods of analysis and actual test results. It was therefore decided to publish the contributions in a RILEM Report, Elfgren et al (1992).

In this paper results are presented from tests of anchor bolts according to the RILEM Round Robin program. Other anchor bolt tests are presented in Elfgren and Ohlsson (1992) and Eligehausen et al (1992).

Fracture and Damage of Concrete and Rock – FDCR-2.
Edited by H.P. Rossmanith. © 1993 Taylor & Francis. ISBN 0 419 18470 8.

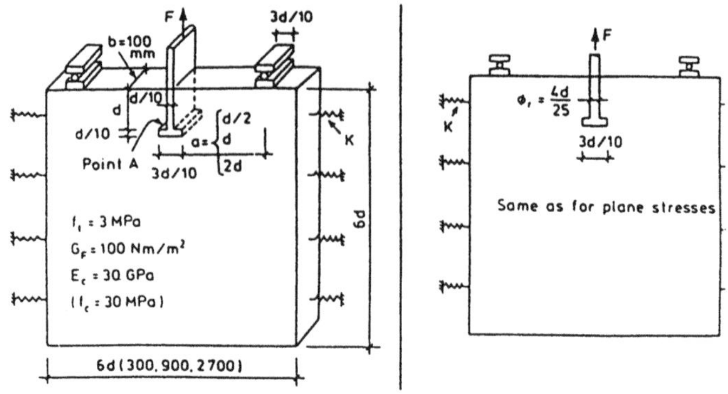

Fig. 1. RILEM Round Robin Analysis and Tests of Anchor Bolts. Proposed geometries for the plain stress and the axisymmetric specimens.

2. Materials and Test set-up

The two test specimens, shown in figure 2, were cast in steel moulds. The geometry of the tests was with small modifications according to the RILEM TC 90-FMA Round Robin Analysis and Tests of Anchor Bolts.

The steel bolts had a diameter of 24 mm. The washer had a thickness of 15 mm and a diameter of 45 mm. The steel quality was 8.8. The bolt and the washer, excecpt on the upper surface of the washer, were treated with form oil before casting, in order to prevent friction between concrete and steel.

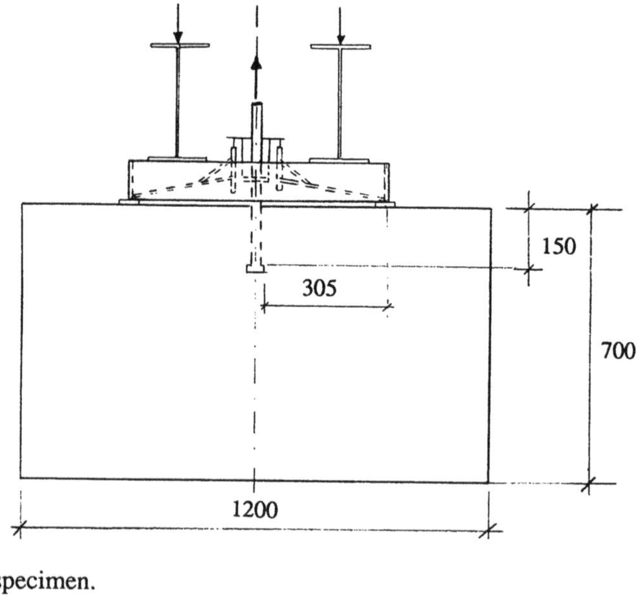

Fig. 2. Test specimen.

The concrete was made of Portland cement and natural aggregates with the following proportions: Cement: 237 kg/m^3, Water: 177 kg/m^3, Aggregate (0-8) mm: 1064 kg/m^3, Aggregate (8-12) mm: 934 kg/m^3, Micro silica: 13 kg/m^3.

Compressive and tensile strengths were tested on cylinders $\phi = 94$ mm and h=94 mm. The compressive strength was 48.2 MPa (mean value of two tests) and the tensile strength 3.7 MPa (mean value of three tests). The age of the concrete was 12 months.

The fracture energy of the concrete was tested with the RILEM three point bending test on concrete beams and was determined to 150.6 Nm/m^2 (mean value of five tests).

The tests were performed in a servohydraulic testing machine with a constant piston velocity. The test set up is shown in figure 3. A steel ring, $\phi_{outer} = 667$ mm, $\phi_{inner} = 643$ mm and h = 100 mm was used as a support on top of the concrete. Below the steel ring, pieces of sheet metal, 130 · 50 · 10 mm^3, were welded to the steel ring, figure 5. Between the test specimen and the pieces of sheet metal was a thin concrete grout. Two stiff steel beams were placed on top of the steel ring, anchoring the test specimen to the floor. Specimen No 1 was tested in the steel mould, made of sheet metal, giving a little extra radial confinement. Specimen No 2 was tested without the steel mould.

Deformations were measured with six LVDT displacement transducers. Three transducers, LVDT A, B and C, measured the vertical displacement of the steel bolt, relative to the supporting steel ring in three directions along the circular steel ring, figure 4. The measuring point on the steel bolt was 70 mm above the concrete surface. Three other transducers, LVDT D, E and F, measured the vertical displacement of the steel ring relative to a point outside the steel ring on the top surface of the concrete, figure 5.

Fig. 3. Test set-up.

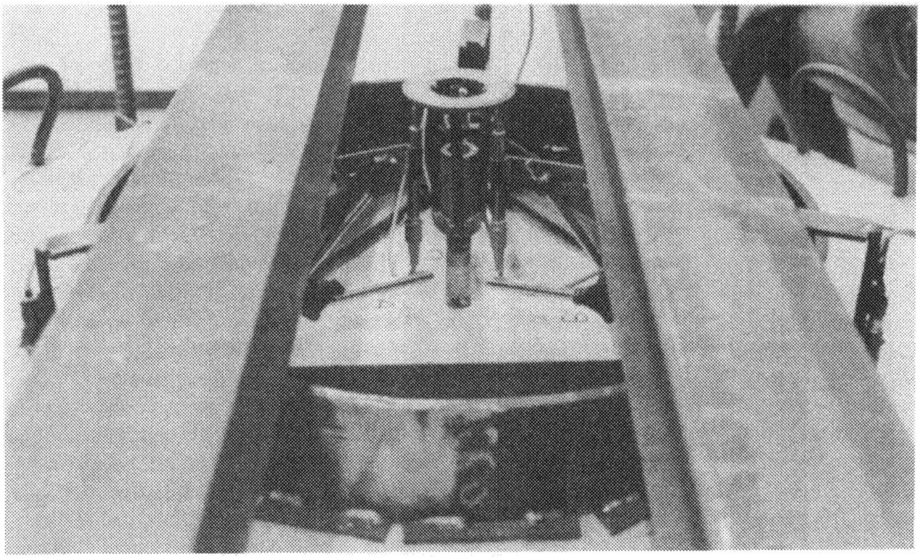

Fig. 4. Deformation measurement, transducer A, B and C.

Fig. 5. Deformation measurement, transducer D.

3. Test results

The load versus deformation curves from the two tests are shown in figures 6-7.

The peak loads in test No 1 and No 2 were 269.1 and 283.0 kN. In test No 1, the range of the displacement transducers was too small to be able to obtain the complete curve. The last part of the curve is therefore vertical.

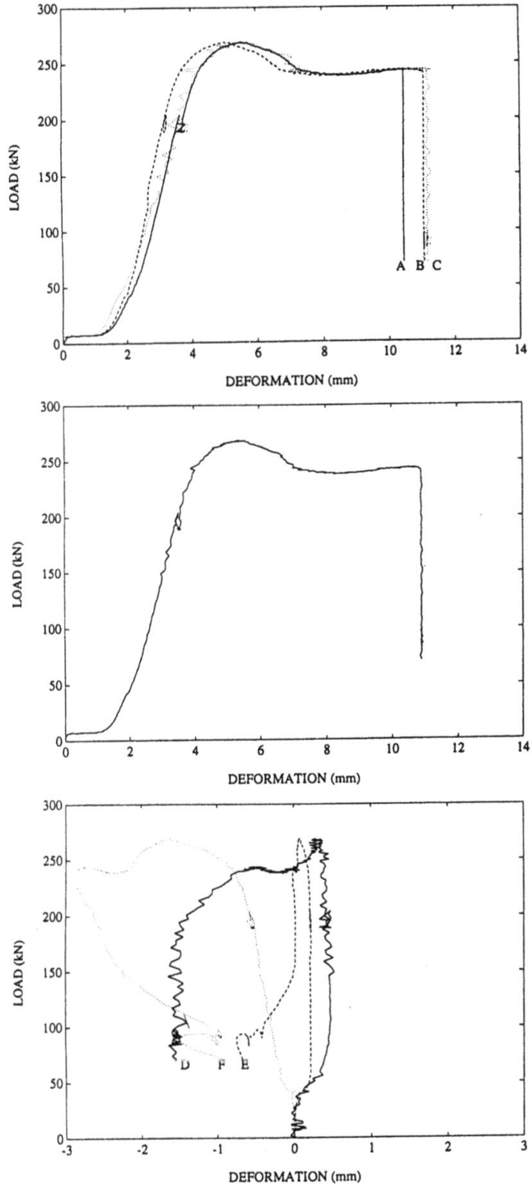

(a)

(b)

(c)

Fig. 6. Test No 1.
(a) Load versus displacement curves for transducers A,B and C showing the anchor bolt displacement relative to the supporting steel ring
(b) Load versus displacement curve. Displacement is mean value of transducers A, B and C.
(c) Load versus displacement curves for transducers D, E, and F showing the vertical displacement of the steel ring relative to a point outside the steel ring on the top surface of the concrete.

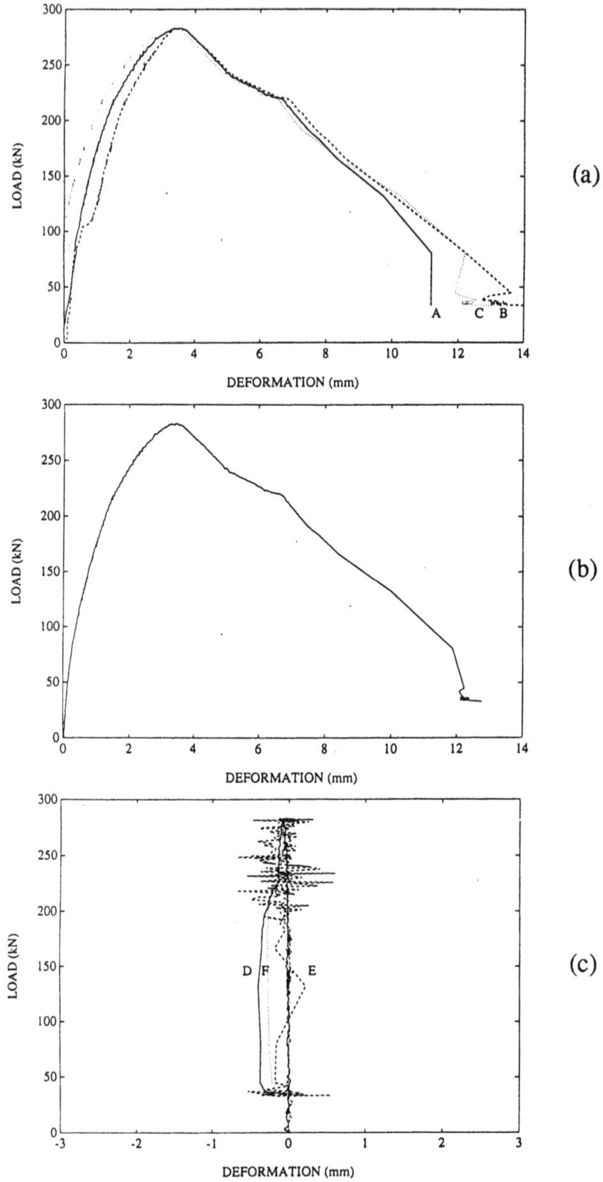

Fig. 7. Test No 2.
(a) Load versus displacement curves for transducers A,B and C showing the anchor bolt displacement relative to the supporting steel ring.
(b) Load versus displacement curve. Displacement is mean value of transducers A, B and C.
(c) Load versus displacement curves for transducers D, E, and F showing the vertical displacement of the steel ring relative to a point outside the steel ring on the top surface of the concrete.

The crack pattern on top of the concrete at the end of the tests are shown in figures 8 and 9.

Radial cracking occurred in both tests. The first radial cracks became visible just after the peak load was reached. In test No 1 the radial cracks grew from the centre all the way out to the edge of the specimen. In test No 2 the radial cracks stopped at the supporting steel ring where a ring of tangential cracks was formed due to local crushing of the concrete.

After the test when the crack pattern had been registered and photographs had been taken the specimen No 2 was reloaded and a concrete cone was pulled up. The concrete cone is shown in figure 10.

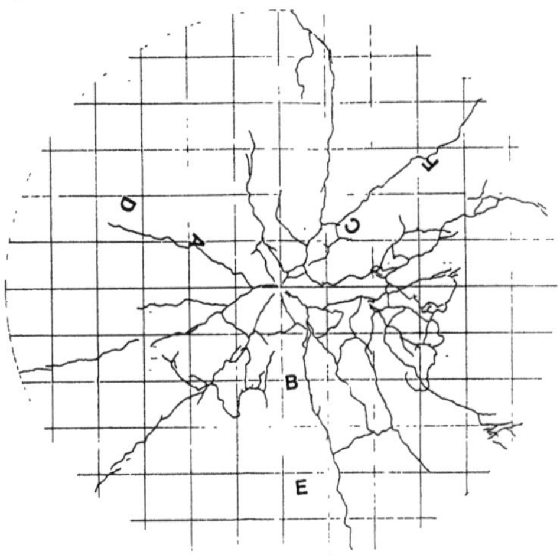

Fig. 8. Test No 1. Crack pattern on top surface.

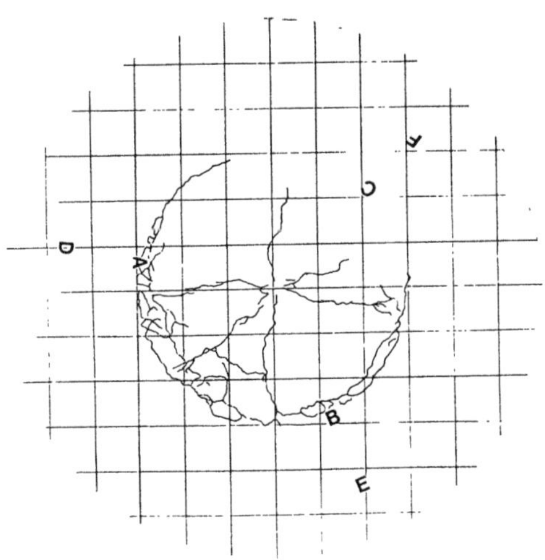

Fig. 9.Test No 2. Crack pattern on top surface.

Fig. 10. Test No 2. Concrete cone.

4. Conclusions

The tests indicates that the failure involves the formation of a concrete cone as well as radial cracking. The failure load is higher than in tests presented by Eligehausen et. al. (1992). The difference can be due to different boundary conditions but should be further investigated.

5. References

Elfgren, L. Eligehausen, R. and Rots J., editors (1992) **Round Robin Analysis and Tests of Anchor bolts**. RILEM Report to be published (1992).

Elfgren, L. and Ohlsson, U. (1992) Anchor Bolts Modelled with Fracture Mechanics. In **Applications of Fracture Mechanics to Concrete Structures.** Edited by A. Carpinteri. Proceedings from the International Workshop on the Applications of Fracture Mechanics to Reinforced Concrete in Torino, Italy, October 6 1990. Elsevier, London, pp. 267-283.

Eligehausen, R. Bouska, P. Cervenka, V. and Pukl, R. (1992) Size Effect of the Concrete Cone Failure Load of Anchor Bolts. In **Fracture Mechanics of Concrete Structures,** edited by Z. P. Bazant. Proceedings from the conference FraMCoS1, held at Beaver Run Resort, Breckenridge, Colorado, USA, June 1-5 1992, Elsevier, London, pp. 517-525.

RILEM (1990) RILEM TC 90 - FMA Fracture Mechanics of Concrete- Applications: Round Robin Analysis of Anchor Bolts - Invitation. In **Materials and Structures**, Vol. 23, January 1990, p. 78.

RILEM (1991) RILEM TC 90 - FMA Fracture Mechanics of Concrete - Applications Round Robin Analysis and Tests of Anchor Bolts - Invitation. Supplement to **RILEM News** 1991:1, 1 p.

56 TESTING OF ANCHOR BOLTS IN PLANE STRESS AND AXISYMMETRIC GEOMETRIES

B.L. KARIHALOO
School of Civil and Mining Engineering, The University of Sydney, Australia

Abstract
This paper describes the results of a series of pull-out tests on bolts anchored in concrete blocks approximating plane stress and axisymmetric conditions. The geometrical and other parameters of the blocks conformed to the requirements of RILEM TC 90-FMA Round Robin testing programme. Tests were conducted without confining the blocks (i.e. $K = 0$). The overall dimensions of the test specimens and the depth of embedment of the anchor bolt were not varied. Likewise, as far as practicable for a given specimen geometry the mix properties were not altered. Thus, the only variable was the distance between the anchor bolt and the support bearing.
Keywords: Anchor Bolts, Pull-out Tests, Plane Stress Geometry, Axisymmetric Geometry, Unconfined Concrete Blocks

1 Introduction

Test data on pull-out of anchor bolts from concrete blocks are scarce. Therefore, following an invitation from the RILEM TC 90-FMA (Fracture Mechanics of Concrete - Applications), a series of pull-out tests were conducted on bolts anchored in concrete blocks approximating plane stress and axisymmetric conditions. The geometrical and other parameters of the concrete blocks conformed very nearly to the values suggested by TC 90-FMA (1990). The tests were conducted without confining the specimens laterally (i.e. $K = 0$). The overall dimensions of the test specimens and the depth of embedment of the anchor bolt (d) were not varied (Figs 1,2). Likewise, as far as practicable all plane stress specimens were made from the same mix, and all axisymmetric specimens were also made from the same, albeit a weaker mix. Thus, the only variable in the series of tests was the distance between the anchor bolt and the support bearing (a in Figs 1,2). The dimensions of the available moulds were such that only three a/d ($=\frac{1}{2}$, 1, 2) ratios were possible in the plane stress configuration and just two ($=\frac{1}{2}$, 1) in the axisymmetric configuration.

The anchor bolts for plane stress specimens were made from mild steel with a

Fracture and Damage of Concrete and Rock – FDCR-2.
Edited by H.P. Rossmanith. ISBN 0 419 18470 8.

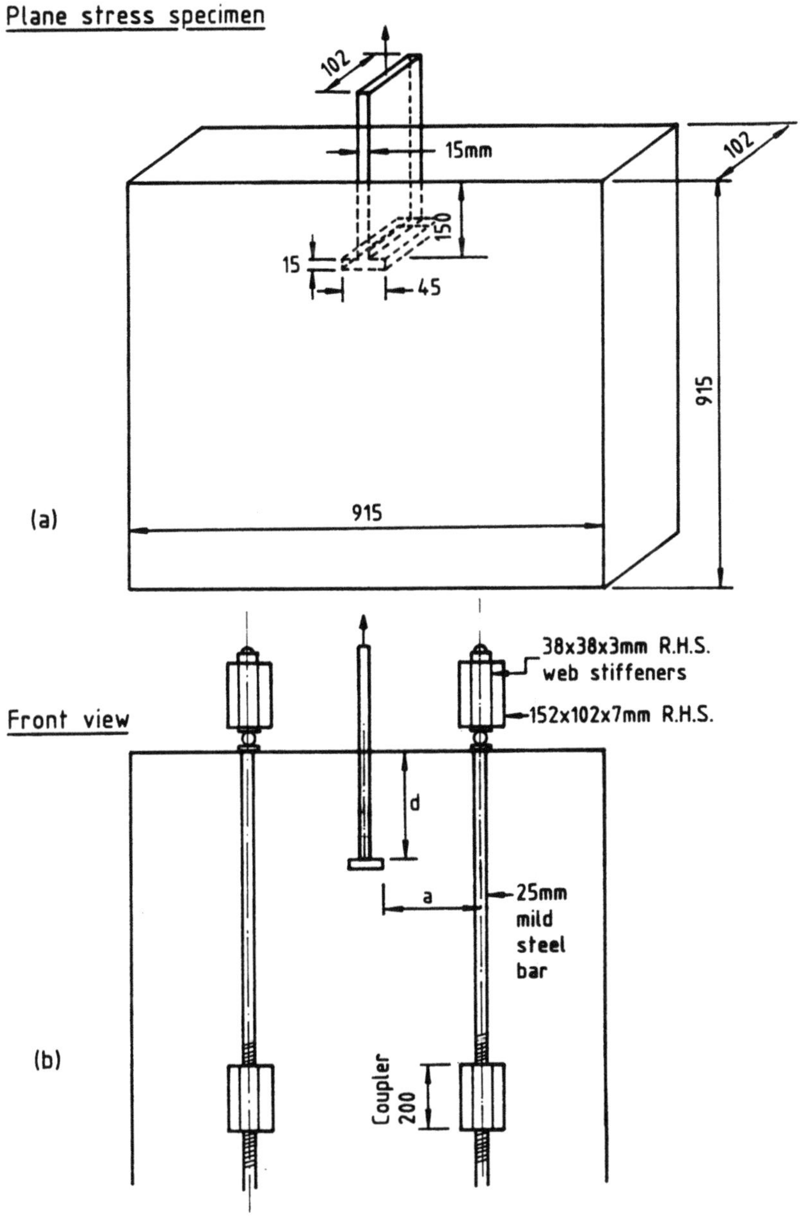

Fig. 1. Plane stress specimen, showing details of reaction frame. The same reaction frame was used for all three values of the ratio a/d.

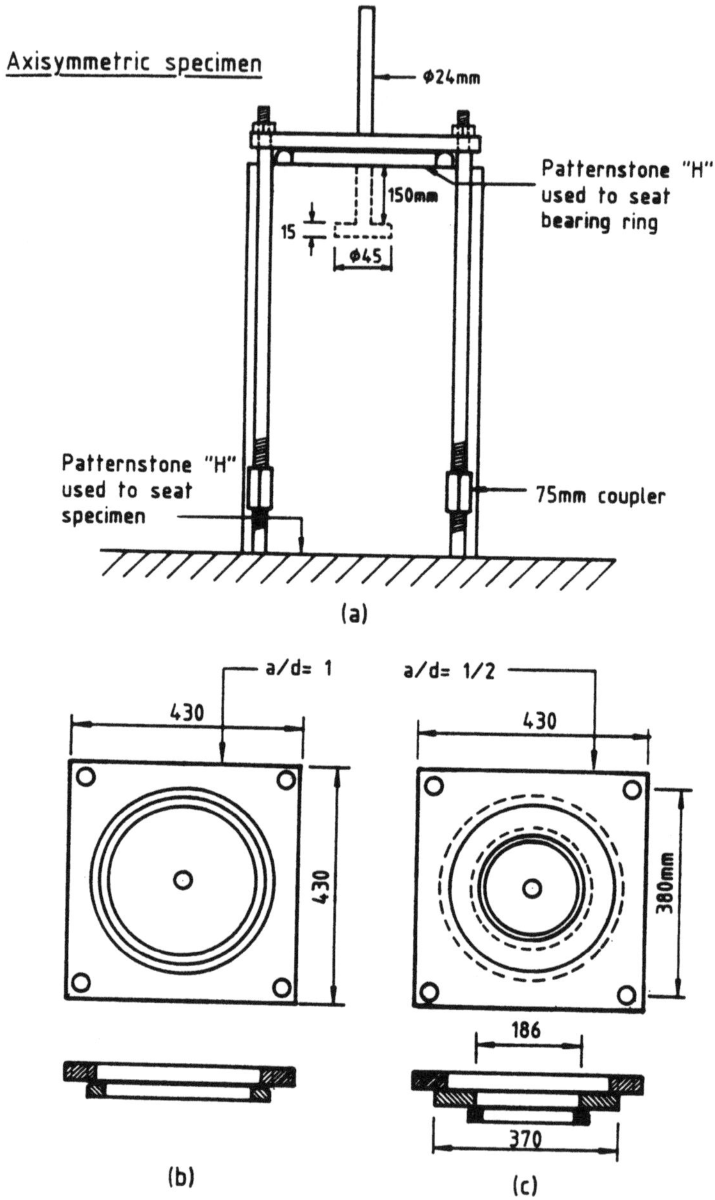

Fig. 2. Axisymmetric specimen, showing details of reaction frame. The transition ring in (c) was used to create $a/d=\frac{1}{2}$ from the bearing ring (b) for $a/d=1$.

yield strength of about 250 MPa, whereas those for axisymmetric specimens were made from high strength steel with a yield strength in excess of 1000 MPa. An earlier attempt at using mild steel anchor bolts for axisymmetric specimens cast from higher strength mix (the same mix as used for making plane stress specimens) proved unsuccessful, in that the anchor bolt failed before being pulled out from the specimen.

2 Mix and specimen preparation

As mentioned above, the mixes used in the two types of specimens were different. Plane stress specimens were made from concrete with the following mix proportions (kg/m^3): 20mm Basalt = 1102, 5mm Basalt = 424, Sydney Sand = 323, Type A Cement = 300, Water = 205, whereas the mix proportions used in the axisymmetric specimens were (kg/m^3): 20mm Basalt = 1052, 5mm Basalt = 445, Sydney Sand = 340, Type A Cement = 281, Water = 215.

The 28-day mean cylinder strength of the mix used in the plane stress specimens was 42.9 MPa (standard deviation 8.4 MPa), whereas the 7-day mean cylinder strength of the mix used in the axisymmetric specimens was 22.6 MPa (standard deviation 4.2 MPa). The indirect tensile strength (28-day) of the first mix was 3.6 MPa (standard deviation 0.3 MPa), whereas that of the second mix was not measured.

All specimens were cast in steel moulds with the anchor embedded in the top surface to the required depth. The concrete was poured into the moulds and compacted using an immersion vibrator. In the case of the axisymmetric specimens, the anchor was held in position by a special locator made to hold it vertical and central in the specimen as the concrete was placed, compacted and finished. The anchor in the plane stress specimen was in contact with the sides of the mould so that it was held in position by simple clamps and checked by a rule and a level. The specimens were stripped after twenty-four hours and placed inside a fog chamber for curing. Cylinders used to determine the compressive strength were also cast from each mix and cured together with the specimens.

3 Test procedure

The plane stress specimens were tested in a servo-hydraulic machine (Dartec, 2000 kN) on the twenty-eighth day of curing, whereas the axisymmetric specimens were tested on the seventh day. This machine is very stiff and therefore unlikely to influence the plot of load-crosshead movement. The special reaction frame arrangements required for testing plane stress and axisymmetric specimens are shown in Figs 1 and 2.

Prior to the application of load, the specimen was centred with respect to the loading grips and levelled. This was achieved by lifting the specimen off the base of the testing machine by means of the anchor, pouring "patternstone" (a variety

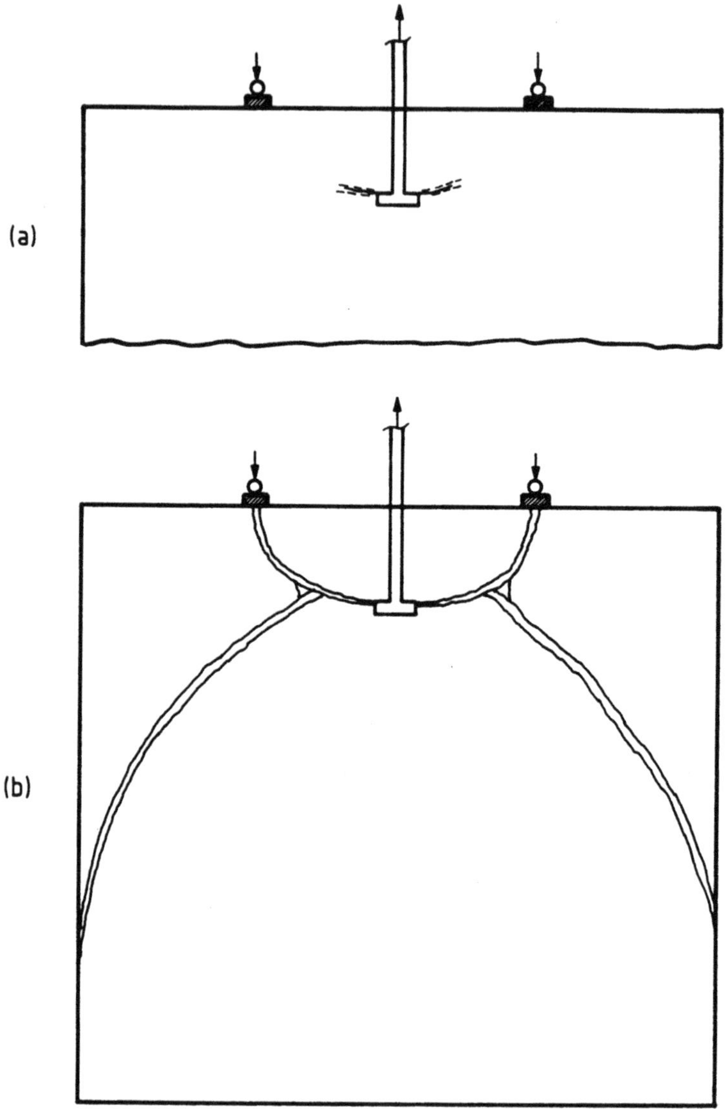

Fig. 3. Typical crack patterns at (a) peak load; (b) termination.

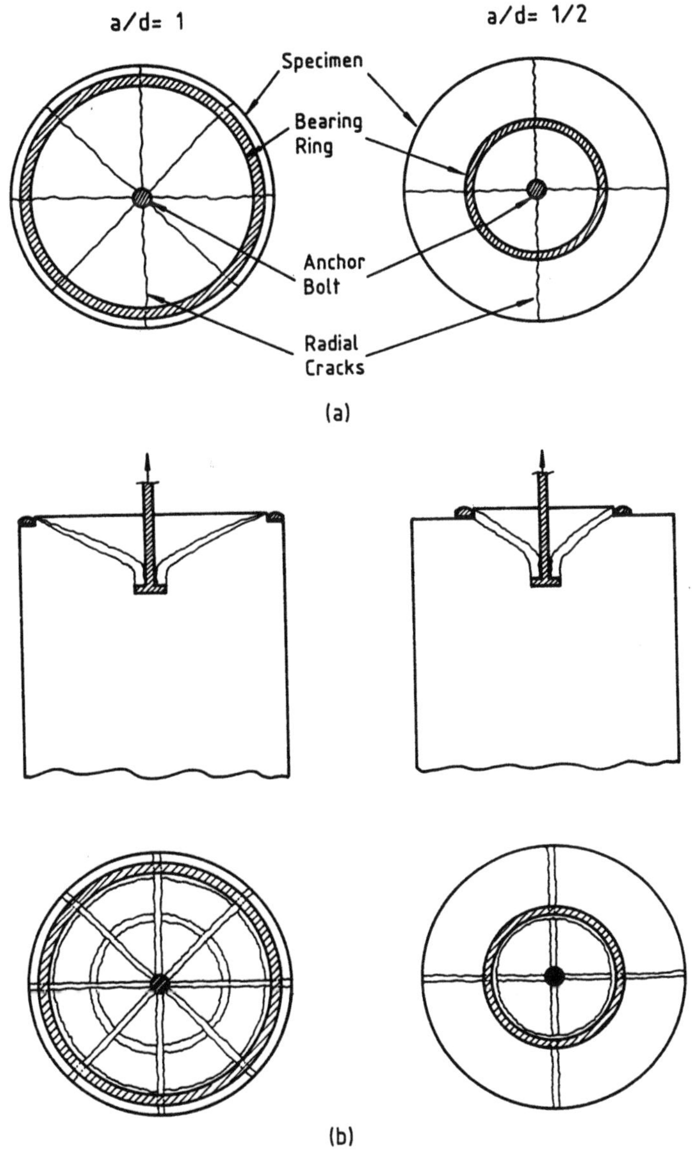

Fig. 4. Typical crack patterns at (a) peak load; (b) termination.

of gypsum plaster) under the specimen and then gently lowering it into the patternstone puddles. The reaction bearings (bearing ring in the case of axisymmetric geometry) were likewise levelled using patternstone. The reaction frame was firmly bolted down in grooves running in the base of the machine.

After the patternstone had set, the anchor bolt was gripped in the machine. Load was applied gradually (cross-head speed 0.014mm/sec for plane stress specimens and 0.025mm/sec for axisymmetric), and a plot of the load against crosshead movement was recorded on an XY-plotter. After the attainment of peak load, the crosshead speed was increased to about 0.05 mm/sec (0.1mm/sec in the case of axisymmetric specimens). The test was terminated when the load dropped to about 5-10% of the peak value.

4 Results and discussion

The mean value and standard deviation (sd) of the peak load and of the crosshead movement at peak load are summarized in Tables 1 and 2 for the plane stress and axisymmetric geometries. Typical crack patterns at peak load and on termination of test are illustrated in Figs 3 and 4. The measured peak loads compare very favourably with the numerical results obtained by Leonard, Rots, de Borst and Feenstra using the DIANA code (1991). To see this, the test results from Table 1 should be compared with their Table 2.3, and the entry corresponding to $a/d = 1$ from Table 2 with their Table 3.4 ($K = 0$). Likewise, it appears that the numerical code is able to predict reasonably accurately the observed crack patterns, especially at the peak load.

In the plane stress specimens the peak load was associated with debonding of the bolt head from concrete and the appearance of cracks at its top edges. These cracks subsequently propagated towards the reaction bearings before bifurcating. One branch continued towards the bearing whereas the second propagated in the opposite direction towards the free surface. The same general cracking pattern was observed for all three values of a/d. It is worth noting that internal cracking was audible even before the attainment of peak load. This kind of internal damage manifests through a nonlinearity in the load-displacement diagram well before the attainment of the peak load. Fig 5 show the load-displacement diagrams of each specimen for two of the three values of a/d.

In the axisymmetric specimens too internal cracking was audible well before the attainment of peak load as is evident from the nonlinearity in the pre-peak load-displacement diagrams (Fig 6). The peak load was associated with the sudden appearance of a number of radial cracks in the top of the specimen; the actual number depended on the a/d ratio. As illustrated schematically in Fig 4, four radial cracks appeared on the top surface when $a/d = 1/2$, but the number increased to eight when $a/d = 1$. For the latter loading geometry, a circumferential crack also appeared at the peak load (Fig 4) in some specimens. The cracks grew wider

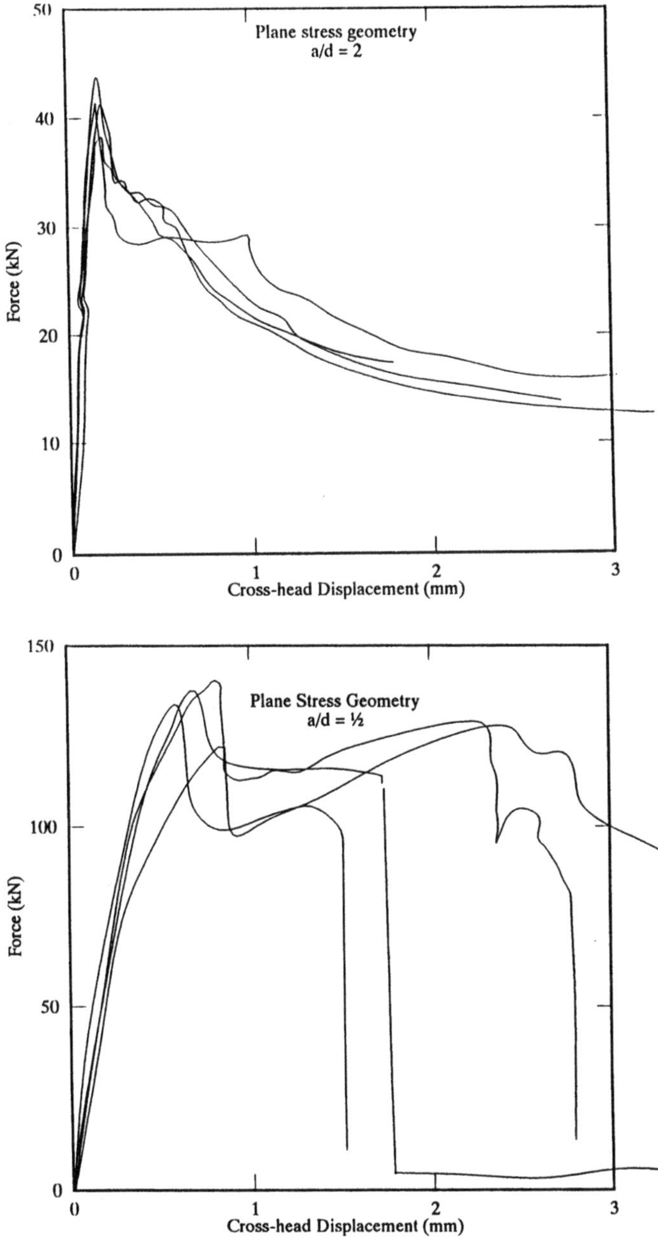

Fig. 5. Load–displacement diagrams of plane stress specimens.

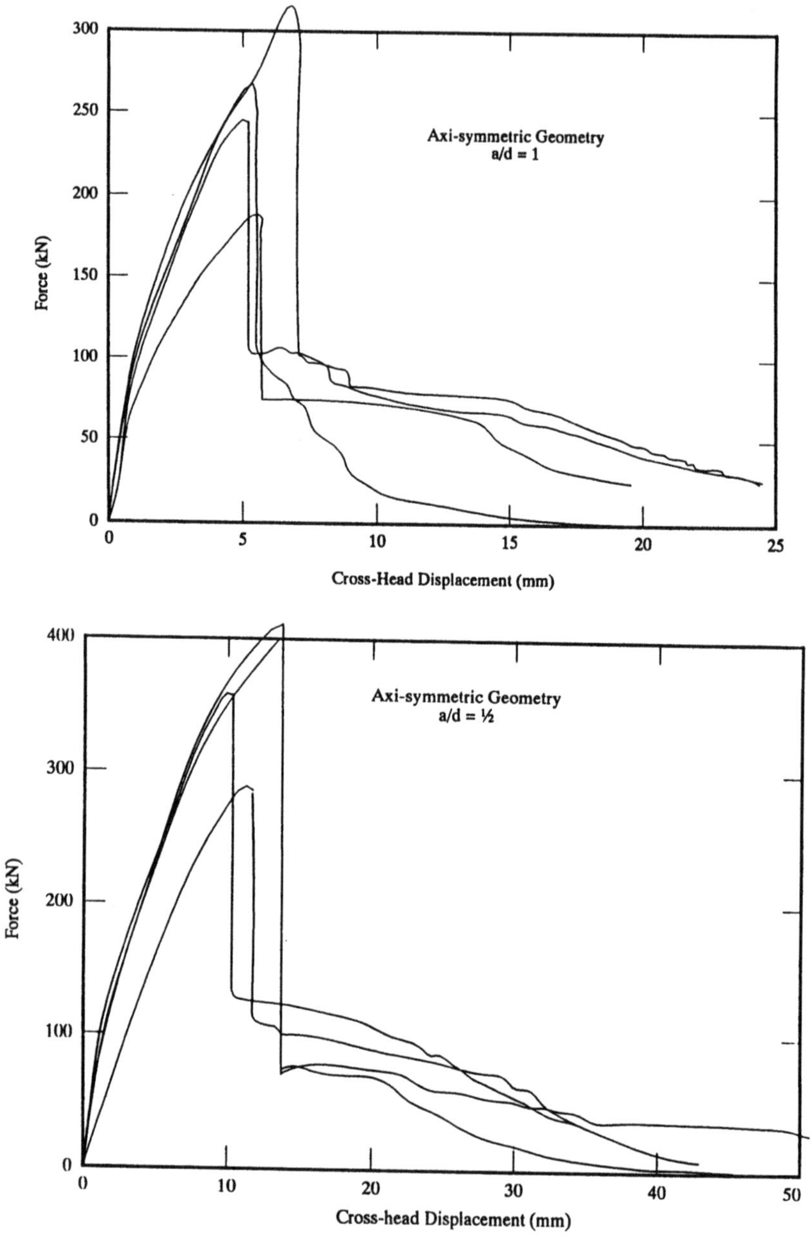

Fig. 6. Load–displacement diagrams of axisymmetric specimens.

on the descending branch of the load-displacement diagram when the anchor bolt was being progressively withdrawn from the specimen. The failure cone invariably disintegrated when the anchor bolt was fully withdrawn.

Table 1. Plane stress specimens

Specimen No	a/d	Peak Load P_{max} (kN)	Displacement δ at P_{max} (mm)	P_{max} (kN) Mean	sd	δ (mm) Mean	sd
1		44.00	0.18				
2		41.76	0.16				
	2			41.36	2.17	0.178	0.013
3		38.75	0.19				
4		40.92	0.18				
1		60.60	0.42				
2		69.80	0.62				
	1			69.24	6.10	0.488	0.109
3		74.70	0.53				
4		71.75	0.38				
1		139.36	0.80				
2		121.43	0.80				
	$\frac{1}{2}$			131.83	7.51	0.698	0.127
3		132.87	0.54				
4		133.67	0.65				

Table 2. Axisymmetric specimens

Specimen No	a/d	Peak Load P_{max} (kN)	Displacement δ at P_{max} (mm)	P_{max} (kN) Mean	sd	δ (mm) Mean	sd
1		188.65	5.50				
2		315.40	6.85				
	1			251.9	52.0	5.76	0.78
3		258.37	5.68				
4		245.19	5.00				
1		293.36	10.50				
2		404.60	13.50				
	$\frac{1}{2}$			367.2	53.9	11.88	1.89
3		409.90	13.50				
4		360.82	10.00				

5 References

Elfgren, L. (1990) Round robin analysis and tests of anchor bolts - invitation. **Mater. Struct.**, 23, 78.

Leonard, H., Rots, J.G., de Borst, R. and Feenstra, P.H. (1991) Contribution to round robin analysis of anchor bolts, Delft, pp 1–21.

57 NUMERICAL ANALYSIS OF ANCHOR BOLTS EMBEDDED IN CONCRETE PLATES

Y. UCHIDA, K. ROKUGO and W. KOYANAGI
Department of Civil Engineering, Gifu University, Japan

Abstract
This contribution deals with the finite element analysis of the pull-out behavior of the anchor bolts embedded in the concrete plates. This is the problem proposed by RILEM TC-90 Committee. Smeared crack approach based on the fictitious crack model is adopted. Numerical results demonstrate that the nominal failure strength of the anchor bolts with different material properties and specimen sizes depend on the ratio of the embedded depth to the characteristic length (d/l_{ch}). The influence of the finite element mesh size on the failure strength of anchor bolts is not serious. In same cases the mesh discretization influences not only the failure strength but also the crack patterns at failure.
Keywords: Anchor bolt, FEM, Smeared crack model, Size effect, Mesh sensitivity

1 Introduction

It is known that the tensile failure of concrete is the dominant mechanism of the pull-out behavior of anchor bolts when the embedded depth is shallow and the supporting points are far from the bolt. Generally, the strengths of concrete members at failure in tension indicate strong size effect. It is important to take account of the tension softening properties of concrete to predict failure strengths or size effect phenomena of those members. In this contribution the analytical results of anchor bolts pull-out problems in plane stress condition are presented. In the finite element analysis the tensile fracture of concrete was modelled using a smeared crack approach based on the fictitious crack model. The specimen geometry, material properties, loading and boundary conditions were in accordance with the Round Robin Analysis and Tests of Anchor Bolts of RILEM TC-90 FMA. The effects of the material properties, the mesh size of finite element and the discretization of mesh on the numerical results were discussed.

2 Material modeling

Before cracking, concrete was modelled as an isotropic and linear elastic material. Once the principal tensile stress exceeds the tensile strength, it was assumed a crack started perpendicular to the direction of the principal stress. After cracking concrete was assumed an orthotropic tension softening material with respect to the axes of orthotropy, where one axis is normal to the crack (tension softening) and the other is parallel with the crack (linear elastic), as shown in Fig.1. The smeared crack

Fracture and Damage of Concrete and Rock – FDCR-2.
Edited by H.P. Rossmanith. © 1993 Taylor & Francis. ISBN 0 419 18470 8.

approach based on the fictitious crack model proposed by Dahlblom et al.(1990) was adopted to define the tensile stress–strain relation of the cracked concrete in the direction normal to the crack. In this model the tensile stress–strain relation in the softening region was composed of the elastic stress–strain relation and the tension softening curve given as the crack opening–cohesive stress relation. The crack opening ω was translated into fracture strain depending on the equivalent length relating to the size and shape of the finite element. The equivalent length is defined as the maximum length of the cracked element in the direction normal to the crack, i.e. the projected length of the element to the crack, as shown in Fig.1.

In this study the bilinear tension softening model proposed by Rokugo et al.(1989) was mainly used, as shown in Fig.2. To examine the influence of the shape of the tension softening model, the linear softening model was also used in the analysis. As shown in Fig.3, a secant type unloading/reloading model was adopted in the softening region.

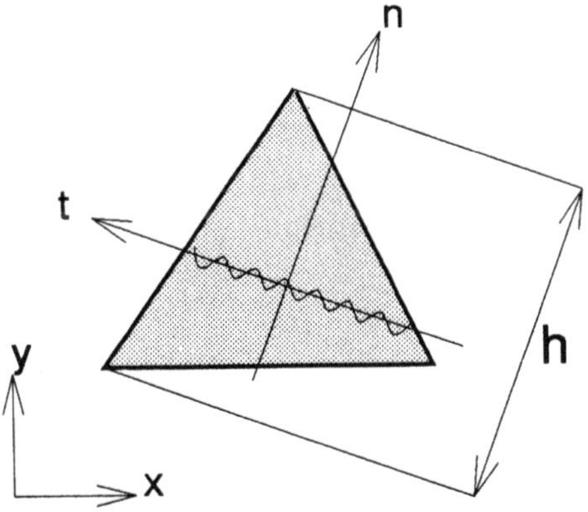

Fig.1 Equvalent length of cracked element.

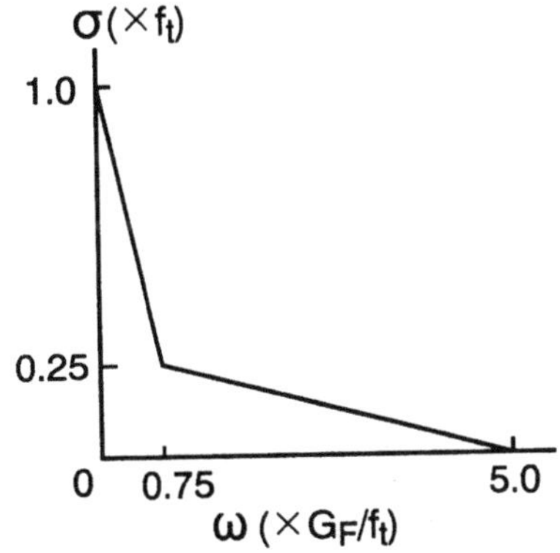

Fig.2 Tension softening curve.

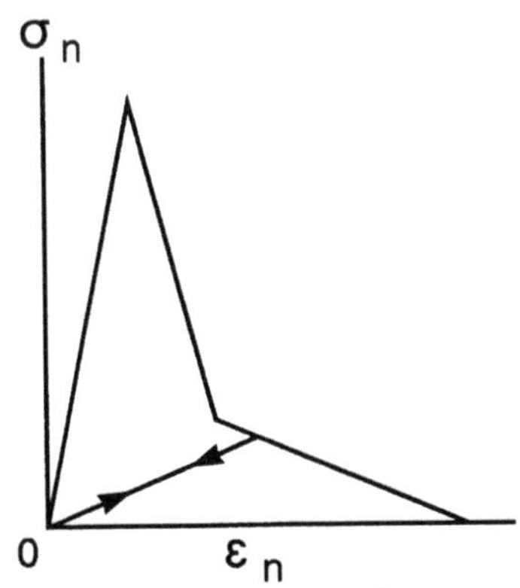

Fig.3 Tensile stress–strain curve.

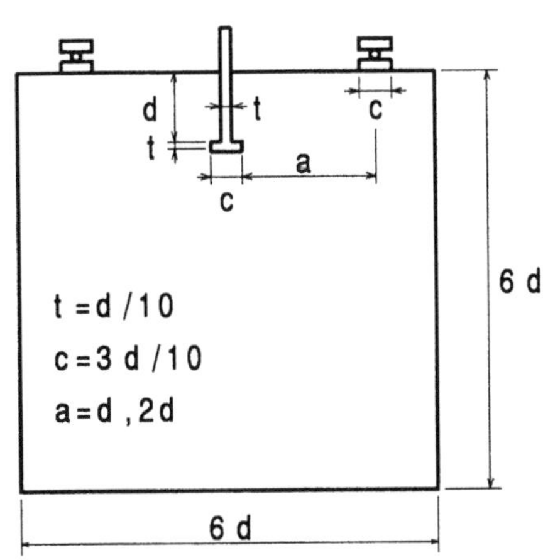

Fig.4 Geometry of pull–out specimens.

Two crack models for cracked concrete were adopted: (1) rotated crack model, in which the principal axes of stresses and strains coincide, (2) fixed crack model, in which the crack direction is fixed after crack initiation.

In case of the rotated crack model, there are no shear displacement nor shear stress on the crack, however the following shear modulus was used in order to enforce co-axiality between principal stresses and strains.

$$\beta G = \frac{\sigma_{11} - \sigma_{22}}{2(\varepsilon_{11} - \varepsilon_{22})} \tag{1}$$

In case of the fixed crack model, if the principal axis rotates, the shear displacement and stress are induced on the crack. Even though a lot of models of shear transfer on a crack have been proposed, most of them are for the crack with relatively large width as visible cracks in reinforced concrete members. For the plain concrete, further studies on mode II or mixed mode fracture are needed, especially with relation to the development of the fracture process zone in front of a macro crack. Therefore, in this study a very simple shear model was adopted for the fixed crack model. The shear stress on the crack was taken as zero and a very low shear retention factor (β=0.001) was chosen to avoid numerical problems.

The material constants used in this study are summarized as follow.

Tensile strength	f_t = 3, 4 MPa
Fracture Energy	G_f = 100, 150, 200 N/m
Young's modulus	E = 30, 40 GPa
Poisson's ratio	ν = 0.2
Tension softening	Bilinear, Linear

(underlined values are recommended in RILEM Proposal)

3 Finite element modelling

The geometry of the specimen is shown in Fig.4. The constant strain triangular elements were used. The half of the specimen was modelled because of symmetry. The bolt was not modelled for finite element, because the concrete was assumed in contact only with the top surface of the bolt head. The supporting plate was modelled included in the specimen. The finite element meshes are given in Fig.5. For the different specimen sizes, the same mesh discretizations were used and the meshes were scaled in proportion to the specimen size. When the distances between the supporting points and the bolt were different, only the supporting plate was moved. Mesh(b) and Mesh(c) in Fig.5 were used to investigate the influence of the mesh dicretization. The meshes were aligned radiately around the bolt head in Mesh(a). The triangles were placed in a bisectional pattern in Mesh(b) and in a cross-diagonal pattern in Mesh(c). To investigate the influence of the mesh size, Mesh(aa) and Mesh(bb) also used. Mesh(aa) and Mesh(bb) were obtained by remeshing Mesh(a) and Mesh(b), respectively, as shown in Fig.6.

The anchor was pulled out by controlling the vertical displacement at the top of the bolt head. To take account of the restraining effect of the bolt head, the horizontal displacement at the top of the bolt head was fixed. The center of the top surface of the supporting plate was supported only in vertically direction. A modified Newton Raphson iteration method was adopted. All solutions were converged with the norm of the residual force vector of 10^{-3} for each loading step.

Totally 30 cases were analyzed. The variables of each case are summarized in Table 1.

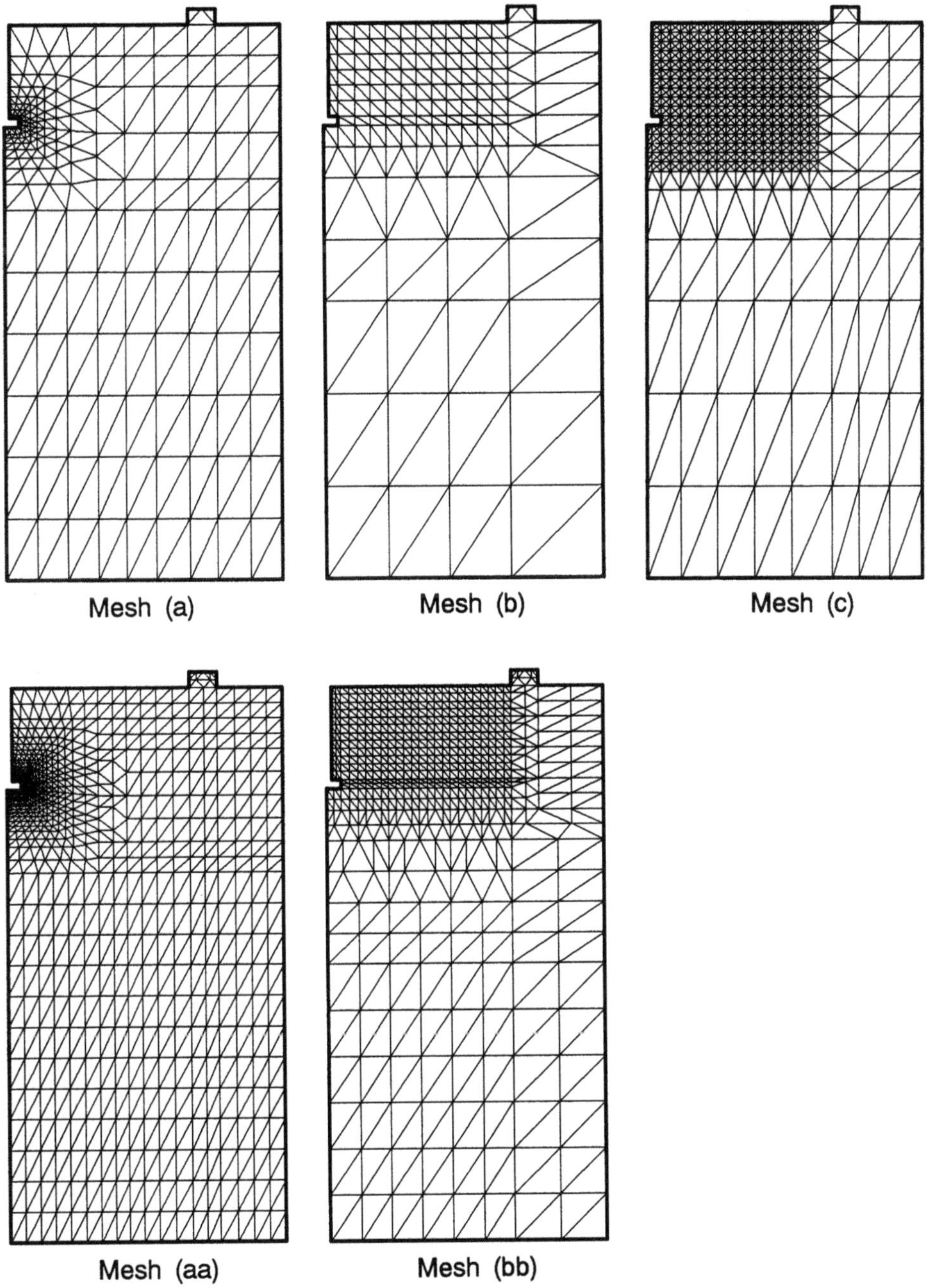

Fig.5 Finite elment meshes.

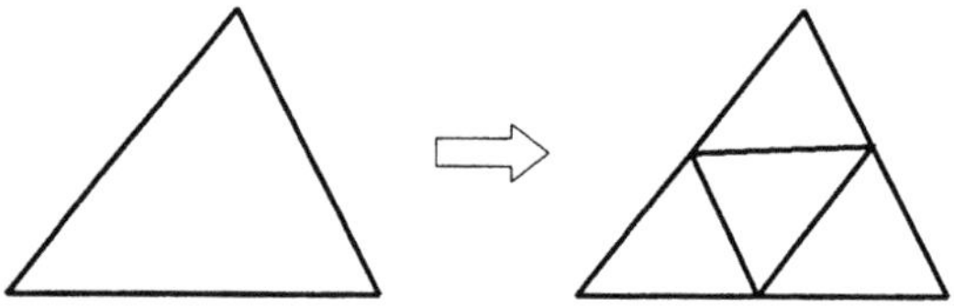

Fig.6 Remeshing of an element.

Table 1 Input data and results of analysis

Case	d (mm)	a	f_t (MPa)	E (GPa)	G_f (N/m)	Tension Softening	Crack Model	Mesh	Max. Load (N/mm)	Disp. (mm)
A	150	2d	3	30	100	bilinear	rotated	(a)	447	0.098
B	↑	↑	4	↑	↑	↑	↑	↑	486	0.098
C	↑	↑	3	40	↑	↑	↑	↑	480	0.138
D	↑	↑	↑	30	150	↑	↑	↑	501	0.192
E	↑	↑	↑	↑	200	↑	↑	↑	538	0.222
F	↑	↑	↑	↑	100	linear	↑	↑	517	0.112
G	↑	↑	↑	↑	↑	bilinear	fixed	↑	459	0.124
H	↑	d	↑	↑	↑	↑	rotated	↑	625	0.134
I	↑	↑	↑	↑	↑	↑	fixed	↑	659	0.160
J	50	2d	↑	↑	↑	↑	rotated	↑	192	0.047
K	↑	↑	↑	↑	↑	↑	fixed	↑	199	0.091
L	450	↑	↑	↑	↑	↑	rotated	↑	886	0.198
M	↑	↑	↑	↑	↑	↑	fixed	↑	937	0.252
N	50	d	↑	↑	↑	↑	rotated	↑	296	0.067
O	↑	↑	↑	↑	↑	↑	fixed	↑	282	0.070
P	450	↑	↑	↑	↑	↑	rotated	↑	1172	0.204
Q	↑	↑	↑	↑	↑	↑	fixed	↑	1246	0.216
R	150	2d	↑	↑	↑	↑	rotated	(aa)	422	0.100
S	↑	↑	↑	↑	↑	↑	fixed	↑	408	0.087
T	↑	d	↑	↑	↑	↑	rotated	↑	570	0.135
U	↑	↑	↑	↑	↑	↑	fixed	↑	590	0.146
V	↑	2d	↑	↑	↑	↑	rotated	(b)	336	0.074
W	↑	↑	↑	↑	↑	↑	fixed	↑	372	0.086
X	↑	↑	↑	↑	↑	↑	rotated	(bb)	321	0.074
Y	50	↑	↑	↑	↑	↑	↑	(b)	161	0.039
Z	450	↑	↑	↑	↑	↑	↑	↑	693	0.142
AA	150	d	↑	↑	↑	↑	↑	↑	541	0.148
AB	↑	↑	↑	↑	↑	↑	fixed	↑	642	0.161
AC	↑	2d	↑	↑	↑	↑	rotated	(c)	388	0.090
AD	↑	d	↑	↑	↑	↑	↑	(c)	571	0.121

(↑: same as above)

4 Results of analysis

4.1 Influence of material properties, crack model and specimen size

The maximum load and the displacement of the upper edge of bolt head at the maximum load are given in Table 1.

Fig.7 and 8 show the correlation between the embedded depth and the maximum load for cases with different material properties, crack models and specimen sizes. The embedded depth d and the maximum load F_{max} were normalized with the characteristic length $l_{ch}(=EG/f_t^2)$ and bdf_t (b:thickness of the specimen), respectively [Elfglen et al.(1992)]. Except for Case F(linear tension softening), the data of Mesh(a) are on one line. As well known, the shape of the tension softening curve has influence on the maximum load of the concrete member and the linear softening makes the maximum load larger as compared to the bilinear softening. This influence was observed in case F. The difference in the crack model (fixed or rotated) had little influence on the maximum load. When $a=2d$ with Mesh(a) the line had a slope of $k≒-0.3$, and $k≒-1/3$ when $a=d$. These results are almost same as the numerical and

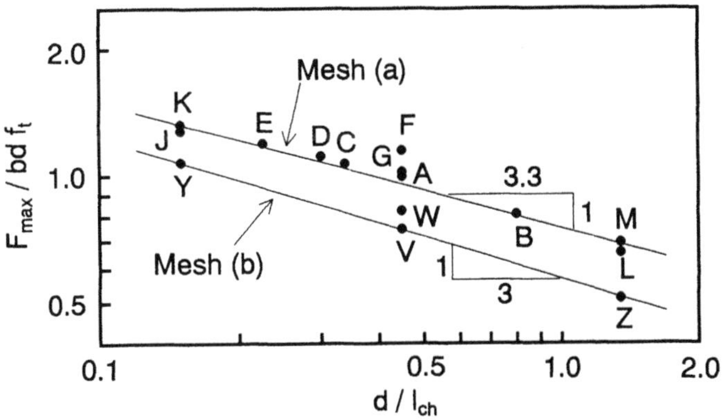

Fig.7 Relations of embedded depth and maximum load ($a=2d$).

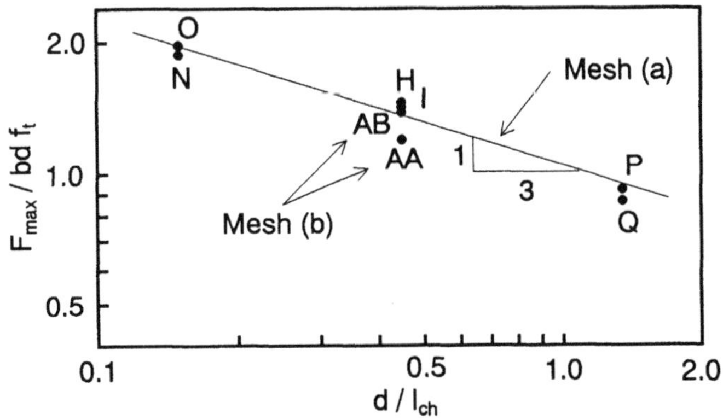

Fig.8 Relations of embedded depth and maximum load ($a=d$).

experimental results of Elfglen et al.(1992). If $k \fallingdotseq -1/3$ is assumed the maximum load is expressed as follow.

$$F_{max} \propto b(EG_f\, f_t\, d^2)^{1/3} \quad (2)$$

As seen from this equation, the maximum load F_{max} is proportional to $d^{2/3}$ and $(EG_f\, f_t\,)^{1/3}$.

4.2 Influence of mesh size and mesh discretization

The Load–Displacement curves showing the influence of the mesh size are given in Fig.9 and Fig.10. The fine mesh gave a little lower maximum load than the coarse mesh, though it was consider that the influence of the mesh size was not serious in the present finite element model. In the finite element analysis the fine mesh has lower elastic stiffness as compared to the coarse mesh. The lower Young's modulus provides the lower maximum load of the anchor bolt, as indicated in equation (2). This would be one of the reasons that the fine mesh provided lower maximum load.

Fig.11 and Fig.12 show the load–displacement curves for three mesh discretizations. Fig.13 shows the cracking patterns at maximum load. In the case of $a=2d$ with the rotated crack model, the mesh discretization had strong influence on the cracking pattern. The maximum load of Mesh(a)(Case A) was about 30% higher than that of Mesh(b)(Case V). Furthermore, there was an apparent difference in the cracking pattern. In case of Mesh(a), the horizontal cracks diffused in the diagonal directions. On the other hand, the cracks localized in a horizontal line in case of Mesh(b). Mesh(c)(Case AC) exhibited an intermediate behavior between Mesh(a) and Mesh(b).

The size effect for Mesh(b) was given in Fig.7. The slope of the curve of Mesh (b) was just $k=-1/3$ and a little steeper than that of Mesh(a).

The development of horizontal cracks in case of $a=2d$ was observed in experiment by Helbling et al.(1991). Cervenka et al.(1991) and several other researchers have also reported the development of horizontal cracks in the analysis. On the other hand, in case of $a=d$ the differences among the load–displacement curves of three meshes(Case H, AA, and AD) were small as shown in Fig.12. However, there was difference in the cracking pattern. The cracks localized in the diagonal direction in case of Mesh(a), and in the horizontal direction in case of Mesh(b). In case of Mesh(c) the diagonal cracks localized and many horizontal cracks also diffused.

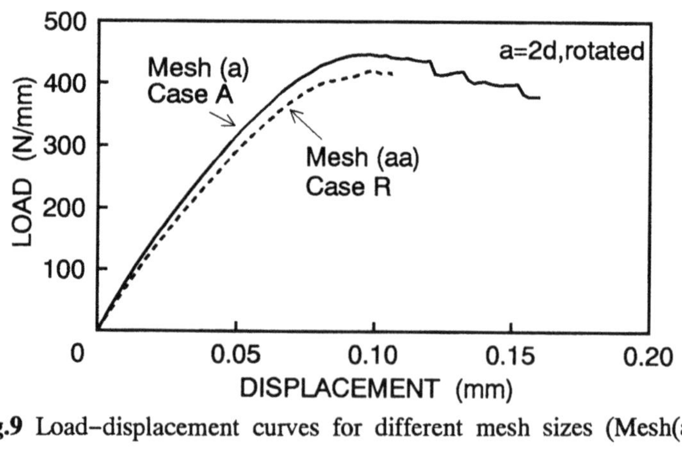

Fig.9 Load–displacement curves for different mesh sizes (Mesh(a),(aa)).

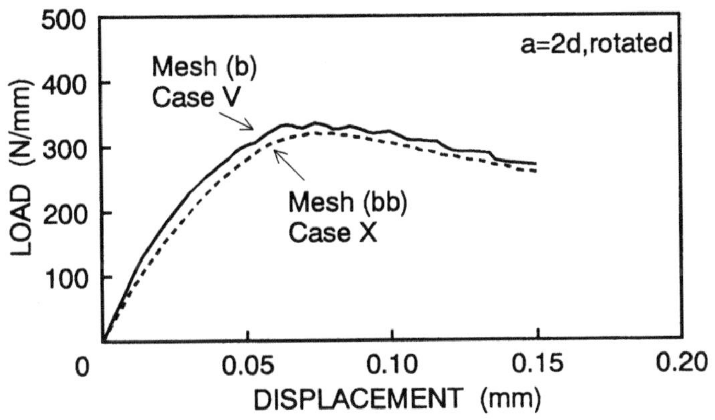

Fig.10 Load–displacement curves for different mesh sizes (Mesh(b),(bb)).

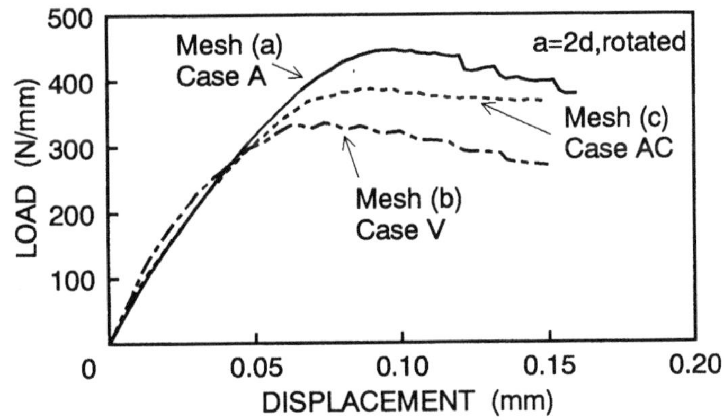

Fig.11 Load–displacement curves for different mesh discretizations ($a=2d$).

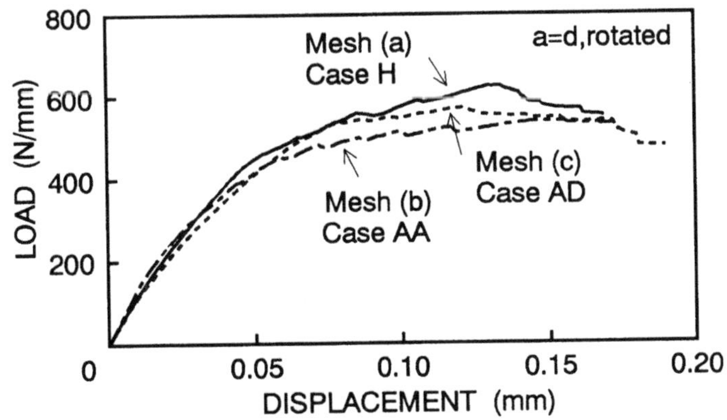

Fig.12 Load–displacement curves for different mesh discretizations ($a=d$).

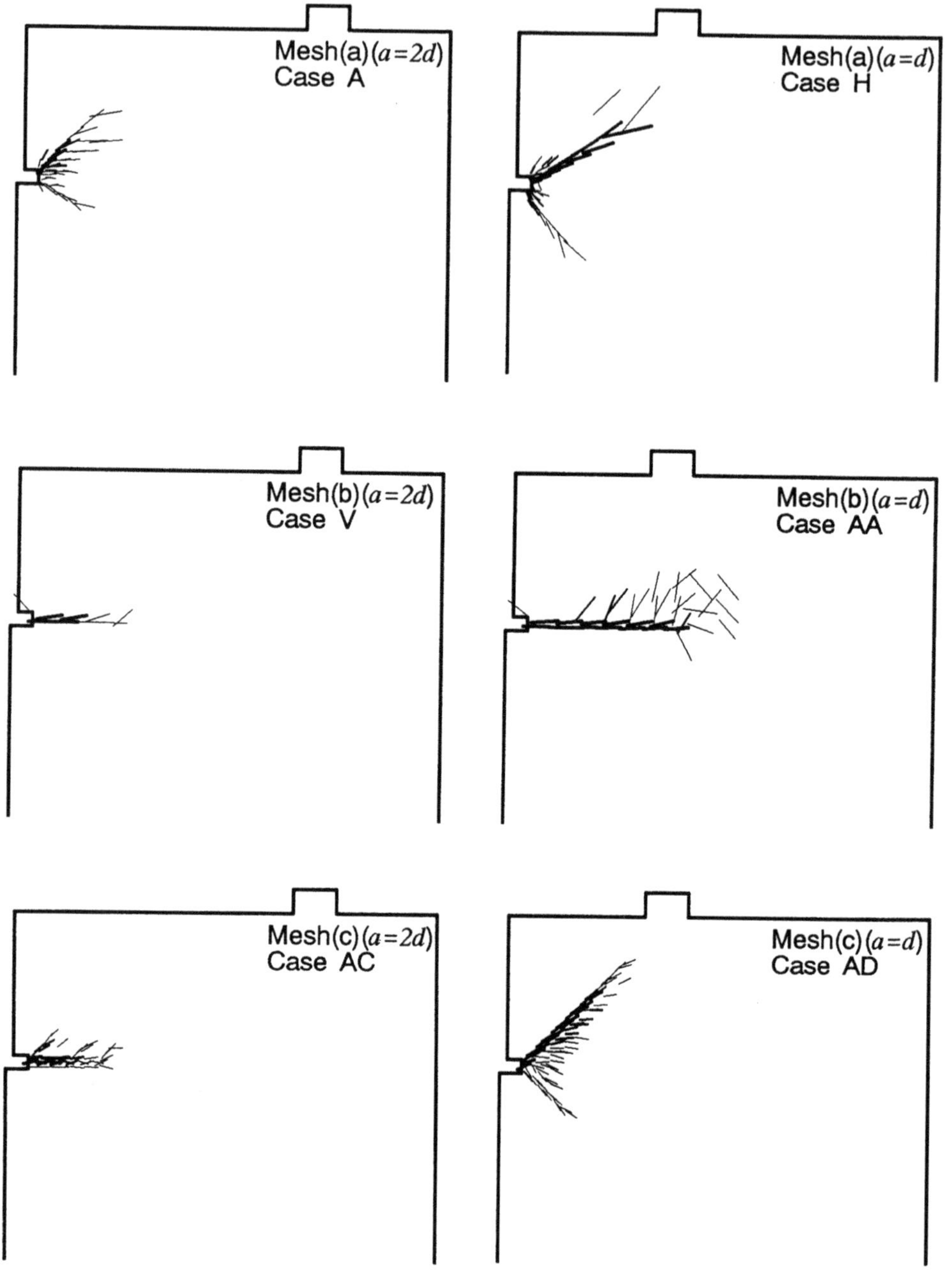

Fig.13 Cracking patterns at maximum load.

5 Conclusions

From the finite element analysis of anchor bolts embedded in concrete plate the following conclusions may be drawn:

(1)The nominal failure strength of anchor bolts (F_{max}/bdf_t) depends on the ratio of the embedded depth to the characteristic length of the concrete (d/l_{ch}). The maximum load F_{max} is proportional to $d^{2/3}$ and $(EG_f f_t)^{1/3}$.

(2)The finite element model used in this study has little mesh size dependency. However, the mesh discretization influences not only the failure strength but also the cracking pattern at failure of the anchor bolt. It is needed to study the mesh sensitivity in the finite element analysis of the localized fracture.

6 References

Cervenka V., Pulk R. and Eligehausen R. (1991) Round robin analysis of anchor bolts 1991, in **Round-Robin Analysis of Anchor Bolts**, Preliminary report, Second edition (com. Elfglen L.).

Dahlblom O. and Ottosen N.S. (1990) Smeared crack analysis using generalized fictitious crack model, **Journal of Engineering Mechanics**, Vol. 116, No. 1, ASCE, pp. 55–76.

Elfgren, L. and Ohlsson, U. (1992) Anchor bolts modelled with fracture mechanics, in **Application of Fracture Mechanics to Reinforced Concrete** (ed. Carpinteri A.), Elsevier Applied Science, London, pp. 267–283.

Helbling A., Alvaredo A.M., and Wittmann F.H. (1991) Round robin test of anchor bolts (Test report No. 1/91), in **Round-Robin Analysis of Anchor Bolts**, Preliminary report, Second edition (com. Elfglen L.).

Rokugo K, Iwasa M., Suzuki T. and Koyanagi W. (1989) Testing methods to determine tensile strain softening curve and fracture energy of concrete, in **Fracture toughness and fracture energy** (eds. Mihashi H., Takahashi H. and Wittmann F.H.), Balkema, Rotterdam, pp. 153–163.

58 ANCHORAGE OF STRANDS IN HOLLOW CORE PRECAST CONCRETE ELEMENTS

K. GYLLTOFT and M. ÅKESSON
Division of Concrete Structures, Chalmers University of Technology, Göteborg, Sweden

Abstract
Anchorage of strands in prestressed hollow core slabs was analysed by a finite element model based on non-linear fracture mechanics. A new material model was developed. The new material model made it possible to simulate the interface behaviour.
Keywords: Anchorage, transmission length, fracture mechanics, finite element analysis, precast concrete, hollow core slab, bond

1 Introduction

Anchorage of strands in prestressed hollow core slabs was analysed by a finite element model based on non-linear fracture mechanics. The purpose of the analysis was to obtain a better understanding of the mechanical behaviour of the hollow core elements by performing parameter studies with the model. In order to simulate the slip between the strand and the surrounding concrete, a new material model has been developed for the interface layer. Cracking in concrete was modelled by a smeared crack model. Until now, some preliminary results have been obtained from the model.

2 Finite element model

2.1 General description

The finite element model is 3-dimensional and the analysis was performed with the program ABAQUS. For the concrete and the strand the standard possibilities for material modelling in ABAQUS was used. The slip behaviour was simulated with an interface layer around the strand for which the new material model was used.

Fracture and Damage of Concrete and Rock – FDCR-2.
Edited by H.P. Rossmanith. © 1993 Taylor & Francis. ISBN 0 419 18470 8.

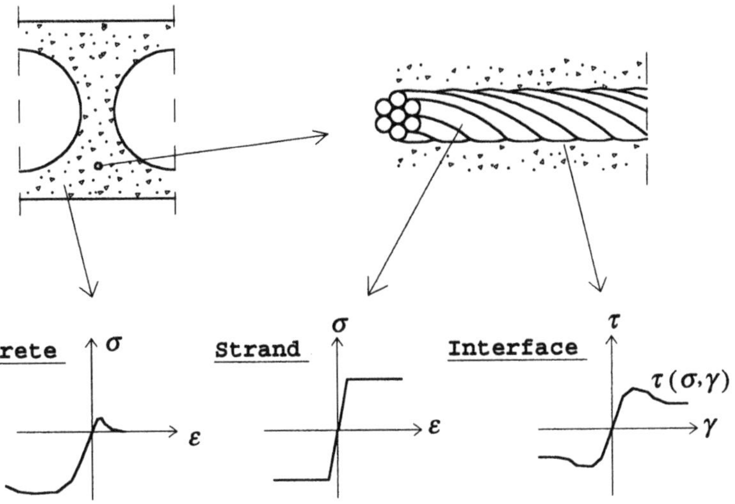

Fig.1. Materials and geometry used in the finite element model.

The constitutive behaviour of the strand was modelled by a linear-elastic perfectly-plastic material model.

The concrete model that was used is based on the "smeared crack" concept. When the stress state is dominantly compressive, the response is modelled by an elastic-plastic theory, using a Drucker-Prager yield surface, an associative flow rule, and an isotropic hardening law. The constitutive behaviour in tension is initially isotropic and then modelled by a Drucker-Prager yield surface (crack detection surface) and an anisotropic hardening law. The damage due to micro cracking is modelled by decreasing modules of elasticity and shear associated with the crack-plane. With the model it is possible to get, at the most, three orthogonal crack-planes in each integration point.

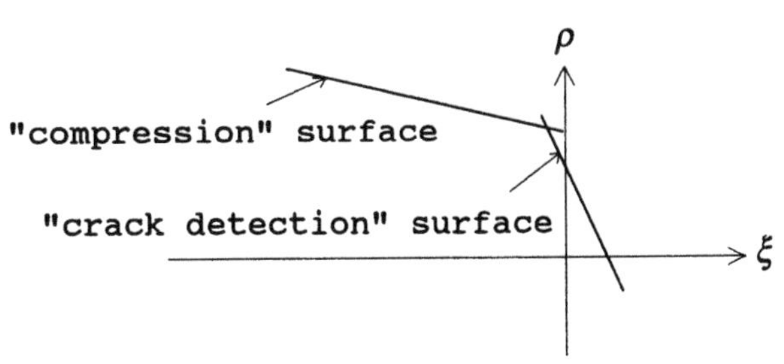

Fig.2. Concrete failure surfaces in the $\xi - \rho$ plane (where ξ is the hydrostatic stress and ρ is the deviatoric stress).

2.2 Interface layer

2.2.1 General

A new material model (with the name BOND) has been developed. This model makes it possible to simulate the slip between the strand and the surrounding concrete. BOND calculates two arbitrary stress components $\sigma_{\perp}$ and $\tau_{||}$ (the other four are set equal to zero), the slip, and the constitutive tangent stiffness, $J_{ij}= \partial\sigma_i/\partial\varepsilon_j$. In the development of the new material model, the first step was to study the reaction forces around an axially loaded steel bar embedded in concrete.

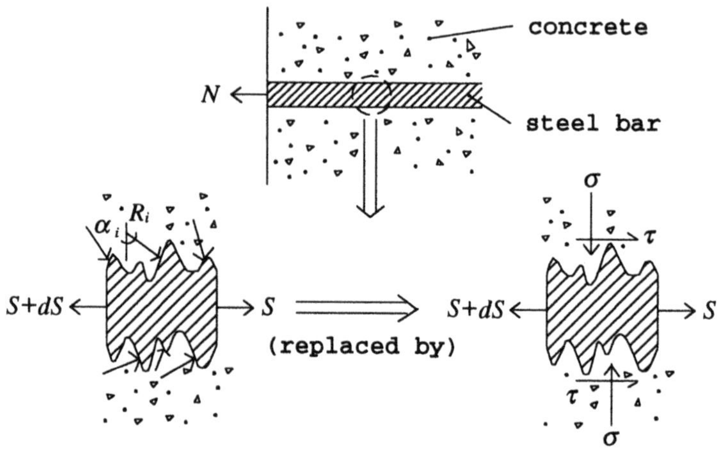

Fig.3. Axially loaded steel bar embedded in concrete with reaction forces.

In order to describe the stresses on a surface around the bar, an interface layer was introduced. The interface has a normal and a shear stiffness. Furthermore it has the special property to expand when it is deformed in shear.

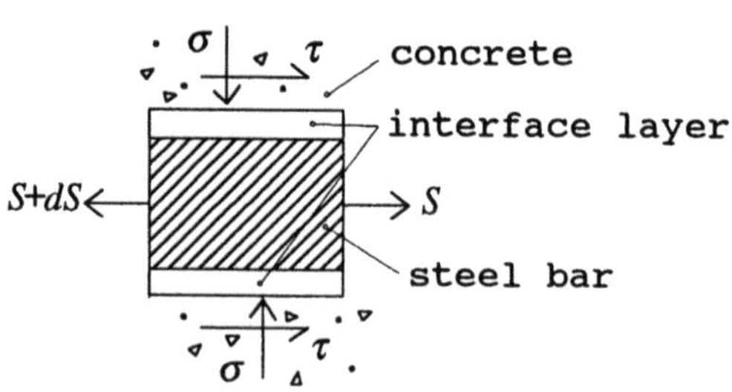

Fig.4. Steel bar element with a surrounding interface layer.

The model is formulated as a conventional elastic-plastic material model with a yield condition, a flow rule and a softening law.

The parameters of the material model are the modulus of elasticity E, the shear modulus G, the adhesion capacity f_{ba}, the coefficient of friction μ and the "expansion-stress" curve (σ_γ, γ). The "expansion-stress" curve is created by curve fitting through the three assumed points $(\sigma_{\gamma 1}, \gamma_1)$, $(\sigma_{\gamma 2}, \gamma_2)$ and $(\sigma_{\gamma 3}, \gamma_3)$, where γ_1 is calculated as $\gamma_1 = \mu \cdot \sigma_{\gamma 1}/G$.

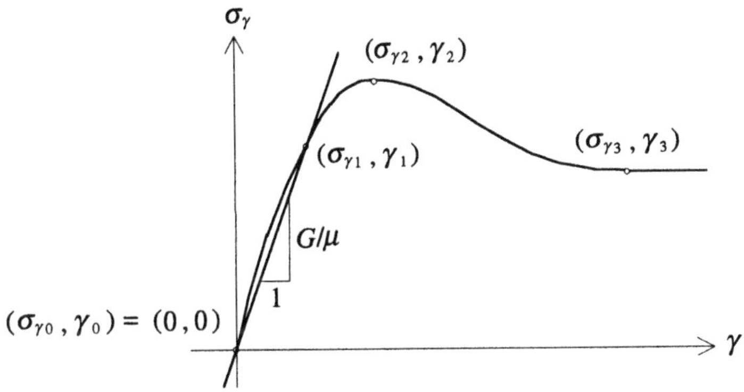

Fig.5. Stress-strain curve with associated parameters describing the expansion stress versus the total shear strain.

2.2.2 Yield condition

The yield condition is described by the Coulomb line

$$F(\sigma, \tau, \gamma^p) = |\tau| + \mu \cdot \sigma - \tau_a(\gamma^p) = 0 \tag{1}$$

where $\tau_a(\gamma^p)$ is a parameter describing adhesion and γ^p is the plastic shear strain.

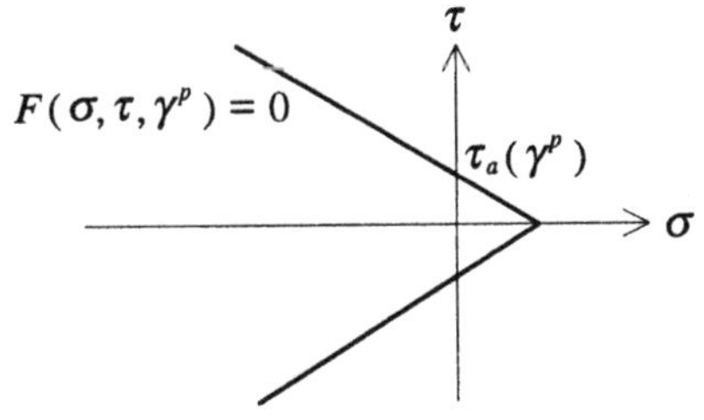

Fig.6. Yield condition.

2.2.3 Flow rule

The flow rule is non-associative and is formulated so

that the "relaxation stress" only affects the "shear trial stress"

$$\underline{d\sigma} = \underline{d\sigma}^e - \underline{d\sigma}^p = [d\tau, d\sigma]^T = [d\tau^e, d\sigma^e]^T - [d\tau^p, 0]^T \tag{2}$$

where $\underline{d\sigma} = \underline{D}^e \cdot \underline{d\varepsilon}^e = \underline{D}^e \cdot \underline{d\varepsilon} - \underline{D}^e \cdot \underline{d\varepsilon}^p$, $\underline{D}^e$ is the elastic tangent stiffness matrix, $\underline{d\varepsilon}^e$, $\underline{d\varepsilon}$ and $\underline{d\varepsilon}^p$ is the elastic-, total- and plastic strain increment vector respectively.

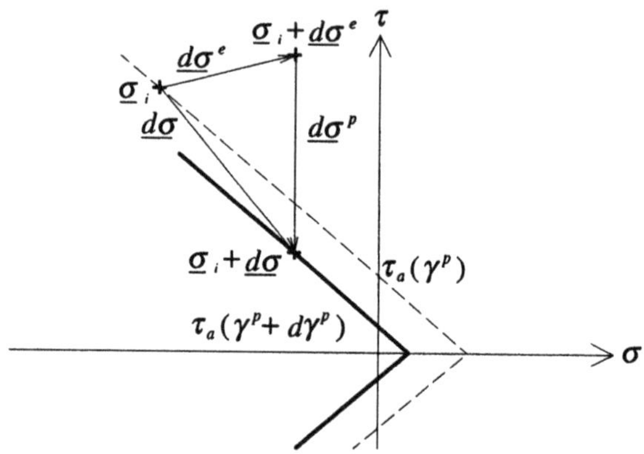

Fig.7. Relation between trial-stress increment, relaxation-stress increment and stress increment.

In matrix notation the flow rule is written as

$$\underline{d\varepsilon}^p = \frac{d\lambda}{((d\sigma_\gamma / d\gamma)^2 + E^2)^{1/2}} \cdot \begin{bmatrix} d\sigma_\gamma / d\gamma \\ E \end{bmatrix} \tag{3}$$

where $d\lambda$ is a plastic multiplier.

2.2.4 Softening law

The softening is determined by the function $\tau_a(\gamma^p)$ which is defined as

$$\tau_a(\gamma^p) = \begin{cases} f_{ba}, \ \gamma^p = 0 \\ f_{ba} - G \cdot \gamma^p / 2, \ 0 < \gamma^p \le 2 \cdot f_{ba} / G \\ 0, \ \gamma^p > 2 \cdot f_{ba} / G \end{cases} \tag{4}$$

When the adhesion strength is reduced to zero in a point, the calculated plastic shear strain corresponds to the slip in that point.

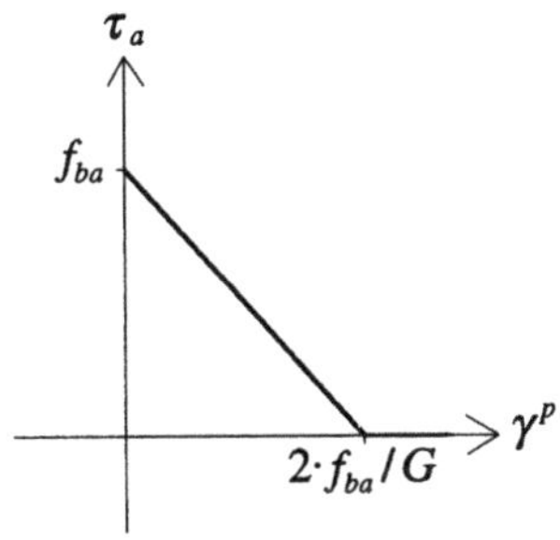

Fig.8. Adhesion strength as a function of the plastic shear strain.

2.2.5 Incremental stress-strain relations

If the updated stress is inside the yield surface (i.e. $F(\sigma_i + d\sigma^e, \tau_i + d\tau^e, \gamma_i^p) \leq 0$) then the response is purely elastic and

$$\begin{bmatrix} d\sigma \\ d\tau \end{bmatrix} = \begin{bmatrix} E & -d\sigma_\gamma / d\gamma \\ 0 & G \end{bmatrix} \cdot \begin{bmatrix} d\varepsilon \\ d\gamma \end{bmatrix} \tag{5}$$

otherwise the deformations are elastic-plastic and

$$\begin{bmatrix} d\sigma \\ d\tau \end{bmatrix} = \begin{bmatrix} E & -d\sigma_\gamma / d\gamma \\ -\mu \cdot E \cdot G / A & B \cdot G / A \end{bmatrix} \cdot \begin{bmatrix} d\varepsilon \\ d\gamma \end{bmatrix} \tag{6}$$

where $A = G + d\tau_a / d\gamma^p$ and $B = (\mu \cdot d\sigma_\gamma / d\gamma + d\tau_a / d\gamma^p)$.

3 Applications

3.1 Introduction

The model has been used to compare results from analyses, with experimental results from a pull-out test of a smooth bar and a pull-out test of a strand. It has also been used to calculate the response of a hollow core slab at release of the prestressing force.

3.2 Pull-out test of a smooth bar

This example concerns pull-out of a hot rolled smooth bar (ϕ16 mm) in a cylindrical concrete body (d = 300 mm, l = 300 mm). The embedment length was 200 mm, the concrete compression strength f_{cc} = 27.5 MPa, the concrete tensile strength f_{ct} = 2.8 MPa and the fracture energy G_f = 100 N/m. N is the pull-out force and s is the slip in the free end.

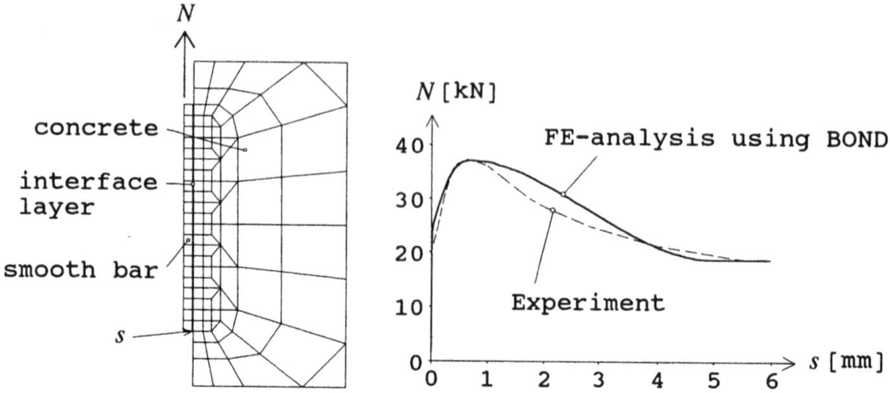

Fig.9. Axisymmetric finite element model and load-slip relation.

3.3 Pull-out test of a strand
This example concerns pull-out of a 7-wire strand (ϕ12.5 mm) in a cubic concrete body (300x300x300 mm). The embedment length was 150 mm, the concrete compression strength f_{cc} = 42.5 MPa, the concrete tensile strength f_{ct} = 3.6 MPa and the fracture energy G_f = 100 N/m. N is the pull-out force and s is the slip in the loaded end.

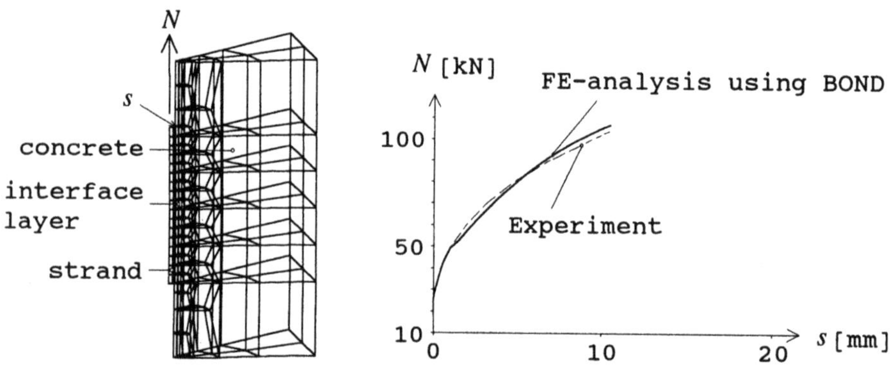

Fig.10. 3-dimensional finite element model and load-slip relation.

3.4 Hollow core slab
This analysis was done in order to examine the response of a hollow core slab with the length l = 2.4 m and the height h = 0.215 m. The initial stress in the 7-wire strand (ϕ 12.7 mm) was σ_{sp0i} MPa, the concrete compression strength f_{cc} = 42.5 MPa, the concrete tensile strength f_{ct} = 3.6 MPa and the fracture energy G_f = 100 N/m.

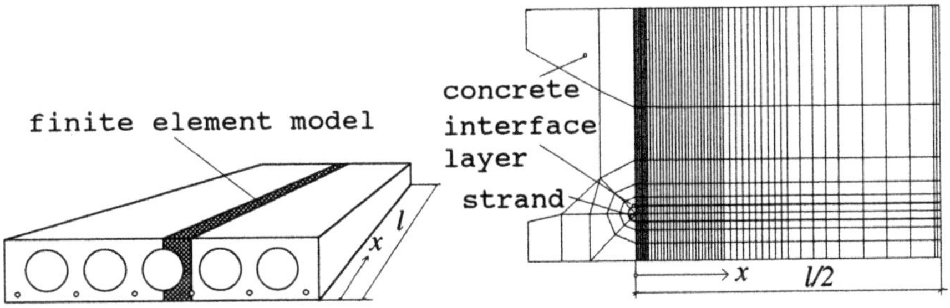

Fig.11. Hollow core element with 3-dimensional finite element model.

The following pictures show the variation along the strand of the slip s (the transmission length is about 0.4 m), the normal stress in the strand σ_{sp0}, the radial compressive stress in the interface layer σ_{rr} and the shear stress in the interface layer τ_{rx}.

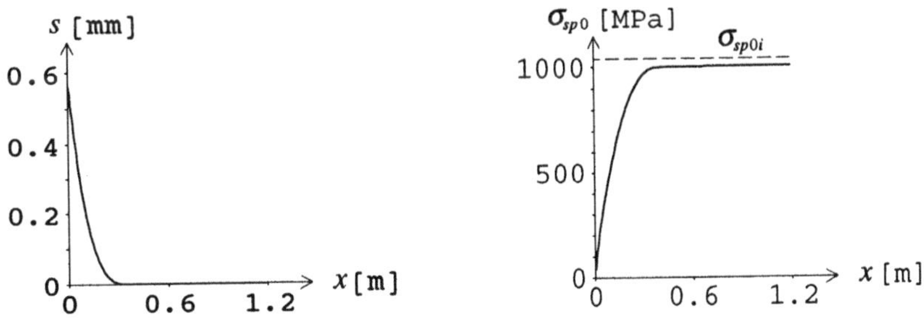

Fig.12. Variation of slip and normal stress along the strand.

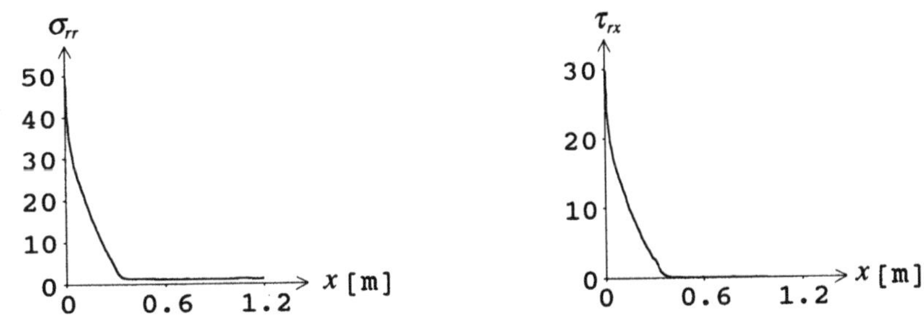

Fig.13. Variation of radial compression stress and shear stress in the interface layer, along the strand.

4 References

Gylltoft, K. (1983) **Fracture mechanics models for fatigue in concrete structures**, Luleå, Sweden, pp. 147-165.

Jokela, J. Lukkariniemi, I. and Tepfers, R. (1981) **Bond of bundled steel strands**, Göteborg, Sweden, pp. 7-15.

59 DESIGN PRINCIPLES AND MONITORING OF ROCK ANCHORS AND BOLTS COMPRISING NON-CORROSIVE GLASS FIBER MATERIALS

M. FAORO
SICOM Ltd, Cologne, Germany

Abstract
During the anchoring of glass fiber reinforced composite (GFRC) rock anchors in the rock or earth-sided grouted elements it has to be taken into consideration that the glass fiber bars due to the specific properties of the material (modulus of elasticity, profile and surface) have - in comparison to conventional anchor steels - a significantly different bond behaviour. On the basis of extensive theoretical and experimental investigations design principles for rock anchors and rock bolts comprising resin bonded glass fiber bars are elaborated and discussed in the present paper. Moreover a new monitoring method on the basis of strain sensors comprising optical fibers is presented and explained.
Keywords: Glass fiber reinforced composite (GFRC), Bond behaviour of GFRC, Design of rock anchors, Monitoring, Strain sensors

1 Introduction

Due to some of their excellent properties high tensile glass fiber composite materials are used more and more often in concrete construction and in the field of earthworks or foundation engineering. Especially because of their high corrosion resistance, their low specific weigth and their high tensile strength anchors comprising this material are very appropriate to be used as permanent anchors. But caused by the specific properties, especially because of the relatively low modulus E and the bond bearing behaviour a bearing behaviour differing from that of conventional steel anchors can be expected. Therefore special considerations have to be made with regard to the dimensioning of the grouted elements.

Starting from a short description of the essential properties of resin bonded glass fiber bars the bond bearing behaviour of same in concrete or cement mortar is presented and discussed in the present contribution. Afterwards an appropriate anchoring device is explained and demonstrated.

Fracture and Damage of Concrete and Rock – FDCR-2.
Edited by H.P. Rossmanith. © 1993 Taylor & Francis. ISBN 0 419 18470 8.

With the aid of these indications design principles for rock anchors or rock bolts are elaborated and explained from theoretical calculations (parameter studies). The used calculation model takes the different modulus of elasticity and the different bond qualities equally into consideration. The essential results of these studies serve as basis for conclusions in the construction practice.

Finally a new type of measuring and monitoring system on the basis of optical fiber sensors is presented.

2 Essential properties and bond behaviour of GFRC-bars

2.1 Composition and properties of the GFRC-bars

The unidirectional glass fiber composite material produced in an industrial way is manufactured in a continous pultrusion process out of a great number of endless glass fibers arranged in a parallel way with composite matrix comprising polyester or epoxy resin. The bar material has a nominal diameter of 7,5 mm and consists of approx. 64.000 individual fibers. With a diameter of the individual fibers in the scale of approx. 25 μm follows a glass fiber percentage of roughly 69 Vol.-% (see fig. 1).

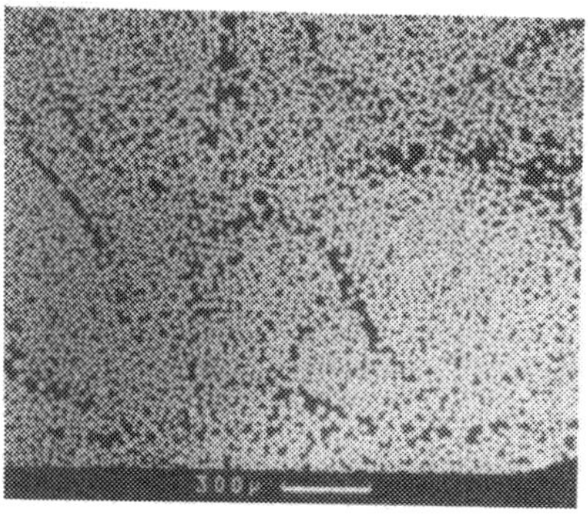

Fig.1: Micro scan cross section image

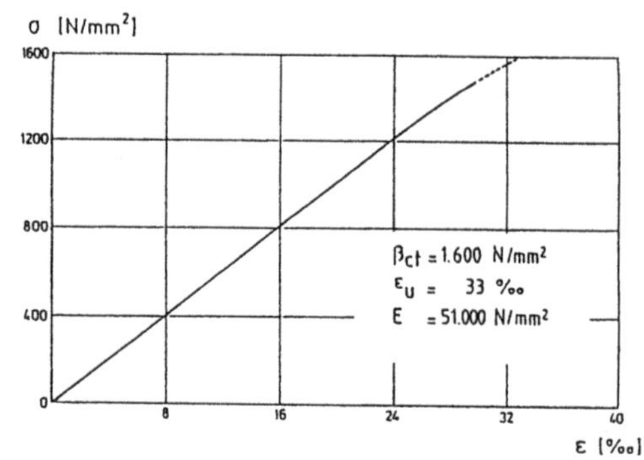

Fig. 2: σ-ϵ-curve of GFRC-bars

The tensile strength and the stress/strain behaviour are of prior interest for the short-term load. With this material the relation between stress and strain can be considered as mainly linear (fig. 2), i.e. the behaviour of the GFRC-elements is almost completely elastic. As can be seen from fig. 2, failure occurs at an ultimate strain of approx. ϵ = 3.3 %. In the working load range, the experimentally found modulus of elasticity of the GRFC-bars amounts to approx. 51.000 N/mm².

A further criterium for the assessment of the applicability of the uni-directional glass fiber composite material is the effect of load duration on the tensile strength. The strength of the material under sustained load comes to approx. 70 % of the short term tensile strength f_{ck}. Caused by the specific structure of the material a very low loss of stress due to relaxation can be recognized under permanent load. During a period of about 50 years the loss of stress amounts only to 3 % of the used initial tensile stress.

Significant effects on the strength are also to be expected if the material is subjected to non-static load. Whereas with prestressing steel the usual and sufficiently accurate stress amplitude at 2 x 10^6 load cycles can be defined as fatigue strength for a certain average stress, this is apparently not permittable for the present GFRC-bars as material failure can also occur with considerably higher load cycles. This fact is taken into consideration in the design concept.

Unidirectioal glass fiber composite bars - in comparison with conventional prestressing steels - do have the same tensile strength but a significantly higher corrosion resistance and a lot of other advantageous properties as follows:

* low own weight
* due to the low modulus of elasticity they have a considerably lower decrease of tension in case of components shortenings due to creep or shrinkage
* high corrosion and chemical resistance
* electromagnetic neutrality

2.2 Bond bearing behaviour of GFRC-bars in concrete

The bond behaviour of the composite bars in both concrete and the injection and grouting mortar is of decisive importance for the transfer of prestressing force to the bars. On technical production reasons the GFRC-bars are wound with chemical fibers which provide the bars with a certain surface roughness and ripple which in turn have a fa-vourable effect on the bonding behaviour in concrete and injection mortar. To increase the resistance to aggressive media, for example alkaline residual moisture from the concrete, and as a protection against defects resulting from mechanical loads and handling on the site, the GFRC-bars are usually also coated with an approx. 0.3 mm thick sheathing comprising melted epoxy resin powder. As a result the surface of the coated bars, and thereby the bonding behaviour in concrete or injection mortar, can be specifically controlled. This coating satisfied expectations - especially concerning density and abrasion.

Moreover there is the possibility to sand the surface of the bars in a not yet hardened state. The individual grains of sand are then embedded in the matrix resin and lead consequently to a significant increase of the roughness of the bar surfaces. But is has to be considered that under these circumstances the required protection through a sheathing or coating does not exist.

In order to study the bond bearing behaviour of GFRC-bars with different surfaces in normal concrete (cement mortar) numerous pull-out-tests with short embeddment lengths have been realized. The result of these kind of tests shows the relation of local bond stress and corresponding relative displacements, the so-called bond laws $\tau = f(s)$. Fig. 3 shows schematically the experimental arrangement.

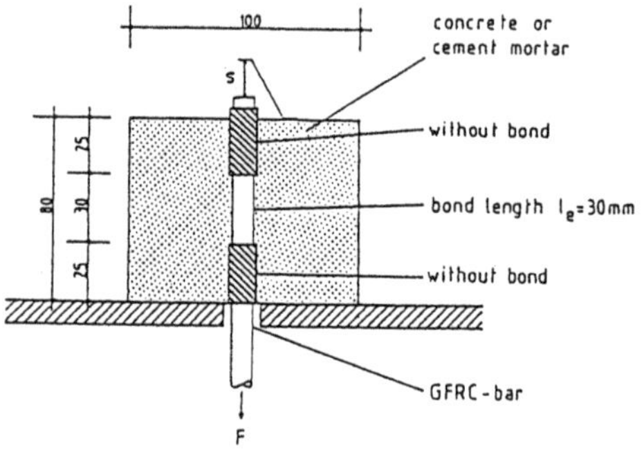

Fig.3: Test set-up of the pull-out tests

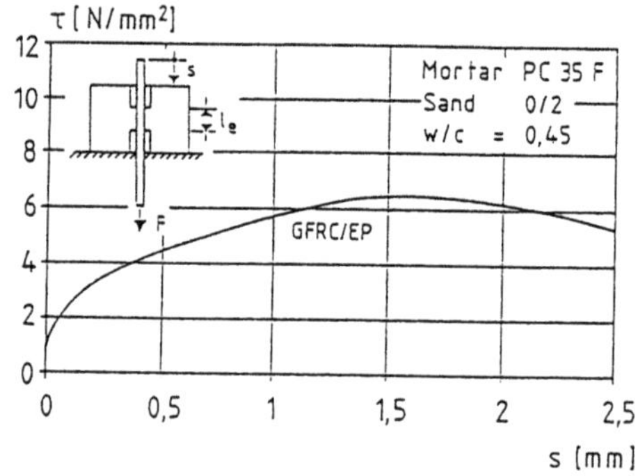

Fig. 4: Bond law for GFRC/EP (Faoro 1992)

Fig.4 shows the experimentally determined bond law for GFRC-bars with epoxy resin powder coating. It is recognizable that the bond stiffness of epoxy powder coated GFRC-bars is considerably lower than that of ribbed steel bars. Also the determined bond strength τ_u is lower. In comparison to conventional reinforcing steels it is evident that the maximum bond stresses are only obtained at relatively high slip values. This can't be recognized especially in the case of sanded GFRC-bars, as shown in fig. 5.

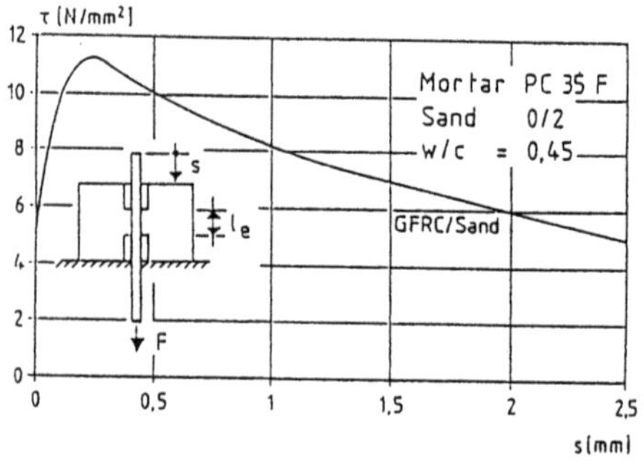

Fig. 5: Bond law for GFRC-bars sanded

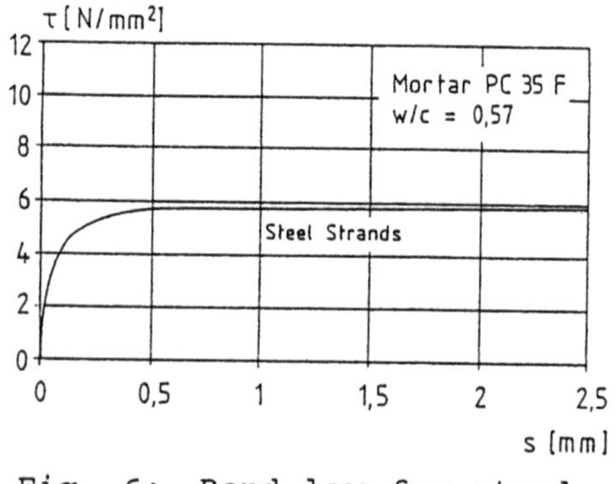

Fig. 6: Bond law for steel strands ϕ 0,6"

In case of sanded bars the bond-stress-slip relations are comparable to those of conventional reinforcing steels (ribbed bars).

Fig. 6 shows a bond law for a bundle consisting of 4 strands with ϕ 0,6" (Trost et.al. 1980) for comparison aims. It is recognizable that the bond stresses - able to be activated -of bundled GFRC-bars (with epoxy powder coating) and strands have approximately the same values. While in the case of steel strands an approx. constant pull-out-resistance can be assumed even when the relative displacements are high, a significantly higher bond strength in case of sanded GFRC-bars has to be taken into account. However the activated bond stresses in this case are highly dependent on the relative displacements (slip) and are approximating a pull-out-resistance of roughly 4 N/mm² for high slip values (s > 2,5 mm).

3 Anchoring of GFRC-bars

An important aspect in respect to practical application of resin bonded glass fiber bars for construction is a secure and economic solution to problems regarding anchorage of the bars. This was recognized at an early stage so that numerous solution proposals can be taken from various publications (compare summary in Faoro 1988). However the solution possiblitities presented are only partly feasible as full utilization of the material strengths in these structures is still only minimal and there are no evidenced test results regarding the long-term and dynamic behaviour of the achorages used.

As an anisotropic composite material, the glass fiber composite bars only withstand transverse compressions to 10 % of their tensile strength.

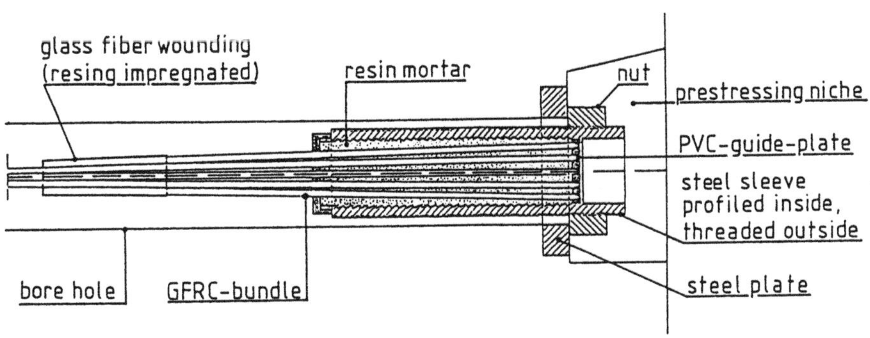

Fig. 7: Grouted anchorage of a bar bundle comprising 19 individual bars in prestressed condition

Therefore completely new solutions were required in the field of anchorage engineering. In addition, the relatively low interlaminary shear strength of the composite material requires a longer anchorage length in comparison with steel prestressing tendons. The grouted tube anchor developed provides a practical solution which also meets the specific material requirements (fig. 7). With this anchorage a "bundle" of up to 19 individual bars are fixed in a profiled steel tube with a purpose-developed resin grouting mortar. Meanwhile extensive theoretical and experimental investigations together with several practical applications in building structures and earthworks have provided impressive substantiation of the serviceability of the anchorage system discussed here.

4 Design principles of GFRC-rock anchors and bolts

4.1 Material anchor heads

GFRC-material has a static fatigue strength of approx. 70 % of the short term tensile strength.

$$f_{clk} = 0.70 \times f_{ck} \qquad (1)$$

f_{clk} - Static fatigue strength of the composite
f_{ck} - Characteristic short term tensile strength of the composite

The admissible continous stress has been fixed according to the reglementations for prestressing steel at 70 % of the static fatigue strength. Therefore the following is valid (see König and Schneider 1992):

$$\sigma_{cw} = 0.70 \times f_{clk} \qquad (2)$$

σ_{cw} - Working stress of the composite

With a characteristic short term tensile strength of approx. f_{ck} = 1.600 N/mm² follows a working stress of approx. σ_{cw} = 800 N/mm².

In case of these considerations it is presupposed that the anchorage of the bar bundle has a high efficiency and that it is able to reach an optimum degree of utilization. The following is demanded:

$$\eta_a = F_u/F_{u,\,cal} \geq 0.95 \qquad (3)$$

η_a - Utilization of the material strength by the anchorage device

Moreover it is taken as granted that also in case of a long term stress and/or static load a similar degree of utilization of the material is given.

4.2 Dimensioning of the grouted elements

4.2.1 General reflections

In the following exclusively the conditions in case of massive rocks are described in order to be able to neglect the yielding of the rock or the soil. This means that the relative displacements between the grouted element and the rock are not taken into consideration, which is also applicable for a solid rock.

The studies are executed on the basis of a theoretical model which has been proved in the practice. Here the distributions of the relative displacements s, the bond stresses τ_c and the bar stresses σ_c of the rock anchors are calculated by the approximate solution of the differential equation of the bond able to be subject to displacements by stepwise integration.

4.2.2 Modelling

The bond laws explained in section 2.2 were described mathematically with the relations indicated by Eligehausen et.al. 1983 and shown in fig. 8.

In case of the used iterative method the non-linear proceedings of $\sigma_c(x)$, $\tau_c(x)$, $v_c(x)$ and $s_c(x)$ are linearized in sections or it is as-sumed that they are running on a constant level. For more details see Faoro 1988.

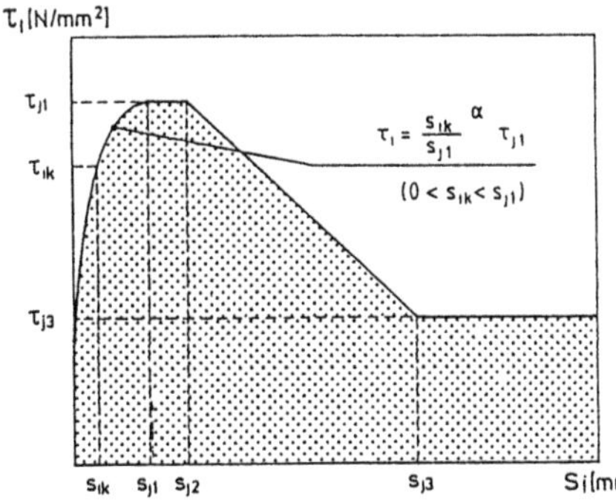

Fig. 8: Analytic form of the bond laws

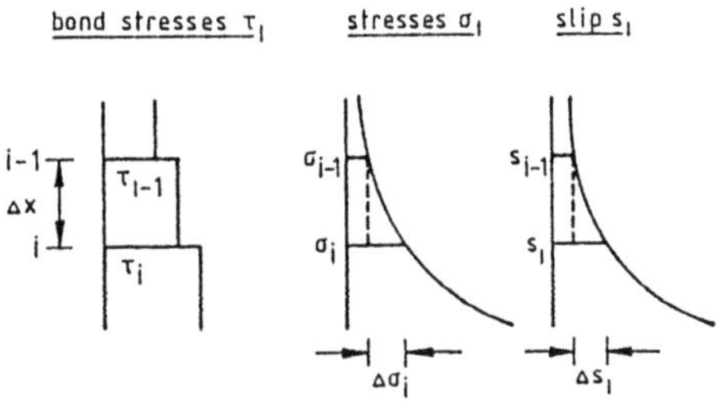

Fig. 9: Theoretical assumptions

4.2.3 Essential results of the parameter studies

The following diagrams show the theoretically determined distribution functions $\sigma_i(x)$, $\tau_i(x)$ and $s_i(x)$ for rock anchors comprising GFRC-material (epoxi powder coated and sanded) and steel strands. The conditions "working load" and "load at rupture" have been examined.

It can be recognized that due to the lower modulus E and the specific bond behaviour, rock anchors comprising GFRC-material with epoxi coating require partly considerably greater transfer lengths of forces than steel strands if stressed highly (ultimate state).

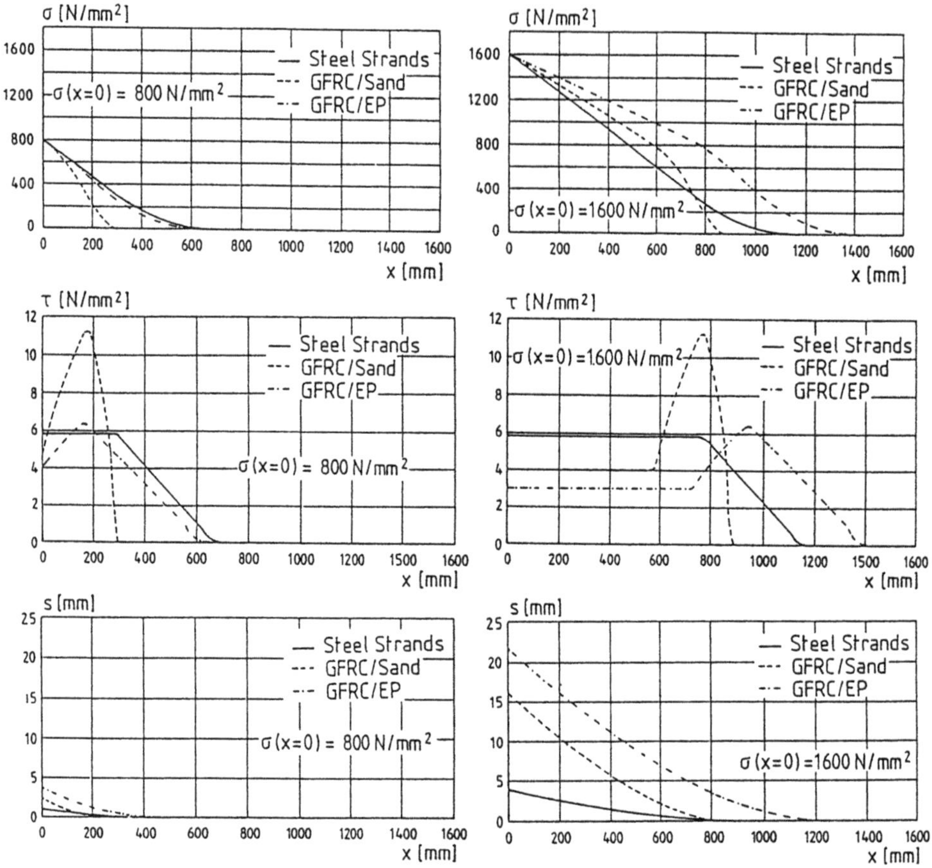

Fig. 10: Distribution functions of $\sigma_i(x)$, $\tau_i(x)$ and $s_i(x)$ for GFRC-steel strands and rock anchors

This disadvantage is made up partly by better bond qualities as e.g. in case of sanded GFRC-bars. Regarding the working stress state a significant difference concerning the load bearing behaviour can't be recognized.

Nevertheless in case of GFRC-rock anchors significantly higher relative displacements s_c between the bars and the grouted mortar always have to be expected. This could lead to eventual friction or abrasion damages. Irrespective of that, the lengths of the grouted elements of GFRC rock anchors and -bolts have to be dimensioned in a generous way.

5 Monitoring of GFRC-rock anchors and bolts

The technical demands on modern load bearing structures in respect to a permanent control and monitoring of the structural condition have increased considerably in the last few years. Therefore thought was given to the permanent monitoring of prestressing reinforcement with the aid of appropriate sensors.

In contrast to metallic materials, the glass fiber composite material opens up the possibility of integration sensor elements during production. The optical fiber sensors used comprise so-called optical fibers (quartz) furnished with a cross-stranded wire coil (fig. 11). The pitch of this coil is so dimensioned that axial tension on the sensor produces mircro-bendings at points of contact between the wires and the optical glass fiber. These microbendings lead to a change of light permeability into the optical fibers which correlates to elongation of the sensor. In addition, local overstretch and points of interference can be located by measurement of the light transmission time. Fig. 12 clarifies the measurement principle.

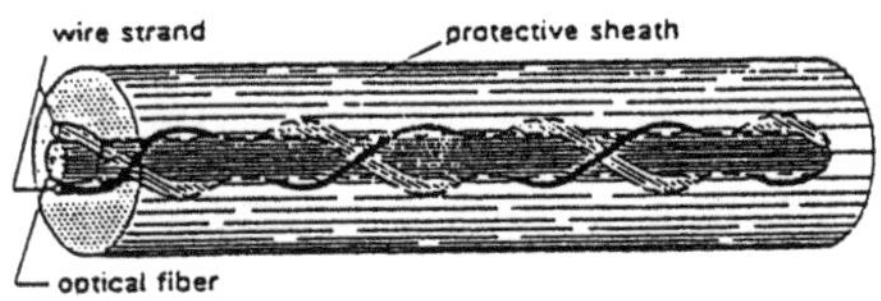

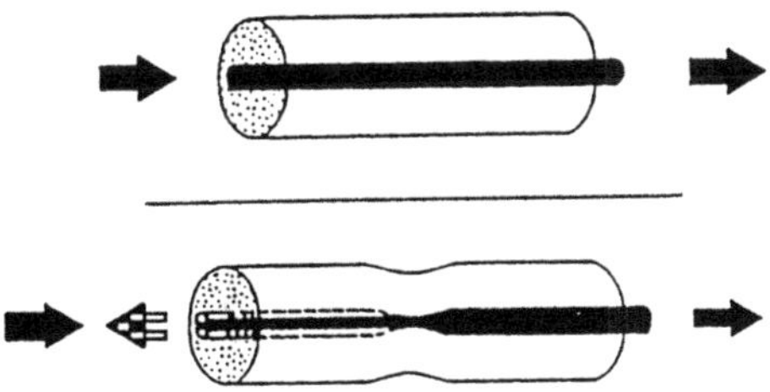

Fig. 11: Strain sensor on an optical fiber basis

Fig. 12: Measurement principle of sensors

6 Conclusions for the practice

Due to the relatively low modulus E of GFRC-rock anchors (E_{steel}/E_{GFRC} = 4) in case of the anchorage by bond in cement greater lengths of the grouted elements are required than in case of conventional rock anchors comprising steel strands. But with increasing bond quality of sanded GFRC-bars the transfer lengths of forces are decreasing so that in this respect there are no significant differences in comparison to steel anchors.

Moreover significantly higher slip values can be expected in case of GFRC-rock anchors. The bearing behaviour of anchors comprising GFRC-material in the soil is being examined at present. This anchor ground let expect - due to the low stiffness - considerably greater relative displacements between the grouted element and the soil. In these cases the bearing behaviour of the soil must be invastigated in detail.

The possiblity to integrate strain sensors already during the manufacture of the fiber composite material opens up very interesting prospects. A permanent monitoring of highly stressed earth and rock anchors or same situated at exposed positions can be realized at any time when required.

7 References

Eligehausen, R.; Popov, E.P. and Bertero, V.V. (1983) Local bond stress slip relationships of deformed bars under generalized excitations, **Report No. UCB/EERC-83/23.** Earthquake Engineering Research Center, College of Engineering, University of California, Berkley: University of California.

Faoro, M. (1988) Zum Tragverhalten kunstharzgebundene Glasfaserstäbe im Bereich von Endverankerungen und Rissen im Beton (On the load bearing behaviour of resin bonded glass fiber bars in the zone of end anchorages and cracks in concrete), in **IWB-Mitteilungen.** - Stuttgart: Fakultät 2, Universität Stuttgart, Dissertation.

Faoro, M. (1992) Bearing and deformation behaviour of structural components with reinforcements comprising resin bonded glass fiber bars and conventional ribbed steel bars. **International Conference on Bond in Concrete**, Riga, Technical University, Riga, Latvia.

König, G. and Schneider K. (1992) Ein Bemessungskonzept für Vorspannung mit Faserverbundwerkstoffen, insbesondere mit Polystal (A design concept for prestressing with fiber composite materials, especially with Polystal) **Forschungskolloquium Faserverbundwerkstoffe im konstruktiven Ingenieurbau und Lichtwellenleitersensorik für die Bauwerksüberwachung.** Technische Universität Braunschweig, Institut für Baustoffe, Massivbau und Brandschutz.

Trost, H.; Cordes, H.; Thormählen, U. and Hagen, H. (1980) Teilweise Vorspannung - Verbundfestigkeit von Spanngliedern und ihre Bedeutung für Rißbildung und Rißbreitenbeschränkung (Partial prestressing - bond strength of prestressing tendons and their influence for crack formation and crack-width limitation) in **Heft 310. Deutscher Ausschuß für Stahlbeton.**- Berlin: Verlag Wilhelm Ernst & Sohn.

60 AE INVERSION ANALYSIS IN PULL-OUT TESTS

M. OHTSU
Department of Civil and Environmental Engineering,
Kumamoto University, Japan

Abstract
Acoustic emission (AE) is recently realized as a technique to identify the fracture process zone. A conventional procedure is the determination of crack locations, based on the arrival time difference. To provide further information on kinematics of internal cracks, a source inversion procedure of AE is proposed. The technique is applied to AE events monitored during a pull-out test of an anchor bolt from concrete block.

Crack kinematics determined are in remarkable agreement with experimental findings. Crack-opening directions of AE events due to tensile cracks are vertical to the final failure surface, while directions of motion due to shear cracks are almost parallel to the failure surface. The shear crack are restrictively observed near the anchorage. It implies that tensile cracks are generated first along the final failure surface and shear slips follow, starting at the bottom of the anchor bolt.
Keywords: Acoustic Emission, Source Inversion, Pull-Out Test.

1 Introduction

Embedded anchors are widespreadly applied to the engineering design for inserts and fixings in concrete structures. The bearing capacity and failure behavior of anchors are often examined by pull-out tests. The pull-out test is also utilized as a means for nondestructive evaluation of concrete (Carino, 1991). Major efforts for research are associated with clarifying the failure process of the pull-out process, because the mechanisms consist of mixed-mode crack propagation.

In fracture mechanics, acoustic emission (AE) is applied to identify the fracture process zone (Min-

Fracture and Damage of Concrete and Rock – FDCR-2.
Edited by H.P. Rossmanith. ISBN 0 419 18470 8.

dess, 1991) and the location of crack front (Chen et al., 1992). The technique applied is the flaw location or the source location (Berthaud et al., 1991; Nomura et al., 1991). Based on the arrival time differences at multi-observation points, crack coordinates are calculated. Because the accuracy of the location technique is limited, the identification of the process zone and the crack front is so far marginal.

Source characterization of AE has been investigated to obtain information on crack kinetics and kinematics (Ohtsu, 1982; Ohtsu and Ono, 1984). Based on these theoretical work, the source location technique is further extended as a source inverse procedure using the moment tensor representation (Ohtsu, 1987; Ohtsu and Ono, 1988; Ohtsu, 1988). The procedure was successfully applied to AE waveforms detected at in situ hydrofracturing tests, where AE sources were classified into tensile cracks and shear cracks and given the direction of crack motion (Ohtsu, 1991).

In the present paper, AE inversion analysis is applied to the pull-out process of a shallow hook anchor.

2 Experiment

A pull-out test was performed by using a concrete block of dimension 1.0 m x 1.0 m x 0.3 m. Compressive strength determined by employing three cylindrical specimens of 10 cm diameter and 20 cm height was 54.3 Mpa at 28-day moisture curing. The velocity of P wave was 4150 m/s, which was determined from traveling time in the cylindrical specimen.

An anchor bolt embedded consists of a washer of 44 mm diameter and 3.2 mm thickness, and a prestressing steel bar of 17 mm diameter. The washer is fixed to the bar by a nut of 27 mm thickness. A head of the bolt embedded is located at 5 cm depth. To decrease the bond between the prestressing bar and surrounding concrete, the steel bar was wrapped up with polyethylene sheet. A set-up is shown in Fig. 1. The steel bar is pulled out by employing a center-hole jack.

During the pull-out process, AE waveforms are recorded, by using six AE sensors attached to the concrete block. Sensors are of resonant type with 150 kHz resonance. Sensor locations are indicated in Fig. 2. Total amplification is 60 dB and the frequency range recorded is 30 kHz to 80 kHz. AE waveforms detected at sensor locations are stored onto a disk through a digital recorder, which converts analog data to digital records at 1 MHz sampling frequency.

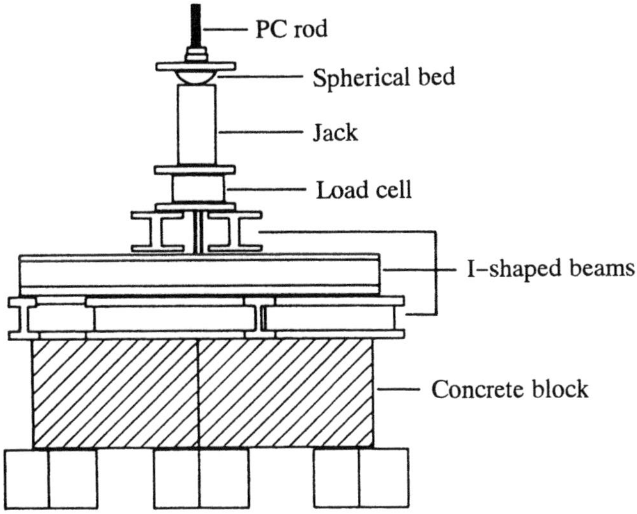

Fig. 1 Experimental set-up of pull-out test.

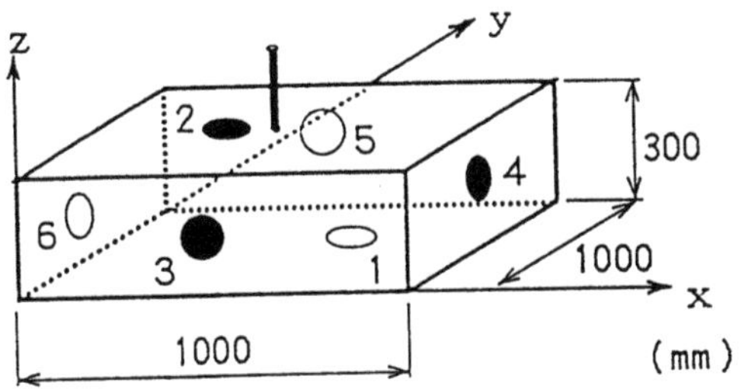

	x	y	z
ch 1	700	480	0
ch 2	410	500	300
ch 3	550	0	120
ch 4	1000	500	150
ch 5	40	100	200
ch 6	0	580	110

Fig. 2 AE sensor locations in the concrete block.

3. Inverse analysis by SiGMA

3.1 Moment tensor

AE waveform $u_i(\mathbf{x},t)$ due to crack displacement $b_k(\mathbf{y},t)$ on crack surface F is theoretically represented, as follows (Ohtsu, 1988);

$$u_i(\mathbf{x},t) = \int_F G_{ip,q}(\mathbf{x},\mathbf{y},t)\, m_{pq} * s(t)dS, \qquad (1)$$

where $G_{ip,q}$ is the spatial derivative of Green's function and source function s(t) represents crack kinetics. m_{pq} is called a moment tensor, and is expressed by using the crack displacement bl_k,

$$m_{pq} = b\ [\lambda l_k n_k \delta_{pq} + \mu(l_p n_q + l_q n_p)]. \quad (2)$$

Here b is the magnitude and l_k is the direction of crack vector. n_k is the normal vector to crack surface F. Thus, moment tensor components contain kinematic information on a crack of crack motion l_k and crack normal n_k. An applicability of eqs. 1 and 2 to actual AE waveforms was already demonstrated by the simulation analysis in the delaminating process of stainless steel overlay (Ohtsu et al., 1987; Yuyama et al., 1988).

To determine moment tensor components from AE waveforms, eq. 1 is simplified, taking into account only the first motion of P wave,

$$u_i(\mathbf{x}) = r_i r_p r_q m_{pq}/(4\pi\rho R v_p^3), \quad (3)$$

where the time dependency is omitted. R is the distance from AE source **y** to observation point **x**, and r_i is its direction cosine. When an AE waveform due crack nucleation is recorded at observation point **x**, AE amplitude of the first motion corresponds to displacement u_i of eq. 3, after the reflection and the system calibration.

In the case that AE waveforms are detected at more than six sensor locations, AE source location procedure could give information on distance R and direction cosine r_i. Eventually eq. 3 leads to a set of linear algebraic equations with six unknown components m_{pq}, because the moment tensor is a symmetric second rank tensor with six independent components. For the analysis, AE measuring system of more than six channels is required.

3.2 Eigenvalue analysis

From eq. 2, it is known that the direction of crack motion and that of crack normal are not readily separable in the moment tensor components. The components of the moment tensor depend on a scalar product (angle) between crack motion and crack normal. As a result, a crack-type classification based on the angle is proposed (Ouyang et al., 1991), although their theoretical treatment is marginal and not generally applicable.

Useful information could be derived from the eigenvalue analysis of the moment tensor. In the case of a shear crack, the two vectors are vertical and the scalar product $l_k n_k$ is equal to zero. Three eigenvalues could be represented as (X, 0, -X). For a tensile crack, the direction of crack normal is identical to that of crack opening. The scalar product $l_k n_k$ is equal to unity. In this case, the eigenval-

ues are decomposed into deviatoric components (Y, -0.5Y, -0.5Y) and hydrostatic mean components (Z, Z, Z). The deviatoric components are named as the compensated linear vector dipole (CLVD) by Knopoff and Randall (1970). Here, AE sources are generally referred to as mixed mode of a tensile crack and a shear crack. Setting the ratio of shear component as X, that of CLVD component as Y, and mean component as Z in the maximum eigenvalue, all eigenvalues are uniquely decomposed, as follow;

$$1.0 = X + Y + Z,$$

The intermediate eigenvalue/the maximum eigenvalue

$$= 0 - 0.5Y + Z,$$

The minimum eigenvalue/the maximum eigenvalue

$$= - X - 0.5Y + Z.$$

After determining the eigenvalues of the moment tensor, the above decomposition is solved and the ratios X:Y:Z are determined. Based on the results of numerical experiments, a simple criterion is proposed for the classification of crack types. When the ratio, X, is bigger than 55 %, AE source is referred to as a shear crack. In the case less than 45 %, AE source is classified into a tensile crack.

In the procedure, the ratios of eigenvalues and the relative values of moment tensor components are necessary. It implies that amplitude of the first motion in eq. 3 should be recorded as relative values. Consequently, the sensor calibration is only needed to compensate the equal sensitivity.

After the classification of cracks, directions of crack motions are obtained from eigenvectors. For the tensile crack, the first eigenvector represents the direction of crack opening. In the case of the shear crack, either the direction of crack motion or crack normal is determined from the sum of the first and the third eigenvectors. If the one direction is determined, the other is easily identified because the two directions are vertical.

3.3 SiGMA procedure

All treatments discussed above are implemented into a computer software. After eq. 3, it is named SiGMA (simplified Green's function for moment tensor analysis). Digital records of AE waveforms are analyzed by the SiGMA procedure, where more than six waveforms for each AE source are necessary. In the first stage, all waveforms for one AE event are shown on the CRT screen. On each waveform, the arrival time and the amplitude of the first motion are read and stored. In the next stage, the location of AE source is computed from the arrival time differences. Then, moment tensor components are determined from the amplitudes of the first motions. In the final stage, the eigenvalue analysis is performed, and crack type and crack orientation are obtained.

4 Results and discussion

In the pull-out test, the shape of failure surface is closely related to the apex angle between an anchor head and reacting supports. For the low apex angle, the failure surface becomes of a conical frustum. In the case of high apex angle, a trumpet-shaped frustum is observed. In the present experiment, the apex angle was bigger than 60°, but the depth of anchor bolt is rather shallow. As a result, the failure surface showed thin trumpet-shaped frustum.

From AE activity, the initiation of circumferential cracking near the edge of the anchor head was observed, prior to reaching the peak load. Then, these circumferential cracks propagated, directing the supports. AE sources detected at this stage are plotted in Fig. 3. Plan and elevation views are shown. The final failure surface is indicated by solid curve in the elevation view. AE sources are plotted at their locations with crack type and the direction of crack motion. Tensile cracks are denoted by arrow symbols. The direction of crack opening corresponds to the direction of the arrow. Shear cracks are denoted by cross symbols. Because either vector n_k or vector l_k is determined from the eigenvectors, two possible vertical directions are indicated by the direction of the cross.

From the plan view, it is observed that tensile cracks are dominantly located at the boundary of AE cluster. It is noted that opening directions of tensile type are radial. The distribution of AE sources suggest mechanisms that tensile cracks are first generated at the final failure surfaces and shear cracks follow. In the elevation view, locations of cracks are along the final failure surface. In respect to tensile cracks, the directions of crack opening are remarkably vertical to the final failure surface.

In contrast, many of shear cracks are dominantly found near the anchor head. It suggests that slip motions are generated around the bottom of the anchor bolt on the crack surfaces nucleated by the tensile cracks.

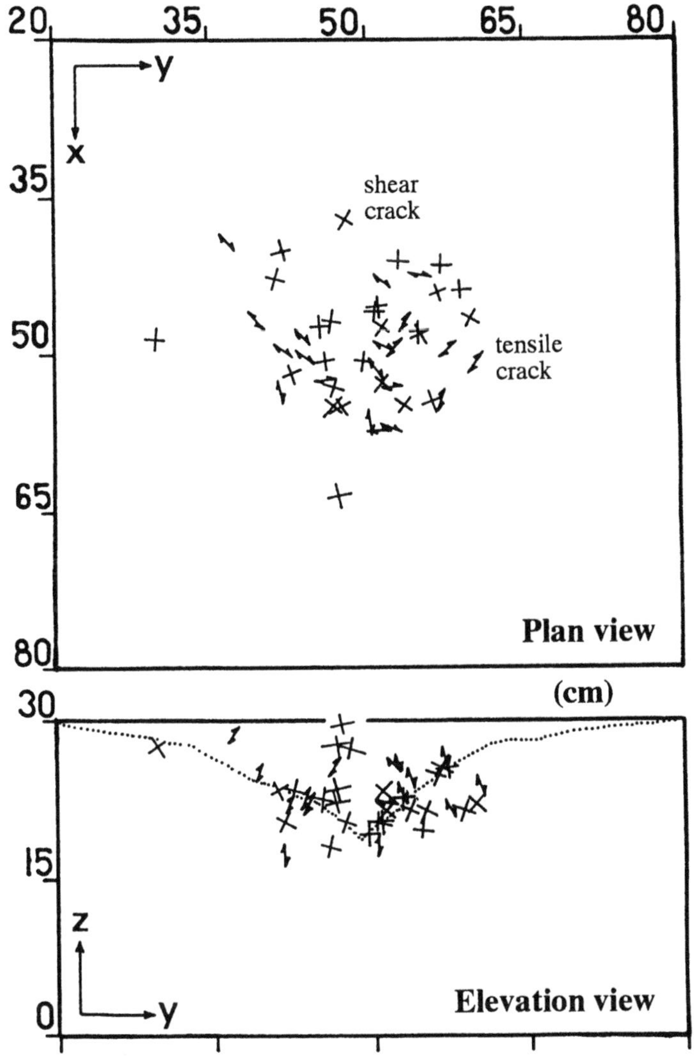

Fig. 3 Plan and elevation views of AE source locations with crack type and crack orientation observed prior to the peak load.

AE sources observed after the peak load are shown in Fig. 4. Shear cracks are observed in the wider region than in Fig. 3. The directions of tensile cracks in the plan view are not necessarily radial. It may imply that tensile cracks are generated due to other mechanisms than circumferential crack propagation, such as crash of matrix and the delamination of

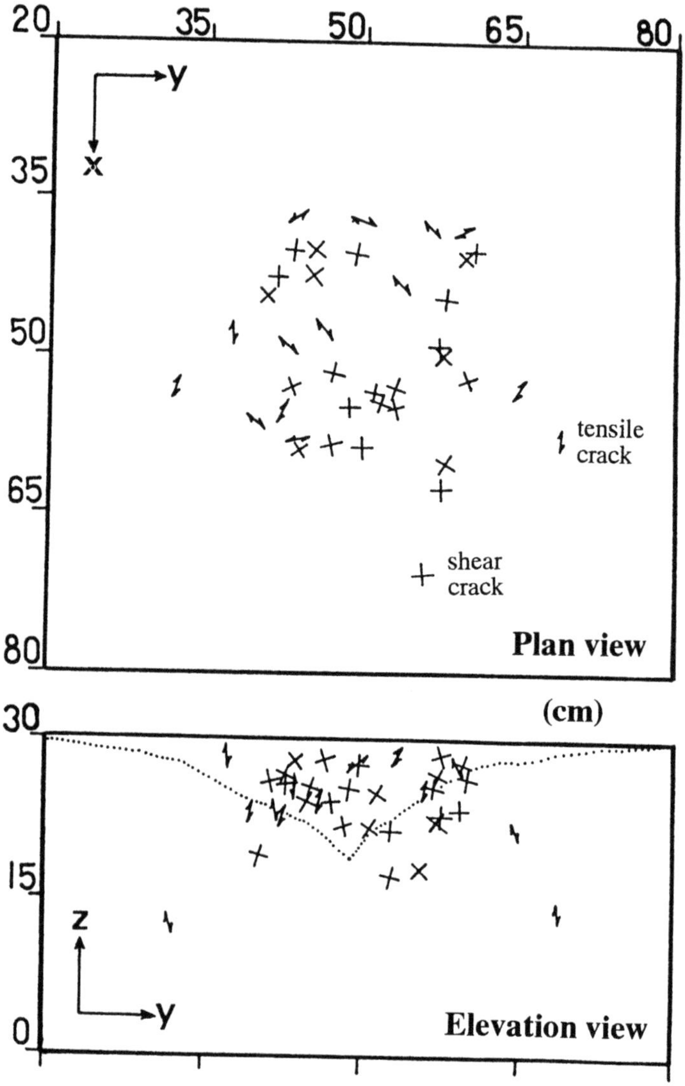

Fig. 4 Plan and elevation views of AE source locations with crack type and crack orientation observed at right after the peak load.

aggregate interface. In the elevation view, the directions of tensile cracks are still vertical to the final failure surface and AE sources are located in the wider cluster. From these results, it is clearly observed that directions of crack motions determined by the AE source inversion are in good agreement with those predicted from the final failure surface.

5 Conclusion

AE source inversion procedure by 'SiGMA software' is applied to the pull-out test of anchor bolt. The following conclusions are stated.
(1) AE sources are located as close as the final failure surfaces. AE sources are classified into tensile cracks and shear cracks. It is demonstrated that both cracks of the tensile and the shear are observed in the pull-out process.
(2) In the pre-peak load, the tensile cracks are mainly observed in the outer boundary of AE cluster. It implies the failure process that the tensile cracks are nucleated first and then the shear cracks follow on the crack surface created.
(3) The opening directions of the tensile cracks are radial from the anchor bolt in the plan view, and are vertical to the final failure surface in the elevation view. It suggests that the final failure surface propagates from the anchor bolt, and is created by coalescing the tensile cracks.
(4) The directions of the shear cracks are almost parallel to the failure surface. Therefore, it is found that the shear cracks follow the tensile cracks, generating slip motion on the crack surface created.

6 Acknowledgement

The pull-out tests were performed in the Department of Civil Engineering, Gifu University. The authors wish to thank Prof. W. Koyanagi, Prof. K. Rokugo and Mr. H. Iwase for their assistance on the experiments.

7 References

Berthaud, Y., Ringot, E. and Schmitt, N. (1991) Experimental Measurements of Localization for Tensile Tests on Concrete, in **Fracture Processes in Concrete, Rock and Ceramics, Vol.1** (eds. J. G. M. van Mier, J. G. Rots and A. Bakker), E & F.N. Spon, London, pp. 41-50.

Carino, N. J.(1991) Pullout Test, in **Handbook on Nondestructive testing of Concrete** (eds. V. M. Malhotra and N. J. Carino), CRC Press, Boca Raton, pp. 39-82.

Chen, H. L., Cheng, C. T. and Chen S. E. (1992), Determination of Fracture Parameters of Mortar and Concrete Beams by Using Acoustic Emission, **Materials Evaluation**, Vol. 50, No. 7, pp. 888-894.

Knopoff, L. and Randall, M. J.(1970),The Compensated

Linear-Vector Dipole: A Possible Mechanism for Deep Earthquakes, **J. Geophysical Research**, Vol. 75, No. 26, pp. 4957-4963.

Mindess, S. (1991) Fracture Process Zone Detection, in **Fracture Mechanics Test Methods for Concrete** (eds. S. P. Shah and A. Carpinteri), Chapman and Hall, London, pp. 231-262.

Nomura, N., Mihashi, H. and Izumi, M. (1991) Properties of Fracture Process Zone and Tension Softening Behavior of Concrete, in **Fracture Processes in Concrete, Rock and Ceramics, Vol. 1** (eds. J. G. M. van Mier, J. G. Rots and A. Bakker), E & F.N. Spon, London, pp. 51-60.

Ohtsu, M. (1982) Source Mechanism and Waveform Analysis of Acoustic Emission in Concrete, **J. Acoustic Emission**, Vol. 1, No. 2, pp. 103-112.

Ohtsu, M. and Ono, K.(1984) A Generalized Theory of Acoustic Emission and Green's Functions in A Half Space, **J. Acoustic Emission**, Vol. 3, No. 1, pp. 27-40.

Ohtsu, M.(1987) Determination of Crack Orientation by Acoustic Emission, **Materials Evaluation**, Vol. 45, No. 9, pp. 1070-1075.

Ohtsu, M., Yuyama, S. and Imanaka, T.(1987) Theoretical Treatment of Acoustic Emission Sources in Microfracturing due to Disbonding, **J. Acoustical Society of America**, Vol. 82, No. 2, pp. 506-512.

Ohtsu, M. and Ono, K. (1988), AE Source Location and Orientation Determination of Tensile Cracks from Surface Observation, **NDT International**, Vol. 21, No. 3, pp. 143-150.

Ohtsu, M. (1988) Source Inversion of Acoustic Emission Waveform," **Proc. of JSCE**, No. 398/I-10, pp. 71-79.

Ohtsu, M. (1991) Simplified Moment Tensor Analysis and Unified Decomposition of Acoustic Emission Source: Application to In Situ Hydrofracturing Test, **J. Geophysical Research**, Vol. 96, No. B4, pp. 6211-6221.

Ouyang, C., Landis, E. and Shah, S. P. (1991), Damage Assessment in Concrete Using Quantitative Acoustic Emission, **J. Engineering Mechanics**, Vol. 117, No. 11, ASCE, pp. 2681-2698.

Yuyama, S., Imanaka, T. and Ohtsu, M.(1988) Quantitative Evaluation of Microfracture due to Disbonding by Waveform Analysis of Acoustic Emission, **J. Acoustical Society of America**, Vol. 83, No. 3, pp. 976-983.

61 ON CONCRETE FATIGUE FRACTURE IN PULL-OUT TESTS

P. BOCCA, E. CADONI and S. VALENTE
Department of Structural Engineering,
Politecnic of Turin, Italy

Abstract
Anchorage fatigue failure has been investigated through pull-out tests carried out on concrete slabs by applying sinusoidal shaped loading cycles to anchor bolts previously embedded in the casting. As is known, in pull-out tests involving a contrast ring of considerable size compared to bolt depth, concrete failure is caused by a tensile stress field localised at the end of the bolt head, as born out by the fact that this is the area where both the main crack and the microcracking zone are initiated. Damage propagation is studied as a function of the number of loading cycles for two concrete types having different fracture energy, $\mathcal{G}_F$, and different Carpinteri's brittleness number, s_E, maximum load intensity being the variable taken into account. Experimental results and the subsequent numerical elaboration have shown that in order to predict anchorage fatigue life in concrete the relationship between the increase in displacement, or the opening of the crack, and the number of loading cycles is definitely preferable than the relationship, commonly used for metals, between crack propagation velocity and the number of cycles.

1 Introduction

Fatigue failure of the bond between steel and concrete with consequent pulling out of the anchorage has been investigated through a series of pull-out tests where steel anchor bolts, embedded in the concrete during casting, are subjected to loading cycles.

The pull-out test has already been used in order to determine concrete compressive strength in situ (Bocca, 1979; 1984), to study the collapse mechanism involved in anchorage pull-out (Ballarini et al., 1986; Bocca et al., 1990b; Elfgren, 1989; Eligehausen and Sawade, 1989; Krenchel and Shah, 1986; Miller and Keer, 1982; Stone and Carino, 1983; Stone and Giza, 1985) and to determine concrete fracture energy in situ (Bocca et al, 1989; Valente, 1991).

When the contrast ring is large compared to bolt depth, the stress field produced by the support reactions turns out to be negligible compared to the stress field close to the stem. In this case, a tensile stress field around the bolt head triggers off the crack and, when static tests are considered, brittle behaviour is observed during collapse. By using smaller diameter contrast rings or different types of anchorage, such as those commonly used in the building industry, a brittle-ductile, or ductile behaviour is observed instead, that is to say, friction or hooping phenomena, due to the presence of reinforcement, or

Fracture and Damage of Concrete and Rock – FDCR-2.
Edited by H.P. Rossmanith. © 1993 Taylor & Francis. ISBN 0 419 18470 8.

interlocking effects, etc. occur alongside with diffused micro and macrocracking that enhance the ductility of the material.

This means that the chosen mechanism meets a twofold need:

- from a theoretical standpoint, in the static domain, it ensures that concrete behaviour will be similar to the fracture behaviour of many metals for which it is possible to control the propagation velocity of a well defined crack;
- the results acquired, when used to predict the fatigue life of other types of anchorages will be very conservative from the standpoint of safety.

With these premises in mind, this investigation was aimed at:

a) ascertaining, through a series of tests, whether or not the brittle behaviour as described above undergoes appreciable variations after a pre-determined number of fatigue cycles, so as to collect useful evidence as to the applicability to concrete of fatigue life prediction laws based on crack propagation velocity;
b) testing bolt fatigue life as a function of the number of load cycles (involving different load levels) for two concrete types with different fracture energy, $\mathcal{G}_F$, and different Carpinteri's brittleness number, s_E.

2 Testing programme

The tests were performed on 28 steel bolts (Fig. 1) embedded, before the casting, in concrete slabs sized 100x100x15 cm.

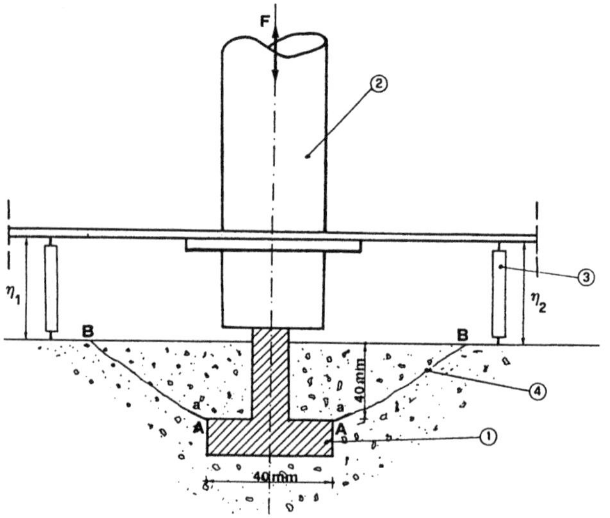

Fig. 1 - Layout of testing equipment.
1. Steel bolt
2. MTS
3. Inductive displ. transducer
4. Crack trajectory

Two types of concrete, having the composition and mechanical properties given in table 1 were tested. Compressive strength, R, was evaluated on cubes (with 16 cm long sides). The determination of the secant modulus, E, and fracture energy, $\mathcal{G}_F$, were performed on 16x16x50 cm prisms and 10x10x84 cm notched prisms, respectively. The next to the last line of the tables gives Carpinteri's brittleness number for structural elements of the two concrete types being considered. It has been determined from static pull-out tests on slabs by considering a conventional characteristic dimension and checked through three-point tests on small-sized notched beams.

Table 1 Concrete composition and Mechanical properties.

MATERIAL	Concrete A	Concrete B
Composition	- Portland cement type 425: 450 kg/m^3 - Alluvial aggregate with maximum diameter = 12 mm: 910 kg/m^3 - Alluvial sand with 0 - 8 mm diameter: 910 kg/m^3 - Plasticizer: 1 kg/m^3	- Portland cement type 425: 400 kg/m^3 - Alluvial sand with 0 - 8 mm diameter: 1700 kg/m^3
Compressive strength, R	51.6 MPa	27.6 MPa
Young's modulus, E	27500 MPa	17200 MPa
Fracture Energy, $\mathcal{G}_F$	119 N/m	39 N/m
Carpinteri's Brittleness number, s_E, for struct. elements	4.4 10^{-2}	3.7 10^{-2}
Carpinteri's Brittleness number, s_E, after N fatigue cycles	4.3 10^{-2}	5.5 10^{-2}

It should be pointed out that before the tests, the slabs and test pieces were kept for 30 days at a temperature of about 20°C at a relative humidity of approx. 65%.

Static pull-out tests were performed by means of an MTS with maximum load of 250 KN by imposing a constant velocity of the load application point of $\dot{\eta} = 5 \cdot 10^{-6}$ m/s. Instantaneous displacement, η, was calculated as the arithmetical mean of the η_1 and η_2 values measured by two inductive transducers placed in a diametrically opposed position with respect to the anchor bolt (Fig. 1).

The measuring points of the transducers on the surface of the slab and on the testing machine were chosen so as to minimize possible displacement errors due to play in the mechanical connection or elastic strains in the materials. A large diameter contrast ring (50 cm) was used so as not to affect the cracking surface, in both static and dynamic tests.

Dynamic tests were carried out by applying a sinusoidal loading cycle, as given from the expression:

$F = F_{min} + F_0 \sin \omega t$, where F_{min} was the same for all bolts and came to 100 daN, while frequency, $\omega/2\pi$, was 1 Hz.

Maximum load, F_{max}, was kept constant throughout the test. Force-displacement diagrams (F, η) were recorded according to the following procedure:

– recording the first loading/unloading cycle by keeping the load increase rate constant;

– recording the (F, η) loading/unloading diagram after N_0, N_1, N_2, ... N_i fatigue cycles up to bolt failure;

– repeating the tests for two different load levels F_{max}/F_p, (F_p being the bolt pull-out force).

Load levels turned out to be, respectively:

– for concrete A : $F_{max}/F_p = 0.8$ (high load level), $F_{max}/F_p = 0.6$ (low load level), $F_{max}/F_p = 0.73$ (medium load level);

– for concrete B : $F_{max}/F_p = 0.6$ (high load level), $F_{max}/F_p = 0.4$ (low load level), $F_{max}/F_p = 0.49$ (medium load level).

Preliminary tests performed on the bolts made it possible to identify, with some approximation, the load value intervals for the two different concrete types outside which it is not reasonable to speak of fatigue, as failure occurs after very few loading cycles or, conversely, fatigue cycles tend to infinity. And namely we get:

– concrete A: $0.5 < F_{max}/F_p < 0.9$

– concrete B: $0.3 < F_{max}/F_p < 0.6$

On some bolts, after N cycles, having reached a pre-determined crack length (Fig. 1), the full (F, η) curve extended to the post-peak phase was determined so as to be able to calculate total fracture energy W_f.

3 Test results and discussion

Figs. 2 and 3 show the force-displacement diagrams (F - η) obtained on A and B type concrete. Dashed line diagrams refer to the static tests performed before subjecting the tests pieces to fatigue cycles, while solid line diagrams have been obtained from static tests after subjecting the bolts to repeated fatigue cycles. The number of load cycles, N, to be performed before the static tests was determined on the basis of the value of permanent strain, η_0, taken conventionally to be roughly equal to that of maximum strain, η_{max}, recorded during the first loading cycle.

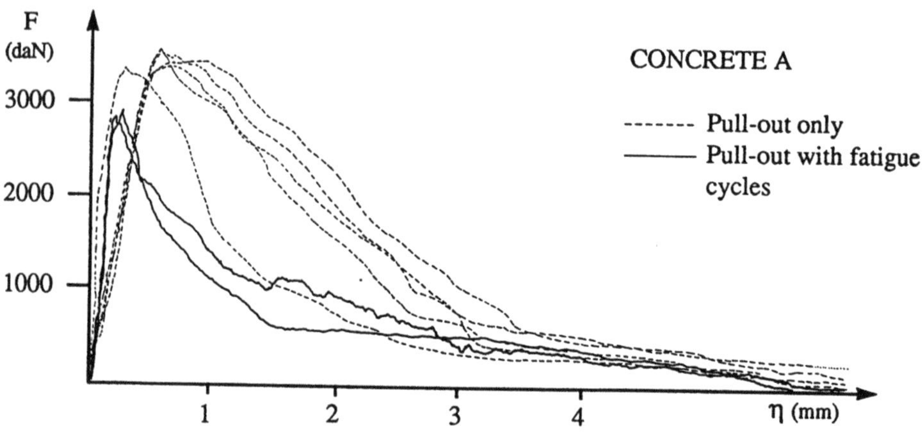

Fig. 2 - Comparison between the force-displacement diagram obtained on the specimens subjected to fatigue and those that were not subjected to fatigue.

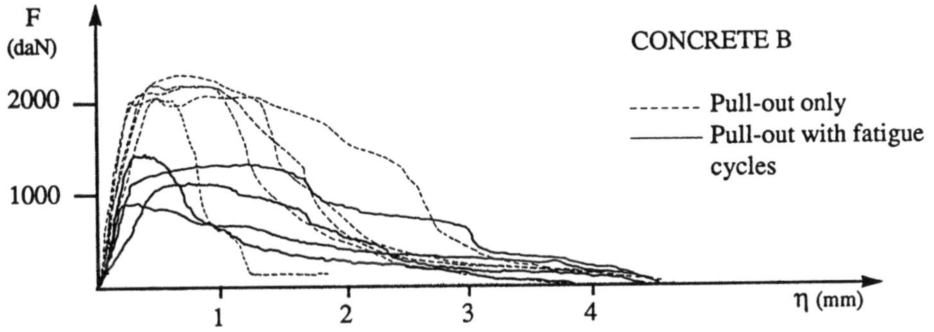

Fig. 3 - Comparison between the force-displacement diagram obtained on the specimens subjected to fatigue and those that were not subjected to fatigue.

Table 2 - Experimental results.

MATERIAL		F_p (daN)	F_{av} (daN)	η_p (mm)	η_{av} (mm)	W_F (daN·mm)	$W_{F,av}$ (daN·mm)
CONCRETE A	Static test without fatigue	3325		0.58		7365	
		3300		0.55		5910	
		3175	3230	0.64	0.57	6317	6377
		3225		0.80		7696	
		3125		0.27		4600	
	Static test after N fatigue cycles	2692	2728	-	-	4937	4791
		2763				4644	
CONCRETE B	Static test without fatigue	1875		0.44		1451	
		2125		0.60		3531	
		1900	1985	0.40	0.57	4308	3073
		2000		1.00		2880	
		2025		0.40		3190	
	Static test after N fatigue cycles	853				2473	
		1258	1140	-	-	2406	2512
		1075				3459	
		1373				1709	

CONCRETE A			CONCRETE B		
bolt no.	F_{max}	N_f	bolt no.	F_{max}	N_f
1	1959	28000	7	1034	2600
2	2521	4000	8	1055	1900
3	1835	67000	9	1051	360000
4	2508	303000	10	937	33700
5	2503	57600	11	824	195000
6	2831	10000			

Table 2 summarises the data obtained from the diagrams, in particular, pull-out force, F_p, displacement at peak load, η_p, and total fracture energy, W_f. On the basis of the diagrams, concrete brittle-ductile behaviour was assessed through Carpinteri's brittleness number as expressed by $s_E = \mathcal{G}_F/(\sigma_u \cdot b)$, where $\mathcal{G}_F$ stands for fracture energy referred to the unit area, σ_u is the ultimate tensile strength, b is the characteristic dimension of the structural element.

From a comparison between three-point tests, pull-out tests and earlier tests (Bocca et al., 1989), it was determined that in concrete:

$$\frac{\mathcal{G}_F}{\sigma_u \cdot b} \equiv \frac{W_f}{F_p \cdot b_0} \tag{1}$$

b_0 being a proportionality constant that depends on the characteristic dimension of the bolt, W_f and F_p being the values of pull-out fracture energy and maximum pull-out force, respectively.

Exp. (1) made it possible to work out the values of s_E contained in the last two lines of table 1. It can be observed that for bolts embedded in concrete A the brittleness number remains essentially the same before and after fatigue testing, while in concrete B, s_E increases to a considerable extent before and after the fatigue cycles, indicating a marked increase in ductility at failure.

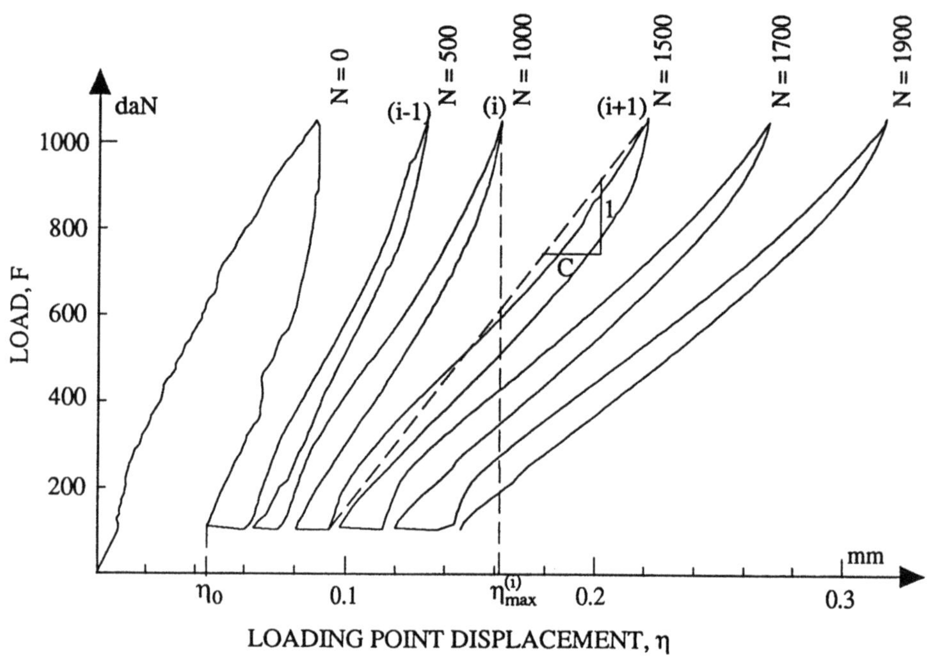

Fig. 4 - Load-displacement curves for different number of cycles.

Fig. 4 illustrates typical F -η curves as recorded during the loading/unloading phase by interrupting the application of the sinusoidal load at different times. The diagram shows the procedure adopted to determine the value of maximum strain, η_{max}, for each cycle and the value of compliance, C, given in the diagrams of figs. 5 - 8. Figs. 5 - 7, which give the number of cycles on a logarithmic scale on the abscissas and the values of η_{max} and C (compliance) on the ordinates, lend themselves to several observations:

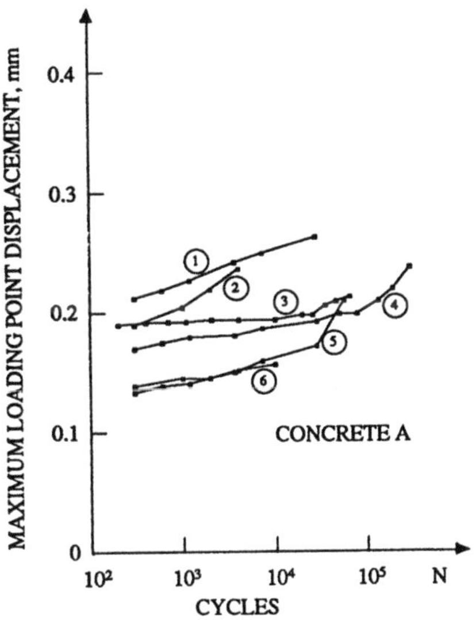

Fig. 5 - Max loading point displacement vs $\log_{10}$ of number of cycles.

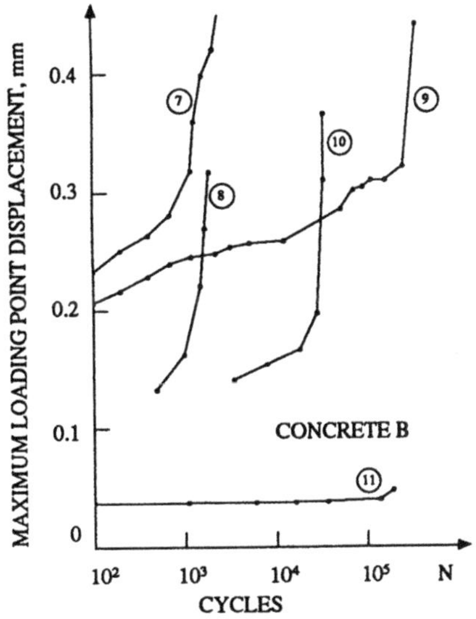

Fig. 6 - Max loading point displacement vs $\log_{10}$ of number of cycles.

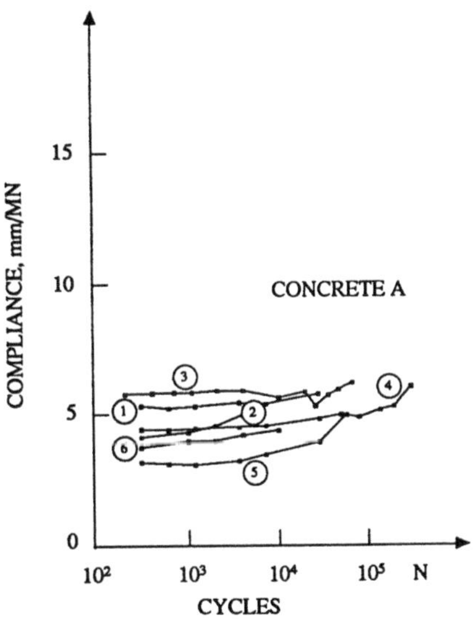

Fig. 7a - Compliance vs $\log_{10}$ number of cycles.

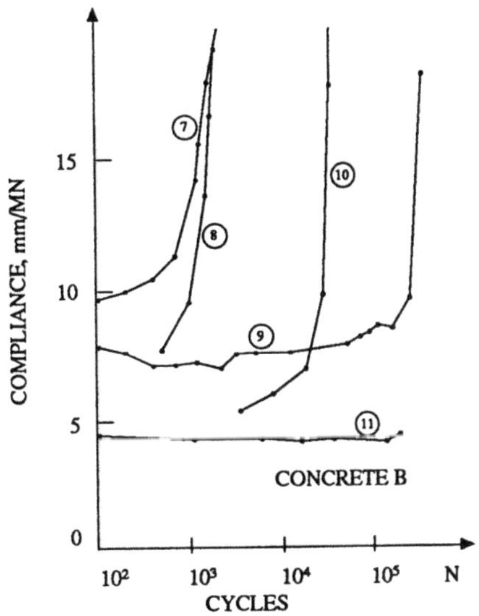

Fig. 7b - Compliance vs $\log_{10}$ number of cycles.

a) with increasing load, fatigue life is obviously seen to decrease, but the results exhibit considerable scatter and therefore we may speak of the phenomenon purely in terms of tendency;

b) fatigue behaviour is remarkably different in the two concrete types: in concrete A, having constant brittleness number, s_E, damage is seen to increase steadily up to failure. In concrete B, with an increasing s_E number and increasing ductility, fatigue damage is seen to increase rapidly in the final stage;
c) The increase in fatigue damage with increasing number of cycles is reflected in a similar manner by the two magnitudes being considered, η_{max} and C; it should be noted, however, that the η_{max}-N curve seems to be more sensitive than the (C - N) curve. In particular, for concrete A the (C - N) curves appear rather flat throughout the loading cycles and hence cannot be used very effectively to predict the residual life of the structural element. This phenomenon, combined with other factors, such as the presence of reinforcement (that will interact in the damaging process) or the onset of widespread cracking at the anchorages (Shah, 1984), make it inadvisable to predict fatigue life through a relationship based on compliance values (or the relationship based on crack propagation velocity which is derived from the former).
To the foregoing considerations, in the case being considered, we should add the practical difficulty of following the evolution of crack propagation at the anchorage.

Fig. 8 shows a normalised version of the curves discussed so far. N_f and η_f stand for the number of cycles and the strain values reached at failure by each bolt. On average, the values of N_f and η_f turn out to be:
– concrete A: $\eta_f = 0.22$ mm $\quad N_f =$ 78.200 cycles
– concrete B: $\eta_f = 0.32$ mm $\quad N_f =$ 118.600 cycles

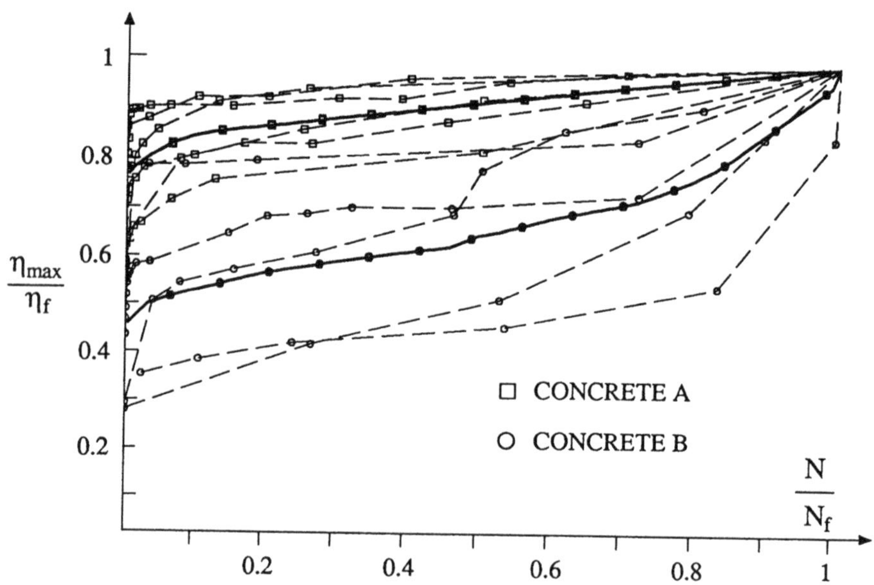

Fig. 8 - Dimensionless loading point displacement versus dimensionless number of cycles.

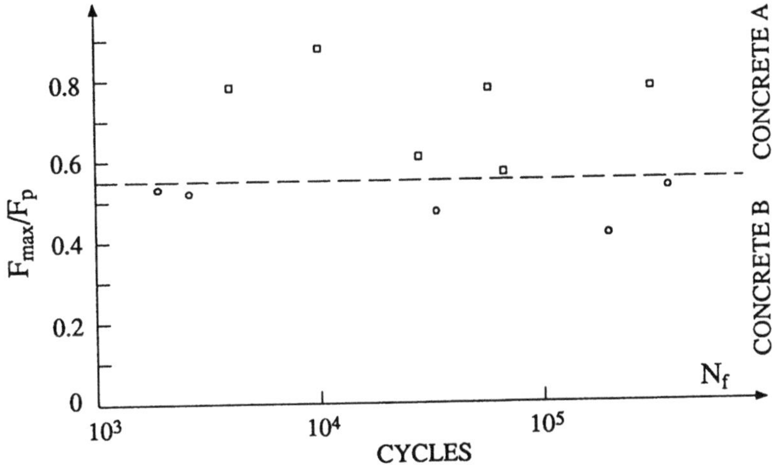

Fig. 9 - Fatigue loading ratio, (F_{max}/F_p), as a function of number of cycles at collapse, (N_f).

Finally, fig. 9 shows that fatigue life, N_f, being the same, specimens manufactured from type A concrete can support a larger fraction of the static collapse load compared to those manufactured from concrete B.

4 Conclusions

Experimental tests have shown that anchorage fatigue life may be predicted more effectively through a relationship based on the increase in the displacement of the load application point, η, as a function of the number of cycles, N, rather than through a relationship based on the crack propagation velocity as a function of the number of cycles. By comparing displacement - number of cycles curves (η, N) and compliance - number of cycles (C, N) curves, in fact, we find that though their behaviour is quite similar, the latter are less sensitive and will not reflect as closely the evolution of damage, entailing a greater possibility of error. This statement, at all events, applies solely to axial-symmetrical forces, i.e., it does not apply to elements subjected to different forces, such as bending.

The tests have shown that the fatigue life of a structural element is significantly affected by concrete brittle-ductile behaviour: though further studies will be necessary, a comparison of the test results obtained reveals a close link between the variation in Carpinteri's brittleness number, s_E, with increasing number of cycles and bolt fatigue life. Load width being the same, in fact, an increase in s_E over time corresponds to a reduction in residual fatigue life.

5 References

Ballarini, R., Shah, S. and Keer, L. (1986). Failure Characteristics of Short Anchor Bolts Embedded in a Brittle Material. In **Proceedings of the Royal Society**, London A 404, pp. 35 - 54.

Bocca, P. (1979). Indagine teorico-sperimentale sulla determinazione della resistenza del calcestruzzo mediante prova di estrazione. **L'industria Italiana del Cemento,** pp. 45-52.

Bocca, P. (1984). The Application of Pull-Out Test to High- Strength Concrete Estimation. **Materials and Structures,** 17, pp. 211 - 216.

Bocca, P., Carpinteri, A. and Valente, S. (1989). Evaluation of Concrete Fracture Energy Through a Pull-out Testing Procedure. In **Proceedings of the International Conference on Recent Developments on the Fracture of Concrete and Rock** (eds. S.P. Shah, S.E. Swartz and B. Barr), Elsevier Applied Science, pp. 347 - 356.

Bocca, P., Carpinteri, A. and Valente, S. (1990a). Size Effects in the Mixed Mode Crack Propagation: Softening and Snap-back Analysis. **Engineering Fracture Mechanics,** 35, pp. 159 - 170.

Bocca, P., Carpinteri, A. and Valente, S. (1990b). Fracture Mechanics Evaluation of Anchorage Bearing Capacity in Concrete. In **Proceedings of the International Workshop on the Application of Fracture Mechanics to Reinforced Concrete** (ed. A. Carpinteri), Elsevier Applied Science, pp. 231 -265.

Carpinteri, A. (1992). **Applications of Fracture Mechanics to Reinforced Concrete,** Elsevier Applied Science.

Elfgren, L. (1989). **Fracture Mechanics of Concrete Structures, Report of the RILEM Committee 90-FMA,** Chapman and Hall.

Eligehausen, R. and Sawade, G. (1989). A Fracture Mechanics based Description of the Pull-out Behaviour of Headed Studs Embedded in Concrete. In **Fracture Mechanics of Concrete Structures, RILEM Report** (ed. L. Elfgren), Chapman an Hall, pp. 281 - 299.

Krenchel, H. and Shah, S.P. (1986). Fracture Analysis of Pull-out Testing. **Materials & Structures,** 18, 108, pp. 439 - 446.

Miller, G. and Keer, L. (1982). Approximate Analytical Model of Anchor Pull-out Test, **Journal of Applied Mechanics (ASME),** 49, pp. 768 - 772.

Shah, S.P. (1984). Predictions of Cumulative Damage for Concrete and Reinforced Concrete. **Material & Structures,** RILEM 17, pp. 65 - 68.

Stone, W.C. and Carino, N.J. (1983). Deformation and Failure in Large-Scale Pull-out Tests, **American Concrete Institute Journal,** pp. 501 - 513.

Stone, W.C. and Giza, B.J. (1985). The Effect of Geometry and Aggregate on the Reliability of the Pull-out Test, **Concrete International,** 2, pp. 27 - 36.

Valente, S. (1991). Analisi della prova di estrazione tramite un approccio di meccanica della frattura. In **Danneggiamento e Diagnosi di Materiali e Strutture** (eds. A. Carpinteri and P. Bocca), Pitagora Editrice, Bologna, pp. 25 - 51.

62 UPLIFT BEHAVIOUR OF PLATE ANCHORS EMBEDDED IN SANDS: EFFECT OF GRAIN SHAPE, RELATIVE DENSITY AND DEPTH

A. BOUAZZA
CNERIB, Souidania, Algeria

Abstract
The paper describes and presents results of an investigation into the effect of relative density, depth of embedment and grain shape on the uplift behaviour of plate anchors subjected to static loads. The investigation was carried out on a 37.5 mm diameter plate anchor embedded in three types of loose and dense sand with similar uniformity, mean grain size diameter but having different shape (rounded, subrounded and subangular respectively). The effect of the above parameters on the ultimate uplift capacity was examined in a series of static tests with varying densities and depth of embedment (2<D/B<5). Stereo-photogrammetric method was also used to assess the effect of the sand grains angularity on the diplacement fields. It was found that in addition to depth of embedment and relative density, the grain shape significantly affected the ultimate uplift capacity and the displacement fields. In conclusion, research in this subject has been extended beyond the parameters of relative density and depth of embedment, by the introduction of a new parameter (grain shape) into the analysis.
Keywords: Plate anchor, Uplift, Sand, Relative density, Grain shape, Depth, Displacement field.

1 Introduction

Different types of anchor are used extensively in offshore and onshore construction, depending on the magnitude and type of loading, type of structure and subsoil conditions. A comprehensive description of different types of anchors

Fracture and Damage of Concrete and Rock – FDCR-2.
Edited by H.P. Rossmanith. © 1993 Taylor & Francis. ISBN 0 419 18470 8.

and their use in the field is reported by Taylor et al (1975) and Hanna (1982).

Plate anchors are among the most common types of anchor used in civil engineering construction as they represent an economical alternative to gravity and other embedded anchors for resisting uplift forces in both a marine and a land environment. Typically, such anchors are used to support transmission towers and to secure submerged pipelines, moorings and cables to the sea bed. Previous research into the uplift capacity of plate anchors embedded in cohesionless soil has shown that two distinct categories of anchors can be identified: shallow anchors and deep anchors. In the former the anchor is installed close to the surface of the soil, and the failure surface in the soil extends from the tip of the anchor to the ground surface with significant surface movements. An increase in depth of embedment results in another type of failure in which the surface does not extend to the surface but instead forms locally around the anchor. This type of failure exemplifies the deep anchor mode. The dimensionless ratio of anchor embedment depth D to anchor diameter B can also be used to classify shallow and deep anchors.

This paper presents the results of model tests on shallow plate anchors and shows the effect on uplift resistance of varying the grain shape, the depth of embedment and the relative density on which the anchor is embedded. The results of this study are compared with the current design methods. This paper shows also the effect of grain shape on the displacement fields.

2 Background

Most of the approaches used to predict anchor uplift resistance are based upon the limit equilibrium concept. The major difference amongst the various methods involves the assumed shape of the slip surface developed within the failing soil mass. Limit equilibrium conditions compute the ultimate anchor uplift capacity (P) as the combined effects of the soil weight contained within the assumed failure surface, the shearing resistance of the soil integrated over the failure surface, and the dead weight of the anchor structure itself. A review of most of the theoretical studies used for evaluation of the ultimate capacity of plate anchors embedded in sand has been given by Bouazza (1990). It is not the intention of this paper to review all pertinent theories, however, but those for plate anchors provided by Meyerhof & Adams (1968), Rowe & Davis (1982), and Frydman & Shiham (1989) will be briefly discussed below since these are the most widely referred to in literature. The methods presented are used in practice in the same way as Terzaghi's (1948) bearing capacity formula, by the use of dimensionless parameters.

Terms used in anchor research are the depth/diameter ratio (D/B) and the breakout factor (Nu); the former is defined as the depth (D) from ground surface to the tip of the anchor, divided by the plate diameter B. The latter is defined as the ultimate pressure on the anchor plate (Pu/A), where A is the anchor area, divided by the effective overburden pressure (γD) where γ is the soil unit weight. Thus, for cohesionless soils:

$$Nu = (Pu/A)/\gamma D \tag{1}$$

The use of a non-dimensional breakout factor allows the reader to compare different theories as well as results from anchor tests with dissimilar anchor sizes, soil densities and embedment depths.

2-1 Meyerhof & Adams' theory(1968)

The above authors proposed a general uplift capacity theory which was complementary to the general bearing capacity theory proposed by Meyerhof. A curved failure surface was adopted extending vertically above the perimeter of the anchor and solved for a strip footing. This theory was then applied to circular and rectangular anchors by the introduction of a shape factor term (s) into the general expression. The ultimate uplift in general shear was computed by summing the shearing resistance from cohesion and passive earth pressure at an angle from a vertical cylindrical surface through the anchor edge and is determined as:

$$P = (\pi/2)\, B \gamma D^2 s\, K \tan \varphi \tag{2}$$

Hence the breakout factor may be written as:

$$Nu = 2\,(D/B)\, s\, K\mu \tan\varphi + 1 \tag{3}$$

where, $K\mu$ is the coefficient of lateral pressure and φ is the angle of internal friction.

2-2 Rowe & Davis' theory(1982)

The above authors described a theoretical investigation on cohesionless soils in which consideration was given to the effect of anchor embedment, friction angle, dilatancy, initial stress state K0 and anchor roughness. Their theoretical appraoch was based on an elastoplastic finite element analysis. The soil was assumed to have a Mohr-Coulomb failure criterion and either an associated or a non associated flow rule. The anchor breakout factor was given as:

$$Nu = F_{\gamma} \; R_{\psi} R_{k} \; R_{r} \qquad (4)$$

Where F_{γ} is a basic anchor capacity factor, and R_{ψ} R_{k} R_{r} are correction factors for the effect of soil dilatency, initial stress and anchor roughness respectively.

2-3 Frydman & Shiham's theory (1989)

Frydman & Shaham (1989) proposed a method based on the work of Vermeer & Sutjiadi (1985) and Murray & Geddes (1987) who considered the equilibrium of a truncated wedge pulled out together with a rectangular plate; in both cases, they suggested the following upper bound to the breakout factor

$$Nu = 1 + D/B \tan \varphi \qquad (5)$$

Analysing data collected from literature and using statistical analysis Frydman & Shaham introduced a shape factor (s) which was subsequently injected in expression (5) to give the following breakout factors:

For dense sand ($D/B < 8$)

$$Nu = 2(1+(4/\pi)^{1/2} D/B \tan \varphi)\ (0.51+2.35 \log(4/\pi)^{1/2} D/B) \qquad (6)$$

For loose sand ($D/B > 1.8$)

$$Nu = 1.5\ (1+(4/\pi)^{1/2} D/B \tan \varphi) \qquad (7)$$

3 Testing program

Model anchor pull out tests were performed in Leighton Buzzard sand, Douglasmuir sand and Ballotini sand which exhibited the engineering properties given in table n°1.

The tests were performed in a steel tank 500 mm diameter and 700 mm, large enough for tests to be carried out without any side effects. The model anchor was made from brass, 37.5 mm diameter and 3 mm thick, with an anchor shaft 4 mm diameter screwed into the brass disc to make the anchor unit. At the start of each test a 75 mm thick layer of sand was compacted into the bottom of the tank. The anchor plate was then placed in position on top of this layer, and the sand rained in layers into the tank from a required height to produce a constant density (Kolbusewski & Jones, 1961). The process of sand laying was repeated until the anchor was buried to the required depth. For the subrounded sand, relative densities (ID) and angle of internal friction (φ) varied between 13% to 85% and 32° to 42.8° respectively; for the subangular sand, they varied between 31% to 76% and 36° to 43.7° respectively whereas for the well rounded sand they varied from 50% to 97% and 30.3° to 34.9° respectively.

The anchor unit was attached to a load cell which was connected to a pneumatic loading device to apply a static loading, under load control (incremental loading). As the applied load approached the failure value, the load increment was reduced to give a more clearly defined load value. The ultimate load was taken as that measured at the last increment before total failure occured (Fadl, 1981). A linear variable displacement transducer recorded the displacement of the anchor throughout the tests and this information, plus the load cell reading, was logged at intervals throughout each test.

Table n° 1 Sand characteristics

	Leighton Buzzard	Douglasmuir	Ballotini
mean grain size D50 (mm)	0.80	0.80	0.80
uniformity	1.80	1.80	1.80
specific gravity	2.65	2.66	2.92
shape	sub-rounded	sub-angular	Well rounded
max porosity (%)	44.2	46.4	42.10
min porosity (%)	33.2	34.95	36.10

A separate special series of tests was carried out to determine the displacement fields at different stages of the pull out in a loose and a dense sand. In the present study semi spatial tests were carried out in a separate tank (600mm X 250mm X 600mm), (B=50 mm) and where the soil was deposited in layers. Stereo-photogrammetry(Butterfield et al,1970, Bouazza, 1991,1992) has been used to assess the effect of relative density and grain shape on the displacement fields.

4 Discussion

A typical effect of relative density is depicted in figure 1 where an anchor with a depth/diameter ratio D/B varying from 2 to 5 is considered. A clear pattern emerges from this figure where it can be seen that the effect of the

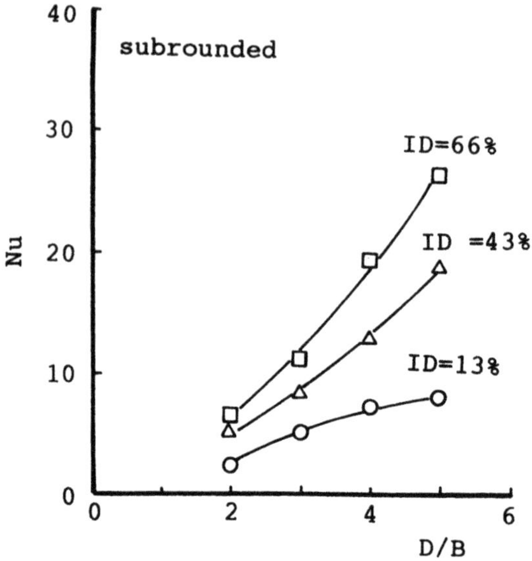

Fig. 1 Effect of relative density (ID) on the breakout factor (Nu) at different depth/diameter ratio D/B.

degree of densification of the soil is to greatly increase the uplift resistance of the anchor. Depending on the depth of embedment, increasing the relative density of the sand from loose to dense condition increases the breakout factor to around 3 times. an increase in relative density (ID) will promote a greater compactness,and hence,a higher intergranular contact among the soil particles. This would mean an increase in the angle of internal friction of the sand surrounding the anchor, thus allowing larger shearing resistance to be developed during loading. Fig.2 shows the effect of relative density on the vertical movement of the anchor, the movement at failure (δ_f)of the anchor surrounded by loose soil being approximately 1.5 times greater than that of the same anchor embedded in dense sand. This was because the shear strength of the sand was rapidly mobilized when the anchor was pulled out of the dense sand. When the sand is denser, the degree of interlocking is high therefore the load from the anchor transferred to the sand only required to breakout the soil mass rather than to first densify it.

Typical load displacement curves at various embedment ratios are shown in figure 3. It is seen that the ultimate uplift capacity increases with the increase in depth of embedment. For these tests rupture planes were clearly observed at the surface. The load displacement diagrams also demonstrate that the displacement (δ) of the majority of the plate anchors at which the ultimate

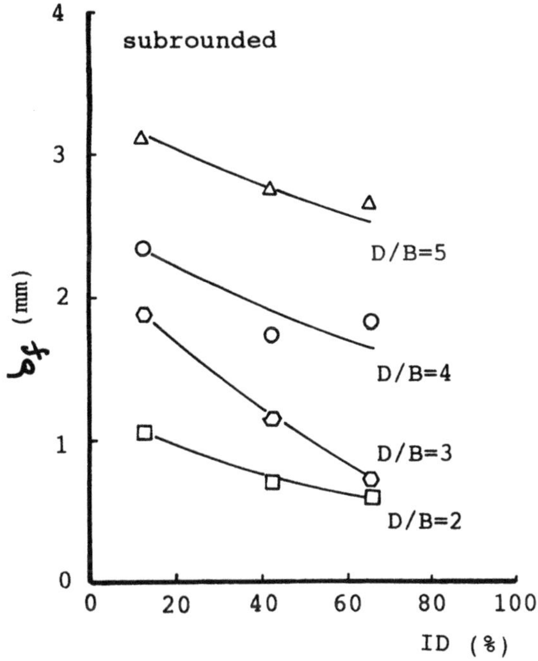

Fig. 2 Typical anchor displacement at failure -relative density relationship.

static resistance occurs is small and ranges from 0.2% to 2.4% of the initial depth of embedment. This indicates that only a small displacement of the anchor is required to mobilize the frictional resistance of the sand.

A comparison of the behaviour of plate anchors in rounded, subrounded and subangular sands is made in figure 4 where the breakout factor (Nu) is plotted against the relative density (ID). The data show a marked increase in the breakout factor with increase of angularity of the particles and relative density, with Ballotini exhibiting the least increase. At a relative density of 65%, Douglasmuir sand with subangular particles registered breakout factors which were higher than the values recorded for the rounded and subrounded sand. This increase in angularity shows also that for example at D/B=2 and a breakout factor (Nu) of say 5 there is a spread in relative density from 30% to 85%. These observations suggest that grain shape can drastically affect the uplift behaviour of a plate anchor.

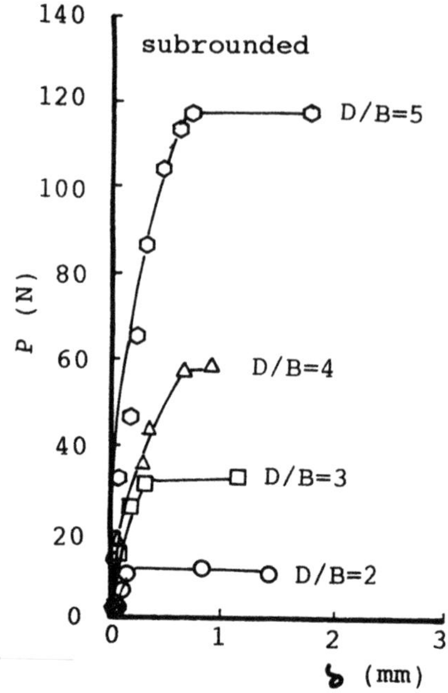

Fig. 3 Effect of anchor depth on load displacement relationships for anchors embedded in dense sand.

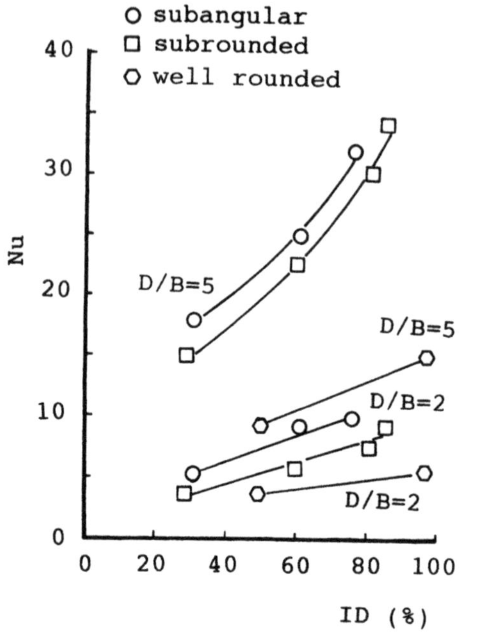

Fig. 4 Breakout factor Nu vs relative density (ID) for shallow anchors: grain shape effect.

The stereo-photogrammetric technique has been applied o monitor the development of the displacement field. Tests were performed at two different densities loose and dense. During each test a sequence of six photographs was taken, the first prior to pulling, the remainder at intervals of 1 mm vertical displacement. Due to the large amount of data accumulated only two sets of tests are herein reported: test n°1 ID=15%, test n°2 ID=75%. Particle movements are presented using the first (no pulling) and last (final pulling) photograph.

In dense sand, figure 5 show that the zone of sand suffering displacement for both materials extends all the way above the plate anchor to reach the ground surface. However, in the case of the subrounded sand this zone radiated from 0.25 X B to 2 X B to the side of the plate anchor whereas in the subangular sand it radiated form 0.25 X B to 2.75 X B. In the loose sand, figure 6, it can be seen that the displacement in the subrounded sand have mainly occured 2 X B above the plate anchor and 0.25 X B laterally whereas in the subangular sand the disturbed area extended 2 X B above the plate and radiated 0.75 X B to its side. It is clear that the observations made in the present experiment study show that the extent of the displaced sand mass at a given depth and relative density is substantially influenced by the grain shape.

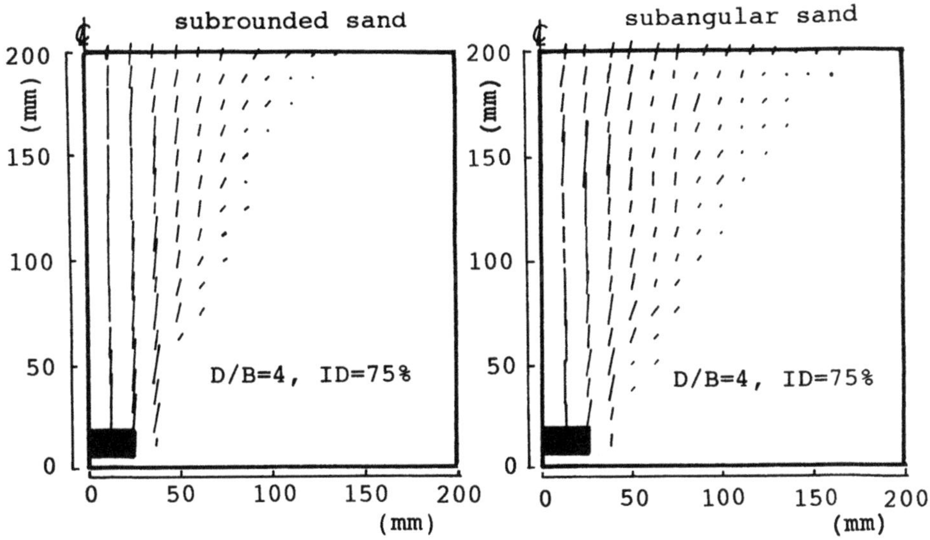

Fig. 5 Displacement fields in dense sands:
a) subrounded, b) subangular.

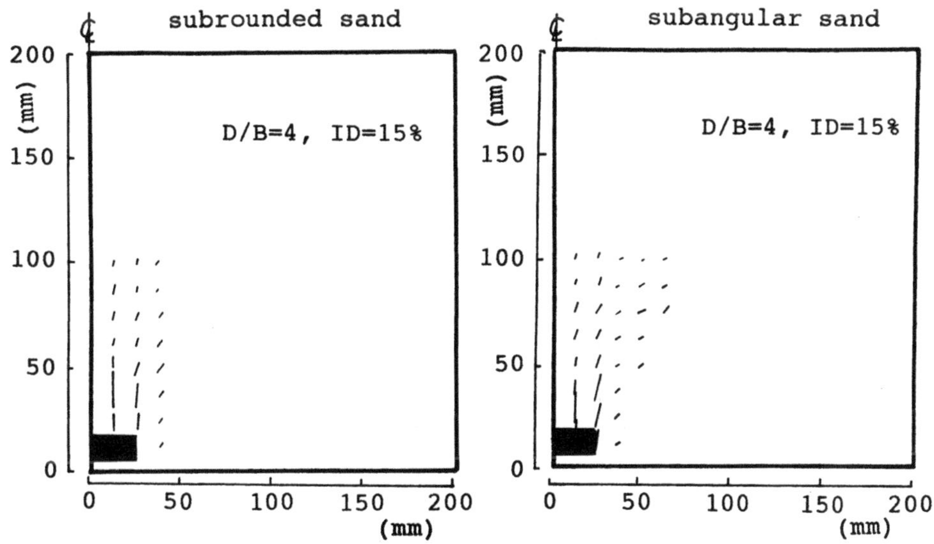

Fig. 6 Displacement fields in loose sands:
a) subrounded, b) subangular.

In order to compare the present experimental results with various existing theories, the experimental breakout factors are shown in figure 7 along the theoretical plots as obtained from the the theories of Meyerhof & Adams, and Frydman & Shiham.

For the subrounded soil the method described by Frydman &

Shiham seems to correlate best with the test data. However, in the case of the subangular sand the same approach appears to overpredict the experimental results. The same observation holds for the Meyerhof & Adams' theory. The other method employed to predict the sand tests is seen to underpredict the breakout factor by as much as 300%. The source of the differences is due to the fact that different type of sands have been tested. The above discussion indicates the importance of grain shape in addition to the relative density and the angle of internal friction when considering different sands.

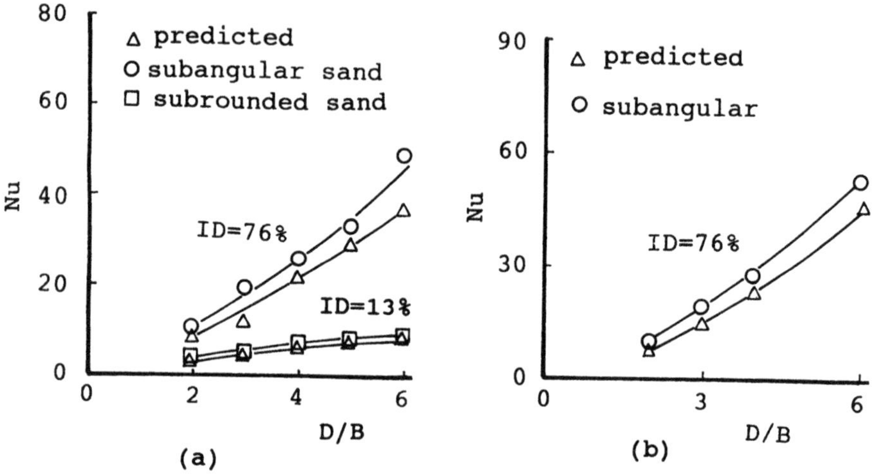

Fig. 7 Comparison of different design theories with present test results:
a) Frydman & Shiham (1989), b) Meyerhof & Adams (1968).

5 Conclusions

Results of laboratory model tests for ultimate uplift capacity of shallow circular plate anchors embedded in cohesionless soils have been presented. Based on the present study, the following conclusions can be drawn.

1- Load displacement behaviour of anchor plates and the breakout factors depend on the depth of embedment and relative density.
2- In sands having similar grading and grain size but different grain shape, the breakout factor and displacement field are influenced by the shape of the sand grain.
3- It was found that the mode of failure and the magnitude and direction of the sand displacements were regulated to a large extent by the sand relative density.

4- The theories developed by various researchers have been compared with present test results. It was observed that discrepancies occured when different sands are taken into account. In the future, great care should be taken when considering different sands and the ID- φ concept.

6 References

Bouazza, A. (1990) Uplift behaviour of circular plate anchors embedded in cohesionless soils. **PhD thesis**, University of Glasgow, Glasgow (U.K.).

Bouazza, A. (1991) Sand deformations around a plate anchor subjected to an uplift load, in **Proc. 9th Asian Conf. on Soil Mechanics & Fdns Engng.**, Bangkok (Thailande), Dec.

Bouazza, A. (1992) The use of the stereo-photogrammetry method in geotechnical engineering work. in **Proc. 17th Int. Society of Photogrammetry & Remote Sensing Congress,** Washington D.C. (U.S.A.), Aug. (in press).

Butterfield, R., Harkness, R.M. & Andrawes, K.Z. (1970) Stereo-photogrammetric method for measuring displacement fields. **Geotechnique**, 20, n°3, pp.308 -314.

Fadl, M.O. (1981) Behaviour of plate anchors. **PhD thesis**, University of Glasgow, Glasgow (U.K.)

Frydman, S. and Shiham, I. (1989) Pull out capacity of slab anchors in sand. **Canadian Geotechnical J.**, 26, n°3, pp. 385-400.

Hanna, T.H. (1982) **Foundations in tension**. Trans. Tech. Publication, n°6.

Kolbuszewski, J.J. and Jones, R.H. (1961) The preparation of sand samples for laboratory testing, in **Proc. Midlands Soil Mechanics Society**, 4, pp. 107-123.

Meyerhof, G.G. and Adams, J.D. (1968) The ultimate uplift capacity of foundations. **Canadian Geotechnical J.**, 5, n°4, pp. 225-244.

Murray, E.J. and Geddes, J.D. (1987) Uplift of anchor plates in sands. **J. Geotechnical Engng. Div.**,113 n°3,pp. 202-215.

Rowe, R.K. and Davis, E.H. (1982) The behaviour of anchor plates in sand. **Geotechnique**, 32,n°1, pp. 25-41.

Taylor, R.J., Jones, D. & Beard, R.M. (1975) **Hand book for uplift resisting anchors**. Naval Civ. Engineering Lab., Port Hueneme, California (U.S.A.).

Terzaghi, K. (1948) **Theoretical soil mechanics**. John Wiley & Sons, New York (U.S.A.).

Vermeer, P.A. and Sutjiadi, W. (1985) The uplift resistance of shallow embedded anchors, **in Proc. 11th Int. Conf. Soil Mechanics & Fdns Engng,** San Fransisco (U.S.A.), 3, pp. 1635-1638.

PART EIGHT

AUSTRIAN INNOVATIVE CONSTRUCTION

63 INTRODUCING AUSTRIAN INNOVATIVE CONSTRUCTION

K.H. BUCHNER
Institute for Economic Development, Vienna, Austria

The Austrian Federal Economic Chamber focuses on the effective representation of the interests of Austrian businessmen and women with respect to legislature government and social partners.

The chamber also offers a comprehensive range of consulting and services to its members and will continue to step up its activities in the area of internationalisation of trade and industry and technology.

The Austrian Federal Economic Chamber itself acts as an umbrella organisation co-ordinating regional and professional economic interests and representing them with respect to the state and organisations furthering the interests of other social groups. The Austrian Federal Economic Chamber and the regional chambers together form the chamber of trade organisation.

An important function of the chamber of trade organisation is to improve the competitiveness of individual firms and the economy as a whole. For this purpose the Institute for Economic Development (known as "WIFI" for short) was established in the regional chambers as well as in the Austrian Federal Economic Chamber.

One main task of the institute is to promote every kind of innovative and technological activities as for instance to facilitate the introduction of new technologies especially in small and medium sized enterprises.

Within this promotion of new technologies the WIFI together with branch organisations - as for instance in our case the building guild and industry also encourages the cooperation between science and economy. So its among our tasks to hold close contacts to university institutes. This science and economy cooperation is also the reason that the conference "FDCR-2" takes place within the rooms of the Austrian Federal Economic Chamber.

It is expected that this conference will give further inputs to the Austrian construction industry for its steadily and ongoing innovative development.

Fracture and Damage of Concrete and Rock – FDCR-2.
Edited by H.P. Rossmanith. © 1993 Taylor & Francis. ISBN 0 419 18470 8.

New systems with classical formwork technologies

V. MITTERHAMMER
Gleitbau Gesellschaft
Salzburg, Austria

1 Videofilm generally showing different systems of Formwork

2 Technical details in connection with examples of executed structures

- slipforming (vertical, tapered, conical)
- jumpforming (vertical, conical)
- special systems

3 Summary and Discussion

As a "Specialist for Specialities" the presentation will comprise examples starting with simple silo-constructions, wind shields for chimneys, piers and pylons for bridges and finally structures as TV-towers, offshore platforms, etc..

Almost any shape and dimension of highrise concrete structures are possible to be formed in an economical and fast way with desired qualities by the presented formwork technologies. In any way no crane will be used to move the forms and therefore this capacity will be available for the transportation of all the construction elements.

Fracture and Damage of Concrete and Rock – FDCR-2.
Edited by H.P. Rossmanith. © 1993 Taylor & Francis. ISBN 0 419 18470 8.

New technologies for electrochemical concrete repair :PAIR

W. KALTENEGGER
Gleitbau Gesellschaft
Salzbug, Austria

1 Theory of Electro - Chemical process
Desalination (Chloride-removal)
Realkalisation

2 Practical Experience
Parking Ground - TU Vienna
Arch-Bridge, Vorarlberg
W.Jauregg Hospital, Linz

3 Experts Research Results

4 Summary and Discussion

Since several years concrete repair volume is increasing fast. Different repair methods were developed. Electro-chemical removal of chlorides and realkalisation has certain advantages in the field of repair systems:

- It solves the problem at its roots
- process is non-intrusive and non-destructive
- no altering of structure in terms of its structural performance or appearance
- significant extending of the useful life of the structure
- environmentally safe procedure

Fracture and Damage of Concrete and Rock – FDCR-2.
Edited by H.P. Rossmanith. © 1993 Taylor & Francis. ISBN 0 419 18470 8.

Transient pressure analysis applied during epoxy grouting of cracks in concrete and rock

G. STADLER
Insond GmbH
Vienna, Austria

The remedial measures to reinstate the integrity of concrete at Kölnbreindam together with an intrigueing concept of additional support of this 200 m high double curvature arch dam, has led to a sizeable grouting program using special expoxies.
Cracks in concrete had to be bonded and watertightened without inducing movement to the structure, or propagating existing cracks and further.

Limits had been set to grouting pressures, quantities and strucktural movements, as is usual for this type of work.

In addition, however grouting at Kölnbreindam has been new technological approaches on this matter.

Designer, Dr. Lombardi, did introduce a new "Intensity" criteria, watching carefully the product of pressure times quantity in each borehole and every packer position.

Further, more information is collected during the grouting process itself. A newly formulated "Continuous Injection Shut-in System" provides "diagnostic" information on: the condition of the crack, flow behaviour and rheological suitability of the grouting fluid.

Numerical investigation of transient pressure curves, registered after shut-in of the grout flow, provide a new type of approximation relating to hydraulically representative crack dimensions.

Thus, design assumptions may in future be verified more easily, and corrections during the process of grouting may be decided on the spot, based on actual field data.

In applaying these new techniques, Kölnbrein grouting shows a stunning 80 % filling ratio for all "accessible" cracks at only 50 % of drilling meters originally planned.

Fracture and Damage of Concrete and Rock – FDCR-2.
Edited by H.P. Rossmanith. © 1993 Taylor & Francis. ISBN 0 419 18470 8.

Underground measurements within numerical design of caverns

G. ZENZ & J. KUMMERER
Tauernplan Consulting GmbH
Salzburg, Austria

Abstract
Numerical design analysis for underground openings depend on a variety of different influences which are in our case reduced to the determination of material parameters, geological situation and In situ stress level. Regarding to the uncertainties of local conditions for determining material parameters, it is very necessary to combine both geological and mechanical experience for deciding which location to use for in-situ large scale tests. During intensive investigations for a power cavern rock mechanical tests, in small as well as in large scale, were carried out.

The gained results with numerical backanalysis of different test layouts led to a set of parameters. For example it was to examine the initial state of stress within the exploration tunnel for biaxial compression test with the measurements of Doorstopper tests in combination with material property assumptions. The parameters are, because of the large scale tests, well prepared for the use within a finite element model of the whole cavern.

For special interest within the calculations for design of a valve chamber was explicitely to consider discrete rock interfaces. The numerical calculations where carried out with the help of contact friction interface elements. With these elements the dilatancy along rock planes, assuming Coulomb friction law, are encountered for. The numerical behaviour and the calculated results are discussed in comparison with a modified continuous nonlinear material model only.

A video tape showing the examination of rock mechanical tests will give an impression and a practical impact on the exploration work during the design stage of a power cavern.

Fracture and Damage of Concrete and Rock – FDCR-2.
Edited by H.P. Rossmanith. ISBN 0 419 18470 8.

FRP-Rockbolts – technology

H. HELBLING
H. Weidmann AG.
Rapperswil, Switzerland

Introduction

Rockbolts have made possible several new construction designs in civil engineering are used mainly as a temporary support measure. FRP-Rockbolts however, not only open up new design possibilities, but also offer consulting and project engineers a highly useful addition to the commonly used steel bolts. They further increase the flexibility of rock bolting, giving a support system which can easily be adjusted to the prevailing rock conditions met during driving. FRP-Rockbolts are best suited in applications that require high tensile strength and the particular characteristics of fiberglass materials. Applied correctly they contribute substantially to the reduction of construction cost.

What are the advantages of FRP-Rockbolts?

CORROSIONS RESISTANCE
FRP-Rockbolts are corrosion resistant and therefore, durable. This eliminates expensive corrosion protection measures. FRP-Rockbolts are the only currently available rockbolting system which can be regarded as permanent, even under difficult conditions with aggresive water and rock deformations. Corrosion is the most pressing material problem worldwide. With capital, downtime and material costs rising, corrosion prevention is high on many industries list of priorities. Every minute, several tons of steel are lost to corrosion. The cost of corrosion protection is enormous, even then however the problem is delayed rather than solved. Conversely, correctly applied FRP-Rockbolts support the construction over its entire life. Because of its anti-corrosion concept, the FRP-Rockbolts Technology is used increasingly in land reinforcement and as ground anchors. Because of the durability of the materials, FRP-Rockbolt can be utilised not only as temporary support, as with regular steel bolts, but also integrated into the

Fracture and Damage of Concrete and Rock – FDCR-2.
Edited by H.P. Rossmanith. © 1993 Taylor & Francis. ISBN 0 419 18470 8.

calculations for the final loadbearing structure.Insitu concrete shells can therefore be reduced or even eliminated. This so-called single-lining tunnel construction method is widely used in Switzerland in military and civil caverns. The final support consists solely of rockbolts and reinforced shotcrete. With the achieved improvements of those materials, single-lining tunnels will further increase in importance. The cost reductions, especially in the planned large Alpine tunnels are ernormous. The construction cost of a railway tunnel for low speed or a highway tunnel with moderate traffic can be reduced by up 50 %. In tunnels that required a structural shell, this second shell is needed mainly as support for the sealing material, but it can also contribute to the bearing of the load.

CUTTABILITY

FRP-Rockbolts can be cut easily and avoid any damage to the equipment used to enlarge pilot tunnels or to remove the ore body. Despite its high tensile strenght, the FRP-Rockbolt System is cuttable. This characteristic is very useful when securing a pilot tunnel which needs to be enlarged to the required profile with a roadheader or TBM. Costs and down time due to damage to the cutter head which occur with regular steel bolts can be eliminated. The time-consuming and dangerous shearing of protruding steel bolts is not required and the spoil material contains no scrap-iron, which could damage or interrupt conveyor and spoil removal systems.

HANDLING

FRP-Rockbolts are flexible and can be set even in the most confined locations in tunneling and mining sites.

FRP-Rockbolts weigh only 25 % of steel bolts, reducing transportation costs and easing handling problems.

FRP-Rockbolts are light and flexible making handling easy and reducing transportation costs. Long bolts can be set even in the narrow and confined spaces of tunneling sites. Expensive scaffolding and lifting equiptment is not required.

These advantages are of great importance in cases of restorations of tunnels when regular traffic must be maintained. The use of rockbolts in mining has been increasing for years. With the introduction of FRP-Rockbolts, the choice of rockbolts has expanded and new applications have been made possible. Today, most FRP-Rockbolts are used in coal mines because they provide the necessary support, but can easily be cut and do not impede the excavation of coal.

MAI Systems – innovative anchor technology

S. GENSHOFER
MAI Systems International
Millstatt, Austria

A short description of the concept behind the MAI Systems product range. The MAI System consists of a well designed range of system components - innovative injection drilling anchors ("MAI-Anchors"), anchor mortar ("MAI-Mortar") and the proven injection pumps for cement grouts and mortar ("MAI-Pump").

The component parts of MAI-Anchors like drill bits, anchor rods, couplings, plates and nuts are available in different dimensions and special qualities, in addition the use and applications of MAI-mortar mixing pumps are fully explained.

The quick and easy installation of the MAI-Anchor is an uncased bore-hole, the injection of the anchor with cement or anchor mortar as well as the ability to prestress the tension section with a torque wrench or a hydraulic press, will be fully explained.

The advantages of the system are described and demonstrated, showing practical examples of applications. A significant benefit is the availability of high-quality and economically priced drill bits.

Different types and dimensions of drill bits make it possible to adjust to the prevailing ground conditions (geological structure) and enables the formation of an uncased bore-hole using a non reclaimable drill bit.

Using the innovative couplings, anchors of practically any length required can be installed. When connected with the anchor rod the couplings are tight enough to enable drilling with air, water or cement slurry as a debris/ cooling medium. The applications are cavity construction (tunnelling, mining shafts), enbankment reinforcement, pilings with or without additional support in difficult ground and rock conditions. The change of load from tension to compression enables the use of MAI-Anchors for pile foundations too.

Fracture and Damage of Concrete and Rock – FDCR-2.
Edited by H.P. Rossmanith. ISBN 0 419 18470 8.

On the necessity of being able to forecast the fracturing of rock in mechanical tunnel boring

E. MIKURA
Universale Bau
Vienna, Austria

In ordered to be able to forecast the headway in tunnelling with tunnel boring machines it is necessary to know the typical failures of each of the rock structures involved. On one hand, it is the rock mass with its structure and texture which is responsible for fractures, whereas the mineral composition, on the other hand, determines the type of failure.

Each type of rock structure requires a different testing method and an analysis of its own. Current analyses and classifications are at present being made for the excavation of San Nicla (Switzerland). As a side effect of these analysis, rock zones with increased liability to fracturing are detected. It is often possible to ascertain the origin of the fractures. Most frequently, they are caused by stress occuring at hte sides of valleys or by recent tectonic events. Weathering symptoms help to define the age of fracture.

Fracture and Damage of Concrete and Rock – FDCR-2.
Edited by H.P. Rossmanith. © 1993 Taylor & Francis. ISBN 0 419 18470 8.

Recycling of construction material residues – recycling technologies

H. Hirnschall
Beyer & Co
Graz, Austria

1 General:

In Austria approximately 20 million tons of construction material residues are originated annually. The Waste Economy legislation will provide the legal framework for these materials. The required measures are stipulated in the decree on the separation of construction material residues and waste.

2 Separation Methods:

2.1 Separation at the construction site

Excavation material can be separated directly at the site. Separation of other material is suggested to be done at the site but is not always possible.

2.2 Separation with preparation plants

The separation with preparation plants covers the following steps: presorting - charging - presieving - magnet sorting - sorting with blower - wet sorting - hand sorting. The general aim of sorting is to get clean mineral material which can be used for further applications.

3 Preparation plants:

3.1 Quarry plants

These can be subdivided into mobile, semimobile and stationary equipment, or according to the technology. An magnetic separation is added when steel reinforced concrete is recycled.

3.2 Sieving plants

These are used for the classification of high quality recycling material.

4 Recycled products:

As far as recycled products conform to the "standard for recycled building material" they can be used as secondary raw material for

- general constructions
- road constructions
- filling material
- aggregates for concrete production or for construction material with bituminous binders.

Fracture and Damage of Concrete and Rock – FDCR-2.
Edited by H.P. Rossmanith. © 1993 Taylor & Francis. ISBN 0 419 18470 8.

AUTHOR INDEX

SUBJECT INDEX

For Product Safety Concerns and Information please contact our EU representative GPSR@taylorandfrancis.com
Taylor & Francis Verlag GmbH, Kaufingerstraße 24, 80331 München, Germany

www.ingramcontent.com/pod-product-compliance
Ingram Content Group UK Ltd.
Pitfield, Milton Keynes, MK11 3LW, UK
UKHW022005180926
475701UK00003B/206
* 9 7 8 0 3 6 7 8 6 3 8 2 1 *